Illustrating the Basic Gra

Vertical Shift Principle

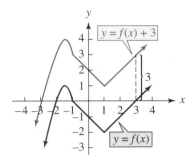

Horizontal Shift Principle

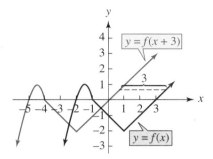

Stretching Principle

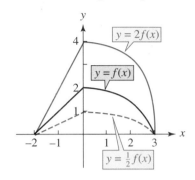

Graphing $-f(x)$

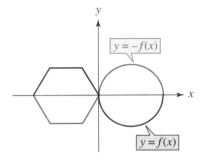

Graphing $f(-x)$

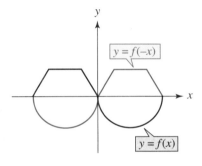

Graphing $f^{-1}(x)$

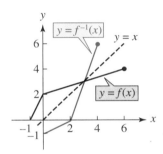

Combining the Graphing Principles to Graph $y = (x + 2)^3 - 8$

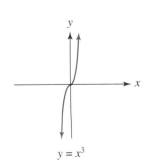

$y = x^3$

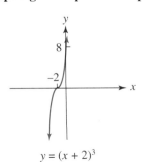

$y = (x + 2)^3$

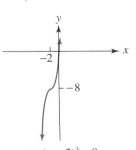

$y = (x + 2)^3 - 8$

COLLEGE ALGEBRA
Understanding Functions

A GRAPHING APPROACH

Arthur Goodman
Queens College

Lewis Hirsch
Rutgers University

THOMSON

BROOKS/COLE

Australia • Canada • Mexico • Singapore • Spain
United Kingdom • United States

THOMSON

™

BROOKS/COLE

Publisher: *Bob Pirtle*

Acquisitions Editor: *Jennifer Huber*

Assistant Editor: *Rebecca Subity*

Editorial Assistant: *Carrie Dodson*

Technology Project Manager: *Christopher Delgado*

Marketing Manager: *Leah Thomson*

Marketing Assistant: *Jessica Perry*

Advertising Project Manager: *Bryan Vann*

Project Manager, Editorial Production: *Andy Marinkovich*

Art Director: *Vernon Boes*

Print/Media Buyer: *Doreen Suruki*

Permissions Editor: *Joohee Lee*

Production Service: *Susan L. Reiland*

Text Designer: *Tani Hasegawa*

Illustrators: *Lori Heckelman, Rolin Graphics*

Copy Editor: *Christine Levesque*

Cover Illustration and Design: *E. Kelly Shoemaker*

Cover Printer: *R.R. Donnelley (Willard)*

Compositor: *Better Graphics, Inc.*

Printer: *R.R. Donnelley (Willard)*

Printed in the United States of America

1 2 3 4 5 6 7   08 07 06 05 04

Library of Congress Control Number: 2003113662

Student Edition: ISBN 0-534-42327-2

Instructor's Edition: ISBN 0-534-42328-0

Brooks/Cole–Thomson Learning
10 Davis Drive
Belmont, CA 94002
USA

Asia
Thomson Learning
5 Shenton Way #01-01
UIC Building
Singapore 068808

Australia/New Zealand
Thomson Learning
102 Dodds Street
Southbank, Victoria 3006
Australia

Canada
Nelson
1120 Birchmount Road
Toronto, Ontario M1K 5G4
Canada

Europe/Middle East/Africa
Thomson Learning
High Holborn House
50/51 Bedford Row
London WC1R 4LR
United Kingdom

Latin America
Thomson Learning
Seneca, 53
Colonia Polanco
11560 Mexico D.F.
Mexico

Spain/Portugal
Paraninfo
Calle/Magallanes, 25
28015 Madrid, Spain

CONTENTS

PREFACE TO THE INSTRUCTOR

For more than a decade, the mathematical community has been focusing on what and how to teach undergraduate mathematics. Motivated by the fact that unacceptably high percentages of students who take college-level mathematics courses do not successfully complete their course sequence, professionals have made various suggestions about how to revamp the undergraduate mathematics curriculum and integrate the new technology into the classroom.

While there are many exciting and innovative ideas being discussed, our experience has been that one of the main factors contributing to students' poor success rate is that they come to mathematics courses without the level of mathematical sophistication necessary to be successful. By this we mean that many students have had insufficient practice in the kinds of thinking required for understanding and applying the concepts of mathematics. Even students whose algebraic skills are adequate often do not know how to apply these skills appropriately. They may have a bag full of tools, and they may even know how to use each tool, but they frequently do not recognize which tools are appropriate to a given situation. This lack of sophistication also manifests itself in the fact that students often think about a problem in only one way—algebraically. Understanding concepts often requires students to view a problem from two additional perspectives: numerically and geometrically (graphically).

In this text we have again made a conscientious effort to afford the student every opportunity to look at questions from a variety of points of view; to take the time to analyze a problem carefully so that he/she clearly understands what is being asked; to reformulate problems in more familiar terms; and to recognize that many mathematical problems require a significant investment in time thinking. We hope that if students are repeatedly exposed to this philosophy, they will come to understand that they should be investing most of their time and effort into conceptualizing and formulating the problem carefully, rather than focusing on the answer. If they learn to ask themselves the right questions, the answers will take care of themselves.

This text is intended for students who want to establish a firm foundation for college-level mathematics courses. While it is assumed that the students who use this book have taken intermediate algebra, it comes as no surprise to those who have taught college algebra that students arrive with many different levels of algebra skills. Many students never really mastered the material when they learned it, while for others there may be a hiatus of several years between their study of algebra and functions. Consequently, the review material in Chapters 1 and 2 is rather detailed. Instructors may choose to go over the review material or to assign it to the students to review on their own, and, therefore, begin the course with Chapter 3. However, particular attention should be paid to Section 2.5 on *Quadratic and Rational Inequalities* and Section 2.6 on *Substitution*. The material in these sections may be new to a number of your students and is used repeatedly throughout the text.

A perusal of the Contents shows that the topics covered and their order are fairly traditional. What we think distinguishes this text is our approach, the ideas we have chosen to emphasize, and some of the special features we have introduced.

This text is replete with figures. In particular, Chapters 3 and 4 alone contain more than 200 figures, so that every idea that benefits from an accompanying figure has one. On many occasions, instead of referring a student to a graph that appeared previously, we repeat the graph to help clarify the discussion.

A great deal of attention has been paid to developing graphing principles and techniques and to encouraging students to familiarize themselves with a catalog of basic graphs and equations. This familiarity will aid students in other mathematical settings by allowing students to concentrate on the new material and not letting them sit and wonder, "Where did that graph come from?"

Pedagogy   Concepts are developed through a series of carefully constructed illustrative examples. Through the course of these examples we compare and contrast related ideas, helping the student to understand the sometimes subtle distinctions among various situations. In addition, every opportunity has been taken to point out how "new" ideas relate to topics the student has already encountered.

Common errors that frequently plague students are highlighted, and perhaps more important, the underlying misconception that leads to these errors is explained.

Graphing Calculators   The graphing calculator is a marvelous tool that the text employs both as a problem solver and as an exploratory tool to anticipate upcoming concepts and to clarify ideas already discussed.

While much of the text can be understood without the aid of a graphing calculator, there are numerous occasions where the text discussion is much richer (and the examples used more interesting) because of the availability and utilization of a graphing calculator.

It is our firm belief that the graphing calculator, employed properly, can substantially enhance a student's understanding of college algebra.

Special Features

1. **Functions and Graphs** In Chapters 3 and 4 we have spent an enormous amount of time in introducing and developing the idea of a function and its graph. Particular attention has been paid to pointing out the connection between the algebraic and graphical interpretations of certain important concepts.

   An entire section (3.6) is devoted to the idea of interpreting graphs: developing the ability to look at the graph of an equation (extracting graphical information) and to recognize the key features that describe the relationship given by the equation (an algebraic relationship).

2. **Different Perspectives** boxes appear wherever there is an opportunity to highlight the connection between the algebraic and graphical interpretation of the same idea. See Section 3.6, for example. In this way the student is encouraged to think about mathematical ideas from more than one point of view.

3. **Applications and Mathematical Modeling** We have made a conscientious effort to include word problems and applications wherever possible, and generally speaking, they are integrated throughout the text. However, there are a few places where applications are concentrated.

   Section 4.3 offers a broad introduction to the idea of mathematical modeling, giving the student opportunities to put the function concept to use in a variety of situations.

   Section 4.4, on quadratic functions, includes optimization problems that carry these ideas a step further. Such problems allow the students to see why we might want to express one quantity as a function of another and present yet another opportunity to see how the algebraic, graphical, and numerical perspectives of each offer particular insight into a problem.

Chapter 6 includes a variety of exercises that illustrate the remarkable range of disciplines in which quantities are related by exponential or logarithmic functions.

4. **Graphing Calculators** The text treats the calculator as a tool students bring to every exercise (just as they bring their algebraic skills). Through experience, and some guidance from illustrative examples, students will learn where and when calculators are necessary or appropriate to use. In addition, there are specific exercises that are designated to be done with a graphing calculator.

5. **Problem Analysis** As mentioned above, we feel very strongly about helping students develop at least the minimum level of mathematical sophistication and maturity necessary for success in higher-level mathematics courses. One step we have made in this direction is to take certain problems and present their solutions in a question-and-answer format, so that students can see some of the thought processes involved in approaching and solving new or unfamiliar problems. In this way we hope the student will develop appropriate problem-solving strategies.

6. **Margin Comments** Throughout the text we have used the margins to add another dimension to the discussion in the text. Recognizing that students are often too passive as they read mathematics, the questions and comments placed in the margin are specifically designed to more actively involve the student in the development presented in the text.

7. **Technology Corners** have been developed to help students appreciate some of the subtler aspects of using a graphing calculator. These are interspersed at various points in the text where appropriate.

Ancillaries

Instructors using this text may obtain the following supplements from the publisher:

For the Instructor

Annotated Instructor's Edition (0-534-42328-0)   This special version of the complete student text contains a Resource Integration Guide and answers printed next to all respective exercises. Graphs, tables, and other answers appear in a special answer section in the back of the text.

Test Bank (0-534-42331-0)   The *Test Bank* includes 8 tests per chapter as well as 3 final exams. The tests are made up of a combination of multiple-choice, free-response, true/false, and fill-in-the-blank questions.

Complete Solutions Manual (0-534-42330-2)   The *Complete Solutions Manual* provides worked solutions to all of the problems in the text.

BCA Instructor Version (0-534-42333-7)   New update 3.0. With a balance of efficiency and high-performance, simplicity and versatility, *Brooks/Cole Assessment* gives you the power to transform the learning and teaching experience. *BCA Instructor Version* is made up of two components, *BCA Testing* and *BCA Tutorial*. *BCA Testing* is a revolutionary, Internet-ready, text-specific testing suite that allows instructors to customize exams and track student progress in an accessible, browser-based format. BCA offers full algorithmic generation of problems and free-response mathematics. *BCA Tutorial* is a text-specific, interactive tutorial software program that is delivered via the Web (at http://bca.brookscole.com) and is offered in both student and instructor versions. Like *BCA Testing*, it is browser-based, making it an intuitive mathematical guide even for students with little technological proficiency. So sophisticated, it's simple, *BCA Tutorial* allows students to work with real math notation in real time, providing instant analysis and feedback. The tracking program built into the instructor version of the software enables instructors to carefully monitor student progress. The complete integration of the testing, tutorial, and course management components simplifies your routine tasks. Results flow automatically to your gradebook and you can easily communicate with individuals, sections, or entire courses.

New! BCA has been updated with a diagnostic tool to assess and place students in any course and provide a personalized study plan. Other new updates include:

- *Manipulate and format your tests* with the new "rich text format" (RTF) conversion tool. RTF conversion allows instructors to open tests in most word processors like Microsoft® Word for further formatting and customization.
- *BCA now offers several new problem types* to provide greater variety and more diverse challenges in your tests.
- *Redesigned Tutorial interface* effectively engages students and helps them learn math concepts faster.

Text-Specific Videotapes (0-534-42337-X)   These text-specific videotape sets, available at no charge to qualified adopters of the text, feature 10- to 20-minute problem-solving lessons that cover each section of every chapter.

WebTutor ToolBox for Blackboard and WebCT
ISBN: 0-534-27489-7 (*Blackboard*) ISBN: 0-534-27488-9 (*WebCT*)
Preloaded with content and available free via PIN code when packaged with this text, **WebTutor ToolBox for Blackboard and WebCT** pairs all the content of this text's rich Book Companion Web Site with all the sophisticated course management functionality of Blackboard and WebCT products. You can assign materials (including online quizzes) and have the results flow AUTOMATICALLY to your gradebook. **ToolBox** is ready to use as soon as you log on—or, you can customize its preloaded content by uploading images and other resources, adding Web links, or creating your own practice materials. Students only have access to student resources on the Web site. Instructors can enter a PIN code for access to password-protected Instructor Resources.

MyCourse 2.1   Ask us about our new free online course builder! Whether you want only the easy-to-use tools to build it or the content to furnish it, Brooks/Cole offers you a simple solution for a custom course Web site that allows you to assign, track, and report on student progress; load your syllabus; and more. Highlights of Version 2.1 include new features that let you import your class roster and grade book, see assignments as students will see them using the "Student Preview Mode," create individual performance grade books, and copy content from an existing course to a new course! Contact your Brooks/Cole representative for details, or visit http://mycourse.thomsonlearning.com

For the Student

Student Solutions Manual (0-534-42329-9)   The *Student Solutions Manual* provides worked solutions to the odd-numbered problems in the text.

Brooks/Cole Mathematics Web Site http://mathematics.brookscole.com   When you adopt a Thomson-Brooks/Cole mathematics text, you and your students will have access to a variety of teaching and learning resources. This Web site features everything from book-specific resources to newsgroups. It's a great way to make teaching and learning an interactive and intriguing experience.

InfoTrac® College Edition is a world-class, online university library that offers the full text of articles from almost 5000 scholarly and popular publications—updated daily and going back as far as 22 years. Both adopters and their students receive unlimited access for four months.

BCA Tutorial Student Version (0-534-42334-5)   New update 3.0. This text-specific, interactive, Web-based tutorial system is browser-based, making it an intuitive mathematical guide, even for students with little technological proficiency. So sophisticated, it's simple, *BCA Tutorial* allows students to work with real math notation in real time, providing instant analysis and feedback. The entire textbook is available in PDF format through BCA Tutorial, as are section-specific video tutorials, unlimited practice problems, and additional student resources such as a glossary, Web links, and more. The tracking program built into the instructor version of the software enables instructors to carefully monitor student progress. *BCA Tutorial Student Version* is also offered on a dual-platform CD-ROM. And, when students get stuck on a particular problem or concept, they need only log on to *vMentor*, accessed through BCA Tutorial, where they can talk (using their own computer microphones) to vMentor tutors who will skillfully guide them through the problem using the interactive whiteboard for illustration.

New! A personalized study plan can be generated by a course-specific diagnostic built into BCA Tutorial. With a personalized study plan, students can focus their time where they need it the most, creating a positive learning environment and paving a pathway to success in their mathematics course.

Interactive Video Skillbuilder CD-ROM (0-534-42332-9)   Think of it as portable office hours! The Interactive Video Skillbuilder CD-ROM contains more than eight hours of video instruction. The problems worked during each video lesson are shown next to the viewing screen so that students can try working them before watching the solution. To help students evaluate their progress, each section contains a 10-question Web quiz (the results of which can be e-mailed to the instructor) and each chapter contains a chapter test, with answers to each problem on each test. Also includes MathCue tutorial and testing software. Keyed to the text, MathCue includes these features:

- MathCue Skill Builder—Presents problems to solve, evaluates answers and tutors students by displaying complete solutions with step-by-step explanations. Students may view partial solutions to get started on a problem, see a continuous record of progress in a session, and back up to review missed problems.
- MathCue Quiz—Generates large numbers of quiz problems keyed to problem types from each section of the book. Students receive immediate feedback after entering their answers.
- MathCue Chapter Test—Provides large numbers of problems keyed to problem types from each chapter. Feedback is given only after the test is completed.
- MathCue Solution Finder—This unique tool allows students to enter their own basic problems and receive step-by-step help as if they were working with a tutor.
- Create customized MathCue sessions—Students work default Skill Builders, Quizzes and Tests or customize sessions to include desired problems from one or more sections or chapters. These customized session setups can be saved and run repeatedly. *Special note for instructors:* Instructors can create targeted, customized MathCue session files to send to students as e-mail attachments for use as make-up assignments, extra credit assignments, specialized review, quizzes, or other purposes.
- Print or e-mail score reports—Score reports for any MathCue session can be printed or sent to instructors via MathCue's secure e-mail score system.

vMentor   An ideal solution for homework help and tutoring. Accessed seamlessly through BCA Tutorial, *vMentor* is an efficient way to provide supplemental assistance that can substantially improve student performance, increase test scores, and enhance technical aptitude. Students will have access, via the Web, to highly qualified tutors with thorough knowledge of our textbooks. When students get stuck on a particular problem or concept, they need only log on to *vMentor*, when they can talk (using their own computer microphones) to *vMentor* tutors who will skillfully guide them through the problem using the whiteboard for illustration. Brooks/Cole also offers vClass, an online virtual classroom environment that is customizable and easy to use. vClass keeps students engaged with full two-way audio, instant messaging, and an interactive whiteboard—all in one intuitive, graphical interface. For information about obtaining a vClass site license, please contact your Thomson·Brooks/Cole representative.

Acknowledgments

The authors sincerely thank the following reviewers for their many thoughtful comments and suggestions during the preparation of this text:

Edward Boylan, Rutgers University

Antonio Castillo, Palo Alto College

Ginny Crisonino, Union County College

Samuel P. Evers, University of Alabama

Susan Fife, Houston Community College

Frances Gulick, University of Maryland

Zenas Hartvigson, University of Colorado at Denver

Lonnie Haas, North Dakota State University

Edward Matzdorff, California State University—Chico

David Tepper, Baruch College

Walter Wang, Baruch College

We also are grateful to our students who were kind enough to allow us to class-test this text in its manuscript form. Their criticisms and input were invaluable.

Obviously, the writing and production of a textbook is a collaborative effort. We must thank the staff at Brooks/Cole: Jennifer Huber, Carrie Dodson, Rebecca Subity, and Andy Marinkovich. Special thanks to Susan Reiland for her editorial skills and Ross Rueger for his proofreading and exercise checking. Of course, the authors are solely responsible for any errors that remain, and we would be most appreciative if they were brought to our attention.

Finally, we would like to thank our wives, Sora and Cindy, respectively, and our families for understanding the enormous amount of time such a project demands.

Arthur Goodman
Lewis Hirsch

This textbook is intended to expose you to many ideas and help you to develop a variety of skills you will find useful in your continuing study of mathematics. Perhaps the most important of these skills is the ability to approach a problem mathematically: read a question, understand what is being asked, and develop a coherent strategy as to how the problem is to be solved. (Doing these three things will not insure your success, but at least it will give you a fighting chance.) As you read the textbook, you will notice that a great deal of emphasis is given to the idea of *really understanding* what a question is saying and what it is asking.

You should always read a mathematics text with pencil and paper in hand. While we have included almost every step in the solutions to the illustrative examples, invariably some steps (often algebraic ones) are left out. You should fill these steps in by yourself.

The Graphing Calculator

A graphing calculator is a very powerful tool for doing calculations quickly and accurately, for exploring graphs, and for solving equations. However, while a graphing calculator should help to facilitate your understanding of mathematics and your ability to solve many problems, it is not a substitute for understanding what you are doing and why. When you sit down to work on a math problem you should have the tools you will need—pencil, paper, and calculator. You need to understand the role the calculator will play in your strategy for solving the problem.

Knowing how and *when* to use a calculator for a problem can make a difference in the accuracy of your answer as well. In this text, unless otherwise specified, all equations require exact answers (answers that are not rounded). Suppose, for example, you need to enter $\frac{33}{7}$ in your calculator before using it in your calculations to solve an equation. If you round $\frac{33}{7}$ to 4.71, you will automatically introduce a rounding error in your calculations. (How much this error is magnified and reflected in the solution will depend on the operations performed on, or with, this number.) Your solution will not be exact. Even if you enter $\frac{33}{7}$ in your calculator as $\boxed{3}\ \boxed{3}\ \boxed{\div}\ \boxed{7}\ \boxed{=}$, your calculator will round the number before performing computations; the result will be a more accurate answer, but it still may not be exact.

The calculator has numerous useful features, many of which will be used in the text. To get the most out of your calculator, you need to become familiar with its various capabilities. You should read your calculator manual carefully to learn how to use the utilities you have available. As you read text material that involves the use of the calculator pay particular attention to the accompanying discussion, which explains what you are trying to accomplish and how the calculator is helping you achieve that goal.

In this text, we often discuss algebraic, numeric, and graphical methods of solving the same problems. Sometimes we ask for a particular method to solve a problem, but often the method is dictated by the kind of solution being requested. Though there are exceptions, generally speaking, exact answers are found algebraically, while rounded answers may be found by either algebraic or graphical methods.

The Calculator and Accuracy of Measurement   Measurements of physical quantities are almost never exact; they are approximations which contain errors. For example, when you use a ruler to measure the length of an object, you know that at some point you have to estimate the measurement to the closest, say tenth, of a unit.

The simplest way to write a number that also indicates the error of measurement involved is by writing fewer or more digits. Recording a measured quantity as 4.2, for example, usually means that we are confident that the estimate is accurate to one decimal place and that its true value lies between 4.15 and 4.25. Recording the same value as 4.20 means we are confident that the value is accurate to two decimal places, and that its true value lies between 4.195 and 4.205. Notice that the precision of measurement is indicated by the number of digits in the recorded value, and that the last digit is always uncertain.

The use of the calculator has made it easy to confuse accuracy of computations with precision of measurement. For example, suppose we want to find the length of the hypotenuse, c, of a right triangle given that the legs measure 16.4 cm and 82.6 cm. By the Pythagorean Theorem, we arrive at the value $c = \sqrt{(16.4)^2 + (82.6)^2}$. Using a calculator (which can display eight digits), we arrive at $c = \mathbf{84.212351}$. This display gives us the impression of a level of precision which is unwarranted. The legs are accurate to only one decimal place, and these measurements are used to compute the hypotenuse. How can the hypotenuse be more precise than either of the two measurements used to compute it? Consequently, for problems involving computations with measurements, our final answer should never have a greater degree of accuracy than any of the measurements used to compute it. For our triangle problem, this means that the best estimate is $c = 84.2$ cm.

Some Remarks about the Notation Used in This Text

Set Notation

A **set** is a well-defined collection of objects. One way to designate a set is to list the members or elements of the set within braces: for example, $A = \{3, 4, 5, 6, 7, 8\}$. Alternatively, we can write a set using *set builder notation*: for example, the set A listed above can be written using set builder notation as:
$$A = \{x \mid 2 < x < 9, x \text{ is an integer}\}.$$
The symbol we use to designate that an object is a member of a particular set is "$\in$". Hence $5 \in A$ is a symbolic way of writing that 5 is a member of the set A. The set B is a **subset** of the set A, written $B \subset A$, if all elements of B are also in A. For example, the set of integers is a subset of the set of real numbers since the real numbers "contain" the integers, or equivalently, every integer is a real number.

We can create new sets from existing sets by the set "operations" of *union* and *intersection*. The **union** of two sets A and B, written $A \cup B$, is made up of the elements in A or B, or in both A and B. For example if $S = \{a, b, c\}$ and $T = \{a, d, f\}$ then $S \cup T = \{a, b, c, d, f\}$. The **intersection** of two sets A and B, written $A \cap B$, is made up of all elements common to both A and B. For the sets S and T defined above, $S \cap T = \{a\}$.

Implication and Equivalence

When we read a statement such as "If he passes the final then he will pass the course" we understand this to mean that one condition (passing the final) guarantees another (passing the course). A statement which indicates that one condition causes or implies another, is called an *implication*. Although implications may appear stated in "if-then" form, often the "if-then" content of the statement may not be explicitly stated and is expected to be understood. For example, the statement "All dogs are animals" can be restated as "If it is a dog, then it is an animal".

The statement "p implies q", symbolically written $p \Rightarrow q$, means if p is *true, then we can conclude that q is true*. We call p the *hypothesis* and q the *conclusion*. For example the statement "All dogs are animals" is a statement equivalent to the implication "If it is a dog, then it is an animal." Notice what it says, but more important, notice what it doesn't say: If it is a dog, then we can conclude that it is an animal, but if it is an animal, we cannot conclude anything about it being a dog. The mathematical statement "$a = b \Rightarrow a^2 = b^2$" means that if two quantities are equal, then we can conclude that their squares are equal. However, this statement tells us nothing about what happens if the squares of two quantities are equal.

Often in mathematics, we would like to assert that two statements are "equivalent." By equivalent we mean that given either statement, we may conclude the other. In other words, using the language of logic, saying p and q are equivalent is saying "$p \Rightarrow q$ AND $q \Rightarrow p$." With this statement notice that if we are given p, we can conclude q AND given q we can conclude p. We generally write this as $p \Leftrightarrow q$, which is read "p *if and only if* q." The "only if" allows us to also conclude that p is true given q is true. Note that $p \Leftrightarrow q$ means that either p and q are both true or p and q are both false (one statement cannot be true when the other is false). For example, we know that if $a = b$ then we conclude $a + c = b + c$; and, if $a + c = b + c$ then we can conclude $a = b$. Therefore the two statements, $a = b$ and $a + c = b + c$ are equivalent to each other and we may write $a = b \Leftrightarrow a + c = b + c$.

Algebra:
The Fundamentals

1

This chapter reviews some of the important and fundamental skills and concepts of algebra. It is not designed to serve as a substitute for a solid prerequisite course in algebra, but instead it is an overview of the basics of "symbolic arithmetic" as well as a demonstration of how the more rigorous or complex algebra problems are actually approached and solved by using the basic fundamental properties.

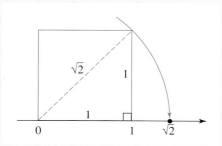

1.1   The Real Numbers

We will begin with some of the basic sets of numbers with which you are already familiar:

The **natural numbers** (counting numbers): $N = \{1, 2, 3, 4, 5, \dots \}$

The **whole numbers**: $W = \{0, 1, 2, 3, 4, \dots \}$

The **integers**: $Z = \{\dots, -3, -2, -1, 0, 1, 2, 3, \dots \}$

The **rational numbers**: $Q = \{\frac{p}{q} \mid p, q \in Z, \text{ and } q \neq 0\}$

We associate the rational numbers with points on the number line (see Figure 1.1). The number associated with a point on the number line is called the **coordinate** of the point.

Figure 1.1

The number associated with a point on the number line is called the **coordinate** of the point.

After associating each rational number with a point on the number line, we find that there are still points that remain unlabeled. The numbers associated with these unlabeled points are called *irrational numbers*. To get a better idea of what the irrational numbers look like, we examine rational numbers in decimal form.

You will recall that certain fractions have a finite decimal representation, such as $\frac{1}{4} = 0.25$, whereas others have an infinite decimal representation, such as $\frac{1}{3} = 0.3333\overline{3}$. (The bar means that the digits under the bar repeat without end.) The first case is called a **terminating decimal** (the decimal ends or at some point is followed by zeros); the second is called a **repeating decimal** (the same group of digits in the decimal is repeated indefinitely).

If a decimal terminates, we can easily recognize it as a rational number. For example, 0.863 is equal to $\frac{863}{1000}$. On the other hand, if the decimal repeats, the process of writing it as a quotient of integers is not quite as straightforward. Example 1 illustrates how we can write the repeating decimal $0.\overline{189189189}$ as a rational number.

If a number can be written as a quotient of two integers, then it is a rational number. Why does this imply that 2.36 is a rational number?

Example 1   Show that $0.\overline{189189189}$ is a rational number.

Solution   To show that a number is rational, we must somehow represent it as a quotient of integers, p/q. We start as follows:

Let $x =$ $0.189189\overline{189}$ We multiply both sides of the equation by 1000. Then

$1000x =$ $189.189189\overline{189}$

Now subtract $x = 0.189189\overline{189}$ from $1000x = 189.189189\overline{189}$:

Many graphing calculators can convert decimals into fractions. On the TI-83 Plus this is done with FRAC in the MATH menu, as illustrated in the following screen.

```
.189189189189189
▶Frac
               7/37
```

$$1000x = 189.189189\overline{189}$$
$$\underline{\quad x = \quad\;\; 0.189189\overline{189}}$$

Notice that the repeating decimal portions of the numbers match up exactly. Thus

$$999x = 189$$

$$x = \frac{189}{999} = \boxed{\frac{7}{37}}$$

The equation $1000x = 189.189189\overline{189}$ is found by multiplying both sides of the equation $x = 0.189189\overline{189}$ by 1000 *(we chose 1000 because it is the power of 10 needed to move the decimal point over and match up the infinitely repeating decimal portions of the numbers).* ∎

Hence, terminating and repeating decimals represent rational numbers. It is a fact that decimals that are both *nonterminating and nonrepeating are* **not** *rational numbers*. In other words, such a decimal cannot be represented as the quotient of two integers. (Exercise 53 outlines how we might prove that a specific number cannot be represented as a quotient of two integers.)

This set of nonrepeating, nonterminating decimals is called the set of **irrational numbers**. For example, π, $\sqrt{2}$, and $\sqrt[5]{3}$ are irrational numbers: We cannot write their exact values as decimals. At best we approximate their decimal values using a table, calculator, or computer and use the symbol $\approx$ to indicate an approximation. For example, $\sqrt{2} \approx 1.414$ ($\sqrt{2}$ is approximately 1.414).

The important thing for us to recognize is that irrational numbers also represent points on the number line (see Exercises 50–52). If we take all the rational numbers together with all the irrational numbers (both positive and negative), we get all the points on the number line. This set is called the set of **real numbers** and is usually designated by the letter $\boldsymbol{R}$:

The real numbers $\boldsymbol{R} = \{x \mid x$ corresponds to a point on the number line $\}$.

Unless stated to the contrary, we will assume that we are always working within the framework of the real number system.

Figure 1.2 illustrates the relationships among the sets discussed previously.

Figure 1.2

Relationships among subsets of real numbers

$N \subset W \subset Z \subset Q \subset R$

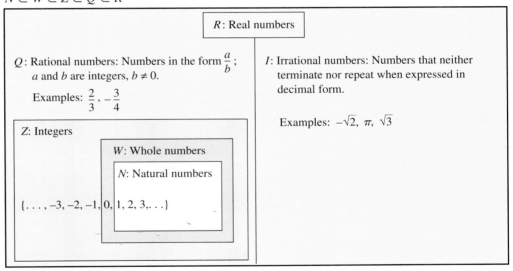

Properties of Real Numbers

The real numbers, along with the operations of addition $(+)$ and multiplication $(\cdot)$, obey the 11 properties listed in the following box. Most of these properties are straightforward and may seem trivial. Nevertheless, we will see that these 11 basic properties are quite powerful in that they are the basis for simplifying algebraic expressions.

The Commutative Properties

1. *For addition:* $a + b = b + a$
2. *For multiplication:* $ab = ba$

The Associative Properties

3. *For addition:* $a + (b + c) = (a + b) + c$
4. *For multiplication:* $a(bc) = (ab)c$

The Distributive Property

5. $a(b + c) = ab + ac$ or $(b + c)a = ba + ca$

Identities

6. *For addition:*   There is a unique real number called the **additive identity**, represented by 0, which has the property that $a + 0 = 0 + a = a$ for all real numbers, a.
7. *For multiplication:*   There is a unique real number called the **multiplicative identity**, represented by 1, which has the property that $a \cdot 1 = 1 \cdot a = a$ for all real numbers, a.

Inverses

8. *For addition:*   Each real number a has a unique **additive inverse**, represented by $-a$, which has the property that

$$a + (-a) = (-a) + a = 0$$

9. *For multiplication:*   Each real number a, except 0, has a unique **multiplicative inverse**, represented by $\frac{1}{a}$, which has the property that

$$a\left(\frac{1}{a}\right) = \left(\frac{1}{a}\right)a = 1$$

Closure Properties

10. *For addition:*   The sum of two real numbers is a real number.
11. *For multiplication:*   The product of two real numbers is a real number.

In the product ab, a and b are called **factors**; in the sum $a + b$, a and b are called **terms**.

When we state that a set is closed under an operation, we mean that when we perform the operation on two elements in the set, the result will be an element in the set. For example, we could start with the set of whole numbers and develop a system of whole numbers that has the associative, commutative, distributive, and identity properties (there are no inverses). This system would be closed under addition and multiplication; that is, sums and products of whole numbers are again whole numbers.

Keep in mind that the real number properties apply not only to numbers and single variables, but to more complex expressions as well. For example, the statement:

$$(2x^2 + 3x - 1)(x + 2) = (2x^2 + 3x - 1)(x) + (2x^2 + 3x - 1)(2)$$

is an application of the distributive property where $2x^2 + 3x - 1$ is distributed over $x + 2$.

Example 2 | If the following statements are true, give the property that the statement illustrates. If the statement is not true for all real numbers, write FALSE.

(a) $\left(\dfrac{3}{2x^2 + 1}\right)\left(\dfrac{2x^2 + 1}{3}\right) = 1$ **(b)** $10[.30(20 - x)] = 600 - 30x$

(c) $5(x + y)z = 5z(x + y)$

Solution | The general approach is to examine what is actually changing (and what hasn't changed) as we move from left to right.

(a) The multiplicative inverse property

(b) If we were to multiply this out using the associative property, first we would get

$$10[.30(20 - x)] = [(10)(.30)](20 - x) = 3[20 - x]$$ Then use the distributive property to get

$$= 60 - 3x$$

We would arrive at the same answer had we applied the distributive property first. The answer is false .

(c) The difference between the left-hand and right-hand expressions is that the positions of z and $x + y$ are interchanged. Hence part (c) illustrates the commutative property of multiplication. ∎

Example 3 | Show that the set of irrational numbers is not closed under multiplication.

Solution | Let's analyze this problem carefully.

What do we need to do?	Show that the set of irrational numbers is not closed under multiplication
How do we start?	First understand what is being asked. This requires knowing what closure is.
What does the statement "the irrationals are closed under multiplication" mean?	It means that the product of two irrational numbers is an irrational number.
How do we show that it is not closed?	We are being asked to find an example of two irrational numbers whose product is not irrational.

In mathematics, when we are asked to show, prove, or justify, we may need to go back to the definitions of the terms that are being used. Often we have to rephrase the problem using a definition.

In this example, we have to ask ourselves what it means for the set of irrational numbers to be closed under multiplication. Reviewing the previous comments on closure, we find it means that the product of two irrational numbers is always an irrational number. Since we want to show that this set is *not* closed under multiplication, we need only find one example of a pair of irrational numbers whose product is *not* irrational. This is called a *counterexample*. Two such numbers are $\sqrt{2}$ and $\sqrt{8}$; they are irrational numbers whose product, $\sqrt{2}\,\sqrt{8} = \sqrt{16} = 4$, is a rational number. So the set of irrational numbers is not closed under multiplication. ∎

Order and the Real Number Line

Numerical order is indicated by the "less than" symbol, $<$, and can be defined by using the number line:

> $a < b$ means that a is to the left of b on the number line.

Hence $3 < \pi$ is the symbolic statement meaning that 3 is to the left of π on the number line, as shown in Figure 1.3.

Figure 1.3

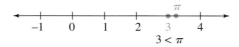

$$3 < \pi$$

Algebraically, the symbol $<$ has the following meaning:

> $a < b$ means that $b - a$ is a positive number.

Hence $3 < 8$, *since* $8 - 3 = 5$ is positive; we can also see that $\sqrt{2} < 2$ *since* $2 - \sqrt{2} \approx 2 - 1.414$ is positive.

The "greater than" symbol is similarly defined: $a > b$ *means that* $a - b$ *is positive*.

The inequality symbol $\leq$ means "less than or equal to"; hence, $6 \leq 6$. Similarly defined, $\geq$ means "greater than or equal to." We usually put a slash through equality and inequality symbols when we want to indicate that the statement is not true; for example $5 \neq 4 - 1$ means "5 is not equal to $4 - 1$."

Inequalities using the $<$ and $>$ symbols are called *strict* inequalities, and inequalities using the $\leq$ and $\geq$ symbols are called *weak* inequalities. By the definition of less than and greater than,

$$a > 0 \text{ means that } a \text{ is positive} \quad \text{and} \quad a < 0 \text{ means that } a \text{ is negative.}$$

Different Perspectives: **Inequalities**

Graphical Description

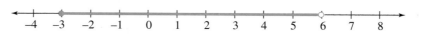

$a < b$ means that a is to the left of b on the number line.

Algebraic Description

$a < b$ means $b - a$ is a positive number.

The double inequality, $a < x < b$, is used to indicate betweenness. For example, $-3 < x < 6$ means that x is between -3 and 6. The double inequality is actually a combination of two inequalities that must be satisfied simultaneously; that is, $a < x < b$ is a combination of the two inequalities $a < x$ **and** $x < b$, where x satisfies *both* inequalities at the same time. Obviously, for the double inequality $a < x < b$ to make sense, a must be less than b. The graph of $-3 \leq x < 6$ is shown in Figure 1.4:

Why doesn't it make sense to write
$-2 < x < -4$ or $2 < x > 5$?

Figure 1.4

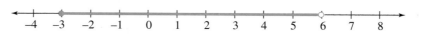

The real numbers satisfy the following property.

The Trichotomy Property

For $a, b \in \mathbf{R}$, one and only one of the following holds:

$$a < b, \quad a > b, \quad \text{or} \quad a = b$$

Interval Notation

By an interval, we mean a section of the number line.

Another way to express sets of numbers described by inequalities is to use interval notation. Interval notation is a convenient and compact way of representing intervals on the number line. We will begin with bounded intervals—that is, intervals that have two endpoints.

We use parentheses to indicate that an endpoint is *not* included and brackets to indicate that an endpoint *is* included. Hence, for $a < b$, we have the following.

Interval Notation: Bounded Intervals

Set Notation		Interval Notation	Line Graph
$\{x \mid a \leq x \leq b\}$	is written as	$[a, b]$, called the *closed interval from a to b.*	
$\{x \mid a < x < b\}$	is written as	(a, b), called the *open interval from a to b.*	
$\{x \mid a \leq x < b\}$	is written as	$[a, b)$, called a *half-open interval:*	
		closed at a and open at b.	
$\{x \mid a < x \leq b\}$	is written as	$(a, b]$, called a *half-open interval:*	
		open at a and closed at b.	

The smaller number is always written to the left of the larger. Unfortunately, the open interval (a, b) uses the same notation as the ordered pair (a, b). It should always be clear, however, from the context of a problem whether we are talking about an interval or an ordered pair.

When expressing unbounded intervals, lines, or half-lines, we use the infinity symbol, ∞ or $-\infty$, with interval notation, as follows.

Interval Notation: Unbounded Intervals

Set Notation		Interval Notation	Line Graph
$\{x \mid x \geq a\}$	is written as	$[a, \infty)$	
$\{x \mid x > a\}$	is written as	(a, ∞)	
$\{x \mid x \leq a\}$	is written as	$(-\infty, a]$	
$\{x \mid x < a\}$	is written as	$(-\infty, a)$	

The symbols ∞ and $-\infty$ do not represent numbers; they are simply symbols to indicate that the interval goes on forever, or increases (or decreases) without bound. Therefore, we always write a parenthesis next to the ∞ symbol.

Remember that whenever we use interval notation, we are working within the framework of the real number system. The heavy line on the graph indicates that all points on the line are included.

Example 4 | Graph the following inequalities on the number line, and express the set using interval notation.

(a) $\{x \mid x > -3\}$ (b) $\{s \mid s \leq 4\}$ (c) $\{x \mid -2 \leq x \leq 6\}$
(d) $\{x \mid -2 < x < 6\}$ (e) $\{t \mid -2 < t \leq 6\}$

Solution

(a) $\{x \mid x > -3\}$ 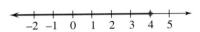   which is $(-3, \infty)$

(b) $\{s \mid s \leq 4\}$ which is $(-\infty, 4]$

(c) $\{x \mid -2 \leq x \leq 6\}$ which is $[-2, 6]$

(d) $\{x \mid -2 < x < 6\}$ which is $(-2, 6)$

(e) $\{t \mid -2 < t \leq 6\}$ which is $(-2, 6]$

Absolute Value

Geometrically, the **absolute value** of a number is its distance from zero on the number line. The absolute value of x is symbolized $|x|$. Hence

$|-4| = 4$ Since -4 is 4 units away from zero on the number line (Figure 1.5)

Figure 1.5

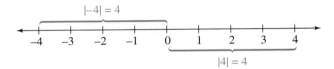

Also,

$|4| = 4$ since 4 is 4 units away from zero on the number line

Algebraically, we define absolute value as follows:

Definition of Absolute Value

$$|x| = \begin{cases} x, & \text{if } x \geq 0 \\ -x, & \text{if } x < 0 \end{cases}$$

A word about this notation: The notation symbolically expresses that the definition of $|x|$ depends on the value of x: If x is nonnegative ($x \geq 0$), then $|x|$ is simply x; on the other hand, if x is negative ($x < 0$), then $|x|$ is $-x$, which makes $|x|$ positive. For example, $|5| = 5$ *since* $5 \geq 0$, and $|-5| = -(-5) = 5$ *since* $-5 < 0$. Consequently, $|x|$ can never be negative.

Example 5 | Write the following without absolute value symbols.
(a) $|\pi - 3|$ (b) $|3 - \pi|$ (c) $|x^6 + 1|$ (d) $|y - 3|$

Solution | According to the definition of absolute value, to evaluate absolute value expressions, we must know whether the expression within absolute values is positive or negative.

(a) Since $\pi \approx 3.14$, then $\pi - 3$ is positive; hence

$$|\pi - 3| = \boxed{\pi - 3} \qquad \text{(since } \pi - 3 \geq 0\text{)}$$

(b) On the other hand, $3 - \pi$ is negative; hence

$$|3 - \pi| = -(3 - \pi) = \boxed{\pi - 3} \qquad \text{(since } 3 - \pi < 0\text{)}$$

Note: $|\pi - 3| = |3 - \pi|$

(c) Although x is a variable, we do know that x^6 must always be nonnegative. Therefore, $x^6 + 1$ must be positive; hence

Why must x^6 be nonnegative?

$$|x^6 + 1| = \boxed{x^6 + 1} \qquad \text{(since } x^6 + 1 > 0\text{)}$$

(d) We cannot determine whether the expression $y - 3$ is positive or negative, so we cannot evaluate this expression. However, we can use the definition of absolute value to rewrite this expression without the absolute value symbols:

$$|y - 3| = y - 3 \quad \text{when } y - 3 \geq 0 \quad \text{(that is, when } y \geq 3\text{)}$$
$$|y - 3| = -(y - 3) = 3 - y \quad \text{when} \quad y - 3 < 0 \quad \text{(that is, when } y < 3\text{)}$$

We can consolidate this as:

$$|y - 3| = \begin{cases} y - 3 & \text{when } y \geq 3 \\ 3 - y & \text{when } y < 3 \end{cases}$$

Distance on the Number Line

We can see by Figure 1.6 that the distance between two points on the real number line can be found using the differences of the coordinates; for example, the distance between the points with coordinates 5 and 2 is $5 - 2 = 3$ units.

Figure 1.6

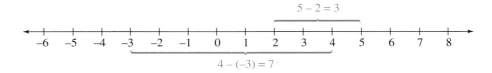

Because we want distance to be positive, we use absolute value to define distance as follows:

The Distance Between Two Points on the Number Line
On the real number line, the distance between points with coordinates a and b is $|a - b|$ or $|b - a|$.

Notice that the distance between 4 and -3 is 7 whether we compute it as $|4 - (-3)|$ or $|-3 - 4|$. The outline of an algebraic proof that $|a - b| = |b - a|$ is discussed in Exercise 54. This definition is consistent with the geometric definition of $|x|$ as the distance from 0 on the number line, since $|x| = |x - 0|$.

1.1 Exercises

In Exercises 1–6, express each infinitely repeating decimal as a quotient of integers, $\dfrac{p}{q}$.

1. $0.22\overline{2}$

2. $0.35\overline{35}$

3. $4.55\overline{5}$

4. $6.23\overline{23}$

5. $8.238\overline{238}$

6. $14.354\overline{354}$

In Exercises 7–16, if a given statement is true, give the property that the statement illustrates. If the statement is not true for all real numbers, write FALSE, and give a counterexample (an example that demonstrates that the statement is not true).

7. $\left(x + \dfrac{1}{2}\right) + \dfrac{2}{3} = x + \left(\dfrac{1}{2} + \dfrac{2}{3}\right)$

8. $y + (5 + x) = y + (x + 5)$

9. $3 + (xy) = (3 + x)(3 + y)$

10. $5a + 0 = 5a$

11. $[3(xy)z] = [(3x)(yz)]$

12. $\left(\dfrac{3}{4} + x\right)1 = \dfrac{3}{4} + x$

13. $(x - y + z)(a + b) = (x - y + z)a + (x - y + z)b$

14. $(x + 4) + [-(x + 4)] = 0$

15. $\dfrac{1}{x^2 + 1} \cdot (x^2 + 1) = 1$

16. $a(bc) = (ab)(ac)$

17. Show that the product of two rational numbers is a rational number. HINT: Start with two rational numbers, $\dfrac{a}{b}$ and $\dfrac{c}{d}$, and find their product.

18. Show that the sum of two rational numbers is a rational number.

19. Is the sum of two irrational numbers always irrational? If not, give a counterexample.

20. Is the difference of two whole numbers always a whole number? If not, give a counterexample.

In Exercises 21–26, graph each set on a real number line.

21. $\{x \mid x < 4\}$

22. $\{x \mid x \geq -5\}$

23. $\{x \mid x > 5\}$

24. $\{x \mid -3 < x \leq 2\}$

25. $\{x \mid -8 < x < -2\}$

26. $\{x \mid -2 \leq x < 4\}$

In Exercises 27–36, graph the set on the number line and express the set using interval notation.

27. $\{x \mid x > 5\}$

28. $\{x \mid x \leq -1\}$

29. $\{x \mid x \geq -5\}$

30. $\{x \mid -3 < x\}$

31. $\{x \mid -8 \leq x < -5\}$

32. $\{x \mid 0 < x \leq 6\}$

33. $\{x \mid -2 \geq x\}$

34. $\{x \mid -3 < x < 4\}$

35. $\{x \mid -9 < x \leq -2\}$

36. $\{x \mid 0 \leq x \leq 6\}$

In Exercises 37–44, write each expression without absolute value symbols.

37. $|3 - 5|$

38. $|\pi - 3.14|$

39. $|\sqrt{2} - 1|$

40. $|1 - \sqrt{2}|$

41. $|x - 5|$

42. $|x + 4|$

43. $|x^2 + 1|$

44. $|x^4 + 3|$

In Exercises 45–48, find the distance on the number line between each pair of points with the given coordinates.

45. 2 and 3

46. 5 and -9

47. -3 and 8

48. -8 and -4

Questions for Thought

49. The *transitive property* of inequalities states that if $a < b$ and $b < c$, then $a < c$. Prove this property. HINT: Determine what each statement means by the algebraic definition of an inequality.

50. Locate $\sqrt{2}$ on the number line following these steps.
   (a) Draw a unit square on the number line such that the base of the square is the line segment starting at 0 and ending at 1.
   (b) The diagonal of the square divides the square into two right triangles, where the diagonal is the hypotenuse of both right triangles. (See figure below.)

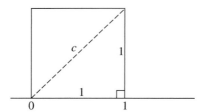

   (c) Using the Pythagorean Theorem from geometry, show that the length of the diagonal of the square is $\sqrt{2}$.

(d) Place the point of a compass on 0 and open it to the length of the diagonal.

(e) Rotate the compass down to the number line to locate $\sqrt{2}$ on the number line. (See figure.)

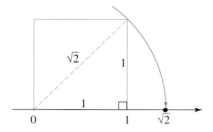

51. Locate $\sqrt{3}$ on the number line using the method described by Exercise 50. HINT: First locate $\sqrt{2}$.

52. Locate $\sqrt{5}$ on the number line using the method described by Exercise 50.

53. The following is an outline of how we might prove that $\sqrt{2}$ is not a rational number.

(a) Show that for every integer n, n^2 is even if and only if n is even.

(b) Assume $\sqrt{2} = \dfrac{m}{n}$, where $\dfrac{m}{n}$ is *reduced to lowest terms.*

(c) Multiply both sides of the preceding equation by n and square each side. Show that this implies that m^2 is even and, therefore, m is even.

(d) If m is even, then we can write $m = 2k$ for some k. Substitute $2k$ for m in the original equation; following a similar approach, show that this implies that n is even.

(e) Why does this contradict the original assumption? What does this mean about the original assumption?

54. Prove that $|a - b| = |b - a|$ for all real numbers a and b. HINT: By the trichotomy property, only one of the following holds: $a = b$, $a < b$, or $a > b$. Check each case by using the definition of $|a - b|$ to determine whether for all real numbers a and b, $|a - b| = |b - a|$.

1.2 Operations with Real Numbers

Subtraction and Division

Subtraction and division are defined in the following way:

Subtraction: $\quad a - b = a + (-b) \qquad$ **Division:** $\quad \dfrac{a}{b} = a\left(\dfrac{1}{b}\right) \quad$ for $b \neq 0$

Thus subtracting b means *adding the additive inverse of b*; dividing by b means *multiplying by the multiplicative inverse of b.* For example,

$$5 - (+8) = 5 + (-8) = -3 \quad \text{and} \quad \frac{6}{2} = 6\left(\frac{1}{2}\right) = 3$$

What happens if we try to divide a nonzero number by 0? Let's assume that $\frac{6}{0} = x$. This means that $x \cdot 0 = 6$. But any number multiplied by 0 is equal to 0. Therefore there is no such number x and division of a nonzero number by zero is *undefined.* On the other hand, if $a \neq 0$, $\frac{0}{a} = 0$, since $0 \cdot a = 0$.

If we try to divide 0 by 0, we run into problems for another reason. Let's suppose we let $\frac{0}{0}$ be equal to some number r. Then this means $r \cdot 0 = 0$. But this is true for all numbers r. This means that any number will work; that is, $\frac{0}{0}$ is not unique. Therefore, we say that $\frac{0}{0}$ is *indeterminate* (there is no unique answer). To summarize:

> If a is any nonzero number, then
>
> $$\frac{a}{0} \text{ is undefined,} \quad \text{whereas} \quad \frac{0}{a} = 0 \quad \text{and} \quad \frac{0}{0} \text{ is indeterminate}$$

Exponents

We define x^n as follows:

> $x^n = x \cdot x \cdot x \cdots x$, *where the factor x occurs n times, and n is a natural number.*

For x^n, x is called the **base** and n is called the **exponent**. A natural number exponent tells how many times x occurs as a factor in the product. x^n is called the **nth power of x**. (Note that $x = x^1$; that is, an unwritten exponent is assumed to be 1.)

To compute or evaluate an expression with a numerical base means to multiply out the expression. For example,

$$3^5 = 3 \cdot 3 \cdot 3 \cdot 3 \cdot 3 = 243$$
$$(-2)^4 = (-2)(-2)(-2)(-2) = +16 \qquad \text{On the other hand,}$$
$$-2^4 = -(2^4) = -(2 \cdot 2 \cdot 2 \cdot 2) = -16$$

x^2 is read "x squared."
x^3 is read "x cubed."

Keep in mind that the exponent applies only to the quantity to its immediate left. In the expression $3 \cdot 2^4$, we raise 2 to the fourth power and then multiply by 3 to get 48. Similarly, in -2^4, we raise 2 to the fourth power and then use the minus sign to get -16.

Multiple Operations

If there is more than one operation to perform in an expression, there must be agreement as to which operation should be performed first: an order of operations. For example, to evaluate the expression $3 + 2 \cdot 4$, we could arrive at the answer 20 or 11, depending on whether we choose to add or multiply first. We agree to the following **order of operations**:

1. Start by performing operations within the grouping symbols, beginning with the innermost grouping.
2. Then evaluate powers (and roots) in any order.
3. Perform multiplication and division, working from left to right.
4. Finally, perform addition and subtraction, working from left to right.

For example, to evaluate the numerical expression $4[-5 - 3(-2)^2]$, we would proceed as follows:

Try evaluating this expression with your calculator for practice.

$$4[-5 - 3(-2)^2] = 4[-5 - 3\,(4)] = 4[-5 - 12] = 4(-17) = -68$$

Substitution

Applications of algebra very often require us to substitute numerical values for variables and then to perform the operations with the substituted values. In other words, we are asked to evaluate an expression given the values for the variables.

Example 1 Given $x_1 = -2$, $y_1 = 5$, $x_2 = 3$, and $y_2 = -7$, evaluate the following:

$$\sqrt{(x_2 - x_1)^2 + (y_2 - y_1)^2}$$

Solution x_1, x_2, y_1, and y_2 are called **subscripted variables**. It is usually a good idea to enclose the values being substituted in parentheses. This helps avoid confusing the original operations. Substitute $x_1 = -2$, $y_1 = 5$, $x_2 = 3$, and $y_2 = -7$ into the expression:

$$\sqrt{(x_2 - x_1)^2 + (y_2 - y_1)^2} = \sqrt{[(3) - (-2)]^2 + [(-7) - (5)]^2}$$ Work inside brackets first

$$= \sqrt{[5]^2 + [-12]^2}$$ Powers under the radical

$$= \sqrt{25 + 144}$$

$$= \sqrt{169} = \boxed{13}$$

∎

Example 2 Using population data from 1980 to 1996 presented in Figure 1.7, we can estimate the total population of the United States by the linear equation $P = 2.364t - 4{,}454$, or by the quadratic equation $P = 0.024826t^2 - 96.344t + 93{,}661.4$, where P is the population in millions in year t.

Figure 1.7

U.S. Population: 1980–1996
[In thousands (180,671 represents 180,671,000).
Estimates as of July 1. Total population includes Armed Forces abroad;
civilian population excludes Armed Forces.]

Year	Total population	Resident population	Civilian population
1980	227,726	227,225	225,621
1981	229,966	229,466	227,818
1982	232,188	231,664	229,995
1983	234,307	233,792	232,097
1984	236,348	235,825	234,110
1985	238,466	237,924	236,219
1986	240,651	240,133	238,412
1987	242,804	242,289	240,550
1988	245,021	244,499	242,817
1989	247,342	246,819	245,131
1990	249,948	249,439	247,798
1991	252,639	252,127	250,517
1992	255,374	254,995	253,410
1993	258,083	257,746	256,273
1994	260,599	260,289	258,877
1995	263,044	262,765	261,414
1996	265,463	265,190	263,904

Source: U.S. Census Bureau, *Current Population Reports*, P25-802 and P25-1095; and "Monthly estimates of the United States population: April 1, 1980 to November 1, 1998"; release date: December 28, 1998; <http://www.census.gov/population/estimates/nation/intfile1-1.txt>.

(a) Compute the total population in 1998 predicted by each equation to the nearest thousand.

(b) The actual total population in 1998 was 270,561,000. Which equation gives the better estimate for the actual population in 1998?

Solution **(a)** We compute the total population of the United States by substituting 1998 into each equation using a calculator. The computation for the linear equation gives us 269.272, which is $\boxed{269{,}272{,}000}$. The computation for the quadratic equation gives us 271.579304, which is $\boxed{271{,}579{,}000}$ rounded to the nearest thousand. See the calculator screen on page 14.

~~~
2.364*1998−4454
            269.272
.024826*1998²−96
.344*1998+93661.
4
      271.579304
~~~

(b) We are interested in which predicted value comes closer to the actual value, so we look at the differences between the predicted values and the actual value. Since it is of no interest whether the predicted estimate is above or below the actual value, we look at the absolute value of the differences. For the linear equation, the difference is $|269{,}272{,}000 - 270{,}561{,}000| = 1{,}289{,}000$. For the quadratic equation we have $|271{,}579{,}000 - 270{,}561{,}000| = 1{,}018{,}000$. We can see that the quadratic equation yields a smaller difference in absolute value and therefore gives a better predicted value. ∎

Fractions and Their Operations

Whereas a rational number is a quotient of integers, $\frac{p}{q}$ where $q \neq 0$, a *fraction* is a quotient of any two numbers $\frac{a}{b}$ where $b \neq 0$. We know from previous experience with fractions that two fractions may look different but may actually be equivalent. Hence we define what we mean by "=" with fractions as follows:

Equivalence of Fractions

$$\frac{a}{b} = \frac{c}{d} \quad \text{if and only if} \quad ad = bc \qquad (b, d \neq 0)$$

Hence $\dfrac{6}{13} = \dfrac{24}{52}$ *because* $6 \cdot 52 = 13 \cdot 24 = 312$.

A fraction reduced to lowest terms or written in simplest form is a fraction that has no factors (other than ± 1) common to both its numerator and denominator. If we require that all fractions be reduced to lowest terms, then, rather than resorting to the definition of equivalence, we can observe by inspection that two fractions are equivalent. Hence, the fundamental principle of fractions is usually more useful:

The Fundamental Principle of Fractions

$$\frac{a \cdot k}{b \cdot k} = \frac{a}{b} \quad b, k \neq 0$$

This principle says that if we multiply or divide the numerator and the denominator of a fraction by the same nonzero expression, we obtain an equivalent fraction.

To reduce a fraction to lowest terms often requires us to factor both numerator and denominator and then use the fundamental principle by dividing out factors common to both the numerator and denominator. For example, $\dfrac{24}{52} = \dfrac{4 \cdot 6}{4 \cdot 13} = \dfrac{6}{13}$.

We define operations with fractions as follows:

Operations with Fractions

Multiplication

$$\left(\frac{a}{b}\right)\left(\frac{c}{d}\right) = \frac{ac}{bd}$$

Division

$$\left(\frac{a}{b}\right) \div \left(\frac{c}{d}\right) = \frac{ad}{bc}$$

Addition

$$\frac{a}{b} + \frac{c}{d} = \frac{ad + bc}{bd}$$

Subtraction

$$\frac{a}{b} - \frac{c}{d} = \frac{ad - bc}{bd}$$

Notice that subtraction is still adding the additive inverse, and division is still multiplying by the reciprocal. For example,

$$\left(\frac{2}{7}\right) \div \left(\frac{5}{8}\right) = \left(\frac{2}{7}\right) \cdot \left(\frac{8}{5}\right) = \frac{2 \cdot 8}{7 \cdot 5} = \frac{16}{35}$$

If the denominators of two factors are identical, we can use the distributive property to find

$$\frac{a}{c} + \frac{b}{c} = \frac{a + b}{c}$$

Hence, an alternative method for combining fractions with different denominators is to express the fractions as equivalent fractions having their least common denominator (LCD) as denominator and then add them as fractions with the same denominator as above. For example,

$$\frac{5}{18} + \frac{11}{12}$$ The LCD is 36, the smallest multiple of both 12 and 18.

$$= \frac{5 \cdot 2}{36} + \frac{11 \cdot 3}{36} = \frac{10}{36} + \frac{33}{36} = \frac{43}{36}$$

Example 3   Evaluate the numerical expression.

$$3\left(-\frac{2}{3}\right)^2 + 4\left(-\frac{2}{3}\right) + 5$$

Solution

We could evaluate this expression on a calculator as follows. Note that the calculator returns a decimal value.

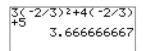

```
3(-2/3)²+4(-2/3)
+5
           3.666666667
```

$$3\left(-\frac{2}{3}\right)^2 + 4\left(-\frac{2}{3}\right) + 5 = 3\left(\frac{4}{9}\right) - \frac{8}{3} + 5 = \frac{4}{3} - \frac{8}{3} + 5 = \frac{4}{3} - \frac{8}{3} + \frac{15}{3}$$

$$= \frac{11}{3}$$

Complex Fractions

Example 4   Write as a simple fraction reduced to lowest terms. $\dfrac{\dfrac{3}{5} + \dfrac{2}{3}}{\dfrac{3}{4} - \dfrac{2}{5}}$

Solution   We offer two methods of solution. For our first method, we treat this as a multiple-operation problem. That is, we treat it as the problem

$$\left(\frac{3}{5} + \frac{2}{3}\right) \div \left(\frac{3}{4} - \frac{2}{5}\right)$$

$$\frac{\dfrac{3}{5} + \dfrac{2}{3}}{\dfrac{3}{4} - \dfrac{2}{5}} = \frac{\dfrac{9 + 10}{15}}{\dfrac{15 - 8}{20}} = \frac{\dfrac{19}{15}}{\dfrac{7}{20}} = \left(\frac{19}{15}\right)\left(\frac{20}{7}\right) = \frac{19 \cdot \overset{4}{20}}{\underset{3}{15} \cdot 7} = \frac{76}{21}$$

An alternative way to clear denominators of fractions within a complex fraction is to apply the fundamental principle by multiplying the numerator and denominator of the complex fraction by the LCD of *all* simple fractions in the complex fraction.

In this case, the LCD of $\frac{3}{5}$, $\frac{2}{3}$, $\frac{3}{4}$, and $\frac{2}{5}$ is 60:

$$\frac{\left(\frac{3}{5} + \frac{2}{3}\right)}{\left(\frac{3}{4} - \frac{2}{5}\right)} \cdot \frac{\frac{60}{1}}{\frac{60}{1}} = \frac{\left(\frac{3}{5}\right)\frac{\overset{12}{60}}{1} + \left(\frac{2}{3}\right)\frac{\overset{20}{60}}{1}}{\left(\frac{3}{4}\right)\frac{\overset{15}{60}}{1} - \left(\frac{2}{5}\right)\frac{\overset{12}{60}}{1}} = \frac{36 + 40}{45 - 24} = \boxed{\frac{76}{21}}$$

Example 5   The *harmonic mean* of n positive numbers, $X_1, X_2, X_3, \ldots, X_n$, is defined as follows:

$$h = \frac{n}{\dfrac{1}{X_1} + \dfrac{1}{X_2} + \dfrac{1}{X_3} + \cdots + \dfrac{1}{X_n}}$$

(a) Find the exact value of the harmonic mean of 4, 6, and 7.

(b) Find the harmonic mean of 4, 6, and 7 rounded to four places using a calculator.

Solution   (a) Since we have three numbers, $n = 3$ and we substitute $X_1 = 4$, $X_2 = 6$, and $X_3 = 7$ in the given formula.

$$h = \frac{n}{\dfrac{1}{X_1} + \dfrac{1}{X_2} + \dfrac{1}{X_3} + \cdots + \dfrac{1}{X_n}}$$

Substitute $X_1 = 4$, $X_2 = 6$, $X_3 = 7$, and $n = 3$ to get

$$= \frac{3}{\dfrac{1}{4} + \dfrac{1}{6} + \dfrac{1}{7}}$$

Multiply the numerator and denominator by the LCD of $\frac{1}{4}$, $\frac{1}{6}$, and $\frac{1}{7}$, which is 84.

$$= \frac{3}{\dfrac{1}{4} + \dfrac{1}{6} + \dfrac{1}{7}} \cdot \frac{\dfrac{84}{1}}{\dfrac{84}{1}}$$

$$= \frac{252}{21 + 14 + 12} = \boxed{\frac{252}{47}}$$

(b) To find the harmonic mean with a graphing calculator, we simply enter the complex fraction as it appears in part (a). Note the use of parentheses to ensure that the divisions are done properly.

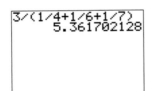

Verify that the fractional answer in part (a) agrees with the decimal answer in part (b).

Rounding the answer to four decimal places, we get $\boxed{5.3617}$.

1.2 Exercises

In Exercises 1–26, evaluate the numerical expressions.

1. $-3 + (-6) - (+4) - (-8)$

2. $-6 + (-2) - (-9) - (-4)$

3. $(-6)(-2)(-3)$

4. $(-5)(-8)(-6)$

5. $-2 - 3.552$

6. $-8 + 5.582$

7. $-4 + 7.29$

8. $-4 - 7.29$

9. $-2[3 - (2 - 5)]$

10. $-6[2 - (5 - 8)]$

11. $2 - (-3)^2$

12. $2(-3)^2$

13. $6 - [4 - (5 - 8)^2]$

14. $9 - \{3 - [6 - 2(9 - 4)^2]\}$

15. $|-6 - 5| - |6 - 5|$

16. $|-4 - 4| - |4 - 4|$

17. $\dfrac{|-9 - 5|}{|-9| - |5|}$

18. $\dfrac{|-6 - 12|}{|-6| - |12|}$

19. $\dfrac{3}{4} - \dfrac{2}{3} + \dfrac{1}{2}$

20. $\dfrac{3}{5}\left(-\dfrac{2}{3}\right) - \dfrac{1}{2}$

21. $\left(-\dfrac{2}{5}\right)^2 - \dfrac{3}{4}$

22. $\left(-\dfrac{3}{4}\right) - \left(\dfrac{2}{3}\right)^2$

23. $6\left(-\dfrac{2}{3}\right)^2 + \left(-\dfrac{2}{3}\right) - 2$

24. $6\left(\dfrac{1}{2}\right)^2 + \left(\dfrac{1}{2}\right) - 2$

25. $3\left(-\dfrac{1}{5}\right)^2 + 2\left(-\dfrac{1}{5}\right) - 3$

26. $2\left(-\dfrac{1}{4}\right)^2 - 3\left(-\dfrac{1}{4}\right) + 8$

In Exercises 27–30, write each as a simple fraction reduced to lowest terms.

27. $\dfrac{3 + \dfrac{3}{5}}{5 - \dfrac{1}{8}}$

28. $\dfrac{4 - \dfrac{2}{3}}{\dfrac{2}{5} - 6}$

29. $\dfrac{\dfrac{2}{3} - \dfrac{1}{2}}{\dfrac{1}{8} + \dfrac{2}{5}}$

30. $\dfrac{\dfrac{3}{5} - \dfrac{1}{2}}{\dfrac{7}{10} - 2}$

In Exercises 31–34, evaluate each expression given $x = -1$ and $y = -2$.

31. $2x^2 - 4y^2$

32. $|x - y| - |x| - |y|$

33. $\dfrac{x^2 - 2xy + y^2}{x - y}$

34. $(x - y)^2 - x^2 - y^2$

35. The *geometric mean* of n positive numbers, $X_1, X_2, X_3, \ldots, X_n$, is defined as follows:

$$g = \sqrt[n]{X_1 \cdot X_2 \cdot X_3 \cdots X_n}$$

Find the geometric mean of 5, 8, 7, 9, 7, 8, and 6 rounded to four places.

36. The *harmonic mean* of n positive numbers, $X_1, X_2, X_3, \ldots, X_n$, is defined in Example 5. Equivalently, we can define the harmonic mean as follows:

$$h = \dfrac{1}{\left[\dfrac{1}{X_1} + \dfrac{1}{X_2} + \dfrac{1}{X_3} + \cdots + \dfrac{1}{X_n}\right]/n}$$

Find the harmonic mean of 5, 8, 7, 9, 7, 8, and 6 both exactly and rounded to four places.

37. Given $s_e = s_y\sqrt{1 - r_{xy}^2}$, if $s_y = 1.25$ and $r_{xy} = 0.4$, find s_e rounded to two places.

38. Given $s_e = s_y\sqrt{1 - r_{xy}^2}$, if $s_y = 2.24$ and $r_{xy} = 0.73$, find s_e rounded to two places.

39. Given

$$Z = \dfrac{Z_{r_1} - Z_{r_2}}{\sqrt{\dfrac{1}{n_1 - 3} + \dfrac{1}{n_2 - 3}}}$$

compute Z to two places for $Z_{r_1} = 0.50$, $Z_{r_2} = 0.32$, $n_1 = 65$, and $n_2 = 83$.

40. Given

$$t = \dfrac{\overline{X} - a}{\dfrac{s_x}{\sqrt{n}}}$$

compute t to two places for $\overline{X} = 90$, $a = 95$, $s_x = 5.2$, and $n = 15$.

41. Given

$$\sigma_r = \sqrt{\dfrac{1 - \rho^2}{n - 1}}$$

compute σ_r to three places for $\rho = 0.5$ and $n = 100$.

42. Given

$$\sigma_r = \sqrt{\dfrac{1 - \rho^2}{n - 1}}$$

compute σ_r to three places for $\rho = 0.67$ and $n = 128$.

43. Based on population data from 1900 to 1990, the resident population of the United States can be estimated by the linear equation $P = 1.92364t - 3,588$, where P is the population in millions at year t.
  (a) Compute the resident population for 1990 predicted by the equation.
  (b) The actual population in 1990 was 249 million. Find the difference between the actual population in 1990 and the population predicted by the equation for 1990.

44. The national health expenditures (the amount of money spent on health care) for the United States from 1980 to 1997 can be estimated by the equation

$$E = 51.73723(t - 1,980) + 212.17$$

where E is the amount spent in billions of dollars during year t.
  (a) Use this equation to determine how much was spent in the United States on health care in 1994, 1995, 1996, and 1997.
  (b) Compute the differences between expenditures for each pair of consecutive years found in part (a), and then compute the percent increase (from the preceding year) for each year from 1995 to 1997.

45. The national health expenditures for hospital care only (the amount of money spent on hospital care) for the United States from 1980 to 1997 can be estimated by the equation

$$E = 16.6333(t - 1,980) + 97.1$$

where E is the amount spent in billions of dollars during year t.
  (a) Use this equation to determine how much was spent in the United States on hospital care in 1994, 1995, 1996, and 1997.
  (b) Compute the differences between hospital care expenditures for each pair of consecutive years found in part (a), and then compute the percent increase (from the preceding year) for each year from 1994 to 1997.

46. The following table from the Bureau of Labor Statistics gives the percentage of females over the age of 16 in the civilian population employed in January of each of the years listed.

Year	Employment rate
1993	53.7%
1994	54.9%
1995	55.5%
1996	55.5%
1997	56.5%
1998	57.1%

Based on these data, a statistician suggests that the following formula can be used to approximate the female employment rate R given year t (in January of the years 1993 through 1998):

$$R = -0.025(x - 1,990)^2 + 0.9(x - 1,990) + 51.4$$

  (a) Use this formula to compute R for each of the years 1993 through 1998.
  (b) Compare the values collected by the Department of Labor with the values obtained from the formula, and comment about the accuracy of the formula.

47. Use the real number properties and the definition of rational addition to prove
$$\frac{a}{c} + \frac{b}{c} = \frac{a+b}{c}.$$

48. Use the *definition of equivalent fractions* to show:
  (a) $\dfrac{1}{\sqrt{2}} = \dfrac{\sqrt{2}}{2}$ **(b)** $\dfrac{5}{\sqrt{3}} = \dfrac{5\sqrt{3}}{3}$

49. Use the *fundamental principle of fractions* to show:
  (a) $\dfrac{1}{\sqrt{2}} = \dfrac{\sqrt{2}}{2}$ **(b)** $\dfrac{5}{\sqrt{3}} = \dfrac{5\sqrt{3}}{3}$

50. Consider the expression $\dfrac{3}{x}$. Discuss the following.
  (a) What happens to the value of $\dfrac{3}{x}$ as x gets larger?
  (b) What happens to the value of $\dfrac{3}{x}$ as x remains positive but gets smaller (closer to 0)?
  (c) What happens to the value of $\dfrac{3}{x}$ as x remains negative but gets closer to 0?
  (d) Can $\dfrac{3}{x}$ ever be zero? Explain your answer.

51. Consider y defined in the following way:
$$y = \begin{cases} x + 1 & \text{if } x < 2 \\ x^2 & \text{if } x \geq 2 \end{cases}$$

This notation indicates that y is defined by two possible rules, dependent on the value of x: If x is less than 2, use $x + 1$ for y; if x is greater than or equal to 2, then use x^2 for y. Hence, if $x = -5$, then, since -5 is less than 2, we use the first rule to find y: $y = x + 1 = -5 + 1 = -4$. If $x = 8$, then, since 8 is greater than (or equal to) 2, we use the second rule to find y: $y = x^2 = (8)^2 = 64$. Find y, defined previously, for $x = 1$ and for $x = 3$.

52. Given
$$y = \begin{cases} x & \text{if } x \geq 0 \\ x - 3 & \text{if } x < 0 \end{cases}$$
find y when **(a)** $x = 4$, **(b)** $x = -3$, **(c)** $x = 0$. HINT: See Exercise 51.

1.3 Polynomials

An **algebraic expression** is an expression obtained by adding, subtracting, multiplying, dividing, and taking roots of constants and/or variables. For example,

$$2x^{-1/2} + 7, \quad \sqrt{3x - 4}, \quad \frac{4x^2 + x - 5}{x + 6}, \quad \text{and} \quad x^3 - 2x + 3$$

are algebraic expressions. In this section we will review a particular type of algebraic expression: polynomial expressions.

Polynomials

A **polynomial in one variable** is an expression of the form

$$a_n x^n + a_{n-1} x^{n-1} + a_{n-2} x^{n-2} + \cdots + a_2 x^2 + a_1 x + a_0, \quad \text{where } a_n \neq 0$$

The a_i's are real numbers, x is a variable, and n is a nonnegative integer called the **degree of the polynomial**. Each expression $a_i x^i$ is called a **term** of the polynomial.

When a polynomial is written with its terms arranged in descending powers, it is said to be in **standard form**. A polynomial such as $5 + 3x - 4x^2$ is written in standard form as $-4x^2 + 3x + 5$. In this form, by the definition, $n = 2$, $a_2 = -4$, $a_1 = 3$, and $a_0 = 5$. Note how the subscripts of a conveniently match the exponents of x.

When a polynomial is written with descending powers, we assume that missing powers of the variable have coefficients of 0. For example, $3x^5 - 2x + 3$ can be rewritten as $3x^5 + 0x^4 + 0x^3 + 0x^2 - 2x + 3$.

A polynomial in more than one variable contains terms such as $ax^m y^n z^s$, where a is real, x, y, and z are variables, and m, n, and s are nonnegative integers.

We may classify polynomials by the number of terms making up the polynomial: A *monomial* is a polynomial consisting of one term, a *binomial* is a polynomial consisting of two terms, and a *trinomial* is a polynomial consisting of three terms.

Besides the number of terms, we can also classify a polynomial by its degree. First, we define the degree of a monomial: The **degree of a monomial** is the sum of the exponents of its variables. For example, $-5x^3 y^4 z$ has degree 8, since $3 + 4 + 1 = 8$ (remember $z = z^1$). The **degree of a polynomial** is the highest degree of any monomial in it. For example, the polynomial $7x^2 y^4 - 3x^7 y^5 z^2$ has degree 14, since the highest degree of any monomial in the polynomial is 14, the degree of the second term; the polynomial $7x^4 - 4x^6$ has degree 6, since the highest-degree term, $-4x^6$, has degree 6.

The degree of a *nonzero* constant is 0 (we can rewrite 4 as $4x^0$). When the number 0 is considered as a polynomial, we call it the **zero polynomial**; the degree of the zero polynomial is undefined.

A third way to classify polynomials is by the number of variables. The following is an example of a polynomial in two variables written in standard form:

$$x^7 y + x^6 + x^3 y^2 + x^2 y^3 + x$$

The terms are arranged in descending degree order with the powers of x in descending order.

Polynomial Operations

Recall that $x^n = x \cdot x \cdot x \cdot x \cdots x$, where the factor x occurs n times. Hence $3xxxxxyyy = 3x^5 y^3$.

Using the definition of exponential notation and the real number properties, we can derive the following rules of natural number exponents:

$$x^n x^m = x^{n+m}, \qquad (x^n)^m = x^{nm}, \quad \text{and} \quad (xy)^n = x^n y^n$$

We use these first few rules of exponents to find products of powers with the same base. Using the associative and commutative properties of multiplication (that is, ignoring order and grouping), we can find the following product:

$$(5x^3y^6)(-6x^4y^2) = (5)(-6)x^3x^4y^6y^2 = -30x^7y^8$$

The distributive property allows us to combine like terms; for example,

$$3x^2y^3 - 8x^2y^3 = (3 - 8)x^2y^3 = -5x^2y^3$$

Addition (and subtraction) of polynomials is simply a matter of removing grouping symbols and combining like terms. For example,

$$(2x^2 - 3) + (x - 4) - (5x^2 - 1) = 2x^2 - 3 + x - 4 - 5x^2 + 1$$
$$= -3x^2 + x - 6 \qquad \text{Notice } -(5x^2 - 1) = -1(5x^2 - 1) = -5x^2 + 1.$$

The distributive property, along with the other real number properties and the first rule of exponents, give us procedures for multiplying polynomials. For example,

$$3x^3y^2(2x^2 - 7y^3) = 3x^3y^2(2x^2) - 3x^3y^2(7y^3) = 6x^5y^2 - 21x^3y^5$$

Keep in mind that variables in equivalent expressions may stand not only for numbers but for other variables, expressions, or polynomials as well. Hence we can apply the distributive property in multiplying $(2x + 5)(x + 3)$ as follows:

$$(2x + 5)(x + 3) = 2x(x + 3) + 5(x + 3)$$
$$(B + C) \cdot A \quad = B \cdot A \quad + C \cdot A$$

Note that we let A stand for the binomial $x + 3$ in applying the distributive property. The problem is still unfinished, for we must now apply the distributive property again and then combine terms:

$$(2x + 5)(x + 3) \qquad \text{Distribute } x + 3.$$
$$= (2x)(x + 3) + (5)(x + 3) \qquad \text{Apply the distributive property again.}$$
$$= (2x)x + (2x)3 + (5)x + (5)3 \qquad \text{Simplify and combine like terms.}$$
$$= 2x^2 + 6x + 5x + 15$$
$$= 2x^2 + 11x + 15$$

To multiply two polynomials, multiply each term of one polynomial by each term of the other.

Example 1 Multiply the following:
(a) $(3a - 2)(2a + 5)$ (b) $(x + y + z)(x + y - z)$

Solution (a) $(3a - 2)(2a + 5) = (3a)(2a + 5) - 2(2a + 5) = 6a^2 + 15a - 4a - 10$
$$= 6a^2 + 11a - 10$$

(b) $(x + y + z)(x + y - z) = x(x + y - z) + y(x + y - z) + z(x + y - z)$
$$= x^2 + xy - xz + xy + y^2 - yz + xz + yz - z^2$$
$$= x^2 + 2xy + y^2 - z^2$$

We can now develop some procedures for multiplying polynomials quickly as long as we keep in mind why the process works. Our emphasis will be on quick procedures for binomial multiplication because such products occur frequently in algebra.

General Forms

1. $(x + a)(x + b) = x^2 + (a + b)x + ab$
2. $(ax + b)(cx + d) = acx^2 + (ad + bc)x + bd$

Example 2   Perform the following operations.
(a) $(x + 3)(x + 5)$ (b) $(x^2y - 3s^2)(x^2y - 4s^2)$

Solution   (a) $(x + 3)(x + 5) = x^2 + (3 + 5)x + 5 \cdot 3$

$$= x^2 + 8x + 15 \qquad \text{Using general form 1}$$

(b) $(x^2y - 3s^2)(x^2y - 4s^2)$ Using general form 2

$$= (x^2y)^2 + (-3 - 4)(x^2y)(s^2) + (-3)(-4)(s^2)^2$$

$$= x^4y^2 - 7x^2ys^2 + 12s^4$$

Special products are specific products of binomials that can be derived from the general forms (which, in turn, are derived from the distributive property).

Special Products

1. $(a + b)(a - b) = a^2 - b^2$ Difference of two squares
2. $(a + b)^2 = a^2 + 2ab + b^2$ Perfect square of a sum
3. $(a - b)^2 = a^2 - 2ab + b^2$ Perfect square of a difference

Special products are important in factoring; in many cases, the quickest way to factor an expression is to recognize it as a special product. In addition, recognizing and using special products can reduce the time needed for multiplication.

Example 3   Perform the following operations.
(a) $(2a - 7b)(2a + 7b)$ (b) $(2a - 7b)^2$ (c) $[(x + y) - 3]^2$
(d) $(x + y - 3)(x + y + 3)$ (e) $2x(x - 4)^2 - (x + 4)(x - 4)$

Solution   The first three problems can be worked by using the general forms or, more slowly, by using the distributive property. The quickest way, however, is to use special products:

(a) $(2a - 7b)(2a + 7b) = (2a)^2 - (7b)^2$ Special product 1: difference of squares

$$= 4a^2 - 49b^2$$

Keep in mind that when you square a binomial, you should get a middle term in the product.

(b) $(2a - 7b)^2 = (2a)^2 - 2(2a)(7b) + (7b)^2$ Special product 3: perfect square of a difference

$$= 4a^2 - 28ab + 49b^2$$

Study the differences between parts (a) and (b); also note their similarities.

(c) $[(x + y) - 3]^2$

This is a square of a trinomial. However, if we treat $x + y$ as the first term, and 3 as the second term, we can use special product 3, the square of the difference between two terms, as follows:

$[(x + y) - 3]^2$

Apply the square of a difference.

$= (x + y)^2 - 2(3)(x + y) + 3^2$

Then apply special product 2 to $(x + y)^2$ and multiply out $2(3)(x + y)$.

$$= x^2 + 2xy + y^2 - 6x - 6y + 9$$

(d) At first glance, $(x + y - 3)(x + y + 3)$ does not seem to be in special product or general form. However, we can regroup within parentheses to get it into special product or general form. This product is a difference of two squares with $x + y$ as the first term and 3 as the second term. We insert parentheses around $x + y$ to help see it more clearly.

$[(x + y) - 3][(x + y) + 3]$

Apply the difference of squares.

$= (x + y)^2 - 3^2$

Then apply special product 2 to $(x + y)^2$.

$$= x^2 + 2xy + y^2 - 9$$

(e) We follow the same order of operations discussed in Section 1.2 (that is, parentheses, exponents, multiplication and division, and, finally, addition and subtraction).

$2x(x - 4)^2 - (x + 4)(x - 4)$

Square the binomial $(x - 4)$ and find the difference of squares $(x + 4)(x - 4)$.

$= 2x(x^2 - 8x + 16) - (x^2 - 16)$

Distribute $2x$ and subtract $x^2 - 16$.

$= 2x^3 - 16x^2 + 32x - x^2 + 16$

$$= 2x^3 - 17x^2 + 32x + 16$$

Note that we multiplied $(x + 4)(x - 4)$ before we subtracted. It is a good habit to retain parentheses to remind us that we are subtracting the entire expression $(x^2 - 16)$. ∎

Example 4 | In terms of x, find the area of the shaded region in Figure 1.8.

Figure 1.8

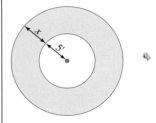

Solution | To find the area of the shaded region in the Figure 1.8, we have to find the difference between the larger circular region and the smaller circular region. The area of a circle is πr^2 where r is the radius; for the smaller circle, the radius is 5 ft and for the larger circle, the radius is $x + 5$. Hence,

$$A_{larger\ circle} - A_{smaller\ circle}$$
$$\pi(x + 5)^2 - \pi(5^2)$$
$$= \pi(x^2 + 10x + 25) - 25\pi$$
$$= \pi x^2 + 10\pi x + 25\pi - 25\pi$$
$$= \boxed{\pi x^2 + 10\pi x}\ \text{sq ft}$$

Example 5

An open box is to be made from a 1-ft by 3-ft rectangular piece of cardboard by cutting out identical squares of length x from each of the corners of the sheet and then folding up the sides on the dashed lines, as illustrated in Figure 1.9. Find the volume of the box in terms of x.

Figure 1.9

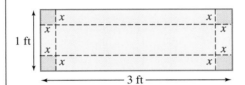

Solution

The volume of a box is given by $V = lwh$, where l = length, w = width, and h = height.

Figure 1.10

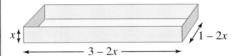

By Figure 1.9, if the cardboard is folded up along the dashed lines, then its height is x. The length of the box is $3 - 2x$, because we are eliminating x from each side of the cardboard. For this same reason, the width of the box is $1 - 2x$ (see Figure 1.10). Hence,

$$V = lwh$$
$$= (3 - 2x)(1 - 2x)x$$
$$= (3 - 8x + 4x^2)x$$
$$= \boxed{3x - 8x^2 + 4x^3}\ \text{cubic feet}$$

Factoring

To factor a polynomial is to rewrite it as a product of polynomials. The distributive property gives us a method for factoring polynomials as well as for multiplying polynomials.

$$\underset{\leftarrow\ Factoring}{\overset{Multiplying \rightarrow}{a(b + c) = ab + ac}}$$

Unless otherwise noted, we will be factoring over the integers; that is, all polynomial factors should have integer coefficients.

The most basic type of factoring is factoring out the greatest common factor. For example, the greatest common monomial factor of $24x^2y^3 - 16xy^3 - 8y^4$ is $8y^3$ because $8y^3$ is the greatest factor common to *all three terms*, $24x^2y^3$, $-16xy^3$, and $-8y^4$. Therefore,

$$24x^2y^3 - 16xy^3 - 8y^4 = (8y^3)(3x^2) + (8y^3)(-2x) + (8y^3)(-y)$$

By the distributive property we get

$$= 8y^3(3x^2 - 2x - y)$$

We can generalize common factoring to more complex expressions, as illustrated next.

Example 6

Factor the following completely:

(a) $3x(y - 4) + 2(y - 4)$ **(b)** $(x + 2)^2 + (x + 2)$

(c) $4x(x - 4)^2 - 2(2x^2 + 1)(x - 4)$

Solution

(a) To factor an expression such as $3x(y - 4) + 2(y - 4)$, we note that $y - 4$ is common to both expressions, $3x(y - 4)$ and $2(y - 4)$, and therefore can be factored out, just as we would factor A from $3xA + 2A$.

$$3x \cdot A \quad + 2 \cdot A \quad = \quad A \cdot (3x + 2)$$

$$3x(y - 4) + 2(y - 4) = \quad (y - 4)(3x + 2)$$

(b) To factor $(x + 2)^2 + (x + 2)$, we note that $x + 2$ is the factor common to both $(x + 2)^2$ and $x + 2$. This is like factoring $A^2 + A$ to get $A(A + 1)$.

$$A^2 \quad + \quad A \quad = \quad A \cdot [\quad A \quad + 1]$$

$$(x + 2)^2 + (x + 2) = (x + 2)[(x + 2) + 1] \qquad \text{Then simplify } [(x + 2) + 1].$$

$$= \quad (x + 2)(x + 3)$$

(c) Again, to factor $4x(x - 4)^2 - 2(2x^2 + 1)(x - 4)$, first observe that $2(x - 4)$ is common to each term. Hence we can factor out $2(x - 4)$, and we are left with $2x(x - 4)$ and $-(2x^2 + 1)$.

$$4x(x - 4)^2 - 2(2x^2 + 1)(x - 4) \qquad \text{Factor out } 2(x - 4).$$

$$= 2(x - 4)[2x(x - 4) - (2x^2 + 1)] \qquad \text{Next, simplify inside the brackets.}$$

$$= 2(x - 4)[2x^2 - 8x - 2x^2 - 1]$$

$$= 2(x - 4)[-8x - 1] \qquad \text{We can factor out } -1 \text{ from } -8x - 1 \text{ to get}$$

$$= \quad -2(x - 4)(8x + 1)$$

The greatest common factor of a polynomial is not always apparent. We often have to take a step or two to put the polynomial in factorable form. First, we may have to group the terms and then factor the groups before it becomes clear what can be factored from the entire expression. For example, in factoring

$$ax + ay + bx + by$$

you can see that there is no common monomial we can factor from *all four terms*. However, if we group the terms by pairs and factor the pairs, we can then factor a binomial from each term as follows:

$$ax + ay + bx + by \qquad \text{Group the pairs together.}$$

$$= ax + ay \quad + \quad bx + by \qquad \text{Then factor each pair.}$$

$$= a(x + y) \quad + \quad b(x + y) \qquad \text{Factor } x + y \text{ from each group.}$$

$$= (x + y)(a + b)$$

Grouping and factoring *parts* of a polynomial to factor the polynomial itself is called *factoring by grouping*.

Example 7

Factor the following completely.

(a) $3xb - 2b + 15x - 10$ (b) $x^3 + 2x^2 - 3x - 6$

Solution

(a) $3xb - 2b + 15x - 10$ There is no factor common to all terms, so group in pairs and factor each pair.

$$= (3xb - 2b) + (15x - 10)$$

$$= b(3x - 2) + 5(3x - 2)$$ Now factor $3x - 2$ from each group.

$$= (3x - 2)(b + 5)$$

(b) When a negative sign appears between the pairs of binomials we intend to group, we occasionally have to factor out a negative factor so that the binomial factors are identical. For example, in factoring $x^3 - 2x^2 - 3x - 6$, we would have to factor out -3 from $-3x - 6$:

$x^3 - 2x^2 - 3x - 6$ First separate the pairs.

$$= x^3 - 2x^2 - 3x - 6$$ Factor out x^2 from the first pair, and -3 from the second pair. [Be careful: Check with multiplication.]

$$= x^2(x + 2) - 3(x + 2)$$ Factor out $x + 2$ to get

$$= (x + 2)(x^2 - 3)$$

Factoring Trinomials

From basic algebra, we should already be familiar with factoring polynomials of the form $Ax^2 + Bx + C$. The simplest cases are those with leading coefficient $A = 1$, that is, when the trinomial is of the form $x^2 + Bx + C$. The product

$$(x + p)(x + q) = x^2 + (p + q)x + pq$$

indicates that if the trinomial $x^2 + Bx + C$ can be factored into two binomials, $(x + p)$ and $(x + q)$, then B is the sum of p and q and C is the product of p and q. Thus, all we need to find are two factors of C that sum to B.

For example, to factor $x^2 + 8x + 12$, we need to find two factors of $+12$ that add up to 8. If we systematically check the factors of 12 ($1 \cdot 12$; $2 \cdot 6$; $3 \cdot 4$), we arrive at $+6$ and $+2$ as the factors of $+12$ that add up to 8. Thus,

$$x^2 + 8x + 12 = (x + 6)(x + 2)$$

On the other hand, factoring the general trinomial $Ax^2 + Bx + C$ requires more trial and error. We saw in the previously that

$$(ax + b)(cx + d) = acx^2 + (ad + bc)x + bd$$

which is obviously more complex than $(x + p)(x + q)$; it is complicated by the coefficients of the x-terms in the binomials. Note the relationships between the constants a, b, c, and d, and the coefficients of the trinomial product.

Therefore, if $Ax^2 + Bx + C$ were to factor into two binomials, A would be the product of the x-coefficients in the binomials ($a \cdot c$), C would be the product of the numerical term coefficients in the binomials ($b \cdot d$), and B would be the result of the interaction ($ad + bc$) of the four coefficents a, b, c, and d.

For example, to factor $2x^2 + 5x + 3$, we would proceed with the following analysis:

The possible factors of 2 are $2 \cdot 1$: These must be the binomial x-term coefficients. The possible factors of 3 are $3 \cdot 1$: These must be the binomial numerical term coefficients. We arrive at two possible answers:

$$(2x + 1)(x + 3) \quad \text{or} \quad (2x + 3)(x + 1)$$

Multiplying these possible factorizations we get:

$$2x^2 + 7x + 3 \quad \text{or} \quad 2x^2 + 5x + 3$$

Note that both first and last terms are identical. The middle term indicates $(2x + 3)(x + 1)$ is the answer. Hence, $2x^2 + 5x + 3 = (2x + 3)(x + 1)$.

Example 8 Factor the following completely.
(a) $3y^3 - 6y^2 - 105y$ (b) $12a^3 + 2a^2 - 4a$

Solution (a) $3y^3 - 6y^2 - 105y$

Don't forget: Always factor the greatest common factor first.

$$= 3y(y^2 - 2y - 35)$$

Next, factor $y^2 - 2y - 35$.

$$= \boxed{3y(y - 7)(y + 5)}$$

(b) $12a^3 + 2a^2 - 4a$

Factor the common monomial, $2a$, first.

$$= 2a(6a^2 + a - 2)$$

Factor $6a^2 + a - 2$ into $(2a - 1)(3a + 2)$.

$$= \boxed{2a(2a - 1)(3a + 2)}$$ ∎

Another approach to factoring trinomials is to *factor by grouping* as described next.

Factoring Trinomials by Grouping

To factor $Ax^2 + Bx + C$ we would:

For example:
To factor $12x^2 - 17x - 5$

1. Find the product AC:

$AC = (12)(-5) = -60$

2. Find two factors of AC that sum to B:

-20 and $+3$ are two factors of -60 which sum to -17.

3. Rewrite the middle term as a sum of terms whose coefficients are the factors found in step 2.

Rewrite $-17x$ as $-20x + 3x$. Thus, $12x^2 - 17x - 5$
$= 12x^2 - 20x + 3x - 5$

4. Factor by grouping.

$= 4x(3x - 5) + 1(3x - 5)$
$= (3x - 5)(4x + 1)$

Although we still have to look for factors (of a number larger than A or C), this process is usually a bit more efficient than the trial-and-error process.

Example 9 Factor $15a^2 - ab - 6b^2$ completely by grouping.

Solution $15a^2 - ab - 6b^2$

The factors of $(15)(-6) = -90$ that add up to -1 are -10 and $+9$; hence, we rewrite the middle term, $-ab$ as $-10ab + 9ab$.

$$= 15a^2 - 10ab + 9ab - 6b^2$$

Factor by grouping to get

$$= 5a(3a - 2b) + 3b(3a - 2b)$$

$$= \boxed{(3a - 2b)(5a + 3b)}$$ ∎

If there are no pairs of factors of AC that add up to the coefficient of the middle term, then the trinomial does not factor into two binomials.

Factoring Using Special Products

If we try to factor $9x^2 + 30x + 25$ by trial and error, it may take a while to arrive at the correct factorization. However, if we recognize this polynomial as a form of a special product, we can cut down our labor a bit.

We again list the special products that we have had so far and add two more.

Special Products

1. $a^2 - b^2 = (a + b)(a - b)$	Difference of two squares
2. $a^2 + 2ab + b^2 = (a + b)^2$	Perfect square of a sum
3. $a^2 - 2ab + b^2 = (a - b)^2$	Perfect square of a difference
4. $a^3 - b^3 = (a - b)(a^2 + ab + b^2)$	Difference of two cubes
5. $a^3 + b^3 = (a + b)(a^2 - ab + b^2)$	Sum of two cubes

The first special product, $a^2 - b^2$, is called the difference of two squares. For example, we could rewrite the binomial $4x^2 - 9y^2$ as the difference of two squares: $(2x)^2 - (3y)^2$, which factors into $(2x - 3y)(2x + 3y)$. Special products 2 and 3 are called perfect squares (just as $36 = 6^2$ is a perfect square).

If a trinomial is factorable, it can be factored by the previous methods discussed in this section. However, if a trinomial can be recognized as a special product, it can be factored quickly and with less effort.

For example, when we see a trinomial such as $16x^2 - 40xy + 25y^2$, there are quite a few possible combinations of factors to check if we use the trial-and-error method or the factoring by grouping method. Our experience with special products, however, makes us suspicious when we see two perfect square terms, $16x^2$ and $25y^2$, in the trinomial. We would immediately check to see whether it is a perfect square by choosing $(4x - 5y)(4x - 5y)$ as the first possible factorization. Multiplication will confirm our suspicion that this is indeed the proper factorization of $16x^2 - 40xy + 25y^2$.

Special products 4 and 5 are called the difference and sum, respectively, of two cubes, which both factor as a product of a binomial and a trinomial.

Note the similarities and differences between factoring the sum and difference of two cubes. Keep in mind $a^3 - b^3$ is *not* the same as $(a - b)^3$ which, when multiplied out, will have middle terms. Also, neither $a^2 - ab + b^2$ nor a $a^2 + ab + b^2$ (the trinomial factors of the sum and difference of cubes) will factor into two binomials. (Don't confuse it with the perfect square $a^2 + 2ab + b^2$, which has binomial factors.)

Example 10 Factor the following completely.
(a) $9x^2 + 30x + 25$ **(b)** $x^4 - y^4$ **(c)** $27x^3 + y^3$ **(d)** $24x^4 - 3x$

Solution **(a)** $9x^2 + 30x + 25 = (3x)^2 + 2(3x)(5) + 5^2$ Perfect square of a sum

$$= (3x + 5)^2$$

(b) $x^4 - y^4$ Rewrite as a difference of two squares.

$$= (x^2)^2 - (y^2)^2$$ Factor: difference of squares.

$$= (x^2 - y^2)(x^2 + y^2)$$ Now factor $x^2 - y^2$.

$$= (x - y)(x + y)(x^2 + y^2)$$

(c) $27x^3 + y^3$ Rewrite as a sum of cubes.

$$= (3x)^3 + y^3$$ Factor: sum of two cubes.

$$= (3x + y)[(3x)^2 - (3x)(y) + y^2]$$ Simplify inside brackets.

$$= (3x + y)(9x^2 - 3xy + y^2)$$

(d) $24x^4 - 3x$ Factor out the common monomial, $3x$, first.

$$= 3x[8x^3 - 1]$$ Now rewrite $8x^3 - 1$ as a difference of cubes.

$$= 3x[(2x)^3 - 1^3]$$ Factor as a difference of cubes.

$$= 3x(2x - 1)[(2x)^2 + (2x)(1) + 1^2]$$ Simplify inside brackets.

$$= 3x(2x - 1)(4x^2 + 2x + 1)$$

We will now apply what we know about factoring special products to more complex expressions. For example, we would factor $(x + y)^2 - (r + s)^2$ by first recognizing that it is a difference of two squares. Hence,

$$A^2 \ - \ B^2 \ = (\ A \ - \ B \ )(\ A \ + \ B \ )$$
$$(x + y)^2 - (r + s)^2 = [(x + y) - (r + s)][(x + y) + (r + s)]$$

Then simplify inside the brackets: $(x + y - r - s)(x + y + r + s)$.

As before, we can factor more complex expressions by recognizing a given form as a special product.

Example 11

Factor completely:

(a) $(x - 3)^2 - 4$ **(b)** $x^3 - x^2 - 4x + 4$ **(c)** $x^2 - 4xy - 4y^2 - 9$

Solution

Try part (a) by multiplying out first then factoring. Which is easier?

(a) $(x - 3)^2 - 4$ is the difference of two squares

$$= (x - 3)^2 - 2^2$$

$$= [(x - 3) - 2][(x - 3) + 2]$$ Special Product 1. Then simplify inside the brackets.

$$= (x - 5)(x - 1)$$

(b) $x^3 - x^2 - 4x + 4$ First separate the pairs.

$$= x^3 - x^2 \ - \ 4x + 4$$ Factor out x^2 from the first pair, and -4 from the second pair. [Be careful: Check with multiplication.]

$$= x^2(x - 1) - 4(x - 1)$$ Factor out $x - 1$ to get

$$= (x^2 - 4)(x - 1)$$ Now factor $x^2 - 4$.

$$= (x - 2)(x + 2)(x - 1)$$

(c) If we tried to factor $x^2 - 4xy + 4y^2 - 9$ by grouping in pairs, we would find no common factors. But suppose we group together the first three terms.

$$x^2 - 4xy + 4y^2 - 9 = (x^2 - 4xy + 4y^2) - 9$$ Note that $x^2 - 4xy + 4y^2$ is the perfect square $(x - 2y)^2$, and $9 = 3^2$.

$$= (x - 2y)^2 - (3)^2$$ A difference of two squares.

$$= [(x - 2y) - 3][(x - 2y) + 3]$$

$$= (x - 2y - 3)(x - 2y + 3)$$

In general, we offer the following advice for factoring polynomials:

1. Always factor out the greatest common factor first.
2. If the polynomial to be factored is a binomial, then it may be a difference of two squares or a sum or difference of two cubes.
3. If the polynomial to be factored is a trinomial, then (1) if two of the three terms are perfect squares, the polynomial may be a perfect square or (2) otherwise the polynomial may be one of the general forms.
4. If the polynomial to be factored consists of four or more terms, then try factoring by grouping.

1.3 Exercises

In Exercises 1–32, perform the operations and express your answers in simplest form.

1. $(3x^2y)(-2xy^2)$
2. $(-7x^2y)(3xy^3)$
3. $(-3xy)^2(5xy)$
4. $(6xy^2)^2(-3x^2y^3)^2$
5. $(2x - 3)(3x + 2)$
6. $(5x - 9)(3x + 2)$
7. $(2x - 7)(3x + 1)$
8. $(2x - 5)(7x + 3)$
9. $(3x - 2)(2x^2 - 3x + 1)$
10. $(2x - 1)(4x^2 + 2x + 1)$
11. $(5x + 3)(25x^2 - 15x + 9)$
12. $(2x + 3)(x^2 - 2x + 1)$
13. $(x - 3)(x - 2)(x + 1)$
14. $(x - 4)(x + 5)(x - 1)$
15. $(x - 3y)^2$
16. $(x - 3y)(x + 3y)$
17. $(x^2 - 7)(x^2 + 7)$
18. $(x^2 + 7)^2$
19. $(2x - 3y)^2$
20. $(2x - 3y)(2x + 3y)$
21. $(x - y + 7)(x - y - 7)$
22. $(2x + 3 - y)(2x + 3 + y)$
23. $(x - y + 2)^2$
24. $(x + y + 3)^2$
25. $(x - 2)^2 - (x - 2)(x + 2)$
26. $(x - 2)(x + 2) - (x - 2)^2$
27. $5y(y - 5)^2 - (y - 5)(y + 5)$
28. $6x(x - 3)(x + 3) - (x - 3)^2$
29. $2(x - 3)^2 - 2x^2$
30. $5(x - 1)^2 - 5x^2$
31. $2(x + h)^2 + 1 - (2x^2 + 1)$
32. $3(x + h)^2 + 2(x + h) - 2 - (3x^2 + 2x - 2)$
33. In terms of x, find the area of the shaded regions given in the figures in parts (a)–(d).

(a)
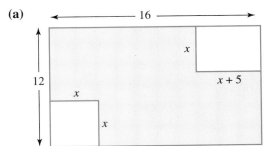

(b)
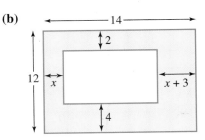

(c)
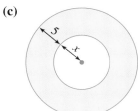

(d)
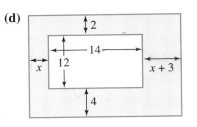

34. An open box is to be made from a 2-ft by 3-ft rectangular piece of cardboard by cutting out identical squares of length x from each of the corners of the sheet and then folding up the sides on the dashed lines (see Example 5). Find the volume of the box in terms of x.

35. An open box is made by cutting squares off the corners of a 2-ft by 3-ft rectangular piece of cardboard and then folding up on the dashed line (see Example 5). If the squares to be cut out have length x, find the *surface area* of the box in terms of x.

36. A rectangular garden is surrounded by a path of uniform width 2 feet. If the length of the garden is twice the width, express the total area of the path and garden in terms of the width of the garden.

In Exercises 37–38, find the area and perimeter of the given figure in terms of x. All the arcs are semicircles.

37.

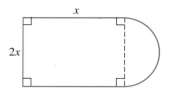

38.

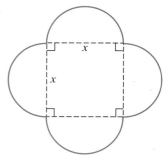

In Exercises 39–40, find the surface area of the rectangular solid in terms of x.

39.

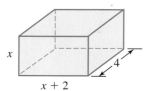

40.

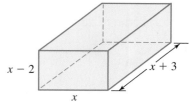

In Exercises 41–70, factor each expression as completely as possible.

41. $x^2 + x - 20$

42. $6x^2 - 7x - 3$

43. $12x^2 + 10x - 12$

44. $3x^3 - 3x^2 - 6x$

45. $3a(b - 2) - (b - 2)$

46. $(x - 3)^2 - (x - 3)$

47. $5x(x - 1)^2 + 10(x - 1)^3$

48. $3x(x - 2) + 2(x - 2)^2$

49. $3x(x - 2)(2x^2 + 1) - 6(x + 3)(2x^2 + 1)$

50. $4x(x - 2)^2(x - 1) + 8(x - 2)(x - 1)$

51. $ax + bx - 2a - 2b$ **52.** $x^3 - 5x - 2x^2 + 10$

53. $x^3 - 5x - 3x^2 + 15$ **54.** $x^3 - 7x - 2x^2 + 14$

55. $x^2 - 9$ **56.** $x^4 - 16$

57. $x^2 - 8x + 16$ **58.** $4x^2 + 4xy + y^2$

59. $8x^3 - 1$ **60.** $27x^3 + 8$

61. $81x^3 - 24$ **62.** $64x - 27xy^3$

63. $x^3 + 3x^2 - 16x - 48$ **64.** $x^3 - 3x^2 - 25x + 75$

65. $x^4 - 10x^2 + 24$ **66.** $2x^3 - 50x + 2x^2 - 50$

67. $x^2 - 2xy + y^2 - 16$ **68.** $x^2 + 4x + 4 - y^2$

69. $a^5 - a^3 - a^2 + 1$ **70.** $x^5 - x^2 - 4x^3 + 4$

71. If a race has n entrants, the number of possible different first-, second-, and third-place finishes (excluding ties) is given by the formula $P = n^3 - 3n^2 + 2n$. Factor this polynomial.

72. The volume of a cylindrical shell (see the accompanying figure) is given by the formula $V = \pi R^2 h - \pi r^2 h$. Factor this polynomial.

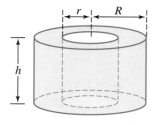

1.4   Rational Expressions

An algebraic fraction is a quotient of two algebraic expressions, $\frac{a}{b}$ (where $b \neq 0$). We define a **rational expression** as a quotient of two polynomials, $\frac{p}{q}$, provided the denominator is not the zero polynomial. (Remember the zero polynomial is simply 0.)

But even if the denominator is not the zero polynomial, we must still be careful about division by zero; a nonzero polynomial can have a value of zero when we substitute certain values for the variable. For example, $\frac{3x - 4}{2x - 1}$ is a rational expression with the nonzero polynomial $2x - 1$ in the denominator. However, because $2x - 1$ is zero when $x = \frac{1}{2}$, the expression $\frac{3x - 4}{2x - 1}$ is not defined for $x = \frac{1}{2}$. For the same reason, the expression $\frac{x + y}{x - y}$ is undefined when $x = y$.

Equivalent Fractions

We already defined what we meant by equivalent fractions in Section 1.2. We also mentioned in that section that since we require fractions to be reduced to lowest terms, we can observe their equivalence by inspection. Reducing fractions to lowest terms requires use of the fundamental principle of fractions:

$$\frac{a \cdot k}{b \cdot k} = \frac{a}{b} \quad (b, k \neq 0)$$

Again, a fraction reduced to lowest terms or written in simplest form is a fraction that has no factors (other than ± 1) common to both its numerator and denominator. This requires us to factor both numerator and denominator and then divide out factors common to the numerator and denominator; for example,

$$\frac{x^2 - y^2}{(x - y)^2} = \frac{(x - y)(x + y)}{(x - y)(x - y)} = \frac{x + y}{x - y} \qquad \text{First factor, then reduce.}$$

Remember, the fundamental principle of fractions allows us to reduce using common **factors**, not terms.

Example 1   Express the following in simplest form: $\dfrac{5(x^2 + 2)^2 - 5x(x^2 + 2)(2x)}{(x^2 + 2)^4}$

Solution   Rather than starting this problem by performing operations to simplify the numerator, we begin by factoring the common factor of $5(x^2 + 2)$ from the numerator:

$$\frac{5(x^2 + 2)^2 - 5x(x^2 + 2)(2x)}{(x^2 + 2)^4} \qquad \text{Factor } 5(x^2 + 2) \text{ from the numerator.}$$

$$= \frac{5(x^2 + 2)[(x^2 + 2) - x(2x)]}{(x^2 + 2)^4} \qquad \text{Now simplify } [(x^2 + 2) - x(2x)].$$

$$= \frac{5(x^2 + 2)(2 - x^2)}{(x^2 + 2)^4} \qquad \text{Reduce by a factor of } (x^2 + 2).$$

$$= \frac{5(2 - x^2)}{(x^2 + 2)^3}$$

Try this example by simplifying the numerator first.

Approaching the problem in this way cuts down the labor of performing polynomial operations in the numerator. More important, however, if you simplify the numerator as a first step, you may not recognize the polynomial in the numerator as being factorable. ∎

Operations with Rational Expressions

We perform arithmetic operations with rational expressions as we would for rational numbers (see page 14).

Example 2 | Perform the operations. Express your answer in simplest form.

(a) $\dfrac{x + 2y}{x - y} \cdot \dfrac{y - x}{x - 2y}$

(b) $\dfrac{a^2 - ab + b^2}{a^2b - ab^2} \div \dfrac{a^3 + b^3}{a^3 + a^2b}$

Solution | (a) $\dfrac{x + 2y}{x - y} \cdot \dfrac{y - x}{x - 2y}$

$$= \dfrac{x + 2y}{x - y} \cdot \dfrac{(-1)(x - y)}{x - 2y}$$

Since $y - x = (-1)(x - y)$
Reduce.

$$= \dfrac{-1(x + 2y)}{x - 2y} \quad \text{or} \quad \dfrac{-x - 2y}{x - 2y}$$

(b) $\dfrac{a^2 - ab + b^2}{a^2b - ab^2} \div \dfrac{a^3 + b^3}{a^3 + a^2b}$

Change the divisor into its reciprocal and multiply.

$$= \dfrac{a^2 - ab + b^2}{a^2b - ab^2} \cdot \dfrac{a^3 + a^2b}{a^3 + b^3}$$

Factor and reduce.

$$= \dfrac{a^2 - ab + b^2}{ab(a - b)} \cdot \dfrac{a^2(a + b)}{(a + b)(a^2 - ab + b^2)} = \dfrac{a}{b(a - b)}$$

∎

Example 3 | Perform the operations: Express your answer in simplest form.

(a) $\dfrac{3}{7x^3y} - \dfrac{5}{3xy^2} + x$

(b) $\dfrac{5x - 10}{x - 4} + \dfrac{3x - 2}{4 - x}$

(c) $-\dfrac{4x(x + 1)}{(x^2 - 2)^3} + \dfrac{1}{(x^2 - 2)^2}$

Solution | (a) $\dfrac{3}{7x^3y} - \dfrac{5}{3xy^2} + x$

The LCD is $7 \cdot 3 \cdot x^3 \cdot y^2 = 21x^3y^2$.
Change to equivalent fractions with $21x^3y^2$ in the denominator.

$$= \dfrac{3(3y)}{7x^3y(3y)} - \dfrac{5(7x^2)}{(3xy^2)(7x^2)} + \dfrac{x(21x^3y^2)}{21x^3y^2}$$

$$= \dfrac{3(3y) - 5(7x^2) + x(21x^3y^2)}{21x^3y^2}$$

Place numerators over the LCD, $21x^3y^2$, and combine.

Why can't this function be reduced further?

$$= \dfrac{9y - 35x^2 + 21x^4y^2}{21x^3y^2}$$

Your last step should be to simplify (reduce) your answer. In this case, the answer is already in simplest form.

(b) $\dfrac{5x - 10}{x - 4} + \dfrac{3x - 2}{4 - x}$

NOTE: Since $x - 4$ is the negative of $4 - x$, we multiply the numerator and denominator of the second fraction by -1.

$$= \frac{5x - 10}{x - 4} + \frac{(3x - 2)(-1)}{(4 - x)(-1)}$$

Now the denominators are the same.

$$= \frac{5x - 10}{x - 4} + \frac{-3x + 2}{x - 4}$$

Combine numerators.

$$= \frac{5x - 10 + (-3x + 2)}{x - 4} = \frac{2x - 8}{x - 4}$$

Factor and reduce.

$$= \frac{2(x - 4)}{x - 4} = \boxed{2}$$

(c) $-\dfrac{4x(x + 1)}{(x^2 - 2)^3} + \dfrac{1}{(x^2 - 2)^2}$ The LCD is $(x^2 - 2)^3$.

Next, rewrite each fraction as an equivalent fraction with the LCD in the denominator.

$$= -\frac{4x(x + 1)}{(x^2 - 2)^3} + \frac{1 \cdot (x^2 - 2)}{(x^2 - 2)^3}$$

Combine numerators; place over the common denominator.

$$= \frac{-4x(x + 1) + (x^2 - 2)}{(x^2 - 2)^3} = \boxed{\frac{-3x^2 - 4x - 2}{(x^2 - 2)^3}}$$

∎

We simplify complex fractions with algebraic expressions in the same way we simplified arithmetic complex fractions, as demonstrated in this example.

Example 4   Express as a simple fraction reduced to lowest terms.

$$\frac{\dfrac{3}{x + h} - \dfrac{3}{x}}{h}$$

Solution   Using method 2 for simplifying complex fractions as discussed in the previous section, we multiply the numerator and denominator by $x(x + h)$, which is the LCD of $\dfrac{3}{x + h}$ and $\dfrac{3}{x}$.

$$\frac{\dfrac{3}{x + h} - \dfrac{3}{x}}{h} = \frac{\left(\dfrac{3}{x + h} - \dfrac{3}{x}\right)}{h} \cdot \frac{x(x + h)}{x(x + h)}$$

Apply the distributive property.

$$= \frac{\left(\dfrac{3}{x + h}\right)\dfrac{x(x + h)}{1} - \left(\dfrac{3}{x}\right)\dfrac{x(x + h)}{1}}{h[x(x + h)]}$$

Reduce where appropriate.

$$= \frac{3x - 3(x + h)}{hx(x + h)}$$

Now simplify the numerator.

$$= \frac{-3h}{hx(x + h)}$$

Reduce by the common factor of h.

$$= \boxed{\frac{-3}{x(x + h)}}$$

∎

1.4 Exercises

In Exercises 1–12, simplify the fraction.

1. $\dfrac{(x-y)^2}{x^2-y^2}$

2. $\dfrac{x^2-x-6}{x^2-9}$

3. $\dfrac{2x^3-4x^2-6x}{4x^3-12x^2}$

4. $\dfrac{3x^3-30x^2+75x}{25x^2-x^4}$

5. $\dfrac{2x^3-2xy^2}{4x^4-8x^3y+4x^2y^2}$

6. $\dfrac{3x^3-24}{5x-10}$

7. $\dfrac{4x(x+2)^2-2x^2(2)(x+2)}{(x+2)^4}$

8. $\dfrac{(x^2-5)(4x)-2x^2(x^2-5)}{(x^2-5)^2}$

9. $\dfrac{3(x+h)+1-(3x+1)}{h}$

10. $\dfrac{(x+h)^2+2(x+h)-3-(x^2+2x-3)}{h}$

11. $\dfrac{(x^2-4)^2(-4)-(-4x)(4x)(x^2-4)}{(x^2-4)^4}$

12. $\dfrac{(x^2-3)^2(-6x)-(-3x^2-9)(4x)(x^2-3)}{(x^2-3)^4}$

In Exercises 13–30, perform the operations and express each answer in simplest form.

13. $\dfrac{6x^2-7x-3}{4x^2-12x+9}\cdot\dfrac{2x-3}{3x^2-5x-2}$

14. $\dfrac{3x-3}{(2x-5)^2}\div\dfrac{2x^2+3x-5}{4x^2-25}$

15. $\dfrac{5x-1}{5x+2}\div\dfrac{25x^2-10x+4}{125x^3+8}$

16. $\dfrac{x^2+2x-3}{2x^3+3x^2-5x}\cdot\dfrac{6x^3+15x^2}{9x^2-9x-108}$

17. $\dfrac{27x^3-8}{9x^2-4}\div(9x^2+6x+4)$

18. $\dfrac{6x^2-8x}{3x^3-x^2-4x}\cdot\dfrac{4x^4-8x^3-12x^2}{2x-6}$

19. $\dfrac{5}{9x^2y^3}-\dfrac{2}{15xy^3}$

20. $\dfrac{2}{3x^2}+\dfrac{4}{5x^3}+x$

21. $\dfrac{x}{x-y}-\dfrac{x}{y}$

22. $\dfrac{2y+10}{3-y}+\dfrac{y+7}{y-3}$

23. $\dfrac{2a+6}{a^2+5a+6}+\dfrac{5a+1}{a+2}$

24. $\dfrac{x}{x-2}+\dfrac{2}{x^2-4x+4}$

25. $\dfrac{3x}{x-1}+\dfrac{x+3}{x-2}+2x-3$

26. $\dfrac{3y-4}{y-5}+\dfrac{4y}{10+3y-y^2}$

27. $\dfrac{x+7}{x+5}-\dfrac{x}{x-3}+\dfrac{16}{x^2+2x-15}$

28. $\dfrac{3x+2}{x^2-2x+1}-\dfrac{7x-3}{x^2-1}+\dfrac{5}{x+1}$

29. $-\dfrac{x^2+1}{(x-2)^2}+\dfrac{2x}{x-2}$

30. $\dfrac{2}{(x^2-3)^2}-\dfrac{8x^2}{(x^2-3)^3}$

In Exercises 31–42, express each as a simple fraction reduced to lowest terms.

31. $\dfrac{1-\frac{1}{x^2}}{\frac{x-1}{x}}$

32. $\dfrac{x+\frac{2}{xy^2}}{\frac{1}{x}+2}$

33. $\dfrac{\frac{1}{x+3}-\frac{1}{x+1}}{2}$

34. $\dfrac{\frac{3}{x-1}-\frac{3}{x-4}}{3}$

35. $\dfrac{\frac{5}{x+h}-\frac{5}{x}}{h}$

36. $\dfrac{\frac{3}{(x+h)^2}-\frac{3}{x^2}}{h}$

37. $\dfrac{\frac{x+h}{x+h+2}-\frac{x}{x+2}}{h}$

38. $\dfrac{\frac{3-(x+h)}{x+h}-\frac{3-x}{x}}{h}$

39. $\dfrac{\frac{1}{y^2}-\frac{4}{x^2}}{\frac{x}{y^2}-\frac{4}{x}}$

40. $\dfrac{1-\frac{8}{x}}{x+6-\frac{16}{x}}$

41. $\dfrac{\frac{3x+1}{x-4}}{\frac{2x+1}{x+2}-\frac{x}{x-4}}$

42. $1-\dfrac{1}{1-\frac{1}{1-\frac{1}{x}}}$

Questions for Thought

43. Describe what happens to the value of the polynomial x^2+2x+3 as x gets larger and larger.

44. Describe what happens to the value of $\dfrac{x+3}{x-4}$ if:

(a) $x>4$, and x gets closer and closer to 4.

(b) $x<4$, and x gets closer and closer to 4.

1.5 Exponents and Radicals

We originally defined x^n as x being taken as a factor n times. Based on this definition, n must be a natural number. It does not make sense for n to be negative or zero. However, we can extend the definition of exponents to include 0 and negative exponents with the following:

Definition of Zero and Negative Exponents

Definition of Zero Exponent	Definition of Negative Exponents
$x^0 = 1 \qquad (x \neq 0)$	$x^{-n} = \dfrac{1}{x^n} \qquad (x \neq 0)$

NOTE: 0^0 is undefined.

As a result of the definition of negative exponents, we have $\dfrac{1}{x^{-n}} = \dfrac{1}{\dfrac{1}{x^n}} = x^n$.

We find that *changing the sign of the exponent of a power changes the power into its reciprocal.* Thus

$$x^{-3} = \frac{1}{x^3} \qquad \frac{1}{x^{-4}} = x^4 \qquad \frac{1}{2^{-2}} = 2^2 = 4 \qquad 10^{-5} = \frac{1}{10^5} = \frac{1}{100,000}$$

Do not confuse the sign of an exponent with the sign of the base. For example, -2 is a number 2 units to the left of zero on the number line, whereas $2^{-1} = \dfrac{1}{2}$ is a positive number, one-half unit to the right of zero, as shown in Figure 1.11. The rules for exponents are listed in the following box.

Figure 1.11

Rules for Integer Exponents

1. $x^n x^m = x^{n+m}$

2. $(x^n)^m = x^{nm}$

3. $(xy)^n = x^n y^n$

4. $\dfrac{x^n}{x^m} = x^{n-m} \qquad (x \neq 0)$

5. $\left(\dfrac{x}{y}\right)^n = \dfrac{x^n}{y^n} \qquad (y \neq 0)$

These rules can be used to shortcut the process of multiplying and dividing expressions involving exponents.

Example 1 Perform the operations and simplify; express each answer as a single simple fraction with positive exponents only.

(a) $\dfrac{(2a^{-2}b)^{-3}}{a^{-4}}$

(b) $\dfrac{a^{-2} - b^{-2}}{a^{-1} + b^{-1}}$

(c) $5(x - 6)^{-1} + 5x(x - 6)^{-2}$

Solution

(a) $\dfrac{(2a^{-2}b)^{-3}}{a^{-4}}$ Use rule 3 and rule 2.

$= \dfrac{2^{-3}a^6 b^{-3}}{a^{-4}}$ Use rule 4.

$= 2^{-3}a^{6-(-4)}b^{-3}$

$= 2^{-3}a^{10}b^{-3}$ Rewrite using positive exponents.

$= \dfrac{a^{10}}{2^3 b^3} = \dfrac{a^{10}}{8b^3}$

(b) Remember that the rules of exponents apply to factors, not terms. We must rewrite each *term* in the numerator and denominator using the definition of negative exponents.

$$\frac{a^{-2} - b^{-2}}{a^{-1} + b^{-1}} = \frac{\dfrac{1}{a^2} - \dfrac{1}{b^2}}{\dfrac{1}{a} + \dfrac{1}{b}}$$

Multiply the numerator and denominator of the complex fraction by a^2b^2. (Why?)

$$= \frac{\left(\dfrac{1}{a^2} - \dfrac{1}{b^2}\right) a^2b^2}{\left(\dfrac{1}{a} + \dfrac{1}{b}\right) a^2b^2} = \frac{b^2 - a^2}{ab^2 + a^2b} = \frac{(b - a)(b + a)}{ab(b + a)} = \boxed{\frac{b - a}{ab}}$$

(c) $5(x - 6)^{-1} + 5x(x - 6)^{-2}$

Rewrite each term using positive exponents.

$$= \frac{5}{x - 6} + \frac{5x}{(x - 6)^2}$$

The LCD is $(x - 6)^2$.

$$= \frac{5(x - 6)}{(x - 6)^2} + \frac{5x}{(x - 6)^2}$$

$$= \frac{5(x - 6) + 5x}{(x - 6)^2} = \boxed{\frac{10x - 30}{(x - 6)^2}}$$

Try part (c) by factoring out $5(x - 6)^{-2}$ from each term.

Scientific Notation

Scientific notation is a way of concisely writing very large or very small numbers. For example, the approximate mass of the earth written in standard decimal form is 5,980,000,000,000,000,000,000,000 kg. Using scientific notation, we can write this number as 5.98×10^{24}.

Definition of Scientific Notation

A number is said to be in **scientific notation** if it has the form $a \times 10^n$ *where* $1 \leq a < 10$ *and n is an integer.*

1.645×10^4 is written in scientific notation with $a = 1.645$ and $n = 4$. 6.8×10^{-5} is written in scientific notation with $a = 6.8$ and $n = -5$.

Converting from scientific to standard notation is straightforward: It is simply a matter of moving the decimal point right or left. For example,

$$1.642 \times 10^5 = 1.642 \times 100{,}000 = 164{,}200$$

$$7.3 \times 10^{-4} = 7.3 \times 0.0001 = 0.00073$$

The middle step may be bypassed by observing that the number of places the decimal point is moved is given by the absolute value of the exponent of 10; the direction the decimal point is moved is determined by the sign of the exponent.

Converting from standard to scientific notation takes a bit more thought, but again it is simply a matter of shifting decimal points right or left, as demonstrated in Example 2.

Example 2   Express the following in scientific notation:
(a) 78,964   **(b)** 0.00751

Solution   **(a)** 78,964

Scientific notation requires the number to have the following form:

$$= 7.8964 \times 10^n$$

What remains is to find the power of 10.

Because we moved the decimal point of 78,964 to the left 4 places, we obviously changed the value of the number. To restore the number to its original value, we multiply it by the power of 10 that will move it 4 places right; this is 10^{+4}. Hence,

$$78,964 = \boxed{7.8964 \times 10^4}$$

(b) 0.00751   In scientific notation, the number must have the following form:

$$= 7.51 \times 10^n$$ Because we moved the decimal point right 3 places, we need to mutiply it by the power of 10 that will move it back 3 places left: 10^{-3}.

$$0.00751 = \boxed{7.51 \times 10^{-3}}$$

In other words, the power of 10 restores the decimal point back to its original position.   ▮

If you tried using a calculator to compute $6{,}500{,}000 \times 950{,}000$, your calculator will probably display 6.175 E 12. When an answer is given in this form, it is usually because the answer has more digits than the calculator can display. 6.175 E 12 is the way the calculator expresses the number 6.175×10^{12}; likewise, a calculator display of 2.3852 E -15 means 2.3852×10^{-15}.

Often computations can be simplified using scientific notation. We first convert all numbers from standard to scientific notation and then perform the computations.

Example 3   If the mass of a particle is 4.3×10^{-7} grams, then is the mass of 5 million of these particles closest to 1 gram, 10 grams, 100 grams, or 1000 grams?

Solution   To find the mass of 5 million of these particles, we simply multiply the mass of one particle by 5 million $= 5{,}000{,}000$ as follows:

$$(4.3 \times 10^{-7})(5{,}000{,}000)$$ Rewrite 5,000,000 using scientific notation.

$$= (4.3 \times 10^{-7})(5 \times 10^6)$$ Separate out powers of 10.

$$= (4.3)(5) \times (10^{-7})(10^6) = 21.5 \times 10^{-1}$$

$$= 2.15 \text{ gram}$$

Of the options given, this is the closest to $\boxed{1 \text{ gram}}$.   ▮

Rational Exponents

Next we extend the definition of exponents even further to include rational number exponents. To do this, we assume that we want the rules for integer exponents also to apply to rational exponents and then use the rules to show us how to define a rational exponent. For example, how do we define $a^{1/2}$? Consider $9^{1/2}$:

If we apply rule 2 and square $9^{1/2}$, we get $(9^{1/2})^2 = 9^{2/2} = 9^1 = 9$.

So $9^{1/2}$ is a number that, when squared, yields 9.

There are two possible answers: 3 and -3, since squaring either number will yield 9. To avoid ambiguity, we define $a^{1/2}$ (called the principal square root of a) as the *nonnegative* quantity that, when squared, yields a. Thus $9^{1/2} = 3$.

We arrive at the definition of $a^{1/3}$ in the same way as we did for $a^{1/2}$: For example, if we cube $8^{1/3}$, we get $(8^{1/3})^3 = 8^{3/3} = 8^1 = 8$. Thus, $8^{1/3}$ is the number that, when cubed, yields 8. Since $2^3 = 8$ we have $8^{1/3} = 2$. Similarly $(-27)^{1/3} = -3$.

Do we have two choices for the value of $8^{1/3}$?

Thus we define $a^{1/3}$ (called the cube root of a) as the quantity that, when cubed, yields a.

We generalize with the following definition for real numbers a.

Definition of a Rational Exponent: $a^{1/n}$

If n is an odd positive integer, then

$$a^{1/n} = b \quad \text{if and only if} \quad b^n = a$$

If n is an even positive integer and $a \geq 0$ then

$$a^{1/n} = |b| \quad \text{if and only if} \quad b^n = a$$

$a^{1/n}$ is called the **principal nth root of a.** Hence, $a^{1/n}$ is the real number (nonnegative when n is even) that, when raised to the nth power, yields a.

Notice that with even roots we have to be careful about two things: (1) we want only the principal (nonnegative) root of the expression, and (2) since raising any real number to an even power will always yield a nonnegative number, the even root of a negative number cannot be a real number. Hence

$$(16)^{1/2} = 4 \quad since \quad 4^2 = 16 \qquad 27^{1/3} = 3 \quad since \quad 3^3 = 27$$

$$(-125)^{1/3} = -5 \quad since \quad (-5)^3 = -125 \qquad (-16)^{1/4} \quad \text{is not a real number}$$

$$\left(\frac{1}{81}\right)^{1/4} = \frac{1}{3} \quad since \quad \left(\frac{1}{3}\right)^4 = \frac{1}{81}$$

Thus far we have defined $a^{1/n}$, where n is a natural number. With some help from the second rule for exponents, we can define the expression $a^{m/n}$, where m and n are natural numbers and $\frac{m}{n}$ is reduced to lowest terms.

Definition of a Rational Exponent: $a^{m/n}$

If $a^{1/n}$ is a real number, then $a^{m/n} = (a^{1/n})^m$ (the nth root of a raised to the mth power).

We can also define negative rational exponents:

$$a^{-m/n} = \frac{1}{a^{m/n}} \qquad (a \neq 0)$$

Now that we have defined rational exponents, we assert that *the rules for integer exponents hold for rational exponents as well, provided the root is a real number*—that is, provided we avoid even roots of negative numbers.

By the exponent rules, we find $(a^{1/n})^m = (a^m)^{1/n}$. Hence we can view $a^{m/n}$ two ways:

If $a^{1/n}$ is a real number, $a^{m/n} = (a^{1/n})^m = (a^m)^{1/n}$.

Example 4   Evaluate the following:

(a) $27^{2/3}$ (b) $36^{-1/2}$ (c) $(-32)^{-3/5}$

Solution   In general, it is easier to find the root *before* raising to a power.

(a) $27^{2/3} = (27^{1/3})^2 = 3^2 = \boxed{9}$

```
27^(2/3)
              9
36^(-1/2)
     .166666667
(-32)^(-3/5)
          -.125
```

(b) $36^{-1/2}$ Begin by using the definition of a negative exponent.

$$= \frac{1}{36^{1/2}} = \frac{1}{6}$$ Note that the sign of the exponent has no effect on the sign of the base.

(c) $(-32)^{-3/5} = \dfrac{1}{(-32)^{3/5}}$ Rewrite using positive exponents.

$$= \frac{1}{[(-32)^{1/5}]^3}$$ Find the root first.

$$= \frac{1}{(-2)^3} = -\frac{1}{8}$$

One point of confusion that frequently arises involves the differences between negative exponents and fractional exponents: Negative exponents involve *reciprocals* of the base, whereas fractional exponents yield *roots* of the base. For example,

$$x^{-4} = \frac{1}{x^4}, \quad \text{whereas} \quad x^{1/4} = \text{the fourth root of } x$$

$$16^{-4} = \frac{1}{16^4} = \frac{1}{65{,}536}, \quad \text{whereas} \quad 16^{1/4} = 2$$

What are the differences among 4^{-2}, $4^{1/2}$, and $4^{-1/2}$?

As we stated earlier, we defined rational exponents so that the rules for integer exponents were also valid for rational exponents. Thus we simplify expressions involving rational exponents by applying the same rules we used for integer exponents. As with integer exponents, when we are asked to simplify an expression, the bases and exponents should appear as few times as possible.

Unless otherwise noted, we assume all variables represent positive real numbers.

Example 5   Perform the operations and simplify. Express your answer using positive exponents only.

(a) $(5a^{1/2})^2$ **(b)** $\left(\dfrac{a^{1/2}a^{-2/3}}{a^{1/4}}\right)^8$ **(c)** $2x(x^2+4)^{3/5}(x^2+4)^{-1/2}$

Solution   Make sure you can justify each step in the solutions.

Note that squaring $a^{1/2}$ does **not** yield $a^{1/4}$.

(a) $(5a^{1/2})^2 = 5^2(a^{1/2})^2 = 5^2 a^{(1/2)\cdot 2} = 25a$

(b) $\left(\dfrac{a^{1/2}a^{-2/3}}{a^{1/4}}\right)^8$ You may choose to simplify inside parentheses first; however, for this problem, if we apply rules 5 and 3 first, the computations with fractions are simpler.

$$= \frac{(a^{1/2})^8(a^{-2/3})^8}{(a^{1/4})^8} = \frac{a^4 a^{-16/3}}{a^2} = a^{4-(16/3)-2} = a^{-10/3} = \frac{1}{a^{10/3}}$$

(c) $2x(x^2+4)^{3/5}(x^2+4)^{-1/2} = 2x(x^2+4)^{\frac{3}{5}-\frac{1}{2}}$

$$= 2x(x^2+4)^{1/10}$$

Example 6 | Perform the operations and simplify the following: $(5a^{1/2} + 3b^{1/2})^2$

Solution | $(5a^{1/2} + 3b^{1/2})^2$ For squaring a binomial, we can use the perfect square special product.

$$= (5a^{1/2})^2 + 2(5a^{1/2})(3b^{1/2}) + (3b^{1/2})^2$$
$$= 5^2 a^{2/2} + 30a^{1/2} b^{1/2} + 3^2 b^{2/2}$$
$$= 25a + 30a^{1/2}b^{1/2} + 9b$$

In basic algebra we are used to factoring polynomials. We can use the same principles of factoring that we learned in algebra to factor expressions with terms containing rational exponents. For example, in factoring $x^3 + x^2$, we know that the greatest common factor is x^2, which is x raised to the smallest exponent appearing in the expression. Hence, $x^3 + x^2 = x^2(x + 1)$.

In the same way, we can factor $x + x^{1/3}$ noting that the greatest common factor is $x^{1/3}$, which is x raised to the *smallest* exponent appearing in the expression. Hence,

$$x + x^{1/3} = x^{1/3}(x^{2/3} + 1)$$ NOTE: $x = x^1 = x^{\frac{2}{3}+\frac{1}{3}} = x^{\frac{2}{3}}x^{\frac{1}{3}}$

Similarly,

$$x^{-1/2} - x^{3/2} = x^{-1/2}(1 - x^2)$$ NOTE: $x^{3/2} = x^{2-\frac{1}{2}} = x^2 x^{-\frac{1}{2}}$

Example 7 | Factor the following: $10x(x - 3)^{1/2} + 5(x - 3)^{3/2}$

Solution | In the expression $10x(x - 3)^{1/2} + 5(x - 3)^{3/2}$, each term has a common factor of $5(x - 3)^{1/2}$. So we can factor out this common factor and simplify the rest of the expression as follows:

$$10x(x - 3)^{1/2} + 5(x - 3)^{3/2}$$

Rewrite each term with $5(x - 3)^{1/2}$ as a factor.

$$= [5(x - 3)^{1/2}][2x] + [5(x - 3)^{1/2}][x - 3]$$

Factor out $5(x - 3)^{1/2}$ from each term.

$$= [5(x - 3)^{1/2}][2x + (x - 3)]$$

Simplify $[2x + (x - 3)]$.

$$= [5(x - 3)^{1/2}][3x - 3]$$

Factor $[3x - 3]$.

$$= [5(x - 3)^{1/2}][3(x - 1)]$$

$$= 15(x - 3)^{1/2}(x - 1)$$

Example 8 | Express as a single simple fraction reduced to lowest terms:

$$2x(x^2 - 2)^{1/2} + x^2(x^2 - 2)^{-1/2}$$

Solution | $2x(x^2 - 2)^{1/2} + x^2(x^2 - 2)^{-1/2}$ Rewrite each term using positive exponents.

$$= 2x(x^2 - 2)^{1/2} + \frac{x^2}{(x^2 - 2)^{1/2}}$$ The LCD is $(x^2 - 2)^{1/2}$.

$$= \frac{2x(x^2 - 2)^{1/2}(x^2 - 2)^{1/2}}{(x^2 - 2)^{1/2}} + \frac{x^2}{(x^2 - 2)^{1/2}}$$

$$= \frac{2x(x^2 - 2) + x^2}{(x^2 - 2)^{1/2}}$$

$$= \frac{2x^3 + x^2 - 4x}{(x^2 - 2)^{1/2}}$$

A second way to approach the same problem is to first factor the expression $x(x^2 - 2)^{-1/2}$ from each term:

$2x(x^2 - 2)^{1/2} + x^2(x^2 - 2)^{-1/2}$

Rewrite each term with $x(x^2 - 2)^{-1/2}$ as a factor.

Note that
$(x^2 - 2)^{-1/2}(x^2 - 2)$
$= (x^2 - 2)^{(-1/2)+1}$
$= (x^2 - 2)^{1/2}$

$= [x(x^2 - 2)^{-1/2}][2(x^2 - 2)] + [x(x^2 - 2)^{-1/2}][x]$

Factor out $x(x^2 - 2)^{-\frac{1}{2}}$.

$= [x(x^2 - 2)^{-1/2}][2(x^2 - 2) + x]$

Now rewrite using positive exponents.

$= \dfrac{x[2(x^2 - 2) + x]}{(x^2 - 2)^{1/2}}$

Simplify the numerator.

$= \dfrac{2x^3 + x^2 - 4x}{(x^2 - 2)^{1/2}}$

∎

Radical Expressions

Radical notation is an alternative way of writing an expression with rational exponents. That is, we define for real numbers a,

Definition of the *n*th Root of *a*

$$\sqrt[n]{a} = a^{1/n} \quad \text{where } n \text{ is a positive integer}$$

$\sqrt[n]{a}$ is also called the **principal *n*th root of *a*.** In $\sqrt[n]{a}$, n is called the **index** of the radical, the symbol $\sqrt{}$, is called the **radical** or **radical sign**, and the expression a under the radical is called the **radicand**.

Since $\sqrt[n]{a} = a^{1/n}$, where n is a positive integer, we have the following.

For a a real number and n a positive *odd* integer: $\sqrt[n]{a} = b$ if and only if $b^n = a$.

For a a real nonnegative number and n a positive *even* integer: $\sqrt[n]{a} = |b|$ if and only if $b^n = a$.

Thus $\sqrt[n]{a}$ is the quantity (nonnegative when n is even) that, when raised to the nth power, yields a.

$$\sqrt[3]{64} = 4 \quad since \quad 4^3 = 64$$
$$\sqrt[5]{-32} = -2 \quad since \quad (-2)^5 = -32$$
$$\sqrt{9} = 3 \quad since \quad 3^2 = 9$$

Note that even though $(-3)^2 = 9$, we are interested only in the principal or nonnegative square root.

$\sqrt[6]{-64}$ is not a real number *since no real number when raised to the sixth power will yield a negative number.*

In general, if the nth root of a exists, we find the following.

For a a real number and n a positive integer,

$$\sqrt[n]{a^n} = \begin{cases} |a|, & \text{if } n \text{ is even} \\ a, & \text{if } n \text{ is odd} \end{cases}$$

For example,
$$\sqrt[3]{5^3} = 5$$
$$\sqrt[4]{(-3)^4} = |-3| = 3 \qquad \text{NOTE:} \quad \sqrt[4]{(-3)^4} = \sqrt[4]{81} = 3$$

Also note that $\left(\sqrt[4]{-3}\right)^4$ is not equal to $\sqrt[4]{(-3)^4}$, since $\sqrt[4]{-3}$ is not a real number.
Using radical notation, the following statements are equivalent:

If $a^{1/n}$ is a real number, then $(a^m)^{1/n} = (a^{1/n})^m$.

If $\sqrt[n]{a}$ is a real number, then $\sqrt[n]{a^m} = \left(\sqrt[n]{a}\right)^m$.

Example 9 | Rewrite as an expression without using radicals:

$$\frac{2x}{\sqrt{x^2 - 5}}$$

Solution | $\dfrac{2x}{\sqrt{x^2 - 5}} = \dfrac{2x}{(x^2 - 5)^{1/2}}$ or $2x(x^2 - 5)^{-1/2}$ ∎

Simplifying Radical Expressions

Now that we have defined radicals, our next step is to determine how to put them in simplified form. Recall that a radical expression is simplified if exponents of factors of the radicand are less than the index and there are no fractions under the radical. Along with the definition of a radical, the following three radical properties will provide us with much of what we need to simplify radicals. These properties of radicals are a consequence of their exponential counterparts.

Properties of Radicals

If $\sqrt[n]{a}$ and $\sqrt[n]{b}$ are real numbers:

1. $\sqrt[n]{ab} = \sqrt[n]{a}\sqrt[n]{b}$ **2.** $\sqrt[n]{\dfrac{a}{b}} = \dfrac{\sqrt[n]{a}}{\sqrt[n]{b}}$ $(b \neq 0)$

How are the properties of radicals related to the rules of exponents?

3. $\sqrt[m]{\sqrt[n]{a}} = \sqrt[mn]{a}$ $(a \geq 0)$

Properties 1 and 2 are actually forms of the third and fifth rules of exponents, and property 3 is a result of applying the second rule of exponents: $(a^{1/n})^{1/m} = a^{1/(mn)}$.

Given the definition of a radical and the three radical properties, we can simplify many radical expressions. Here are a few examples of how we would use the properties to simplify radicals.

Example 10 | Express the following in simplest radical form.

(a) $\sqrt{25x^4 y^2}$ (b) $\sqrt[3]{\dfrac{27}{y^{18}}}$ (c) $\sqrt[3]{24}$ (d) $\sqrt[3]{\sqrt[4]{2x^2}}$ (e) $\sqrt[3]{x^3 + y^3}$

Solution | (a) $\sqrt{25x^4 y^2}$

We stated earlier that all variables represent positive real numbers. Otherwise, $\sqrt{y^2} = |y|$, yet we can write $\sqrt{x^4} = x^2$ rather than $|x^2|$. Why?

$= \sqrt{25}\sqrt{x^4}\sqrt{y^2}$ Apply property 1.

Since $5^2 = 25$, $(x^2)^2 = x^4$, and $(y)^2 = y^2$, we have

$= 5x^2 y$

(b) $\sqrt[3]{\dfrac{27}{y^{18}}} = \dfrac{\sqrt[3]{27}}{\sqrt[3]{y^{18}}} = \boxed{\dfrac{3}{y^6}}$

(c) $\sqrt[3]{24} = \sqrt[3]{8 \cdot 3} = \sqrt[3]{8}\sqrt[3]{3} = \boxed{2\sqrt[3]{3}}$

(d) $\sqrt[3]{\sqrt[4]{2x^2}} = \boxed{\sqrt[12]{2x^2}}$

(e) $\sqrt[3]{x^3 + y^3}$ $\boxed{\text{cannot be simplified}}$ (Why not?)

Operations with Radicals

The properties of radicals can also be used to simplify products and quotients of radicals. When the first two properties of radicals are read from right to left, they state that *if the indices are the same*, the product (quotient) of radicals is the radical of the product (quotient).

Example 11 Perform the operations and simplify.

(a) $\sqrt{3a^2b}\ \sqrt{6ab^3}$ **(b)** $\left(ab\sqrt[3]{2a^2b}\right)\left(2a\sqrt[3]{4ab}\right)$

Solution **(a)** We could simplify each radical first, but after we multiply, it may be necessary to simplify again. It is often better to multiply radicands first.

$$\sqrt{3a^2b}\ \sqrt{6ab^3} = \sqrt{(3a^2b)(6ab^3)}$$
$$= \sqrt{18a^3b^4} \qquad \text{Then simplify.}$$
$$= \sqrt{9a^2b^4}\sqrt{2a}$$
$$= \boxed{3ab^2\sqrt{2a}}$$

(b) Because this is all multiplication, we use the associative and commutative properties to reorder and regroup variables, and then use radical property 1 to multiply the radicals.

$$\left(ab\sqrt[3]{2a^2b}\right)\left(2a\sqrt[3]{4ab}\right) = (ab)(2a)\sqrt[3]{(2a^2b)(4ab)}$$
$$= 2a^2b\sqrt[3]{8a^3b^2} \qquad \text{Then simplify the radical.}$$
$$= 2a^2b\left(\sqrt[3]{8a^3}\ \sqrt[3]{b^2}\right)$$
$$= 2a^2b\left(2a\sqrt[3]{b^2}\right) \qquad \text{Multiply the expressions outside the radical.}$$
$$= \boxed{4a^3b\sqrt[3]{b^2}}$$

Example 12 Express the following in simplest radical form: $\dfrac{2x\sqrt{24xy^3}}{5x^2y\sqrt{12xy}}$

Solution | We could simplify each radical first, but after simplifying, multiplying, and reducing, we may have to simplify again. Because this expression is a quotient, we can use property 2 to collect all radicands under one radical.

$$\frac{2x\sqrt{24xy^3}}{5x^2y\sqrt{12xy}}$$

$$= \frac{2x}{5x^2y}\sqrt{\frac{24xy^3}{12xy}} \qquad \text{Then reduce fractions inside and outside the radicals.}$$

$$= \frac{2}{5xy}\sqrt{2y^2} \qquad \text{Now simplify the radical expression.}$$

$$= \frac{2}{5xy}\left(y\sqrt{2}\right) \qquad \text{Then reduce again.}$$

$$\boxed{= \frac{2\sqrt{2}}{5x}}$$

The first two properties allow us to multiply and divide radical expressions with the same index. If we need to express a product or quotient of radical expressions with different indices under a single radical, we can convert to fractional exponents, as demonstrated next.

Example 13 | Express $\sqrt{x}\sqrt[3]{x^2}$ as a simplified radical.

Solution | We begin by rewriting the given expression using fractional exponents:

Alternatively, we can rewrite $x^{7/6}$ as $x^{6/6} \cdot x^{1/6} = x \cdot x^{1/6}$ and then rewrite this product using radical notation.

$$\sqrt{x}\sqrt[3]{x^2} = x^{1/2}x^{2/3} = x^{\frac{1}{2}+\frac{2}{3}} = x^{7/6} = \sqrt[6]{x^7} = \sqrt[6]{x^6}\sqrt[6]{x} = \boxed{x\sqrt[6]{x}}$$

It is important for you to be comfortable working with expressions involving fractional exponents or their equivalent radical forms.

The distributive property allows us to combine terms with radical factors. For example, we can combine $3\sqrt{2} + 4\sqrt{2}$ as follows:

$$3\sqrt{2} + 4\sqrt{2} = (3 + 4)\sqrt{2} = 7\sqrt{2}$$

Example 14 | Simplify $\sqrt{27} - \sqrt{81} + \sqrt{12}$.

Solution | Simplify each term first.

$$\sqrt{27} - \sqrt{81} + \sqrt{12} = \sqrt{9}\sqrt{3} - \sqrt{81} + \sqrt{4}\sqrt{3} = 3\sqrt{3} - 9 + 2\sqrt{3}$$
$$\text{Then combine where possible.}$$

$$\boxed{= 5\sqrt{3} - 9}$$

We cannot combine $5\sqrt{3} - 9$, just as we cannot combine $5x - 9$.

As with multiplying polynomials, we may use the distributive property or special products to multiply expressions with more than one radical term.

Example 15   Perform the indicated operations.

(a) $\left(2\sqrt{x} - 3\sqrt{y}\right)\left(2\sqrt{x} + 3\sqrt{y}\right)$ (b) $\left(\sqrt{x+y}\right)^2 - \left(\sqrt{x} + \sqrt{y}\right)^2$

Solution   (a) $\left(2\sqrt{x} - 3\sqrt{y}\right)\left(2\sqrt{x} + 3\sqrt{y}\right)$ This is a difference of squares.

$$= \left(2\sqrt{x}\right)^2 - \left(3\sqrt{y}\right)^2 \qquad \text{Square each term.}$$

$$= 2^2\left(\sqrt{x}\right)^2 - 3^2\left(\sqrt{y}\right)^2$$

$$= \boxed{4x - 9y}$$

(b) $\left(\sqrt{x+y}\right)^2 - \left(\sqrt{x} + \sqrt{y}\right)^2$

$$= x + y - \left[\left(\sqrt{x}\right)^2 + 2\sqrt{x}\sqrt{y} + \left(\sqrt{y}\right)^2\right] \qquad \begin{array}{l}\text{Note the differences in the} \\ \text{way we handle } (\sqrt{x+y})^2 \\ \text{and } (\sqrt{x}+\sqrt{y})^2.\end{array}$$

$$= x + y - \left[x + 2\sqrt{xy} + y\right]$$

$$= x + y - x - 2\sqrt{xy} - y = \boxed{-2\sqrt{xy}}$$ ∎

How would you write $3x - 5y$ as a product using the difference of squares? The difference of cubes?

Example 15(a) illustrates that any expression containing a difference can be factored if we lift our restriction of requiring factors to be polynomials with integer coefficients.

Rationalizing Radical Expressions

Sometimes fractions involving radical expressions are easier to work with if the radical is eliminated from the numerator (or denominator); this is called *rationalizing the numerator (or denominator)*. This can be done using the fundamental principle of fractions.

Example 16   Rationalize the following:

(a) the denominator of $\dfrac{\sqrt{2}}{\sqrt{5}}$ (b) the numerator of $\dfrac{\sqrt{2}}{\sqrt{5}}$

Solution   (a) We apply the fundamental principle of fractions and multiply the numerator and denominator by $\sqrt{5}$. We choose $\sqrt{5}$ because multiplying the denominator by $\sqrt{5}$ will give us the square root of a perfect square, which in turn yields a rational expression. Thus

$$\frac{\sqrt{2}}{\sqrt{5}} = \frac{\sqrt{2}}{\sqrt{5}} \frac{\sqrt{5}}{\sqrt{5}} = \frac{\sqrt{10}}{\sqrt{5^2}} = \boxed{\frac{\sqrt{10}}{5}}$$

(b) To eliminate the radical from the numerator, we use the fundamental principle of fractions and multiply the numerator and denominator by $\sqrt{2}$:

$$\frac{\sqrt{2}}{\sqrt{5}} = \frac{\sqrt{2}}{\sqrt{5}} \frac{\sqrt{2}}{\sqrt{2}} = \boxed{\frac{2}{\sqrt{10}}}$$ ∎

In rationalizing the denominator, we try to multiply the numerator and the denominator of the fraction by *the expression that will make the denominator the nth root of a perfect nth power.*

Example 17

Rationalize the denominator of $\dfrac{1}{\sqrt[3]{2x^2}}$.

Solution

What principle allows you to "eliminate the radical in the denominator" when you rationalize the denominator?

To change $\sqrt[3]{2x^2}$ into the cube root of a perfect cube, we can multiply the numerator and denominator by $\sqrt[3]{4x}$. Then $\sqrt[3]{2x^2}\sqrt[3]{4x} = \sqrt[3]{8x^3} = 2x$.

$$\frac{1}{\sqrt[3]{2x^2}} = \frac{1}{\sqrt[3]{2x^2}} \cdot \frac{\sqrt[3]{4x}}{\sqrt[3]{4x}} = \frac{\sqrt[3]{4x}}{\sqrt[3]{8x^3}} = \boxed{\frac{\sqrt[3]{4x}}{2x}}$$ ∎

In Example 16 we rationalized the numerator or denominator of a fraction with a single radical term. If the numerator or denominator of a fraction contains two terms and involves square roots, we exploit the difference of two squares and multiply the numerator and denominator by a conjugate of the radical expression to be rationalized, as illustrated in the next two examples.

*Recall that a **conjugate** of $a + b$ is $a - b$.*

Example 18

Rationalize the denominator of $\dfrac{2}{3 + \sqrt{5}}$.

Solution

$\dfrac{2}{3 + \sqrt{5}}$

Multiply the numerator and the denominator by $3 - \sqrt{5}$, a conjugate of $3 + \sqrt{5}$.

$$= \frac{2}{3 + \sqrt{5}} \cdot \frac{3 - \sqrt{5}}{3 - \sqrt{5}}$$

The denominator is a difference of squares.

$$= \frac{2(3 - \sqrt{5})}{(3)^2 - (\sqrt{5})^2}$$

Don't multiply out the numerator yet.

$$= \frac{2(3 - \sqrt{5})}{9 - 5} = \frac{\overset{1}{\cancel{2}}(3 - \sqrt{5})}{\underset{2}{\cancel{4}}} = \boxed{\frac{3 - \sqrt{5}}{2}}$$ ∎

Example 19

Rationalize the numerator and simplify: $\dfrac{\sqrt{x} + 2}{x - 4}$

Solution

We can approach this problem in two ways. First, we can rationalize by multiplying the numerator and denominator by $\sqrt{x} - 2$:

$$\frac{\sqrt{x} + 2}{x - 4} = \frac{\sqrt{x} + 2}{x - 4} \cdot \frac{\sqrt{x} - 2}{\sqrt{x} - 2}$$

$$= \frac{(\sqrt{x})^2 - 2^2}{(x - 4)(\sqrt{x} - 2)}$$

Don't multiply out the denominator yet.

$$= \frac{x - 4}{(x - 4)(\sqrt{x} - 2)}$$

Now reduce.

$$= \boxed{\frac{1}{\sqrt{x} - 2}}$$

A second way to arrive at the same result is to rewrite the denominator as a difference of squares. Noting that each term of the binomial in the denominator (x and 4) is the square of each term of the binomial in the numerator ($\sqrt{x}$ and 2), we can factor the denominator as follows:

$$\frac{\sqrt{x} + 2}{x - 4} = \frac{\sqrt{x} + 2}{(\sqrt{x} + 2)(\sqrt{x} - 2)} = \frac{1}{\sqrt{x} - 2}$$ ∎

1.5 Exercises

In Exercises 1–8, perform the operations and simplify. Express answers with positive exponents only.

1. $\dfrac{(x^2y)^2(x^3y)^2}{(xy)^4}$

2. $\dfrac{(3a^2b)^2(2ab)^3}{6a^2b^3}$

3. $\dfrac{(3x^{-1}y^{-2})^{-3}(x^2y^{-1})^3}{(9x^{-2}y)^{-2}}$

4. $\dfrac{4^{-2}16^{-1}}{2^{-4}}$

5. $\dfrac{x^{-1}+2}{x+2}$

6. $\dfrac{x^{-1}+2}{(x+2)^{-1}}$

7. $\dfrac{x^{-1}+y^{-1}}{xy^{-1}}$

8. $\dfrac{x^{-2}+y^{-2}}{(x+y)^{-2}}$

9. Express 7,500,000,000 using scientific notation.

10. Express 0.000000653 using scientific notation.

11. The sun is approximately 93 million miles from Earth. How long does it take light, which travels at approximately 186,000 miles per second, to reach us from the sun?

12. Scientists estimate that sun converts about 700 million tons of hydrogen to helium every second. If the sun contains about 1.49×10^{27} tons of hydrogen, approximately how many years will it take to use up the sun's supply of hydrogen?

In Exercises 13–16, evaluate the expression.

13. $32^{-3/5}$

14. $\left(\dfrac{1}{8}\right)^{2/3}$

15. $\left(-\dfrac{8}{27}\right)^{-2/3}$

16. $\dfrac{9^{-1/2}81^{1/4}}{27^{-2/3}3^{-2}}$

In Exercises 17–24, perform the operations and express in simplest form with positive exponents only.

17. $\dfrac{x^{1/2}y^{2/3}}{x^{-1/3}}$

18. $\dfrac{x^{1/2}x^{-2/3}}{x^{-3/4}}$

19. $(x^2+1)^{-1/2}(x^2+1)^{3/5}$

20. $\dfrac{(x^2+1)^{1/3}}{(x^2+1)^{-2/3}}$

21. $(x^{1/2}+2)^2$

22. $(x^{1/3}+y^{1/3})(x^{2/3}-x^{1/3}y^{1/3}+y^{2/3})$

23. $(x^{1/2}+2)(x^{1/2}-2)$

24. $4x(x^{-1/2}+1)^2$

In Exercises 25–30, express as a single fraction with positive exponents only.

25. $3(x+1)^{-1}-3x(x+1)^{-2}$

26. $2x(x+1)^{-2}-6(x+1)^{-1}$

27. $4x^3(x^2+1)^{-1/2}+4x(x^2+1)^{1/2}$

28. $2x^2(x^2-2)^{-2/3}+(x^2-2)^{1/3}$

29. $\dfrac{3x^3}{2}(x^3-3)^{-1/2}+(x^3-3)^{1/2}$

30. $\dfrac{2x^2}{3}(x^2-4)^{-2/3}+(x^2-4)^{1/3}$

In Exercises 31–32, rewrite as an expression without radicals.

31. $\dfrac{3}{x\sqrt{x^3-3}}$

32. $\dfrac{3x^2}{\sqrt{x^2-1}}$

In Exercises 33–38, express in simplest radical form.

33. $\sqrt{24x^4y^6}$

34. $\sqrt[3]{24x^4y^6}$

35. $\sqrt{\dfrac{16x^8}{9}}$

36. $\sqrt{\dfrac{24x^8}{9x^3}}$

37. $\sqrt[4]{x^4-y^4}$

38. $\sqrt[3]{(x-y)^3}$

In Exercises 39–44, perform the operations and express your answer in simplest radical form.

39. $\sqrt[4]{54y^2}\,\sqrt[4]{48y^4}$

40. $\sqrt[3]{20ab^5}\,\sqrt[3]{50b^4}$

41. $\dfrac{\sqrt[4]{x^2y^{17}}}{\sqrt[4]{x^{14}y}}$

42. $\dfrac{\sqrt[3]{a^{13}b^2}}{\sqrt[3]{a^4b^5}}$

43. $\sqrt{3x^3}\,\sqrt[3]{81x^5}$

44. $\sqrt{\sqrt[3]{5x^2}}$

In Exercises 45–48, rationalize the denominator and express your answer in simplest radical form.

45. $\dfrac{5}{\sqrt{3x^5}}$

46. $\dfrac{5}{\sqrt[3]{3x^5}}$

47. $\dfrac{1}{\sqrt{x}-4}$

48. $\dfrac{a-b}{\sqrt{a}-\sqrt{b}}$

In Exercises 49–54, rationalize the numerator and simplify the fraction where possible.

49. $\dfrac{\sqrt{x^2-3x+2}}{x-1}$

50. $\dfrac{\sqrt{5}}{5x+25}$

51. $\dfrac{\sqrt{x}-3}{x-9}$

52. $\dfrac{\sqrt{x}+2}{x^2-3x-4}$

53. $\dfrac{\sqrt{x+h}-\sqrt{x}}{h}$

54. $\dfrac{\sqrt{x+h+3}-\sqrt{x+3}}{h}$

In Exercises 55–74, perform the operations and express your answer in simplest radical form; rationalize denominators where possible.

55. $(3xy^2\sqrt{x^2y})(2x\sqrt{18xy^2})$

56. $(3a\sqrt[3]{2b^4})(2a^2\sqrt[3]{4b^2})$

57. $\sqrt[3]{\dfrac{81x^2y^4}{2x^3y}}$

58. $\sqrt[3]{\dfrac{64a^2b^5}{9a^4b^2}}$

59. $\dfrac{xy^2\sqrt{24x^2}}{x\sqrt{8x^5y}}$

60. $\dfrac{5x^2\sqrt{a^3b^2}}{4y^2\sqrt{a^7b^3}}$

61. $\dfrac{\sqrt{x^2-2x+1}}{\sqrt{x-1}}$

62. $\dfrac{\sqrt{(x^2-2)^3}}{\sqrt{x^2-2}}$

63. $\sqrt{27}-3\sqrt{18}+6\sqrt{12}$

64. $2\sqrt{32}-\sqrt{8}-3\sqrt{2}$

65. $(2\sqrt{3}-5)(2\sqrt{3}+2)$

66. $(3\sqrt{x}-2)(2\sqrt{x}+5)$

67. $(\sqrt{a}-5)(\sqrt{a}+5)$

68. $\left(\sqrt{x-5}\right)^2$

69. $\left(\sqrt{x}-3\right)^2-\left(\sqrt{x-3}\right)^2$

70. $\left(\sqrt{x-6}\right)^2-\left(\sqrt{x}-6\right)^2$

71. $\dfrac{\sqrt{2}}{\sqrt{7}-\sqrt{5}}$

72. $\dfrac{20+\sqrt{60}}{4}$

73. $\dfrac{12}{\sqrt{6}-2}-\dfrac{36}{\sqrt{6}}$

74. $\dfrac{20}{\sqrt{7}+3}+\dfrac{28}{\sqrt{7}}$

In Exercises 75–80, express as a single simple fraction.

75. $2x-\dfrac{1}{\sqrt{x^2-2}}$

76. $5x+\dfrac{2x}{\sqrt{x+3}}$

77. $3\sqrt{x^2-2}-\dfrac{2}{\sqrt{x^2-2}}$

78. $3x\sqrt{x^2+4}+\dfrac{5x}{\sqrt{x^2+4}}$

79. $\dfrac{\sqrt{\dfrac{2}{x+h}}-\sqrt{\dfrac{2}{x}}}{h}$

80. $\dfrac{\dfrac{1}{\sqrt{x+h+1}}-\dfrac{1}{\sqrt{x+1}}}{h}$

1.6   The Complex Numbers

When we solve the equation $3x+5=7$, we get $x=\dfrac{2}{3}$ as a solution. If we were restricted to the set of integers, then this equation would have no solutions. However, if we are allowed to use the set of rational numbers, then the solution $x=\dfrac{2}{3}$ is available. Similarly, to solve the equation $x^2=3$, we have to "go beyond" the rationals and define irrationals (and hence the real numbers) to obtain the solutions $x=\sqrt{3}$ and $x=-\sqrt{3}$.

It may seem as though our system is complete, but there are polynomial equations we are still unable to solve. For example, the equation $x^2=-5$ has no real number solution, because no real number, when squared, yields a negative number.

To obtain the solutions to such equations, we define a system that goes beyond the real numbers. We begin by defining i.

The imaginary unit, i, is defined by
$$i=\sqrt{-1}$$

Given this definition, we have:

$$i$$
$$i^2 = -1$$
$$i^3 = i^2 i = (-1)i = -i$$
$$i^4 = i^2 i^2 = (-1)(-1) = +1$$

If we continue, $i^5 = i^4 i = (1)i = i$, $i^6 = i^4 i^2 = (1)(-1) = -1, \ldots$. We find that this cycle repeats itself after i^4, so that any power of i can be written as i, -1, $-i$, or 1.

Example 1 | Rewrite as ± 1 or $\pm i$:   **(a)** i^{39} **(b)** i^{26}

Solution | Because $i^4 = 1$, it would be convenient to factor the largest perfect fourth power of i.

(a) $i^{39} = i^{36} i^3$ 36 is the largest multiple of 4 in 39; hence i^{36} is the greatest perfect 4th power factor of i^{39}. We rewrite i^{36} as a perfect 4th power.

$$= (i^4)^9 i^3 \quad \text{Since } i^4 = 1 \text{, we get}$$

$$= (1)^9 i^3 = i^3 = -i.$$

Hence $i^{39} = \boxed{-i}$.

Actually, we find that $i^s = i^r$, where r is the remainder when s is divided by 4, and then rewrite the expression as $\pm i$ or ± 1.

(b) $i^{26} = (i^4)^6 i^2$

$$= 1^6 i^2$$

$$= i^2 = \boxed{-1}$$

We note the following: $(i\sqrt{a})^2 = i^2(\sqrt{a})^2 = (-1)a = -a$

Thus we see that $i\sqrt{a}$ is a number whose square is $-a$. Therefore, for $a > 0$ we have the following:

> If a is a real number and $a > 0$, then
> $$\sqrt{-a} = i\sqrt{a}$$

Using i, we can now translate square roots with negative radicands as follows:

$$\sqrt{-4} = i\sqrt{4} = i(2) = 2i$$
$$\sqrt{-\frac{1}{16}} = i\sqrt{\frac{1}{16}} = i\left(\frac{1}{4}\right) = \frac{1}{4}i$$

We will use i, the imaginary unit, to define a new type of number—a *complex number*.

Definition of a Complex Number

A **complex number** is a number that can be written in the form $a + bi$, where a and b are real numbers and i is the imaginary unit. a is called the **real part** of $a + bi$, and b is called the **imaginary part** of $a + bi$.

For example, in the complex number $3 + 4i$, 3 is the real part and 4 is the imaginary part.

If a is a real number, then we can rewrite it as the complex number $a + 0i$. Because we can put any real number into complex form (with the imaginary part equal to zero), we conclude that all real numbers are complex numbers. In other words, the set of real numbers is a subset of the complex numbers.

If we designate the set of complex numbers as C, then $R \subset C$.

If a nonzero complex number does not have a real part (real part is zero), then we say that the number is pure imaginary. For example, $3i$, $-4i$, and $i\sqrt{5}$ are pure imaginary numbers.

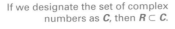

Example 2 | Put the following in $a + bi$ form.

(a) $5 + \sqrt{-16}$ (b) -5 (c) $\dfrac{6 - \sqrt{-3}}{2}$

Solution | (a) $5 + \sqrt{-16} = \boxed{5 + 4i}$ Since $\sqrt{-16} = i\sqrt{16} = 4i$

(b) $-5 = \boxed{-5 + 0i}$

(c) $\dfrac{6 - \sqrt{-3}}{2} = \dfrac{6 - i\sqrt{3}}{2} = \dfrac{6}{2} - \dfrac{\sqrt{3}}{2}i = \boxed{3 - \dfrac{\sqrt{3}}{2}i}$ The real part is 3; the imaginary part is $-\dfrac{\sqrt{3}}{2}$.

Two complex numbers are equal if their real parts are identical *and* their imaginary parts are identical. Algebraically, we state this as follows.

Equivalence of Complex Numbers
$a + bi = c + di$ if and only if $a = c$ and $b = d$

Now that we have defined complex numbers, our next step is to examine operations on complex numbers.

Addition and Subtraction of Complex Numbers

Addition and subtraction of complex numbers are relatively straightforward.

Addition of complex numbers:
$$(a + bi) + (c + di) = (a + c) + (b + d)i$$

Subtraction of complex numbers:
$$(a + bi) - (c + di) = (a - c) + (b - d)i$$

This says that the real part of the sum (difference) is the sum (difference) of the real parts and the imaginary part of the sum (difference) is the sum (difference) of the imaginary parts.

Example 3 | Perform the operations.

(a) $(3 + 4i) + (5 - 6i)$ (b) $(6 + 9i) - (3 - 2i)$

Solution   **(a)** We could use the definition of addition and subtraction:

$$(3 + 4i) + (5 - 6i) = (3 + 5) + (4 - 6)i = \boxed{8 - 2i}$$

or we could treat i as a variable and combine terms:

$$(3 + 4i) + (5 - 6i) = 3 + 4i + 5 - 6i = \boxed{8 - 2i}$$

(b) $(6 + 9i) - (3 - 2i) = (6 - 3) + [9 - (-2)]i$

$$= \boxed{3 + 11i} \qquad \text{By using the definition} \qquad ▮$$

Products of Complex Numbers

We can treat a complex number as a binomial and multiply two complex numbers using binomial multiplication:

$$(a + bi)(c + di) = ac + (ad + bc)i + bdi^2 \qquad \text{General form 2}$$
$$= ac + (ad + bc)i + bd(-1) \qquad \text{Since } i^2 = -1$$
$$= ac - bd + (ad + bc)i \qquad \begin{array}{l}\text{Rearrange terms so that the}\\\text{real parts are together and the}\\\text{imaginary parts are together.}\end{array}$$

Thus we formulate the rule for multiplying complex numbers.

Product of Complex Number
$$(a + bi)(c + di) = (ac - bd) + (ad + bc)i$$

It is probably not very useful to memorize this rule as a special product at this point. It is better just to multiply two complex numbers as binomials, substitute -1 for i^2, and then combine the real parts and imaginary parts to conform to $a + bi$ form.

Example 4   Perform the operations.
(a) $(3 + 2i)(4 - 5i)$ **(b)** $(3 - 7i)^2$ **(c)** $(5 - 2i)(5 + 2i)$
(d) $(3 + i)^2 - 2(3 + i)$

Solution   **(a)** $(3 + 2i)(4 - 5i)$

$$= 12 + (8 - 15)i - 10i^2 \qquad \text{Since } i^2 = -1, -10i^2 = -10(-1) = +10$$

$$= 12 - 7i + 10 = \boxed{22 - 7i}$$

(b) $(3 - 7i)^2 = (3)^2 - 2(3)(7i) + (7i)^2 \qquad \begin{array}{l}\text{Special product: a perfect square}\\\text{of a difference}\end{array}$

$$= 9 - 42i + 49i^2 \qquad \text{Since } 49i^2 = 49(-1) = -49$$

$$= 9 - 42i - 49 = \boxed{-40 - 42i}$$

(c) $(5 - 2i)(5 + 2i) = 5^2 - (2i)^2 \qquad \text{Special product: the difference of squares}$

$$= 25 - 4i^2 \qquad \text{Since } -4i^2 = -4(-1) = +4$$

$$= 25 + 4 = \boxed{29} \qquad \text{Note that the result is a real number.}$$

(d) $(3 + i)^2 - 2(3 + i) \qquad \text{Powers first: square } (3 + i).$

$$= 9 + 6i + i^2 - 2(3 + i)$$

$$= 9 + 6i + i^2 - 6 - 2i$$

$$= 9 + 6i - 1 - 6 - 2i = \boxed{2 + 4i} \qquad ▮$$

Observe that in part (c) of Example 4, the product of two complex numbers $5 - 2i$ and $5 + 2i$ yields the real number 29. The two numbers $5 - 2i$ and $5 + 2i$ are *complex conjugates* of each other. A **complex conjugate** of $a + bi$ is $a - bi$ (and conversely, a complex conjugate of $a - bi$ is $a + bi$). In general, *the product of two complex numbers that are conjugates of each other yields a real number:*

$$(a + bi)(a - bi) = a^2 - (bi)^2 \qquad \text{Difference of squares}$$
$$= a^2 - b^2 i^2 \qquad \text{Since } i^2 = -1, -b^2 i^2 = -b^2(-1) = b^2.$$
$$= a^2 + b^2 \qquad \text{Note that } a^2 + b^2 \text{ is a real number.}$$

We will use this result to help us find certain quotients of complex numbers.

Quotients of Complex Numbers

Our goal is to express a quotient of complex numbers in the form $a + bi$. If we have a quotient of a complex number and a real number, where the real number is the divisor, then expressing the quotient in the form $a + bi$ is similar to dividing a polynomial by a monomial. For example,

$$\frac{6 - 5i}{2} = \frac{6}{2} - \frac{5}{2}i = 3 - \frac{5}{2}i$$

Keeping in mind that $i = \sqrt{-1}$ is a radical expression, we will find the rationalizing techniques we used with quotients of radicals useful when working with quotients of complex numbers when the divisor has an imaginary part.

Example 5 | Express the following in the form $a + bi$.

(a) $\dfrac{4 + 3i}{2i}$ (b) $\dfrac{7 - 4i}{2 + i}$

Solution | (a) We will treat this expression in the same manner as we would if we were rationalizing a denominator: Multiply the numerator and denominator by i.

$$\frac{4 + 3i}{2i} = \frac{(4 + 3i)}{2i} \cdot \frac{i}{i} = \frac{4i + 3i^2}{2i^2} \qquad \text{And, since } i^2 = -1,$$

$$= \frac{4i + 3(-1)}{2(-1)} = \frac{-3 + 4i}{-2} = \boxed{\frac{3}{2} - 2i}$$

(b) Our goal is to get a real number in the denominator so that we may express our result in the form $a + bi$. We proceed as follows:

$$\frac{7 - 4i}{2 + i} \qquad \text{Multiply the numerator and denominator by a conjugate of the denominator, which is } 2 - i.$$

$$= \frac{(7 - 4i)(2 - i)}{(2 + i)(2 - i)}$$

$$= \frac{14 - 7i - 8i + 4i^2}{2^2 - i^2} \qquad \text{Since } i^2 = -1, \text{ we get}$$

$$= \frac{14 - 7i - 8i - 4}{4 + 1} \qquad \text{Note that the denominator is now a real number.}$$

$$= \frac{10 - 15i}{5} \qquad \text{Then rewrite in complex number form.}$$

$$= \frac{10}{5} - \frac{15}{5}i = \boxed{2 - 3i}$$

We have defined complex numbers and their operations. Complex numbers, together with the arithmetic operations as we have defined them, obey the same properties as the real numbers (associative, commutative, distributive, inverses, etc.) and form a system called the complex number system, designated by C.

We have extended the natural numbers to the whole numbers, the whole numbers to the integers, the integers to the rationals, the rationals to the reals, and finally, the reals to the complex numbers, C (see Figure 1.12):

Figure 1.12

The complex number system

$$N \subset W \subset Z \subset Q \subset R \subset C$$

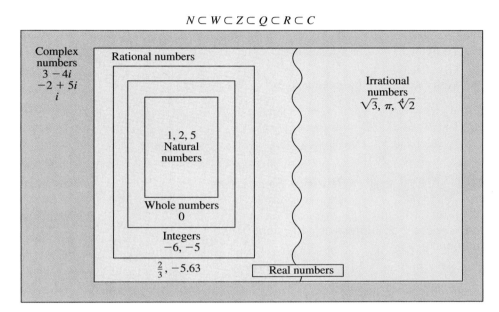

In Section 2.3 we will see that any second-degree equation has its solution in C, the complex number system. In Chapter 5 we will see that any polynomial equation has all its solutions in C.

1.6 Exercises

In Exercises 1–6, express each as ± 1 or $\pm i$.

1. i^{16} **2.** i^{35}

3. i^{100} **4.** i^{43}

5. i^{4n+3} (n an integer)   **6.** i^{4n+1} (n an integer)

In Exercises 7–10, write in $a + bi$ form.

7. $3 + \sqrt{-16}$ **8.** $2 - \sqrt{-12}$

9. $5 - \sqrt{-\dfrac{1}{12}}$ **10.** $\dfrac{2 + \sqrt{-12}}{2}$

In Exercises 11–32, perform the operations and write the results in $a + bi$ form.

11. $(3 + 2i) + (6 - 8i)$ **12.** $(7 + 2i) + (3 - 5i)$

13. $(8 - 2i) + (9 + 2i)$ **14.** $(7 + 6i) - (7 - 6i)$

15. $(2 - 3i)(5 + i)$ **16.** $(5 + 3i)(2 - i)$

17. $(5 - 6i)(5 + 6i)$ **18.** $(3 - 8i)(3 + 8i)$

19. $(2 - 3i)^2$ **20.** $(5 - i)^2$

21. $\dfrac{2 + 8i}{2}$ **22.** $\dfrac{5 - 15i}{3}$

23. $\dfrac{2 + 3i}{i}$ **24.** $\dfrac{6 - 2i}{2i}$

25. $\dfrac{3}{2 + i}$ **26.** $\dfrac{4}{3 - i}$

27. $\dfrac{i}{2 - i}$ **28.** $\dfrac{i}{2 + i}$

29. $\dfrac{i + 1}{i - 1}$ **30.** $\dfrac{i - 1}{i + 1}$

31. $\dfrac{6 - 2i}{5 + 2i}$

32. $\dfrac{1 + i}{3 + 2i}$

In Exercises 33–36, perform the operations and write the results in $a + bi$ form.

33. $(2 + i)^2 - 4(2 + i)$

34. $2(1 - i)^2 - 4(1 - i)$

35. $(2 - 2i)^2 + 2(2 + i)$

36. $(1 - 3i)^2 - 2(1 + 3i)$

37. Verify that $3 + 2i$ is a solution to $x^2 - 6x + 13 = 0$.

38. Verify that $2 - i$ is a solution to $3x^2 - 12x = -15$.

39. Express i^{-1} in $a + bi$ form.

40. Express i^{-2} in $a + bi$ form.

Question for Thought

41. The number i allows us to represent even roots of negative numbers other than square roots. Show how we can represent $\sqrt[6]{-1}$ using i. HINT: $\sqrt[6]{x} = x^{1/6} = (x^{1/3})^{1/2} = \sqrt{\sqrt[3]{x}}$. How about $\sqrt[10]{-1}$? $\sqrt[14]{-1}$?

Chapter 1   *Review Exercises*

In Exercises 1–2, express as a quotient of two integers.

1. $8.252\overline{525}$

2. $72.47247\overline{247}$

In Exercises 3–4, state the real number property being illustrated.

3. $5 - (x + 4) = 5 - (4 + x)$

4. $(x + 3)(x^2 - 3) = x(x^2 - 3) + 3(x^2 - 3)$

In Exercises 5–8, graph the set on the real number line and express the set using interval notation.

5. $\{x \mid x \geq -4\}$

6. $\{x \mid -8 < x \leq -1\}$

7. $\{x \mid x < 3\}$

8. $\{x \mid x \geq 0\}$

In Exercises 9–10, write each expression without absolute value symbols.

9. $|1 - \sqrt{5}|$

10. $|-7x^2 - 3|$

In Exercises 11–16, evaluate the numerical expressions.

11. $6 - \{4 - [5 - 2(3 - 9)]\}$

12. $|-5| - |-9| - |-5 - 9|$

13. $2\left(\dfrac{2}{3}\right)^2 - 4\left(\dfrac{2}{3}\right) + 6$

14. $5\left(-\dfrac{1}{3}\right)^2 - 2\left(-\dfrac{1}{3}\right) - 4$

15. $\dfrac{3 - \dfrac{1}{2}}{\dfrac{2}{5} - 1}$

16. $\dfrac{\dfrac{1}{5} - \dfrac{2}{3}}{1 - \dfrac{3}{4}}$

In Exercises 17–26, perform the operations and express your answer in simplest form.

17. $(-2xy)^2(3xy)^3$

18. $3xy^2(2xy^2 - 3y^2)$

19. $(x - 3)(2x + 7)$

20. $(3x - 4)(2x - 1)$

21. $(5a^3 - b)(5a^3 + b)$

22. $(5a^3 - b)^2$

23. $(3x - 1)(9x^2 + 3x + 1)$

24. $(1 + 2x)(1 - 4x + 16x^2)$

25. $2x(x - 2)^2 - (x + 2)(x - 2)$

26. $(x - 2)^2(x - 3)^2$

In Exercises 27–36, factor as completely as possible.

27. $x^2 - 2x - 15$

28. $6x^2 - 11x - 10$

29. $3y(x - 2)^2 + 5(x - 2)$

30. $2x(x + 8)^2 - 2x^2(x + 8)^3$

31. $25x^2 - 30x + 9$

32. $a^4 - 81$

33. $a^3 + a^2 - 16a - 16$

34. $2x^3 + 6x^2 - 18x - 54$

35. $a^5 - a^3 + a^2 - 1$

36. $a^2 + 4a + 4 - x^2$

In Exercises 37–44, perform the operations and express your answer in simplest form.

37. $\dfrac{4x^2 - 12xy + 9y^2}{2x^2 - xy - 3y^2} \cdot \dfrac{3x + y}{6x^2 - 7xy - 3y^2}$

38. $\dfrac{2x^2}{x^2 + 3x + 9} \div \dfrac{2x^3 - 18x}{x^3 - 27}$

39. $\dfrac{3x}{x - 2} + \dfrac{1}{x}$

40. $\dfrac{2x - 3}{x + 3} - \dfrac{x - 1}{x - 3}$

41. $\dfrac{3x}{(x - 1)^2} - \dfrac{2}{x - 1}$

42. $\dfrac{2x}{x^2 - 1} + \dfrac{2}{x^2 - 2x + 1}$

43. $\dfrac{\dfrac{1}{x} + \dfrac{6}{y}}{\dfrac{1}{x^2} - \dfrac{36}{y^2}}$

44. $\dfrac{\dfrac{3y - 1}{2y + 1} + \dfrac{2y}{4y^2 - 1}}{\dfrac{3}{2y - 1} + 3}$

In Exercises 45–52, perform the operations and express your answer in simplest form with positive exponents only.

45. $(xy^2)^{-2}(xy^{-2})^{-3}$

46. $(3x^{-1}y^{-2})^{-1}(2x^{-2}y)^{-3}$

47. $\dfrac{x^{-1}y^{-2}}{xy^{-2}}$

48. $\dfrac{x^{-1} + y^{-2}}{xy^{-2}}$

49. $\dfrac{3x^{-2}y^{-2}}{x^{-1}y}$

50. $\dfrac{3x^{-1} - y^{-2}}{x^{-1} + y}$

51. $\dfrac{x^{2/3}x^{-1/4}}{x^{-2/5}}$

52. $\dfrac{2x(x^2 + 4)^{-3/2}}{(x^2 + 4)^{-1/2}}$

In Exercises 53–54, evaluate the number.

53. $(-64)^{-2/3}$

54. $81^{-3/4}$

In Exercises 55–56, express as a single simplified fraction with positive exponents only.

55. $2x(x^2 + 1)^{1/2} + x^2\left(\dfrac{1}{2}\right)(x^2 + 1)^{-3/2}$

56. $(x^2 - 1)^{1/3} + \dfrac{2x^2}{3}(x^2 - 1)^{-2/3}$

In Exercises 57–58, rewrite the expression using rational exponents.

57. $3x^2\sqrt{x^2 + 1}$

58. $4x^5\sqrt[3]{x^2 - 1}$

In Exercises 59–60, rewrite the expression using radical notation.

59. $2x(x + 1)^{1/2} + x^2(x^2 + 1)^{-3/2}$

60. $(x^2 - 1)^{1/3} + \dfrac{2x^2}{3}(x^2 - 1)^{-2/3}$

In Exercises 61–62, rewrite in simplest radical form.

61. $\sqrt[3]{48x^{12}y^{15}}$

62. $\sqrt{\dfrac{48x^{12}y^{15}}{16x^4y^8}}$

In Exercises 63–64, express in simplest radical form with the denominator rationalized.

63. $\dfrac{5x}{\sqrt{x^5}}$

64. $\dfrac{\sqrt{8}}{\sqrt{6} - \sqrt{2}}$

In Exercises 65–74, perform the operations and express in simplest radical form with the denominator rationalized.

65. $\sqrt[3]{48a^2b}\,\sqrt[3]{12b}$

66. $\left(2x^3\sqrt{x^2}\right)\left(3x^2\sqrt{x^2y^4}\right)$

67. $\dfrac{\sqrt{28}}{\sqrt{63}}$

68. $\dfrac{\left(xy\sqrt{2xy}\right)\left(3x\sqrt{y}\right)}{\sqrt{4x^3}}$

69. $\left(\sqrt{2x} - 2\right)^2$

70. $\left(\sqrt{2x} - 2\right)\left(\sqrt{2x} + 2\right)$

71. $\left(\sqrt{a} - 4\right)^2 - \left(\sqrt{a} - 4\right)^2$

72. $\dfrac{\sqrt{x^2 + 2x - 3}}{\sqrt{x + 3}}$

73. $\dfrac{\sqrt{8}}{\sqrt{11} - \sqrt{3}}$

74. $\dfrac{x^2 - 7x - 18}{\sqrt{x + 3}}$

In Exercises 75–76, express as a single fraction.

75. $2 - \dfrac{2}{\sqrt{x^2 - 4}}$

76. $5\sqrt{3x^2 - 1} - \dfrac{5}{\sqrt{3x^2 - 1}}$

77. Two common units of measurements in physics and chemistry are the *angstrom* (written Å), which is 10^{-8} cm, and the *micron* (written μ), which is 10^{-4} cm. How many angstroms are there in a micron?

78. An astronomical unit AU is approximately 150,000,000 km. If the planet Pluto is 39.44 AU from the sun, what is the distance from Pluto to the sun in km?

In Exercises 79–84, perform the operations and express the result in $a + bi$ form.

79. $(3 + 4i) + (2 - 5i)$

80. $(6 - 3i) - (2 + 5i)$

81. $(5 + i)^2$

82. $(3 - 2i)(3 + 2i)$

83. $\dfrac{2 + 3i}{2i}$

84. $\dfrac{6 + 3i}{2 - 3i}$

Chapter 1   Practice Test

1. State the real number property illustrated by
$$(x^2 + 3)\left(\frac{1}{x^2 + 3}\right) = 1.$$

2. Graph the sets on the real number line and express the sets using interval notation.

   (a) $\{x \mid x \le -2\}$ (b) $\{x \mid -6 < x \le -2\}$

3. Write without absolute value symbols:
$$|1 - \sqrt{5}|$$

4. Evaluate the numerical expression: $\dfrac{\dfrac{2}{5} - 3}{1 - \dfrac{3}{2}}$

5. Perform the operations and express your answer in simplest form.

   (a) $(3a^3 - 2b)(3a^3 + 2b)$

   (b) $3x(x - 5)^2 - (x + 5)(x - 5)$

6. Factor as completely as possible.

   (a) $10x^2 + x - 2$

   (b) $3y(y + 3)^2 + 15(y + 3)$

   (c) $x^3 + 2x^2 - 9x - 18$

7. Perform the operations. Express your answer in simplest form with positive exponents only.

   (a) $\dfrac{3}{x + 1} - \dfrac{2x}{(x + 1)^2}$

   (b) $\dfrac{\dfrac{1}{b} - \dfrac{5}{a}}{\dfrac{1}{b^2} - \dfrac{25}{a^2}}$

   (c) $\dfrac{x^{-2} + y^{-2}}{(x + y)^{-2}}$

   (d) $\dfrac{3x^2(x^3 - 2)^{-1/3}}{(x^3 - 2)^{2/3}}$

8. Rewrite each in simplest radical form. Rationalize denominators where possible.

   (a) $\sqrt{\dfrac{32x^9y^7}{18x^4y^8}}$

   (b) $\dfrac{5x^2}{\sqrt{2x}}$

   (c) $(3x - \sqrt{2})^2$

   (d) $\dfrac{\sqrt{9}}{\sqrt{10} - \sqrt{7}}$

9. Perform the operations and express in $a + bi$ form.

   (a) $(3 - 5i)^2$

   (b) $\dfrac{2 - 3i}{3 + i}$

Equations and Inequalities

2

Undoubtedly the primary goal in studying algebra is to develop the ability to solve different types of equations and inequalities. In this chapter we review a variety of algebraic methods used to solve different types of equations and inequalities. In addition, to apply these algebraic skills we begin a continuing process of examining how to represent real-life situations mathematically.

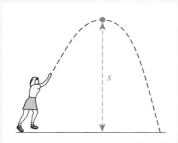

2.1   First-Degree Equations and Inequalities in One Variable

First-Degree Equations

An equation is a symbolic statement of equality. That is, rather than writing "Twice a number is four less than the number," we write $2x = x - 4$. Our goal is to find the solution to a given equation. By **solution** we mean the value or values of the variable that make the algebraic statement true.

In some cases, we may have an equation that is always true. For example, $x + 6 = x + 8 - 2$ is *true for all values of the variable for which it is defined.* We call such equations **identities**. On the other hand, there are equations for which there is no solution, such as $x + 2 = x + 1$. Such an equation is called a **contradiction**. Equations that are true for some values of the variable and false for other values are called **conditional**. The conditional equation $3x - 2 = 5x - 1$ is true when x is $-\frac{1}{2}$ and false when x is any other value.

Definition of a First-Degree Equation

A **first-degree equation in one variable** is an equation that can be put in the form $ax + b = 0$, where a and b are constants and $a \neq 0$.

In other words, a first-degree equation is an equation in which the highest degree of the variable is 1.

Equations that have identical solutions are called **equivalent equations**. The equations

$$3x - [5 + 2(x - 1)] = 2x + 1$$
$$x - 3 = 2x + 1$$
$$x = -4$$

are equivalent since they each have exactly the same solution, -4. Our goal is to take an equation and, with the help of a few properties, gradually change the given equation into an equivalent *obvious equation,* an equation of the form $x = a$ (or $a = x$), where x is the variable for which we are solving.

The **substitution property of equality:** If two quantities are equal, then one quantity can be substituted for the other.

The substitution property states that if $a = b$, then a can be used to replace b and vice versa. Hence, the statement "if $a = b$, then $a + c = b + c$" is simply a particular application of the substitution property, as are the other properties cited in the following box.

Properties of Equality

For $a, b, c \in R$, if $a = b$, then

1. $a + c = b + c$ Addition property
2. $ac = bc$ Multiplication property
3. $a - c = b - c$ Subtraction property
4. $\dfrac{a}{c} = \dfrac{b}{c}$ $(c \neq 0)$ Division property

Because subtraction and division are defined respectively in terms of addition and multiplication, the third and fourth properties are included for completeness and not really necessary (and all four properties are really specific applications of the substitution property). These properties imply the following:

> Adding (subtracting) the same quantity to (from) both sides of an equation will produce an equivalent equation.
>
> Multiplying (dividing) both sides of an equation by the same *nonzero* quantity will produce an equivalent equation.

Now we have two (or four) ways to transform an equation into an equivalent equation. Our goal is to apply the preceding properties to a given equation until we have reduced it to an obvious equation.

In general, our strategy is as follows.

1. First, simplify each side of an equation as much as possible. If there are fractions in the equation, use the multiplication property to "clear the denominators."
2. Using the addition (and/or subtraction) property, collect all terms containing the varible for which we are solving on one side of the equation.
3. Use the division (or multiplication) property to isolate the variable and get the obvious equation.
4. Finally, check all solutions in the original equation.

Example 1 Solve for x.

(a) $820x = 10x + 30(50 - x)$

(b) $\dfrac{3}{x-1} + 5 = \dfrac{4-x}{x-1}$

(c) $\dfrac{8}{x-5} - \dfrac{5}{x+3} = \dfrac{3x+49}{x^2-2x-15}$

Solution (a) $820x = 10x + 30(50 - x)$ Simplify the right-hand side.

$820x = 10x + 1500 - 30x$

$820x = 1500 - 20x$ Apply the addition property (add $20x$ to each side of the equation).

$840x = 1500$ Now use the division property (divide each side of the equation by 840).

$$x = \frac{1500}{840} = \boxed{\frac{25}{14}}$$

Remember to check by substituting $\dfrac{25}{14}$ for x in the original equation.

(b) $\dfrac{3}{x-1} + 5 = \dfrac{4-x}{x-1}$ Multiply each side of the equation by $x-1$ to clear the denominators.

$\dfrac{3}{x-1}(x-1) + 5(x-1) = \dfrac{4-x}{x-1}(x-1)$ Multiply.

$3 + 5(x-1) = 4 - x$ Simplify the left-hand side.

$5x - 2 = 4 - x$ Apply equality properties.

$6x = 6$

$x = 1$

It seems as though we have a solution to this equation. However, if we check this solution by substituting the value 1 for x in the original equation, we find the fractions are undefined. Thus, the solution we arrived at by applying the properties does not work. Did we make a mistake? No, when we multiply both sides by $x - 1$, we must be sure that $x - 1 \neq 0$ to arrive at an equivalent equation. This means that x cannot equal 1; hence, the equation has no solutions .

Why do we exclude 5 and −3 in Example 1(c)?

(c) $\dfrac{8}{x - 5} - \dfrac{5}{x + 3} = \dfrac{3x + 49}{x^2 - 2x - 15}$ Multiply both sides of the equation by the LCD of all the fractions. Since $x^2 - 2x - 15 = (x - 5)(x + 3)$, the LCD is $(x - 5)(x + 3)$. (NOTE: $x \neq 5$, $x \neq -3$)

$$\dfrac{8}{x - 5}(x - 5)(x + 3) - \dfrac{5}{x + 3}(x - 5)(x + 3) = \dfrac{3x + 49}{(x - 5)(x + 3)}(x - 5)(x + 3)$$

Reduce.

$$8(x + 3) - 5(x - 5) = 3x + 49 \qquad \text{Simplify the left-hand side.}$$

$$3x + 49 = 3x + 49$$

At this point we notice that this equation is always true. If we start applying the properties of equality, we would end up with an equation such as $0 = 0$ or $49 = 49$. Since these statements are always true *and since they are equivalent to the original equation*, the first equation must be true for all reals *except* 5 and −3. ∎

We employ the same strategies for solving literal equations (equations involving more than one letter).

Example 2   Solve the following for the indicated variable.

(a) $A = \dfrac{1}{2}h(b_1 + b_2)$ (for b_1)   (b) $y = \dfrac{x - 1}{2x + 3}$ (for x)

(c) $\dfrac{1}{R} = \dfrac{1}{R_1} + \dfrac{1}{R_2}$ (for R_1)

Solution   (a) $A = \dfrac{1}{2}h(b_1 + b_2)$ (for b_1)   Multiply both sides by 2 to eliminate denominators.

$$2A = h(b_1 + b_2) \qquad \text{Multiply out the right-hand side.}$$

$$2A = hb_1 + hb_2 \qquad \text{Isolate the variable you are solving for } (b_1).$$

$$2A - hb_2 = hb_1 \qquad \text{Divide both sides by the multiplier of } b_1.$$

$$\boxed{\dfrac{2A - hb_2}{h} = b_1}$$

(b) $y = \dfrac{x - 1}{2x + 3}$ (for x)   Multiply both sides by $2x + 3$.

$$y(2x + 3) = x - 1 \qquad \text{Simplify the left-hand side.}$$

$$2xy + 3y = x - 1 \qquad \text{Now isolate the } x\text{-terms (collect all } x\text{-terms on one side and all non-}x\text{-terms on the other side).}$$

$$2xy - x = -3y - 1 \qquad \text{Now factor } x \text{ from the left-hand side.}$$

$$(2y - 1)x = -3y - 1 \qquad \text{Finally, divide both sides by the multiplier of } x: 2y - 1.$$

$$\boxed{x = \dfrac{-3y - 1}{2y - 1}} \qquad \text{NOTE: } y \neq \dfrac{1}{2}.$$

(c) $\dfrac{1}{R} = \dfrac{1}{R_1} + \dfrac{1}{R_2}$ (for R_1)   Multiply both sides by the LCD: RR_1R_2.

$$\frac{1}{R}(RR_1R_2) = \frac{1}{R_1}(RR_1R_2) + \frac{1}{R_2}(RR_1R_2)$$

$$R_1R_2 = RR_2 + RR_1 \qquad \text{Isolate } R_1.$$

$$R_1R_2 - RR_1 = RR_2 \qquad \text{Factor out } R_1 \text{ from the left-hand side.}$$

$$R_1(R_2 - R) = RR_2 \qquad \text{Divide both sides by } R_2 - R.$$

$$R_1 = \frac{RR_2}{R_2 - R}$$

Does it make sense for R_2 to equal R?

Example 3 Given the formula $P = \dfrac{kV}{T}$, where k is a fixed constant, what happens to T when P is doubled and V is tripled?

Solution Since we want to examine what happens to T when P is doubled and V is tripled, let's first solve for T:

$$P = \frac{kV}{T} \qquad \text{Solve for } T \text{ to get}$$

$$T = \frac{kV}{P} \qquad \text{Doubling } P \text{ means we use } 2P \text{ instead of } P \text{ and tripling } V \text{ means using } 3V$$
$$\text{instead of } V \text{ in the formula } T = \frac{kV}{P}. \text{ Hence the new } T, \text{ which we call } T', \text{ is}$$

$$T' = \frac{k[3V]}{[2P]}$$

$$= \frac{3}{2}\frac{kV}{P} \qquad \text{And since } \frac{kV}{P} = T,$$

$$T' = \frac{3}{2}T$$

Thus, doubling P and tripling V causes T to be multiplied by $\dfrac{3}{2}$.

Example 4 A stock loses 30% of its previous day's value on September 21. The next day, September 22, it gains 30% of its previous day's value.
(a) If you have $2,400 worth of that stock on September 20, how much is it worth at the end of the day on September 22?
(b) If you have x worth of that stock on September 20, how much is it worth at the end of the day on September 22

Solution (a) A 30% loss means that on September 21, the stock is worth 70% of its value from September 20 (subtract the percent loss from 100%). To find the percent of an amount we multiply the value by the decimal equivalent of the percent, 70%, which is 0.70. Hence the stock valued at $2,400 on September 20 is worth $(0.70)(2,400) = \$1,680$ on September 21.

> A 30% gain will not offset a 30% loss. Keep in mind the differences between percent and actual values: The actual value of a 30% change is dependent on the previous value, and the previous values are different on September 20 and September 21.

On September 22, the stock gains 30% *of its September 21 value*. This means the stock is worth 130% of its September 21 value (add the percent gain to 100%). We convert 130% to a decimal and multiply it by the new value, $1,680, to get $(1.30)(1,680) = \$2,184$.

Hence if the stock was worth $2,400 on September 21, then the value of the stock on September 22 is $2,184.

(b) A 30% loss means that on September 21, the stock is worth 70% of its value from September 20. The find the percent of an amount we multiply the value by the decimal equivalent of the percent, 70%, which is 0.70. Hence the stock valued at $x on September 20 is worth $0.70x$ on September 21.

On September 22, the stock gains 30% *of its September 21 value*. This means the stock is worth 130% of its September 21 value. We convert 130% to a decimal and multiply it by the new value, $0.70x$, to get

Value on September 22 = 130% of the value on September 21

$$= (1.3)(0.70x)$$
$$= 0.91x$$

Hence if the stock was worth $x on September 21, then the value of the stock on September 22 is $0.91x.$ ∎

Most real-life applications of mathematics involve translating a problem into mathematical language and solving it mathematically. Students from arithmetic through calculus and higher mathematics find applications to be the most difficult skill to master in mathematics. The main difficulty lies in the ability to develop or build a model that describes the verbal problem in mathematical terms. This is called *building a mathematical model*: a mathematical equation or expression that describes the problem.

Very often the process of creating a mathematical model requires that we borrow from other disciplines or apply our previous experiences and our common sense: We may have to remember previously seen relationships or formulas. The subtle change of a word or two in a problem may change the entire approach or model. Although we cannot be very specific with our advice, we can offer the following outline as an aid to solving verbal problems.

General Problem-Solving Strategy

1. *Read the problem carefully*, as many times as necessary to understand (or visualize) what the problem is saying and what it is asking.
2. *Use diagrams* whenever possible. It usually makes the given information clearer. If you are drawing a diagram, make sure you label it correctly with what you are given and what you are trying to find.
3. *Find the underlying relationship or formula* relevant to the given problem. Ask whether there is some underlying relationship or formula that you need to know. Does the diagram yield the relationship? If not, do the words of the problem give you the relationship?
4. Clearly *identify the unknown quantity* (or quantities) in the problem, and label it (them) using one variable.
5. By using the underlying formula or relationship in the problem, *write an equation* involving the unknown quantity (or quantities).
6. *Solve* the equation.
7. *Answer the question.* (Make sure you have answered the question that was asked.)
8. *Check* the answer(s) in the original words of the problem. Does it make sense?

Example 5   A computer discount store held an end-of-summer sale on two types of computers. They collected $41,800 on the sale of 58 computers. If one type sold for $600 and the other type sold for $850, how many of each type were sold?

Solution | If we let x = the number of \$600 computers sold, then $58 - x$ = the number of \$850 computers sold (since 58 computers were sold all together).

Our equation involves the amount of money collected on the sale of each type of computer. (That is, the *value* of the computers sold). We compute the amount as follows:

Amount collected from sale of \$600 computers	+	Amount collected from sale of \$850 computers	=	Total amount collected
$\left(\begin{array}{c}\text{No. of \$600}\\\text{computers sold}\end{array}\right)\left(\begin{array}{c}\text{Cost of a \$600}\\\text{computer}\end{array}\right)$	+	$\left(\begin{array}{c}\text{No. of \$850}\\\text{computers sold}\end{array}\right)\left(\begin{array}{c}\text{Cost of an \$850}\\\text{computer}\end{array}\right)$	=	\$41,800
$x \quad \cdot \quad 600$	+	$(58 - x) \quad \cdot \quad 850$	=	$41,800$

Our equation is $600x + 850(58 - x) = 41{,}800$, which yields $x = 30$. Hence there were

$$x = 30 \text{ computers sold at \$600 and } 58 - 30 = 28 \text{ computers sold at \$850.}$$

Example 6 | Art can process 200 forms per hour, and Lew can process 150 forms per hour. How long would it take to process 900 forms working together if Art starts $\frac{1}{2}$ hour after Lew begins?

Solution | The underlying relationship is that the quantity Q of work done is equal to the product of the rate R at which the work is done and the time T doing the work, or $Q = RT$. If Art can process 200 forms in 1 hr, then he can process $200 \times 2 = 400$ forms in 2 hr, $200 \times 3 = 600$ forms in 3 hr, etc.

We can let t be the number of hours Lew works. Then Lew processes $150t$ forms in t hours. On the other hand, Art starts $\frac{1}{2}$ hour later, so his time working is $t - \frac{1}{2}$ hours; therefore, he processes $200\left(t - \frac{1}{2}\right)$ forms.

Our equation expresses the contribution made by each one.

$$\underbrace{150t}_{\substack{\text{Number of forms}\\\text{processed by Lew}}} + \underbrace{200(t - 1/2)}_{\substack{\text{Number of forms}\\\text{processed by Art}}} = \underbrace{900}_{\substack{\text{Total number of}\\\text{forms to be processed}}}$$

This yields

$$350t - 100 = 900.$$

Solve for t to get $t = 2\frac{6}{7}$ hours to process 900 forms.

Example 7 | If it takes Laura 3 hours to mow a lawn and Maria 2 hours to mow the same lawn, how long would it take for them to mow the lawn working together?

Solution | This is the same type of rate problem, but the rates are stated slightly differently than in Example 6. To say that it takes 3 hours to mow a lawn means Laura's rate is $\frac{1}{3}$ of a lawn per hour. Notice $\frac{1}{3}$ of a lawn per hour means in 2 hours, $\frac{2}{3}$ of the lawn is completed; in 3 hours, $\frac{3}{3}$, or 1 entire lawn is completed, etc. Now that we have their rates in familiar form, we can use the same approach we did in Example 6.

If t is the time it takes for them to mow the lawn working together, then Laura can do $\frac{1}{3}t$ of the lawn in t hours, and Maria can do $\frac{1}{2}t$ of the lawn in t hours. So together, in t hours they do $\frac{1}{3}t + \frac{1}{2}t$ of the lawn. Since we want to determine how long it takes to complete 1 lawn, our equation is as follows:

$$\underbrace{\frac{1}{3}t}_{\substack{\text{Part of lawn}\\\text{completed by Laura}}} + \underbrace{\frac{1}{2}t}_{\substack{\text{Part of lawn}\\\text{completed by Maria}}} = \underbrace{1}_{\substack{\text{One complete}\\\text{lawn}}}$$

Solving for t, we get $\boxed{\dfrac{6}{5} \text{ hours or 1 hour 12 minutes}}$.

First-Degree Inequalities

Definition of a First-Degree Inequality

A **first-degree inequality** is an inequality that can be put in the form $ax + b < 0$, where a and b are constants with $a \neq 0$. (The $<$ symbol can be replaced with $>$, $\leq$, or $\geq$.)

As with equations, solving inequalities means to find the values of the variable that make the algebraic statement true. As with equations, we may have an inequality that is *true for all values of the variable*, such as $x + 6 < x + 8$. We call these inequalities **identities**. An inequality for which there is no solution, for example, $x + 5 < x + 3$, is called a **contradiction**. Inequalities that are true for some values of the variable and false for other values are called **conditional.**

Two inequalities are *equivalent* if they have identical solutions. Our goal will be to take more complicated inequalities such as $2 - x - 3 < 5 - 2x$ and reduce them to equivalent "obvious inequalities" such as $x < 6$. For double inequalities, our goal will be to isolate x in the middle member, for example, $-3 < x \leq 2$. For this we will need the properties of inequalities listed in the box.

Keep in mind that although conditional first-degree equations have single solutions, the solutions to conditional first-degree inequalities are infinite sets.

Properties of Inequalities

For a, b, and $c \in R$, if $a < b$ then

1. $a + c < b + c$

2. $a - c < b - c$

3. $ac < bc$ when c is positive; $ac > bc$ when c is negative

4. $\dfrac{a}{c} < \dfrac{b}{c}$ when c is positive; $\dfrac{a}{c} > \dfrac{b}{c}$ when c is negative

(Again the $<$ symbol can be replaced with $>$, $\geq$, or $\leq$.) We can verbally summarize these properties as follows:

To produce an equivalent inequality, we may add (subtract) the same quantity to (from) both sides of an inequality, or multiply (divide) both sides of an inequality by the same *positive* quantity. On the other hand, we must reverse the inequality symbol to produce an equivalent inequality if we multiply (divide) both sides of an inequality by the same *negative* quantity.

Example 8 Solve for the variable. Express your answer using interval notation.

(a) $5x + 8(20 - x) \geq 2(x - 5)$ **(b)** $0 \leq \dfrac{5}{9}(F - 32) \leq 100$

(c) $-2 \leq \dfrac{5}{3} - 3x < 5$

Solution **(a)** $5x + 8(20 - x) \geq 2(x - 5)$ Simplify each side.

$\qquad 5x + 160 - 8x \geq 2x - 10$

$\qquad\qquad 160 - 3x \geq 2x - 10$ Now apply the inequality properties.

$\qquad\qquad\qquad -5x \geq -170$ Divide both sides by -5.

$\qquad\qquad\qquad\qquad x \leq 34$ Note the inequality symbol is reversed.

Using interval notation, the answer is $(-\infty, 34]$.

(b) $0 \leq \dfrac{5}{9}(F - 32) \leq 100$

This double inequality is actually two inequalities that must be satisfied simultaneously:

$$0 \leq \frac{5}{9}(F - 32) \quad \textbf{and} \quad \frac{5}{9}(F - 32) \leq 100$$

Solve Example 7(b) using two inequalities

We could break the double inequality up into two inequalities, $0 \leq \dfrac{5}{9}(F - 32)$ and $\dfrac{5}{9}(F - 32) \leq 100$, solve them, and rewrite the solution as a double inequality. But we can shorten the solution by isolating F in the middle member as follows:

$$0 \leq \frac{5}{9}(F - 32) \leq 100 \qquad \text{Multiply each member by } \tfrac{9}{5}.$$

$$\frac{9}{5}(0) \leq \frac{9}{5} \cdot \frac{5}{9}(F - 32) \leq \frac{9}{5}(100)$$

$$0 \leq F - 32 \leq 180 \qquad \text{Add 32 to each member.}$$

$$32 \leq F \leq 212 \quad \text{or, using interval notation,} \quad \boxed{[32, 212]}$$

(c) $-2 \leq \dfrac{5}{3} - 3x < 5$ Multiply each member by 3.

$\qquad -6 \leq 5 - 9x < 15$ Subtract 5 from each member.

$\qquad -11 \leq -9x < 10$ Divide each member by -9.

$\qquad \dfrac{11}{9} \geq x > -\dfrac{10}{9}$ Note we reversed the inequality symbols. The preferred form of the answer is

$$-\frac{10}{9} < x \leq \frac{11}{9} \quad \text{or, using interval notation,} \quad \boxed{\left(-\frac{10}{9}, \frac{11}{9}\right]} \qquad \blacksquare$$

Example 9 A solution of 10% (by volume) alcohol must be mixed together with a solution of 25% alcohol to make 24 liters of a solution that is at least 15% but no more than 20% alcohol. How much of the 10% alcohol can be used to produce a mixture with an alcohol content within the given limits?

Solution | This problem is similar in structure to Example 5 in this section. In Example 5 we differentiated quantity (how many) from value (the cost of each), but in this problem we differentiate the quantity (amount) of the solution from the pure alcohol content of each solution. For example, 40 gallons of a 25% solution of alcohol would contain $0.25(40) = 10$ gallons of pure alcohol.

We develop an *inequality* reflecting the fact that the amounts of pure alcohol in each of the solutions to be mixed should sum to the amount of pure alcohol in the final mixture.

If we let $x =$ the amount of 10% solution, then $24 - x =$ amount of 25% solution. Our inequality is

$$0.15(24) \quad \leq \quad 0.10x \quad + \quad 0.25(24 - x) \quad \leq \quad 0.20(24)$$

Total amount of alcohol in 24 liters of a 15% solution	Amount of alcohol in the 10% solution	Amount of alcohol in the 25% solution	Total amount of alcohol in 24 liters of a 20% solution

$$0.15(24) \quad \leq \quad 0.10x + 0.25(24 - x) \quad \leq 0.20(24)$$

Clear decimals: Multiply all three members by 100.

$$(100)\,0.15(24) \leq (100)[0.10x + 0.25(24 - x)] \leq (100)[0.20(24)]$$
$$15(24) \leq \quad 10x + 25(24 - x) \quad \leq 20(24)$$

Simplify each member to get

$$360 \leq \quad 600 - 15x \quad \leq 480$$

Solving for x, we get

$$16 \geq x \geq 8 \quad \text{or equivalently} \quad 8 \leq x \leq 16$$

Hence between 8 and 16 liters of 10% alcohol can be used to produce mixtures within the desired limits. ∎

2.1 Exercises

In Exercises 1–22, solve for the given variable. For inequalities, express your answer using interval notation.

1. $2 - 3(x - 4) = 2(x - 1)$

2. $3x - [2 + 3(2 - x)] = 5 - (3 - x)$

3. $4x + \dfrac{2}{3} \leq 2x - (3x + 1)$

4. $5x - 2 > 3x - \left(x - \dfrac{1}{5}\right)$

5. $5\{y - [2 - (y - 3)]\} > y - 2$

6. $2 - \{a - 6[a - (4 - a)]\} \leq 3(a + 2)$

7. $\dfrac{2}{3}x - 5 = \dfrac{3}{2}x + 4(x - 1)$

8. $\dfrac{3}{4}(2x - 3) = \dfrac{2}{3}x + 5$

9. $-\dfrac{7}{y} + 1 = -13$

10. $\dfrac{2}{x + 3} - 4 = 8$

11. $\dfrac{3x + 1}{2} - \dfrac{1}{3} < 1$

12. $\dfrac{5x - 2}{3} \geq \dfrac{x + 3}{4}$

13. $-\dfrac{7}{y} + 3 = 3$

14. $\dfrac{3}{a - 2} + 5 = \dfrac{1 + a}{a - 2}$

15. $\dfrac{2}{x + 3} + 4 = \dfrac{5 - x}{x + 3}$

16. $\dfrac{7}{a-1} + 4 = \dfrac{a+6}{a-1}$

17. $\dfrac{4a+1}{a^2-a-6} = \dfrac{2}{a-3} + \dfrac{5}{a+2}$

18. $\dfrac{5}{y+3} + \dfrac{2}{y} = \dfrac{y-12}{y^2+3y}$

19. $\dfrac{5}{2x+1} + \dfrac{3}{2x-1} = \dfrac{22}{4x^2-1}$

20. $\dfrac{1}{3x+4} + \dfrac{8}{9x^2-16} = \dfrac{1}{3x-4}$

21. $\dfrac{6}{3x+5} - \dfrac{2}{x-4} = \dfrac{10}{3x^2-7x-20}$

22. $\dfrac{6}{x^2-3x} = \dfrac{12}{x} + \dfrac{1}{x-3}$

In Exercises 23–36, solve for the given variable.

23. $3x + 2y - 4 = 5x - 3y + 2$ for y

24. $\dfrac{x}{3} + \dfrac{y}{4} = \dfrac{x}{2} + 3$ for x

25. $S = 2LH + 2LW + 2WH$ for W

26. $ax + b = cx + d$ for x

27. $A = \dfrac{1}{2}h(b_1 + b_2)$ for h

28. $A = \dfrac{1}{2}h(b_1 + b_2)$ for b_1

29. $(3x - 2)(2y - 1) = 0$ for y

30. $(3x - 2)(2y - 1) = 0$ for x

31. $\dfrac{x-\mu}{s} < 1.96$ for s $(s > 0)$

32. $\dfrac{x-\mu}{s} < 1.96$ for μ $(s > 0)$

33. $\dfrac{1}{f} = \dfrac{1}{f_1} + \dfrac{1}{f_2}$ for f

34. $\dfrac{1}{R} = \dfrac{1}{R_1} + \dfrac{1}{R_2} + \dfrac{1}{R_3}$ for R_2

35. $y = \dfrac{3x-2}{x}$ for x

36. $y = \dfrac{2x+3}{5x-1}$ for x

In Exercises 37–44, solve for x. Graph your answer on the number line.

37. $-4 < 3x - 2 < 5$

38. $-5 \le 2x - 5 < 8$

39. $0 < 5 - 2x \le 4$

40. $-1 \le 3 - \dfrac{1}{2}x \le 2$

41. $-\dfrac{3}{2} < \dfrac{1}{3} - x < 2$

42. $0 \le \dfrac{2-3x}{5} < \dfrac{1}{2}$

43. $2x - 3 < 4 < 3x - 1$

44. $5x - 2 < 5 \le 2x - 1$

45. If $s = kt^2$, where k is a fixed constant, what happens to s if t is tripled?

46. If $V = \dfrac{kT}{P}$, where k is a fixed constant, what happens to P if V is halved?

47. If $E = \dfrac{k}{d^2}$, where k is a fixed constant, what happens to E when d is tripled?

48. If $F = \dfrac{km_1m_2}{d^2}$, where k is a fixed constant, what happens to F when d is doubled and m_1 is halved?

49. A stock loses 50% of its previous day's value on October 29. The next day, October 30, it gains 50% of its previous day's value.
  (a) If you have $2,800 worth of that stock on October 28, how much is it worth at the end of the day on October 30?
  (b) If you have x worth of that stock on October 28, how much is it worth at the end of the day on October 30?

50. Suppose you bought stock A at 28 ($28 per share) and stock B at 77 ($77 per share). The next day stock A falls 5 points to 23 and stock B rises 5 points to 82.
  (a) If you originally bought $2,000 worth of stock A and $2,000 worth of stock B, how much are your stocks worth the day after you bought them?
  (b) If you originally bought x worth of stock A and y worth of stock B, how much are your stocks worth the day after you bought them?
  (c) If you originally bought x worth of stock A and x worth of stock B, how much are your stocks worth the day after you bought them?

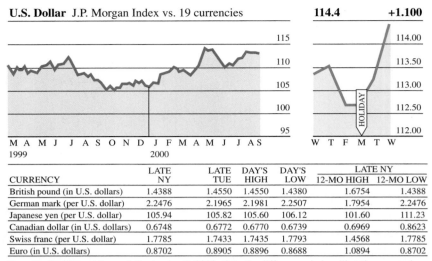

U.S. Dollar J.P. Morgan Index vs. 19 currencies

| CURRENCY | LATE NY | LATE TUE | DAY'S HIGH | DAY'S LOW | LATE NY | |
					12-MO HIGH	12-MO LOW
British pound (in U.S. dollars)	1.4388	1.4550	1.4550	1.4380	1.6754	1.4388
German mark (per U.S. dollar)	2.2476	2.1965	2.1981	2.2507	1.7954	2.2476
Japanese yen (per U.S. dollar)	105.94	105.82	105.60	106.12	101.60	111.23
Canadian dollar (in U.S. dollars)	0.6748	0.6772	0.6770	0.6739	0.6969	0.8623
Swiss franc (per U.S. dollar)	1.7785	1.7433	1.7435	1.7793	1.4568	1.7785
Euro (in U.S. dollars)	0.8702	0.8905	0.8896	0.8688	1.0894	0.8702

From the *Wall Street Journal*, September 7, 2000. Note the differences in handling the conversions when currencies are given *in* U.S. dollars versus *per* U.S. dollar.

For Exercises 51–54, refer to the chart and table above.

51. On September 7, the euro (European dollar) was worth $0.8702 (U.S. dollars). Express the value of *x* euros in U.S. dollars. Use the model to estimate the value of 300 euros in U.S. dollars.

52. On September 7, the euro (European dollar) was worth $0.8702 (U.S. dollars). Express the value of *x* U.S. dollars in terms of euros. Use the model to estimate the value of $300 in euros.

53. On September 7, the British pound was worth $1.4388 (U.S. dollars). Express the value of *x* British pounds in terms of U.S. dollars. Use the model to estimate the value of 400 British pounds in U.S. dollars.

54. On September 2, the U.S. dollar was worth $105.94 Japanese yen. Express the value of *x* U.S. dollars in terms of the Japanese yen. Use the model to estimate the value of $200 in yen.

55. A truck carries a load of 50 boxes; some are 20-lb boxes, and the rest are 25-lb boxes. If the total weight of all boxes is 1175 lb, how many of each type are there?

56. A merchant wishes to purchase a shipment of clock radios. Simple AM models cost $25 each, whereas AM/FM models cost $30 each. In addition, there is a delivery charge of $70. If he spends $700 on 24 clock radios, how many of each type did he buy?

57. Orchestra seats to a certain Broadway show are $48 each and balcony seats are $28 each. If a theater club spends $2328 on the purchase of 56 seats, how many orchestra seats were purchased?

58. A plumber charges $22 per hour for her time and $13 per hour for her assistant's time. On a certain job the assistant works alone for 2 hours doing preparatory work; then the plumber and her assistant complete the job together. If the total bill for the job was $271, how many hours does the plumber work?

59. A manager needs 7200 copies of a document. A new duplicating machine can make 70 copies per minute, and an older model can make 50 copies per minute. The older machine begins making copies but breaks down before completing the job and is replaced by the new machine, which completes the job. If the total time for the job is 1 hour and 50 minutes, how many copies did the older machine make?

60. An experienced worker can process 60 items per hour, whereas a new worker can process 30 items per hour. How many hours will it take to complete a job of processing 500 items if the new worker begins the job 3 hours before being joined by the experienced worker?

61. How many ounces of a 20% solution of alcohol (by volume) must be mixed with 5 ounces of a 50% solution of alcohol to get a mixture of 30% alcohol?

62. Tamara's radiator has a 3-gallon capacity. Her radiator is filled to capacity with a 30% mixture of antifreeze and water. She drains off some of the old antifreeze and refills the radiator to capacity with pure water to get a 20% mixture. How much did she drain off?

63. Cindy wants to invest $8,000 in two money-market certificates; one yields 9% interest per year and the other higher-risk certificate yields 14% per year. If she needs to receive an income of at least $890 a year from her two investments, what is the *minimum* amount she should invest in the higher-risk certificate to get her desired income?

64. Sheldon's math teacher assigns course grades in the following way: Two exams are given, each worth 20% of the course grade; quizzes and homework combined together are worth 20% of the course grade; and the final exam is worth 40% of the course grade. Sheldon received a grade of 85 on the first exam, 65 on the second exam, and had a quiz and homework combined average of 72. What is the minimum score he must get on the final for him to receive a grade of at least 80 for the course?

65. Pipe A can fill a pool with water in 3 days, and pipe B can fill the same pool in 2 days. If both pipes were used, how long would it take to fill the pool?

66. Pipes A and B together can fill the pool in 2 days. If pipe A alone can fill the pool in 5 days, how long would it take to fill the pool with pipe B alone?

67. When a bathtub faucet is turned on (and the drain is shut), it can fill a tub in 10 minutes: when the drain is open (and the faucet is turned off), a full tub can be emptied in 15 minutes. How long would it take the tub to fill if the faucet were turned on with the drain left open?

68. How long would it take for the bathtub of Exercise 67 to fill if the tub could empty in 6 minutes when the drain is open?

Questions for Thought

69. If $\dfrac{1}{a} < 0$, what can be said about a?

70. When does $\dfrac{1}{a}$ equal 0? Explain your answer.

71. A student was given the inequality: $\dfrac{3}{x-2} > 4$.

The first step the student took in solving this inequality was to transform it into $3 > 4(x-2)$. Explain what the student did wrong.

2.2   Absolute Value Equations and Inequalities

Absolute Value Equations

Consider the equation $|x| = 3$. Recall from Section 1.1 that we visualize $|x|$ as the distance from x to 0 on the number line. Hence $|x| = 3$ means that x is 3 units away from 0 on the number line.

Given this definition, we can see by Figure 2.1 that there are two possible answers: $x = -3$ or $x = 3$, since both -3 and 3 are three units away from 0.

Figure 2.1

The meaning of $|x| = 3$

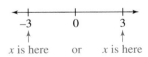

x is here   or   *x* is here

Example 1 | Solve the following: $|4x - 7| = 5$

Solution | $4x - 7$ must be 5 units away from 0 on the number line, as pictured in Figure 2.2.

Figure 2.2

The meaning of $|4x - 7| = 5$

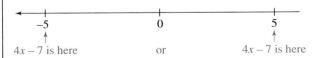

$4x - 7$ is here   or $4x - 7$ is here

Therefore,

$$4x - 7 = -5 \quad \text{or} \quad 4x - 7 = 5 \qquad \text{We solve each equation to get}$$

Hence, $x = \dfrac{1}{2}$ or $x = 3$

Thus we see algebraically that

> For $a \geq 0$,
>
> $|u| = a$ is equivalent to $u = -a$ or $u = a$

Example 2 | Solve the following: $|2x - 8| + 4 = 10$

Solution |
$$|2x - 8| + 4 = 10$$
First we put the equation in the form $|u| = a$ by isolating the absolute value on one side of the equation.

$$|2x - 8| = 6$$
This means

$$2x - 8 = -6 \quad \text{or} \quad 2x - 8 = 6$$
Solve each equation to get

$$x = 1 \quad \text{or} \quad x = 7$$

Example 3 | Solve the following: $|2x - 5| = |3x - 5|$

Solution | For the absolute value of two expressions to be equal, they must be the same distance from 0 on the number line. This can happen either if the two expressions are equal or if they are negatives of each other. (If $|a| = |b|$, then either $a = b$ or $a = -b$.) Hence

$$2x - 5 = 3x - 5 \quad \text{or} \quad 2x - 5 = -(3x - 5)$$
Solve each equation to get

$$x = 0 \quad \text{or} \quad x = 2$$

Absolute Value Inequalities

To solve absolute value inequalities we again refer back to the geometric definition of absolute value. We first note that $|x| < 5$ means that x must lie within 5 units of 0 on the number line. Hence, by Figure 2.3, we see that x must lie between -5 and 5, or $-5 < x < 5$.

Figure 2.3
The meaning of $|x| < 5$

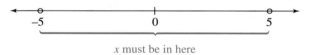

x must be in here

Example 4 | Solve for x: $|5x - 3| < 8$

Solution | $5x - 3$ must be less than 8 units away from 0 on the number line, as pictured in Figure 2.4.

Figure 2.4
The meaning of $|5x - 3| < 8$

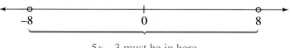

$5x - 3$ must be in here

Therefore, $-8 < 5x - 3 < 8$ Solving this double inequality, we get

$$-1 < x < \frac{11}{5}$$

Using interval notation, the solution is $(-1, 11/5)$.

The procedure for solving $|x| > a$ is different from the procedure for solving $|x| < a$. This is why it is important for you to be able to visualize the meaning of absolute value.

For example, $|x| > 5$ means that x is *more* than 5 units from 0 on the number line. By Figure 2.5, we see that the solutions lie out at the ends of the number line. We must use two inequalities to describe the solution set: $x < -5$ or $x > 5$. (We cannot use the word "and" between the sets because that would imply that x must satisfy both conditions at the same time, which is clearly impossible.)

Figure 2.5
The meaning of $|x| > 5$

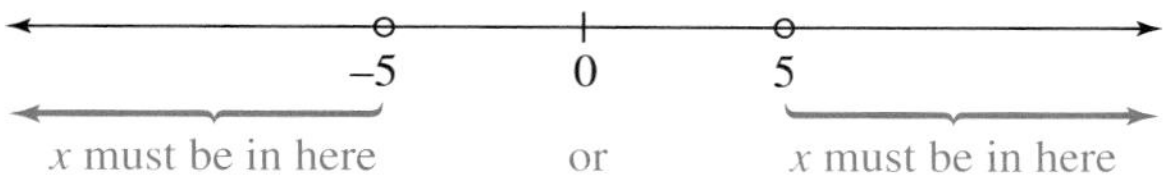

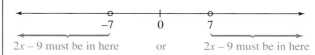

x must be in here   or x must be in here

Example 5

Solve for x: $|2x - 9| > 7$

Solution

$2x - 9$ must be more than 7 units away from 0 on the number line, as pictured in Figure 2.6.

Figure 2.6
The meaning of $|2x - 9| > 7$

$2x - 9$ must be in here   or $2x - 9$ must be in here

Therefore,

$$2x - 9 < -7 \quad \text{or} \quad 2x - 9 > 7 \qquad \text{We solve each inequality to get}$$
$$x < 1 \quad \text{or} \quad x > 8$$

The notation $(-\infty, 1) \cup (8, \infty)$ means x must be in the interval $(-\infty, 1)$ **or** in the interval $(8, \infty)$.

Using interval notation, we can write the solution as: $(-\infty, 1) \cup (8, \infty)$.   ∎

Algebraically, we can summarize solving absolute value inequalities as follows.

For $a > 0$,

$$|u| < a \quad \text{is equivalent to} \quad -a < u < a$$
$$|u| > a \quad \text{is equivalent to} \quad u < -a \quad \text{or} \quad u > a$$

Example 6

Solve the following. Express the answers using interval notation.

(a) $\left| \dfrac{3 - 7x}{2} \right| \le 12$ **(b)** $\left| \dfrac{2}{3}x + 1 \right| \ge 5$ **(c)** $|9x + 1| + 5 < 1$

Solution

(a) $\left| \dfrac{3 - 7x}{2} \right| \le 12$ This translates into

$$-12 \le \frac{3 - 7x}{2} \le 12 \qquad \text{Multiply each member by 2 to get}$$

$$-24 \le 3 - 7x \le 24 \qquad \text{Add } -3 \text{ to each member.}$$

$$-27 \le -7x \le 21 \qquad \text{Then divide each member by } -7.$$

$$\frac{27}{7} \ge x \ge -3 \qquad \text{Notice the inequality symbols are reversed.}$$

Using interval notation, we get $\left[-3, \dfrac{27}{7} \right]$.

(b) $\left|\dfrac{2}{3}x + 1\right| \geq 5$ This translates into

$\dfrac{2}{3}x + 1 \leq -5$ or $\dfrac{2}{3}x + 1 \geq 5$ Solve each inequality:

$2x + 3 \leq -15$ or $2x + 3 \geq 15$

$x \leq -9$ or $x \geq 6$ which is $(-\infty, -9] \cup [6, \infty)$.

(c) $|9x + 1| + 5 < 1$ First isolate the absolute value by adding -5 to each side of the inequality.

$|9x + 1| < -4$

At this point we can stop; the inequality states that a quantity in absolute value is less than a negative number. Since absolute values are always nonnegative, this is impossible, and so the answer is no solution . ∎

Different Perspectives: Absolute Value Equations and Inequalities

Graphical Description

The absolute value of x, $|x|$, is its distance from 0 on the number line. For $a > 0$,

$|x| = a$ means that x is
a units from 0 on the number line:

$|x| < a$ means x is less than
a units from 0 on the number line:

$|x| > a$ means that x is more than
a units from 0 on the number line:

Algebraic Description

(In the following, the symbol $\Leftrightarrow$ means "is equivalent to.") For $a > 0$,

$|x| = a \Leftrightarrow x = a$ or $x = -a$

$|x| < a \Leftrightarrow -a < x < a$

$|x| > a \Leftrightarrow x < -a$ or $x > a$

We complete this section with a few more absolute value properties.

More Absolute Value Properties

1. $|a| \geq 0$

2. $|ab| = |a||b|$

3. $\left|\dfrac{a}{b}\right| = \dfrac{|a|}{|b|}$

4. $|a|^2 = |a^2| = a^2$

5. $|a + b| \leq |a| + |b|$ (The triangle inequality)

6. $|a| - |b| \leq |a - b|$

Example 7   Using the preceding properties, show that we can write $\left|\dfrac{-2x^2 - 3}{3}\right|$ as a single expression without absolute value symbols.

Solution $\left|\dfrac{-2x^2 - 3}{3}\right| = \dfrac{|-2x^2 - 3|}{|3|}$ By property 3

$= \dfrac{|(-1)(2x^2 + 3)|}{|3|}$ Factoring out -1 from $-2x^2 - 3$

$= \dfrac{|-1||2x^2 + 3|}{|3|}$ By property 2

At this point we know that $|-1| = 1$ and $|3| = 3$; since $2x^2 + 3$ is always positive (why?), we conclude

$$\left| \frac{-2x^2 - 3}{3} \right| = \frac{2x^2 + 3}{3}$$

∎

Example 8 An economic forecasting corporation predicts that in the next year, the price of oil will stay within 5% of its current price. According to this prediction, what will the price range be for a barrel of oil if the current price for a barrel is $30?

Solution The company is predicting that the difference between the current price and the future price should remain within 5% of its current value. Letting F = the future price, we can write this as an absolute value inequality:

We could just as well have used $|F - 30|$.

|Current price $- F| \le 5\%$ of the current price   Substitute the current price of $30.

$$|30 - F| \le 0.05(30)$$

$$|30 - F| \le 1.50$$

We start by rewriting the inequality without the absolute value symbols:

$$-1.50 \le 30 - F \le 1.50 \qquad \text{Subtract 30 from each member.}$$

$$-31.50 \le \quad -F \quad \le -28.50 \qquad \text{Divide each member by } -1.$$

$$31.50 \ge \quad F \quad \ge 28.50 \qquad \text{Rewrite the double inequality in standard form.}$$

$$28.50 \le \quad F \quad \le 31.50$$

Hence the actual price range predicted for the following year is between $28.50 and $31.50.

∎

2.2 Exercises

In Exercises 1–34, solve the absolute value equations or inequalities. Express the solutions to the inequalities using interval notation.

1. $|x| = 12$

2. $|x| = -4$

3. $|x| < 9$

4. $|x| \ge 9$

5. $|x - 8| \le 2$

6. $|x + 3| > 5$

7. $|2x - 4| = 7$

8. $|5 - 3x| = 8$

9. $|5 + 2x| < 3$

10. $|7 + 3x| \ge 4$

11. $\left| x - \frac{2}{3} \right| > \frac{2}{3}$

12. $\left| x - \frac{1}{5} \right| \le \frac{3}{5}$

13. $|3 - 2x| < 3$

14. $|4 - 3x| > 5$

15. $\left| \frac{x}{3} + 1 \right| = \frac{3}{4}$

16. $\left| 1 - \frac{x}{5} \right| = \frac{4}{3}$

17. $\left| \frac{x}{2} - \frac{2}{3} \right| > \frac{1}{2}$

18. $\left| \frac{x}{5} + \frac{1}{2} \right| < \frac{3}{5}$

19. $\left| \frac{x}{2} - \frac{x}{3} \right| \ge \frac{1}{2}$

20. $\left| \frac{x}{5} + \frac{x}{2} \right| < \frac{1}{2}$

21. $\left| \frac{2x - 3}{2} \right| = 4$

22. $\left| \frac{1 - 3x}{5} \right| < 6$

23. $\left| \frac{5 - x}{5} \right| = \frac{1}{2}$

24. $\left| \frac{3 - 5x}{2} \right| \le \frac{1}{2}$

25. $|3x - 2| = |-5|$

26. $|2 - 7x| = |-2|$

27. $|3x - 4| = |x|$

28. $|2x - 3| = |x - 5|$

29. $|3x - 2| + 3 = 6$

30. $|1 - 5x| - 4 = 7$

31. $|5 - 3x| - 4 < 8$

32. $|2 - 3x| + 5 > 9$

33. $|3x - 2| + 3 \le 1$

34. $|1 - 5x| + 4 \ge 2$

In Exercises 35–42, rewrite the following as a single expression without absolute value symbols, *if possible*.

35. $|x^2 + 8|$

36. $|3x^2 + 10|$

37. $|-4x^2 - 9|$

38. $|-5 - 3x^4|$

39. $|3x^3 + 3|$

40. $|2x^2 - 5|$

41. $\left| \dfrac{-5 - x^6}{4} \right|$

42. $\left| \dfrac{7x^4 + 5}{-3} \right|$

43. Prove that if $\left| \dfrac{1}{a} \right| < \left| \dfrac{1}{b} \right|$ and $a, b \neq 0$, then $|b| < |a|$.

44. The Yadav Heating and Plumbing Company subcontracts out the manufacture of thermostats to another company. The Yadav Company requires the subcontractor to produce thermostats accurate to 0.75°F. It tests the thermostats by putting them in a room and taking a reading when the temperature is a true value of 40°F, 50°F, and 80°F. What is the allowable range of thermostat readings for each temperature?

45. A weather forecaster predicts the day's high temperature and claims that she is always accurate to within 5°F. What is the acceptable range of temperature values for a forecast high of 79°F?

46. A political pollster predicts that in 2 days, the mayor of a certain city will win with 64% of the votes. If the pollster's projected margin of error for this prediction is 6 percentage points, give the range of values of the mayor's proportion of votes predicted by this pollster.

47. In a certain state, it was predicted that next year's sales tax should not change by more than 8% of the current sales tax. Currently, the sales tax in this state is 6%. Based on this prediction, what is the range of possible values of the sales tax for next year?

Question for Thought

48. Prove $|x| - |y| \leq |x - y|$. HINT: Start with the triangle inequality (property 5). Substitute $x - y$ for a and y for b.

2.3   Quadratic Equations

Quadratic Equations

A **second-degree equation** is a polynomial equation in which the highest degree of the variable is 2. In particular, a second-degree equation in one unknown is called a **quadratic equation**. We define the **standard form** of a quadratic equation as $Ax^2 + Bx + C = 0$, where $A \neq 0$.

As with all other equations, the solutions for quadratic equations are values of the variable that make the equation a true statement. The solutions of $Ax^2 + Bx + C = 0$ are also called the *roots of the polynomial equation* $Ax^2 + Bx + C = 0$. We will find the following property of real numbers particularly useful.

> **The Zero-Product Rule**
>
> If $a \cdot b = 0$, then $a = 0$ or $b = 0$.

In solving the equation $Ax^2 + Bx + C = 0$, if the polynomial $Ax^2 + Bx + C$ can be factored, then we can use the zero-product rule to reduce the problem to that of solving two linear equations. For example, to solve the equation $6x^2 - x - 2 = 0$, we can factor the left-hand side to get $(3x - 2)(2x + 1) = 0$. Hence we can conclude that $3x - 2 = 0$ or $2x + 1 = 0$, which yields $x = \dfrac{2}{3}$ or $x = -\dfrac{1}{2}$.

Another method is to apply the Square Root Theorem.

> **The Square Root Theorem**
>
> If $u^2 = d$, then $u = \pm\sqrt{d}$.

This theorem is usually applied to solve quadratic equations that have no first-degree term. For example, in solving the equation $x^2 = 3$, we can apply this theorem to get $x = \pm\sqrt{3}$.

Example 1

Solve the following.

(a) $4x^2 + 10x = 6$

(b) $\dfrac{y}{y-5} - \dfrac{3}{y+1} = \dfrac{30}{y^2 - 4y - 5}$

(c) $5x^2 - 6 = 8$

(d) $(x-2)^2 = 6$

Throughout this section, unless otherwise mentioned, the check is left to the student.

Solution

(a)
$$4x^2 + 10x = 6 \quad \text{Put into standard form.}$$
$$4x^2 + 10x - 6 = 0 \quad \text{Factor the left-hand side.}$$
$$2(2x - 1)(x + 3) = 0 \quad \text{Hence we have}$$
$$2x - 1 = 0 \quad \text{or} \quad x + 3 = 0 \quad \text{Solving each linear equation, we get}$$

We ignored the factor of 2 when we applied the zero-product rule. What allows us to do that?

$$x = \frac{1}{2} \quad \text{or} \quad x = -3$$

(b) $\dfrac{y}{y-5} - \dfrac{3}{y+1} = \dfrac{30}{y^2 - 4y - 5}$

Noting that $y^2 - 4y - 5 = (y-5)(y+1)$, we multiply both sides by the LCD of all the fractions: $(y-5)(y+1)$.

$$\frac{y}{y-5}(y-5)(y+1) - \frac{3}{y+1}(y-5)(y+1) = \frac{30}{(y-5)(y+1)}(y-5)(y+1)$$

This yields

$$y(y+1) - 3(y-5) = 30 \quad \text{Simplify each side and put into standard form.}$$
$$y^2 - 2y - 15 = 0 \quad \text{Factor.}$$
$$(y-5)(y+3) = 0 \quad \text{This yields}$$
$$y = 5 \quad \text{or} \quad y = -3$$

In checking the solutions, we find that we must eliminate $y = 5$ as a possible solution, because it leads to an undefined expression (denominators are equal to 0). Thus one of the "solutions" we arrived at by applying the properties of equality does not work. Did we make a mistake? No; we are multiplying both sides of the equation by $y - 5$, which is 0 when $y = 5$. Recall that multiplying both sides of an equation by 0 does not necessarily give us an equivalent equation. Hence the only solution is $y = -3$.

(c) We note that there is no first-degree term, so our approach will be to apply the Square Root Theorem.

$$5x^2 - 6 = 8 \quad \text{Isolate } x^2 \text{ on the left-hand side before applying the Square Root Theorem.}$$
$$5x^2 = 14$$
$$x^2 = \frac{14}{5} \quad \text{Applying the Square Root Theorem, we get}$$

$$x = \pm\sqrt{\frac{14}{5}} \quad \text{or (with the denominator rationalized)} \quad x = \pm\frac{\sqrt{70}}{5}$$

(d) We could multiply out the left-hand side and put it in standard form; however, we would find that the quadratic expression in standard form does not factor with integer coefficients. Since it is in the form of a squared quantity equal to a number, we will apply the Square Root Theorem first.

$$(x - 2)^2 = 6 \qquad \text{Apply the Square Root Theorem.}$$
$$x - 2 = \pm\sqrt{6} \qquad \text{Isolate } x \text{ (add 2 to both sides).}$$
$$x = \boxed{2 \pm \sqrt{6}}$$

Try solving $(x - 2)^2 = 6$ by multiplying out the left-hand side first and using the factoring method.

Completing the Square

Part (d) of Example 1 illustrates that if we can construct a perfect square binomial from a quadratic equation (i.e., get the equation in the form $(x + p)^2 = d$), then we can apply the Square Root Theorem and solve for x to get $x = -p \pm \sqrt{d}$.

The method of constructing a perfect square is called **completing the square**. It is based on the fact that in multiplying out the perfect square $(x + p)^2$, with p a constant, we get

$$(x + p)^2 = x^2 + 2px + p^2$$

Notice the relationship between the constant term, p^2, and the coefficient of the middle term, $2p$: The constant term is the square of half the coefficient of the middle term, or

$$\left[\frac{1}{2}(2p)\right]^2 = p^2$$

Suppose we started out with the equation $x^2 + 6x - 5 = 3$.

$$x^2 + 6x - 5 = 3 \qquad \text{Let's rewrite this equation as}$$
$$x^2 + 6x \quad\;\; = 8 \qquad \text{Then we make the left-hand side a perfect}$$
$$\text{square by adding } \left[\frac{1}{2}(6)\right]^2 = 9 \text{ to both sides of the}$$
$$\text{equation to get}$$
$$x^2 + 6x + 9 = 8 + 9 \qquad \text{Next, we rewrite this equation as}$$
$$(x + 3)^2 = 17 \qquad \text{We then apply the Square Root Theorem to get}$$
$$x + 3 = \pm\sqrt{17} \qquad \text{Finally, we isolate } x.$$
$$x = -3 \pm \sqrt{17}$$

Example 2 | Solve by completing the square: $2x^2 - 8x + 4 = 6$

Solution | We note that the leading coefficient (in this case, the coefficient of x^2) is *not* 1, and completing the square is based on squaring a binomial with the resultant leading coefficient being 1. Thus our first step is to make the leading coefficient 1.

$$2x^2 - 8x + 4 = 6 \qquad \text{Divide both sides by 2, the coefficient of } x^2.$$
$$x^2 - 4x + 2 = 3 \qquad \text{Isolate the constant term on the right-hand side. (This}$$
$$\text{step makes it clearer to see how we complete the}$$
$$\text{square on the left-hand side.)}$$
$$x^2 - 4x \quad\;\; = 1 \qquad \text{Take half the coefficient of the middle term and square it}$$
$$\left(\left[\frac{1}{2}(-4)\right]^2 = 4\right); \text{ we add 4 to both sides of the equation.}$$
$$x^2 - 4x + 4 = 1 + 4 \qquad \text{Factor the left-hand side and simplify the right-hand side.}$$
$$(x - 2)^2 = 5 \qquad \text{Solve for } x \text{ using the Square Root Theorem.}$$
$$x - 2 = \pm\sqrt{5} \qquad \text{Isolate } x.$$
$$x = \boxed{2 \pm \sqrt{5}}$$

Unlike the factoring method, all quadratic equations can be solved by completing the square. Next, we will complete the square for the general quadratic equation: $Ax^2 + Bx + C = 0,\ A > 0$.

$$Ax^2 + Bx + C = 0$$

Divide both sides of the equation by the leading coefficient A.

$$x^2 + \frac{B}{A}x + \frac{C}{A} = 0$$

Isolate the constant term on the right-hand side.

$$x^2 + \frac{B}{A}x \qquad = -\frac{C}{A}$$

Take half the coefficient of the middle term, square it, $\left[\frac{1}{2}\left(\frac{B}{A}\right)\right]^2 = \frac{B^2}{4A^2}$, and add it to both sides of the equation.

$$x^2 + \frac{B}{A}x + \frac{B^2}{4A^2} = \frac{B^2}{4A^2} - \frac{C}{A}$$

Factor the left-hand side and simplify the right-hand side.

$$\left(x + \frac{B}{2A}\right)^2 = \frac{B^2}{4A^2} - \frac{C}{A} = \frac{B^2 - 4AC}{4A^2}$$

Apply the Square Root Theorem.

$$x + \frac{B}{2A} = \pm\sqrt{\frac{B^2 - 4AC}{4A^2}}$$

Isolate x.

$$x = -\frac{B}{2A} \pm \sqrt{\frac{B^2 - 4AC}{4A^2}}$$

Simplify the right-hand side (recall $A > 0$; hence, $\sqrt{4A^2} = 2A$).

$$= -\frac{B}{2A} \pm \frac{\sqrt{B^2 - 4AC}}{2A}$$

Solve for x by completing the square for the equation $Ax^2 + Bx + C = 0$ for $A < 0$.

$$x = \frac{-B \pm \sqrt{B^2 - 4AC}}{2A}$$

Similarly, we can demonstrate the same result for $A < 0$. Hence, we have

The Quadratic Formula

If $Ax^2 + Bx + C = 0$ and $A \neq 0$, then

$$x = \frac{-B \pm \sqrt{B^2 - 4AC}}{2A}$$

Example 3 Solve the following using the quadratic formula: $2x^2 - 2x = -8$

Solution Before using the quadratic formula, we put the equation in standard form and identify A, B, and C as follows: $2x^2 - 2x + 8 = 0$. Hence $A = 2$, $B = -2$, and $C = 8$. By the quadratic formula, we get

$$x = \frac{-(-2) \pm \sqrt{(-2)^2 - 4(2)(8)}}{2(2)} = \frac{2 \pm \sqrt{-60}}{4}$$

$$= \frac{2 \pm 2i\sqrt{15}}{4}$$

Factor and reduce.

$$= \frac{2(1 \pm i\sqrt{15})}{4} = \frac{1 \pm i\sqrt{15}}{2}$$

There are no real solutions.

In solving quadratic equations in general, we advise that unless the equation is already in the form $(x + p)^2 = d$, or $(x + a)(x + b) = 0$, you should put it in standard form. If there is no first-degree term, use the Square Root Theorem; otherwise try factoring. If the expression does not factor easily, then either use the quadratic formula or complete the square.

Example 4 Jenna throws a ball into the air. The equation $s = -16t^2 + 46t + 6$ gives the height of the ball, s (in feet), above the ground t seconds after she throws it. (See Figure 2.7.)
(a) How far above the ground is the ball exactly $\frac{3}{4}$ of a second after she throws it?
(b) How long will it take for the ball to hit the ground?

Figure 2.7

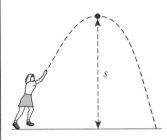

Solution **(a)** Since the ball is s feet above the ground in t seconds, to find its height above the ground we simply substitute $\frac{3}{4}$ for t in the given equation:

$$s = -16t^2 + 46t + 6$$

$$s = -16\left(\frac{3}{4}\right)^2 + 46\left(\frac{3}{4}\right) + 6 = -16\left(\frac{9}{16}\right) + 46\left(\frac{3}{4}\right) + 6 = \frac{63}{2}$$

$$= \boxed{31\frac{1}{2} \text{ feet}}$$

(b) Since the ball is s feet above the ground in t seconds, we note that when the ball hits the ground, $s = 0$. Hence we are looking for t when $s = 0$:

$$s = 0 = -16t^2 + 46t + 6 \qquad \text{Solve for } t \text{ by factoring.}$$

$$0 = -2(8t + 1)(t - 3) \qquad \text{Hence}$$

$$t = -\frac{1}{8} \quad \text{or} \quad x = 3 \qquad \text{Eliminate the negative answer. (Why?)}$$

It takes 3 seconds for the ball to hit the ground .

Example 5 A 20-ft by 55-ft rectangular swimming pool is surrounded by a concrete walkway of uniform width. If the area of the concrete walkway is 400 sq ft, find the width of the walkway.

Solution We can draw a diagram of the pool and walkway, labeling the uniform width of the concrete walkway x, as pictured in Figure 2.8.

Figure 2.8

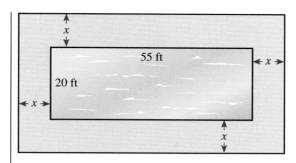

Looking at Figure 2.8, we can see that the length of the outer rectangle is $55 + x + x = 55 + 2x$ feet, and the width of the outer rectangle is $20 + x + x = 20 + 2x$ feet. Therefore, the area of the outer rectangle (the pool *and* walkway) is given by $(55 + 2x)(20 + 2x)$. Since the area of the pool is $20 \times 55 = 1100$ sq ft, and the area of the walkway is given as 400 sq ft, the diagram shows us that

$$\underbrace{1100}_{\text{Area of pool}} + \underbrace{400}_{\text{Area of walkway}} = \underbrace{(55 + 2x)(20 + 2x)}_{\text{Area of outer rectangle}}$$

Hence our equation is

$1100 + 400 = (55 + 2x)(20 + 2x)$	Which becomes
$1500 = 1100 + 150x + 4x^2$	And is put into standard form as
$0 = 4x^2 + 150x - 400$	Solve by factoring:
$0 = 2(2x - 5)(x + 40)$	Hence
$x = \dfrac{5}{2}$ or $x = -40$	Eliminate the negative answer.

The width of the walkway is $\dfrac{5}{2} = 2\dfrac{1}{2}$ feet .

The Discriminant

Let us further examine the quadratic formula; in particular, let's look at the radical portion $\sqrt{B^2 - 4AC}$. Notice that we often arrive at two solutions for the equation $Ax^2 + Bx + C = 0$:

$$x = \frac{-B + \sqrt{B^2 - 4AC}}{2A} \quad \text{and} \quad \frac{-B - \sqrt{B^2 - 4AC}}{2A}$$

So if $B^2 - 4AC = 0$, we would only have one solution: $x = -\dfrac{B}{2A}$. If $B^2 - 4AC < 0$, we would have two distinct solutions but the solutions would be complex. On the other hand, if $B^2 - 4AC > 0$, then the solutions would be distinct and real. The expression $B^2 - 4AC$ is called the **discriminant** of the equation $Ax^2 + Bx + C = 0$.

Example 6

Use the discriminant to describe the roots of $5x^2 - 3x = -4$.

Solution

$5x^2 - 3x = -4.$	First put the equation in standard form.
$5x^2 - 3x + 4 = 0$	Identify A, B, and C: $A = 5$, $B = -3$, and $C = 4$. Now calculate the discriminant:
$B^2 - 4AC = (-3)^2 - 4(5)(4) = -71$	Since the discriminant is negative, the roots are distinct and complex .

Occasionally we will be interested in constructing an equation given its solutions.

Suppose we wanted to find an equation with the solution 5. $x = 5$ is the simplest, or, $x - 5 = 0$. In fact, there are an infinite number of equations with this solution since we can add, subtract, multiply, and divide both sides of any of these equations by any (nonzero) constant and come up with a different looking equivalent equation. In general, we can construct an equation with solutions $r_1, r_2, r_3, \ldots$, and r_n by forming the product: $k(x - r_1)(x - r_2)(x - r_3) \cdots (x - r_n) = 0$, where $k \neq 0$ is any real number.

Example 7   Find an equation with solutions -3 and 4.

Solution   The simplest equation to use given the solutions -3 and 4 is $[x - (-3)][x - 4] = 0$, which can be simplified to $(x + 3)(x - 4) = 0$, or $x^2 - x - 12 = 0$. Another solution is $5(x + 3)(x - 4) = 0$. In fact, any equation of the form $k(x + 3)(x - 4) = 0$, where k is a real nonzero number, is an equation with the given solutions.   ∎

2.3 Exercises

In Exercises 1–6, solve for the variable by factoring or using the Square Root Theorem.

1. $0 = y^2 - 65$

2. $x^2 + 7 = 2$

3. $2x^2 - 7x = 15$

4. $2a^2 - 12 = -5a$

5. $x - 3 = \dfrac{1}{x + 3}$

6. $\dfrac{3}{x + 4} + \dfrac{5}{x + 2} = 6$

In Exercises 7–10, solve by completing the square.

7. $x^2 + 2x - 4 = 0$

8. $2a^2 + 4a - 3 = 0$

9. $\dfrac{1}{a - 5} + \dfrac{3}{a + 2} = 4$

10. $3a^2 - 6a + 5 = 0$

In Exercises 11–24, solve by any algebraic method.

11. $(x + 5)(x - 7) = 3$

12. $(t + 3)(t - 4) = t(t + 2)$

13. $(a + 1)(a - 3) = (3a + 1)(a - 2)$

14. $(2y + 3)(y + 5) = (y - 1)(y + 4)$

15. $(z + 3)(z - 1) = (z - 2)^2$

16. $(w - 3)^2 = (w + 4)(w - 2)$

17. $\dfrac{3}{x - 2} = x$

18. $\dfrac{1}{x + 2} = x - 4$

19. $\dfrac{x}{x + 2} - \dfrac{3}{x} = \dfrac{x + 1}{x}$

20. $\dfrac{3}{x - 2} + \dfrac{7}{x + 2} = \dfrac{x + 1}{x - 2}$

21. $\dfrac{x + 1}{3} = \dfrac{-5}{x + 5}$

22. $\dfrac{2y}{y - 2} = \dfrac{y + 3}{y}$

23. $0.35x - \dfrac{4}{x} = 0.8$

24. $0.001x^2 - 2x = 0.1$

25. Given $s_e = s_y \sqrt{1 - r_{xy}^2}$, if $s_e = 1.24$ and $r_{xy} = 0.63$, determine s_y to two places.

26. Given $s_e = s_y \sqrt{1 - r_{xy}^2}$, if $s_e = 1.72$ and $s_y = 3.73$, find r_{xy} to two places.

27. Given $Z = \dfrac{Z_{r_1} - Z_{r_2}}{\sqrt{\dfrac{1}{n_1 - 3} + \dfrac{1}{n_2 - 3}}}$ determine Z_{r_1} to two places, for $Z = 0.8$, $Z_{r_2} = 0.52$, $n_1 = 83$, and $n_2 = 68$.

28. Given $Z = \dfrac{Z_{r_1} - Z_{r_2}}{\sqrt{\dfrac{1}{n_1 - 3} + \dfrac{1}{n_2 - 3}}}$, find Z_{r_2} to two places, for $Z = 1.2$, $Z_{r_1} = 0.85$, $n_1 = 14$, and $n_2 = 30$.

29. Given $t = \dfrac{\overline{X} - a}{\dfrac{s_x}{\sqrt{n}}}$, determine a to one place, for $t = 10$, $\overline{X} = 70$, $s_x = 1.6$, and $n = 80$.

30. Given $t = \dfrac{\overline{X} - a}{\dfrac{s_x}{\sqrt{n}}}$, determine $\overline{X}$ to two places for $a = 65$, $t = 1.3$, $s_x = 3.4$, and $n = 50$.

In Exercises 31–36, use the discriminant to describe the roots of the equations.

31. $x^2 - 2x + 5 = 0$ **32.** $x^2 - 2x - 5 = 0$

33. $2x^2 - 7x = 15$ **34.** $2a^2 - 12 = -5a$

35. $x^2 = -12$ **36.** $5x^2 = 3x + 4$

In Exercises 37–42, find an equation with the given roots.

37. $5, 3$ **38.** $-2, 8$

39. $7, 0$ **40.** $-6, -4$

41. $i, -i$ **42.** $1 - i, 1 + i$

43. The sum of a number and its reciprocal is $\dfrac{13}{6}$. Find the number.

44. The difference between a number and its reciprocal is $\dfrac{8}{3}$. Find the number.

45. The product of two numbers is 5. If their sum is $\dfrac{9}{2}$, find the numbers.

46. The product of two numbers is 6. If their sum is $\dfrac{11}{2}$, find the numbers.

47. The length of a rectangle is 5 more than twice its width. If its area is 75 square feet, find its dimensions.

48. The length of a rectangle is 1 less than twice its width. If its area is 45 square feet, find its dimensions.

49. Find the length of the side of a square if its area is 8 square feet.

50. Find the length of the side of a square if its area is a square feet.

51. Find the area of a square if its diagonal is 8 inches.

52. Find the area of a square if its diagonal is a inches.

53. A 21-ft by 21-ft square swimming pool is surrounded by a path of uniform width. If the area of the path is 184 square feet, find the width of the path.

54. An 8-in. by 10-in. picture is surrounded by a frame of uniform width. If the area of the frame is 40 square inches, find the width of the frame.

55. A circular garden is surrounded by a path of uniform width. If the path has area 57π square feet and the radius of the garden is 8 feet, find the width of the path.

56. A circular garden is surrounded by a path of uniform width. If the path has area 200π square feet and the radius of the garden is 10 feet, find the width of the path to the nearest foot.

57. Jose drops a ball straight down from a building. The equation $s = -16t^2 + 160$ gives the distance, s, in feet the ball is above the ground t seconds after he drops it.
   (a) How high above the ground is the ball at $t = 2$ seconds?
   (b) How long does it take for the ball to hit the ground?

58. Samantha throws a ball straight up into the air off a building. The equation $s = -16t^2 + 48t + 28$ gives the distance, s, in feet the ball is above the ground t seconds after she throws it up.
   (a) How high above the ground is the ball at $t = 3/2$ seconds?
   (b) How long does it take for the ball to hit the ground?

59. Verify that $2 + i$ is a solution to $2x^2 - 8x + 10 = 0$.

60. Verify that $2 + \sqrt{3}$ is a solution to $x^2 - 4x + 1 = 0$.

2.4 Radical Equations: Equations Reducible to Quadratic Form

Radical Equations

The multiplicative property of equality yields the following theorem:

Theorem 2.1

If $a = b$, then $a^n = b^n$.

If we want to solve an equation containing a radical or fractional exponent such as $x^{1/3} = 5$ or $\sqrt[3]{x} = 5$, then we can apply Theorem 2.1 by raising each side of the equation to the third power:

$$x^{1/3} = 5 \qquad \text{Raise each side to the third power (the power that will yield } x^1\text{).}$$
$$(x^{1/3})^3 = (5)^3 \qquad \text{By the second rule of exponents, } (x^{1/3})^3 = x^{(1/3)\cdot 3} = x^1.$$
$$x = 5^3 = 125$$

We should note some things about using this theorem before we continue. First, using this theorem (i.e., raising each side of an equation to the same power) will *not* necessarily yield an equivalent equation, as was true with the other equality properties. This theorem guarantees only that solutions to the first equation will show up as solutions to the transformed equation.

For example, if we start with the equation $x = 2$, we observe that there is only one solution to this equation, 2. However, if we square each side of this equation, we will arrive at $x^2 = 4$, which has two solutions: $x = +2$ and $x = -2$. By squaring both sides of the equation, we picked up a solution to the second equation that was not a solution to the first. This extra solution is called an *extraneous solution*. Thus, we must always *check the possible solutions in the original equation to ensure we have not picked up extraneous solutions.*

Example 1   Solve the following.
(a) $\sqrt{3x} - 2 = 5$
(b) $(x - 2)^{1/5} = 3$
(c) $\sqrt{2x + 1} - \sqrt{4x} = -1$
(d) $x^{-5/2} = 32$

Solution   (a) $\sqrt{3x} - 2 = 5$

If we tried squaring both sides now, and did it properly, the new equation would then be $3x - 4\sqrt{3x} + 4 = 25$. (Note that we still have a radical in the transformed equation.) Your first step should be to isolate the radical, and then apply the theorem as follows:

$$\sqrt{3x} - 2 = 5 \qquad \text{Isolate } \sqrt{3x}.$$
$$\sqrt{3x} = 7 \qquad \text{Then square both sides of the equation}$$
$$(\sqrt{3x})^2 = 7^2 \qquad \text{to get}$$
$$3x = 49 \qquad \text{Now isolate } x.$$

$$x = \boxed{\dfrac{49}{3}}$$

You must check your answer in the original equation. In this case, the answer checks.

(b) $(x - 2)^{1/5} = 3 \qquad \text{Raise each side to the fifth power}$
$$[(x - 2)^{1/5}]^5 = 3^5 \qquad \text{to get}$$
$$x - 2 = 243 \qquad \text{Hence}$$

$$x = \boxed{245}$$

Again, if you check, you will find that the answer is a solution to the original equation.

(c) $\sqrt{2x+1} - \sqrt{4x} = -1$

Squaring both sides at this point will produce the radical $\sqrt{4x(2x-1)}$ in the middle term. It is usually easier to isolate the more complicated radical first, then square both sides, as demonstrated here:

$$\sqrt{2x+1} - \sqrt{4x} = -1$$
  Isolate $\sqrt{2x+1}$ on the left-hand side.

$$\sqrt{2x+1} = \sqrt{4x} - 1$$
  Now square both sides.

$$(\sqrt{2x+1})^2 = (\sqrt{4x}-1)^2$$

$$2x + 1 = 4x - 2\sqrt{4x} + 1$$
  (Note the differences between how the right-hand side and left-hand side are squared.) Isolate $\sqrt{4x}$.

$$-2x = -2\sqrt{4x}$$
  Divide both sides by -2.

$$x = \sqrt{4x}$$
  Now square both sides of the equation.

$$x^2 = 4x$$
  A quadratic equation: Put into standard form.

$$x^2 - 4x = 0$$
  Solve by factoring.

$$x(x-4) = 0$$
  Hence

$$x = 0 \quad \text{or} \quad x = 4$$

If we check both answers in the original equation, we find that 0 does not work, but 4 does. The answer is $\boxed{4}$.

(d) $x^{-5/2} = 32$

As long as we are careful, we can apply the power theorem with n as a rational exponent as well.

$$x^{-5/2} = 32 \quad \text{Raise each side to the } -2/5 \text{ power:}$$

$$[x^{-5/2}]^{-2/5} = 32^{-2/5}$$

$$x = 32^{-2/5}$$

$$= \frac{1}{32^{2/5}} = \frac{1}{[32^{1/5}]^2} = \frac{1}{[2]^2} = \boxed{\frac{1}{4}} \quad \text{Again, check your solution.}$$

Equations Reducible to Quadratic Form

The equation $x^4 + x^2 - 6 = 0$ is not a quadratic equation; however, we can use the methods discussed in this section to solve such an equation.

You may see that $x^4 + x^2 - 6 = 0$ factors into $(x^2 - 2)(x^2 + 3) = 0$, which means we can set each factor equal to zero to get $x^2 - 2 = 0$ and $x^2 + 3 = 0$. Solving each equation by applying the Square Root Theorem, we get $x = \pm\sqrt{2}$ and $x = \pm i\sqrt{3}$.

If you don't see that it factors in this way, we can use the method of **substitution of variables** as follows:

Let $u = x^2$ then $u^2 = (x^2)^2 = x^4$. By substituting u for x^2 and u^2 for x^4 in $x^4 + x^2 - 6 = 0$,

$$x^4 + x^2 - 6 = 0 \quad \text{Becomes}$$

$$u^2 + u - 6 = 0 \quad \text{We recognize } u^2 + u - 6 = 0 \text{ as a quadratic equation, which can be solved by factoring.}$$

$$(u+3)(u-2) = 0 \quad \text{Hence}$$

$$u = -3 \quad \text{or} \quad u = 2 \quad \text{But we must solve for the original variable, } x. \text{ We substitute back } x^2 \text{ for } u \text{ and get}$$

$$x^2 = -3 \quad \text{or} \quad x^2 = 2 \quad \text{Which yields}$$

$$x = \pm i\sqrt{3} \quad \text{or} \quad x = \pm\sqrt{2}$$

The pattern to watch for is that, ignoring the coefficients, the higher-degree expression should be the square of the lower-degree expression. (The reverse is true if the exponents are negative.)

Example 2 | Solve the following: $4x^{2/3} - 9x^{1/3} + 2 = 0$

Solution | $4x^{2/3} - 9x^{1/3} + 2 = 0$

Note that $x^{2/3}$ is the square of $x^{1/3}$. If we let $u = x^{1/3}$, then $u^2 = (x^{1/3})^2 = x^{2/3}$.

Substitute $u = x^{1/3}$, and $u^2 = x^{2/3}$ to get

$$4u^2 - 9u + 2 = 0$$ Solve by factoring.

$$(4u - 1)(u - 2) = 0$$ Thus

$$u = 1/4 \quad \text{or} \quad u = 2$$ Substitute back $x^{1/3}$ for u to get.

$$x^{1/3} = \frac{1}{4} \quad \text{or} \quad x^{1/3} = 2$$ Cube both sides of each equation to get

$$(x^{1/3})^3 = \left(\frac{1}{4}\right)^3 \quad \text{or} \quad (x^{1/3})^3 = (2)^3$$

$$x = \frac{1}{64} \quad \text{or} \quad x = 8$$

We can apply the same procedure to solve second-degree literal equations as well.

Example 3 | Solve the following for y:

$$3x^2 - 2xy = 5y^2$$

Solution | We note that because the equation contains both a first-degree term and a second-degree term of the variable we are solving for, y, we cannot use the Square Root Theorem. We will factor:

$$3x^2 - 2xy = 5y^2$$ Put into standard form

$$3x^2 - 2xy - 5y^2 = 0$$ and factor the left-hand side.

$$(3x - 5y)(x + y) = 0$$ Hence we have

$$3x - 5y = 0 \quad \text{or} \quad x + y = 0$$ We solve each first-degree literal equation for y

$$3x = 5y \quad \text{or} \quad y = -x$$ to get:

$$y = \frac{3x}{5} \quad \text{or} \quad y = -x$$

Example 4 | Solve the following for r:

$$s_e = s_y \sqrt{1 - r^2} \quad (\text{assume } s_y > 0)$$

Solution | $s_e = s_y\sqrt{1 - r^2}$ Square each side.

$$(s_e)^2 = (s_y)^2(1 - r^2)$$ Multiply out the right-hand side.

$$(s_e)^2 = (s_y)^2 - (s_y)^2 r^2$$ Isolate r^2.

$$(s_y)^2 r^2 = (s_y)^2 - (s_e)^2$$

$$r^2 = \frac{(s_y)^2 - (s_e)^2}{(s_y)^2}$$ Apply the Square Root Theorem.

$$r = \pm\sqrt{\frac{(s_y)^2 - (s_e)^2}{(s_y)^2}} = \pm\frac{\sqrt{(s_y)^2 - (s_e)^2}}{s_y}$$

Example 5   Solve the following: $x^3 + 4x^2 - x - 4 = 0$

Solution

$$x^3 + 4x^2 - x - 4 = 0 \quad \text{Factor the left-hand side by grouping.}$$

$$x^2(x + 4) - (x + 4) = 0 \quad \text{Factor } x + 4 \text{ from the left-hand side.}$$

$$(x^2 - 1)(x + 4) = 0 \quad \text{We can still factor } x^2 - 1.$$

$$(x - 1)(x + 1)(x + 4) = 0 \quad \text{Set each factor equal to 0 and get}$$

$$x = 1, -1, \quad \text{or} \quad -4$$

Example 6   Solve the following: $\dfrac{(x - 5)^{1/2}(x + 3)^{1/5}}{(x - 2)^{1/3}} = 0$

Solution   We are being asked to determine when this fraction is equal to 0. Provided we keep in mind that $x \neq 2$ (that is, the denominator cannot be 0), we can start to solve this equation by multiplying both sides of the equation by $(x - 2)^{1/3}$. Alternatively, we can simply recognize that the only way a fraction can be equal to 0 is if its numerator is 0 (and its denominator is not 0). Hence if $x \neq 2$,

$$\dfrac{(x - 5)^{1/2}(x + 3)^{1/5}}{(x - 2)^{1/3}} = 0 \quad \text{Becomes}$$

Show that if $q \neq 0$, then the rational expression $\dfrac{p}{q} = 0$ if and only if $p = 0$.

$$(x - 5)^{1/2}(x + 3)^{1/5} = 0 \quad \begin{array}{l}\text{Since the product is equal to 0, we set each}\\\text{factor to 0 to get}\end{array}$$

$$(x - 5)^{1/2} = 0 \quad \text{or} \quad (x + 3)^{1/5} = 0 \quad \text{Which yields}$$

$$x = 5 \quad \text{or} \quad x = -3$$

2.4 Exercises

1. Given $s_e = s_y\sqrt{1 - r_{xy}^2}$, if $s_y = 1.04$ and $s_e = 0.90$, find r_{xy} to two places.

2. Given $t = \dfrac{\overline{X} - a}{\dfrac{s_x}{\sqrt{n}}}$, find the nearest whole number n for $t = -4.2$, $a = 70$, $\overline{X} = 66$, and $s_x = 3.4$.

3. Given $t = \dfrac{\overline{X} - a}{\dfrac{s_x}{\sqrt{n}}}$, find the nearest whole number n for $t = 2.1$, $a = 85$, $\overline{X} = 90$, and $s_x = 5.8$.

4. Given $Z = \dfrac{Z_{r_1} - Z_{r_2}}{\sqrt{\dfrac{1}{n_1 - 3} + \dfrac{1}{n_2 - 3}}}$, find the nearest whole number n_1 for $Z = 0.8$, $Z_{r_1} = 0.62$, $Z_{r_2} = 0.52$, and $n_2 = 83$.

In Exercises 5–40, solve for the variable.

5. $\sqrt{a} - 8 = 14$

6. $5\sqrt{2x} - 7 = 8$

7. $\sqrt{x + 4} = 9$

8. $3\sqrt{x} = \sqrt{3x}$

9. $\sqrt{x - 2} = x - 4$

10. $\sqrt{3x + 1} + 3 = x$

11. $\sqrt{7x + 1} - 2\sqrt{x} = 2$

12. $\sqrt{3s + 4} - \sqrt{s} = 2$

13. $x^4 - 17x^2 + 16 = 0$

14. $x^3 - 2x^2 - 15x = 0$

15. $x^{2/3} - 6x^{1/3} - 7 = 0$

16. $2x^{1/2} - 9x^{1/4} + 4 = 0$

17. $x^{1/2} + 8x^{1/4} + 7 = 0$

18. $x^{-2} - 2x^{-1} - 15 = 0$

19. $2a^{-1/2} - 9a^{-1/4} = -4$

20. $x^{-2/3} - 3x^{-1/3} = 10$ **21.** $\sqrt{a} - \sqrt[4]{a} - 6 = 0$

22. $(a - 4)^2 - 3(a - 4) - 10 = 0$

23. $\left(x + \dfrac{12}{x}\right)^2 - 15\left(x + \dfrac{12}{x}\right) + 56 = 0$

24. $x^{2/3} = 9$ **25.** $a^{3/4} + 2 = 5$

26. $x^{5/3} + 4 = 6$ **27.** $(x - 3)^{2/3} = 9$

28. $(y + 2)^{3/5} = 3$ **29.** $y^{-3/4} = 3$

30. $x^{-2/3} + 5 = 12$

31. $y^4 - 7y^2 - 18 = 0$

32. $x^4 - 81 = 0$

33. $a^4 - 35a^2 - 36 = 0$

34. $a^4 - 15a^2 - 16 = 0$

35. $x^3 - 9x + 4x^2 - 36 = 0$

36. $a^3 + a^2 - 4a - 4 = 0$

37. $\dfrac{(x - 1)^2}{(x - 8)^3} = 0$

38. $\dfrac{(x - 3)^5(x + 6)^3}{(x - 5)^2} = 0$

39. $\dfrac{(x - 3)^{1/3}}{(x + 7)^{2/3}} = 0$

40. $\dfrac{(x + 1)^{1/5}(x - 6)^{2/3}}{(x - 1)^{1/2}} = 0$

In Exercises 41–46, solve for the given variable.

41. $K = \dfrac{2gm}{s^2}$ for s $(s > 0)$

42. $V = \dfrac{2}{3}\pi r^2$ for r $(r > 0)$

43. $3a^2 + 2b = 5b - 2a^2x - 2$ for a

44. $3x^2 + 8 = 5b - 2\pi x^2 - 2$ for x

45. $\sigma_r = \sqrt{\dfrac{1 - \rho^2}{n - 1}}$ for ρ $(\rho > 0)$

46. $x^2 - 6xy + 5y^2 = 0$ for y

Question for Thought

47. Use the multiplication property of equality to prove the following: If $a = b$, then $a^n = b^n$.

2.5   Quadratic and Rational Inequalities

Quadratic Inequalities

A quadratic inequality is in **standard form** if it is in the form $Ax^2 + Bx + C < 0$. (We can replace $<$ with $>$, $\leq$, or $\geq$.)

If we keep in mind that $u > 0$ means u is positive, then solving an inequality such as $2x^2 + 5x - 3 > 0$ means we are interested in finding the values of x that will make $2x^2 + 5x - 3$ positive. Or, since $2x^2 + 5x - 3 = (2x - 1)(x + 3)$, we are looking for values of x that make $(2x - 1)(x + 3)$ positive.

Although we will refer to the real number line, our approach in solving quadratic inequalities will be primarily algebraic. After putting the inequality in standard form, we will determine the sign of each factor of the expression for various values of x. Then we determine the solution by examining the sign of the product. This process is called a *sign analysis*.

Let's return to the problem $2x^2 + 5x - 3 > 0$: This translates into finding values of x that make $(2x - 1)(x + 3)$ positive. For $(2x - 1)(x + 3)$ to be positive, the factors must be either both positive or both negative. To determine when this happens, we first find the values of x for which $(2x - 1)(x + 3)$ is equal to 0; we call these the *cut points* of $(2x - 1)(x + 3)$. The cut points are $\dfrac{1}{2}$ and -3.

Next we draw a number line and examine the sign of each factor as x takes on various values on the number line, especially around the cut points $\frac{1}{2}$ and -3.

Figure 2.9

We determine the sign of each factor by substituting a value from each interval. For instance, if we substitute $x = -5$ into $x + 3$ we get -2 and therefore the factor $x + 3$ is negative to the left of -3.

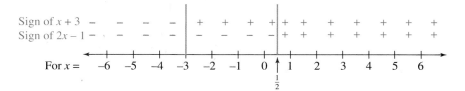

Figure 2.9 illustrates that the factor $x + 3$ is negative when $x < -3$, and positive when $x > -3$. It also shows that $2x - 1$ is negative for $x < \frac{1}{2}$ and positive for $x > \frac{1}{2}$. Looking at the figure and the signs of the factors, we easily observe that the *product* of the two factors is positive when $x < -3$ or $x > \frac{1}{2}$. See Figure 2.10.

Figure 2.10

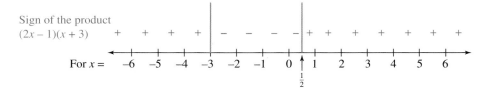

Hence $2x^2 + 5x - 3 > 0$ (is positive) when x is in either of the two intervals $(-\infty, -3)$ or $(\frac{1}{2}, \infty)$, so the solution is $(-\infty, -3) \cup (\frac{1}{2}, \infty)$.

We can shortcut this analysis by observing that (1) the cut points of the inequalities will break up the number line into intervals, and (2) most importantly, the sign of the product does not change *within* an interval (excluding the cut points); that is, *if the expression is positive (or negative) for one value within the interval, it is positive (or negative) for all values within the interval.*

Hence, if we want to find the sign of the product within each interval, all we need to do is test one value within the interval. For example, we found that the cut points for the expression $2x^2 + 5x - 3 = (2x - 1)(x + 3)$ are -3 and $\frac{1}{2}$. This breaks up the number line into three intervals, as shown in Figure 2.11.

Figure 2.11

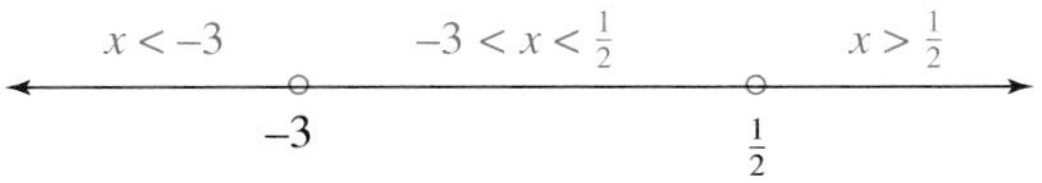

Review the preceding sign analysis and think about why the product never changes its sign within an interval.

$$x < -3 \qquad -3 < x < \tfrac{1}{2} \qquad x > \tfrac{1}{2}$$

We choose any value in the leftmost interval (i.e., any value less than -3), say, $x = -10$. When $x = -10$, the expression $(2x - 1)(x + 3)$ becomes $(-21)(-7)$, which is positive, and therefore $(2x - 1)(x + 3)$ is positive for *all* values of x in the interval $(-\infty, -3)$.

We choose any value in the middle interval, $(-3, \frac{1}{2})$, such as $x = 0$, and note that when $x = 0$ the expression $(2x - 1)(x + 3)$ becomes $(-1)(3)$, which is negative; therefore, $(2x - 1)(x + 3)$ is negative for *all* values of x in $(-3, \frac{1}{2})$.

Finally, choose any value in the interval $(\frac{1}{2}, \infty)$, such as $x = 100$, and note that for $x = 100$, the expression $(2x - 1)(x + 3)$ is positive; therefore, $(2x - 1)(x + 3)$ is positive for *all* values of x in $(\frac{1}{2}, \infty)$. (Note we are not interested in the exact value but only in the *sign* of the evaluated expression.)

Hence the solution is $(-\infty, -3) \cup (\frac{1}{2}, \infty)$ because $2x^2 + 5x - 3$ is positive when x lies in these intervals. (On the other hand, the identical analysis also yields the solution to $2x^2 + 5x - 3 < 0$. That is, $2x^2 + 5x - 3$ is negative when x is in the interval $(-3, \frac{1}{2})$.)

We can apply the same approach to solving higher-degree polynomial inequalities.

Example 1 Solve the following: $x^4 \geq 9x^2$

Solution

$$x^4 \geq 9x^2 \qquad \text{Put the inequality in standard form.}$$
$$x^4 - 9x^2 \geq 0 \qquad \text{Factor.}$$
$$x^2(x+3)(x-3) \geq 0 \qquad \begin{array}{l}\text{Find the cut points. Set } x^2, x+3, \text{ and } x-3 \text{ equal to 0.}\\\text{The cut points are 0, } -3 \text{ and 3.}\end{array}$$

We plot the cut points on the number line in Figure 2.12, choose any value within each interval determined by the cut points, and check the sign of the product. We used $x = -10$, $x = -2$, $x = 2$, and $x = 10$ as test values for the four intervals and recorded the sign of $x^2(x+3)(x-3)$ on each interval in Figure 2.12.

Figure 2.12

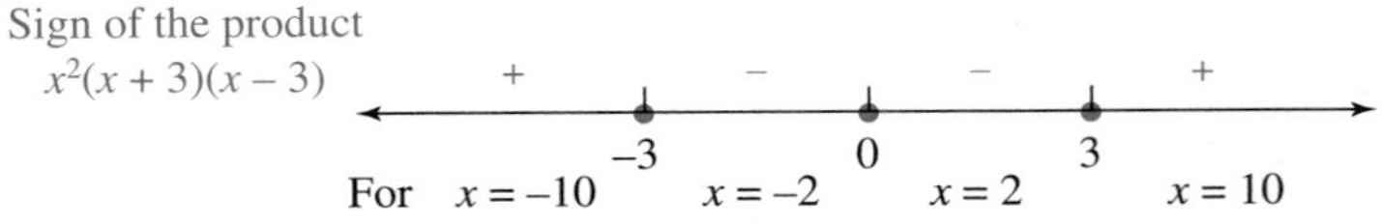

Sign of the product
$x^2(x+3)(x-3)$

$$+ \qquad - \qquad - \qquad +$$
$$-3 \qquad 0 \qquad 3$$
For $x=-10 \qquad x=-2 \qquad x=2 \qquad x=10$

Thus the solution to the original inequality is $(-\infty, -3] \cup [3, \infty)$. Note that the endpoints -3 and 3 are included since they do satisfy the inequality.

Example 2 Solve the following: $x^2 - 2x - 2 < 0$

Solution

Because we cannot factor $x^2 - 2x - 2$, we use the quadratic formula to find that its roots are $1 \pm \sqrt{3}$. This gives the cut points for the polynomial $x^2 - 2x - 2$. We use the sign analysis in Figure 2.13 with the test points given. NOTE: $1 + \sqrt{3} \approx 2.7$ and $1 - \sqrt{3} \approx -0.7$.

Figure 2.13

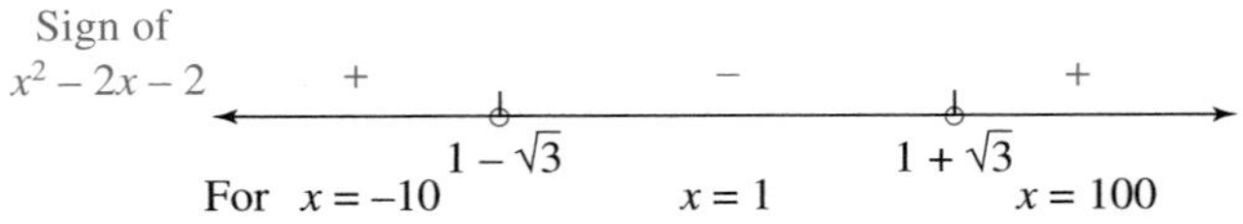

Sign of
$x^2 - 2x - 2$

$$+ \qquad - \qquad +$$
$$1 - \sqrt{3} \qquad 1 + \sqrt{3}$$
For $x=-10 \qquad x=1 \qquad x=100$

Substituting the test values -10, 1, and 100 for x in the expression $x^2 - 2x - 2$, we find that $x^2 - 2x - 2$ is negative only when x is within the interval $(1 - \sqrt{3}, 1 + \sqrt{3})$.

Example 3 For what values of x is $\sqrt{5x - 2x^2}$ a real number?

Solution

We first note that $\sqrt{5x - 2x^2}$ is real when the expression under the radical is nonnegative. Hence, the question reduces to solving the quadratic inequality $5x - 2x^2 \geq 0$.

$$5x - 2x^2 \geq 0 \qquad \text{Factor.}$$
$$x(5 - 2x) \geq 0 \qquad \text{The cut points are 0 and } \frac{5}{2}. \text{ (See Figure 2.14.)}$$

Figure 2.14

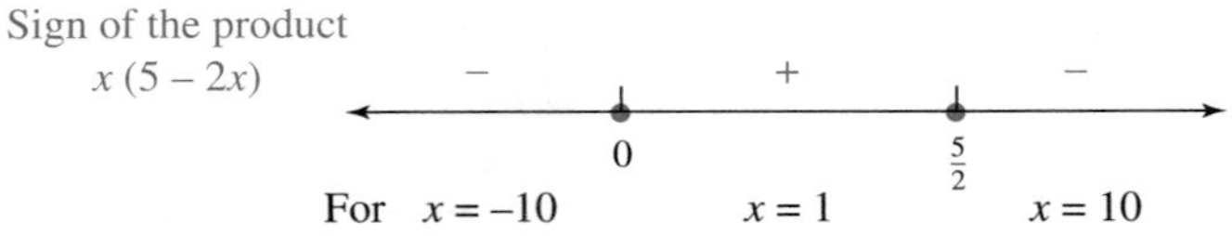

Sign of the product
$x(5-2x)$

$$- \qquad + \qquad -$$
$$0 \qquad \frac{5}{2}$$
For $x=-10 \qquad x=1 \qquad x=10$

Thus $\sqrt{5x - 2x^2}$ is real when $0 \leq x \leq \frac{5}{2}$. Note the cut points are included.

Rational Inequalities

We can apply this same method in solving rational inequalities such as

$$\frac{x - 4}{x - 3} > 0 \quad \text{or} \quad \frac{2x - 1}{x - 5} \leq 2$$

These inequalities are different from the linear inequalities with constants in the denominator. In attempting to solve $\frac{x - 4}{x - 3} < 0$, our first inclination might be to clear the denominator by multiplying both sides of the inequality by $x - 3$. With *equations,* this approach is appropriate provided we keep in mind $x \neq 3$. With inequalities, however, we need to know whether the multiplier, $x - 3$, is positive or negative to be able to determine whether the inequality symbol should be reversed. Although we could reason out the answer this way on a case-by-case basis, it is easier to approach this problem in a manner similar to that used for quadratic inequalities, as illustrated in the next example.

Example 4   Solve the following: **(a)** $\dfrac{x - 4}{x - 3} < 0$ **(b)** $\dfrac{2x - 1}{x - 5} \leq 2$

Solution   **(a)** To solve $\dfrac{x - 4}{x - 3} < 0$, we must find the values of x that make $\dfrac{x - 4}{x - 3}$ negative. For this quotient to be negative, the numerator and denominator must have opposite signs. *We now define the cut points of a rational expression to be the value(s) where the numerator is* 0 *or where the denominator is* 0.

$$\frac{x - 4}{x - 3} < 0$$ The cut points are 4 (where the numerator is 0) and 3 (where the denominator is 0). We plot the cut points on the number line and perform the sign analysis to get Figure 2.15.

Figure 2.15

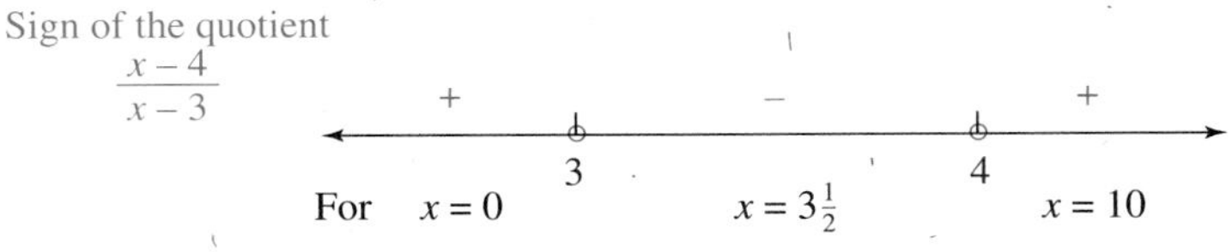

Sign of the quotient
$\dfrac{x - 4}{x - 3}$

$$\begin{array}{ccccc} & + & & - & & + \\ \hline & & 3 & & 4 & \end{array}$$

For $x = 0$ $x = 3\frac{1}{2}$ $x = 10$

The solution is $(3, 4)$.

(b) We must first put this expression in the form $R \leq 0$, where R is a rational expression, before we can apply the sign analysis:

$$\frac{2x - 1}{x - 5} \leq 2$$ Get 0 on the right-hand side.

$$\frac{2x - 1}{x - 5} - 2 \leq 0$$ Express the left-hand side as a single fraction.

$$\frac{2x - 1}{x - 5} - \frac{2(x - 5)}{x - 5} \leq 0$$

$$\frac{2x - 1 - 2(x - 5)}{x - 5} \leq 0$$ Which becomes

$$\frac{9}{x - 5} \leq 0$$ There is only one cut point: 5. See Figure 2.16.

Figure 2.16

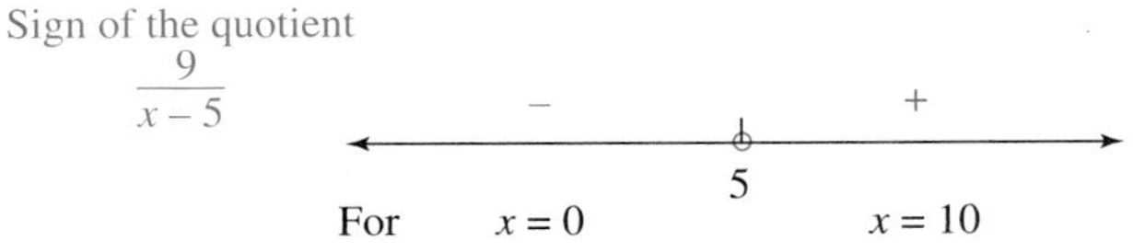

Sign of the quotient
$\dfrac{9}{x - 5}$

$$\begin{array}{ccc} & - & & + \\ \hline & & 5 & \end{array}$$

For $x = 0$ $x = 10$

We exclude $x = 5$. (Why?) Hence, the answer is $(-\infty, 5)$.

2.5 Exercises

In Exercises 1–50, solve the inequalities. Express your answer using interval notation.

1. $x^2 + 2x - 24 > 0$ **2.** $x^2 + 3x - 10 < 0$

3. $2x^2 - 3x - 2 < 0$ **4.** $5x^2 - 14x - 3 > 0$

5. $x^2 - 5x \le 24$ **6.** $x^2 - 10 \ge -1$

7. $3x - 2x^2 \le -5$ **8.** $13x \le 5 - 6x^2$

9. $15x^2 - 2 \le 7x$ **10.** $6y^2 + 1 \le 5y$

11. $8x^2 - 6 > -8x$ **12.** $6y^2 - 12 > -21y$

13. $9x^2 - 6x + 1 \ge 0$ **14.** $10x + 1 \le -25x^2$

15. $x^2 - 2x > -1$ **16.** $x^2 + 4x + 4 < 0$

17. $y^2 > 1$ **18.** $x^2 \le 16$

19. $x^3 + 4x^2 - x - 4 < 0$ **20.** $x^4 - 16 > 0$

21. $x^3 + 2x^2 - 4x - 8 \ge 0$

22. $x^3 + x^2 - 9x - 9 > 0$

23. $3 + 3x - x^2 > 0$ **24.** $x^2 - 2x - 2 < 0$

25. $2x^2 - x - 2 \ge 0$ **26.** $1 + 3x - 2x^2 \le 0$

27. $\dfrac{5}{x + 1} < 0$ **28.** $\dfrac{3}{x - 4} \le 0$

29. $\dfrac{2x - 1}{x + 5} \ge 0$ **30.** $\dfrac{x - 7}{x + 7} > 0$

31. $\dfrac{x - 2}{x + 1} < 0$ **32.** $\dfrac{2x + 3}{5 - x} \le 0$

33. $\dfrac{5x - 1}{2 - x} \ge 0$ **34.** $\dfrac{7x + 3}{x - 1} > 0$

35. $\dfrac{-2x}{x - 1} \le 0$ **36.** $\dfrac{-4x}{x + 1} \ge 0$

37. $\dfrac{3x}{x - 1} < 1$ **38.** $\dfrac{4}{x - 1} > 3$

39. $\dfrac{x - 1}{x + 1} \ge 2$ **40.** $\dfrac{2x - 3}{x - 3} \le 1$

41. $\dfrac{x - 2}{(x + 1)(x - 3)} \ge 0$ **42.** $\dfrac{x + 5}{(1 - x)(2x + 3)} \le 0$

43. $\dfrac{2x}{x^2 - 1} \le 0$ **44.** $\dfrac{5x}{x^2 - 16} \ge 0$

45. $\dfrac{9 - x^2}{x + 5} < 0$

46. $\dfrac{4 - x^2}{x - 3} > 0$

47. $\dfrac{-2x}{x^2 + 1} \le 0$

48. $\dfrac{5x}{x^2 + 16} \ge 0$

49. $\dfrac{a^2 - 1}{a^2 - 16} > 0$

50. $\dfrac{x^2 - 2x - 3}{x^2 + x - 2} < 0$

In Exercises 51–54, given that $|a| < |b| \Leftrightarrow a^2 < b^2$, use this property to solve the following absolute value inequalities. Express your answer using interval notation.

51. $|2x - 1| < |x|$ **52.** $|3x - 1| > |x|$

53. $|x - 3| < |x + 2|$ **54.** $|2x + 1| > |x + 3|$

55. A biologist finds that the size of the population of a certain species of marine life is related to the temperature of the water at a particular location in the following way:

$$N = 1,000(-T^2 + 42T - 320)$$

where N is the size of the population and T is the water temperature in degrees Celsius. What must the water temperature range be to support a population size of at least 96,000?

56. The concentration of a certain drug in the bloodstream varies with time in the following way:

$$C = \dfrac{3t}{t^2 + t + 1}$$

where C is the concentration of the drug in the blood stream in milligram/liter t hours after it is taken orally. It was determined that the drug is effective if the concentration is at least 0.6 milligram/liter. During what time interval is the drug effective?

2.6   Substitution

Until now we have discussed single equations and inequalities as conditions placed on one variable. On occasion, we may have to deal with more than one equation in which conditions are placed on one or more variables.

Example 1 | If the perimeter of a square is 32 inches, find its area.

Solution | We begin with a diagram, Figure 2.17, and the formula for the area of a square, which is $A = s^2$, where s is the length of a side. This formula requires that we know s, the length of a side, to compute the area. We are not given the length of the side, but we are given the perimeter. Hence we find the formula for the perimeter of a square and see whether we can somehow use that information to find the length of the side. The formula for perimeter is $P = 4s$.

To find s, given the perimeter, we substitute $P = 32$ in the formula for perimeter and solve for s:

$$P = 4s \quad \Rightarrow \quad 32 = 4s \quad \Rightarrow \quad s = 8 \text{ inches}$$

Now that we have $s = 8$, we can find the area of the square by substituting $s = 8$ in the formula for area, $A = s^2$:

$$A = s^2 \quad \Rightarrow \quad A = 8^2 \quad \Rightarrow \quad A = 64 \text{ sq in.}$$

Hence the area of the square is 64 sq in.   ∎

Figure 2.17

Suppose we are given a task that requires us continuously to find the area of a square given its perimeter. Rather than continuing through the two-step process described in Example 1, it would be more convenient if we had a single formula yielding a direct relationship between the area of a square and its perimeter. We may state this problem as "Find the area of a square given its perimeter" or "Express the area of a square in terms of its perimeter." We will demonstrate that the steps taken in the process of developing this formula are not much different from the approach taken in Example 1.

Example 2 | If the perimeter of a square is P, find its area, A, in terms of P.

Solution | Again, our first response is to begin with the formula for the area of a square, which is $A = s^2$, where s is the length of a side.

This formula requires that we have s, the length of a side, to compute the area. We are not given the length of the side, but we are given the perimeter. Hence we find a formula that relates the perimeter to the length of a side. This formula is $P = 4s$.

To find s given P means to solve for s in terms of P. Hence

$$P = 4s \quad \Rightarrow \quad s = \frac{P}{4}$$

Now that we have s (in terms of P), we can find the area of the square by substituting $s = \frac{P}{4}$ in the formula for area, $A = s^2$:

$$A = s^2 \qquad \text{Substitute } s = \frac{P}{4}.$$

$$A = \left(\frac{P}{4}\right)^2 = \frac{P^2}{16}$$

Hence $A = \dfrac{P^2}{16}$.

Notice that if $P = 32$, then $A = \dfrac{32^2}{16} = 64$, as we found in Example 1.   ∎

In finding the area of a square, A, in terms of its perimeter, P, we had to solve explicitly for the side, s, in $P = 4s$, and then substitute $\frac{P}{4}$ for s in the formula $A = s^2$. This method of replacing a variable by an expression involving another is called *substitution*.

Example 3   Find the area, A, of a square in terms of the length of its diagonal, d.

Solution   We approach this as if we were given a value for the length of the diagonal and had to find the area. Again, our first response is to begin with the formula for the area of a square, which is $A = s^2$, where s is the length of a side.

The formula requires that we have s, the length of a side, to compute the area. We are not given the length of the side, but we are given the length of its diagonal as d. Hence we need to find a relationship between the side s and the diagonal d. We can draw a picture of a square with a diagonal, labeled d, and side labeled s. See Figure 2.18.

We note that a right triangle is formed with the diagonal as the hypotenuse, and by the Pythagorean Theorem, the relationship between its sides and diagonal is

$$s^2 + s^2 = d^2 \qquad \text{which simplifies to} \qquad 2s^2 = d^2$$

We still need s to find A, so we solve for s in terms of d:

$$2s^2 = d^2$$

$$s^2 = \frac{d^2}{2} \qquad \text{We can stop here.}$$

Rather than continue our attempt to solve for s, we note that we have A in terms of s^2. Since we now have s^2 in terms of d, we can find the area of the square by substituting $s^2 = \frac{d^2}{2}$ in the formula for area, $A = s^2$:

$$A = s^2 \quad \text{Substitute } s^2 = \frac{d^2}{2}.$$

$$A = \frac{d^2}{2}$$

Hence $A = \dfrac{d^2}{2}$. ∎

Figure 2.18

Example 4   A boat is anchored in the middle of a still lake when its fuel tank begins to leak, causing a circular slick on its surface (see Figure 2.19). If the radius of the slick is growing at a constant rate of 2 ft/min, how much area will the slick cover **(a)** 1 hour from the time it started? **(b)** t minutes from the time it started?

Figure 2.19

Solution   **(a)** In this example we begin with the area of a circle, which is given by $A = \pi r^2$. To find the area we need to find the *radius*. We are given that the radius is growing at 2 ft/min. Using the formula *distance* = *rate* × *time*, we can determine that at the end of 1 hour (60 minutes), the radius is $2 \times 60 = 120$ feet.

Since $r = 120$, the area of the slick after 1 hour is

$$A = \pi r^2$$
$$= \pi(120)^2$$
$$= 14{,}400\pi \approx 45{,}240 \text{ sq ft}$$

(b) Again, the area of a circle is given by $A = \pi r^2$. We need to find the radius. We are given that the radius is growing at 2 ft/min. Using the distance formula, we can determine that the radius is $r = 2t$. See Figure 2.20.

The area of the slick after t minutes is

$$A = \pi r^2$$
$$= \pi(2t)^2$$
$$= 4\pi t^2$$

Hence $A = 4\pi t^2$ sq ft.

Figure 2.20

■

Example 5 | A farmer wants to enclose a rectangular field with a fence that costs $9 per foot. The field is to be four times as long as it is wide, and have an area of A square feet. Let C be the cost of enclosing the field with a fence. Find a formula that expresses the cost C in terms of A.

Solution | We begin with a diagram (see Figure 2.21).

Figure 2.21

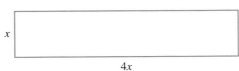

Since we are told that the length of the rectangle is four times its width, we have labeled the width as x and the length as $4x$.

The cost of enclosing the field with a fence is found by multiplying the number of feet of fencing required by the price per foot. The number of feet of fencing required is the perimeter of the rectangle, which is $2x + 2(4x) = 10x$. Therefore the total cost of the fence is

$$C = (\$9 \text{ per foot}) \cdot (10x \text{ feet}) = 90x \text{ dollars}$$

This gives us the cost of the fence in terms of x, but we are asked for the cost in terms of A, the area.

We now look for a connection between x and the area A. The area of the rectangle is $x \cdot 4x$, so that we have $A = 4x^2$. Solving this equation for x, we get

$$A = 4x^2 \implies x^2 = \frac{A}{4} \implies x = \pm\sqrt{\frac{A}{4}} = \frac{\pm\sqrt{A}}{2} \qquad \begin{array}{l}\text{Since } x \text{ must be positive,}\\ \text{we reject the negative}\\ \text{solution, and so}\end{array}$$

$$x = \frac{\sqrt{A}}{2}$$

Substituting for x in the cost equation obtained above, we have

$$C = 90x \qquad \text{Substitute } x = \frac{\sqrt{A}}{2}.$$
$$= 90\left(\frac{\sqrt{A}}{2}\right) = 45\sqrt{A}$$

Thus the cost of enclosing the field is $45\sqrt{A}$ dollars.

■

Example 6   Given that the volume of a sphere is found by $V = \dfrac{4}{3}\pi r^3$, where r is its radius, find the volume of a sphere in terms of its diameter.

Solution   We start out with the volume of the sphere given in terms of its radius as $V = \dfrac{4}{3}\pi r^3$.

Ask yourself: How would you find the volume given the diameter is 10?

Because we want to find the volume in terms of the diameter, we need to find a relationship between the radius and the diameter of a sphere. We already know that if d is the diameter and r is the radius, then

$$d = 2r \quad \Rightarrow \quad r = \frac{d}{2}$$

Hence, given the diameter, we can find its radius. Now we substitute for the radius in the formula for volume as follows:

$$V = \frac{4}{3}\pi r^3 \qquad \text{Now substitute } \tfrac{d}{2} \text{ for } r.$$

$$= \frac{4}{3}\pi\left(\frac{d}{2}\right)^3 \qquad \text{Simplify}$$

$$= \frac{4\pi d^3}{24} \qquad \text{Hence } \boxed{V = \dfrac{\pi d^3}{6}}.$$

2.6 Exercises

1. Find the area of a square if **(a)** its perimeter is 84 inches and **(b)** its perimeter is x inches.

2. Find the perimeter of a square if **(a)** its area is 90 sq in. and **(b)** its area is A sq in.

3. Find the area of a square if its diagonal is 5 feet.

4. Find the area of a square if its diagonal is x feet.

5. Find the diagonal of a square if its area is 84 sq cm.

6. Express the diagonal of a square in terms of its area, A.

7. Find the diagonal of a square in terms of its perimeter, P.

8. Find the perimeter of a square if its diagonal is 12 feet.

9. Express the perimeter of a square in terms of its diagonal, d.

10. Find the perimeter of a square if its diagonal is 4 inches.

11. Find the volume of a sphere given its diameter is 8 feet.

12. Find the surface area of a sphere given its diameter is 8 feet.

13. Express the surface area of a sphere in terms of its diameter, d.

14. Express the surface area of a sphere in terms of its volume, V.

Use the following figure for Exercises 15–18. The circle is inscribed in the square.

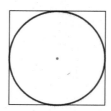

15. If the radius of the circle is 5 inches, what is the area of the square?

16. If the side of the square is 8 inches, find the area of the circle.

17. If the radius of the circle is r inches, express the area of the square in terms of r.

18. If the side of the square is s inches, express the area of the circle in terms of s.

19. The radius of a circle is growing at a constant rate of 3 in./sec. **(a)** What is the area of the circle after 2 seconds (starting at $r = 0$)? **(b)** What is the area of the circle after t seconds (starting at $r = 0$)?

20. The area of a circle is growing at a constant rate of 10 sq in./sec. **(a)** What is the radius of the circle after 2 seconds (starting at $A = 0$)? **(b)** What is the radius of the circle after t seconds (starting at $A = 0$)?

21. The radius of a snowball is growing at a constant rate of 3 in./min. What is the volume of the snowball after 5 minutes (starting at $r = 0$)?

22. The volume of a snowball is growing at a constant rate of 8 cu in./min. What is the radius of the snowball after 5 minutes (starting at $V = 0$)?

23. The radius of a snowball is growing at a constant rate of 3 in./min. What is the volume of the snowball after t minutes (starting at $r = 0$)?

24. The volume of a snowball is growing at a constant rate of 8 cu in./min. What is the radius of the snowball after t minutes (starting at $V = 0$)?

25. A farmer wants to enclose a rectangular field with fencing materials; the fencing material costs $8 per foot. If the perimeter of the field is 500 feet, how much would it cost to enclose the field with fencing?

26. A farmer wants to enclose a rectangular field with fencing materials; the fencing material costs $8 per foot. If the perimeter of the field is P feet, how much would it cost (in terms of P) to enclose the field with fencing?

27. A farmer wants to enclose a square field with fencing materials; the fencing material costs $8 per foot. If the area of the field is 200 sq ft, how much would it cost to enclose the field with fencing?

28. A farmer wants to enclose a square field with fencing materials; the fencing material costs x dollars per foot. If the area of the field is 1,000 sq ft, how much would it cost (in terms of x) to enclose the field with fencing?

29. A gardener wants to enclose a circular garden with fencing materials; the fencing material costs $2 per foot. If the radius of the garden is 5 feet, how much would it cost to enclose the garden with fencing?

30. A gardener wants to enclose a circular garden with fencing materials; the fencing material costs $3 per foot. If the radius of the garden is r feet, how much would it cost (in terms of r) to enclose the garden with fencing?

31. A gardener wants to enclose a circular garden with fencing materials; the fencing material costs $3 per foot. If the area of the garden is 90 sq ft, how much would it cost to enclose the garden with fencing?

32. A gardener wants to enclose a circular garden with fencing materials; the fencing material costs $3 per foot. If the area of the garden is A square feet, how much would it cost (in terms of A) to enclose the garden with fencing?

33. Two boats leave a harbor at the same time at right angles to each other. If one boat is traveling at 30 mi/hr and the other is traveling at 40 mi/hr, how far away are the boats from each other after 2 hours?

34. Two boats leave a harbor at the same time at right angles to each other. If one boat is traveling at 30 mi/hr and the other is traveling at 40 mi/hr, how far away are the boats from each other after t hours?

Chapter 2 *Review Exercises*

In Exercises 1–8, solve for the variable. For inequalities, express your answer using interval notation.

1. $2 - \{3 - [2(1 - x)]\} = 5x + 3$

2. $3 - 2[2 - (x - 6)] = 2x + 8$

3. $3x - \dfrac{2}{5}x \geq -2x + 3(1 - x)$

4. $\dfrac{3x - 2}{5} - 2 < \dfrac{2x - 3}{2}$

5. $\dfrac{4}{x^2 - 2x} - \dfrac{3}{2x} = \dfrac{17}{6x}$

6. $\dfrac{3x}{x - 1} - 2 = \dfrac{3}{x - 1}$

7. $2 < 5 - 3x \leq \dfrac{7}{3}$

8. $-2 \leq \dfrac{5}{2}x + 3 \leq 4$

In Exercises 9–10, solve for the given variable.

9. $C = \dfrac{5}{9}(F - 32)$ for F

10. $y = \dfrac{x - 9}{3x + 1}$ for x

11. Raju has 2 liters of a 60% solution of alcohol. What is the minimum amount of water he should add to the solution to have a solution that is less than 40% alcohol?

12. It takes Collin 3 hours to paint a room and it takes Mary $2\frac{1}{2}$ hours to paint the same room. How long would it take for them to paint the room working together?

In Exercises 13–20, solve for the given variable. For inequalities, express your answer using interval notation.

13. $|7x - 2| = 9$

14. $|2x - 3| = 4$

15. $|5 - 3x| < 6$

16. $|5 - 3x| \geq 6$

17. $\left|\dfrac{1 - 4x}{3}\right| \geq 5$

18. $\left|1 - \dfrac{4}{3}x\right| < 3$

19. $|x - 1| = |5x + 3|$

20. $|x - 2| = |x + 2|$

In Exercises 21–22, solve by completing the square.

21. $x^2 - x = 8$

22. $2t^2 = 8t + 10$

In Exercises 23–32, solve by any algebraic method.

23. $3x^2 + 8 = 23$

24. $6x^2 = 2 - x$

25. $\dfrac{3}{x - 2} + \dfrac{2}{x - 1} = \dfrac{3}{2}$

26. $\dfrac{1}{x} + x = 2$

27. $\dfrac{6}{x^2 - 2x - 3} + \dfrac{x}{x - 3} = 3$

28. $\dfrac{(x - 3)^{1/3}(x + 1)^{1/2}}{(x - 2)^{1/5}} = 0$

29. $\sqrt{3x + 1} + 1 = x$

30. $(x - 3)^{-1/3} = 4$

31. $2x^{2/3} - x^{1/3} = 3$

32. $\sqrt{5x + 1} - 1 = \sqrt{3x}$

In Exercises 33–34, solve for the given variable.

33. $5x^2 + 7y^2 = 9$ for x

34. $15a^2 + 7ab - 2b^2 = 0$ for a

In Exercises 35–42, solve the inequalities and express your answers using interval notation.

35. $3x^2 - 14x - 5 > 0$

36. $2x^2 \geq -5x - 3$

37. $x^2 - 4x + 4 < 0$

38. $x^3 + 2x^2 - x - 2 \leq 0$

39. $\dfrac{3x - 2}{x + 1} > 0$

40. $\dfrac{3 - 2x}{x - 5} \leq 0$

41. $\dfrac{3x - 1}{x} \leq 2$

42. $\dfrac{x}{3x + 1} < 5$

43. Carlos throws a ball straight up into the air off a building. The equation $s = -16t^2 + 80t + 44$ gives the distance, s, in feet the ball is above the ground t seconds after he tosses it up.
   (a) How high above the ground is the ball at $t = 2$ seconds?
   (b) How long does it take for the ball to hit the ground?

44. A circular garden is surrounded by a path of uniform width. If the path has area 44π sq ft and the radius of the garden is 10 feet, find the width of the path.

45. The radius of a circle is growing at a constant rate of 5 in./sec. What is the area of the circle after t minutes (starting at $r = 0$)?

46. A gardener wants to enclose a circular garden with fencing materials; the fencing material costs $3 per foot. If the radius of the garden is r feet, how much would it cost (in terms of r) to enclose the garden with fencing?

Chapter 2   Practice Test

In Exercises 1–14, solve for the variable. For inequalities, express your answers using interval notation.

1. $4 - [3(2 - x)] = x + 3$

2. $\dfrac{x - 1}{3} - 1 < \dfrac{x + 3}{2}$

3. $|7 + 2x| = 7$

4. $\left|\dfrac{2}{3}x - 2\right| \leq 5$

5. $|7 - 2x| > 4$

6. $|2 - 3x| = |3x - 2|$

7. $(2x - 1)(x + 2) = 5x + 2$

8. $3x^2 - 2x = 5 - 2x$

9. $\dfrac{2}{x - 2} + \dfrac{3}{x + 2} = \dfrac{5}{x^2 - 4}$

10. $\sqrt{x - 5} + 4 = x - 1$

11. $2x^2 \leq x + 3$

12. $(x - 1)(x - 2) > 2x - 2$

13. $\dfrac{x - 2}{2x + 1} > 0$

14. $\dfrac{x - 5}{x} \geq 2$

15. Ken can process 200 forms in 3 hours and Kim can process the same 200 forms in $2\frac{1}{3}$ hours. How long would it take them to process the 200 forms working together?

16. Sandy throws a ball up into the air. The equation $s = -16t^2 + 40t + 96$ gives the distance, s, in feet the ball is above the ground t seconds after she tosses it up. How long does it take for the ball to hit the ground?

17. A farmer wants to enclose a circular field with fencing materials. The fencing material costs $6 per foot. If the area of the field is A square feet, how much would it cost (in terms of A) to enclose the field with fencing?

Functions and Graphs: Part 1

3

The concept of a function is one of the most important ideas in mathematics. It is an idea that unifies many different branches of mathematics and allows a variety of other disciplines to be "mathematized." We often hear or read statements such as "On a hot summer day, a city's electrical power consumption is a function of the temperature" or "Life insurance rates are a function of a person's age." We understand such statements to mean that there is a relationship between one of the quantities and the other.

In this chapter we describe different types of mathematical relationships and define precisely what we mean by a function. In addition, we begin a discussion of how to graph various types of equations and functions that will continue throughout the text.

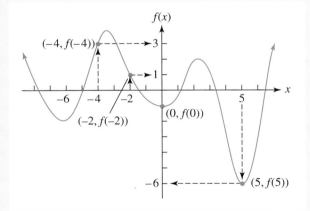

3.1 The Cartesian Coordinate System: Graphing Straight Lines and Circles

Consider the following situation:

> As an employee in a sales department you are offered the option of receiving your salary two possible ways: Either you can be paid a straight commission of 9% of your gross sales, or you can receive a base pay of $240 per week plus a commission of 6% of your gross sales. Which salary plan should you choose?

We can write an equation that represents the weekly salary s earned under each plan in terms of the gross weekly sales g:

$s = 0.09g$ The weekly salary under the 9% straight commission plan.

$s = 240 + 0.06g$ The weekly salary under the base pay plus 6% commission plan.

To make an informed choice, we could compare the two plans for various amounts of weekly sales. However, if we had a picture of these two equations, it would be much easier to compare them. This is done in Section 7.1.

Each of these equations is a first-degree equation in two variables. We now turn our attention to the graphs of such equations. When we solve a first-degree equation in one variable, we are trying to find all values of the variable that satisfy a certain condition. Every equation is, in fact, a condition. For example, when we solve an equation such as $3x - 2 = 11$, we are looking for all numbers that satisfy the condition that "two less than three times a number is 11." An equation in one variable imposes a condition on one unknown quantity.

Similarly, a first-degree equation in *two variables* imposes a condition on two unknown quantities. For example, when we solve an equation of the form $y = 3x - 6$, we are looking for all *pairs* of numbers x and y that satisfy the condition that "the y-value is 6 less than three times the x-value." Keep in mind that a single solution to the equation $y = 3x - 6$ consists of *two* numbers—an x-value and a y-value. Thus the pair of numbers $x = 5, y = 9$ is one solution to this equation.

In fact, we can generate infinitely many solutions to the equation $y = 3x - 6$ by simply choosing an x-value and then finding the corresponding y-value that makes the pair of numbers satisfy the equation.

For example, if we let $x = -1$, then $y = 3(-1) - 6 = -9$. Thus $x = -1$, $y = -9$ is *one* solution to this equation.

Table 3.1 contains some of the solutions to the equation $y = 3x - 6$. Since we cannot list the infinitely many pairs of numbers x and y that satisfy this equation, how can we exhibit all the solutions? One answer lies in using the Cartesian (rectangular) coordinate system invented by the French philosopher and mathematician René Descartes (1596–1650).

Table 3.1

x	$y = 3x - 6$	y
–2	$y = 3(-2) - 6$	–12
–1	$y = 3(-1) - 6$	–9
0	$y = 3(0) - 6$	–6
1	$y = 3(1) - 6$	–3
2	$y = 3(2) - 6$	0

Figure 3.1 illustrates the familiar rectangular coordinate system. The horizontal number line is called the *x*-axis; the vertical number line is called the *y*-axis. Together they are called the **coordinate axes**, and their common zero point is called the **origin**.

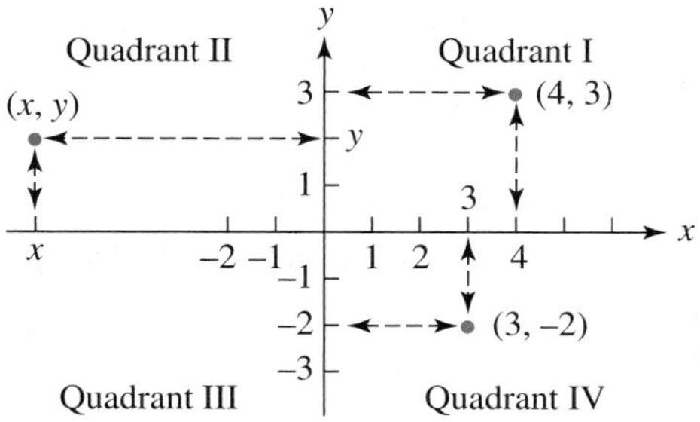

Figure 3.1

Finding the ordered pair (*x*, *y*) corresponding to a point in a rectangular coordinate system

As shown in Figure 3.1, the coordinate axes divide the plane into four parts, which are called **quadrants**. They are numbered from I to IV in counterclockwise order, starting in the upper right-hand quadrant. *Note that the points **on** the coordinate axes are not considered as being **in** any of the quadrants.*

Figure 3.1 also illustrates how every ordered pair is associated with a point, and how every point is associated with an ordered pair. The *x*-coordinate is found by *projecting* the point vertically to the *x*-axis and the *y*-coordinate is found by projecting the point horizontally to the *y*-axis.

Thus we have a one-to-one correspondence between the set of points in the plane and the set of ordered pairs of real numbers; that is, every point in the plane is assigned a unique pair of real numbers, and every pair of real numbers is assigned a unique point in the plane. For this reason we frequently refer to the point (*x*, *y*) rather than the ordered pair (*x*, *y*).

Keeping the meaning of an ordered pair in mind, we recognize that all points on the *x*-axis are of the form (*x*, 0). (Why?) Similarly, all points on the *y*-axis are of the form (0, *y*). (Why?)

With this coordinate system in hand, we can return to the question, How can we exhibit the solution set to the equation $y = 3x - 6$? If we look back at the pairs of numbers (*x*, *y*) appearing in Table 3.1 that satisfy the equation $y = 3x - 6$ and plot these points in a coordinate plane, we see that the points seem to lie on a straight line. See Figure 3.2. By drawing a straight line through the points, we are saying two things: First, every ordered pair that satisfies the equation is a point on the line and second, every point on the line corresponds to an ordered pair that satisfies the equation. Often we will simply say that *the point satisfies the equation*. We will find the following definition useful.

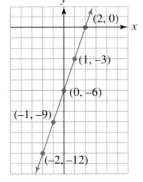

Figure 3.2

Ordered pairs that satisfy $y = 3x - 6$

Definition of a Graph

The **graph** of an equation is the set of all points whose ordered pairs satisfy the equation.

Thus the straight line in Figure 3.2 is the graph of the equation $y = 3x - 6$. In fact, the graph of any first-degree equation in two variables is a straight line. We state this formally in the following theorem.

Theorem 3.1

The graph of an equation of the form $Ax + By = C$ (where A and B are not both equal to zero) is a straight line.

For this reason a first-degree equation is often called a **linear equation**. The form $Ax + By = C$ is called the *general form* for the equation of a line. As we continue with our discussion of straight lines, we will see that the converse of this theorem is also true; that is, a straight line graph has an equation of the form $Ax + By = C$.

Keep in mind that we have not proven this theorem. (The fact that the points we graphed *seemed* to fall on a straight line does not prove that all the points that satisfy the equation will.) The proofs of this theorem and its converse will be the substance of much of our discussion in the first three sections of this chapter.

Different Perspectives: The Graph of an Equation

An equation and its graph offer two ways of looking at a relationship between two variables.

Algebraic Description

An equation such as $y = 2x - 5$ describes a relationship in which the y-value is 5 less than twice the x-value. Ordered pairs such as $(2, -1)$, $(4, 3)$, and $(0, -5)$ satisfy this relationship and hence are solutions to the equation.

Graphical Description

The graph of an equation gives a pictorial representation of the relationship. The following is the graph of the equation $y = 2x - 5$.

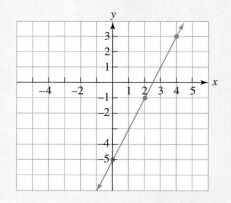

Every point (x, y) on the graph also gives a pair of numbers that satisfies the equation and every pair of numbers that satisfies the equation corresponds to a point on the graph. We can see that the points $(2, -1)$, $(4, 3)$, and $(0, -5)$ are on the graph and by substituting we can check that they satisfy the equation $y = 2x - 5$.

Using a graphing calculator, we demonstrate that the same relationship can be represented by an equation, a graph, and a table of values. It is important that you understand and be able to get information from any of the three representations, for often one may be more accessible or more accurate than the other.

For example, Figure 3.3(a) shows the graph of the equation $y = 2x + 6$ in the standard $[-10, 10]$ by $[-10, 10]$ window. (When referring to a calculator window, the first interval indicates Xmin and Xmax; the second interval indicates Ymin and Ymax.) These are the minimum and maximum x- and y-values in the window.

Figure 3.3

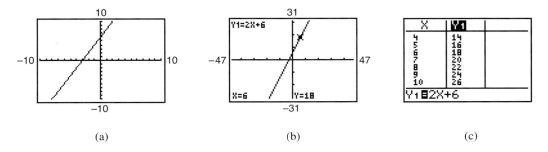

(a)   (b)   (c)

The ⏢TRACE⏢ key allows us to identify points on a graph. Notice that for each value of x, we are given a corresponding y-value that lies on the line and therefore also satisfies the equation $y = 2x + 6$. See Figure 3.3(b). Some calculators, such as the TI-83 Plus, have a TABLE option, which allows us to examine a table of x- and y-values for this equation. (See Figure 3.3(c).)

As we proceed through the text, we will point out that the form of a particular equation often allows us to recognize what its graph is and, in so doing, to identify particular features of the graph that make it easier to draw. For instance, Theorem 3.1 tells us that the graph of a first-degree equation in two variables is a straight line. Therefore, if we want to graph such an equation, we need find only two points that are on the line and then draw the line that passes through these two points.

Throughout our work in graphing there are certain points on the graph to which we want to pay particular attention.

Definition of x- and y-intercepts

The **x-intercepts** of a graph are the x-values of the points where the graph crosses the x-axis.

The **y-intercepts** of a graph are the y-values of the points where the graph crosses the y-axis.

Look back at Figure 3.2. The graph of $y = 3x - 6$ crosses the x-axis at $(2, 0)$ and so has an x-intercept of 2. The graph crosses the y-axis at $(0, -6)$ and so has a y-intercept of -6.

Note that since the x-intercepts correspond to points on the x-axis (and every point on the x-axis has a y-coordinate of 0), x-intercepts can be found by substituting $y = 0$ into the equation and solving for x; similarly, y-intercepts can be found by substituting $x = 0$ into the equation and solving for y. See Figure 3.4. *Whenever possible (and practical), we label the x- and y-intercepts of a graph.*

Figure 3.4

Finding the x- and y-intercepts of a line

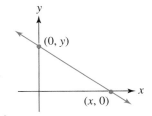

Let's illustrate how we can use the intercepts to graph a linear equation.

Example 1 | Sketch the graphs: (a) $5y - 2x = 12$ (b) $y = -3x$

Solution | (a) We recognize $5y - 2x = 12$ as a first-degree equation in two variables, and so its graph is a straight line. Any two points on the line will be sufficient to allow us to draw the graph. We will find the intercepts, since they are often easy to compute. To find the x-intercept, we set $y = 0$ and solve for x. In other words, we let $y = 0$ in $5y - 2x = 12$ and get $5(0) - 2x = 12 \Rightarrow x = -6$. Thus the x-intercept is -6, which means that the line crosses the x-axis at $(-6, 0)$.

Similarly, to find the y-intercept, let $x = 0$ in $5y - 2x = 12$ and get $y = \dfrac{12}{5}$, which means that the line crosses the y-axis at $\left(0, \dfrac{12}{5}\right)$.

To sketch the graph of $5y - 2x = 12$, we plot the points corresponding to the intercepts and draw the line passing through them. See Figure 3.5.

Figure 3.5

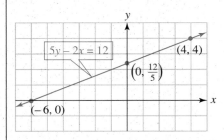

Because any two points we find determine a line, it is generally a good idea to find a third point as a check. For example, we may choose an x-value, say, $x = 4$, and find the corresponding y-value:

$$5y - 2(4) = 12, \quad \text{which gives} \quad 5y = 20 \quad \text{and so} \quad y = 4$$

Therefore the point $(4, 4)$ should be on the line, and it is. (See Figure 3.5.)

(b) As in part (a), we recognize that the graph of $y = -3x$ is a straight line. We begin by finding the intercepts. We let $y = 0$ in $y = -3x$ and get $x = 0$. The x-intercept is 0, which means the line goes through $(0, 0)$. Since the line goes through the origin, we automatically know that the y-intercept is also 0.

The intercepts give us only *one* point. To sketch the graph, we must find an additional point. We do this by arbitrarily choosing a value for x (or y) and solving for the other variable.

$$y = -3x \qquad \text{We choose } x = 2.$$
$$y = -3(2) = -6 \qquad \text{Thus the point } (2, -6) \text{ is on the line.}$$

The graph appears in Figure 3.6.

Figure 3.6

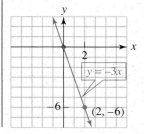

Different Perspectives: Intercepts

Consider the graphical and algebraic interpretations of the x- and y-intercepts of the graph of $3x - 2y = 6$.

Graphical Interpretation

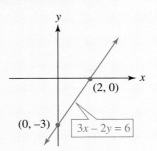

The line crosses the x-axis at the point (2, 0).
The line crosses the y-axis at the point (0, −3).

→ The x-intercept is 2.

→ The y-intercept is −3.

Algebraic Interpretation

To find the x-intercept, set $y = 0$ and solve for x:

$$3x - 2y = 6$$
$$3x - 2(0) = 6$$
$$3x = 6$$
$$x = 2$$

To find the y-intercept, set $x = 0$ and solve for y:

$$3x - 2y = 6$$
$$3(0) - 2y = 6$$
$$-2y = 6$$
$$y = -3$$

Example 2 Graph the equation $0.34y - 2.95x = 14.8$ using a graphing calculator, and find the intercepts accurate to two decimal places. Check your answers algebraically.

Solution If we want to use a graphing calculator to graph an equation relating x and y, we must first solve the equation explicitly for y. Hence the equation $0.34y - 2.95x = 14.8$ becomes $y = \frac{2.95}{0.34}x + \frac{14.8}{0.34}$. The graph of this equation using a standard $[-10, 10]$ by $[-10, 10]$ window is shown in Figure 3.7(a).

Figure 3.7

To get the standard window on the TI-83 Plus, press

ZOOM [6:ZStandard]

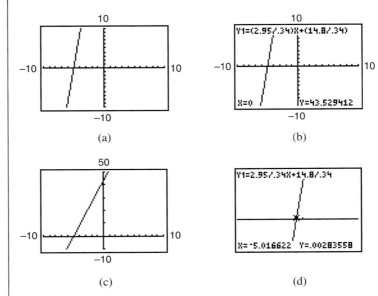

In this window you cannot see where the graph crosses the y-axis. However, we can press the TRACE key to display x- and y-values *on* the graph, as illustrated by the point X=0 and Y=43.529412 in Figure 3.7(b). Hence, the graph crosses the y-axis at approximately 43.5. We reset the window to $[-10, 10]$ by $[-10, 50]$, to ensure that we see both intercepts of the graph (see Figure 3.7(c)). After zooming in a few times (using the $\boxed{\text{ZOOM}}$ key) near the x-intercept, we find that the x-intercept is -5.02 (to two decimal places). See Figure 3.7(d).

Algebraically, to find the y-intercept, set $x = 0$ and solve for y:

$$0.34y - 2.95x = 14.8 \Rightarrow 0.34y - 2.95(0) = 14.8 \Rightarrow y = \frac{14.8}{0.34} = 43.53$$
to two decimal places

To find the x-intercept, set $y = 0$ and solve for x:

$$0.34y - 2.95x = 14.8 \Rightarrow 0.34(0) - 2.95x = 14.8 \Rightarrow x = -\frac{14.8}{2.95} = -5.02$$
to two decimal places

These algebraic values agree with the graphical values we obtained previously. ∎

Example 3 Sketch the graphs of the following equations in a rectangular coordinate system:
(a) $x = -4$ (b) $y = 2$

Solution (a) Had we been asked to graph the equation $x = -4$ on a number line, we would have drawn the graph shown in Figure 3.8.

Figure 3.8

The graph of $x = -4$ on a number line

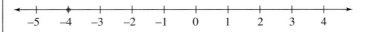

However, in this example we are being asked to graph in a rectangular co-ordinate system. The equation $x = -4$ is in the form $Ax + By = C$, with $A = 1$, $B = 0$, and $C = -4$:

$$Ax + By = C$$
$$1x + 0y = -4$$

As mentioned before, an equation can be viewed as a condition that x and y must satisfy. The equation $x = -4$ imposes the condition that the x-coordinate of any point satisfying the equation must be -4. Since y does not appear in the equation $x = -4$, there is *no* condition on y. In other words, x must be equal to -4, but y can be any real number. This gives a line parallel to and 4 units to the left of the y-axis. The graph appears in Figure 3.9(a).

Figure 3.9

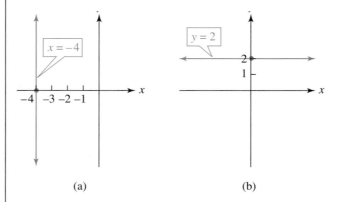

(a)   (b)

(b) Similarly, the equation $y = 2$ fits the general form $Ax + By = C$ with $A = 0$, $B = 1$, and $C = 2$. The equation $y = 2$ imposes the condition that the y-coordinate of any point on the line must be 2, while x can be any real number. This gives a line parallel to and 2 units above the x-axis. The graph appears in Figure 3.9(b). ∎

In general, the graph of an equation of the form $x = h$ is a straight line parallel to the y-axis and passing through the point $(h, 0)$. The graph of an equation of the form $y = k$ is a straight line parallel to the x-axis and passing through the point $(0, k)$.

Many real-life relationships give rise to linear equations.

Example 4   A cellular phone company charges $7 per week to rent a phone and $0.25 per minute of airtime.
(a) Express the cost C per week (in dollars) in terms of m, the total number of minutes of airtime.
(b) Use the equation obained in part (a) to compute the cost during a week in which there are 87 minutes of airtime.
(c) If your weekly charge was $23.25, how many minutes of airtime were used?

Solution   (a) To compute the weekly cost, we add the rental fee of $7 and the airtime fee that is computed by multiplying the number of minutes of airtime m by the cost per minute, which is $0.25. Thus we have

$$C = 7 + 0.25m$$

(b) If there are 87 minutes of airtime, then we substitute $m = 87$ in the equation obtained in part (a).

$$C = 7 + 0.25m \qquad \text{Substitute } m = 87.$$
$$= 7 + 0.25(87) = \boxed{28.75}$$

Thus the cost for a week in which 87 minutes of airtime is used would be $28.75.
(c) To find the number of minutes of airtime that generates a weekly charge of $23.25, we substitute $C = 23.25$ into the equation obtained in part (a) and then solve for m.

$$C = 7 + 0.25m \qquad \text{Substitute } C = 23.25.$$
$$23.25 = 7 + 0.25m \qquad \text{Solve for } m.$$
$$= \frac{16.25}{0.25} = \boxed{65}$$

Thus 65 minutes of airtime generates a weekly charge of $23.25.

The Distance and Midpoint Formulas

Although we tend to take the Cartesian coordinate system rather for granted, it is important to recognize what a revolutionary idea it was (remember that it is only about 350 years old). Having a coordinate system allows us to approach geometric problems from an algebraic point of view. The ideas we are now going to discuss are part of a branch of mathematics called *analytic geometry.*

For example, the distance between two points and the midpoint of a line segment are both inherently geometric ideas; nevertheless, with the aid of a coordinate system we can derive an algebraic formula for each.

Example 5   Find the distance between the given pair of points:
(a) $(-4, 3)$ and $(2, 3)$
(b) $(4, -2)$ and $(4, 5)$

Solution | We plot the given pairs of points in Figure 3.10.

(a) Since the points $(-4, 3)$ and $(2, 3)$ lie on the same horizontal line, the distance between them is the same as the distance between the points -4 and 2 on the x-axis. As we saw in Chapter 1, the distance between two points a and b on a number line is $|a - b|$. Therefore, the distance between $(-4, 3)$ and $(2, 3)$ is $|2 - (-4)| = \boxed{6}$.

(b) Similarly, the points $(4, -2)$ and $(4, 5)$ lie on the same vertical line, and so the distance between them is the absolute value of the difference of their y-coordinates. Therefore, the distance between $(4, -2)$ and $(4, 5)$ is $|5 - (-2)| = \boxed{7}$.

Figure 3.10

Finding horizontal and vertical distances

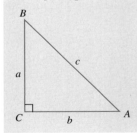

In the previous example we could simply have found the distances by counting horizontal or vertical units. We chose to take a somewhat more formal approach so that we could recognize that the distance between the points (x_1, y_1) and (x_2, y_1), which lie on the same horizontal line, is $|x_2 - x_1|$, and the distance between the points (x_1, y_1) and (x_1, y_2), which lie on the same vertical line, is $|y_2 - y_1|$.

To derive a formula for the distance between *any* two points, we will also need the Pythagorean Theorem.

The Pythagorean Theorem and Its Converse

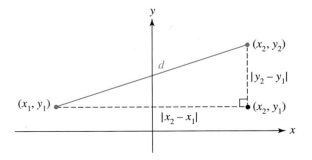

In right triangle ABC labeled as indicated, $a^2 + b^2 = c^2$. The converse of this is also true. That is, if $a^2 + b^2 = c^2$, then $\triangle ABC$ is right triangle with c as hypotenuse.

Now suppose we want to find the distance d between any two points (x_1, y_1) and (x_2, y_2). We can drop perpendiculars toward the x- and y-axes, as indicated in Figure 3.11, forming a right triangle whose third vertex is (x_2, y_1).

Why is this third vertex (x_2, y_1)?

Figure 3.11

Finding the distance between two points

Since the points (x_1, y_1) and (x_2, y_1) lie on the same horizontal line, the length of the horizontal side of the triangle is $|x_2 - x_1|$. Similarly, the length of the vertical side of the triangle is $|y_2 - y_1|$. Therefore, by the Pythagorean Theorem we have

$$d^2 = |x_2 - x_1|^2 + |y_2 - y_1|^2$$

Since the distance, d, must be positive, we take the positive square root to get

$$d = \sqrt{|x_2 - x_1|^2 + |y_2 - y_1|^2}$$

We leave it as an exercise for the reader to verify that $|x_2 - x_1|^2 = (x_2 - x_1)^2$ and similarly for the y's.

$$d = \sqrt{(x_2 - x_1)^2 + (y_2 - y_1)^2}$$

We have thus derived the distance formula.

The Distance Formula

The distance between the points (x_1, y_1) and (x_2, y_2) in the Cartesian plane is

$$d = \sqrt{(x_2 - x_1)^2 + (y_2 - y_1)^2}$$

Example 6 Find the distance between the points $(3, -1)$ and $(-1, 4)$.

Solution Even though this example can be done without a diagram, we strongly urge you to draw a diagram to accompany your solution whenever possible. We plot the points in Figure 3.12. We arbitrarily let $(x_1, y_1) = (3, -1)$ and $(x_2, y_2) = (-1, 4)$. Applying the distance formula, we get

$$d = \sqrt{(x_2 - x_1)^2 + (y_2 - y_1)^2}$$
$$d = \sqrt{(-1 - 3)^2 + [4 - (-1)]^2} = \sqrt{(-4)^2 + 5^2} = \sqrt{16 + 25}$$

$$= \boxed{\sqrt{41}} \approx 6.4$$

Figure 3.12

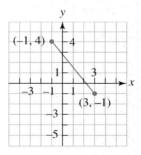

Sometimes we need to know the midpoint of a line segment. In Figure 3.13 we have drawn a line segment joining the points $P(x_1, y_1)$ and $Q(x_2, y_2)$.

Figure 3.13

Finding the midpoint of a line segment

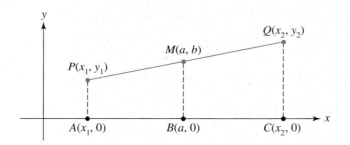

Why do A, B, and C have coordinates $(x_1, 0)$, $(a, 0)$, and $(x_2, 0)$, respectively?

Let $M(a, b)$ be the midpoint of line segment $\overline{PQ}$. We have drawn dashed lines through P, $M(a, b)$, and Q parallel to the y-axis and labeled the points $A(x_1, 0)$, $B(a, 0)$, and $C(x_2, 0)$ on the x-axis.

It is a basic fact from geometry that if parallel lines intercept equal segments on one line, then they intercept equal line segments on any line. In other words, if $\overline{PM} = \overline{MQ}$, then it must follow that $\overline{AB} = \overline{BC}$, which means that B is the midpoint of $\overline{AC}$. But since $\overline{AC}$ is on the x-axis, its midpoint is just the average of the x-coordinates. Therefore, $a = \dfrac{x_1 + x_2}{2}$. Similarly, $b = \dfrac{y_1 + y_2}{2}$. We have thus shown the following.

The Midpoint Formula

The midpoint M of the line segment joining the points $P(x_1, y_1)$ and $Q(x_2, y_2)$ is

$$M\left(\frac{x_1 + x_2}{2}, \frac{y_1 + y_2}{2}\right)$$

Example 7   Find the midpoint of the line segment joining the points $(-3, 5)$ and $(6, 3)$.

Solution   Using the midpoint formula, we find the midpoint to be

$$\left(\frac{-3 + 6}{2}, \frac{5 + 3}{2}\right) = \left(\frac{3}{2}, 4\right)$$

See Figure 3.14.

Figure 3.14

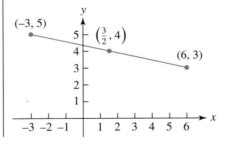

Circles

A **circle** is defined to be the set of all points in a plane whose distance from a fixed point is constant. The fixed point is called the center, C, and the constant distance from the center to the circle is called the radius, r (where $r > 0$).

Let's place a circle of radius r on the Cartesian plane and center it at the point (h, k). Pick a point in the plane and call it (x, y). See Figure 3.15.

Figure 3.15

A circle with center (h, k) and radius r

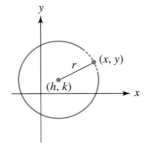

The definition of a circle tells us that in order for (x, y) to be on the circle, the distance from the center, (h, k), to (x, y) must be r. By the distance formula, we have

$$\sqrt{(x - h)^2 + (y - k)^2} = r$$

For convenience we eliminate the radical by squaring both sides of this equation to get the following.

The Standard Form for the Equation of a Circle

$$(x - h)^2 + (y - k)^2 = r^2$$

is the equation of a circle with center (h, k) and radius r.

This equation is the **standard form for the equation of a circle**. The center and radius are all that is needed to describe or graph a circle.

Example 8 | Find the equation of a circle with center $(2, -5)$ and radius 6.

Solution | Looking at the standard form, since the center is $(2, -5)$, we have $h = 2$ and $k = -5$; since the radius is 6, $r = 6$.

$$(x - h)^2 + (y - k)^2 = r^2 \qquad \text{Substitute } h = 2, k = -5, \text{ and } r = 6.$$
$$(x - 2)^2 + (y - (-5))^2 = 6^2$$

Hence, the equation of the circle is

$$(x - 2)^2 + (y + 5)^2 = 36 \qquad \text{which, when multiplied out, is}$$

$$x^2 + y^2 - 4x + 10y - 7 = 0$$

Although both forms of the answer are acceptable, the standard form for the equation of the circle has the significant advantage of making the center and radius easily recognizable. ∎

Example 9 | Sketch the graph of the following equations:
(a) $(x + 3)^2 + (y - 4)^2 = 8$ (b) $x^2 + y^2 = 9$

Solution | (a) We recognize that the given equation is in the standard form for the equation of a circle, $(x - h)^2 + (y - k)^2 = r^2$, and so we can simply read off the values h, k, and r, but we must be careful of the signs.

$$x - h = x + 3 \implies -h = 3 \implies h = -3$$
$$y - k = y - 4 \implies k = 4$$
$$r^2 = 8 \implies r = \sqrt{8} = 2\sqrt{2}$$

Thus the center of the circle is $(-3, 4)$; the radius is $2\sqrt{2} \approx 2.8$. With this information, we can easily sketch the graph of the circle, which appears in Figure 3.16(a) on page 110.

Figure 3.16

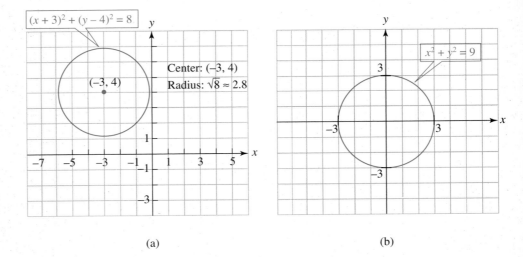

(a)

(b)

(b) The equation $x^2 + y^2 = 9$ is also in standard form. (It can be thought of as $(x - 0)^2 + (y - 0)^2 = 3^2$.) Consequently, the center is $(0, 0)$ and the radius is 3. The graph appears in Figure 3.16(b). ∎

Suppose the equation in part (a) of Example 9 had not been given to us in standard form, where identifying the center and radius of a circle is straightforward, but rather in its multiplied-out form, $x^2 + y^2 + 6x - 8y + 17 = 0$. Would we recognize that this is the equation of a circle? How do we find the center and radius?

If the equation $(x - h)^2 + (y - k)^2 = r^2$ is multiplied out, we get an equation of the form $x^2 - 2hx + h^2 + y^2 - 2ky + k^2 = r^2$ (remember that h, k, and r are just real numbers). The key feature of this equation is that it is a *second-degree equation in which x^2 and y^2 both appear with coefficient* 1. Thus if we have a second-degree equation in x and y in which the coefficients of x^2 and y^2 are both 1 or can both be made 1, we should recognize that it is the equation of a circle.

So if we are given an equation such as $2x^2 + 2y^2 - 8x + 16y = 10$, we should recognize that it is the equation of a circle, and our goal would be to put it in standard form so that we may read off the center and radius.

The standard form of a circle involves perfect squares; this suggests that the technique of completing the square discussed in Chapter 2 would be useful.

Example 10   Find the center and radius of the following: $2x^2 + 2y^2 - 8x + 16y = 10$

Solution   Since the coefficients of x^2 any y^2 are equal, we recognize that this is the equation of a circle. We find the center and radius by completing the square. We begin by making the coefficient of the squared terms equal to 1:

$$2x^2 + 2y^2 - 8x + 16y = 10 \quad \text{Divide each side by 2.}$$

$$x^2 + y^2 - 4x + 8y = 5 \quad \text{Group terms as shown.}$$

$$(x^2 - 4x \quad) + (y^2 + 8y \quad) = 5 \quad \text{Complete the square for each quadratic}$$

expression: $\left[\frac{1}{2}(-4)\right]^2 = 4$; $\left[\frac{1}{2}(8)\right]^2 = 16$.

Add both numbers to both sides of the equation.

$$(x^2 - 4x + \mathbf{4}) + (y^2 + 8y + \mathbf{16}) = 5 + \mathbf{4} + \mathbf{16}$$

Rewrite the quadratic expressions in perfect square form.

$$(x - 2)^2 + (y + 4)^2 = 25$$

Thus, we have a circle with   center $(2, -4)$ and radius $\sqrt{25} = 5$. ∎

As we proceed through the text, one of our major goals is to build a catalog of basic equations and their graphs—that is, a catalog of equations whose graphs we recognize at a glance. Based on our work thus far, our catalog contains the graphs of straight lines and circles. If we encounter the equation $x^2 + y^2 = 9$, we should immediately recognize that its graph is a circle with center $(0, 0)$ and radius 3.

Example 11   Find the equation of a circle whose diameter has endpoints $(2, 3)$ and $(-4, 7)$.

Solution   We draw a diagram (Figure 3.17) so that we may visualize what is given and what needs to be found. Let's analyze this problem carefully in order to develop a strategy for the solution.

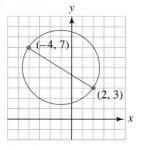

Figure 3.17

The circle with diameter having endpoints $(2, 3)$ and $(-4, 7)$

What do we need to find?	The equation of the circle with the given diameter
What is needed to find an equation of a circle?	The center and the radius
What information is given in the problem?	The diameter's endpoints
How can I restate the problem in simpler terms?	Find the center and radius of the circle given the endpoints of a diameter
What additional knowledge or information do I need to solve this simpler problem?	That the center of the circle is the midpoint of the diameter (requiring the midpoint formula) and that the radius of a circle is the distance from the center of the circle to any point on the circle, including the given endpoints of a diameter (requiring the distance formula)

To locate the center, we find the midpoint between $(2, 3)$ and $(-4, 7)$, which is

$$\left(\frac{2 + (-4)}{2}, \frac{3 + 7}{2}\right) = (-1, 5)$$

Therefore, the center of the circle is $(-1, 5)$.

To find the radius, we use the distance formula to find the distance between the center, $(-1, 5)$, and one of the endpoints of the diameter; we use the endpoint $(2, 3)$:

Can you find the radius using only the endpoints of the diameter?

$$r = \sqrt{(-1 - 2)^2 + (5 - 3)^2} = \sqrt{9 + 4} = \sqrt{13}$$ Hence the radius is $\sqrt{13}$.

Having found that the center is $(-1, 5)$ and the radius is $\sqrt{13}$, we can write the equation of the circle:

$$(x + 1)^2 + (y - 5)^2 = (\sqrt{13})^2 \quad \text{or} \quad \boxed{(x + 1)^2 + (y - 5)^2 = 13} \quad \blacksquare$$

Two of the major questions on which we will focus our attention throughout the text are:

Question 1:   Given an equation, how do we find its graph?

Question 2:   Given a graph, how do we find its equation?

In this section we have discussed question 1 as it pertains to straight lines and circles and question 2 as it pertains to circles. That is, given the equations $3x + 5y = 8$ and $x^2 + y^2 = 9$, we recognize that the first is a straight line and the second is a circle, *and* because of this recognition we can sketch their graphs fairly easily. Conversely, if we specify the center and radius of a circle (that is, we know its graph), we can write its equation. In the next two sections we address question 2 as it pertains to straight lines: How do we obtain an equation of a line whose graph we already know?

3.1 Exercises

In Exercises 1–14, find the *x*- and *y*-intercepts of the given equation.

1. $5y - 4x = 20$

2. $7x - 2y = 14$

3. $x + 3y = 6$

4. $y + 2x = 8$

5. $3x - 8y = 16$

6. $7y + 5x = 10$

7. $2x + 9y + 6 = 0$

8. $6x - 5y + 3 = 0$

9. $x = -3$

10. $y - 5 = 0$

11. $6y = 5x + 8$

12. $4x = 3y - 10$

13. $2x = 3y$

14. $7x - 2y = 0$

In Exercises 15–32, sketch the graph of the given equation in a rectangular coordinate system. Label the intercepts.

15. $7y - 3x = 21$

16. $5x - 4y + 20 = 0$

17. $y = 2x - 8$

18. $y = -3x - 6$

19. $\dfrac{x}{3} + \dfrac{y}{2} = 2$

20. $\dfrac{y}{4} - \dfrac{x}{5} = 1$

21. $\dfrac{3y}{2} - x = 6$

22. $\dfrac{5x}{6} + y = 10$

23. $5x + 2y - 9 = 0$

24. $6y - 3x = 10$

25. $y = \dfrac{3}{2}$

26. $x = -2$

27. $y = 3x$

28. $y = 3$

29. $5x - 4 = 0$

30. $5x - 4y = 0$

31. $y = \dfrac{2}{5}x + 4$

32. $y = -\dfrac{3}{4}x - 6$

In Exercises 33–40, use a graphing calculator to find the intercepts of the graphs of the equations. Round your answers to two places, and check your answers algebraically.

33. $y = -2x + 7$

34. $y = \dfrac{3}{5}x - \dfrac{2}{3}$

35. $y = \dfrac{0.02x + 5}{3}$

36. $y = \dfrac{1}{3} + \dfrac{7}{4}x$

37. $2.45x = 0.5y - 1$

38. $0.26x - 1.5y = 11.2$

39. $\dfrac{4.4x + 3y}{3} = 2.8$

40. $\dfrac{5x}{4} - \dfrac{4y}{5} = 1$

41. Using *d* for the vertical axis and *t* for the horizontal axis, sketch the graph of the equation $d = 5t$.

42. Using *D* for the vertical axis and *p* for the horizontal axis, sketch the graph of the equation $D = -20p + 160$.

43. Using *s* for the vertical axis and *t* for the horizontal axis, sketch the graph of the equation $s = 0.5t + 15$.

44. Using *V* for the vertical axis and *p* for the horizontal axis, sketch the graph of the equation $V = 200 - 8p$.

45. An electronics discount store wants to use up a credit of $9,110 with its supplier to order a shipment of VCRs and TVs. Each VCR costs $95 and each TV costs $125.
  (a) Let *v* represent the cost of one VCR and *t* represent the cost of one TV. Write an equation that reflects the given situation.
  (b) Sketch the graph of this relationship. Be sure to label the coordinate axes clearly.
  (c) If 30 VCRs are ordered, use the equation you obtained in part (a) to find the number of TVs.

46. A computer store budgets $16,400 to buy computers and laser printers. Each computer costs $650 and each printer costs $235.
  (a) Let *c* represent the cost of one computer and *p* represent the cost of one printer. Write an equation that reflects the given situation.
  (b) Sketch the graph of this relationship. Be sure to label the coordinate axes clearly.
  (c) If the shipment contained 18 computers, use the equation you obtained in part (a) to find the number of printers.

47. Dana is working as a clerk in a bank where she inspects documents. It takes her 9 minutes to check a loan application and 15 minutes to check a credit report. On a certain day she is scheduled to spend 6 hours inspecting documents.
  (a) Let *a* represent the number of loan applications and *c* represent the number of credit reports. Write an equation that reflects the given situation.
  (b) Sketch the graph of this relationship. Be sure to label the coordinate axes clearly.

(c) If Mary inspected 30 loan applications, use the equation you obtained in part (a) to find the number of credit reports she can check.

48. A seamstress can hem a skirt in 10 minutes and cuff a pair of pants in 15 minutes. On a certain day she has allotted 300 minutes to do these tasks.
(a) Let s represent the number of skirts she hems and p represent the number of pairs of slacks she cuffs. Write an equation that reflects the given situation.
(b) Sketch the graph of this relationship. Be sure to label the coordinate axes clearly.
(c) If she hems 18 skirts, use the equation you obtained in part (a) to find the number of pairs of pants she cuffs.

In Exercises 49–54, find the length and the midpoint of the line segment joining the two points.

49. $A(-3, 4)$ and $B(1, -1)$

50. $P(2, -3)$ and $Q(3, 6)$

51. $R(1, 5)$ and $S(-1, -4)$ **52.** $E(0, 0)$ and $F(2, 4)$

53. $C(0, 0)$ and $D(a, a)$ **54.** $T(0, 0)$ and $U(a, 2a)$

55. Draw the triangle ABC with vertices $A(-1, 0)$, $B(4, 0)$, and $C(4, 6)$. What is the area of triangle ABC ?

56. Draw the rectangle $ABCD$ with vertices $A(5, 0)$, $B(5, 4)$, $C(-3, 4)$, and $D(-3, 0)$. What is the area of rectangle $ABCD$?

57. Use the converse of the Pythagorean Theorem to prove that the three points $A(2, 1)$, $B(7, 2)$, and $C(5, -1)$ are the vertices of a right triangle. What is the area of $\triangle ABC$?

58. Which of the three points $(1, 5)$, $(2, 4)$, and $(3, 3)$ is the farthest from the origin? Would the answer be the same if we interchanged the x- and y-coordinates of each point?

59. Find the value(s) of w so that the points $(0, 3)$ and $(6, w)$ are 10 units apart.

60. Find the value(s) of t so that the points $(3, -2)$ and $(1, t)$ are 4 units apart.

61. Can you find a point $(x, 4)$ that is 2 units from the point $(5, 1)$? Explain.

62. Can you find a point $(-2, y)$ that is 1 unit from the origin? Explain.

In Exercises 63–66, write an equation of the circle with the given center, C, and radius, r.

63. $C = (2, 3); r = 3$

64. $C = (7, -4); r = 5$

65. $C = \left(\dfrac{1}{2}, 4\right); r = 6$

66. $C = \left(-\dfrac{3}{4}, -2\right); r = \sqrt{7}$

In Exercises 67–74, identify the center and radius of the given circle.

67. $(x - 3)^2 + (y - 2)^2 = 16$

68. $\left(x - \dfrac{1}{2}\right)^2 + (y + 3)^2 = 24$

69. $x^2 + y^2 = 16$

70. $x^2 + (y + 2)^2 = 72$

71. $x^2 + y^2 - 6x - 10y = -9$

72. $2x^2 + 2y^2 - 8x + 12y + 8 = 0$

73. $3x^2 + 3y^2 + 18y = 0$

74. $5x^2 - 20x + 5y^2 = 5$

75. Sketch the graph of $(x - 2)^2 + (y + 3)^2 = 4$.

76. Sketch the graph of $x^2 + y^2 + 6x + 33 = 10y$.

77. Find an equation of the circle with a diameter having endpoints $(-2, 8)$ and $(4, -5)$.

78. Find an equation of the circle with a diameter having endpoints $(-3, 5)$ and $(-4, 0)$.

79. Find an equation of the circle passing through the point $(2, 6)$, with center $(3, -5)$.

80. Find an equation of the circle passing through the point $(3, -2)$, with center $(2, 5)$.

81. Find the circumference of the circle passing through the point $(-3, 4)$, with center $(5, 2)$.

82. Find the area of the circle passing through the point $(3, -2)$, with center $(5, 2)$.

83. Find an equation of the circle tangent to the x-axis, with center $(3, -2)$.

84. Find an equation of the circle tangent to the y-axis, with center $(3, -2)$.

85. Find an equation of the circle tangent to the x-axis at $(3, 0)$ and tangent to the y-axis at $(0, -3)$.

86. Find an equation of all circles with radius 6 and tangent to both the x- and y-axes. HINT: Check each quadrant.

87. The circle with center $(0, 0)$ and radius 1 is called the *unit circle*.
(a) Write an equation of the unit circle.
(b) Determine which of the following points are on the unit circle.

$$\left(\dfrac{3}{5}, -\dfrac{4}{5}\right), \quad \left(-\dfrac{\sqrt{3}}{2}, \dfrac{1}{2}\right), \quad \left(\dfrac{2}{3}, \dfrac{3}{5}\right)$$

Questions for Thought

88. Does the equation $\dfrac{x}{4} + \dfrac{y}{5} = 2$ fit the form of a

first-degree equation in two variables, that is, $Ax + By = C$? If so, what are the values A, B, and C? Are the values A, B, and C unique? Explain.

89. Use the distance formula to prove that the

point $M\left(\dfrac{x_1 + x_2}{2}, \dfrac{y_1 + y_2}{2}\right)$ is equidistant from

the points $P(x_1, y_1)$ and $Q(x_2, y_2)$.

(a) Does this prove that the point

$M\left(\dfrac{x_1 + x_2}{2}, \dfrac{y_1 + y_2}{2}\right)$ is the midpoint of the

line segment connecting the points $P(x_1, y_1)$ and $Q(x_2, y_2)$? Explain.

(b) Can you use the distance formula to prove that M is the midpoint of the segment $\overline{PQ}$? How?

(c) Do you think this proof of the midpoint formula is easier or harder than that offered in the text's presentation?

90. Define x- and y-intercepts in two ways:
   (a) In terms of the graph of an equation
   (b) In terms of an equation of a graph

91. How could the distance formula be used to prove that the three points $P(-3, 4)$, $Q(2, -1)$, and $R(3, -2)$ are collinear (lie on the same line)?

92. We have been tacitly assuming that we always use units of the same length along the x- and y-axes; however, this is not necessarily the case. Describe how you would choose units along the coordinate axes to sketch the graph of $y = 0.01x + 2$.

93. In our discussion of the derivation of the midpoint formula, we found the x-coordinate of the midpoint can be found by averaging the two x-coordinates. Suppose $x_1 < x_2$; show that we get the same result if instead we add one-half the distance between x_1 and x_2 to x_1.

94. In the derivation of the distance formula we stated that $|x_2 - x_1|^2 = (x_2 - x_1)^2$. Prove this fact. HINT: Recall that $|x| = x$ if $x \geq 0$ or $-x$ if $x < 0$.

3.2   Slope

In the previous section, we raised the two general questions: Given an equation, how do we find its graph? Given a graph, how do we find its equation? In the last section, we answered the first question for first-degree equations in two variables. We will repeat these questions for a variety of graphs and equations as we proceed through this book. In this section we begin to answer the second question when the graph is a straight line. The ideas we develop in this section play a pivotal role in calculus as well.

Keep in mind that, as we pointed out before, an equation is a condition that all points on the graph must satisfy. Once we find such a condition, we obtain an equation by translating this condition mathematically.

Suppose we are given any nonvertical line L. We choose any four distinct points on L and label the points $P(x_1, y_1)$, $Q(x_2, y_2)$, $S(x_3, y_3)$, and $T(x_4, y_4)$. We have also formed right triangles PQR and STU. See Figure 3.18.

Figure 3.18

Triangles PQR and STU are similar.

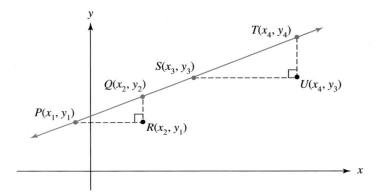

Since $\overline{PR}$ and $\overline{SU}$ are parallel, as are $\overline{QR}$ and $\overline{TU}$, we have $\angle QPR \cong \angle TSU$ (they are corresponding angles formed by parallel lines), and $\angle R$ and $\angle U$ are both right angles. Consequently, triangles PQR and STU are similar triangles; therefore, their corresponding sides are in proportion. That is,

NOTE: $|\overline{QR}|$ means the length of line segment QR.

$$\frac{|\overline{QR}|}{|\overline{PR}|} = \frac{|\overline{TU}|}{|\overline{SU}|}$$ Hence

$$\frac{y_2 - y_1}{x_2 - x_1} = \frac{y_4 - y_3}{x_4 - x_3}$$

Will the ratio $\frac{y_2 - y_1}{x_2 - x_1}$ be constant if the points (x_1, y_1) and (x_2, y_2) are on a graph that is not a straight line?

In other words, because triangles PQR and STU are similar, whenever we move from one point on a nonvertical line to another point on the line, the ratio of the change in y-coordinates to the change in x-coordinates remains *constant* for each line. This is exactly what we are looking for—a condition that all points on the line must satisfy.

We will actually derive an equation for a line from this condition in the next section. The remainder of this section is devoted to amplifying this idea that the ratio of the change in y to the change in x is constant. We begin with the following definition.

Definition of Slope

Let $P_1(x_1, y_1)$ and $P_2(x_2, y_2)$ be any two distinct points on a nonvertical line L. The **slope** of the line L, denoted by m, is given by

$$m = \frac{y_2 - y_1}{x_2 - x_1} = \frac{\text{change in } y}{\text{change in } x}$$

Note that, based on the preceding discussion, the slope of a nonvertical line is well defined. That is, every nonvertical line has a unique slope. Regardless of which two points we choose, the ratio of the change in y to the change in x will be constant.

Example 1   Find the slope of the line segment joining the two points:
(a) $P(-1, 2)$ and $Q(3, 8)$ (b) $P(-1, 2)$ and $S(3, -1)$
(c) $P(-1, 2)$ and $R(4, 2)$

Solution   Although it is not always necessary, it is generally a good idea to draw a diagram to help visualize the given information. See Figure 3.19.

Figure 3.19

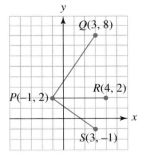

(a) Using the points $P(-1, 2)$ and $Q(3, 8)$ and the formula for the slope of a line, we compute the slope of the line passing through P and Q as

It does not matter which point is first and which is second, as long as we are consistent for both the x- and y-coordinates.

$$m = \frac{8 - 2}{3 - (-1)} = \frac{6}{4} = \frac{3}{2} \quad \text{or} \quad m = \frac{2 - 8}{-1 - 3} = \frac{-6}{-4} = \frac{3}{2}$$

Thus the slope is $m = \dfrac{3}{2}$. Remember to subtract the x-coordinates in the same order as the y-coordinates.

(b) Using the points $P(-1, 2)$ and $S(3, -1)$, we compute the slope of the line passing through P and S as

$$m = \frac{-1 - 2}{3 - (-1)} = \frac{-3}{4} = -\frac{3}{4}$$

(c) Using the points $P(-1, 2)$ and $R(4, 2)$, we compute the slope of the line passing through P and R as

$$m = \frac{2 - 2}{4 - (-1)} = \frac{0}{5} = 0$$

What does this number, the slope, tell us about a line? We begin with the following ground rule.

Whenever we describe a graph, we describe it as we move from *left to right*. In other words, we describe it for increasing values of x. (See Figure 3.20.)

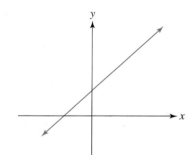

This line is rising.
This means the y-values are increasing as we move from left to right on the line.

(a)

This line is falling.
This means the y-values are decreasing as we move from left to right on the line.

(b)

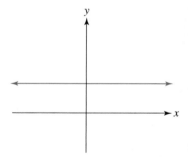

This line is neither rising nor falling.
This means the y-values are constant as we move from left to right on the line.

(c)

Figure 3.20

The slope is actually a rate of change. As we move from one point on a line to another, the slope tells us how much y is changing as compared to how much x is changing. What is important to recognize is that the slope of a line is merely a number that gives us information about a line's "steepness" and its direction.

For instance, if we look back at Figure 3.19, the line through P and Q has a slope of $\dfrac{3}{2}$, which means that as we move from left to right on the line, the ratio of the change in y to the change in x is 3 to 2. Therefore, to get from one point on the line to another point on the line, a change of 2 units in the x-coordinate is accompanied by a change of 3 units in the y-coordinate. Thus the line through P and Q, which has a *positive* slope, rises (goes up) as we move from left to right.

The line through P and S in Figure 3.19 has a slope of $\dfrac{-3}{4}$, which means that as we move from left to right on the line, a 4-unit change in the x-coordinate is accompanied by a -3-unit change in the y-coordinate (meaning the y-coordinate goes down 3 units). Thus the line through P and S, which has a *negative* slope, falls (goes down) as we move from left to right. In other words, on a line with positive slope, the y-values increase as we move from left to right, whereas on a line with negative slope the y-values decrease as we move from left to right.

What does the slope actually tell us about a line?

Figure 3.21 illustrates lines with various slopes passing through the point (3, 2). Note that the larger the slope in absolute value, the steeper the line.

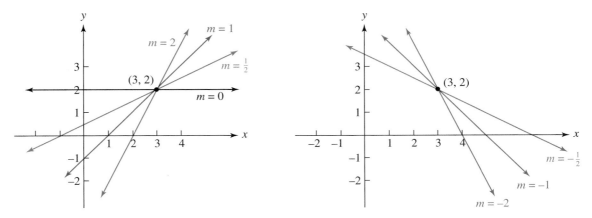

Figure 3.21
Lines with various slopes passing through (3, 2)

In our definition of the slope, note that we specified the line L to be nonvertical. Why? If we try to compute the slope of a vertical line such as the one passing through the points (2, 1) and (2, 5), we get

$$m = \frac{5 - 1}{2 - 2} = \frac{4}{0} \quad \text{which is undefined}$$

Thus *the slope of a vertical line is undefined.*

Be careful not to confuse a line that has slope 0 and is horizontal with a line whose slope is undefined and is vertical. See Figure 3.22.

Figure 3.22

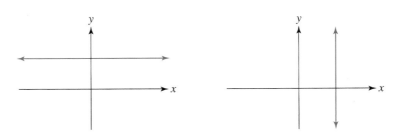

A horizontal line; its slope is 0

(a)

A vertical line; its slope is undefined

(b)

How does a line of slope 2 compare to a line of slope $\frac{1}{2}$?

How does a line of slope 2 compare to a line of slope -2?

Summary of Slopes

1. A line with positive slope rises as we move from left to right.
2. A line with negative slope falls as we move from left to right.
3. A line with zero slope is horizontal.
4. A line with undefined slope is vertical.

Example 2 | Sketch the graph of the line with slope $-\frac{1}{2}$ passing through the point $(3, 2)$.

Solution | To graph the line, we need to have two points on the line. We are given one point; how do we find a second point? The given slope of $-\frac{1}{2}$ can be thought of as $\frac{-1}{2}$ or $\frac{1}{-2}$. If we think of the slope as $\frac{-1}{2}$, then a 2-unit change in x is accompanied by a -1-unit change in y; that is, to get from one point on the line to another, we move 2 units to the right and 1 unit down. Alternatively, viewing the slope as $\frac{1}{-2}$, we would move 2 units to the left and 1 unit up. Either way, we end up with the same line. See Figure 3.23.

Remember that the slope is the change in y divided by the change in x.

Figure 3.23

The graph of the line with slope $-\frac{1}{2}$ passing through $(3, 2)$

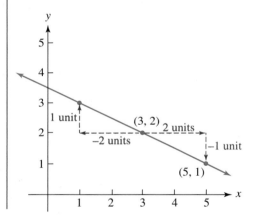

What is most important to recognize here is that once we specify one point and a slope, the line is completely determined.

A point and a slope determine a line.

Example 3 | An engineer has specified that sewage pipe for a certain building must have a 3% drop in grade.
(a) How much vertical clearance must a builder allow for a sewage pipe that is to carry waste from a point in a building to a point in the street that is 300 feet away in a horizontal direction?
(b) How long does this pipe have to be (to the nearest tenth of a foot)?

Solution **(a)** Figure 3.24 illustrates the given situation.

Figure 3.24

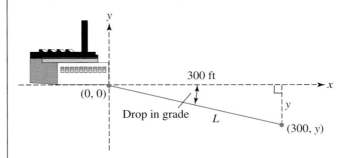

(0, 0)

300 ft

Drop in grade   L

(300, y)

To simplify the computation, we have set up the situation as if the building end of the pipe is situated at the origin. As the diagram illustrates, we need the horizontal distance (the *x*-coordinate) to be 300 feet and we want to find the vertical clearance (the *y*-coordinate) that will give a 3% drop in grade. A 3% drop in grade (or downgrade) means that the ratio of the vertical distance to the horizontal distance is -0.03. In effect, we require that the slope of the line segment *L* be -0.03. We therefore have

$$\frac{y - 0}{300 - 0} = -0.03 \qquad \text{Now we solve for } y.$$

$$y = -0.03(300) = \boxed{-9}$$

Therefore, the builder must set the pipe so that the street end of the pipe is 9 feet lower than the building end.

(b) In Figure 3.24 we have labeled the length of the pipe *L*. Since we have a right triangle, we can use the Pythagorean Theorem to find the length of *L*:

$$L^2 = 9^2 + 300^2$$

$$L^2 = 90{,}081 \qquad \text{Find the square root of 90,081 on a calculator.}$$

$$\boxed{L = 300.1 \text{ ft}} \qquad \text{Rounded to the nearest tenth}$$ ∎

Example 4 Suppose that the UBC Cable Company had 3.4 million cable subscribers at the end of 1996 and that during the next decade the number of subscribers is expected to increase by 1.2 million subscribers per year.

(a) Letting $x = 0, 1, 2, 3, \ldots$ represent 1996, 1997, 1998, $\ldots$ and letting *y* represent the number of subscribers (in millions) at the end of year *x*, write an equation relating *x* and *y*.

(b) Sketch the graph of the equation obtained in part (a).

(c) Find the slope of the line graphed in part (b) and relate it to the equation obtained in part (a).

Solution **(a)** Since we are starting with 3.4 million subscribers and adding 1.2 million subscribers per year, in year *x* the company would have

$$y = 3.4 + 1.2x \quad \text{million subscribers}$$

(b) We recognize that the equation found in part (a) is a first-degree equation and so its graph is a straight line. Thus we need only find two points on the line. By choosing $x = 0$ and $x = 1$, we find that the points $(0, 3.4)$ and $(1, 4.6)$ are on the line. The graph appears in Figure 3.25 on page 120.

Figure 3.25

The graph of $y = 3.4 + 1.2x$

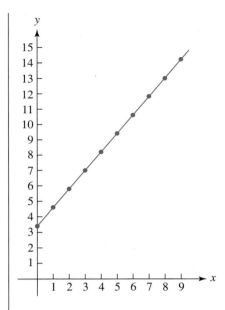

(c) We can use the same two points $(0, 3.4)$ and $(1, 4.6)$ to compute the slope. To understand the significance of the slope, let's include the meaning of the variables:

$$m = \frac{\text{change in } y}{\text{change in } x} = \frac{\text{change in number of subscribers}}{\text{change in years}}$$

$$= \frac{4.6 - 3.4}{1 - 0} = 1.2 \text{ million subscribers per year}$$

Note that the slope is exactly the same as the rate at which the number of subscribers is changing. This idea of the slope of a line being interpreted as a rate of change will be discussed further. ∎

   Throughout your study of mathematics, parallel and perpendicular lines have played a prominent role, and they are special with regard to their slopes as well.

Theorem 3.2

Let L_1 and L_2 be nonvertical lines with slopes m_1 and m_2, respectively. Then

1. L_1 and L_2 are parallel if and only if their slopes are equal, that is, $m_1 = m_2$.

2. L_1 and L_2 are perpendicular if and only if their slopes are negative reciprocals of each other, that is,

$$m_2 = -\frac{1}{m_1} \qquad \text{or equivalently} \qquad m_1 \cdot m_2 = -1$$

Proof   Part 1 of this theorem seems quite plausible. The fact that lines are parallel suggests that they have the same steepness, which in turn means that they have the same slope. A formal proof of this is outlined in Exercise 53 at the end of this section.

Part 2 of this theorem is not nearly as intuitive and we offer a formal proof here. For the sake of simplicity, let's assume that lines L_1 and L_2 intersect at the origin. (A similar proof works in the general case.) Let P and Q be, respectively, the points where the lines L_1 and L_2 intersect the line $x = 1$, as indicated in Figure 3.26.

Figure 3.26

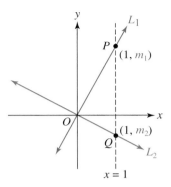

Since the slope of L_1 is $m_1 = \dfrac{m_1}{1}$, as x changes by 1 unit, y changes by m_1 units, and so the coordinates of P are $(1, m_1)$. Similarly, the coordinates of Q are $(1, m_2)$.

Now, L_1 is perpendicular to L_2 if and only if $\angle POQ$ is a right angle. Applying the Pythagorean Theorem, $\angle POQ$ is a right angle if and only if

$$|\overline{PQ}|^2 = |\overline{OP}|^2 + |\overline{OQ}|^2$$

Applying the distance formula, we get

$$\left(\sqrt{(1-1)^2 + (m_1 - m_2)^2} \right)^2 = \left(\sqrt{(1-0)^2 + (m_1 - 0)^2} \right)^2 + \left(\sqrt{(1-0)^2 + (m_2 - 0)^2} \right)^2$$

$$\left(\sqrt{(m_1 - m_2)^2} \right)^2 = \left(\sqrt{1^2 + (m_1)^2} \right)^2 + \left(\sqrt{1^2 + (m_2)^2} \right)^2$$

$$(m_1 - m_2)^2 = 1 + (m_1)^2 + 1 + (m_2)^2$$

$$(m_1)^2 - 2m_1m_2 + (m_2)^2 = (m_1)^2 + (m_2)^2 + 2 \qquad \text{Simplifying this equation, we get}$$

$$-2m_1m_2 = 2$$

$$m_1m_2 = -1$$

as required, which proves part 2 of the theorem in the case where the two lines intersect at the origin.

Example 5   Find the value of t so that the line passing through the points $A(2, t)$ and $B(5, -2)$ is parallel to the line passing through the points $C(-6, 3)$ and $D(0, -4)$.

Solution   A diagram will help us visualize exactly what this example is asking. In Figure 3.27 (page 122) we have plotted the points A, B, C, and D. Note that, although we do not know the exact location of the point $A(2, t)$, we do know that it must fall on the line $x = 2$. Let's analyze this problem carefully in order to develop a strategy for the solution.

Figure 3.27

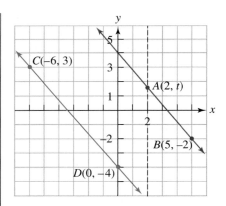

What do we need to find? (We use the diagram to help in this analysis.)	The value of t, the second coordinate of the point A
What information do we have about point A?	We know that the coordinates of A are $(2, t)$ and that the line passing through A and B is parallel to the line passing through C and D.
What do we know about parallel lines?	We know that parallel lines have the same slope.
How do I use the slopes to find the coordinates of the point?	Find the slope of each line by using the definition of slope: $m = \dfrac{y_2 - y_1}{x_2 - x_1}$. The fact that the lines are to be parallel tells us that the slopes must be equal.

Knowing that the slopes must be equal allows us to write and solve an equation involving t:

$$m_{\overline{AB}} = m_{\overline{CD}}$$

$$\frac{t - (-2)}{2 - 5} = \frac{3 - (-4)}{-6 - 0}$$

$$\frac{t + 2}{-3} = \frac{7}{-6} \qquad \text{Multiply both sides of the equation by } -3.$$

$$t + 2 = \frac{7}{2}$$

$$t = \frac{3}{2} \qquad \text{Does this value agree with Figure 3.27?}$$

3.2 Exercises

In Exercises 1–6, sketch the line through the given points and compute its slope.

1. $(2, 1)$ and $(5, 6)$ **2.** $(-3, 2)$ and $(4, -1)$

3. $(0, 3)$ and $(3, 0)$ **4.** $(-4, 0)$ and $(0, -4)$

5. $(-2, -1)$ and $(1, -3)$ **6.** $(-3, -4)$ and $(1, 5)$

In Exercises 7–20, find the slope of the line passing through the given points.

7. $(2, 7)$ and $(4, 10)$ **8.** $(-1, 5)$ and $(2, -3)$

9. $(-4, -3)$ and $(2, 1)$ **10.** $(-2, 6)$ and $(-4, -8)$

11. $(-4, 2)$ and $(6, 2)$ **12.** $(-1, 3)$ and $(-1, -5)$

13. $\left(\dfrac{1}{2}, \dfrac{3}{5}\right)$ and $\left(\dfrac{3}{4}, \dfrac{2}{3}\right)$

14. $\left(\dfrac{1}{6}, \dfrac{1}{4}\right)$ and $\left(-\dfrac{1}{2}, 2\right)$

15. $(2, \sqrt{3})$ and $(4, \sqrt{27})$

16. $(\sqrt{8}, 4)$ and $(\sqrt{18}, -2)$

17. (a, a^2) and (b, b^2) $(a \neq b)$

18. (b, a^2) and (a, b^2) $(a \neq b)$

19. (r, s) and $(r + s, 2s)$ $(s \neq 0)$

20. $(-p, n)$ and $(2p, 5n)$ $(p \neq 0)$

In Exercises 21–26, sketch the graph of the line passing through the given point and having the indicated slope.

21. $(-1, 2), m = 3$

22. $(2, -3), m = -2$

23. $(4, 0), m = -\dfrac{2}{3}$

24. $(0, -3), m = \dfrac{1}{4}$

25. $(-5, 1), m = 0$

26. $(-5, 1), m$ is undefined

27. The straight line segment in the accompanying diagram represents the relationship between the number of calories burned during a brisk walk, c, and the duration of the walk in minutes, m.

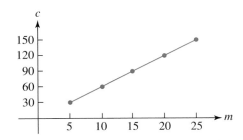

(a) How many calories are burned during a brisk walk that lasts 20 minutes?

(b) Use the graph to determine the slope of the line.

(c) How would you interpret the slope of this line?

28. The straight line segment in the accompanying diagram represents the relationship between a manufac-

turer's cost per radio c (in dollars) and the number r of radios manufactured (in thousands).

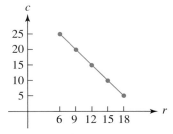

(a) What is the cost per radio if 12,000 radios are manufactured?

(b) Use the graph to determine the slope of the line.

(c) How would you interpret the slope of this line?

29. The straight line segment in the accompanying diagram represents the relationship between the number of gallons of gasoline used, g, and the distance traveled in miles, m, during a 400-mile trip.

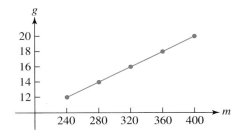

(a) How many gallons are used if the car travels 360 miles?

(b) How many miles can the car travel using 14 gallons?

(c) Compute the car's gas consumption in miles per gallon using the results of parts (a) and (b). Are your results the same? Should they be?

(d) Use the graph to determine the slope of the line.

(e) How would you interpret the slope of this line?

30. The straight line segment in the accompanying diagram represents the relationship between the number of miles m and the number of hours h that a car travels during a 300-mile trip.

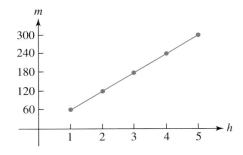

(a) How many miles does the car travel in 2 hours?

(b) How many hours does it take for the car to travel 300 miles?

(c) Compute the car's average speed using the results of parts (a) and (b). Are the results the same? Should they be?

(d) Use the graph to determine the slope of the line.

(e) How would you interpret the slope of this line?

31. Suppose that Jane works part-time making deliveries for a caterer. She gets paid a base salary of $65 per day plus $20 for each delivery she makes that day.

(a) Let d represent the number of deliveries she makes and let A represent the amount she earns for each day that she works. Write an equation relating d and A.

(b) Sketch the graph of the equation obtained in part (a) representing d along the horizontal axis.

(c) Find the slope of the line graphed in part (a) and relate it to the equation obtained in part (a).

32. Suppose that a swimming pool has 1000 gallons of water in it when a valve that drains 35 gallons of water from the pool each minute is opened.

(a) Let m represent the number of minutes after the valve is opened and let N represent the number of gallons of water in the pool. Write an equation relating m and N.

(b) Sketch the graph of the equation obtained in part (a) representing m along the horizontal axis.

(c) Find the slope of the line graphed in part (b) and relate it to the equation obtained in part (a).

In Exercises 33–38, determine whether the line passing through the points P_1 and P_2 is parallel or perpendicular (or neither) to the line passing through the points P_3 and P_4.

33. $P_1(2, 1)$, $P_2(4, 3)$, $P_3(-2, -1)$, $P_4(-4, -3)$

34. $P_1(-1, 4)$, $P_2(3, 2)$, $P_3(2, -3)$, $P_4(3, -1)$

35. $P_1(-3, -2)$, $P_2(-1, 1)$, $P_3(5, 4)$, $P_4(7, 1)$

36. $P_1(2, 9)$, $P_2(5, 10)$, $P_3(-4, -5)$, $P_4(-1, -6)$

37. $P_1(1, 4)$, $P_2(-3, 4)$, $P_3(-2, 7)$, $P_4(-2, -3)$

38. $P_1(1, 2)$, $P_2(3, 6)$, $P_3(-6, -7)$, $P_4(-3, -1)$

39. Find the value of c so that the line passing through the points $(2, 5)$ and $(-4, c)$ has slope $-\dfrac{1}{2}$.

40. Find the value of a so that the line passing through the points $(-2, 4)$ and $(a, 1)$ has slope $\dfrac{2}{3}$.

41. Find the value of t so that the line through the points $(0, t)$ and $(t, -1)$ is parallel to the line through the points $(1, 2)$ and $(2, -3)$.

42. Find the value of c so that the line through the points $(c, 1)$ and $(1, c)$ is perpendicular to the line through the points $(-2, 5)$ and $(3, -4)$.

43. Find the value(s) of h so that the line through the points $(h, 1)$ and $(-1, h)$ is perpendicular to the line through the points $(7, h)$ and $(h, 2)$.

44. Find the value(s) of t so that the line through the points (t, t) and $(3, 4)$ is parallel to the line through the points $(-9, t)$ and $(t, 9)$.

45. Prove that the points $P(-3, 0)$, $Q(1, 2)$, and $R(3, -2)$ are the vertices of a right triangle.

46. Prove that the points $A(0, -4)$, $B(4, -2)$, $C(5, 4)$, and $D(1, 2)$ are the vertices of a parallelogram.

47. Prove that the points $P(-3, -2)$, $Q(1, 4)$, $R(-2, 4)$, and $S(-4, 1)$ are the vertices of a trapezoid.

48. Prove that the points $A(-2, -5)$, $B(2, -4)$, $C(1, 0)$, and $D(-3, -1)$ are the vertices of a square.

49. Let $ABCD$ be the parallelogram with vertices $A(0, 0)$, $B(4, 0)$, $C(5, 2)$, and $D(1, 2)$. Prove that the quadrilateral formed by joining the midpoints of the sides of $ABCD$ is also a parallelogram.

50. Let $ABCD$ be the square with vertices $A(0, 0)$, $B(6, 0)$, $C(6, 6)$, and $D(0, 6)$. Prove that the quadrilateral formed by joining the midpoints of the sides of $ABCD$ is also a square.

51. The accompanying figure illustrates four lines with slopes m_1, m_2, m_3, and m_4. List these slopes in increasing order.

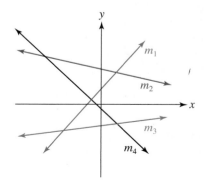

52. The accompanying figure illustrates three lines with equations as indicated.
  (a) Show that the values b_1, b_2, and b_3 are the y-intercepts of their respective lines.
  (b) List the slopes in decreasing order and the y-intercepts in increasing order.

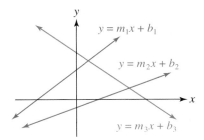

Questions for Thought

53. In this exercise we outline a proof of part 1 of Theorem 3.2, which says that two nonvertical lines are parallel if and only if their slopes are equal.

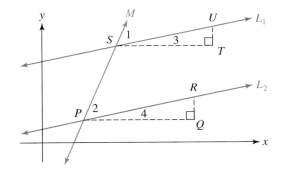

Let L_1 and L_2 be two nonvertical lines with slopes m_1 and m_2, respectively, crossed by line M, as in the accompanying figure. We have drawn $\overline{ST}$ and $\overline{PQ}$ parallel to the x-axis, and $\overline{UT}$ and $\overline{RQ}$ parallel to the y-axis. Justify each of the following statements.
  (a) Since $\overline{PQ}$ is parallel to $\overline{ST}$, $\angle MST \cong \angle SPQ$.
  (b) L_1 is parallel to L_2 if and only if $\angle 1 \cong \angle 2$.
  (c) L_1 is parallel to L_2 if and only if $\angle 3 \cong \angle 4$.

(d) Since $\angle T$ and $\angle Q$ are right angles, L_1 is parallel to L_2 if and only if $\triangle STU$ is similar to $\triangle PQR$.
  (e) L_1 is parallel to L_2 if and only if
$$\frac{|\overline{UT}|}{|\overline{ST}|} = \frac{|\overline{RQ}|}{|\overline{PQ}|}$$
  (f) L_1 is parallel to L_2 if and only if $m_1 = m_2$.

54. How would you describe a line whose slope is positive? Negative? Zero? Undefined?

55. How can the idea of slope be used to prove that the three points $(-6, -4)$, $(2, -2)$, and $(6, -1)$ are collinear?

56. What happens to our visualization of a line with slope 3 if we don't insist that the units along both axes be the same? Sketch the graph of a line passing through the point $(2, 3)$ with slope 3 if the units along the x-axis are twice as large as the units along the y-axis. Now draw the graph again if the units along the y-axis are twice as large as the units along the x-axis.

57. Prove part 2 of Theorem 3.2 in the case where the two lines intersect at the point (a, b).

58. As was mentioned before, the slope of a line can be thought of as a rate of change. Suppose that the height h, in meters, of an object above the ground after t seconds is given by the equation $h = 3t + 2$ for $t \geq 0$. Thus when $t = 4$, $h = 14$, meaning that the object is 14 meters above the ground after 4 seconds.
  (a) How far has the object traveled from time $t = 5$ to time $t = 8$?
  (b) What is the average speed of the object for these 3 seconds? Remember that average speed is computed by dividing the distance traveled by the time elapsed.
  (c) Repeat parts (a) and (b) as t changes from time $t = 10$ to time $t = 15$.
  (d) Letting t be the horizontal axis and h be the vertical axis, sketch the graph of $h = 3t + 2$ for $t \geq 0$. What is the slope of this line?
  (e) How are the average speed of the object and the slope of the line related?

3.3   Equations of a Line

We are now ready to answer the question raised earlier: Given a line, how do we find its equation? Suppose we are given a line L with slope m that passes through the point (x_1, y_1), as indicated in Figure 3.28.

Figure 3.28

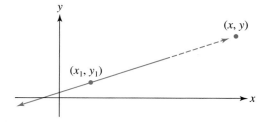

If (x, y) is any other point, then the condition it must satisfy to be on the line is that the slope of the line as determined by the points (x_1, y_1) and (x, y) must be m. If the slope is m, then the point is on the line, and if the slope is not m, then the point is not on the line.

Algebraically, we translate this condition as

$$\frac{y - y_1}{x - x_1} = m \qquad \text{for } x \neq x_1$$

Multiplying both sides of the equation by $x - x_1$, we get the following.

Point–Slope Form for an Equation of a Straight Line

An equation of the line with slope m passing through the point (x_1, y_1) is

$$y - y_1 = m(x - x_1)$$

Keep in mind that in the **point–slope form**, (x_1, y_1) denotes the *given* point and (x, y) denotes any other point on the line.

Example 1   Write an equation of the line with slope $\dfrac{2}{3}$ that passes through the point $(-2, 5)$.

Solution   We are given exactly the information necessary to use the point–slope form for an equation of a line. The given point $(-2, 5)$ corresponds to (x_1, y_1) and the slope is $m = \dfrac{2}{3}$.

$y - y_1 = m(x - x_1)$ Substitute $(-2, 5)$ for (x_1, y_1) and $\dfrac{2}{3}$ for m.

$y - 5 = \dfrac{2}{3}[x - (-2)]$ and so an equation of the line is $\boxed{y - 5 = \dfrac{2}{3}(x + 2)}$

For the time being, we leave the answer in this form. We discuss the different possible forms for the answer in Example 3. ∎

Calculator Exploration

Use a graphing calculator to:

1. Graph the following on the same set of coordinate axes in the standard $[-10, 10]$ by $[-10, 10]$ window: $y = 2x$, $y = 3x$, $y = 5x$. What can you conclude about how m affects the graph of the equation $y = mx$?

2. Graph the following on the same set of coordinate axes in the standard $[-10, 10]$ by $[-10, 10]$ window: $y = 3x - 2$, $y = 3x$, $y = 3x + 4$. What can you conclude about how b affects the graph of the equation $y = 3x + b$?

The point–slope form allows us to write an equation of a line given its slope and *any* point on the line. Let's see what happens if the given point happens to correspond to the y-intercept of the line; that is, suppose the line has slope m and a y-intercept of b, which means that the line passes through the point $(0, b)$.

Applying the point–slope form, we get

$$y - y_1 = m(x - x_1) \qquad \text{Substitute } (0, b) \text{ for } (x_1, y_1).$$
$$y - b = m(x - 0)$$
$$y - b = mx$$

This equation is usually solved explicitly for y, giving the following.

The Slope–Intercept Form for an Equation of a Straight Line

An equation of the line with slope m and y-intercept b is

$$y = mx + b$$

Example 2 Write an equation of the line with slope -2 passing through the point $(0, -3)$.

Solution Since the line passes through the point $(0, -3)$, we know that the y-intercept is -3. Therefore, from the given information, we can apply the slope–intercept form with $m = -2$ and $b = -3$:

$$y = -2x - 3$$

Example 3 Write an equation of the line passing through the points $(1, 2)$ and $(3, -5)$.

Solution This example illustrates that in writing an equation of a line (as long as the line is not vertical), we always have the choice of using either the point–slope or the slope–intercept form. In either case we need to compute the slope of the given line:

$$m = \frac{-5 - 2}{3 - 1} = \frac{-7}{2}$$

If we choose to use the point–slope form, we may use either of the given points.

Using the point $(1, 2)$, we get $y - 2 = -\dfrac{7}{2}(x - 1)$.

Using the point $(3, -5)$, we get $y + 5 = -\dfrac{7}{2}(x - 3)$.

If we choose to use the slope–intercept form, we proceed as follows.

$$y = mx + b$$

We substitute $m = -\dfrac{7}{2}$. We need to find the value of b.

$$y = -\frac{7}{2}x + b$$

Since we know the line passes through the points (1, 2) and (3, −5), either of these points satisfies the equation. We choose to substitute (1, 2) and solve for b.

$$2 = -\frac{7}{2}(1) + b \quad\Rightarrow\quad b = \frac{11}{2}$$

Therefore, we get $y = -\dfrac{7}{2}x + \dfrac{11}{2}$.

Show that the first two answers are equivalent to the third by putting them in the form $y = mx + b$.

Although these three answers may look different, they are, in fact, equivalent. If we take the first two answers and put them in slope–intercept form, we get the third answer. In fact, one of the advantages of the slope–intercept form is that everyone's answer looks the same. Even though we have just seen that using the slope––intercept form can involve some extra computation, for most situations the slope–intercept form of the equation is the most useful and is preferred.

What are the advantages and disadvantages of using the slope–intercept form?

One of the most useful features of the slope–intercept form is that when an equation of a line is written in the form $y = mx + b$, the slope of the line is readily identified as the coefficient of x.

Example 4   Find the slope of the line whose equation is:
(a) $y = -4x + 7$ (b) $3x - 4y = 18$

Solution   (a) By comparing the slope–intercept form $y = mx + b$ with $y = -4x + 7$, we can simply read off the slope (as well as the y-intercept).

$$
\begin{array}{ccc}
y = & mx & + b \\
\downarrow & \downarrow & \downarrow \\
y = & -4x & + 7
\end{array}
$$

Therefore, the slope is -4.

(b) We may read the slope of a line from its equation, provided that the equation is *exactly* in slope–intercept form.

$$3x - 4y = 18$$ We solve this equation explicitly for y.

$$-4y = -3x + 18 \quad\Rightarrow\quad y = \frac{3}{4}x - \frac{9}{2}$$

Compute the slope in part (b) by finding two points on the line and using the definition of slope. Which method is easier?

Therefore, the slope is $\dfrac{3}{4}$.

Example 5   Write an equation of the line that passes through the point (4, 0) and is perpendicular to the line whose equation is $5y - 3x = 15$.

Solution   Again, it is a good idea to draw a diagram to help us visualize what the example is asking. We plot the point (4, 0) and sketch the line whose equation is $5y - 3x = 15$.

See Figure 3.29. We have drawn a dashed line perpendicular to $5y - 3x = 15$, which passes through the point $(4, 0)$. It is the equation of this dashed line that we seek.

Figure 3.29

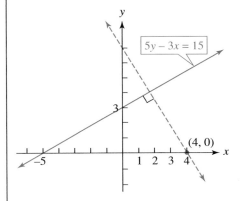

Let's analyze this problem carefully to develop a strategy for the solution.

What do we need to find?   An equation of a line

What is needed to find an equation of a line?   A point on the line and the slope of the line

What information is given in the problem?   The point is given, along with an equation for a line perpendicular to the line whose equation we want to find.

How do we find the slope of a line given the equation of a line perpendicular to it?   The slopes of perpendicular lines are negative reciprocals of each other. If we find the slope of the perpendicular line, we compute its negative reciprocal and we have the slope of the line we want to find.

How do we find the slope of the line whose equation is given?   Put the given equation into slope–intercept form.

Based on this analysis, we begin by finding the slope of the line whose equation is given:

$$5y - 3x = 15 \implies 5y = 3x + 15 \implies y = \frac{3}{5}x + 3$$

So the slope of the given line is $\frac{3}{5}$.

Therefore, the slope of the perpendicular line is $-\frac{5}{3}$.

Now that we have the point $(4, 0)$ and slope $-\frac{5}{3}$, we can use the point–slope form to get

$$y - 0 = -\frac{5}{3}(x - 4)$$

and so our final answer for an equation of the perpendicular line is

$$y = -\frac{5}{3}x + \frac{20}{3}.$$

See the Technology Corner at the end of this section. It deals with how perpendicular lines appear on a graphing calculator.

Example 6 | Write an equation of the line passing through the given pair of points.
(a) $(4, 3)$ and $(-2, 3)$ **(b)** $(2, -3)$ and $(2, 5)$

Solution | **(a)** The slope of the line is $m = \dfrac{3-3}{4-(-2)} = \dfrac{0}{6} = 0$. Using the point–slope form

with the point $(4, 3)$, we get

$$y - 3 = 0(x - 4) \quad \Rightarrow \quad y - 3 = 0 \quad \Rightarrow \quad \boxed{y = 3}$$

Alternatively, we could have recognized from the given points that the line passing through them is horizontal, 3 units above the x-axis; as we saw in Section 3.1, an equation of this line is $y = 3$. (See Figure 3.30.)

(b) If we attempt to compute the slope of the line passing through $(2, -3)$ and $(2, 5)$, we get

$$m = \frac{5-(-3)}{2-2} = \frac{8}{0} \qquad \text{which is undefined}$$

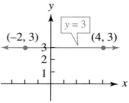

Figure 3.30

Figure 3.31

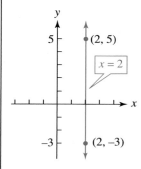

Consequently we cannot use either the point–slope or the slope–intercept form, since either form requires the line to have a slope. (This situation explains why the discussion leading up to the derivation of both forms specified a non-vertical line L.) Once we recognize that these points determine a vertical line 2 units to the right of the y-axis (see Figure 3.31), we know the equation of this line is $\boxed{x = 2}$.   ∎

Can the equation of a vertical line be put in slope–intercept form?

Example 7 | Find an equation of the line that is the perpendicular bisector of the line segment joining the points $P(-3, 1)$ and $Q(4, 2)$.

Solution | Figure 3.32 shows $\overline{PQ}$ with its perpendicular bisector indicated by the dashed line. Let's analyze this problem carefully to develop a strategy for the solution.

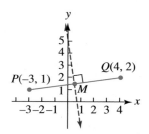

Figure 3.32

What do we need to find? (Use the diagram to help in this analysis.)	An equation of a line
What is needed to find an equation of a line?	A point on the line and the slope of the line
What information is given in the problem?	The line we need to find is the perpendicular bisector of a line segment joining two given points.
What is the perpendicular bisector?	It is a line that divides the given line segment into two equal parts and is perpendicular to the line segment.

How does the given information help us to find the point?	Since the perpendicular bisector divides the line into two equal parts, it must pass through the midpoint of the line segment.
How does the given information help us to find the slope?	Slopes of perpendicular lines are negative reciprocals of each other. If we find the slope of the given line segment, we compute its negative reciprocal and we have the slope of the line we seek.
How can we restate the problem in simpler terms?	Find the midpoint (by the midpoint formula) and the slope of the perpendicular bisector by taking the negative reciprocal of the slope of the line segment.

Based on this analysis, we begin by using the midpoint formula to find the midpoint of $\overline{PQ}$:

$$\text{Midpoint of } \overline{PQ} = \left(\frac{-3+4}{2}, \frac{1+2}{2}\right) = \left(\frac{1}{2}, \frac{3}{2}\right)$$

We find the slope of $\overline{PQ}$ by using the two given points:

$$m_{\overline{PQ}} = \frac{2-1}{4-(-3)} = \frac{1}{7}$$

Therefore, the slope of a perpendicular line will be -7, and we have the following as an equation of the perpendicular bisector of $\overline{PQ}$:

$$y - \frac{3}{2} = -7\left(x - \frac{1}{2}\right) \quad \text{or} \quad y = -7x + 5$$

As we continue through the text, we will see that straight lines and their equations are used in a wide variety of applications.

Example 8

Suppose that a shoe store finds that it sells 23 pairs of shoes per day at $30 per pair and 20 pairs of shoes per day at $36 per pair. Assuming a linear relationship exists between P, the price of a pair of shoes, and N, the number of pairs sold, predict the number of pairs of shoes sold per day at $40 per pair.

Solution

We may express the given information as ordered pairs. We are free to express the ordered pairs with P as the first coordinate and N as the second coordinate or vice versa. Reading through the example, it seems reasonable to assume that the number of pairs of shoes sold *depends* on the price. In sketching the graph of a relationship between two quantities, it is generally accepted that the dependent quantity (in this case the number of pairs of shoes sold) is taken as the vertical axis and the quantity on which it depends (in this case the price) is taken as the horizontal axis. Consequently, we will write the ordered pairs in the form (P, N).

The given information gives us the ordered pairs $(30, 23)$ and $(36, 20)$. Since we are told that the relationship between price and number of pairs sold is linear, we compute the slope of the line:

$$m = \frac{23-20}{30-36} = \frac{3}{-6} = -\frac{1}{2}$$

We can use the point–slope form with the point $(36, 20)$ to write an equation describing the relationship between P and N:

$$N - 20 = -\frac{1}{2}(P - 36)$$

Now we may substitute $P = 40$ to determine the number of pairs of shoes sold when the price is \$40:

Is there an advantage to transforming the equation $N - 20 = -\frac{1}{2}(P - 36)$ into slope–intercept form?

$$N - 20 = -\frac{1}{2}(P - 36) \qquad \text{Substitute } P = 40.$$

$$N - 20 = -\frac{1}{2}(40 - 36)$$

$$N - 20 = -2$$

$$N = 18$$

Thus, assuming a linear relationship, the store will sell 18 pairs of shoes per day at \$40 per pair. Figure 3.33 illustrates the linear relationship between the price and number of pairs of shoes sold. Several aspects of this graph are worth noting. First, the graph appears only in the first quadrant, because it makes no sense for either P (the price per pair of shoes) or N (the number of pairs of shoes sold per day) to be negative. Second, N can realistically take on only positive integer values (you can't sell a fraction of a pair of shoes), and so the graph should just be a discrete set of points. Nevertheless, to analyze the graph, it is often useful to draw the entire line segment.

Figure 3.33

The linear relationship between the price and the number of pairs of shoes sold

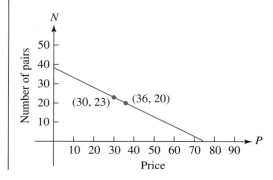

One final comment: In this section we have proven that the equation of a straight line is a first-degree equation in x and y. We still have not proven Theorem 3.1 (originally stated in Section 3.1), which states that the graph of an equation of the form $Ax + By = C$ (with both A and B not equal to zero) is a straight line. The proof of this theorem is outlined in Exercise 86.

Example 9   Figure 3.34 shows a graph of the relationship between the daily cost, C, of electricity to keep a particular home at a constant $72°$, and the average daily outdoor

Figure 3.34

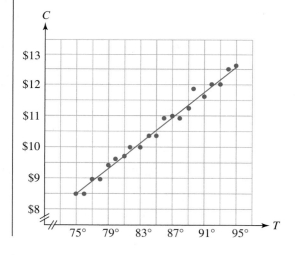

temperature, T, in July. The horizontal axis represents the average daily temperature and the vertical axis represents the daily cost of electricity. For the temperature range given in the figure, the actual data points are plotted. These data points can be approximated by the line $C = 0.21T - 7.30$. What does the slope of the line tell you about this relationship?

Solution

The slope gives the ratio of vertical change to horizontal change. Since the vertical axis is cost and the horizontal axis is temperature, the slope is the ratio of cost to temperature, in this case, the cost (in dollars) per degree. The slope of this line is 0.21, which means that to keep the home at a constant temperature of 72°, each 1-degree increase of the outdoor temperature increases the cost by $0.21 per day. Of course, this relationship holds only when the average outdoor temperature is between 75° and 95°. ∎

Technology Corner

Figure TC.1 shows the graphs of $y = \frac{1}{3}x - 6$ and $y = -3x + 6$ using he standard $[-10, 10]$ by $[-10, 10]$ window.

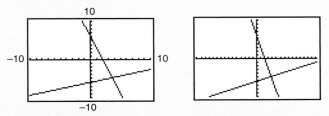

Figure TC.1   **Figure TC.2**

Although the lines should be perpendicular (because their slopes are negative reciprocals), they do not appear to be perpendicular in this viewing window. This is because units along the x-axis are a different size from those along the y-axis. Most calculators have a key or setting that allows you to correct this problem automatically.

Figure TC.2 shows how the same graphs would look on the T1-83 Plus using the ZSquare selection in the [ZOOM] menu. Check the new window settings in the ZSquare window and you will see that the x-settings are about 1.5 times the y-settings, which makes the units along both axes the same size.

Figure TC.3 illustrates how the circle $x^2 + y^2 = 9$ looks in the ZStandard window and in the ZSquare window.

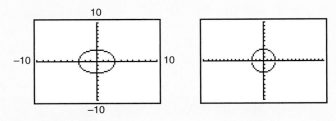

Figure TC.3

3.3 Exercises

In Exercises 1–4, write an equation of the line satisfying the given conditions and sketch its graph.

1. The line has slope 3 and passes through the point $(-1, 2)$.

2. The line has slope $-\dfrac{1}{4}$ and passes through the point $(0, 0)$.

3. The line has slope $\dfrac{2}{5}$ and passes through the point $(-2, -3)$.

4. The line has slope -3 and passes through the point $(0, 2)$.

In Exercises 5–22, write an equation of the line satisfying the given conditions.

5. Line L has slope 4 and passes through the point $(-3, 4)$.

6. Line L has slope $\dfrac{3}{5}$ and passes through the point $(-2, -1)$.

7. Line R passes through the points $(2, 6)$ and $(8, 3)$.

8. Line T passes through the points $(-2, -5)$ and $(-6, -7)$.

9. Line F passes through the origin and has slope 4.

10. Line G passes through the origin and has slope $-\dfrac{1}{4}$.

11. Line L has slope $-\dfrac{2}{3}$ and a y-intercept of 4.

12. Line L has slope $\dfrac{1}{4}$ and an x-intercept of -2.

13. Line R has an x-intercept of 5 and a y-intercept of -1.

14. Line Q has a y-intercept of 3 and an x-intercept of -2.

15. Line V is vertical and passes through the point $(-3, 4)$.

16. Line H is horizontal and passes through the point $(-3, 4)$.

17. Line T has slope $-\dfrac{1}{3}$ and crosses the y-axis at $y = 5$.

18. Line R has slope $\dfrac{2}{7}$ and crosses the x-axis at $x = -5$.

19. Line F has no x-intercept and passes through the point $(3, 5)$.

20. Line G has no y-intercept and passes through the point $(3, 5)$.

21. Line L passes through the points $(2, -5)$ and $(-3, -5)$.

22. Line L passes through the points $(-4, 6)$ and $(-4, -1)$.

In Exercises 23–30, find the slope of the line whose equation is given.

23. $y = -2x + 7$

24. $y = \dfrac{3}{5}x - \dfrac{2}{3}$

25. $y = \dfrac{2x + 5}{3}$

26. $y = \dfrac{1}{3} + \dfrac{7}{4}x$

27. $x = 5y - 1$

28. $6x - 5y = 10$

29. $\dfrac{4x + 3y}{3} = 2$

30. $\dfrac{5x}{4} - \dfrac{4y}{5} = 1$

In Exercises 31–44, write an equation of the line satisfying the given conditions.

31. Line S passes through the point $(1, -5)$ and is parallel to $y = 4x - 7$.

32. Line S passes through the point $(-2, 3)$ and is perpendicular to $y = -4x + 1$.

33. Line M passes through the point $(4, 7)$ and is perpendicular to $y = -\dfrac{3}{4}x + 9$.

34. Line T passes through the point $(5, -5)$ and is parallel to $y = \dfrac{4}{5}x + 2$.

35. Line S passes through the point $(5, 0)$ and is parallel to $5y - 4x = 9$.

36. Line R passes through the point $(0, 5)$ and is perpendicular to $7x + 4y = 1$.

37. Line M passes through the point $(7, 4)$ and is perpendicular to $6x - 7y = 6$.

38. Line T passes through the point $(-3, 3)$ and is parallel to $10y - 8x = 7$.

39. Line L passes through the point $(0, 0)$ and is perpendicular to $y = x$.

40. Line L passes through the point $(0, 0)$ and is parallel to $x + y = 8$.

41. Line L passes through the point $(-4, 5)$ and is parallel to the x-axis.

42. Line L passes through the point $(-4, 5)$ and is parallel to the y-axis.

43. Line T is perpendicular to $2x + 7y = 14$ and has the same y-intercept.

44. Line T is parallel to $6y - 4x = 9$ and has the same x-intercept.

In Exercises 45–52, use a graphing calculator to graph the two given equations on the same set of coordinate axes. Identify the slopes for each and discuss the similarities and differences between the two graphs.

45. $y = x, y = 2x$

46. $y = 2x, y = 3x$

47. $y = x - 3, y = 2x - 3$

48. $y = 2x - 1, y = 3x - 1$

49. $y = x, y = -x$

50. $y = 2x, y = -2x$

51. $y = 5x - 2, y = -5x - 2$

52. $y = 3x + 1, y = -3x + 1$

In Exercises 53–56, use a graphing calculator to graph the three given equations on the same set of coordinate axes. Identify the y-intercepts for each and discuss the similarities and differences among the three graphs.

53. $y = x, y = x - 3, y = x + 3$

54. $y = 2x, y = 2x - 3, y = 2x + 3$

55. $y = 5x, y = 5x - 1, y = 5x + 2$

56. $y = -2x + 3, y = -2x - 2, y = -2x - 4$

57. Find an equation of the perpendicular bisector of the line segment joining $(-4, 3)$ and $(1, 0)$.

58. Find an equation of the perpendicular bisector of the line segment joining the centers of the circles $x^2 + y^2 - 4y = 1$ and $x^2 + 3x + y^2 = 5$.

59. Find an equation of the perpendicular bisector of the line segment joining the points where the line $5y - 3x = 2$ crosses the x-axis and the y-axis.

60. (a) Find an equation of the perpendicular bisector of the line segment joining $(1, 2)$ and $(5, 8)$ using the method outlined in Example 7.
(b) Find an equation of this perpendicular bisector by using the distance formula and verify that you get the same result as in (a). HINT: The perpendicular bisector of a line segment is the line containing all points equidistant from both endpoints.

61. A car rental company charges $29 per day plus $0.30 per mile. Write an equation relating the cost, C, of renting a car for a day to the number of miles, m, driven.

62. A car rental company charges $39 per day plus $0.22 per mile and also requires customers to purchase a collision damage waiver at a cost of $8 per day. Write an equation relating the cost, C, of renting a car for 3 days to the number of miles, m, driven.

63. Suppose that a manufacturer determines that there is a linear relationship between P, the profit earned, and x, the number of items produced. The profit is $600 on 50 items and $750 on 65 items.
(a) Write an equation relating x and P.
(b) What would the expected profit be if 90 items were produced?

64. A quality control inspector finds that there is a relationship between D, the number of defective items produced, and h, the number of overtime hours put in by the employees. When the employees put in a total of 120 overtime hours, 80 defective items were found; when a total of 70 overtime hours were put in, 60 defective items were found. If the inspector suspects that the relationship between h and D is linear, how many defective items would she expect to find if 90 overtime hours are put in?

65. In physiology, a jogger's heart rate, N, in beats per minute is related linearly to the jogger's speed, s. A jogger's heartbeat is 80 beats per minute at a speed of 15 ft/sec and 85 beats per minute at a speed of 18 ft/sec.
(a) Write an equation expressing N in terms of s.
(b) Using the equation obtained in (a), predict the jogger's heart rate at a speed of 25 ft/sec.

66. An assembly-line worker receives an hourly salary of $13.50. In addition, the worker receives a piecework commission of $0.65 per unit produced.
(a) Write an equation relating the hourly wages, W, and the number of items produced per hour, x.
(b) Use the equation obtained in part (a) to determine the hourly wage if the worker produces 5 units per hour.
(c) Use the equation obtained in part (a) to find the number of units per hour a worker must produce if he wants to have an hourly wage of at least $22.

67. According to the U.S. Census Bureau the metropolitan Los Angeles area had the following populations in the years listed.

Year	1990	2000
Population (in millions)	14.5	16.3

(a) Use the given data to write a linear relationship between the year and the population of the metropolitan Los Angeles area.
(b) Use the equation obtained in part (a) to predict the population in 2010.

68. The following table contains the percentage of foreign-born citizens living in the United States in the years listed as reported by the U.S. Census Bureau.

Year	1990	2000
Percentage of foreign-born citizens	8.0	10.4

(a) Use the given data to write a linear relationship between the year and the percentage of foreign-born citizens.

(b) Use the equation obtained in part (a) to predict the percentage of foreign-born citizens in 2010.

69. The following table indicates the number of people living below the poverty level in the United States as reported by the Department of Commerce for the years listed.

Year	1998	1999
Numer of people below poverty level (in millions)	34.5	32.3

(a) Use the given data to write a linear relationship between the year and the number of people below the poverty level.

(b) Use the equation obtained in part (a) to predict the number of people below the poverty level in 2005.

70. The following table indicates the median price of a single-family home in the United States as reported by the Department of Commerce for the years listed.

Year	1999	2000
Price (in thousands)	104.9	110.1

(a) Use the given data to write a linear relationship between the year and the median price of a single-family home.

(b) Use the equation obtained in part (a) to predict the median price of a home in 2007.

71. A business buys a piece of machinery for $8500. Assuming that the machinery depreciates linearly to a value of zero dollars in 12 years, write an equation expressing the value, V, of the machinery after t years.

72. A small company buys a computer system for $30,000 and, for tax purposes, uses the linear depreciation method to arrive at a salvage value of $1800 after 10 years. Write an equation describing the value, V, of the computer system after n years.

73. A car rental agency buys a car for $12,375 and incurs expenses of $8.25 per day to operate the vehicle. It then rents the car for $49.95 per day.

(a) Write an equation for the total cost, C, to buy and operate the car for d days.

(b) Write an equation for the total income, I, earned if this car is rented for d days.

(c) The profit the rental company earns on this car is obtained by computing $I - C$. The *break-even point* for a business is that point at which the income and costs are equal (i.e., the profit is 0). Find the number of days this car must be rented for the company to break even.

(d) Sketch the graphs of the equations for C and I on the same coordinate system and give a graphical description of the break-even point.

74. In economics, a *demand equation* describes the relationship between the price, p, of an item and the number of items, d, that can be sold at that price. The *supply equation* describes the relationship between the projected selling price, p, of an item, and the number of items, s, that the manufacturer is willing to supply at that price.

(a) Would you expect the demand, d, to increase or decrease as the price, p, increases? Explain.

(b) Would you expect the supply price, s, to increase or decrease as price, p, increases? Explain.

(c) A clothing manufacturer estimates that for a certain swimsuit, the demand equation is given by $d = \dfrac{100 - p}{0.05}$ and the supply equation is given by $s = \dfrac{p - 70}{0.03}$. Do d and s as given by these equations agree with your answers to parts (a) and (b)?

(d) Sketch the graphs of the demand equation and the supply equation on the same coordinate system. (You will need to label the vertical axis as both d and s.)

(e) Using the graphs obtained in part (d) and given the fact that p is the price and that d and s are the number of items produced, what is the possible range of values that makes sense for p, d, and s?

(f) The price at which the supply and demand are equal is called the *equilibrium price*. Use the equations given in part (c) to find the equilibrium price for these swimsuits.

75. An audio equipment company knows that it can sell 275 portable cassette players each month at a price of $62.50. Based on survey information, the company also believes that for each $2.50 increase in price, 10 fewer units will be sold each month.

(a) Let N denote the number of cassette players sold each month and P be the price per unit. Assuming a linear relationship between the number of players sold and the price per player, write an equation for N in terms of P.

(b) Using the equation obtained in part (a), determine how many players will be sold at a price of $75.

(c) At what price would the company sell 240 players per month?

76. The cost of producing an item usually consists of two components: *fixed costs*, such as the cost of machinery, insurance, and taxes, and *cost per item*, such as the cost of raw material, labor, or utilities. Suppose the cost, C, of producing x items is approximately given by the equation

$$C = 20x + 120$$

(a) What is the fixed cost? HINT: What is the cost to produce 0 items?

(b) An important quantity in economics, known as the *marginal cost*, is approximately equal to the cost of producing *one* additional item—that is, the difference between the cost of producing x items and $x + 1$ items. Compute the marginal cost for this cost equation.

(c) How much would it cost to produce 250 items?

77. Two popular systems for measuring temperature are the Celsius scale, C, and the Fahrenheit scale, F. Water freezes at $0°C$ and $32°F$; water boils at $100°C$ and $212°F$. Given that the Celsius and Fahrenheit scales are linearly related, find an equation that gives the Celsius temperature in terms of the Fahrenheit temperature. What is the Celsius equivalent to $98.6°F$?

78. Use the information given (and obtained) in Exercise 77 to write an equation that gives the Fahrenheit temperature in terms of the Celsius temperature.

79. Anthropologists often extrapolate the appearance of a human being from the parts of a skeleton that they uncover. For example, it has been found that there is a linear relationship between the length, f, of the femur (thigh bone), and the height, h, of the human from whom it came. Suppose that it is known that a femur of length 47.5 cm corresponds to a height of 177.8 cm and that a femur of length 39.1 cm corresponds to a height of 146.3 cm.

(a) Write an equation expressing the height in terms of the length of the femur.

(b) Rounding to the nearest tenth, what height would correspond to a femur of length 52 cm?

80. From elementary geometry we know that a tangent line drawn to a circle is perpendicular to a radius drawn to the point of tangency. Use this fact to write an equation of the tangent line to the circle $x^2 + y^2 = 25$ at the point $(-3, 4)$.

81. Sketch the graph of the circle $x^2 + y^2 = 169$ and write an equation of the tangent line to this circle at the point $(5, -12)$. (See Exercise 80.)

82. Sketch the graph of $y = x - 2$, and use the graph to answer the following questions.

(a) For what values of x is the graph above the x-axis?

(b) For what values of x is $x - 2 > 0$?

(c) For what values of x is $y > 0$?

83. Sketch the graph of $y = 2x + 3$, and use the graph to answer the following questions.

(a) For what values of x is the graph below the x-axis?

(b) For what values of x is $2x + 3 < 0$?

(c) For what values of x is $y < 0$?

84. A company manufactures monitors. The accompanying figure shows a graph of the relationship between the wholesale price of a monitor, p, in dollars, and the company's daily revenue, R, in thousands of dollars. The horizontal axis represents the wholesale price of a monitor and the vertical axis represents the daily revenue. For the price range given in the figure, the actual data points are plotted. These data points can be approximated by the line $P = 1700p - 130,000$. What does the slope of the line tell you about this relationship?

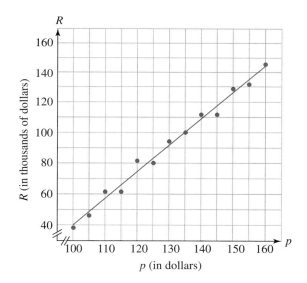

Questions for Thought

85. Given the line with equation $5y - 7x = 11$, describe two ways to find its slope. Which method is easier?

86. In this exercise we outline a proof of the fact that the graph of an equation of the form $Ax + By = C$, where $B \neq 0$, is a straight line. (Note that if $B = 0$, the equation becomes $Ax = C$, which we know to have a vertical line as its graph.)

Our strategy is to show that *any* two ordered pairs (x_1, y_1) and (x_2, y_2) will give the same slope. It then follows that the graph is a straight line. We can proceed as follows: First show that if (x_1, y_1) is a point satisfying $Ax + By = C$, then $y_1 = \dfrac{C - Ax_1}{B}$, which gives the ordered pair $\left(x_1, \dfrac{C - Ax_1}{B}\right)$ on the graph. Similarly, for (x_2, y_2) we get $\left(x_2, \dfrac{C - Ax_2}{B}\right)$ on the

graph. Now show that the slope determined by two such points is independent of x_1 and x_2.

87. There is yet another form for a first-degree equation in two variables: Prove that the equation of a line passing through $(a, 0)$ and $(0, b)$, where $a \neq 0$, $b \neq 0$, can be written in the form

$$\frac{x}{a} + \frac{y}{b} = 1$$

Why do you think this form is called the *intercept form*?

88. Consider the following 10 data points:

$(-1, -4.8)$, $(0, -2)$, $(0.5, -0.4)$, $(1, 1)$, $(1.5, 2.7)$, $(2, 3.8)$, $(2.4, 5.5)$, $(3.1, 7.2)$, $(3.7, 8.4)$, $(4, 10)$

Draw a straight line that, in your opinion, lies as close as possible to as many of these data points as possible. Estimate the equation of this line.

3.4   Relations and Functions

Our everyday lives are filled with situations in which we encounter relationships between two sets. For example,

To each taxpayer in the United States, there corresponds a Social Security number.

To each automobile, there corresponds a license plate number.

To each circle, there corresponds a circumference.

To each number, there corresponds its square.

In order to apply mathematics to a variety of disciplines, we must make the idea of a "relationship" between two sets mathematically precise. Let's begin with the following definition:

Definition of a Relation

A **relation** is a correspondence between two sets so that to each member of the first set (called the **domain**) there corresponds one or more members of the second set (called the **range**).

Let's consider the relation described by the following table, in which we associate each person listed with his or her telephone number.

Domain	Range
Sarah $\longrightarrow$	297-4419
John $\searrow$	
Ginger $\longrightarrow$	348-7743
Lamar $\longrightarrow$	459-8810
$\searrow$	459-8811

Note that Lamar has two telephone numbers, whereas the number 348-7743 is associated with two people, John and Ginger.

Let's look at a few more examples of relations.

Relation A   **Relation B**   **Relation S**

$u \longrightarrow 4$ $u \searrow 4$ $u \longrightarrow 4$

$v \longrightarrow 7$ $v \searrow 7$ $v \nearrow 7$

$w \longrightarrow 11$ $w \longrightarrow 11$ $w \rightrightarrows 11$

In relation A, u is assigned 4, v is assigned 7, and w is assigned 11.

In relation B, u is assigned 7 and v and w are assigned 11.

In relation S, u is assigned 4, v is assigned 4, and w is assigned 7 and 11.

In all three relations, the domain is $D = \{u, v, w\}$. In relations A and S, the range is $R = \{4, 7, 11\}$; however, in relation B, the range is $R = \{7, 11\}$.

In fact, any type of correspondence gives us a relation as long as each element in the domain is assigned at least one element in the range.

An alternative way of describing a relation is by using ordered pair notation. For example, we may write the correspondences in relation A as

Relation A   **Ordered Pair Notation**

$u \longrightarrow 4$ $(u, 4)$

$v \longrightarrow 7$ $(v, 7)$

$w \longrightarrow 11$ $(w, 11)$

Thus we may write relation A as the set of ordered pairs $A: \{(u, 4), (v, 7), (w, 11)\}$.

When we use ordered pair notation, we will assume (unless stated otherwise) that the first coordinate is a member of the domain and the second coordinate is the associated member of the range.

Any set of ordered pairs describes a relation. The set of all first (x-) coordinates constitutes the domain, and the set of all second (y-)coordinates constitutes the range. The ordered pairs themselves describe the correspondence.

We can rewrite relations B and S as follows:

relation B: $\{(u, 7), (v, 11), (w, 11)\}$

relation S: $\{(u, 4), (v, 4), (w, 7), (w, 11)\}$

Just as with the arrow notation, each set of ordered pairs defines a relation. Thus we can also define a relation as follows.

Definition of a Relation

A **relation** is a set of ordered pairs (x, y). The set of x-values is called the *domain* and the set of y-values is called the *range*.

For example, the set of ordered pairs $R = \{(8, 2), (6, -3), (5, 7), (5, -3)\}$ is the relation between the sets $\{5, 6, 8\}$ and $\{2, -3, 7\}$, where $\{5, 6, 8\}$ is the domain and $\{2, -3, 7\}$ is the range; 8 is assigned 2, 6 is assigned -3, 5 is assigned 7, and 5 is also assigned -3.

If the domain and/or range of a relation is infinite, we cannot list each element assignment, so instead we use set-builder notation to describe the relation. The situation we will encounter most frequently is that of a relation defined by an equation or formula. For example,

Recall that *R* is the symbol for the set of real numbers.

$$M = \{(x, y) | y = 2x - 3, x, y \in \mathbf{R}\}$$

is a relation for which the range value is 3 less than twice the domain value. It is understood that x represents domain values and y represents range values. Hence $(0, -3)$,

$(0.5, -2)$, and $(-2, -7)$ are examples of ordered pairs that are part of the assignment. Often we drop the set-builder notation and simply write that the relation is described by $y = 2x - 3$.

Since a relation can be described as a set of ordered pairs or an equation in two variables, we can use the rectangular coordinate system to specify the relation as well. In Figure 3.35 we have drawn a graph and labeled some of the points on the graph. Since every graph is a set of ordered pairs, it is automatically a relation; *the points on the graph themselves give the relation.*

Figure 3.35

The graph of a relation

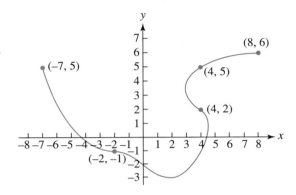

By examining the graph, we determine several things. First, each point on the graph tells us which y-value is assigned to that x-value. For example, $(8, 6)$, $(4, 5)$, $(4, 2)$, and $(-2, -1)$ are on the graph, which means that 6 is assigned to 8, 5 and 2 are both assigned to 4, and -1 is assigned to -2.

Different Perspectives: **Representing Relations**

The following are four different ways of representing the same relation.

Algebraic Description

We can represent a relation by an equation:

$y = 8 - x^2$ for $x = -3, -1, 0, 2, 3$

Table Description

We can represent a relation by a table of values:

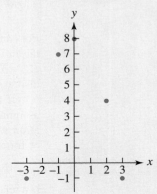

Ordered Pair Description

We can represent a relation by a set of ordered pairs:

$\{(-3, -1), (-1, 7), (0, 8), (2, 4), (3, -1)\}$

Graphical Description

We can represent a relation by a graph:

Figure 3.36

A typical telephone keypad

Functions

Let's consider a real-life relation described by the part of a telephone keypad illustrated in Figure 3.36.

It is very common today to see a business advertise its phone number by saying something like, "To get an over-the-phone price quotation on a new car, dial BUY CARS."

Anyone wanting to call this number can easily decode BUY CARS as the number 289-2277. This is because the telephone keypad sets up a correspondence between the set of letters of the alphabet and the set of numbers {2, 3, 4, 5, 6, 7, 8, 9}. The correspondence between the two sets is that indicated by the telephone keypad— that is, the letters A, B, C correspond to the number 2; the letters D, E, F correspond to the number 3, etc.

What is important to note is that although the telephone keypad sets up a correspondence between letters and numbers, there is a qualitative difference between the letter → number correspondence and the reverse number → letter correspondence: Each letter corresponds to exactly one number, but each number corresponds to more than one letter. Consequently, a phone number such as 438-2253 cannot be *uniquely* encoded into words. The number 438-2253 can be interpreted as GET CAKE or IF U BAKE.

$$
\begin{array}{ccccccc}
4 & 3 & 8 & 2 & 2 & 5 & 3 \\
\downarrow & \downarrow & \downarrow & \downarrow & \downarrow & \downarrow & \downarrow \\
G & E & T & C & A & K & E
\end{array}
\quad \text{or} \quad
\begin{array}{ccccccc}
4 & 3 & 8 & 2 & 2 & 5 & 3 \\
\downarrow & \downarrow & \downarrow & \downarrow & \downarrow & \downarrow & \downarrow \\
I & F & U & B & A & K & E
\end{array}
$$

Mathematically, it is important for us to distinguish among relations that assign a *unique* range element to each domain element (such as the letter → number correspondence on the telephone) and those that do not.

Definition of a Function

> A **function** is a relation in which each element of the domain corresponds to **exactly one element** of the range.

Note that every function is a relation, but not every relation is a function.

Looking back at relations *A*, *B*, and *S*, which we rewrite here,

$$A: \{(u, 4), (v, 7), (w, 11)\} \qquad B: \{(u, 7), (v, 11), (w, 11)\}$$
$$S: \{(u, 4), (v, 4), (w, 7), (w, 11)\}$$

A is a function, since each element in the domain, {*u*, *v*, *w*}, is assigned only one element in the range.

On the other hand, *S* is not a function—in *S*, both 7 and 11 are assigned to *w*.

If two different x's are assigned to the same y, why is B still a function?

In *B*, even though the range element 11 is assigned to two elements of the domain, *v* and *w*, it is still a function since each element in the domain, {*u*, *v*, *w*}, is assigned only *one element of the range*, {7, 11}: *u* is assigned only one value, 7; *v* is assigned only one value, 11; and *w* is assigned only one value, 11.

Recalling the telephone relation, we have that the letter → number assignment is a function (a phrase gives you only one number), but the number → letter assignment is not a function (a number may give you more than one possible phrase).

A function may also be viewed as a "machine" into which you put an element of the domain, *x*, and out of which comes the associated element in the range, *y*. Figure 3.37(a) on page 142 illustrates the idea of a general function machine, whereas Figure 3.37(b) illustrates the function $y = x^2 + 4$.

Figure 3.37

Viewing a function as a machine

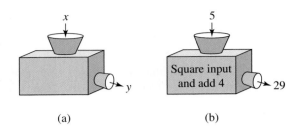

(a)   (b)

Hence we consider x as the input and its associated y as the output. Consider the function $\{(2, 4), (3, 4), (5, 8)\}$. If we put 2 into the machine, out comes 4; if we put in 3, out comes 4; and if we put in 5, out comes 8.

If we try to define a function machine by the relation $\{(2, 5), (3, 8), (3, 9)\}$, then we run into trouble when we put in 3; we cannot be sure whether 8 or 9 will come out. Because two outputs can occur from one input, $\{(2, 5), (3, 8), (3, 9)\}$ is not a function.

Example 1   Determine whether the following relations are functions.
(a) $\{(5, -2), (3, 5), (3, 7)\}$ (b) $\{(2, 4), (3, 4), (6, -4)\}$

Solution

Some of the keys on a calculator are good examples of functions. We press a key such as $\boxed{\sqrt{}}$, enter an input (pick a number), and the calculator displays a unique output.

(a) Since the domain element 3 is assigned two different values in the range, 5 and 7, it is not a function.

(b) Each element in the domain, $\{2, 3, 6\}$, is assigned exactly one value in the range; 2 is assigned only 4, 3 is assigned only 4, and 6 is assigned only -4. Therefore it is a function.   ∎

As with relations, we can describe a function with an equation. For example,

$$y = 2x + 1$$

is a function, since each x will produce only one y.

We will find it helpful to use the following vocabulary: The **independent variable** refers to the variable representing possible values in the domain, and the **dependent variable** refers to the variable representing possible values in the range. Thus in our usual ordered pair notation (x, y), x is the independent variable and y is the dependent variable.

It is very helpful to view functions within the framework of "pick x (a value in the domain) and determine y (the corresponding value in the range)." For the function $y = 2x + 1$, if we choose $x = 4$, then $y = 2(4) + 1 = 9$, so $(4, 9)$ is an assignment described by this function. In this way we can generate as many ordered pairs as we like. The question that remains is, What is the domain of $y = 2x + 1$? To answer this question, we need to set up the following ground rule.

> The **natural domain** of a function is the set of real numbers for which the equation or formula is defined *and* that produces real number values in the range. Such a function is called a *real-valued function of a real variable*. In other words, the function must have a real input and a real output. Unless otherwise specified, a function is assumed to have its *natural domain*.

Thus the domain of $y = 2x + 1$ is the set of all real numbers, since there is no restriction on what real number we may choose for x. The range is also the set of all real numbers. On the other hand, the domain of $y = \dfrac{1}{x + 2}$ is all reals *except* -2, since $x = -2$ produces an undefined value for y. In general, the range of a function is more difficult to identify than the domain; we have more to say about the range a bit later in this section.

To summarize, then, for our purposes a function is a relationship between x and y for which to each real value of x in the domain there corresponds exactly one real value of y in the range.

Example 2   Determine whether the following equations determine y as a function of x; if so, find the domain.

(a) $y = -3x + 5$ (b) $y = \dfrac{2x}{3x - 5}$ (c) $y = x^2$

(d) $y^2 = x$ (e) $y = \dfrac{4}{x^2 + 1}$

Solution   (a) To determine whether $y = -3x + 5$ gives y as a function of x, we need to know whether each x-value uniquely determines a y-value. THINK: Pick x, get exactly one y. Looking at the equation $y = -3x + 5$, we can see that once x is chosen, we multiply it by -3 and then add 5. Thus for each x there is a unique y. Therefore, $y = -3x + 5$ is a function.

As for its domain, we can see that there is no restriction on the values we may choose for x. (Is there any real number you can't multiply by -3 and then add 5 to the result?) Therefore, the domain is the set of all real numbers.

(b) Looking at the equation $y = \dfrac{2x}{3x - 5}$ carefully, we can see that each x-value uniquely determines a y-value. (One x-value cannot produce two different y-values.) Therefore, $y = \dfrac{2x}{3x - 5}$ is a function.

As for its domain, we ask ourselves, Are there any values of x that must be excluded? Since $y = \dfrac{2x}{3x - 5}$ is a fractional expression, we must exclude any value of x that makes the denominator equal to zero. We must have

$$3x - 5 \neq 0 \quad \Rightarrow \quad x \neq \frac{5}{3}$$

Therefore, the domain consists of all real numbers except for $\dfrac{5}{3}$. The domain is $\{x \mid x \neq \frac{5}{3}\}$.

(c) For the equation $y = x^2$, if we choose $x = 3$, we get $y = 9$, and if we choose $x = -3$, we also get $y = 9$. However, this poses no problem as far as $y = x^2$ being a function is concerned. Each x is assigned exactly one y-value, its square. Be careful! The definition of a function requires that each x-value in the domain is assigned exactly one y-value, *not* necessarily that every x-value have a different y-value. Therefore, $y = x^2$ is a function.

Since we can square any number, the domain is the set of all real numbers.

(d) For the equation $y^2 = x$, if we choose $x = 9$ we get $y^2 = 9$, which gives $y = \pm 3$. In other words, there are *two* y-values associated with $x = 9$. Therefore, $y^2 = x$ is not a function.

Look very carefully at parts (c) and (d) to make sure you see why $y = x^2$ is a function of x, but $y^2 = x$ is not.

(e) Examining the equation, we can see that each x-value uniquely determines a y-value and so $y = \dfrac{4}{x^2 + 1}$ is a function.

In trying to determine the domain of $y = \dfrac{4}{x^2 + 1}$, we are again looking at a fractional expression, and so we must again exclude any values that make the denominator equal to zero. In other words, we require

$$x^2 + 1 \neq 0$$
$$x^2 \neq -1 \qquad \text{This statement is true for all real numbers.}$$

In other words, there are no real values of x that make the denominator equal to zero, and so no values of x need to be excluded. Therefore, the domain is the set of all real numbers. ∎

Example 3 | Find the domain of the function $y = \sqrt{3x - x^2}$.

Solution | Let's analyze this problem carefully so that we can clearly understand what is being asked and develop a strategy for the solution.

What do we need to find?	The domain of the function $y = \sqrt{3x - x^2}$
What is a domain?	The real values of x that make y defined and real
When is y defined and real?	When the expression under the radical is nonnegative
How can I restate the problem in simpler terms?	Since nonnegative means greater than or equal to zero, we solve the inequality $3x - x^2 \geq 0$.

Because an even root of a negative number is undefined in the real number system, we need x to satisfy the inequality

$$3x - x^2 \geq 0 \qquad \text{This is a quadratic inequality; we solve it by doing a sign analysis.}$$
$$x(3 - x) \geq 0 \qquad \text{Sign of } 3x - x^2 \quad - - - - - - \mid + + + + + + + + \mid - - - - - -$$
$$\qquad\qquad\qquad\qquad\qquad\qquad\qquad\qquad 0 \qquad\qquad 3$$

Doing a sign analysis was discussed in Section 2.5.

Since we want $3x - x^2 = x(3 - x)$ to be nonnegative, the sign analysis shows us that the domain is

$$\{x \mid 0 \leq x \leq 3\} \qquad \text{or, using interval notation,} \qquad [0, 3]$$ ∎

A comment is in order here. Students are sometimes heard remarking about a problem like Example 3, "I know how to solve a quadratic inequality, but that is not what the problem asked." Well, in fact, the problem did not directly ask that a quadratic inequality be solved. A problem such as this obliges the reader to reformulate the question into more familiar terms. Our analysis of what the question was asking and what condition x must satisfy led us to recognize that the domain could be found by doing a sign analysis. Analyzing a question and breaking it down into smaller, more digestible, parts is an idea we will use many times in this book.

As with relations, we can describe functions by graphs. For example, in Figure 3.38, we have drawn a graph and labeled some of the points on the graph. The graph represents a relation—but is it a function? Since the points (4, 1), (4, 2), and (4, 5) are on the graph, there is a value, $x = 4$, that has three y-values associated with it. But a function can assign only one *y*-value for each *x*. Therefore, we can see that this relation is *not* a function.

Figure 3.38

The graph of a relation that is not a function

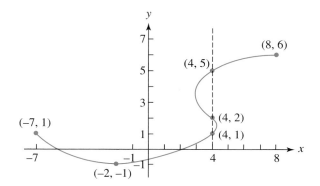

We can generalize this last comment as follows. We can visually determine whether a graph is the graph of a function by using a simple procedure called the **vertical line test**.

The Vertical Line Test

If any vertical line intersects a graph in more than one point, then the graph is not the graph of a function.

Alternatively, the vertical line test says that any vertical line may intersect the graph of a function in at most one point.

Different Perspectives: **Functions**

Consider the graphical and algebraic descriptions of a function.

Graphical Description

This *is* a function:

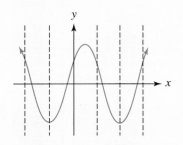

Any vertical line crosses the graph at most once; therefore, each *x*-value is assigned exactly one *y*-value.

This is *not* a function:

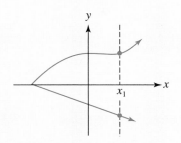

Note that two different *y*-values are assigned to x_1; this violates the definition of a function.

Algebraic Description

If a relation assigns more than one *y*-value to *any x*-value, then it is *not* a function.

Consider $2y^2 - 8 = x$. If $x = 10$, we have

$$2y^2 - 8 = 10$$
$$2y^2 = 18$$
$$y^2 = 9$$
$$y = \pm 3$$

Since two *y*-values correspond to $x = 10$, the equation $2y^2 - 8 = x$ does not define *y* as a function of *x*.

Consider the function defined by the graph in Figure 3.39. We can identify the domain and range of this function by its graph. We note the following: Every x-value between 2 and 10 has a y-value associated with it, whereas x-values to the left of 2 and to the right of 10 have no y-values associated with them. Hence, the domain is the set of all numbers between 2 and 10, inclusive. Alternatively, we can project the points on the graph up or down to the x-axis to see which x-values are in the domain. See Figure 3.40(a).

Figure 3.39

The graph of a function

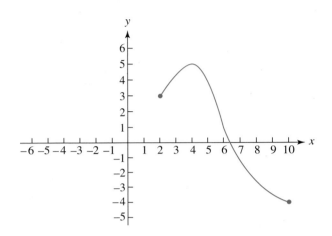

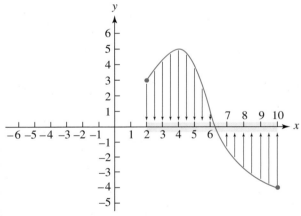

(a) To get the domain we project the points on the graph vertically to the x-axis.
The domain is $\{x \mid 2 \le x \le 10\}$

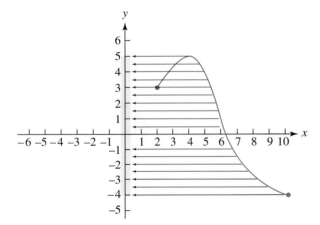

(b) To get the range we project the points on the graph horizontally to the y-axis.
The range is $\{y \mid -4 \le y \le 5\}$

Figure 3.40

As for the range, we can see that the smallest y-coordinate on the graph is -4, and the largest y-coordinate on the graph is 5, and every y-value in between is also the y-coordinate of some point on the graph. Therefore, we can see that the range is the set of numbers between -4 and 5, inclusive. The range can also be seen by projecting the points on the graph left or right to the y-axis. See Figure 3.40(b). When we read the domain and range of a function from its graph, we say that we are finding the domain and range by inspection.

Example 4   Given the graph in Figure 3.41, determine whether it is the graph of a function, and find its domain and range.

Solution   Using the vertical line test, we imagine a vertical line moving across the graph and we can see that any vertical line will cross the graph at most once. Therefore, this is the graph of a function.

In looking for the domain, we project the graph vertically to the x-axis to see which x-values have y-values associated with them. Looking at the graph carefully, we can see that each x-value between -8 and -3 (including -8 but excluding -3) and each x-value between 1 and 4, inclusive, has a y-value associated with it. Therefore, the domain is

$$\{x \mid -8 \le x < -3 \text{ or } 1 \le x \le 4\} \quad \text{or, equivalently,} \quad [-8, -3) \cup [1, 4]$$

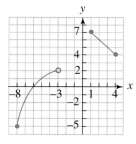

Figure 3.41

In looking for the range, we project the graph horizontally to the y-axis to see which y-values have x-values associated with them. Again looking at the graph carefully, we can see that the range includes each y-value between -5 and 2 (including -5, excluding 2) and each y-value between 4 and 7, inclusive. Therefore, the range is

$$\{y \mid -5 \le y < 2 \text{ or } 4 \le y \le 7\} \quad \text{or, equivalently,} \quad [-5, 2) \cup [4, 7] \quad \blacksquare$$

Mathematical Models

To apply mathematical procedures to real-life situations it is often helpful to have a mathematical formula that allows us to compute or approximate the actual data values. Our concern here is not how this "formula" is derived, but rather how it can be used.

For example, the following table contains data reported by the U.S. Census Bureau on the population density of the state of California (in number of people per square mile) for the years listed.

Year	1930	1960	1980	1990	2000
Population density, D	36.2	100.4	151.4	190.8	217.2

Based on this information, a data analyst suggests that the formula

$$D = 0.012x^2 + 1.04x - 6.03$$

where x represents the number of years since 1900, can be used to approximate the population density.

An equation like this, which represents the density as a function of x, is called a **mathematical model** of the population density.

Let's first see how accurate this model is. For example, if we calculate the value of D when $x = 60$ (meaning the population density in 1960), we obtain $D = 0.012(60^2) + 1.04(60) - 6.03 = 99.57$, which agrees quite well with the actual value of 100.4.

One reason for creating such a mathematical model is to predict the population density in the future. We will continue to discuss a variety of mathematical models throughout this text.

In the next section we will continue our discussion of functions.

3.4 Exercises

In Exercises 1–6, the given arrow diagram defines a relation. Translate the diagram into ordered pair notation. Find the domain and range of the relation. Is the relation a function?

1. $A \longrightarrow 10$
$B \longrightarrow 20$
$C \longrightarrow 30$

2. 1
$3 \searrow a$
$5 \longrightarrow b$
$7 \longrightarrow c$

3. $6 \longrightarrow A$
$10 \longrightarrow B$
$15 \longrightarrow C$
$19 \longrightarrow D$

4. $J \qquad 21$
$K \times 22$
$L \longrightarrow 23$
$M \times 24$
$N \qquad 25$

5. A
$B \searrow$
$C \longrightarrow 7$
$D \nearrow$

6. A
$7 \nearrow B$
$\searrow C$
D

In Exercises 7–12, determine the domain and range of the given relation. Is the relation a function?

7. $\{(-4, -3), (2, -5), (4, 6), (2, 0)\}$

8. $\left\{(8, -2), \left(6, -\dfrac{3}{2}\right), (-1, 5)\right\}$

9. $\{(-\sqrt{3}, 3), (-1, 1), (0, 0), (1, 1), (\sqrt{3}, 3)\}$

10. $\left\{\left(-\dfrac{1}{2}, \dfrac{1}{16}\right), (-1, 1), \left(\dfrac{1}{3}, \dfrac{1}{81}\right)\right\}$

11. $\{(0, 5), (1, 5), (2, 5), (3, 5), (4, 5), (5, 5)\}$

12. $\{(5, 0), (5, 1), (5, 2), (5, 3), (5, 4), (5, 5)\}$

In Exercises 13–42, find the domain of the given relation. Does the given equation describe y as a function of x?

13. $y = -3x + 2$

14. $y = x^2 + 3$

15. $y = \dfrac{1}{x - 2}$

16. $y = \dfrac{x + 4}{x + 8}$

17. $y = x^2 - 3x - 4$

18. $y = \dfrac{1}{x^2 - 3x - 4}$

19. $y = \sqrt{x + 2}$

20. $y = \dfrac{1}{\sqrt{x - 4}}$

21. $y = \dfrac{2x - 5}{3x + 4}$

22. $y = \dfrac{6x}{x^2 - x - 1}$

23. $y = \dfrac{|x|}{x}$

24. $y = \dfrac{-4}{x^3 - 9x^2}$

25. $y = \dfrac{5}{\sqrt{x^2 - 9}}$

26. $y = \sqrt{x^2 - 6x + 5}$

27. $y = \sqrt{x^2 + 16}$

28. $y = \sqrt[3]{x - 8}$

29. $y = \sqrt[5]{x^2 - 2x - 3}$

30. $y = \sqrt[4]{x^2 - 2x + 1}$

31. $y = \sqrt{8 - 5x}$

32. $y = 8 - \sqrt{5x}$

33. $y = \sqrt{|x - 4|}$

34. $y = \dfrac{1}{x^2}$

35. $y = \dfrac{\sqrt{x + 5}}{x - 3}$

36. $y = \dfrac{\sqrt{x - 3}}{x + 2}$

37. $y = \dfrac{7}{\sqrt{3x - 2x^2}}$

38. $y = \dfrac{3x}{x^2 + x + 1}$

39. $y = \dfrac{x^2 - 3x - 10}{x - 5}$

40. $y = \dfrac{x^3 - 5x^2}{x}$

41. $y = \dfrac{2x - 8}{x^2 - 16}$

42. $y = \dfrac{3x + 15}{x^2 + 3}$

In Exercises 43–48, determine whether the given graph is the graph of a function.

43.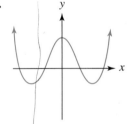

44.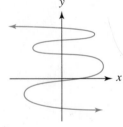

45.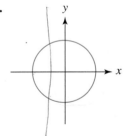

46.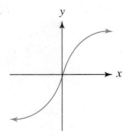

47.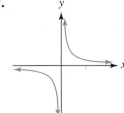

48.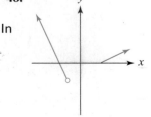

Exercises 49–54, use the given graph to determine the domain and range of the relation or function.

49.

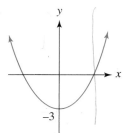

50.

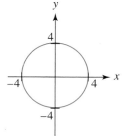

51.

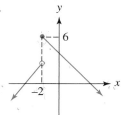

52.

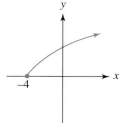

53.

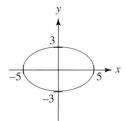

54.

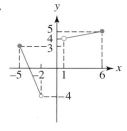

55. Suppose a car rental company charges a rate of $22 per day plus $0.11 per mile. Express the cost of a 5-day car rental as a function of the number of miles driven, m.

56. A truck driver drives from point A due east at 50 mi/hr for h hours and then drives due north at 60 mi/hr for another h hours until reaching point B. Express the distance, d, between points A and B as a function of h.

57. A jogger runs in a road race for a total of 6 hours. She runs at an average speed of 8 km/hr for t hours and the rest of the time at an average speed of 10 km/hr. Express the total distance, d, the jogger runs as a function of t.

58. A farmer has 100 feet of fencing, which is to be cut into two pieces. The first piece, of length x feet, will be erected in the shape of a square, and the remainder of the fencing will also be erected in the shape of a square. Express the total area enclosed by the two squares as a function of x.

59. The Consumer Price Index is a measure of inflation obtained by comparing the price of a basic "shopping cart" of goods during some previous time period (called the base period) and today. The following table contains data from the U.S. Department of Labor on the annual Consumer Price Index (CPI) for the years listed for New York, New Jersey, Pennsylvania, and Connecticut. (The base period for these data is 1982–1984.)

Year	1996	1997	1998	1999	2000
CPI	166.9	170.8	173.6	177.0	182.5

Based on these data, a researcher suggests that the Consumer Price Index can be expressed as the function $C = 0.27n^2 + 2.65n + 167.2$, where n represents the number of years after 1996 ($n = 0$ is 1996, $n = 1$ is 1997, . . .).

(a) Use this function to calculate the CPI for the years given in the table and compare the functional values with the actual data.

(b) Use the function to predict the CPI for 2001. (The actual value was 187.1.)

60. Using the same data from Exercise 59, another government agency suggested the function $C = 0.26n^3 - 1.33n^2 + 4.95n + 166.9$ for the CPI.

(a) Use this function to calculate the CPI for the years given in the table and compare the functional values with the actual data.

(b) Does this function or that of Exercise 59 better match the data?

61. The following table contains data from the U.S. Department of Labor on the annual Consumer Price Index (CPI) for the years listed for midwestern urban consumers. (The base period for these data is 1982–1984.)

Year	1996	1997	1998	1999	2000
CPI	153.0	156.7	159.3	162.7	168.3

Based on these data, a researcher suggests that the Consumer Price Index can be expressed as the function $C = 3.66n - 7152.68$, where n represents the year.

(a) Use this function to calculate the CPI for the years given in the table and compare the functional values with the actual data.

(b) Use the function to predict the CPI for 2001. (The actual value was 172.8.)

62. Using the same data from Exercise 61, another government economist suggested the function $C = 0.329x^2 - 1310.39x + 1,304,950$ for the CPI.

(a) Use this function to calculate the CPI for the years given in the table and compare the functional values with the actual data.

(b) Does this function or that of Exercise 61 better match the data?

(c) Which function is easier to use to compute the CPI?

63. The Dow Jones Industrial Average (DJIA) is a measure of strength of the stock market. The following table lists the high and low values of the DJIA for the years listed.

Year	1996	1997	1998	1999	2000
DJIA Low	5032.94	6391.69	7539.07	9120.67	9796.03
DJIA High	6560.91	8259.31	9374.27	11497.12	11722.98

A stock analyst suggests that the low annual value of the DJIA can be obtained from the function $L = -57.9n^3 + 280.2n^2 + 993.9n + 5061.3$, and high annual value can be obtained from the function $H = -109.5n^3 + 518.4n^2 + 968.3n + 6625.1$, where n represents the number of years after 1996 ($n = 0$ is 1996, $n = 1$ is 1997, . . .).

(a) Use the function L to calculate the low DJIA for the years given in the table and compare the functional values with the actual data.

(b) For which years does the mathematical model come the closest to and the farthest from the actual low value?

(c) Use the model to compute the low DJIA in 2001. (The actual low was 7926.9.)

(d) Use the function H to calculate the high DJIA for the years given in the table and compare the functional values with the actual data.

(e) For which years does the mathematical model come the closest to and the farthest from the actual high value?

(f) Use the model to compute the high DJIA in 2001. (The actual high was 11,436.4.)

64. The following table lists the population density for the state of Texas (in people per square mile) for the years listed as reported by the Bureau of the Census.

Year	1930	1960	1980	1990	2000
Population density, P	22.1	36.4	54.3	64.9	79.6

Based on these data, a statistician suggests that the population density for Texas can be modeled by the function $P = 0.0087y^2 - 33.47y + 32215.9$, where y represents the year.

(a) Use this function to calculate the population density for Texas for the years given in the table and compare the functional values with the actual data.

(b) For which years does the mathematical model come the closest to and the farthest from the actual value?

Questions for Thought

65. State in words what makes a relation a function.

66. The following arrow diagrams define two functions we call F and G: F has domain A and range B; G has domain S and range T. Suppose all the arrows are reversed. Which, if any, of the two new arrow diagrams would be functions? Explain.

67. How would you use a graphing calculator to sketch the graph of the relation $x^2 + y^2 = 4$?

3.5   Function Notation

In the last section we saw that equations such as $y = 3x + 4$ and $y = x^2 - 1$ define y as a function of x. Because these equations are solved explicitly for y, we can say that y is dependent on the value of x. Hence x is called the independent variable ("pick x"), and y is called the dependent variable ("get y"). However, there are many occasions when we wish to state outright or emphasize that y is a function of x. We do this by using a special **function notation**: $f(x)$.

Definition of f(x)

$f(x)$ is the value the function f assigns to x.

When we want to indicate that y is a function of x, we write

f(x) is just another name for y.

$$y = f(x) \qquad f(x) \text{ is read "} f \text{ of } x\text{."}$$

$f(x)$ *is merely another name for the dependent variable.* Thus we can describe an expression such as $x^2 - 1$ as being a function of x by writing

$$f(x) = x^2 - 1 \qquad \text{The expression } x^2 - 1 \text{ is a function of } x.$$

(In fact, many times we write both y and $f(x)$ and write $y = f(x) = x^2 - 1$.)

*It is very important to recognize that the parentheses in $y = f(x)$ are **not** being used to indicate multiplication. Rather, the parentheses are being used to specify the independent variable.*

Most times we will use the letters f, g, h, F, G, and H for functions.

Function notation is a very useful shorthand for substituting values. For example, if $f(x) = x^2 - 1$, instead of asking, "What is the functional value associated with $x = 4$?" all we need to write is "What is $f(4)$?" To compute $f(4)$ we proceed as follows:

$$f(x) = x^2 - 1 \qquad \text{To find } f(4), \text{ we substitute } x = 4 \text{ in } x^2 - 1.$$
$$f(4) = (4)^2 - 1 = 15 \qquad f(4) \text{ is the value of } f(x) \text{ when 4 is substituted for } x.$$

It is useful to think of x as the *input* and $f(x)$ as the *output*. Thus in computing $f(4)$, **4 is the input and $f(4) = 15$ is the output.**

Input

$\downarrow$

Remember that f(x) does not mean f times x.

$$f(4) = 15$$

$\uparrow$

Output

Example 1 Given the function $g(x) = \dfrac{1}{3}x^3 + 2x - 4$, find each of the following:

(a) $g(-3)$ **(b)** $g\left(\dfrac{1}{2}\right)$ **(c)** $g(t)$

Solution Since x is the independent variable, $g(x) = \dfrac{1}{3}x^3 + 2x - 4$ tells us that the rule of assignment for the function g in this example is "take the input, cube it and multiply by $\dfrac{1}{3}$, and then add 2 times the input and subtract 4." We will follow this rule in all parts of the example, regardless of what the input is.

(a) $g(x) = \dfrac{1}{3}x^3 + 2x - 4$ To find $g(-3)$, we replace all occurrences of x with -3.

$$g(-3) = \frac{1}{3}(-3)^3 + 2(-3) - 4$$

$$g(-3) = -19$$

Remember that $g(-3) = -19$ means that when $x = -3$, $g(x) = -19$.

We can also evaluate functions directly on a calculator. Some calculators have a "TABLE" option that displays several values of x (and y). The screen below shows a table from a TI-83 Plus calculator for the values sought in Example 1(a) and 1(b). (Note that decimal values in tables may be rounded.)

X	Y1
-3	-19
.5	-2.958

Y1 ■ (1/3)X^3+2X−4

Verify that $-71/24$ is equal to -2.958 rounded to three decimal places.

(b) $g(x) = \dfrac{1}{3}x^3 + 2x - 4$ To find $g\left(\dfrac{1}{2}\right)$, we replace x with $\dfrac{1}{2}$.

$$g\left(\frac{1}{2}\right) = \frac{1}{3}\left(\frac{1}{2}\right)^3 + 2\left(\frac{1}{2}\right) - 4$$

$$= \frac{1}{3}\left(\frac{1}{8}\right) + 1 - 4$$

$$g\left(\frac{1}{2}\right) = -\frac{71}{24}$$

(c) $g(x) = \dfrac{1}{3}x^3 + 2x - 4$ To find $g(t)$, we replace all occurrences of x with t.

$$g(t) = \frac{1}{3}(t)^3 + 2(t) - 4 \qquad \text{Thus}$$

$$g(t) = \frac{1}{3}t^3 + 2t - 4$$

Example 2 Given the function $f(x) = 3x^2 - 4x + 5$, find each of the following and simplify:

(a) $f\left(\dfrac{1}{t}\right)$ **(b)** $\dfrac{1}{f(t)}$ **(c)** $f(a + 2)$ **(d)** $f(a) + 2$ **(e)** $f(a) + f(2)$

(f) $f(2x)$ **(g)** $2f(x)$

Solution   **(a)** $f(x) = 3x^2 - 4x + 5$ To find $f\left(\dfrac{1}{t}\right)$, we replace x with $\dfrac{1}{t}$.

$$f\left(\frac{1}{t}\right) = 3\left(\frac{1}{t}\right)^2 - 4\left(\frac{1}{t}\right) + 5 = 3\left(\frac{1}{t^2}\right) - \left(\frac{4}{t}\right) + 5 \qquad \text{Thus}$$

$$f\left(\frac{1}{t}\right) = \frac{3}{t^2} - \frac{4}{t} + 5 = \frac{3 - 4t + 5t^2}{t^2}$$

(b) To find $\dfrac{1}{f(t)}$, we first find $f(t)$ and then take its reciprocal.

Since $f(t) = 3t^2 - 4t + 5$, we have $\dfrac{1}{f(t)} = \dfrac{1}{3t^2 - 4t + 5}$.

Note the difference between parts (a) and (b). In part (a) we choose a value t and use its reciprocal as the input, whereas in part (b) we choose a value t and then take the reciprocal of its output.

(c) $f(x) = 3x^2 - 4x + 5$ Keep in mind that x represents the input to the function. It is sometimes helpful to view the function as
$f(\ \ ) = 3(\ \ )^2 - 4(\ \ ) + 5$.
Whatever we substitute into the parentheses on the left (which is being called x), we also substitute on the right. Therefore, we replace all occurrences of x with $a + 2$ and simplify.

$$f(a + 2) = 3(a + 2)^2 - 4(a + 2) + 5$$

$$= 3(a^2 + 4a + 4) - 4a - 8 + 5$$

$$= 3a^2 + 12a + 12 - 4a - 8 + 5 \qquad \text{Thus}$$

$$f(a + 2) = 3a^2 + 8a + 9$$

(d) $f(a) + 2$ means we are to add 2 to $f(a)$. First find $f(a)$:

$$f(a) = 3a^2 - 4a + 5 \qquad \text{And so}$$

$$f(a) + 2 = \underbrace{3a^2 - 4a + 5}_{f(a)} + 2 \qquad \text{Thus}$$

$$f(a) + 2 = 3a^2 - 4a + 7$$

Note the difference between parts (c) and (d). In part (c), we are adding 2 to a and using that as the *input*, whereas in part (d), we are adding 2 to the *output* $f(a)$. Thus there is no reason to expect $f(a + 2)$ and $f(a) + 2$ to be equal, and as we just saw they are not equal.

(e) $f(a) + f(2)$ is asking us to add $f(a)$ to $f(2)$, which means we first compute $f(a)$ and $f(2)$:

$$f(a) + f(2) = \underbrace{(3a^2 - 4a + 5)}_{f(a)} + \underbrace{(3 \cdot 2^2 - 4 \cdot 2 + 5)}_{f(2)}$$

$$= (3a^2 - 4a + 5) + (9) \qquad \text{Hence,}$$

$$f(a) + f(2) = 3a^2 - 4a + 14$$

It is important to recognize that in general, $f(a + b) \neq f(a) + f(b)$

Note the differences between parts (c) and (e). We cannot simply add $f(a)$ to $f(2)$ to get $f(a + 2)$.

(f) $f(x) = 3x^2 - 4x + 5$ Again don't take x literally; x stands for the input. To compute $f(2x)$, $2x$ becomes the input; that is, we replace all occurrences of x with $2x$.

$$f(2x) = 3(2x)^2 - 4(2x) + 5 \qquad \text{Thus}$$

$$f(2x) = 12x^2 - 8x + 5$$

(g) $2f(x)$ means that we multiply $f(x)$ by 2.

$$f(x) = 3x^2 - 4x + 5 \qquad \text{First find } f(x); \text{ then multiply by 2.}$$
$$2f(x) = 2(3x^2 - 4x + 5) \qquad \text{Thus}$$

$$2f(x) = 6x^2 - 8x + 10$$

How are $f(2x)$ and $2f(x)$ different from each other?

Note the difference between parts (f) and (g). In part (f), $f(2x)$, we are doubling x and using that as the input, but in part (g), $2f(x)$, we are doubling the output. Again there is no reason to expect the two results to be equal. ∎

Example 3   Given the function $H(x) = \dfrac{x + 3}{x - 4}$, find $H\left(\dfrac{1}{x + 1}\right)$.

Solution

$$H(x) = \frac{x + 3}{x - 4} \qquad \text{To find } H\left(\frac{1}{x + 1}\right) \text{ we replace each occurrence of } x \text{ by } \frac{1}{x + 1}.$$

$$H\left(\frac{1}{x + 1}\right) = \frac{\dfrac{1}{x + 1} + 3}{\dfrac{1}{x + 1} - 4} \qquad \text{Simplify the complex fraction.}$$

$$= \frac{\left(\dfrac{1}{x + 1} + 3\right)(x + 1)}{\left(\dfrac{1}{x + 1} - 4\right)(x + 1)} = \frac{1 + 3x + 3}{1 - 4x - 4} = \frac{3x + 4}{-4x - 3}$$

∎

Example 4   Given $f(x) = x^2 - x + 1$, find $\dfrac{f(x + 3) - f(x)}{3}$.

Solution   An expression of the form $\dfrac{f(x + h) - f(x)}{h}$ is called a *difference quotient* for $f(x)$. The difference quotient plays a pivotal role in calculus. We hint at the significance of the difference quotient in Exercise 80 but restrict our attention here to simplifying difference quotients algebraically, as we do in this example.

We begin by finding $f(x + 3)$ for $f(x) = x^2 - x + 1$:

$$f(x + 3) = (x + 3)^2 - (x + 3) + 1$$
$$= x^2 + 6x + 9 - x - 3 + 1$$
$$= x^2 + 5x + 7$$

Therefore,

> **Don't forget the parentheses around $f(x)$.**

$$\frac{f(x + 3) - f(x)}{3} = \frac{\overbrace{x^2 + 5x + 7}^{f(x+3)} - \overbrace{(x^2 - x + 1)}^{f(x)}}{3} = \frac{x^2 + 5x + 7 - x^2 + x - 1}{3}$$

$$= \frac{6x + 6}{3} = \frac{6(x + 1)}{3} = \boxed{2(x + 1)} \qquad \blacksquare$$

Split Functions

Although all the functions we have examined thus far in this section have had one rule of assignment for all members of their domain, this need not be the case.

For example, consider the following function $f(x)$:

$$f(x) = \begin{cases} x^2 + 3 & \text{if } -3 \le x < 2 \\ 5x - 4 & \text{if } x \ge 2 \end{cases}$$

This means that we are to use the rule $f(x) = x^2 + 3$ for any x in the interval $[-3, 2)$, but we are to use the rule $f(x) = 5x - 4$ for any x in the interval $[2, \infty)$.

If we want to find $f(-1)$, we *first* note that -1 is between -3 and 2, and so we use the top rule. Thus $f(-1) = (-1)^2 + 3 = 4$.

If we want to find $f(2)$, we note that 2 is greater than or equal to 2, and so we use the bottom rule. Thus $f(2) = 5(2) - 4 = 6$.

If we want to find $f(-4)$, we note that there is no rule for $f(x)$ when x is less than -3. Therefore, -4 is not in the domain and so there is no value associated with it.

Such a function, which has more than one rule of assignment, depending on the input value, is called a **split function**. The *domain of a split function* is the set of x-values for which there is a rule of assignment. Thus the preceding split function $f(x)$ has as its domain $\{x \mid x \ge -3\}$.

> A split function is also called a piecewise defined function.

Example 5   Given the function $G(x) = \begin{cases} x - 1 & \text{if } -5 \le x < -1 \\ x^3 & \text{if } -1 \le x < 2 \\ 6 & \text{if } x > 2 \end{cases}$

Find: **(a)** $G(0)$ **(b)** $G(5)$ **(c)** $G(2)$ **(d)** $G(-5)$

Solution   **(a)** Since 0 lies between -1 and 2, we use the middle rule to get $G(0) = 0^3 = 0$. Thus $\boxed{G(0) = 0}$.

(b) Since 5 is greater than 2, we use the third rule to get $\boxed{G(5) = 6}$. Note that the third rule will assign the value 6 to any x-value greater than 2.

(c) To find $G(2)$ we note that $x = 2$ does not fall into any of the three categories. $x = 2$ is not in the domain of $G(x)$ and so $G(2)$ is undefined .

(d) To find $G(-5)$ we use the first rule and get $G(-5) = -5 - 1 = -6$. Thus $G(-5) = -6$. ∎

Example 6

Because of the pressure to complete a project, an employer offers to pay its employees $12 for each hour of regular time per week, and triple-time for each hour of overtime. The first 35 hours worked per week are considered regular time, and all hours in excess of 35 hours per week are considered overtime.

(a) Express an employee's weekly income, I, from this project as a function of h, the number of hours worked.

(b) Use this function to compute the weekly income if the employee works 35 hours, 42 hours, or 48 hours per week on this project.

Solution

(a) Let's examine how we would compute an employee's weekly income. If the employee works 35 hours or less, we simply multiply the number of hours worked by $12 per hour. If the employee works more than 35 hours, he or she would earn $12 per hour for the first 35 hours plus $36 per hour (triple-time) for all hours in excess of 35 per week.

For instance, if an employee works for 38 hours per week, he or she will earn $12 per hour for the first 35 hours and $36 per hour for the additional 3 hours. The weekly income would be $I = 12(35) + 36(38 - 35) = 420 + 36(3) = \$528$.

Similarly, if the employee works h hours per week, the formula we use to compute the salary depends on *whether h is greater than* 35 *or not*. We need two different rules to determine the salary, depending on how many hours are worked. This suggests that we can use a split function to define I as a function of h, the number of hours the employee works per week:

$$I = I(h) = \begin{cases} 12h & \text{if } 0 \le h \le 35 \\ 36(h - 35) + 420 & \text{if } h > 35 \end{cases}$$

In the second part of this split function, we multiply $36 per hour by $(h - 35)$, which is the number of hours *above* 35 that the employee works, and add to that $420 ($= 35 \times 12$), which is the income earned for the first 35 hours.

(b) To compute the weekly income for an employee who works 35, 42, or 48 hours, we must substitute the number of hours into the applicable rule of the split function.

If the employee works 35 hours, we use the first rule to get

$$I(35) = 12(35) = 420$$

If the employee works 42 hours, we use the second rule to get

$$I(42) = 36(42 - 35) + 420 = 36(7) + 420 = 252 + 420 = 672$$

If the employee works 48 hours, we use the second rule to get

$$I(48) = 36(48 - 35) + 420 = 36(13) + 420 = 468 + 420 = 888$$

Therefore, we see that if the employee works 35 hours, he or she earns $420; 42 hours, $672; and 48 hours, $888. ∎

Thus far, our attention has been focused on understanding and using function notation from an algebraic point of view. Let's now examine the graphical interpretation of $f(x)$ notation.

Interpreting $f(x)$ from a Graph

Let's carefully examine the graph in Figure 3.42(a) and review what it means for a point (x, y) to lie on a graph and how, given the graph of a relationship, we can determine y given x.

We can see in Figure 3.42(a) that the point $(5, -6)$ is on the graph. This means that the graph passes through a point that is 5 horizontal units to the right of the y-axis and 6 vertical units below the x-axis. If we need to find the y-value associated with a particular x-value, we start at the x-value on the x-axis, project vertically upward or downward to the graph, and then project horizontally to the y-axis and read this y-value. Hence if $x = 3$, then we see that for this graph, $y = 1$. Notice that at $x = 3$, the "height" of the graph (the vertical distance from the x-axis) is 1; likewise, if $x = -4$, $y = 3$.

The x-coordinate gives the horizontal distance from the y-axis, and the y-coordinate gives the vertical distance from the x-axis.

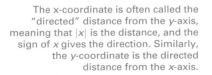

The x-coordinate is often called the "directed" distance from the y-axis, meaning that $|x|$ is the distance, and the sign of x gives the direction. Similarly, the y-coordinate is the directed distance from the x-axis.

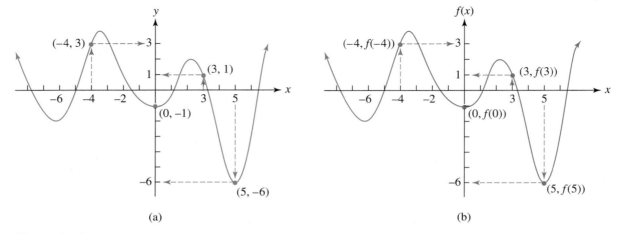

(a)   (b)

Figure 3.42

Returning to the graph in Figure 3.42(a), using function notation, we can label the graph $y = f(x)$. Then, instead of writing $y = 3$ when $x = -4$, we can write $f(-4) = 3$. Similarly, $y = -1$ when $x = 0$ can be written as $f(0) = -1$. Figure 3.42(b) repeats Figure 3.42(a) with the vertical axis labeled $f(x)$ instead of y.

Referring to Figure 3.42(b), let's summarize the equivalence of these alternatives. See Table 3.2.

Table 3.2

(x, y) Notation	$y = f(x)$ Notation	$(x, f(x))$ Notation
$(-4, 3)$	$3 = f(-4)$	$(-4, f(-4))$
$(3, 1)$	$1 = f(3)$	$(3, f(3))$
$(0, -1)$	$-1 = f(0)$	$(0, f(0))$

Again, $f(x)$ is the y-value at x, and gives the "height" of the graph (the vertical distance above or below the x-axis) at x. To be more precise, $f(x)$ is the directed distance of the graph from the x-axis at x. Hence, $f(a)$ is the directed distance of the graph from the x-axis at $x = a$; the directed distance of the graph from the x-axis at $x = b$ is $f(b)$. See Figure 3.43.

Figure 3.43

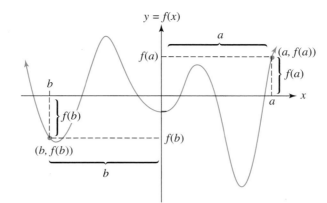

Example 7   Use the function defined by the graph in Figure 3.44 to find each of the following.
(a) $f(6)$ (b) $f(-3)$ (c) $f(0)$ (d) For what values of x is $f(x) = 0$?
(e) Which is larger, $f(-3)$ or $f(3)$?

Figure 3.44

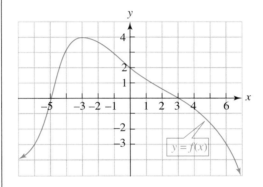

Solution   (a) To find $f(6)$, we must find the y-value that corresponds to $x = 6$. Looking at the graph, we can see that when $x = 6$, the corresponding y-value is $y = -3$. Therefore $f(6) = -3$.

(b) Similarly, from the graph we see that when $x = -3$, the corresponding y-value is $y = 4$. Therefore $f(-3) = 4$.

(c) To find $f(0)$, we must determine where the graph crosses the y-axis (because when $x = 0$ the point must be on the *y-axis*). From the graph we can see that $y = 2$ when $x = 0$, hence $f(0) = 2$.

(d) We are looking for those x-values for which $y = 0$ [remember that $f(x)$ is just another name for y]. But $y = 0$ means that the point must be on the x-axis. Therefore, from the graph we see that $f(-5) = 0$ and $f(3) = 0$, and so the x-values that make $f(x) = 0$ are -5 and 3 .

(e) Although it is true that 3 is larger than -3, this question asks you to compare the y-values or f-values, not the x-values. We cannot answer this question until we determine what the function f actually does to x. We see on the graph that when $x = -3$, the corresponding y value is $y = 4$; hence $f(-3) = 4$. Similarly, the y-value corresponding to $x = 3$ is $y = 0$; thus $f(3) = 0$. So $f(-3) = 4$ is larger than $f(3) = 0$.

∎

At the beginning of this section we learned that, algebraically, $f(a + 3)$ and $f(a) + 3$ are not the same. Keeping in mind that $f(a)$ is the "height" of the graph at a, we can geometrically demonstrate the differences between the two quantities. See Figure 3.45 on page 158.

Figure 3.45

Comparing $f(a) + 3$ and $f(a + 3)$

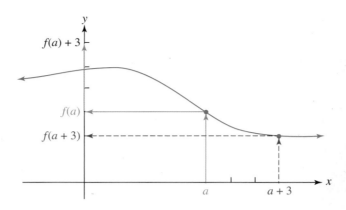

Notice that finding $f(a) + 3$ requires us first to find $f(a)$. To find $f(a)$, we first locate a on the x-axis, move vertically up until we intersect the graph, and then project horizontally onto the y-axis. To find $f(a) + 3$, we move vertically up three more units. This is equivalent to using a as the input and then adding 3 to the output, $f(a)$.

On the other hand, finding $f(a + 3)$ requires us first to find $a + 3$, move vertically up until we intersect the graph, and then project horizontally onto the y-axis. This is the same as using $a + 3$ as the input and then finding the output, $f(a + 3)$. Notice the difference between the two expressions in their values as well as in the procedures used to find each expression.

Different Perspectives: **Function Notation**

Consider the graphical and algebraic interpretation of function notation. Let's consider the function $y = f(x) = x^2 - x - 6$, whose graph appears below.

Graphical Description

Algebraic Description

To find $f(4)$ we substitute 4 for x in $f(x) = x^2 - x - 6$:

$$f(4) = 4^2 - 4 - 6 = 16 - 4 - 6 = 6$$

Hence $f(4) = 6$.

To find $f(4)$ *using the graph,* we are looking for the y-value that corresponds to $x = 4$. From the graph we can see that when $x = 4$, the corresponding y-value is $y = 6$. Hence $f(4) = 6$.

3.5 Exercises

In Exercises 1–20, let $f(x) = 5x - 2$,
$g(x) = 3x^2 - 4x + 1$, and $h(x) = \sqrt{4x - 3}$. Find:

1. $f(6)$
2. $g(-5)$
3. $h(10)$
4. $f(-2)$
5. $g(8)$
6. $h(0)$
7. $f\left(\dfrac{1}{3}\right)$
8. $g\left(-\dfrac{1}{2}\right)$
9. $f(x + 2)$
10. $f(x) + 2$
11. $g(x - 1)$
12. $g(x) - g(1)$
13. $h(x^2)$
14. $[h(x)]^2$
15. $g(3x)$
16. $3g(x)$
17. $f(4x + 7)$
18. $4f(x) + 7$
19. $g(x + h)$
20. $f(x - a)$

In Exercises 21–40, given $F(x) = \dfrac{x}{x + 1}$ and
$G(t) = \dfrac{1}{\sqrt{t - 1}}$, find:

21. $F(9)$
22. $F(-2)$
23. $G(15)$
24. $G(6)$
25. $F\left(\dfrac{1}{2}\right)$
26. $G(1)$
27. $G(t^2)$
28. $F\left(\dfrac{1}{x}\right)$
29. $F(-1)$
30. $F(x^2)$
31. $F(x + 1)$
32. $G(t - 1)$
33. $5G(6)$
34. $G(30)$
35. $F(40)$
36. $8F(5)$
37. $F\left(\dfrac{a}{3}\right)$
38. $G(t^2 + 1)$
39. $F(x - 1)$
40. $F\left(\dfrac{a}{a + 1}\right)$

In Exercises 41–60, let $f(x) = 4x - 3$, $g(x) = \dfrac{1}{x}$, and
$h(x) = x^2 - x$. Find (and simplify):

41. $f(5x + 7)$
42. $5f(x) + 7$
43. $g(x - a)$
44. $f(x) + f(a)$
45. $g(x) - g(a)$
46. $f(x) + a$
47. $g(x) - a$
48. $f(x + a)$
49. $f(4)h(4)$
50. $f(1)g(2)h(-3)$
51. $f[h(4)]$
52. $h[f(4)]$
53. $g[h(x)]$
54. $h[g(x)]$
55. $h(kx)$
56. $kh(x)$
57. $f(g[h(3)])$
58. $f(3)g(3)h(3)$
59. $h\left(\dfrac{1}{x}\right)$
60. $\dfrac{1}{h(x)}$

In Exercises 61–66, let $f(x) = 2 - 3x$ and
$g(x) = x^2 - 3x + 2$. Find (and simplify):

61. $\dfrac{g(x) - g(5)}{x - 5}$
62. $\dfrac{f(x) - f(2)}{x - 2}$
63. $\dfrac{f(x + 3) - f(x)}{3}$
64. $\dfrac{g(x + 4) - g(x)}{4}$
65. $\dfrac{g(x + h) - g(x)}{h}$
66. $\dfrac{f(x + h) - f(x)}{h}$

67. Given $f(x) = \begin{cases} 3x - 5 & \text{if } x < 1 \\ x^2 - 1 & \text{if } x \geq 1 \end{cases}$

    Find: **(a)** $f(-3)$ **(b)** $f(1)$ **(c)** $f(6)$

68. Given $g(x) = \begin{cases} x^2 - 3x + 2 & \text{if } x \leq -5 \\ x + 1 & \text{if } -5 < x < 2 \\ 8 & \text{if } x > 2 \end{cases}$

    Find: **(a)** $g(-4)$ **(b)** $g(2)$ **(c)** $g(-6)$ **(d)** $g(4)$

69. Given $h(s) = \begin{cases} |2s - 1| & \text{if } 0 \leq s \leq 3 \\ \dfrac{s + 1}{s - 1} & \text{if } 3 < s \leq 7 \end{cases}$

    Find: **(a)** $h(0)$ **(b)** $h(3)$ **(c)** $h(5)$ **(d)** $h(8)$

70. Given $f(x) = \begin{cases} 3x - 5 & \text{if } -2 \leq x < 1 \\ x^2 - 4 & \text{if } 1 < x \end{cases}$

    Find the domain of $f(x)$.

71. If an object is thrown upward from ground level with an initial velocity of 50 ft/sec, then its height, h, after t seconds is given by the equation

    $$h(t) = 50t - 16t^2$$

    (a) Find the height of the object after 1, 2, and 3 seconds.
    (b) How many seconds does it take for the object to hit the ground?

72. If an object is dropped from a window 600 feet above ground, then its height, h, after t seconds is given by the equation

    $$h(t) = 600 - 16t^2$$

    (a) Find the height of the object after 1, 2, and 3 seconds.
    (b) Approximately how many seconds does it take for the object to hit the ground?

73. Use the following graph of $y = g(x)$ to find
    (a) $g(-5)$ **(b)** $g(-2)$ **(c)** $g(0)$ **(d)** $g(1)$
    (e) $g(3)$ **(f)** For what value(s) of x is $g(x) = 0$?
    (g) How many solutions are there to the equation
    $g(x) = -1$?

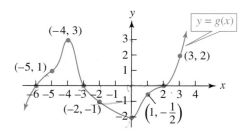

74. Use the following graph of $y = h(x)$ to find
    (a) $h(-2)$ **(b)** $h(6)$
    (c) $h(0)$ **(d)** $h(4)$
    (e) $h(-4)$
    (f) For what value(s) of x is $h(x) = 0$?
    (g) Based on the graph, what is the domain of $h(x)$?
    (h) Based on the graph, what is the range of $h(x)$?

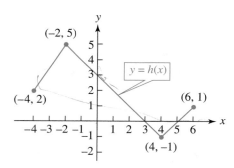

75. Each month, a local utility company charges a flat fee of $7 for the first 35 kilowatt-hours or less, and $0.12 for each kilowatt-hour above 35. Write an equation describing the relationship between the dollar amount, A, of an electric bill and the total number of kilowatt-hours, h, used by a customer each month. Sketch the graph of this equation.

76. Each month, a local phone company charges a flat fee of $11.60 for the first 80 message units and 10.6¢ for each message unit above 80. Write an equation describing the relationship between the dollar amount, A, of a phone bill and the total number of message units, m, used by a customer each month. Sketch the graph of this equation.

77. The number of items a company must produce and sell so that its costs are equal to its revenue (and so its profit is 0) is called the *break-even point*. A manufacturer of trash cans determines that the cost C of producing x cans is given by the function $C = C(x) = 5.75x + 12375$, and the revenue R earned on the sale of x cans is given by the function $R = R(x) = 14x$. Determine the break-even point.

78. A manufacturer of power screwdrivers determines that the cost C of producing s screwdrivers is given by the function $C = C(s) = 14.65s + 37125$, and the revenue R earned on the sale of x cans is given by the function $R = R(s) = 29.50s$. Determine the break-even point.

Questions for Thought

79. Given $f(x) = 5x + 7$, discuss what is *wrong* (if anything) with each of the following:
    (a) $f(x + 4) \overset{?}{=} 5x + 7 + 4 \overset{?}{=} 5x + 11$
    (b) $f(x + 4) \overset{?}{=} (5x + 7)(x + 4) \overset{?}{=} 5x^2 + 27x + 28$
    (c) $f(2x) \overset{?}{=} 2(5x + 7) \overset{?}{=} 10x + 14$
    (d) $f(x^2) \overset{?}{=} (5x + 7)^2 \overset{?}{=} 25x^2 + 70x + 49$

80. Suppose $y = f(x) = 3x - 6$.
    (a) Find $f(9)$ and $f(4)$, and explain why this means the points $(9, 21)$ and $(4, 6)$ lie on the graph of $f(x)$.
    (b) Compute $\dfrac{f(9) - f(4)}{9 - 4}$. What is the significance of this difference quotient?
    (c) Compute the difference quotient $\dfrac{f(x + h) - f(x)}{h}$ for $y = f(x) = mx + b$. What is the significance of the difference quotient for a function whose graph is a straight line?

3.6 Relating Functions to Their Graphs

Suppose we were managing a factory and found that the cost of manufacturing items, $C(x)$, and the number of items produced daily by our factory, x, are related in the following way: $C(x) = 10,000(-3x^2 + 2x + 10)$. As managers, it would be our goal to understand this relationship or know the "behavior of this function," perhaps in an effort to minimize the cost, $C(x)$. Suppose we are medical researchers who find that the decrease in a person's blood pressure, $D(x)$, is related to the amount, x, of a particular drug taken by the person as follows: $D(x) = \frac{1}{3}x^2(10 - x)$. If we want to know what happens to blood pressure as we change the dosage (perhaps to find the appropriate dosage for a particular patient), it is important that we understand this mathematical relationship. In this section we will review how equations are related to graphs. Our goal is to better understand how variables are related in an equation by picturing the relationship with a graph.

Suppose you are given the equation $y = x^2 - 5x + 6$. How would you describe the relationship between x and y? What do you think happens to y as x changes? Does y always increase as x increases? Can y be negative, or is y positive for all values of x?

We could substitute values for x and see what happens to y, as in Table 3.3. However, a table may not give a clear picture of the relationship between x and y; if we miss some values for x, we may miss important values for y. For example, we may miss the minimum or maximum y-values.

Table 3.3

x	-1	0	1	2	3	4
y	12	6	2	0	0	2

We learn how to sketch the graphs of functions such as $y = x^2 - 5x + 6$ in Section 4.4.

It would be nice to have a "snapshot" of this relationship—a picture that tells us at a glance how the variables are related. This picture is what a graph is. A portion of the graph of the equation $y = x^2 - 5x + 6$ appears in Figure 3.46.

Figure 3.46

A portion of the graph of the equation $y = x^2 - 5x + 6$

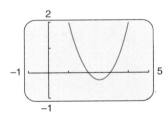

Some of the things the graph allows us to see are that y can be negative, and that Table 3.3 did not capture the minimum y-value. We will continue to discuss how to use a graph to analyze the behavior of a function in this section and throughout this text.

Table 3.4 lists the high value of IBM stock for each month from March through September 2002.

Table 3.4

Month	March	April	May	June	July	August	September
High value of IBM	108.50	102.86	85.69	78.11	73.50	82.49	75.60

Figure 3.47 on page 162 is a graph of the stock data in Table 3.4.

The graph makes it easier to see several things: the general downward trend is obvious; the greatest monthly decline in stock value occurred from April to May; and

Figure 3.47

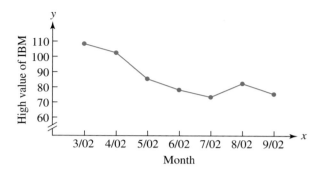

there was a one-month increase in stock value from July to August. Although having the data is useful, having the graph adds significantly to our understanding of the situation.

Let's begin by reemphasizing the difference between "finding $f(5)$" and "finding x such that $f(x) = 5$."

If we are given only the graph of $y = f(x)$, such as the one shown in Figure 3.48, then $f(5)$ is the y-coordinate on the graph when $x = 5$. Since we see that $y = 2$ when $x = 5$, we have $f(5) = 2$. On the other hand, solving $f(x) = 5$ for x means finding the x-coordinate (s) when $y = 5$. In this case, since $x = -3$ when $y = -5$, $x = -3$ is the solution to $f(x) = 5$.

Figure 3.48

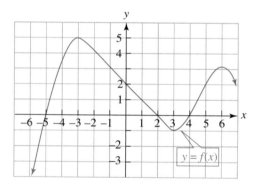

Given the graph, the processes of finding the answers to these "different" questions are similar. However, given the function $f(x)$ stated as a formula or equation, the processes used to answer these questions algebraically are different, as we see in the next example.

Example 1   Given the function $f(x) = x^2 - 2x - 1$,
(a) Find $f(5)$.   (b) Solve the equation $f(x) = 5$.

Solution   (a) To find $f(5)$, we evaluate $f(x)$ at $x = 5$:

$$f(x) = x^2 - 2x - 1 \qquad \text{Substitute } x = 5, \text{ and evaluate.}$$
$$f(5) = (5)^2 - 2(5) - 1 = 14$$

Hence $f(5) = 14$.

(b) To solve the equation $f(x) = 5$, we set $f(x)$ equal to 5 and solve for x:

$$5 = x^2 - 2x - 1$$ Put into standard form.

$$0 = x^2 - 2x - 6$$ Use the quadratic formula:
$$x = \frac{-b \pm \sqrt{b^2 - 4ac}}{2a} \text{ with}$$
$a = 1$, $b = -2$, and $c = -6$.

$$x = \frac{-(-2) \pm \sqrt{(-2)^2 - 4(1)(-6)}}{2(1)}$$

$$x = \frac{2 \pm \sqrt{28}}{2}$$ Simplify.

$$x = \boxed{1 \pm \sqrt{7}}$$

$$x = -1.645751 \quad \text{and} \quad x = 3.645751$$ Accurate to six places   ∎

We could use a graphing calculator to answer the questions in Example 1(b). To solve the equation $f(x) = 5$, we would enter and graph the equation $y = x^2 - 2x - 1$ as shown using the standard $[-10, 10]$ by $[-10, 10]$ window in Figure 3.49(a). We use the $\boxed{\text{TRACE}}$ and $\boxed{\text{ZOOM}}$ keys to find x when $y = 5$. The decimal answers we arrive at using the graphing calculator are X=-1.645612 and X=3.6469415 (see Figures 3.49(b) and (c)), which are close to the six-place decimal approximations to the answers we found algebraically. On the other hand, we could also find $f(5)$ using the $\boxed{\text{TRACE}}$ and $\boxed{\text{ZOOM}}$ keys on a graphing calculator to find y when $x = 5$. But the algebraic or numeric method is more efficient for this purpose.

See Example 2 for another way to use the graphing calculator to solve such an equation.

Figure 3.49

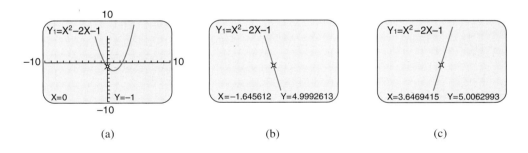

(a)   (b)   (c)

Solving $f(x) = k$

On a graph, to find the values of x where $f(x) = k$ (k a constant), we look for the point *on the graph of* $y = f(x)$ where the directed distance from the x-axis is k units. To find the point, we start on the y-axis at $y = k$, and move horizontally until we intersect the graph of $y = f(x)$. The x-coordinate of this point is the value we are seeking. See Figure 3.50(a) on page 164.

Recall that $y = k$ is a horizontal line $|k|$ units from the x-axis. Hence, another useful way to view this problem is that we are seeking to find the points where the line $y = k$ intersects the graph of $y = f(x)$. See Figure 3.50(b). (There may be more than one point of intersection.)

Hence, we can solve the equation $f(x) = k$ using a graphing calculator by entering two equations: $y = f(x)$ and $y = k$, and reading off the x-coordinate(s) of their intersection(s).

Figure 3.50

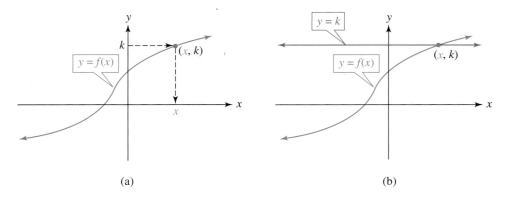

(a)   (b)

Example 2   Solve for x: $x^4 - x + 1 = 5$ accurate to four decimal places.

Solution   We have no algebraic formula or technique available to solve the equation $x^4 - x + 1 = 5$. In such a situation, where we have no algebraic method to solve an equation, we can take advantage of the technology that we do have available to us. We solve this equation using a graphing calculator.

In this example, if we let $y = x^4 - x + 1$, we want the x-coordinate of the point on the graph that has y-coordinate 5. We could use the TRACE and ZOOM keys until we find Y=5, and read off the X value associated with Y=5 as we did in Example 1. Instead, we first enter Y1=X^4-X+1, and then enter Y2=5 and graph both using the window as shown in Figure 3.51(a):

Figure 3.51

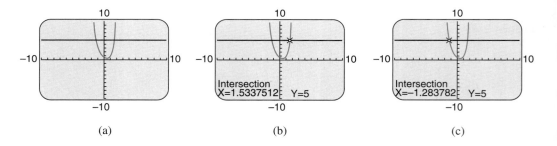

(a)   (b)   (c)

From Figure 3.51(a) we can see that there are at least two points of intersection and therefore at least two solutions to the equation. Most calculators have an intersect menu item that allows you to find the point(s) of intersection quickly with a high degree of accuracy. The TI-83 Plus intersect function is in the CALC menu. Their intersections are shown in Figures 3.51(b) and 3.51(c). Accurate to 4 places, the answers are X=-1.2838 and X=-1.5338.

We mentioned previously that there were "at least" two points of intersection. We cannot say that there are *only* two points of intersection at this time because we are limited by the window we have chosen; we only see a small part of this graph. It is possible that there are other points of intersection outside this window. It turns out that, for this example, there *are* only two points of intersection.

We need to recognize that the graphing calculator is a tool we can use to help us extract some information about a graph, but the graphing calculator alone may not give us a complete picture. It is up to us to use it in conjunction with our knowledge of the function under consideration to obtain a more complete picture of the graph. This is an idea we will return to throughout this text. ∎

Different Perspectives: **Solving $f(x) = k$**

Consider the graphical and algebraic approaches to solving $x^2 - x = 6$.

Graphical Description

We can define $y = x^2 - x$ and $y = 6$, graph the two functions, and look for their intersection.

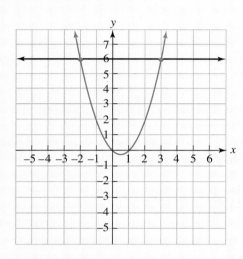

We see that the two functions intersect at $(-2, 6)$ and $(3, 6)$, which means $x^2 - x$ is 6 when $x = -2$ and 3. Therefore, the solutions to the equation $x^2 - x = 6$ are $x = -2$ and $x = 3$.

Thus the following two statements are equivalent:

The graphs of $y = x^2 - x$ and $y = 6$ intersect when $x = -2$ and $x = 3$.

Algebraic Description

Algebraically, we can solve the equation $x^2 - x = 6$ by factoring:

$$x^2 - x = 6 \qquad \text{Put into standard form.}$$
$$x^2 - x - 6 = 0 \qquad \text{Factor.}$$
$$(x + 2)(x - 3) = 0$$
$$x + 2 = 0 \quad \text{and} \quad x - 3 = 0 \quad \text{Hence}$$
$$x = -2, x = 3$$

$\Longleftrightarrow$ $x = -2$ and $x = 3$ are the solutions to the equation $x^2 - x = 6$.

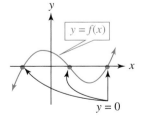

Figure 3.52

The solutions to $f(x) = 0$

The Zeros of a Function

The solutions to the equation $f(x) = 0$ are found by noting the x-values at which the height of the graph of $f(x)$ is 0. We can restate this by saying that we are looking for the values of x where the graph of $f(x)$ intersects the horizontal line $y = 0$ (the x-axis). See Figure 3.52.

Definition of a Zero of a Function

Given a function $y = f(x)$, a solution to the equation $f(x) = 0$ is called a **zero** of the function.

A zero of a function is defined algebraically—it is a solution to the functional equation $y = f(x) = 0$, that is, the value of x that makes $f(x) = 0$. On the other hand, we have defined an x-intercept of a graph as the x-coordinate of a point where the graph crosses the x-axis. Since an x-intercept corresponds to a y-coordinate of 0, each x-intercept of the graph corresponds to a zero of the function. Thus we have a direct relationship between an algebraic concept (a solution to an equation) and a graphical one (a graph crossing the *x*-axis). See Figure 3.53.

Figure 3.53

x_1 is an *x*-intercept, or, equivalently, $f(x_1) = 0$, and so x_1 is a zero of $f(x)$.

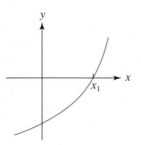

Returning to Figure 3.52, the arrows that appear at the end of the graph indicate that the graph continues in the direction of the arrows, and so Figure 3.52 illustrates all the zeros of $f(x)$.

Using function notation, the *x*-intercepts of a function are the solutions to the equation $f(x) = 0$; the *y*-intercept of a function is $f(0)$.

Example 3 Use a graphing calculator to estimate the zero(s) of $f(x) = 5x^3 - 2x - 1$, rounded to three decimal places.

Solution We begin by entering the function Y1=5X^3-2X-1 in the window, as shown in Figure 3.54(a):

Figure 3.54

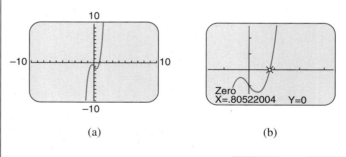

(a)   (b)

We could find the zeros by using the $\boxed{\text{TRACE}}$ and $\boxed{\text{ZOOM}}$ keys as close as possible to where $y = 0$. Alternatively, many calculators have a menu choice that finds the zero directly: For the TI-83 Plus, this is called zero in the CALC menu. Figure 3.54(b) shows that the zero for $f(x)$ is X=.80522004, which is 0.805 accurate to three places.

Interpreting the Graph of $f(x)$: General Trends

Consider the graph of a function $f(x)$ pictured in Figure 3.55. The picture tells us a few things about the relationship between *x* and $y = f(x)$. One thing we can see immediately is that the graph does not cross the *y*-axis. This tells us that 0 is not in the domain of this function.

Let's see what happens when *x* is positive (staying to the right of the *y*-axis). Notice that as *x* gets extremely large—that is, as we move out farther and farther to the right, which we denote as $x \to +\infty$, the values of $f(x)$ (the *y*-values) remain positive (above the *x*-axis) but get closer and closer to 0. On the other hand, as *x* gets closer to 0 *from the right* (*x* is still positive), the values of $f(x)$ (the *y*-values) get extremely large, which we denote as $f(x) \to +\infty$.

Figure 3.55

The graph of a function $y = f(x)$

Different Perspectives: **The Zeros of a Function**

Consider the graphical and algebraic description of the zeros of a function. Let's examine the function $y = f(x) = (x + 3)(x - 1)(x - 2)$, whose graph is shown here.

Graphical Description

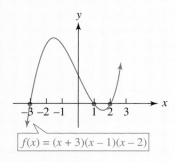

$f(x) = (x + 3)(x - 1)(x - 2)$

The zeros of a function are the x-intercepts of its graph. In this example, the zeros of $f(x) = (x + 3)(x - 1)(x - 2)$ are -3, 1, and 2.

Algebraic Description

The zeros of a function are the values of x for which $f(x) = 0$. For example, to find the zeros of $f(x) = (x + 3)(x - 1)(x - 2)$, we solve the equation $(x + 3)(x - 1)(x - 2) = 0$ for x, which gives $x = -3$, 1, and 2.

To the left of the y-axis (when x is negative), we see that as we move out farther and farther to the left, which we denote as $x \to -\infty$, the values of $f(x)$ get closer and closer to 0 but remain negative. On the other hand, as x approaches 0 *from the left* (x remains negative), the values of $f(x)$ get smaller and smaller, which we denote as $f(x) \to -\infty$.

Examining the graph of $f(x)$ often reveals the general trends of a function for large and small values of x.

Interpreting the Graph of $f(x)$: Inequalities

In the preceding case of $f(x)$, we discussed what happens to y when x was positive or negative. A graph can also tell us when—that is, for what values of x—the function is negative or positive.

Let's examine $y = f(x) = x^2 - 7x + 6$, whose graph appears in Figure 3.56. Look carefully at the graph of $y = f(x) = x^2 - 7x + 6$. Notice that it tells us the relationship between x and y at a glance.

Figure 3.56

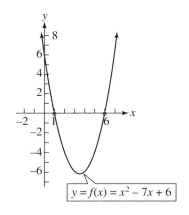

$y = f(x) = x^2 - 7x + 6$

We know that when the graph of any function is below the x-axis, y *is negative,* that is, $y < 0$. For this function, we can see that the graph dips below the x-axis when x is between 1 and 6; see Figure 3.57(a). (Notice that we talk about the behavior of y as x *changes.*)

Figure 3.57

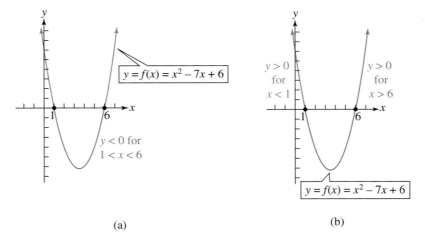

(a)   (b)

Hence, the graph of $y = f(x)$ dips below the x-axis when $1 < x < 6$.   Which means that

$$y \text{ is negative or } y < 0 \text{ when } 1 < x < 6.$$

Since $y = x^2 - 7x + 6$, we have

$$x^2 - 7x + 6 < 0 \text{ when } 1 < x < 6.$$

Thus $1 < x < 6$ is the solution to the inequality $x^2 - 7x + 6 < 0$.

A similar analysis of the graph in Figure 3.57(b) tells us that $x^2 - 7x + 6$ is greater than 0 for $x < 1$ and for $x > 6$.

In Figure 3.58 we have indicated on the x-axis those values of y for which the function $y = f(x)$ is positive (above the x-axis) or negative (below the x-axis). The result is the same as if we had done a sign analysis.

Graph the function $y = x^2 - 7x + 6$ on a graphing calculator using the standard $[-10, 10]$ by $[-10, 10]$ window, and $\boxed{\text{TRACE}}$ the function to the right. As you move the cursor, observe the sign of Y for various values of X, until X is well past 6. When the cursor is below the x-axis, y is negative; when the cursor is above the x-axis, y is positive.

Figure 3.58

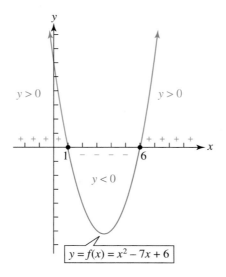

This gives us another method for solving the inequality $f(x) > 0$ or $f(x) < 0$.

To solve the inequality $f(x) > 0$, we graph the function $y = f(x)$ and determine when (for which intervals of x) the graph is *above* the x-axis.

To solve the inequality $f(x) < 0$, we graph the function $y = f(x)$ and determine when (for which intervals of x) the graph is *below* the x-axis.

Example 4   Use a graphing calculator to solve the inequality $x^5 - 4x^2 > -2$, rounded to two decimal places.

Solution   We first rewrite the inequality as $x^5 - 4x^2 + 2 > 0$. Then graph the function $f(x) = x^5 - 4x^2 + 2$, and note the x-intervals where the graph of $f(x)$ lies above the x-axis. Figure 3.59 shows the graph of $f(x)$.

We identify the zeros (accurate to two decimal places), and note that $f(x) > 0$, or the graph of $f(x)$ is above the x-axis, when x is on $(-0.68, 0.75) \cup (1.45, \infty)$. Hence the solution (accurate to two places) is $(-0.68, 0.75) \cup (1.45, \infty)$. ∎

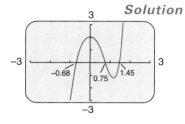

Figure 3.59

Different Perspectives: Solving Inequalities

Graphical Description

We are given the graph of
$y = f(x) = (x + 2)(x - 1)(x - 4)$. Solving the inequality $f(x) > 0$ means we are looking for values of x where the graph of $y = f(x)$ is above the x-axis. Based on the graph, $f(x) > 0$ for x on $(-2, 1) \cup (4, \infty)$.

Algebraic Description

Solving the inequality
$(x + 2)(x - 1)(x - 4) > 0$ means we are seeking the values of x for which $(x + 2)(x - 1)(x - 4)$ is positive; we usually utilize a sign analysis. The cut points are -2, 1, and 4.

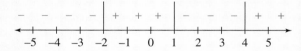

We see that $(x + 2)(x - 1)(x - 4)$ is positive for x on $(-2, 1) \cup (4, \infty)$.

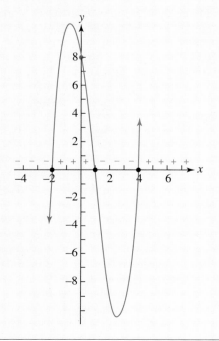

Increasing and Decreasing Functions

When we talk about a function increasing or decreasing, we are referring to the value of $f(x)$ as *x gets larger*. This is consistent with our previous agreement to describe the graph of a function *as we move from left to right*. Hence, when we say that $f(x)$ is an increasing function, we mean that as x increases, $f(x)$ always increases. Geometrically, this means as we move left to right, the graph rises. On the other hand, when we say that $f(x)$ is a decreasing function, we mean that as x increases, $f(x)$ decreases. (Our frame of reference for x is always for increasing values of x.) Geometrically, this means as we move left to right, the graph falls.

When we say that a function $f(x)$ is increasing or decreasing, are we talking about *x*-values or *y*-values?

Algebraically, we define increasing and decreasing functions as follows.

> ### Definition of Increasing and Decreasing Functions
> A function $y = f(x)$ is said to be an **increasing function** on an interval I if and only if for x_1 and x_2 in the interval, $x_1 < x_2$ implies that $f(x_1) < f(x_2)$. A function $y = f(x)$ is said to be a **decreasing function** on an interval I if and only if for x_1 and x_2 in the interval, $x_1 < x_2$ implies that $f(x_1) > f(x_2)$.

Figure 3.60

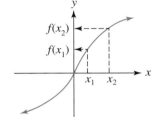

(a) $f(x)$ is increasing: If $x_1 < x_2$, then $f(x_1) < f(x_2)$.

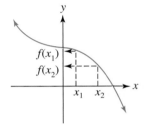

(b) $f(x)$ is decreasing: If $x_1 < x_2$, then $f(x_1) > f(x_2)$.

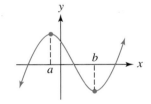

Figure 3.61

This function is increasing on the intervals $(-\infty, a]$ and $[b, \infty)$ and decreasing on the interval $[a, b]$.

The functions illustrated in Figure 3.60 are always increasing or always decreasing, but many functions do not fall into this category: They could be constant functions such as $f(x) = 3$ (which neither increase nor decrease), or perhaps they increase on some intervals and decrease on other intervals. We may be interested in where these functions increase and where they decrease—that is, **in what intervals of *x*** is $f(x)$ increasing and decreasing. See Figure 3.61.

Referring to the graph of $f(x) = x^2 - x - 6$ (see Figure 3.62), we can see that y is decreasing (gets smaller) on the interval $\left(-\infty, \frac{1}{2}\right]$. (Again, notice we talk about the behavior of y for values of x.) We also see that y is increasing on the interval $\left[\frac{1}{2}, \infty\right)$.

Figure 3.62

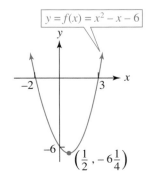

$y = f(x) = x^2 - x - 6$

Notice that at the point $\left(\dfrac{1}{2}, -6\dfrac{1}{4}\right)$ on the graph of $y = x^2 - x - 6$ in Figure 3.62, the function changes from decreasing to increasing. A point on the graph of a function where it changes from decreasing to increasing (or vice versa) is called a **turning point** of the graph.

Example 5 Use the graph of $f(x)$ in Figure 3.63 to identify where $f(x)$ is increasing, where $f(x)$ is decreasing, and the turning points.

Figure 3.63

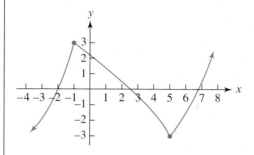

Solution Finding where $f(x)$ is increasing means that we want the x intervals where the graph of $f(x)$ is rising as we move from left to right. We can see from the graph in Figure 3.63 that $f(x)$ increases when x is in the intervals $(-\infty, -1] \cup [5, \infty)$.

Be sure you understand the use of the notation: The graph increases in the INTERVAL $[-1, 5]$; a turning POINT of $f(x)$ is $(-1, 3)$.

On the other hand, finding where $f(x)$ decreases means locating the interval(s) of x where the graph of $f(x)$ falls *as we move from left to right*. Thus we can see that $f(x)$ is decreasing in the interval $[-1, 5]$. The *turning points* are at $(-1, 3)$ and at $(5, -3)$. ∎

The y-value of a turning point at which the graph changes from increasing to decreasing is called a **relative maximum**; the y-value of a turning point at which the graph changes from decreasing to increasing is called a **relative minimum**; see Figure 3.64.

Figure 3.64

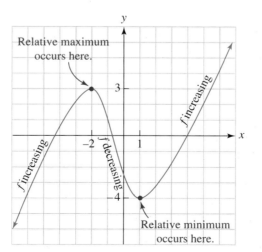

A relative maximum or minimum is also called a local maximum or minimun.

In Figure 3.64, the turning point $(-2, 3)$ gives the relative maximum of $f(x)$ as 3, and the point $(1, -4)$ gives the relative minimum of $f(x)$ as -4.

Example 6 Suppose that an automotive research institute collects the data in the following table, which indicates a certain car's gas mileage g (in miles per gallon) at specific speeds s (in miles per hour).

s	10	20	30	40	50	60
g	17.5	22.8	27.7	30.8	32.5	32.1

Using these data, a researcher conjectures that the gas mileage and speed are related by the equation $g = 10 + 0.8s - 0.007s^2$.
(a) Determine how well the proposed equation agrees with the observed data.
(b) Graph this equation and describe how the speed of the car and its miles per gallon are related.
(c) Approximate (to the nearest mile per hour) the speed that maximizes the car's mileage, and determine this maximum mpg rate (to the nearest tenth).

Solution **(a)** We can compute values of g using the equation with the given values s, and compare these projected values of g with the observed values of g. The TI-83 Plus can compute the values obtained from the equation quickly using the TABLE function. The results appear in the table below.

X	Y1	
10	17.3	
20	23.2	
30	27.7	
40	30.8	
50	32.5	
60	32.8	
70	31.7	

Y1■10+.8X−.007X²

Comparing these results obtained from the equation with the observed values appearing in the original data table, we can see that the mpg rates obtained from the equation for speeds of 30, 40, and 50 mph are exactly the same as the observed values. At speeds of 10, 20, and 60 mph, the values given by the equation are very close to the actual observed values. Thus we can say that, at least for these speeds, the equation seems to predict the corresponding miles per gallon rate very well.
(b) If we graph the function Y1=10+.8X-.007X² in the standard $[-10, 10]$ by $[-10, 10]$ window, we get the graph shown in Figure 3.65(a). After zooming out a few times and adjusting the window settings, we see that the function has an umbrella shape (called a parabola), with a single maximum turning point (Figure 3.65(b)).

Figure 3.65

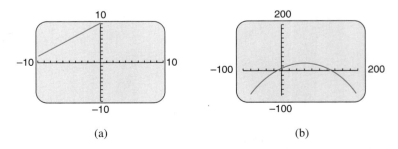

(a)   (b)

The graph indicates that as the speed increases from 10 mph, the miles per gallon rate increases up to a maximum value. When the speed surpasses a certain value (which corresponds to the maximum mpg rate), the miles per gallon rate *decreases* as the speed increases.

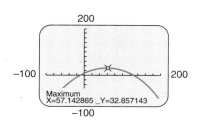

Figure 3.66

(c) We could approximate the maximum mpg rate by using the $\boxed{\text{TRACE}}$ and $\boxed{\text{ZOOM}}$ keys, as we did in the previous example; however, most graphing calculators have the ability to find the maximum value. In the TI-83 Plus we can find the maximum value using `maximum` in the CALC menu.

The graph appears in Figure 3.66, along with the maximum point occurring at $x = 57.1$ and $y = 32.9$ (accurate to one decimal place), which means that the speed that maximizes the car's mileage is 57.1 miles per hour, and the gas mileage at this speed is 32.9 miles per gallon. ∎

We will continue this discussion of increasing and decreasing functions in a later chapter.

As we have demonstrated in this section, a picture of an equation in the form of its graph yields a great deal of information about the relationship between the two variables in the equation. By graphing an equation, we can better understand the nature of the relationship, the general and local trends, and the important values for an equation. In the next section, we continue to examine the relationships expressed by equations by studying their graphs.

3.6 Exercises

1. Use the following graph of $y = f(x)$ to find
   (a) $f(5)$ **(b)** $f(-3)$ **(c)** $f(0)$ **(d)** $f(-1)$
   (e) What are the zeros of $f(x)$?
   (f) Find the domain and range of $f(x)$.

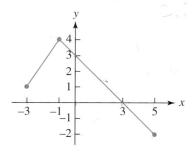

2. Use the following graph of $y = g(x)$ to find
   (a) $g(-2)$ **(b)** $g(2)$ **(c)** $g(-4)$ **(d)** $g(6)$
   (e) What are the zeros of $g(x)$?
   (f) Find the domain and range of $g(x)$.

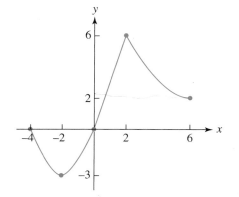

3. Use the following graph of $y = F(x)$ to find
   (a) $F(-7)$ **(b)** $F(-5)$
   (c) $F(-2)$ **(d)** $F(1) + F(4)$
   (e) $F(1 + 4)$
   (f) Based on the graph, what is the domain of $F(x)$?
   (g) Based on the graph, what is the range of $F(x)$?

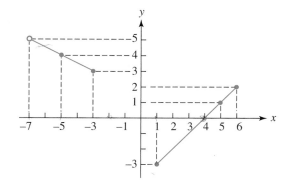

4. Use the following graph (page 162) of $y = g(x)$ to find
   (a) $g(6)$
   (b) $g(-3)$
   (c) $g(3)$
   (d) $g(-1) + g(1)$
   (e) $g(-1 + 1)$
   (f) On what intervals is $g(x)$ constant?
   (g) Based on the graph, what is the domain of $g(x)$?
   (h) Based on the graph, what is the range of $g(x)$?

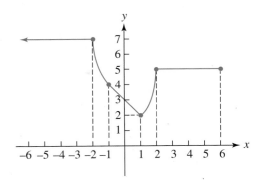

5. (a) Use the following graph of $f(x)$ to estimate its zeros.

(b) Given that this is the graph of $f(x) = x^2 - 2x - 4$, find the zeros of $f(x)$ algebraically and approximate the algebraic answers to the nearest tenth.

(c) What are the domain and range of $f(x)$?

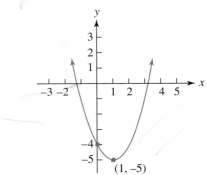

6. (a) Use the following graph of $g(x)$ to estimate its zeros.

(b) Given the graph of $g(x) = -x^2 - 4x + 3$ below, find the zeros of $g(x)$ algebraically and approximate the algebraic answers to the nearest tenth.

(c) What are the domain and range of $g(x)$?

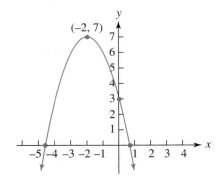

7. (a) Use the graph of $g(x)$ to estimate its zeros.

(b) Given that this is the graph of $g(x) = -x^2 + 3x + 1$, find the zeros of $g(x)$

algebraically and approximate the algebraic answers to the nearest tenth.

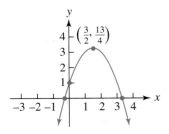

8. (a) Use the following graph of $h(x)$ to estimate its zeros.

(b) Given that this is the graph of $h(x) = x^2 + 5x + 3$, find the zeros of $h(x)$ algebraically and approximate the algebraic answers to the nearest tenth.

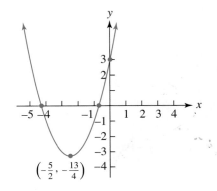

9. (a) Use the following graph of $F(x)$ to estimate its zeros.

(b) Given that this is the graph of $F(x) = x^3 - x^2 - x$, find the zeros of $F(x)$ algebraically and approximate the algebraic answers to the nearest tenth.

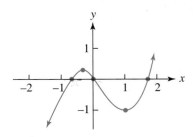

10. (a) Use the following graph of $G(x)$ to estimate its zeros.

(b) Given the graph of $G(x) = -x^3 - 2x^2 + x$, find the zeros of $G(x)$ algebraically and approximate the algebraic answers to the nearest tenth.

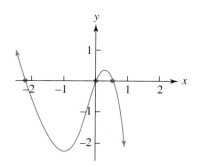

11. Use the following graph of $y = F(x)$ to answer each question.
  (a) What happens to $F(x)$ as $x \to \infty$?
  (b) What happens to $F(x)$ as $x \to -\infty$?
  (c) What happens to $F(x)$ as $x \to 3$?

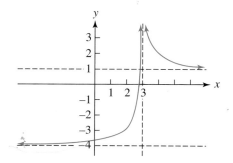

12. Use the following graph of $y = G(x)$ to determine
  (a) What happens to $G(x)$ as $x \to \infty$?
  (b) What happens to $G(x)$ as $x \to -\infty$?
  (c) What happens to $G(x)$ as $x \to -2$?

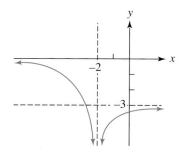

13. Use the following graph of $y = f(x)$ to determine
  (a) The intervals where $f(x)$ is increasing.
  (b) The intervals where $f(x)$ is decreasing.
  (c) The intervals where $f(x)$ is nonnegative.

(d) The intervals where $f(x)$ is positive.
(e) The intervals where $f(x)$ is negative.
(f) Find the relative maximum and minimum values.

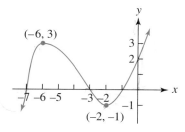

14. Use the following graph of $y = g(x)$ to determine
  (a) The intervals where $g(x)$ is increasing.
  (b) The intervals where $g(x)$ is decreasing.
  (c) The intervals where $g(x)$ is constant.
  (d) The intervals where $g(x)$ is positive.
  (e) The intervals where $g(x)$ is negative.

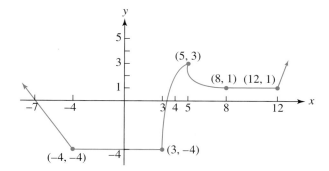

15. Use the following graph of $y = f(x)$ to determine
  (a) The intervals where $f(x)$ is increasing.
  (b) The intervals where $f(x)$ is decreasing.
  (c) The intervals where $f(x)$ is nonnegative.
  (d) The intervals where $f(x)$ is positive.
  (e) The intervals where $f(x)$ is negative.

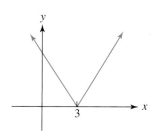

16. Use the following graph of $y = g(x)$ to determine
  (a) The intervals where $g(x)$ is increasing.
  (b) The intervals where $g(x)$ is decreasing.

(c) The intervals where $g(x)$ is nonnegative.
(d) The intervals where $g(x)$ is positive.
(e) The intervals where $g(x)$ is negative.

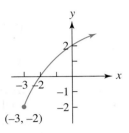

$(-3, -2)$

17. Use the following graph of $y = f(x)$ to determine
    (a) The intervals where $f(x)$ is increasing.
    (b) The intervals where $f(x)$ is decreasing.
    (c) The intervals where $f(x)$ is constant.
    (d) The intervals where $f(x)$ is positive.
    (e) The intervals where $f(x)$ is negative.

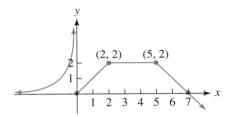

$(2, 2)$ $(5, 2)$

18. Use the following graph of $y = g(x)$ to determine
    (a) The intervals where $g(x)$ is increasing.
    (b) The intervals where $g(x)$ is decreasing.
    (c) The intervals where $g(x)$ is nonnegative.
    (d) The intervals where $g(x)$ is positive.
    (e) The intervals where $g(x)$ is negative.

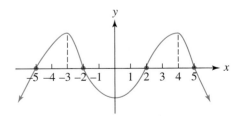

Answer Exercises 19–26 using the graphs in the accompanying figure.

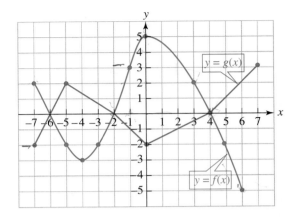

$y = g(x)$

$y = f(x)$

19. Find
    (a) $f(-1)$ (b) $f(0)$ (c) $f(4)$
    (d) For what value(s) of x is $f(x) = -2$?
    (e) For how many value(s) of x is $f(x) = 4$?
    (f) Identify the relative minimum value(s) of $f(x)$.

20. Find
    (a) $g(-1)$ (b) $g(0)$ (c) $g(4)$
    (d) For what value(s) of x is $g(x) = -2$?
    (e) For what value(s) of x is $g(x) = 4$?
    (f) Identify the relative maximum values(s) of $g(x)$.

21. (a) How many zeros does $f(x)$ have?
    (b) For how many values of x is $f(x) = 2$?
    (c) Find the solutions to $f(x) = 6$.
    (d) Identify the relative maximum values(s) of $f(x)$.

22. (a) How many zeros does $g(x)$ have?
    (b) For how many values of x is $g(x) = 2$?
    (c) Find the solutions to $g(x) = 6$.
    (d) Identify the relative minimum values(s) of $g(x)$.

23. (a) For what values of x is $f(x)$ positive?
    (b) For what values of x is $f(x) < 0$?
    (c) Find the solutions to $f(x) \geq 0$.

24. (a) For what values of x is $g(x)$ negative?
    (b) For what values of x is $g(x) \geq 0$?
    (c) Find the solutions to $g(x) \leq 0$.

25. (a) Find $f(3) - g(-7)$.
    (b) For what values of x is $f(x) > g(x)$?
    (c) For what values of x is $g(x) - f(x)$ nonnegative?

26. (a) Find $g(-6) - f(-1)$.
    (b) For what values of x is $f(x) = g(x)$?
    (c) For what values of x is $g(x) > f(x)$?

Answer Exercises 27–34 using the graphs in the following figure.

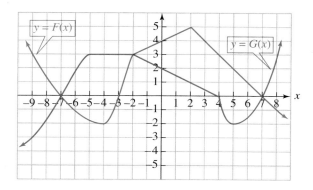

27. Find
   (a) $F(-2)$
   (b) $F(0)$
   (c) $F(2)$
   (d) For what value(s) of x is $F(x) = 3$?
   (e) For what value(s) of x is $F(x) = 0$?
   (f) Identify the relative minimum value(s) of $F(x)$.

28. Find
   (a) $G(2)$
   (b) $G(0)$
   (c) $G(5)$
   (d) For what value(s) of x is $G(x) = 3$?
   (e) For what value(s) of x is $G(x) = -3$?
   (f) Identify the relative maximum values(s) of $G(x)$.

29. **(a)** How many zeros does $F(x)$ have?
   (b) For how many values of x is $F(x) = 2$?
   (c) How many solutions are there to $F(x) = -6$?
   (d) Identify the relative maximum value(s) of $F(x)$.

30. **(a)** How many zeros does $G(x)$ have?
   (b) For how many values of x is $G(x) = 2$?
   (c) How many solutions are there to $G(x) = 6$?
   (d) Identify the relative minimum value(s) of $G(x)$.

31. **(a)** For what values of x is $F(x)$ positive?
   (b) For what values of x is $F(x) < 0$?
   (c) Find the solutions to $F(x) \geq 0$.

32. **(a)** For what values of x is $G(x)$ negative?
   (b) For what values of x is $G(x) \geq 0$?
   (c) Find the solutions to $G(x) \leq 0$.

33. **(a)** Find $F(-4) - G(-4)$.
   (b) For what values of x is $F(x) > G(x)$?
   (c) For what values of x is $G(x) - F(x)$ nonnegative?

34. **(a)** Find $G(7) - F(5)$.
   (b) For what values of x is $F(x) = G(x)$?
   (c) For what values of x is $G(x) > F(x)$?

Graphs are often used in many disciplines. They allow us to visualize relationships between various quantities. The next several exercises illustrate this idea.

35. Ecology   A Russian biologist, G. F. Gause, formulated the principle of *competitive exclusion,* which states that in any given biological community only one species can occupy any given ecological niche for an extended period of time. (An ecological niche refers to an organism's position in the structure of the ecosystem.) In an attempt to support his hypothesis, Gause conducted a number of laboratory experiments. In one such experiment, Gause used the laboratory cultures of two species of paramecium, *Paramecium aurelia* and *Paramecium caudatum.* When the two species were grown under identical conditions in separate containers, *P. aurelia* grew much faster than *P. caudatum.* When the two species were grown together, *P. aurelia* rapidly outmultiplied the *P. caudatum,* which soon died out. The accompanying figure illustrates the results of this experiment. The horizontal t-axis indicates the number of days that have elapsed since the culture was started. The vertical P-axis indicates the number of paramecia present. Thus a point on one of the graphs indicates how many of that type of paramecium are present after t days. The solid graphs indicate the behavior of the mixed cultures, whereas the dotted graphs indicate the behavior of each type of paramecium alone.
   (a) Describe what happens over time to each species when grown alone.
   (b) Describe what happens over time to each species when they are grown together.
   (c) Explain how the graph of the data supports or refutes Gause's principle of competitive exclusion.

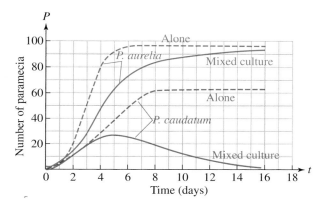

Results of Gause's experiment with two species of paramecium

36. Medicine   The accompanying graph illustrates the relationship between the dosage, d, of a particular drug and its effect on the heart rate, H, in a female monkey. What is the minimum dosage that will maintain a normal heartbeat of 72 beats per minute?

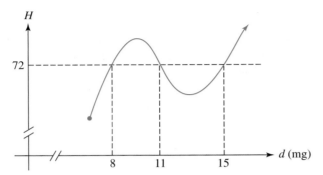

Relationship between drug dosage, d, and heart rate, h

37. Business   The following graph illustrates the relationship between the profit, P, in thousands of dollars a company earns and the number of line machines in operation, x.
(a) What number of machines maximizes the profit?
(b) What is the maximum profit?

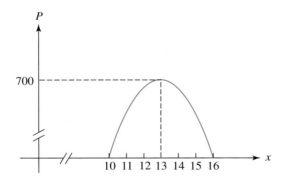

The relationship between profit, P, and the number of machines in operation, X

38. Psychology   The accompanying figure contains the graph of data collected in an experiment to measure how deviant verbal behavior can be influenced by using reinforcing techniques and the withholding of social attention. The solid graph indicates how the number of psychotic verbal responses (measured along the vertical axis) changes over a 36-day period (measured along the horizontal axis) as the psychotic responses are reinforced during the first 18 days and the neutral responses are reinforced during the last 18 days. The dotted graph gives the

same information for the number of neutral verbal responses.
(a) Describe what happens to the number of psychotic verbal responses as they are reinforced while neutral verbal responses are not.
(b) Describe what happens to the number of psychotic verbal responses as they are not reinforced while neutral verbal responses are reinforced.

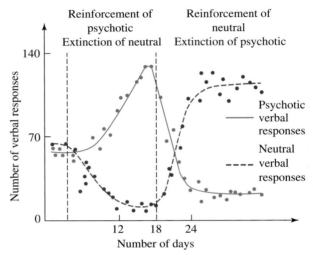

Incidence of psychotic and neutral verbal behavior as affected by social reinforcement techniques

In Exercises 39–44 use a graphing calculator to obtain the graph of the given function and use the graph to determine (a) through (d):
(a) the zeros of the function
(b) the intervals where the function is positive
(c) the intervals where the function is negative
(d) the relative maximum and minimum values of the function

Round off to the nearest hundredth where necessary.

39. $f(x) = x^2 - x - 6$ **40.** $f(x) = -x^2 - 2x + 8$

41. $y = \dfrac{x^2 - 1}{x^2 + 1}$ **42.** $y = \dfrac{-1}{x^2 - 2}$

43. $y = x^3 - 3x + 1$ **44.** $y = x^4 - 2x^3 - x + 1$

In Exercises 45–54, use a graphing calculator to solve the given equations and inequalities. Round your answers to two places.

45. $x^3 - 2x + 5 = 0$ **46.** $x^3 - 2x + 5 = 2$
47. $x^3 - 2x + 5 = 4$ **48.** $x^4 + x^2 + 2 = 1$
49. $x^4 + x^2 + 2 = 3$ **50.** $x^4 - 2x^2 - 3 > 0$
51. $x^4 - 2x^2 \le 3$ **52.** $x^3 - 5x^2 \ge x$
53. $x^3 - 5x^2 > 2 - x$ **54.** $x^5 - 4x^2 < 2x - 4$

3.7 Introduction to Graph Sketching: Symmetry

Figure 3.67

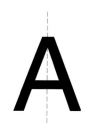

Figure 3.68

Figure 3.69

A graph that exhibits *y*-axis symmetry

Our main goal in this section and throughout the next chapter is to develop the idea that certain *geometric* characteristics of a graph can be determined by examining the *algebraic* characteristics of its equation and vice versa.

Let us now look at a geometric characteristic of a particular graph and see whether we can describe it algebraically. Consider the letter **A** shown in Figure 3.67.

We have drawn a vertical line through the letter and note that if we were to fold the letter along the dotted line, the two halves would coincide. Alternatively, we can say that if a mirror were placed on the vertical line through the **A**, the left half of the letter would be the reflection of the right half and vice versa, as illustrated in Figure 3.68. We say that the letter **A** is **symmetric** about the vertical line through its center, which is called the **axis of symmetry**.

Let's consider the graph in Figure 3.69. We observe that if we imagine a mirror on the *y*-axis, then that portion of the graph to the left of the *y*-axis is the reflection of that portion of the graph to the right of the *y*-axis (and vice versa). Such a graph is said to be **symmetric with respect to the *y*-axis**.

How can we describe this symmetry algebraically? If we look carefully at the graph in Figure 3.69 we can see that the points $(5, 3)$ and $(-5, 3)$ are both on the graph. Similarly, we also have the following pairs of points on the graph: $(4, 0)$ and $(-4, 0)$, $(3, -2)$ and $(-3, -2)$, and $(-2, 0)$ and $(2, 0)$.

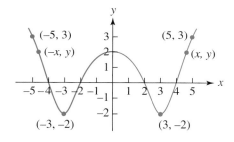

We can state this more concisely: In general, if a graph exhibits *y*-axis symmetry, then whenever a point (x, y) is on the graph, so is $(-x, y)$. This is an algebraic description of *y*-axis symmetry. In fact, since it is so concise, we generally use this algebraic description to define *y*-axis symmetry (which we will do in a moment).

Figure 3.70 illustrates two additional types of symmetry. The graph in Figure 3.70(a) is symmetric with respect to the *x*-axis. Notice that for every point *y*

Figure 3.70

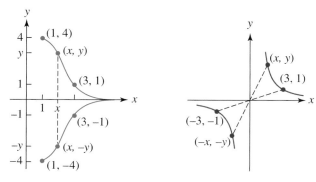

(a) A graph exhibiting
x-axis symmetry

(b) A graph exhibiting
origin symmetry

vertical units on one side of the x-axis, there corresponds another point y vertical units on the other side of the x-axis with the same x-coordinate. Again, we can state this more concisely: Whenever a point (x, y) is on this graph, so is the point $(x, -y)$.

Although we can easily describe symmetry with respect to a line such as the x- or y-axis, symmetry with respect to a point is a bit more difficult to describe. We cover symmetry with respect to the origin here because it does come up frequently in our discussion of functions.

The graph in Figure 3.70(b) is symmetric with respect to the origin. This type of symmetry is called origin symmetry because the origin is the midpoint of the line segment joining the two points (x, y) and $(-x, -y)$. Note that whenever a point (x, y) is on this graph, so is the point $(-x, -y)$.

We thus give the following definitions.

> A graph that exhibits origin symmetry is unchanged if it is rotated 180° about the origin.

Definition of Symmetry

A graph is **symmetric with respect to the y-axis** if whenever (x, y) is on the graph, $(-x, y)$ is also on the graph.

A graph is **symmetric with respect to the x-axis** if whenever (x, y) is on the graph, $(x, -y)$ is also on the graph.

A graph is **symmetric with respect to the origin** if whenever (x, y) is on the graph, $(-x, -y)$ is also on the graph.

If a graph of an equation is symmetric with respect to the y-axis, then by definition, if (x, y) is on the graph, $(-x, y)$ is also on the graph. Hence, both (x, y) and $(-x, y)$ satisfy the equation of the graph. This means that in the equation of the graph, both x and $-x$ will yield the same y, or in other words, *replacing x by $-x$ yields the same equation.* This is a convenient test for y-axis symmetry. In the same way, we can use these definitions to establish the other algebraic tests for symmetry.

Tests for Symmetry

1. The graph of an equation will exhibit **y-axis symmetry** if replacing x by $-x$ yields an equivalent equation.

2. The graph of an equation will exhibit **x-axis symmetry** if replacing y by $-y$ yields an equivalent equation.

3. The graph of an equation will exhibit **origin symmetry** if replacing x by $-x$ and y by $-y$ yields an equivalent equation.

Example 1   Determine what types of symmetry (if any) the graphs of the following equations will exhibit.

(a) $y = x^3$ (b) $y = x^2 - 4$ (c) $x^2 + y^2 = 4$

Solution   We can check each of these equations for symmetry by applying the symmetry tests just described. We do this by first replacing x by $-x$ in the original equation and seeing if we obtain an equivalent equation; then we do the same after replacing y by $-y$; finally, we replace x by $-x$ *and* y by $-y$ and see if we obtain an equivalent equation.

Keep in mind that a graph can exhibit more than one type of symmetry, so that we need to apply the various tests individually. However, if a graph exhibits any two of these symmetries, then it must necessarily exhibit the third symmetry as well. For example, if a graph exhibits both y-axis and x-axis symmetry, then it must necessarily exhibit origin symmetry, because reflecting a portion of a graph about the y-axis and then about the x-axis is equivalent to a reflection through the origin. See Figure 3.71.

Figure 3.71

x-axis and y-axis symmetry implies origin symmetry.

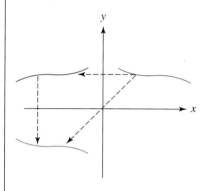

Note that we are answering the geometric questions about symmetry by examining each equation algebraically. Each of the equations in this example is of a type that we are going to discuss in much greater detail later in the text. Even though we have not yet discussed how to obtain the graphs of these equations, we include each of their graphs so that we may visually verify the symmetry that we have deduced algebraically.

(a) **Check for y-axis symmetry**

$y = x^3$ To check for y-axis symmetry, we replace x by $-x$.

$y = (-x)^3 = -x^3$ This is not equivalent to $y = x^3$. Therefore, the graph of $y = x^3$ does not have y-axis symmetry.

Check for x-axis symmetry

$y = x^3$ To check for x-axis symmetry, we replace y by $-y$.

$-y = x^3$ This is not equivalent to $y = x^3$. Therefore, the graph of $y = x^3$ does not have x-axis symmetry.

Check for origin symmetry

$y = x^3$ To check for origin symmetry, we replace x by $-x$ and y by $-y$.

$-y = (-x)^3$

$-y = -x^3$ This is equivalent to $y = x^3$. Therefore, the graph of $y = x^3$ does have origin symmetry.

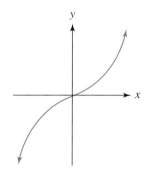

Figure 3.72

The graph of $y = x^3$ exhibits origin symmetry.

The graph of $y = x^3$ appears in Figure 3.72.

(b) **Check for y-axis symmetry**

$y = x^2 - 4$ We replace x by $-x$.

$y = (-x)^2 - 4 = x^2 - 4$ This is the original equation. Therefore, the graph of $y = x^2 - 4$ does have y-axis symmetry.

Check for x-axis symmetry

$y = x^2 - 4$ We replace y by $-y$.

$-y = x^2 - 4$ This is not equivalent to $y = x^2 - 4$. Therefore, the graph of $y = x^2 - 4$ does not have x-axis symmetry.

Check for origin symmetry

$y = x^2 - 4$ We replace x by $-x$ and y by $-y$ to check for origin symmetry.

$-y = (-x)^2 - 4$

$-y = x^2 - 4$ This is not equivalent to $y = x^2 - 4$. Therefore, the graph of $y = x^2 - 4$ does not have origin symmetry.

Actually, once we know that the graph of $y = x^2 - 4$ has y-axis symmetry but not x-axis symmetry, it cannot have origin symmetry, and so the test for origin symmetry was unnecessary. Keep in mind that it *is* possible for a graph to have neither x- nor y-axis symmetry but to be symmetric with respect to the origin. The graph of $y = x^2 - 4$ appears in Figure 3.73.

(c) It is fairly easy to recognize that if we replace x by $-x$ or y by $-y$ in the equation $x^2 + y^2 = 4$, we will obtain an equivalent equation. Thus the equation $x^2 + y^2 = 4$ exhibits y-axis, x-axis, and, necessarily, origin symmetry. As we saw in Section 3.1 the graph of $x^2 + y^2 = 4$ is a circle with center $(0, 0)$ and radius 2; it appears in Figure 3.74.

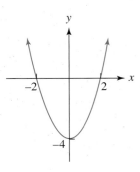

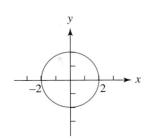

Figure 3.73
The graph of $y = x^2 - 4$ exhibits y-axis symmetry.

Figure 3.74
The graph of $x^2 + y^2 = 4$ exhibits all three symmetries.

Example 2   Find the point symmetric to the point $(-2, 3)$ with respect to the y-axis, x-axis, and origin.

Solution   The point symmetric to the point $(-2, 3)$ with respect to the y-axis has the same "height" as $(-2, 3)$; hence, its y-coordinate is 3. It is the same distance from the x-axis, only on the opposite side of the x-axis; hence, its x-coordinate is 2. The point is $(2, 3)$.

The point symmetric to the point $(-2, 3)$ with respect to the x-axis has the same x-coordinate as $(-2, 3)$, which is -2. It is the same distance from the y-axis, only on the opposite side of the y-axis; hence, its y-coordinate is -3. The point is $(-2, -3)$.

The point symmetric to the point $(-2, 3)$ with respect to the origin is the point with opposite x- and y-coordinates, which is $(2, -3)$. See Figure 3.75.

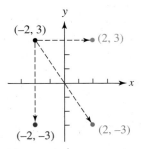

Figure 3.75

Up to this point, our discussion of symmetry has been about graphs of equations in general. If we restrict our attention to the graphs of functions, we can reformulate some of our results about symmetry.

For example, a graph has y-axis symmetry if the same y-value corresponds to both x and $-x$. Using function notation, we can restate this as: If $f(x) = y$, then $f(-x) = y$, or, more concisely, $f(x) = f(-x)$. A function exhibits origin symmetry if opposite y-values correspond to opposite x-values. Again using function notation: If $f(x) = y$, then $f(-x) = -y$. Noting that $f(-x) = -y = -f(x)$, we have $f(-x) = -f(x)$. Using function notation, we restate the symmetry definitions given earlier, as in the box at the top of page 184.

Different Perspectives: Symmetry

Graphical Description

y-axis symmetry:
A graph exhibits y-axis symmetry if the graph is unchanged when it is reflected about the y-axis.

Algebraic Description

y-axis symmetry:
A graph will exhibit y-axis symmetry if the equation of the graph is unchanged when x is replaced by $-x$.

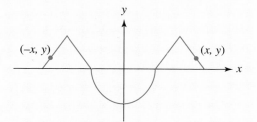

This graph exhibits y-axis symmetry.

x-axis symmetry:
A graph exhibits x-axis symmetry if the graph is unchanged when it is reflected about the x-axis.

x-axis symmetry:
A graph will exhibit x-axis symmetry if the equation of the graph is unchanged when y is replaced by $-y$.

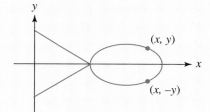

This graph exhibits x-axis symmetry.

Origin symmetry:
A graph exhibits origin symmetry if the graph is unchanged when it is rotated 180° about the origin.

Origin symmetry:
A graph will exhibit origin symmetry if the equation of the graph is unchanged when x is replaced by $-x$ and y is replaced by $-y$.

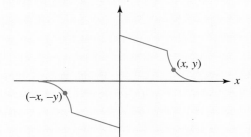

This graph exhibits origin symmetry.

Why don't we talk about *x*-axis
symmetry for the graph of a function?

> The graph of a function $y = f(x)$ has y-axis symmetry if $f(-x) = f(x)$.
> The graph of a function $y = f(x)$ has origin symmetry if $f(-x) = -f(x)$.

Thus to test a function for y-axis or origin symmetry, we examine $f(-x)$. If $f(-x) = f(x)$, then the graph of $y = f(x)$ has y-axis symmetry; if $f(-x) = -f(x)$, then the graph of $y = f(x)$ has origin symmetry.

Example 3   Discuss the symmetry of the following functions:

(a) $y = f(x) = 2x^3 - 3x$ (b) $y = g(x) = -x^4 + 5x^2 - 6$
(c) $y = h(x) = 3x^2 - 6x$

Solution   To determine whether a function has y-axis or origin symmetry, we compare $f(-x)$ with $f(x)$. If they are the same, then the graph of $y = f(x)$ has y-axis symmetry, whereas if they are opposites, then the graph of $y = f(x)$ has origin symmetry.

(a) For $y = f(x) = 2x^3 - 3x$, we have

$$f(-x) = 2(-x)^3 - 3(-x) = -2x^3 + 3x = -(2x^3 - 3x) = -f(x)$$

Therefore, the graph of $f(x)$ has origin symmetry.

(b) For $y = g(x) = -x^4 + 5x^2 - 6$, we have

$$g(-x) = -(-x)^4 + 5(-x)^2 - 6 = -x^4 + 5x^2 - 6 = g(x)$$

Therefore, the graph of $g(x)$ has y-axis symmetry.

(c) For $y = h(x) = 3x^2 - 6x$, we have

$$h(-x) = 3(-x)^2 - 6(-x) = 3x^2 + 6x$$

which is equal to neither $h(x)$ nor $-h(x)$; therefore, the graph of $h(x)$ has neither y-axis symmetry nor origin symmetry.   ∎

Why do you think an odd function is
called odd? Why do you think an even
function is called even?

A function that exhibits y-axis symmetry is often called an *even function*, whereas one that exhibits origin symmetry is called an *odd* function. By looking back at the results of the last example, can you see why this terminology is used? (See also Exercise 42.)

In the next chapter we will continue our discussion of the relationship between the algebraic characteristics of an equation and the geometric characteristics of its graph.

3.7 Exercises

In Exercises 1–8, determine whether the given graph has y-axis symmetry, x-axis symmetry, origin symmetry, or none of these symmetries.

1.

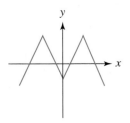

2.

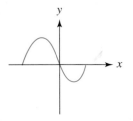

3.

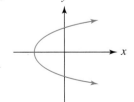

4.

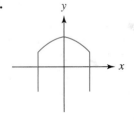

5.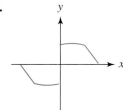

6.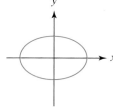

7.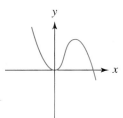

8.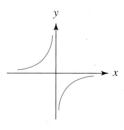

In Exercises 9–12, you are given a portion of a graph. Complete the given graph if it exhibits **(a)** y-axis symmetry and **(b)** origin symmetry.

9.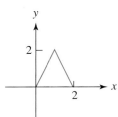

10.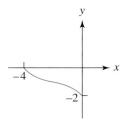

11.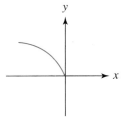

12.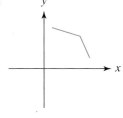

In Exercises 13–16, the endpoints of line segment $\overline{PQ}$ are given. Sketch the reflection of $\overline{PQ}$ about the **(a)** y-axis, **(b)** x-axis, and **(c)** origin.

13. $P(2, 1)$ and $Q(4, 6)$

14. $P(-2, -3)$ and $Q(0, 2)$

15. $P(3, 0)$ and $Q(5, -1)$

16. $P(-4, 0)$ and $Q(0, -4)$

In Exercises 17–34, determine whether the graph of the given function has y-axis symmetry, origin symmetry, or neither.

17. $y = f(x) = 4x^2 - 1$

18. $y = g(x) = -2x^3 + 16$

19. $y = h(x) = x^5 - 2x$

20. $y = f(x) = 6 + x^2 - x^4$

21. $y = f(x) = 2x^2 - 3x + 1$

22. $y = h(x) = 7$

23. $y = F(x) = \dfrac{5}{x}$

24. $y = G(x) = \dfrac{2}{x + 1}$

25. $y = F(x) = -\dfrac{x^4}{x^2 - 1}$

26. $y = G(x) = \dfrac{x^2 + 9}{x}$

27. $y = F(x) = \dfrac{x^3}{x^2 - 1}$

28. $y = G(x) = -\dfrac{x^2 + 9}{x^2}$

29. $y = H(x) = |x|$

30. $y = H(x) = |x - 3|$

31. $y = f(x) = \dfrac{x}{4} - \dfrac{x}{3}$

32. $y = G(x) = \dfrac{5}{x} + \dfrac{2}{x^2}$

33. $y = h(x) = \dfrac{x - 3}{4 - x}$

34. $y = F(x) = \dfrac{x^4 + 1}{2x^2 - 8}$

In Exercises 35–40, use a graphing calculator to determine whether the graph exhibits y-axis symmetry, origin symmetry, or neither.

35. $y = x^3 + x$

36. $y = \dfrac{4}{x + 2}$

37. $y = x^4 - x^2 - 4$

38. $y = 3x - 6$

39. $y = \dfrac{x - 1}{x^2 + 1}$

40. $y = \dfrac{x}{x^2 - 5}$

Questions for Thought

41. We have described the vertical line test for determining whether a graph defines y as a function of x. Suppose we want the graph to also define x as a function of y. What special characteristic must the graph have?

42. As mentioned in this section, a function for which $f(-x) = f(x)$ is called an *even function*, whereas a function for which $f(-x) = -f(x)$ is called an *odd function*. Thus an even function exhibits y-axis symmetry, and an odd function exhibits origin symmetry.
   (a) Show that if $f(x)$ is a polynomial made up exclusively of odd powers of x, then it is an odd function.
   (b) Show that if $f(x)$ is a polynomial made up exclusively of even powers of x, then it is an even function.
   (c) Show that if $f(x)$ is a polynomial made up of both even and odd powers of x, then it is neither an even nor an odd function.
   (d) By examining $f(x) = \sqrt[3]{x}$, show that a function can be odd without having odd powers.

43. **(a)** Suppose $f(x)$ and $g(x)$ are even. What can be said about $f(x) + g(x)$?
   (b) Suppose $f(x)$ and $g(x)$ are odd. What can be said about $f(x) + g(x)$?
   (c) Suppose $f(x)$ is even and $g(x)$ is odd. What can be said about $f(x) + g(x)$?

After completing this chapter you should:

1. Be able to graph straight lines in a rectangular coordinate system. (Section 3.1)

2. Be able to compute the slope of a line. (Section 3.2)
   The slope is the ratio of the change in y to the change in x as we move from one point to another point along the graph. For a line, this ratio is constant.
   For example:
   To find the slope of the line passing through the points $(-3, 2)$ and $(1, -4)$, we use the formula for the slope of a nonvertical line, which is $m = \dfrac{y_2 - y_1}{x_2 - x_1}$, and get

   $$m = \frac{-4 - 2}{1 - (-3)} = \frac{-6}{4} = \frac{-3}{2}$$

3. Understand the significance of the slope of a line. (Section 3.2)

   Lines with positive slope rise as we move from left to right, and lines with negative slope fall as we move from left to right. A line with zero slope is horizontal, whereas a line whose slope is undefined is vertical.

4. Know that two nonvertical lines are parallel if and only if their slopes are equal, and are perpendicular if and only if their slopes are negative reciprocals. (Section 3.2)

5. Be able to write an equation of a line satisfying certain conditions. (Section 3.3)
   A line is determined once we know one point on the line and the slope.
   For example:
   Write an equation of a line passing through the point $(4, -5)$ that is perpendicular to the line whose equation is $3x + 4y = 7$.
   Solution:
   The slope of the line whose equation is $3x + 4y = 7$ can most easily be found by getting this equation into slope–intercept form:

   $$3x + 4y = 7 \quad \Rightarrow \quad y = -\frac{3}{4}x + \frac{7}{4} \qquad \text{Thus the slope of the given line is } -\frac{3}{4}.$$

   The slope of a line perpendicular to the given line will therefore be $\dfrac{4}{3}$. We can write the equation of the perpendicular line by using the point–slope form for the equation of a line, which gives $y - (-5) = \dfrac{4}{3}(x - 4)$; the final answer is usually written as

   $$y = \frac{4}{3}x - \frac{31}{3}$$

6. Recognize and graph equations of circles. (Section 3.1)
   The standard form of the equation of a circle with its center (h, k) and its radius r is $(x - h)^2 + (y - k)^2 = r^2$. Knowing the center and the radius of the circle makes it simple to graph the circle.
   For example:
   Find the center and radius of the circle whose equation is $(x + 2)^2 + (y - 5)^2 = 15$.
   Solution:
   Comparing the given equation to the standard form $(x - h)^2 + (y - k)^2 = r^2$, we can see that $h = -2$, $k = 5$, and $r = \sqrt{15}$. Therefore, the center is $(-2, 5)$ and the radius is $\sqrt{15}$.

7. Understand the definition of a function and be able to find its domain. (Section 3.4)

A function is a relationship between x and y for which to each real value of x in the domain, there corresponds exactly one real value of y in the range.

For example:

Find the domain of the function $y = \dfrac{\sqrt{2x + 5}}{x - 3}$.

Solution:

The domain of this function consists of all real numbers x for which the expression

$\dfrac{\sqrt{2x + 5}}{x - 3}$ is defined and a real number. Consequently, we require that

$$2x + 5 \geq 0 \quad \Rightarrow \quad x \geq -\frac{5}{2}$$

and

$$x - 3 \neq 0 \quad \Rightarrow \quad x \neq 3$$

Therefore, the domain is $\left\{ x \mid x \geq -\dfrac{5}{2}, x \neq 3 \right\}$, or in interval notation

$$\left[-\frac{5}{2}, 3 \right) \cup (3, \infty).$$

8. Use function notation to compute functional values. (Section 3.5)

When using function notation, it is useful to think of x as the input and $f(x)$ as the output.

For example:

Given $f(x) = -x^2 - 3x + 2$, find **(a)** $f(-6)$ and **(b)** $f(x + 5)$.

Solution:

(a) $f(x) = -x^2 - 3x + 2$ To find $f(-6)$, we replace x by -6.

$f(-6) = -(-6)^2 - 3(-6) + 2$

$f(-6) = -16$

(b) $f(x) = -x^2 - 3x + 2$ To find $f(x + 5)$ we replace x by $x + 5$.

$f(x + 5) = -(x + 5)^2 - 3(x + 5) + 2$

$\quad\quad\quad = -(x^2 + 10x + 25) - 3x - 15 + 2$

$\quad\quad\quad = -x^2 - 10x - 25 - 3x - 15 + 2$

$f(x + 5) = -x^2 - 13x - 38$

9. Be able to read information about a function from its graph. (Section 3.6)

For example:

Consider the graph of $f(x)$ in Figure 3.76. From the graph of $f(x)$, we can see, among other things, that $f(3) = -2$; $f(0) = 3$; $f(x)$ has zeros at -3, -1, and 2; $f(x)$ is positive on the interval $(-\infty, -3) \cup (-1, 2)$; and $f(x)$ is increasing on the interval $[-2, 0]$. There are two turning points, occurring at $(0, 3)$ and $(-2, -2)$. The relative maximum is 3, and the relative minimum is -2.

10. Solve an equation and inequality using a graph or a graphing calculator. (Section 3.6)

For example:

There are several ways we can solve the equation $x^4 - 4x^2 - 2x = -4$ using a graphing calculator: We will do it here by graphing the equations

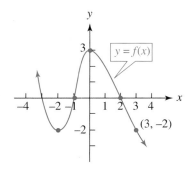

Figure 3.76

Y1=X^4-4X^2-2X and Y2=-4, and see where they intersect. Figure 3.77(a) shows the graphs. Note that the graphs intersect twice. Figures 3.77(b) and 3.77(c) show the coordinates of the intersections. The *x*-coordinates of their intersections are the solutions to the equation $x^4 - 4x^2 - 2x = -4$. Hence the solutions are (rounded to two places) $x = 0.84$ and $x = 2$.

Figure 3.77

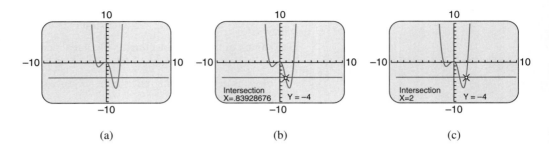

(a)   (b)   (c)

11. Recognize when a graph exhibits *y*-axis, *x*-axis, or origin symmetry. (Section 3.7) See Figure 3.78.

Figure 3.78

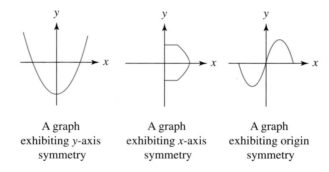

A graph   A graph   A graph
exhibiting *y*-axis   exhibiting *x*-axis   exhibiting origin
symmetry   symmetry   symmetry

12. Be able to use the algebraic formula of a function to determine whether its graph will exhibit *y*-axis or origin symmetry. (Section 3.7)
A function for which $f(-x) = f(x)$ has a graph that exhibits *y*-axis symmetry.
A function for which $f(-x) = -f(x)$ has a graph that exhibits origin symmetry.
For example:
For $f(x) = x^3 - 4x$, we have

$$f(-x) = (-x)^3 - 4(-x)$$
$$= -x^3 + 4x = -(x^3 - 4x) = -f(x)$$

and so the graph of $f(x) = x^3 - 4x$ exhibits origin symmetry. The graph appears in Figure 3.79.

Figure 3.79

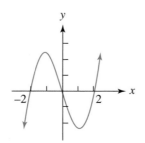

Chapter 3   Review Exercises

In Exercises 1–18, sketch the graph of the given equation on a rectangular coordinate system.

1. $4x - 5y = 20$

2. $y = 2x - 8$

3. $3x + 7y + 14 = 0$

4. $y = -5x$

5. $y = -5$

6. $\dfrac{x}{2} - \dfrac{y}{3} = 4$

7. $3x = 4$

8. $3x = 4y$

9. $x^2 + y^2 = 9$

10. $x + y = 9$

11. $(x - 3)^2 + (y + 4)^2 = 9$

12. $x^2 + y^2 - 6y = 1$

13. $5x + 5y = 20$

14. $5x^2 + 5y^2 = 20$

15. $x^2 - 3x + y^2 + 2y = 2$

16. $2x^2 + 2y^2 - 8x + 4y = 0$

17. $x = -2$

18. $y = \dfrac{3}{2}$

In Exercises 19–22, use the given graph to determine (a) whether it is the graph of a function, (b) its domain, and (c) its range.

19.

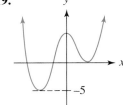

20.

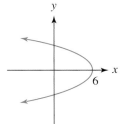

21.

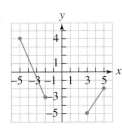

22.

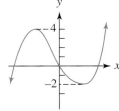

23. Given the equation $y = x^4$, what is y when $x = 2$? When $x = -2$? Does $y = x^4$ define y as a function of x? Explain.

24. Given the equation $y^4 = x$, what is y when $x = 16$? Does $y^4 = x$ define y as a function of x? Explain.

25. Write an equation of the line passing through the points $(-3, 4)$ and $(2, -5)$.

26. Write an equation of the line that passes through the point $(2, 0)$ and is perpendicular to the line whose equation is $4x - 7y = 2$.

27. Find the value(s) of t so that the line passing through the points $(1, -4)$ and $(t, 6)$ is parallel to the line passing through the points $(-3, -1)$ and $(5, t)$.

28. Prove that the points $P(-1, 3)$, $Q(3, -2)$, and $R(13, 6)$ are the vertices of a right triangle.

29. Write an equation of the circle with center $(-4, 1)$ and radius 7.

30. Find the center and radius of the circle with equation $x^2 + (y + 2)^2 = 20$.

31. Find the center and radius of the circle with equation $x^2 + 8x + y^2 - 2y = 8$.

32. Sketch the graph of $x^2 - 4x + y^2 - 6y + 9 = 0$.

33. Write an equation of the circle with center $(-5, 4)$ and radius $\dfrac{1}{2}$.

34. Write an equation of the line tangent to the circle $x^2 + 10x + y^2 = 33$ at the point $(2, -3)$.

35. Write an equation of the circle with a diameter having endpoints $(1, -4)$ and $(0, 5)$.

36. Write an equation of the circle with a radius of 7 and with a center at the point where the line $3y - \dfrac{1}{2}x = 4$ crosses the y-axis.

In Exercises 37–44, find the domain of the given function.

37. $f(x) = \sqrt{5 - 3x}$

38. $g(x) = \sqrt[3]{5 - 3x}$

39. $h(x) = \dfrac{x - 1}{x^2 - 3x - 4}$

40. $F(x) = \dfrac{5}{x^2 + 9}$

41. $G(x) = \sqrt{6x - x^2}$

42. $f(x) = \dfrac{\sqrt{x + 6}}{x - 2}$

43. $F(x) = \dfrac{-2}{\sqrt{x + 4}}$

44. $H(x) = 4x^3 - 5x^2 - x + 7$

In Exercises 45–56, use the given function to find the requested values (if possible).

45. $f(x) = -x^2 + 4x - 1$

$f(-3), \ f(2x), \ 2f(x)$

46. $g(x) = \sqrt{7 - 2x}$

$g(-9), \quad g(0), \quad g\left(\dfrac{1}{2}\right)$

47. $h(x) = \sqrt{5}$

$h(2), \quad h(10), \quad h(-3)$

48. $f(x) = 9 - 4x^2$

$f(x + 3), \quad f(x) + 3$

49. $g(t) = \dfrac{2t - 1}{t + 5}; \quad g\left(\dfrac{1}{2}\right), \quad g(-5), \quad g(t + 1)$

50. $H(r) = \dfrac{r}{4r^2 + 1}; \quad H\left(-\dfrac{1}{2}\right), \quad H\left(\dfrac{1}{r}\right), \quad H(r + 1)$

51. $f(x) = \begin{cases} -2x + 7 & \text{if } x \le -4 \\ x^2 - 1 & \text{if } -1 \le x \le 8 \end{cases}$

$f(0), \quad f(-1), \quad f(-5)$

52. $f(x) = \begin{cases} -x & \text{if } x \le -2 \\ x^2 & \text{if } -2 < x < 4 \\ \sqrt{x} & \text{if } 4 \le x \end{cases}$

$f(3), \quad f(4), \quad f(-2)$

53. $f(x) = 5x^2 - 6x + 3; \quad \dfrac{f(x - 4) - f(x)}{4}$

54. $g(u) = 9 - 7u; \quad \dfrac{g(u + h) - g(u)}{h}$

55. $f(t) = \dfrac{-3}{t}; \quad \dfrac{f(t + h) - f(t)}{h}$

56. $h(x) = \dfrac{x + 1}{x - 1}; \quad \dfrac{h(x - 3) - h(x)}{3}$

57. Use the graph of $y = f(x)$ in the figure to answer the following.

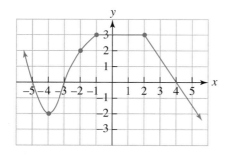

(a) Find $f(-4), f(-2), f(0)$, and $f(1)$.
(b) What are the zeros of $f(x)$?

(c) On what intervals is $f(x) > 0$?
(d) On what intervals is $f(x)$ negative?
(e) On what intervals is $f(x)$ increasing?
(f) On what intervals is $f(x)$ decreasing?
(g) On what intervals is $f(x)$ constant?
(h) Identify the relative minimum.

In Exercises 58–61, examine the given graph to determine whether it exhibits *y*-axis symmetry, *x*-axis symmetry, origin symmetry, or none of these.

58.   **59.**

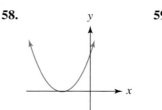   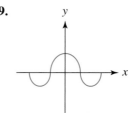

60.   **61.**

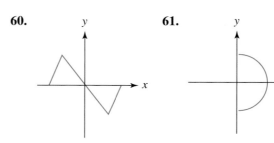

In Exercises 62–67, determine whether the graph of the given function exhibits *y*-axis symmetry, origin symmetry, or neither.

62. $f(x) = \dfrac{-3x}{x^2 + 1}$ **63.** $f(x) = 2x^2 - 5x$

64. $g(x) = \dfrac{x + 4}{x - 4}$

65. $h(x) = 7x - x^3$

66. $F(x) = \sqrt{9 - x^2}$

67. $G(x) = \dfrac{x^2 + 4}{x}$

68. An oil company charges $1.32 per gallon for each of the first 150 gallons of home heating oil, and $1.21 for each gallon above 150. Express the cost, C, of an oil delivery as a function of g, the number of gallons of oil purchased.

69. A student notes that she received a grade of 76 on her first math test after studying 3 hours for the test, and a grade of 85 on her second math test after studying 5 hours. Assuming a linear relationship between her grades on her math tests and the number of hours she studies, what would her grade be if she studied 6 hours for her third math test?

70. A computer consultant charges \$75 for an initial consultation lasting up to 1 hour, and \$95 per hour for each additional hour devoted to a project. Express the consultant's charge, C, as a function of h, the total number of hours of the consultant's time required.

71. At a certain time, the sides of a square are 6 cm long and increasing at the rate of 2.5 cm/min. Express the area of the square t minutes later as a function of t.

In Exercises 72–75, use a graphing calculator to solve the following equations and inequalities. Round your answers to two decimal places.

72. $x^4 - x^2 - 2x = 0$

73. $x^3 - 6x - 5 = -2$

74. $x^5 - 6x^2 + 3 > 0$

75. $x^4 - 2x^2 \le 3x + 1$

Chapter 3   *Practice Test*

1. Given $f(x) = -2x^2 - 5x + 3$, find
   (a) $f(-4)$ **(b)** $f(x + 4)$ **(c)** $f(4x)$

   (d) $f(x^2)$ **(e)** $\dfrac{f(x + h) - f(x)}{h}$

2. Given $g(t) = \dfrac{2t + 1}{t - 2}$, find each of the following and simplify.

   (a) $g\left(\dfrac{2}{3}\right)$ **(b)** $g\left(\dfrac{1}{t}\right)$

3. Given $F(x) = \dfrac{x}{x - 3}$, find the difference quotient

   $\dfrac{F(x + 5) - F(x)}{5}$ and simplify.

4. Sketch the graph of $5x - 3y = 20$. Label the intercepts.

5. Write an equation of the line that crosses the x-axis at 4 and the y-axis at -3.

6. Write an equation of the line passing through the point $(-2, -5)$ that is perpendicular to the line whose equation is $3x - 7y + 10 = 0$.

7. Sketch the graph of the equation $x^2 - 2x + y^2 = 0$. Is it the graph of a function? Explain.

8. Write an equation of the circle with diameter whose endpoints are $(-3, 4)$ and $(1, -3)$.

9. Use the graph of $y = f(x)$ in the accompanying figure to answer the following.

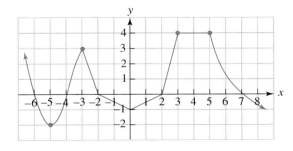

   (a) Find $f(-5), f(-3), f(0)$, and $f(4)$.
   (b) What are the zeros of $f(x)$?
   (c) On what intervals is $f(x)$ positive?
   (d) On what intervals is $f(x) < 0$?
   (e) On what intervals is $f(x)$ increasing?
   (f) On what intervals is $f(x)$ decreasing?
   (g) On what intervals is $f(x)$ constant?
   (h) Identify the relative maximum.

10. The accompanying figure contains a portion of a graph. Complete the remainder of the graph if the graph exhibits
   (a) y-axis symmetry
   (b) origin symmetry
   (c) x-axis symmetry

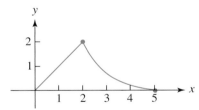

11. Determine whether the graphs of the following functions exhibit y-axis symmetry, origin symmetry, or neither.

   (a) $f(x) = \dfrac{x}{x + 5}$

   (b) $F(x) = 3x^2 - x^4$

   (c) $h(x) = x - \dfrac{1}{x}$

12. The width of a rectangle is w and the length is 3 less than 4 times the width. A fence costing \$3.50 per foot is to be erected around the rectangle. Express the cost, C, of the fence as a function of w.

13. Use a graphing calculator to solve the following. Round your answers to two places.
   (a) $x^3 - x^2 - 3x - 1 = 0$
   (b) $x^4 - x^2 - 1 > 3$

4

Functions and Graphs: Part II

In this chapter we expand upon many of the ideas regarding functions
that were introduced in the previous chapter. In the first two sections
we continue our discussion of graph sketching by describing some
basic graphing principles. In Section 4.3 we begin looking at how we
can apply the idea of functions to model real-life situations. In
Section 4.4 we discuss quadratic functions. Section 4.5 introduces the
algebra of functions—that is, how we can produce new and often more
complex functions from simpler ones. Section 4.6 concludes the
chapter with a discussion of equations that not only define y as a
function of x but also define x as a function of y. Such a function is said
to have an inverse.

Many of the ideas we introduce and develop in this chapter will be used
repeatedly throughout the text as we continue to encounter a wide variety
of special functions.

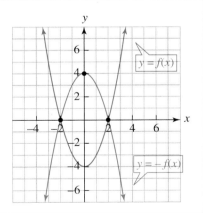

4.1   Basic Graphing Principles

As mentioned previously, one of our main goals is to become thoroughly familiar with a large number of basic graphs. We do this in two ways. First, we learn to recognize the graphs of certain basic functions. As we encounter these basic graphs, we record them in a catalog. (This catalog is similar to the inside front covers of this textbook.) This catalog will contain graphs that should be recognized immediately. Second, we establish a set of basic graphing principles that will allow us to easily graph variations of these basic graphs. We begin this process here and it continues throughout the text.

To demonstrate the various graphing techniques, it is extremely helpful to illustrate how these ideas work with one particular function first. Consequently, we digress for just a moment to determine the graph of the function $y = f(x) = x^2$, which we use as the springboard for our discussion.

Example 1 | Sketch the graph of $y = f(x) = x^2$.

Solution | In general, we try to obtain a graph by analyzing its equation rather than by plotting points. However, since we have to start somewhere, we compute some values to get an idea of what *this* graph looks like. See Table 4.1.

Table 4.1

x	-3	-2	-1	$-\frac{1}{2}$	0	$\frac{1}{2}$	1	2	3
$y = f(x) = x^2$	$(-3)^2$	$(-2)^2$	$(-1)^2$	$(-\frac{1}{2})^2$	$(0)^2$	$(\frac{1}{2})^2$	$(1)^2$	$(2)^2$	$(3)^2$
y	9	4	1	$\frac{1}{4}$	0	$\frac{1}{4}$	1	4	9

Using these points, we draw a smooth curve through them, obtaining the graph in Figure 4.1.* Based on this graph we can see that the domain is the set of all real numbers and the range is the set of all nonnegative real numbers. ∎

We will refer to the graph of $y = f(x) = x^2$ as the **basic parabola**, and it is the next entry in our catalog of basic graphs. The lowest (or highest) point of a parabola is called the **vertex** of the parabola, and the vertical line passing through the vertex is called the **axis of symmetry**. (In Sections 4.4 and 8.2 we engage in a much more detailed discussion of parabolas.)

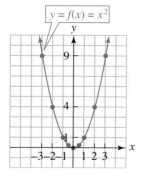

Figure 4.1

*The question of the exact shape of the graph—that is, how to connect the points we have found— is a very important one. To answer this question precisely, one generally needs to have some of the techniques of calculus available.

Calculator Exploration

Using a graphing calculator to graph the following in the standard $[-10, 10]$ by $[-10, 10]$ window.

1. Graph the following on the same set of coordinate axes:

$$y = 0.1x^2 \quad y = 0.5x^2 \quad y = x^2 \quad y = 2x^2 \quad y = 3x^2$$

Describe how changing the coefficient a in $y = ax^2$ affects the graph of $y = ax^2$.

2. Graph the following on the same set of coordinate axes:

$$y = 0.5(x^3 - 3x + 1) \quad y = x^3 - 3x + 1 \quad y = 2(x^3 - 3x + 1)$$

Describe how changing the coefficient a in $y = a(x^3 - 3x + 1)$ affects the graph of $y = x^3 - 3x + 1$.

Suppose we now consider the graph of $y = 2x^2$. Rather than compute a table of values, we recognize that for each x-value, except for $x = 0$, the y-value on the graph of $y = 2x^2$ will be twice as large as the y-value on the graph of $y = x^2$. This gives us a narrower parabola that is somewhat "sharper" at the vertex. The graph appears in Figure 4.2.

Figure 4.2

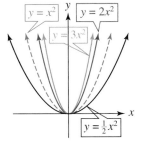

Similarly, the graph of $y = 3x^2$ will be even narrower; its graph also appears in Figure 4.2.

For a given value of x, the y-value on the graph of $y = \frac{1}{2}x^2$ is one-half the y-value on the graph of $y = x^2$. Thus the graph of $y = \frac{1}{2}x^2$ will be wider than the graph of $y = x^2$ and a bit flatter at the vertex. See Figure 4.2. If $a > 0$ (and $a \neq 1$), we say that the graph of $y = ax^2$ is obtained by *stretching* the graph of $y = x^2$.

More generally, the graph of $2f(x)$ is obtained by deflecting the graph of $y = f(x)$ farther away from the x-axis, whereas the graph of $\frac{1}{2}f(x)$ is obtained by pulling the graph of $y = f(x)$ closer to the x-axis. In many (but not all) cases, this can be described by saying that the graph of $af(x)$ is wider or narrower than the graph of $f(x)$, depending on the size of a.

The Stretching Principle
(for $a > 0, a \neq 1$)

The graph of $y = af(x)$ can be obtained by stretching the graph of $y = f(x)$. In other words, the graph of $y = af(x)$ will have the same basic shape as the graph of $y = f(x)$. If $a > 1$, the graph of $y = af(x)$ is obtained by deflecting the graph of $f(x)$ away from the x-axis. If $0 < a < 1$, the graph of $y = af(x)$ is obtained by pulling the graph of $f(x)$ toward the x-axis.

Figure 4.3 illustrates the stretching principle.

Figure 4.3

How does the value of $3f(4)$ compare with the value of $f(4)$?

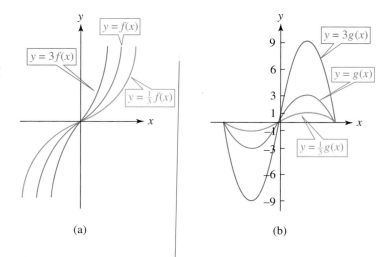

(a)   (b)

Calculator Exploration

Use a graphing calculator to graph the following in the standard $[-10, 10]$ by $[-10, 10]$ window.

1. Graph the following on the same set of coordinate axes:

   $$y = x^2, \quad y = x^2 + 2, \quad y = x^2 - 3$$

   What can you conclude about the effect on the graph of $y = x^2$ of adding a constant to x^2?

2. Graph the following on the same set of coordinate axes:

   $$y = x^5, \quad y = x^5 + 3, \quad y = x^5 - 2$$

   What can you conclude about the effect on the graph of $y = x^5$ of adding a constant to x^5?

Let's use our knowledge of the graph of $y = x^2$ to graph other functions.

Example 2 | Use the graph of $y = f(x) = x^2$ to sketch the following graphs.
(a) $y = g(x) = x^2 + 3$ (b) $y = h(x) = x^2 - 4$

Solution | (a) Although we could complete a table similar to Table 4.1 for $y = x^2 + 3$, there is a much more efficient way to determine its graph. Let us compare the y-values obtained from the equations $y = x^2$ and $y = x^2 + 3$. For each value of x, the associated y-value obtained from $y = x^2 + 3$ is 3 more than the y-value obtained from $y = x^2$. In other words, for a particular x-value, the point on the graph of $y = x^2 + 3$ will be 3 units above the point on the graph of $y = x^2$. Increasing the y-value on a graph by 3 moves that point up 3 units. To obtain the graph of $y = x^2 + 3$, we shift the graph of $y = x^2$ up 3 units. The graph of $y = g(x) = x^2 + 3$ appears in Figure 4.4. The graph of $y = f(x) = x^2$ is also included as indicated by the dashed graph.

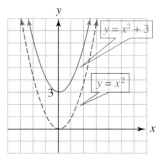

Figure 4.4

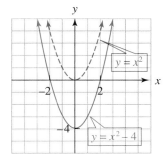

Figure 4.5

(b) A similar analysis tells us that for each x-value, the y-value in $y = x^2 - 4$ is 4 less than the y-value in $y = x^2$. Decreasing the y-value on a graph by 4 units moves that point 4 units down. Therefore, the graph of $y = h(x) = x^2 - 4$ can be obtained by shifting the graph of $y = f(x) = x^2$ down 4 units. The graph of $y = h(x) = x^2 - 4$ appears in Figure 4.5.

The graph clearly indicates that there are x-intercepts, which we can find by setting $f(x) = 0$. (We could have said that we are finding the x-intercepts by setting $y = 0$. We are purposely using y and $f(x)$ interchangeably to remind ourselves that the two represent the same quantity.)

$$y = f(x) = x^2 - 4 \qquad \text{Set } f(x) = 0.$$
$$0 = x^2 - 4 \;\Rightarrow\; x^2 = 4 \;\Rightarrow\; x = \pm 2$$

These x-intercepts appear in Figure 4.5. The y-intercept is found by shifting the original y-intercept down 4 units. We could also have found the y-intercept algebraically by setting $x = 0$ and solving for y, that is, by finding $f(0)$. ∎

Let's now apply these same ideas to a slightly more general situation.

Example 3   Given the graph of $y = f(x)$ shown in Figure 4.6, sketch the following graphs.
(a) $y = f(x) - 3$ **(b)** $y = f(x) + 2$

Figure 4.6

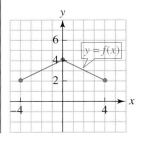

Solution | Using the ideas presented in the last example, we recognize that for a particular x-value, each y-value on the graph $y = f(x) - 3$ will be 3 less than the y-value on the graph of $y = f(x)$, whereas for a particular x-value, each y-value on the graph of $y = f(x) + 2$ will be 2 more than the y-value on the graph of $y = f(x)$. Therefore, to obtain the graph of $y = f(x) - 3$, we shift the original graph of $y = f(x)$ down 3 units, and to obtain the graph of $y = f(x) + 2$, we shift the original graph up 2 units. The graphs appear in Figure 4.7(a) and (b). Again the original function appears as the dashed graph.

Figure 4.7

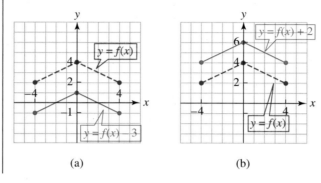

(a)   (b)

We can generalize the results of the last two examples into the following principle of graphing.

In general, how does the value of
$f(2) - 5$ compare with the
value of $f(2)$?

The Vertical Shift Principle for Graphs
(for $c > 0$)

To Obtain the Graph of:	Shift the Graph of $y = f(x)$:
$y = f(x) + c$	c units upward
$y = f(x) - c$	c units downward

Different Perspectives: The Vertical Shift Principle

Graphical Description

To obtain the graph of $y = f(x) + 2$, shift the graph of $y = f(x)$ up 2 units.

To obtain the graph of $y = f(x) - 3$, shift the graph of $y = f(x)$ down 3 units.

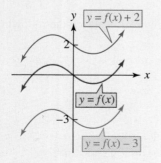

Algebraic/Numeric Description

For each value of x, the value of $y = f(x) + 2$ is 2 greater than the value of $f(x)$. If $f(4) = 5$, then $f(4) + 2 = 5 + 2 = 7$.

For each value of x, the value of $y = f(x) - 3$ is 3 less than the value of $f(x)$. If $f(4) = 5$, than $f(4) - 3 = 5 - 3 = 2$.

On the T1-83 Plus calculator, the list of functions Y1, Y2, . . . may be accessed by choosing Function in the ⎡VARS⎤ Y-VARS menu.

We can use a graphing calculator to demonstrate what happens to a graph when we add a constant to a function. The window below shows three functions. The first function is Y1=X²−3X. The second function is Y2=Y1+4, which means Y2 is 4 more than Y1. The third function is Y3=Y1−4, which means Y3 is 4 less than Y1. The graphs of the three functions are shown below; note that the graphs Y1, Y2, and Y3 have the same exact shape except Y2 is 4 units above Y1, and Y3 is 4 units below Y1.

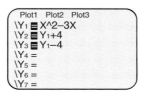
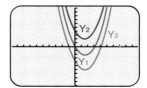

In Section 3.5 we examined some split functions. Let's look at how we might graph a split function.

Example 4 | Given the function

$$f(x) = \begin{cases} x + 2 & \text{if } x < -1 \\ x^2 - 7 & \text{if } -1 \le x \le 4 \end{cases}$$

(a) Find $f(-4)$.   (b) Find $f(3)$.   (c) Find $f(-1)$.   (d) Find $f(6)$.
(e) Sketch the graph of $f(x)$.

Solution | (a) Remember—which rule we use depends on the value of x. To find $f(-4)$ we first note that $x = -4$ satisfies the top condition for x ($x < -1$), and therefore we use the top rule of $f(x)$ to compute $f(-4)$:

$$f(-4) = (-4) + 2 = \boxed{-2}$$ We use the top rule, $x + 2$, to compute $f(-4)$ since $-4 < -1$.

(b) To find $f(3)$ we first note that $x = 3$ satisfies the bottom condition, $-1 \le x \le 4$, and therefore we use the bottom rule of $f(x)$ to compute $f(3)$:

$$f(3) = (3)^2 - 7 = \boxed{2}$$ We use the bottom rule, $x^2 - 7$, to compute $f(3)$ since $-1 \le 3 \le 4$.

(c) $f(-1) = (-1)^2 - 7 = \boxed{-6}$ We use the bottom rule, $x^2 - 7$, to compute $f(-1)$ since $-1 \le -1 \le 4$.

(d) $f(6)$ is undefined. This function if defined only for values for x that are less than or equal to 4. Alternatively, we can say that 6 is not in the domain of the function.

(e) To graph the function $f(x)$, notice that we are actually piecing together portions of two functions, $y = x + 2$ and $y = x^2 - 7$. The first rule gives the line $y = x + 2$, but it is only for values of x less than -1. Hence we "cut off" the portion of this line for values of x greater than (or equal to) -1. [We cut on the line $x = -1$, as shown in Figure 4.8(a) on page 200.]

Figure 4.8

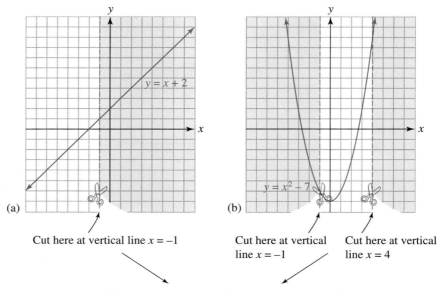

(a)

(b)

Cut here at vertical line $x = -1$

Cut here at vertical line $x = -1$

Cut here at vertical line $x = 4$

Discard shaded regions. Piece together the remaining portions to get the graph shown in Figure 4.9.

The second rule gives the parabola $y = x^2 - 7$, but it is only for values of x between (and including) -1 and 4. Hence, we "cut off" the portions of the line for values of x less than -1 and greater than 4. [We cut on the lines $x = -1$ and $x = 4$, as shown in Figure 4.8(b).] We put the two pieces together as shown in Figure 4.9.

Figure 4.9

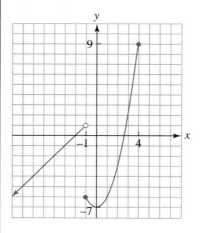

See the Technology Corner at the end of this section to learn how to graph split functions on a graphing calculator.

Note that we put an empty circle at the right end of the line (the portion of the graph that is $y = x + 2$). This occurs where $x = -1$, and indicates that the point where $x = -1$ on the *line* portion of the graph is excluded. Remember, the graph of $f(x)$ is a line only for values of x strictly less than -1.

We put a filled-in circle on the left end of the parabola (the portion of the graph that is $y = x^2 - 7$). This occurs where $x = -1$, and indicates that the point where $x = -1$ on the *parabola* portion of the graph is included. The filled-in circle at the right end of the parabola (at $x = 4$) indicates that this point on the parabola is included as well.

Example 5 Sketch the graph of $y = f(x) = |x|$.

Solution We offer two approaches. The first approach begins by recalling the algebraic definition of absolute value, which is

$$y = f(x) = |x| = \begin{cases} x & \text{if } x \geq 0 \\ -x & \text{if } x < 0 \end{cases}$$

We are, in effect, being asked to graph a split function. The graph of $y = |x|$ will look like the graph of $y = x$ for $x \geq 0$ and will look like the graph of $y = -x$ for $x < 0$. (The graphs of $y = x$ and $y = -x$ are very basic and you should be familiar with them by now.) The graphs appear in Figure 4.10(a) and (b). The graph of $y = f(x) = |x|$ appears in Figure 4.10(c).

Figure 4.10

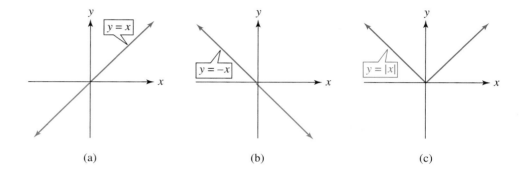

(a)   (b)   (c)

In the second approach we obtain the graph of $y = f(x) = |x|$ by looking at the "underlying" graph—that is, the graph of the function without the absolute value—and then analyzing the effect that the absolute value has on the underlying graph. If we drop the absolute value, we get the equation $y = x$, whose graph appears in Figure 4.10(a). How do the y-values on the graph of $y = |x|$ compare with the y-values on the underlying graph $y = x$? For those x-values where x is negative on the underlying graph, the absolute value makes it positive.

Thus we recognize that the absolute value has the following effect on the underlying graph: that portion of the underlying graph on or above the x-axis is unaffected by the absolute value, whereas that portion of the underlying graph below the x-axis is reflected above the x-axis.

Figure 4.11

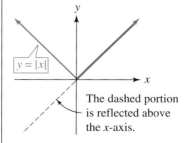

The dashed portion is reflected above the x-axis.

This approach to the graph of $y = f(x) = |x|$ appears in Figure 4.11. Note that the underlying graph of $y = x$ is shown as a dashed line.

The graph of $y = |x|$ should be added to your catalog of basic graphs. ∎

Calculator Exploration

Use a graphing calculator to graph the following in the standard $[-10, 10]$ by $[-10, 10]$ window.

1. Graph the function $y = x^2 - 5$. Then clear the screen and graph $y = |x^2 - 5|$. Observe the differences between the two graphs.

2. Graph the function $y = x^3 - 3x + 1$. Then clear the screen and graph $y = |x^3 - 3x + 1|$. Observe the differences between the two graphs.

3. What can you conclude about the relationship between the graph of $f(x)$ and the graph of $|f(x)|$?

The second approach used in the last example can be used to answer the following more general question. Suppose we know the graph of $y = f(x)$ (which we shall call the underlying graph); how can we obtain the graph of $y = |f(x)|$? Wherever $f(x)$ is nonnegative, the absolute value will leave it unchanged, whereas wherever $f(x)$ is negative, the absolute value will make it positive. Keeping in mind that $f(x)$ is just another name for y, that y being positive means that the point on the graph is above the x-axis, and y being negative means that the point on the graph is below the x-axis, absolute value has the following effect on the underlying graph.

Principle for Graphing $y = |f(x)|$

To obtain the graph of $y = |f(x)|$, begin with the graph of the underlying function $y = f(x)$.

1. Leave that portion of the underlying graph on or above the x-axis unchanged, and

2. Take that portion of the underlying graph below the x-axis and reflect it above the x-axis.

Example 6

How does the distance from the x-axis to the graph of $f(x)$ at $x = 5$ compare to the distance from the x-axis to the graph of $|f(x)|$ at $x = 5$?

Given the graph of $y = f(x)$ in Figure 4.12, sketch the graph of $y = |f(x)|$.

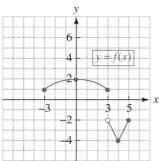

Figure 4.12

Solution

Applying the principle for graphing $y = |f(x)|$, that portion of the graph on the interval $[-3, 3]$ is left alone because it is on or above the x-axis, whereas that portion of the graph below the x-axis on the interval $(3, 5]$ is reflected above the x-axis. The graph appears in Figure 4.13.

Figure 4.13

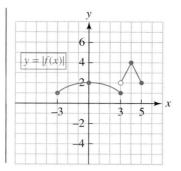

Example 7 Using the graph of $y = |x|$, sketch the graphs of
(a) $y = f(x) = |x - 2|$
(b) $y = f(x) = |x + 3|$

Solution Let's examine the functions from a slightly different point of view, which will lead us to yet another graphing principle.

(a) Looking at the function $y = |x|$, we recognize that y cannot take on negative values. In fact, the smallest y can be is zero, which will occur when the input, x, is zero. This is mirrored in the graph by the fact that the lowest point on the graph of $y = |x|$ is the origin; that is, $y = 0$ when $x = 0$.

If we now look at the function $y = f(x) = |x - 2|$, we again recognize that y cannot be negative. The smallest y can be is zero, and this will happen when $x - 2 = 0$, or when the input is $x = 2$. Thus the lowest point on the graph of $y = |x - 2|$ will be $(2, 0)$.

Similarly, $|x|$ will be equal to 5 when $x = 5$ or -5, whereas $|x - 2|$ will be equal to 5 when $x = 7$ or $x = -3$. Note that 7 is 2 units to the right of 5 and -3 is 2 units to the right of -5. This suggests that the y-values on the graph of $y = |x - 2|$ will be the same as the y-values on the graph of $y = |x|$ but will occur 2 units farther to the right.

Thus the graph of $y = |x - 2|$ can be obtained by shifting the graph of $y = |x|$ two units to the *right*. The graph appears in Figure 4.14. Note that the solid black graph is that of $y = |x|$.

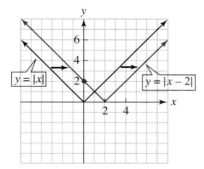

Figure 4.14

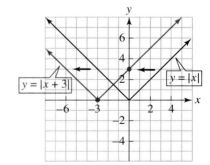

Figure 4.15

(b) A similar analysis for $y = f(x) = |x + 3|$ tells us that the lowest point on its graph will occur when $x + 3 = 0$, that is, when the input x is -3. This suggests that the graph of $y = f(x) = |x + 3|$ can be obtained by shifting the graph of $y = |x|$ horizontally 3 units to the *left*. The graph appears in Figure 4.15.

Calculator Exploration

Use a graphing calculator to graph the following in the standard $[-10, 10]$ by $[-10, 10]$ window.

1. Graph the following on the same set of coordinate axes:

$$y = x^2, \quad y = (x - 1)^2, \quad y = (x + 3)^2$$

What can you conclude about the effect on the graph of $y = x^2$ of adding a constant to x before squaring?

2. Graph the following on the same set of coordinate axes:

$$y = x^3, \quad y = (x - 1)^3, \quad y = (x + 3)^3$$

What can you conclude about the effect on the graph of $y = x^3$ of adding a constant to x before cubing?

Example 8   Sketch the graph of $y = f(x) = (x - 3)^2$.

Solution   Using the same idea as in the previous example, we recognize that since the lowest point of $y = x^2$ occurs when $x = 0$, the lowest point of $y = (x - 3)^2$ occurs when $x - 3 = 0$, or $x = 3$. Thus the graph of $y = (x - 3)^2$ can be obtained by shifting the graph of $y = x^2$ horizontally 3 units to the right. The graph appears in Figure 4.16.

Figure 4.16

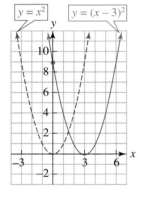

We can generalize the result of the last two examples to the following principle of graphing, illustrated in Figure 4.17.

Figure 4.17

The Horizontal Shift Principle for Graphs
(for $c > 0$)

To Obtain the Graph of:	Shift the Graph of $y = f(x)$:
$y = f(x + c)$	c units to the left
$y = f(x - c)$	c units to the right

In Figure 4.17, note that the "height" of $f(x - 4)$ at x is the same as the "height" of $f(x)$ at $x - 4$.

Be careful when applying the horizontal shift principle. If we have the graph of $y = f(x)$ and we want the graph of $y = f(x + 3)$, there is a natural tendency to expect to shift the graph 3 units to the right, which is *incorrect*. As our analysis has just shown us and as the horizontal shift principle describes, we should shift the graph 3 units *to the left*.

In general, is it true that $f(5) > f(3)$?

Different Perspectives: The Horizontal Shift Principle

Graphical Description

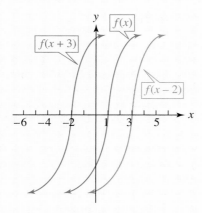

To obtain the graph of $f(x + 3)$, shift the graph of $f(x)$ to the left 3 units.

To obtain the graph of $f(x - 2)$, shift the graph of $f(x)$ to the right 2 units.

Algebraic Description

Consider the value of $f(x)$ at $x = 1$, which is $f(1)$. If we want $f(x + 3)$ to have this same value then we could have $x + 3 = 1 \implies x = -2$. Thus the value of $f(x)$ at $x = 1$ is equal to the value of $f(x + 3)$ at $x = -2$ (three units *to the left* of 1).

If we want $f(x - 2)$ to have the value $f(1)$ then we could have $x - 2 = 1 \implies x = 3$. Thus the value of $f(x)$ at $x = 1$ is equal to the value of $f(x - 2)$ at $x = 3$ (two units *to the right* of 1).

We can use a graphing calculator to demonstrate what happens to a graph when we add a constant to the input of a function. The TI-83 Plus window below shows three functions: the first function is $Y1=X^3-2X$. The second function is $Y2=Y1(X+4)$ (in the TI-83 Plus, we can use function notation with the Y variables as we would use with f). The third function is $Y3=Y1(X-6)$. The graphs of the three functions are shown below; note that the graphs $Y1$, $Y2$, and $Y3$ have the same exact shape except $Y2$ is 4 units to the left of $Y1$, and $Y3$ is 6 units to the right of $Y1$.

We can compare the value of $f(3)$ with $f(3) + 2$. Can we, in general, compare the value of $f(3)$ with $f(3 + 2)$?

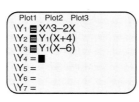

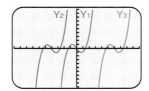

Let's apply this graphing principle in the next example.

Example 9 Given the graph of $y = f(x)$ in Figure 4.18, sketch each graph.
(a) $y = f(x + 4)$ **(b)** $y = f(x - 2)$

Figure 4.18

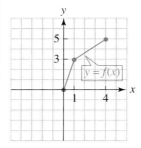

Solution According to the horizontal shift principle, the graph of $f(x + 4)$ is obtained by shifting the given graph of $y = f(x)$ four units to the left, and the graph of $f(x - 2)$ is obtained by shifting the given graph 2 units to the right. The graphs appear in Figure 4.19(a) and (b). Again, the dashed graph is that of the original $f(x)$.

Figure 4.19

Using the horizontal shift principle

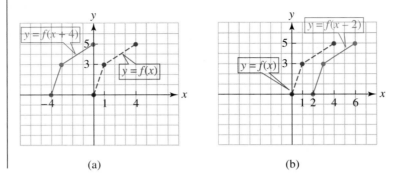

(a)   (b)

Example 10 Sketch the graph of
(a) $g(x) = (x - 2)^2 + 4$ (b) $h(x) = |x + 2| - 3$

Solution (a) The key idea here is recognizing that we can obtain this graph by shifting the basic parabola horizontally and vertically. If we let $f(x) = x^2$, then $f(x - 2) + 4 = (x - 2)^2 + 4 = g(x)$. Hence the graph of $g(x)$ is obtained by taking the basic parabola $f(x) = x^2$ and shifting it 2 units to the right and 4 units up. See Figure 4.20(a).

(b) The key to this problem is to recognize that we can obtain this graph by shifting the graph of the absolute value function horizontally and vertically. If we let $f(x) = |x|$, then $f(x + 2) - 3 = |x + 2| - 3 = h(x)$. Hence $h(x)$ is obtained by taking the absolute value function and shifting it 2 units to the left and 3 units down. The graph appears in Figure 4.20(b).

Figure 4.20

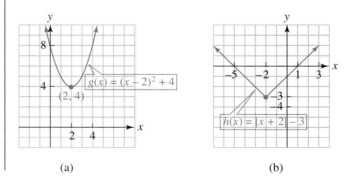

(a)   (b)

In the next section, we continue the development of graphing principles.

Technology Corner

Let's explain some techniques for graphing a split function on a graphing calculator. The keys and menus described here pertain to the T1-83 Plus calculator.

1. To enter a split function, we must use some of the relational and logical connectors in the [TEST] menu.

   If the calculator sees the statement $X > 3$, the calculator assigns the value 1 to this statement if it is true and the value 0 if it is false. Similarly, if the calculator sees a statement such as $X > 2$ and $X < 6$, the calculator assigns the value 1 if this statement is true (meaning that X must be greater than 2 **and** also less than 6) and the value 0 if the statement is false (meaning that either X is not greater than 2 or X is not less than 6).

   Using this information, let's use a graphing calculator to graph the following split function.

$$f(x) = \begin{cases} 2 - x & \text{if } x < 0 \\ x^2 + 2 & \text{if } x \geq 0 \end{cases}$$

We would type in this function $f(x)$ as follows:

   $Y_1 = (2 - X)(X < 0) + (X^2 + 2)(X \geq 0)$

   Before we explain why this function is actually the same as the split function $f(x)$, Figure TC.4 shows that we have entered this function as Y_1 and the graph of this split function.

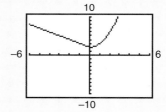

Figure TC.4

   To enter this function, we type the expression as usual. When we need to insert one of these TEST operations, we use the [2ND] [TEST] menu either under TEST or under LOGIC to select an inequality symbol or the "and" connector, which is automatically inserted at the cursor position in Y_1.

   Let's explain how the expression for Y_1 makes it the split function $f(x)$. If $x < 0$, the statement $(X < 0)$ is true and the statement $X \geq 0$ is false, so the calculator assigns a value of 1 to the expression $(X < 0)$ and a value of 0 to the expression $(X \geq 0)$. Therefore, for a value of x less than 0 we have

$$Y_1 = (2 - x)(x < 0) \quad + (x^2 + 2)(x \geq 0) \qquad \text{becomes}$$
$$= (2 - x)(\ 1\ ) \quad + (x^2 + 2)(\ 0\ )$$
$$= 2 - x$$

(*continued*)

whereas for a value of x greater than or equal to 0 we have

$$
\begin{aligned}
Y_1 &= (2 - x)(x < 0) \quad + (x^2 + 2)(x \geq 0) \qquad \text{becomes} \\
&= (2 - x)(\;0\;) \quad + (x^2 + 2)(\;1\;) \\
&= \qquad\qquad\qquad x^2 + 2
\end{aligned}
$$

This agrees exactly with our description of $f(x)$ as a split function.

Using these testing statements allows us to multiply part of the expression by 0 or 1 for certain x-values, thus creating a split function.

2. A graphing calculator has certain quirks that we need to be aware of if we want the graphs we get to agree with what we expect to see.

If we graph the function $y = x + 2$ on the restricted domain $1 \leq x \leq 4$, we would draw the graph in Figure TC.5.

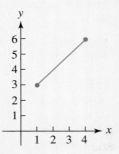

Figure TC.5   The graph of $y = x + 2$ on the restricted domain $1 \leq x \leq 4$

To graph this on a graphing calculator, we might follow the technique we just illustrated above and enter

```
Y₁=(X+2)(1 ≤ X and X ≤ 4)
```

Recall that the idea here was to multiply $(x + 2)$ by 1 for those x's that satisfy the inequality, and to multiply $(x + 2)$ by 0 for those x's that do not. If we graph this on the calculator, we obtain the graph in Figure TC.6.

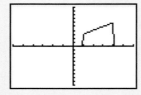

Figure TC.6
The graph of $Y_1=(X+2)(1 ≤ X$ and $X ≤ 4)$

The reason we have unexpected "vertical pieces" in the graph is that the definition of Y_1 makes the function $(x + 2)$ for x between 1 and 4, and 0 elsewhere in the window. In other words, for x's in the window that are less than 1 or greater than 4, y is equal to zero, and so the graph is the x-axis itself. Since the calculator is in "connect mode," it attempts to connect the portions of the graph, and so it draws the vertical lines as "connectors."

However, instead of constructing Y_1 as we did, we can construct the following:

```
Y₁=(X+2)/(1 ≤ X and X ≤ 4)
```

Graphing this on the calculator gives the graph in Figure TC.7.

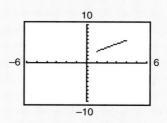

Figure TC.7
The graph of $Y_1 = (X+2)/(1 \leq X \text{ and } X \leq 4)$

Now the calculator is dividing by 1 or 0 rather than multiplying by 1 or 0. When the calculator tries to divide by 0 (for those x's not between 1 and 4), the function is *undefined* so there is no graph at all for those x's and consequently no graph to connect to from the line segment.

4.1 Exercises

1. Given the following graph of $y = f(x)$, sketch the graphs of $y = f(x) - 3$ and $y = f(x - 3)$ on the same coordinate system.

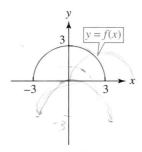

2. Given the following graph of $y = G(x)$, sketch the graphs of $y = G(x) + 2$ and $y = G(x + 2)$ on the same coordinate system.

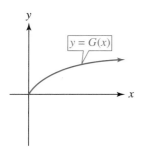

3. Given the following graph of $y = h(x)$, sketch the graphs of $y = h(x) - 1$ and $y = h(x) + 1$ on the same coordinate system.

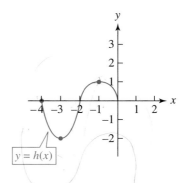

4. Given the following graph of $y = g(x)$, sketch the graphs of $y = g(x + 3)$ and $y = g(x) + 3$ on the same coordinate system.

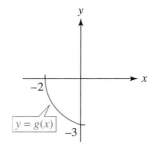

5. Given the following graph of $y = f(x)$, sketch the graphs of $y = f(x - 1) + 2$ and $y = |f(x)|$.

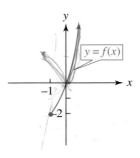

6. Given the following graph of $y = g(x)$, sketch the graphs of $y = g(x + 1) - 2$ and $y = |g(x)|$.

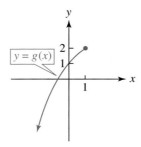

7. Given the following graph of $y = h(x)$, sketch the graph of $y = h(x + 2) + 3$.

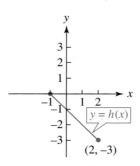

8. Given the following graph of $y = f(x)$, sketch the graph of $y = f(x - 2) - 3$.

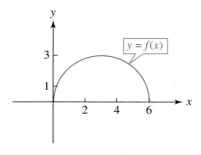

Use the following graph to answer Exercises 9–12.

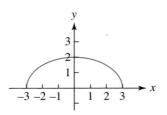

9. If the graph in the figure is the graph of $y = f(x) + 2$, sketch the graph of $y = f(x)$.

10. If the graph in the figure is the graph of $y = f(x + 3)$, sketch the graph of $y = f(x)$.

11. If the graph in the figure is the graph of $y = f(x - 2)$, sketch the graph of $y = f(x)$.

12. If the graph in the figure is the graph of $y = f(x) - 3$, sketch the graph of $y = f(x)$.

In Exercises 13–36, sketch the graph of the given function. Label the intercepts.

13. $y = f(x) = x^2 + 2$

14. $y = f(x) = (x + 2)^2$

15. $y = g(x) = (x - 1)^2$

16. $y = g(x) = x^2 - 1$

17. $y = h(x) = |2x - 5|$

18. $y = h(x) = |2x| - 5$

19. $y = f(x) = |x + 4|$

20. $y = f(x) = |x| + 4$

21. $y = F(x) = (x + 1)^2 - 4$

22. $y = G(x) = (x - 1)^2 + 4$

23. $y = G(x) = (x - 2)^2 - 3$

24. $y = H(x) = (x + 2)^2 + 3$

25. $y = g(x) = |x| - 3$

26. $y = g(x) = |x - 3|$

27. $y = h(x) = |2x|$

28. $y = h(x) = 2|x|$

29. $y = F(x) = |x - 1| + 2$

30. $y = G(x) = |x + 2| - 1$

31. $y = H(x) = |-x + 2|$

32. $y = F(x) = |-x| + 2$

33. $y = f(x) = -|x| + 5$

34. $y = g(x) = |-x + 5|$

35. $y = |x^2 - 4|$

36. $y = |x^2 - 9|$

37. Use the following graph of $y = f(x)$ to find
(a) $f(-4)$ (b) $f(-1)$ (c) $f(0)$ (d) $f(2)$ (e) $f(5)$
(f) For what values, if any, is $f(x) = 2$?
(g) For what values, if any, is $f(x) = 1$?
(h) For what values, if any, is $f(x) = -1$?
(i) For what values, if any, is $f(x) = 4$?

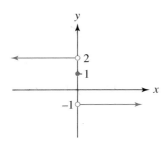

38. Use the graph of $y = F(x)$ to find
(a) $F(6)$ (b) $F(-5)$ (c) $F(1)$
(d) How many zeros does $F(x)$ have?
(e) How many solutions are there to the equation $F(x) = 3$?
(f) Based on the graph, what is the domain of $F(x)$?
(g) Based on the graph, what is the range of $F(x)$?

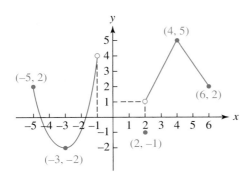

In Exercises 39–50, sketch the graph of the given function and find its domain and range.

39. $f(x) = \begin{cases} 3x - 6 & \text{if } x \le 1 \\ 1 - x & \text{if } x > 1 \end{cases}$

40. $g(x) = \begin{cases} 2x + 5 & \text{if } x < -3 \\ x & \text{if } x \ge -3 \end{cases}$

41. $f(x) = \begin{cases} x - 4 & \text{if } -2 \le x < 2 \\ 4 & \text{if } 2 \le x < 5 \end{cases}$

42. $g(x) = \begin{cases} -3 & \text{if } 0 \le x < 4 \\ -2x & \text{if } 4 < x \le 6 \end{cases}$

43. $f(x) = \begin{cases} x + 6 & \text{if } x < -6 \\ x & \text{if } -6 \le x < 6 \\ 6 - x & \text{if } x > 6 \end{cases}$

44. $f(x) = \begin{cases} 2 & \text{if } x < -3 \\ |x| & \text{if } -2 \le x < 2 \\ 2 & \text{if } x > 2 \end{cases}$

45. $f(x) = \begin{cases} -4 & \text{if } x < -4 \\ |x| & \text{if } -4 \le x \le 4 \\ 4 & \text{if } x > 4 \end{cases}$

46. $f(x) = \begin{cases} 2 & \text{if } x < -2 \\ -x & \text{if } -2 \le x \le 2 \\ -2 & \text{if } x > 2 \end{cases}$

47. $F(x) = \begin{cases} 0 & \text{if } x < 0 \\ \dfrac{7}{30} & \text{if } 0 \le x < 1 \\ \dfrac{9}{15} & \text{if } 1 \le x < 2 \\ 1 & \text{if } x \ge 2 \end{cases}$

48. $G(x) = \begin{cases} 0 & \text{if } x < 0 \\ \dfrac{4}{7} & \text{if } 0 \le x < 3 \\ \dfrac{1}{14} & \text{if } 3 \le x < 5 \\ 1 & \text{if } x \ge 5 \end{cases}$

49. $G(x) = \begin{cases} \dfrac{1}{5} & \text{if } 1 < x < 5 \\ 0 & \text{elsewhere} \end{cases}$

50. $F(x) = \begin{cases} 2 & \text{if } -3 \le x \le 3 \\ 4 & \text{elsewhere} \end{cases}$

Questions for Thought

51. Analyze the function $y = f(x) = -x^2$ and obtain its graph by using your knowledge of the graph of $y = x^2$.

52. Analyze the function $y = f(x) = -|x|$ and obtain its graph by using your knowledge of the graph of $y = |x|$.

53. Based on Exercises 51 and 52, can you describe how to obtain the graph of $y = -f(x)$ from the graph of $y = f(x)$?

54. Describe how you would use the graph of $y = x^2$ to obtain the graph of $y = (x - 3)^2 + 5$.

4.2   More Graphing Principles: Types of Functions

By building on the ideas introduced in the last section, we can complete our repertoire of graphing principles.

Calculator Exploration

Use a graphing calculator to graph the following in the standard $[-10, 10]$ by $[-10, 10]$ window.

1. Graph the following on the same set of coordinate axes:

$$y = x^2 + 2 \qquad\qquad y = -(x^2 + 2)$$

2. Graph the following on the same set of coordinate axes:

$$y = x^3 - 1 \qquad\qquad y = -(x^3 - 1)$$

3. What would you conjecture is the relationship between the graph of $f(x)$ and the graph of $-f(x)$?

Example 1 | Sketch the graph of $y = f(x) = -x^2$.

Solution | Again, rather than plot points we try to analyze this function in light of what we already know about the graph of $y = x^2$. We recognize that for each x-value, the y-value obtained from $y = -x^2$ will be the negative of the y-value obtained from $y = x^2$. All the positive y-values on the graph of $y = x^2$ become negative. Therefore, the graph of $y = -x^2$ can be obtained by turning the graph of $y = x^2$ "upside down." The graph appears in Figure 4.21.

Figure 4.21

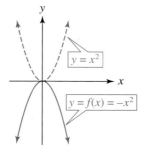

Example 2 | Sketch the graphs of $y = f(x) = 2x + 6$ and $y = g(x) = -(2x + 6)$ on the same coordinate system. How would you describe the relationships between $f(x)$ and $g(x)$ and their graphs?

Solution | We recognize that the graphs of $f(x)$ and $g(x)$ are straight lines. (Why?) We find the intercepts for each line; the graphs appear in Figure 4.22.

We note that $y = f(x) = 2x + 6$ and $y = g(x) = -(2x + 6)$ are negatives of each other. This means that the y-values on the graph of $g(x)$ are the opposites of the y-values on the graph of $f(x)$. In other words, wherever the graph of $f(x)$ is above the

Figure 4.22

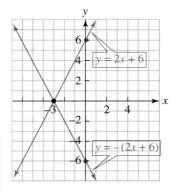

x-axis, the graph of $g(x)$ is below the x-axis, and vice versa. We say that the graph of $g(x)$ is obtained by *reflecting the graph of $f(x)$ about the x-axis.* ∎

We can generalize the results of the last two examples into the following principle of graphing.

Principle for Graphing y = − f(x)

To obtain the graph of $y = -f(x)$, reflect the graph of $y = f(x)$ about the x-axis.

Example 3   Given the following graph of $y = f(x)$, sketch the graph of $y = -f(x)$.

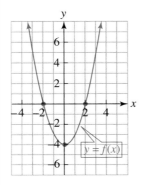

Solution   The graphing principles tell us that to get the graph of $y = -f(x)$, we reflect the entire graph of $f(x)$ about the x-axis. The graph appears in Figure 4.23. The graph of the original $y = f(x)$ appears as the dashed curve.

Figure 4.23

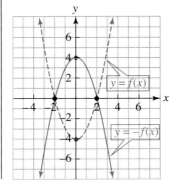

Recall that in $y = Ax^2$, the coefficient A stretches the graph of $y = x^2$. Based on the discussion above, we can see that for the parabola $y = Ax^2$, if $A > 0$, then the parabola opens upward; on the other hand, if $A < 0$, the parabola opens downward.

Example 4 | Sketch the graph of $y = 2 - |x|$.

Solution | We can rewrite the equation as $y = -|x| + 2$ and employ the graphing principles to the graph of $y = |x|$. We start with the graph of $y = |x|$ (Figure 4.24(a)). The negative sign tells us to reflect the graph of $y = |x|$ about the x-axis (Figure 4.24(b)) and the $+2$ tells us then to shift the graph up two units (Figure 4.24(c)).

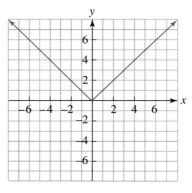

(a) The graph of $y = |x|$

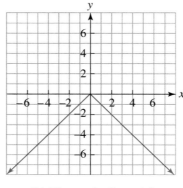

(b) The graph of $y = -|x|$

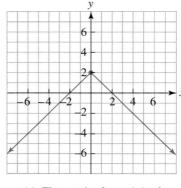

(c) The graph of $y = -|x| + 2$

Figure 4.24
The graph of $y = 2 - |x|$

The x-intercepts are found by setting $y = 0$.

$$0 = 2 - |x| \implies |x| = 2 \implies x = \pm 2$$

In Section 4.7 we saw that a function for which $f(-x) = f(x)$ is symmetric with respect to the y-axis, meaning that the portion of the graph to the left of the y-axis is obtained by reflecting the portion of the graph to the right of the y-axis about the y-axis (and vice versa).

We can generalize this further. Suppose we know what the graph of $y = f(x)$ looks like and we want to sketch the graph of $y = f(-x)$. If the graph has y-axis symmetry, then the graph of $f(-x)$ will be identical to the graph of $f(x)$—but what if the graph does not have y-axis symmetry?

Calculator Exploration

Use a graphing calculator to graph the following. (Try several viewing windows to help in your analysis.)

1. Graph the following on the same set of coordinate axes:
$$f(x) = x^3 + 1 \qquad f(-x) = -x^3 + 1$$

2. Graph the following on the same set of coordinate axes:
$$f(x) = x^5 + 3x^2 \qquad f(-x) = -x^5 + 3x^2$$

3. What can you conclude about the relationship between the graph of $f(x)$ and the graph of $f(-x)$?

Let's suppose $x \neq 0$ so that x and $-x$ are opposites: If x is to the right of the y-axis, then $-x$ is to the left of the y-axis, and vice versa. By replacing x with $-x$ in $f(x)$, we end up on the other side of the y-axis *but at the same height as $f(x)$*. Figure 4.25(a) and (b) illustrate the graph of a function $y = f(x)$ and the graph of $y = f(-x)$.

Figure 4.25

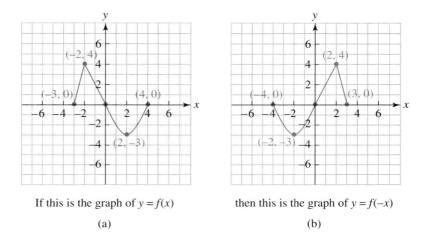

If this is the graph of $y = f(x)$

(a)

then this is the graph of $y = f(-x)$

(b)

Pay particular attention to the points that have been labeled. Note that since the point $(2, -3)$ is on the graph of $y = f(x)$, the point $(-2, -3)$ is on the graph of $y = f(-x)$. In other words, since $f(2) = -3$ for $y = f(x)$, then for $y = f(-x)$ and $x = -2$ we have $f(-x) = f(-(-2)) = f(2) = -3$. The same is true for the other points on the graph.

Verbally, we can describe this by saying that what happens on the right side of the graph of $y = f(x)$ happens on the left side of the graph of $y = f(-x)$, and vice versa. More precisely, the graph of $y = f(-x)$ is obtained by reflecting the graph of $y = f(x)$ about the y-axis.

Based on this discussion, we can formulate the following graphing principle.

Principle for Graphing $y = f(-x)$
To obtain the graph of $y = f(-x)$, reflect the graph of $y = f(x)$ about the y-axis.

Keep in mind that in order to use this graphing principle, we must know what the graph of $y = f(x)$ looks like.

Example 5 Given the graph of $y = f(x)$ in Figure 4.26, sketch the graph of $y = f(-x)$.

Figure 4.26

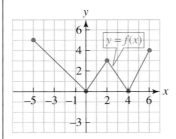

Solution Using the graphing principle for $y = f(-x)$, we reflect the given graph about the y-axis to obtain the graph in Figure 4.27 on page 216.

Figure 4.27

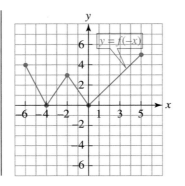

If the graph of $f(x)$ has y-axis symmetry, what is the relationship between the graphs of $f(x)$ and $f(-x)$?

Example 6

Given the graph of $y = f(x)$ in Figure 4.28, sketch each graph.
(a) $y = -f(x)$ **(b)** $y = f(-x)$ **(c)** $y = |f(x)|$

Figure 4.28

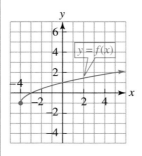

Solution

It is very important to apply the graphing principles carefully. To obtain the graph of $y = -f(x)$, we reflect the original graph about the x-axis. See Figure 4.29(a). To obtain the graph of $y = f(-x)$, we reflect the original graph about the y-axis. See Figure 4.29(b). To obtain the graph of $y = |f(x)|$, we reflect the portion of the graph below the x-axis above the x-axis and leave the portion of the graph above the x-axis alone. See Figure 4.29(c).

Figure 4.29

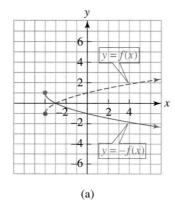

(a)

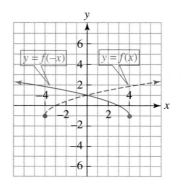

(b)

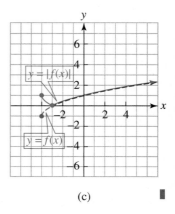

(c)

For ease of reference we summarize the various graphing principles here, with an illustration of each.

Summary of Graphing Principles

The Stretching Principle

(for $a > 0$, $a \neq 1$)

The graph of $y = af(x)$ can be obtained by stretching the graph of $y = f(x)$. In other words, the graph of $y = af(x)$ will have the same basic shape as the graph of $y = f(x)$. If $a > 1$, the graph of $y = af(x)$ is obtained by deflecting the graph of $f(x)$ away from the x-axis. If $0 < a < 1$, the graph of $y = af(x)$ is obtained by pulling the graph of $f(x)$ toward the x-axis.

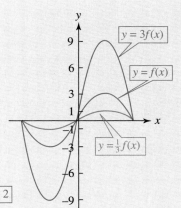

The Vertical Shift Principle for Graphs

(for $c > 0$)

To Obtain the Graph of:	Shift the Graph of $y = f(x)$:
$y = f(x) + c$	c units upward
$y = f(x) - c$	c units downward

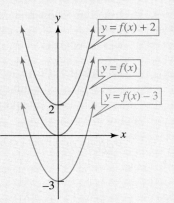

The Horizontal Shift Principle for Graphs

(for $c > 0$)

To Obtain the Graph of:	Shift the Graph of $y = f(x)$:
$y = f(x + c)$	c units to the left
$y = f(x - c)$	c units to the right

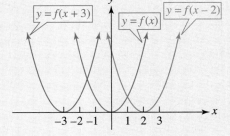

Principle for Graphing $y = |f(x)|$

To obtain the graph of $y = |f(x)|$, begin with the graph of the underlying function $y = f(x)$.

1. Leave that portion of the underlying graph on or above the x-axis unchanged.

2. Take that portion of the underlying graph that is below the x-axis and reflect it above the x-axis.

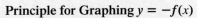

Principle for Graphing $y = -f(x)$

To obtain the graph of $y = -f(x)$, reflect the graph of $f(x)$ about the x-axis.

Principle for Graphing $y = f(-x)$

To obtain the graph of $y = f(-x)$, reflect the graph of $y = f(x)$ about the y-axis.

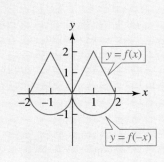

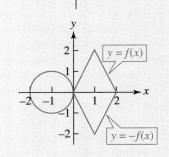

Types of Functions

As we continue to be exposed to a variety of functions and their graphs, it is useful to categorize them for ease of reference.

Constant Functions   A function of the form $y = f(x) = K$, where K is a constant, is called a **constant function**. If $y = f(x) = 6$, then this function assigns a y-value of 6 to every input x. Thus $f(2) = 6, f(-9) = 6, f(0) = 6, f\left(\dfrac{1}{5}\right) = 6$, and so on. Recall from Section 3.2 that the graph of a function of the form $y = K$ is a horizontal line.

Linear Functions   A function of the form $y = f(x) = mx + b$ is called a **linear function**. The name is obviously due to the fact that the graph of a function of the form $y = mx + b$ is a straight line. (Note that a constant function is a linear function with $m = 0$.)

Quadratic Functions   A function of the form $y = f(x) = ax^2 + bx + c$ with $a \neq 0$ is called a **quadratic function**. (If $a = 0$, then the function is $f(x) = bx + c$, which is a linear function.) We discuss the graphs of quadratic functions in great detail in Section 4.4.

Polynomial Functions   Constant, linear, and quadratic functions are all particular examples of a more general class of functions.

> **Definition of a Polynomial Function**
>
> A **polynomial function** is a function of the form
>
> $$y = p(x) = a_n x^n + a_{n-1} x^{n-1} + a_{n-2} x^{n-2} + \cdots + a_2 x^2 + a_1 x + a_0,$$
>
> where $a_n \neq 0$
>
> Each a_i is assumed to be a real number, and n is a nonnegative integer. Such a polynomial function is said to be of degree n.

For example, $f(x) = 6x^3 - 5x + 1$ is an example of a polynomial function of degree 3.

Rational Functions   A function that can be expressed as the quotient of two polynomials is called a **rational function**. $f(x) = \dfrac{x - 1}{x^2 + 1}$ and $g(x) = \dfrac{1}{x - 5}$ are examples of rational functions.

Note that every polynomial function, $p(x)$, is also a rational function, since it can be written as $\dfrac{p(x)}{1}$. We discuss polynomial and rational functions in Chapter 5.

Some other functions we have already discussed or will discuss are $f(x) = |x|$, the absolute value function; $f(x) = \sqrt{x} = x^{1/2}$, a radical function; $f(x) = 2^x$, an exponential function; and logarithmic functions.

We conclude this section with a different type of function. It is worthwhile remembering the *any* rule that assigns a single y-value to each x-value is a bona fide function. Let's consider the following function.

> **Definition of the Greatest Integer Function**
>
> The **greatest integer function**, written $f(x) = [x]$, means the greatest integer less than or equal to x.

Thus, for example,

$$[6.4] = 6 \qquad \text{Because 6 is the greatest integer less than or equal to 6.4}$$
$$[5] = 5 \qquad \text{Because 5 is the greatest integer less than or equal to 5}$$
$$[-2.8] = -3 \qquad \text{Because } -3 \text{ is the greatest integer less than or equal to } -2.8$$

In terms of the number line, the greatest integer function assigns to each x the closest integer at or to the left of x.

Example 7 | Sketch the graph of $y = f(x) = [x]$.

Solution | Since the rule for the greatest integer function is a bit unusual, its graph will be somewhat different from what we have seen thus far.

For instance,

When $0 \leq x < 1$, $y = [x]$ will be equal to 0.

When $1 \leq x < 2$, $y = [x]$ will be equal to 1.

When $2 \leq x < 3$, $y = [x]$ will be equal to 2.

Similarly,

When $-3 \leq x < -2$, $y = [x]$ will be equal to -3.

When $-2 \leq x < -1$, $y = [x]$ will be equal to -2.

When $-1 \leq x < 0$, $y = [x]$ will be equal to -1.

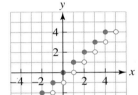

Figure 4.30

On each interval of the form $[n, n + 1)$, where n is an integer, $y = [x]$ will be constant and equal to n. Thus the graph consists of a series of horizontal line segments. A representative portion of the graph appears in Figure 4.30. This type of graph, which consists of a series of horizontal line segments, is called a **step function**. ∎

In the next section we see how the idea of a function can be applied to describe real-life situations.

4.2 Exercises

1. Given the following graph of $y = f(x)$, sketch the graphs of $y = -f(x)$ and $y = f(-x)$.

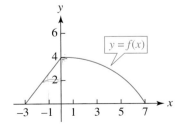

2. Given the following graph of $y = g(x)$, sketch the graphs of $y = g(-x)$ and $y = -g(x)$.

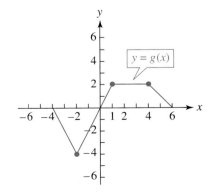

3. Given the following graph of $y = h(x)$, sketch the graphs of $y = h(x) - 2$ and $y = h(x - 2)$.

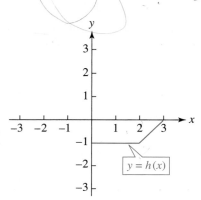

4. Given the following graph of $y = g(x)$, sketch the graphs of $y = g(x + 3)$ and $y = -g(x + 3)$.

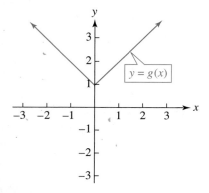

5. Given the following graph of $y = f(x)$, sketch the graphs of $y = f(x - 1)$ and $y = f(1 - x)$.   HINT: To get the graph of $f(1 - x)$, reflect the graph of $f(x + 1)$ about the y-axis. Explanation: Replacing x by $-x$ in $f(x + 1)$ gives $f(-x + 1) = f(1 - x)$ and causes a reflection about the y-axis.

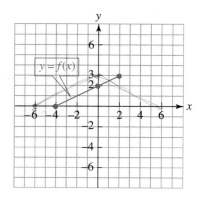

6. Given the following graph of $y = g(x)$, sketch the graphs of $y = g(x - 1)$ and $y = g(1 - x)$. See the hint in Exercise 5.

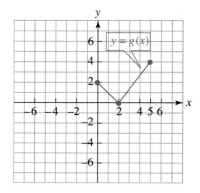

In Exercises 7–12, use the following graph of $y = f(x)$ to obtain the requested graphs.

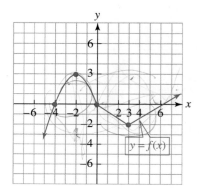

7. $y = f(-x)$ and $y = -f(x)$
8. $y = |f(x)|$ and $y = |f(-x)|$
9. $y = -f(x) + 2$ and $y = -f(x + 2)$
10. $y = -f(x - 2)$ and $y = |f(x - 2)|$
11. $y = f(-x) - 3$ and $y = f(-x - 3)$
12. $y = f(-x) + 3$ and $y = -f(-x) + 3$

In Exercises 13–28, sketch the graph of the given function. Label the intercepts.

13. $y = f(x) = x^2 + 4$
14. $y = f(x) = (x + 4)^2$
15. $y = f(x) = -x^2 + 4$
16. $y = f(x) = -(x + 4)^2$
17. $y = g(x) = |x - 5|$

18. $y = g(x) = |x| - 5$
19. $y = g(x) = 3 - |x|$
20. $y = g(x) = |3 - x|$
21. $y = f(x) = (x + 1)^2 + 1$
22. $y = f(x) = -(x - 1)^2 - 1$
23. $y = h(x) = |x - 9|$
24. $y = h(x) = ||x| - 3|$
25. $y = F(x) = -(x + 1)^2$
26. $y = G(x) = |16 - x|$
27. $f(x) = \begin{cases} |x| - 4 & \text{if } -4 \le x \le 4 \\ 16 - x^2 & \text{elsewhere} \end{cases}$
28. $f(x) = \begin{cases} 2 - |x| & \text{if } -2 \le x \le 2 \\ |x| - 2 & \text{elsewhere} \end{cases}$

Use the following figure in Exercises 29–32.

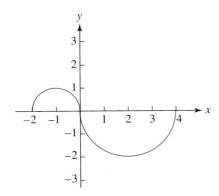

29. If the graph is the graph of $y = f(-x)$, sketch the graph of $y = f(x)$.
30. If the graph is the graph of $y = -f(x)$, sketch the graph of $y = f(x)$.
31. If the graph is the graph of $y = -f(x) + 1$, sketch the graph of $y = f(x)$.
32. If the graph is the graph of $y = f(-x + 1)$, sketch the graph of $y = f(x)$.

In Exercises 33–44, describe in words what you would do to the graph of $y = f(x)$ to obtain the graph of the given function.

33. $f(x) - 4$
34. $f(x - 4)$
35. $f(-x)$
36. $-f(x)$
37. $2 - f(x)$
38. $-f(-x)$
39. $-f(x) + 3$
40. $f(-x) - 3$

41. $f(x + 3) + 4$
42. $f(x - 2) - 5$
43. $f(2 - x)$
44. $f(1 - x) + 1$

In Exercises 45–48, use $[x]$, the greatest integer function, to evaluate each expression.

45. $[6]$
46. $[-8]$
47. $[-2.9]$
48. $[6.7]$

In Exercises 49–52, sketch the graph of the given function.

49. $y = f(x) = [x] + 1$
50. $y = f(x) = [x + 1]$
51. $y = f(x) = [2x]$
52. $y = f(x) = 2[x]$

53. Suppose that the cost of a telephone call between Atlanta and Houston is $0.80 for the first minute and $0.50 for each additional minute or part thereof. We can use the greatest integer function to express the cost, C, of such a phone call, as a function of the length, m, of the phone call in minutes:

$$C = 0.30 - 0.50[-m] \qquad \text{for } m > 0$$

Sketch the graph of this function for a phone call that lasts no more than 5 minutes.

54. Suppose that an overnight delivery service charges $12.50 for the first two pounds and $3 for each additional pound or part thereof. Use the greatest integer function to express the cost, C, of overnight delivery of a package weighing p pounds, and sketch the graph of this function for $0 < p \le 6$.

55. In computer science the function $f(x) = \dfrac{[10x + 0.5]}{10}$ is common. Compute each of the following and determine what operation this function performs.
(a) $f(7)$ (b) $f(2.4)$ (c) $f(5.67)$ (d) $f(8.932)$
(e) $f(-6.48)$

56. In computer science the function $f(x) = [x + 0.5]$ is common. Compute each of the following and determine what operation this function performs.
(a) $f(9)$ (b) $f(2.9)$ (c) $f(6.14)$ (d) $f(7.562)$
(e) $f(-7.29)$

Question for Thought

57. Describe how you would graph $y = a(x - h)^2 + k$.

4.3   Extracting Functions from Real-Life Situations

Let us consider the following problem:

Suppose that a person wants to build a single-story rectangular summer cottage containing 150 square meters of floor space. Suppose further that local building codes require that both the length and width of the building be at least 6 meters. In order to conserve energy, how should the dimensions of the cottage be chosen to minimize the perimeter of the building?

Figure 4.31 illustrates the cottage. We have labeled the length *l* and the width *w*.

Figure 4.31

Problems of this sort, where we are trying to maximize or minimize a certain quantity, are called **optimization problems**.

In the next section, we will learn techniques for solving a particular class of optimization problems. However, before these techniques can be applied to such a real-life situation, it is first necessary to write a function that represents the quantity in which we are interested. (In this problem it is the perimeter of the cottage that is to be minimized.) This section is devoted to the process of extracting a particular function from given information. We expand on the ideas first introduced in Section 2.6.

Example 1   Express the perimeter of the cottage just described as a function of its width, *w*, and find its domain.

Solution   The objective is to express the perimeter of the house as a function of *w*. We have the perimeter, *P*, given by $P = 2w + 2l$. To express the perimeter as a function of one variable, we need to use the fact that the area is to be 150 sq meters:

$$A = lw = 150 \qquad \text{Solve for } l.$$

$$l = \frac{150}{w}$$

We can now substitute for *l* in the perimeter equation, which gives us the perimeter as a function of *w*:

$$P = 2w + 2l \qquad \text{Substitute } l = \frac{150}{w}.$$

$$P = P(w) = 2w + 2\left(\frac{150}{w}\right) = 2w + \frac{300}{w}$$

All that remains for us to do is to determine the domain for the function $P(w)$. We are told that the length and width must be at least 6 meters, so that $w \geq 6$. Since $l \geq 6$ and the area is to be 150 sq meters, we have

Why does $l \geq 6$ imply that $\dfrac{150}{l} \leq \dfrac{150}{6}$?

$$lw = 150 \quad \Rightarrow \quad w = \frac{150}{l} \leq \frac{150}{6} = 25$$

Therefore, the perimeter, P, of the cottage as a function of its width, w, is

$$P = P(w) = 2w + \frac{300}{w} \qquad \text{for } 6 \le w \le 25$$

∎

The remainder of this section is devoted to examining a variety of examples requiring us to extract a function from a realistic situation. As we proceed, we will pause along the way to outline an approach to this type of problem.

Example 2   An oil rig is leaking oil into the ocean, creating a circular slick of oil on the ocean surface (Figure 4.32). Suppose the diameter is growing at a constant rate of 8 meters/min. Express the area of the slick as a function of t, the time in minutes since the leak began.

Solution

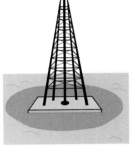

Figure 4.32

What do we need to find?

The area A of the circular oil slick in terms of t, the number of minutes since the leak began

How do we find the area of the slick?

Since the oil slick is circular, we use the formula for the area of a circle, $A = \pi r^2$, where r is the radius of the circle.

We have the area in terms of r, its radius. How do we express the area in terms of t?

We can express the radius in terms of the diameter, and we are given the diameter in terms of the time. In other words, since the radius is half the diameter, we may substitute $r = \dfrac{d}{2}$ into the formula for A:

$$A = \pi r^2 \qquad \text{Substitute } r = \tfrac{d}{2}.$$

$$= \pi \left(\frac{d}{2}\right)^2 = \frac{\pi d^2}{4}$$

We are told that the diameter is growing at a rate of 8 meters/min. If t is the time in minutes since the leak began, we can express the diameter, d, in terms of t by $d = 8t$ (distance = rate × time) and substitute into the formula for A:

$$A = \frac{\pi d^2}{4} \qquad \text{Substitute } d = 8t.$$

$$= \frac{\pi (8t)^2}{4} = 16\pi t^2 \text{ sq meters}$$

Thus the area of the slick as a function of t is $A = A(t) = 16\pi t^2$. The statement of the problem does not tell us how long this oil leak will last, so we assume that the domain of the function is $t \ge 0$.

$$A(t) = 16\pi t^2 \text{ sq meters} \qquad \text{for } t \ge 0$$

∎

Example 3   A rectangular area of 2500 sq ft is to be fenced off. Two opposite sides of the fencing will cost \$3 per foot; the remaining two sides cost \$2 per foot. If x represents the length of the sides requiring the more expensive fencing, express the total cost of the fencing as a function of x.

Solution | *What do we need to find?* | The total cost of the fencing in terms of x, the length of the sides requiring the more expensive fencing

How do we find the total cost?

The cost of fencing is found by multiplying the length of fence by the cost per unit length. Since there are two types of fencing, we need to identify the length of fence at each price. We draw a diagram of the rectangle and label the expensive sides x and the inexpensive sides y.

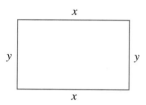

The diagram indicates that we have $2x$ feet of the more expensive fencing; hence the total cost for the \$3 fencing is $3(2x) = 6x$. Since we have $2y$ feet of the \$2 fencing, the total cost of the less expensive fencing is $2(2y) = 4y$. Hence the total cost for all the fencing is $C = 6x + 4y$ dollars.

The cost is expressed in terms of two variables. How do we express the cost in terms of x only?

We need to find a relationship that allows us to express y in terms of x. We know that the area of the rectangle is 2500 sq ft; hence we can write $A = 2500 = xy$. We solve this equation for y to get $y = \dfrac{2500}{x}$ and substitute into the cost equation:

$$C = 6x + 4y \qquad \text{Substitute } y = \frac{2500}{x}$$

$$= 6x + 4\left(\frac{2500}{x}\right) = 6x + \frac{10{,}000}{x}$$

Since x represents the length of a side of the rectangle, the domain for this function is $x > 0$. Thus the answer is

$$C(x) = 6x + \frac{10{,}000}{x} \qquad \text{for } x > 0$$

Example 4 | Use Figure 4.33 to express the area of $\triangle ABC$ as a function of x.

Figure 4.33

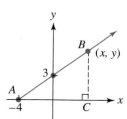

Solution First let's recognize that the question makes sense. A point $B(x, y)$ is chosen on the line passing through the points $A(-4, 0)$ and $(0, 3)$ and a perpendicular is dropped from the point (x, y) to the x-axis, forming $\triangle ABC$. To each point (x, y), there corresponds one triangle and hence one area. Thus to each x, there corresponds exactly one area of $\triangle ABC$, and so the area of the triangle is a function of x.

We begin with the formula for the area, A, of a triangle, $A = \frac{1}{2}bh$. (We often call this the *generic formula*—that is, since we are looking for the area of a triangle we use the formula for the area of *any* triangle.) Next we use the given figure to apply the generic formula to this particular example. We note that the distance from the origin to point A is 4 units and the distance between the origin and point C is x units, so that the length of base AC is $x + 4$. The height BC of the triangle is y. See Figure 4.34.

Figure 4.34

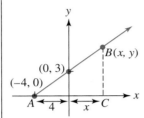

Thus the area formula for this triangle becomes

$$A = \frac{1}{2}(x + 4)y$$

The example asks us to express the area as a function of x. All that remains is to find a relationship that allows us to express y in terms of x. Since the point (x, y) is on the line passing through the points $(-4, 0)$ and $(0, 3)$, the equation of this line gives us the relationship we seek. To get the equation of the line, we first find its slope to be

$$m = \frac{3 - 0}{0 - (-4)} = \frac{3}{4}$$

Since the y-intercept of the line is 3, we have $y = \frac{3}{4}x + 3$ as an equation of the line. Substituting this into the area formula $A = \frac{1}{2}(x + 4)\,y$, we get

$$A = A(x) = \frac{1}{2}(x + 4)\left(\frac{3}{4}x + 3\right) = \frac{3}{8}x^2 + 3x + 6$$

Based on the given diagram, the point (x, y) is to be chosen on the half-line through $(-4, 0)$ and $(0, 3)$ and to the right of the point $(-4, 0)$, and so the domain of $A(x)$ is the set of real numbers greater than -4. Thus the answer is

$$A(x) = \frac{3}{8}x^2 + 3x + 6 \qquad \text{for } x > -4$$

∎

Example 5 Suppose that the surface area of a closed right circular cylinder is 30π sq cm. Express the volume of the cylinder as a function of its radius r.

Solution To answer this question we need to know the formulas for the surface area, S, and volume, V, of a closed right circular cylinder. See Figure 4.35 on page 226. Thus we see that the formula for the volume depends on both r and h. To express the volume as a function of r alone, we need a relationship between r and h.

$S = 2\pi rh + 2\pi r^2$

$V = \pi r^2 h$

Figure 4.35

A closed right circular cylinder with the formulas for its surface area and volume.

The fact that we are told that the surface area is 30π sq cm allows us to use the surface area formula to solve for h in terms of r:

$$S = 2\pi rh + 2\pi r^2$$
$$30\pi = 2\pi rh + 2\pi r^2 \qquad \text{Divide both sides by } 2\pi \text{ and solve for } h.$$
$$15 = rh + r^2$$
$$\frac{15 - r^2}{r} = h$$

Now we can substitute for h in the volume formula:

$$V = \pi r^2 h \qquad \text{We substitute for } h.$$
$$V = V(r) = \pi r^2 \left(\frac{15 - r^2}{r}\right) \qquad \text{Thus the volume as a function of } r \text{ is}$$

$$V = V(r) = 15\pi r - \pi r^3$$

What is the domain of $V(r)$? Clearly, since r represents the radius of the cylinder, r must be positive. However, the height h must also be positive and so from the equation $h = \frac{15 - r^2}{r}$, we require that $15 - r^2 > 0$. Solving this quadratic inequality, we find that the domain of $V(r)$ (which we denote as D_V) is $D_V = \{r \mid 0 < r < \sqrt{15}\}$. ∎

Example 6   Suppose that a computer manufacturing company determines that it can sell 2000 computers at a price of $750 and that for each $25 that the price is increased, 40 fewer computers will be sold. Let n represent the number of $25 increases in the price. Express the total revenue from computer sales as a function of n.

Solution   The revenue is computed by multiplying the number of computers sold by the price per computer.

revenue = (no. of computers sold)(price per computer)

This is the *generic formula*. In a situation like this, it may be helpful to set up a table with some numerical values for n.

n	Price per computer	No. of computers sold	Revenue
0	750	2000	750(2000)
1	750 + 25	2000 − 40	(750 + 25) (2000 − 40)
2	750 + 2(25)	2000 − 2(40)	(750 + 2(25)) (2000 − 2(40))
3	750 + 3(25)	2000 − 3(40)	(750 + 3(25)) (2000 − 3(40))
⋮	⋮	⋮	⋮
n	750 + n(25)	2000 − n(40)	(750 + n(25)) (2000 − n(40))

Thus if we call the revenue function $R(n)$, we have

$$R(n) = (750 + 25n)(2000 - 40n) = -1000n^2 + 20,000n + 1,500,000$$

Since the number of computers sold is $2000 - 40n$ and this number cannot be negative, we need $2000 - 40n \geq 0 \Rightarrow n \leq 50$. Thus the domain of $R(n)$ is $\{n \mid 0 \leq n \leq 50\}$. ∎

Example 7 | Figure 4.36 illustrates a rectangle inscribed in a circle. Express the area of rectangle $EFGH$ as a function of radius, r, of the circle.

Figure 4.36

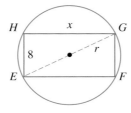

Solution | Since we are looking for the area, A, of the rectangle, the generic formula for this example is $A = bh$, where b and h represent the rectangle's base and height, respectively. In this example this formula becomes $A = 8x$.

All that remains is for us to find a relationship between x and r. It is a basic fact from geometry that since angle EHG is a right angle, the diagonal EG of the rectangle is a diameter of the circle. And so we may label the diagonal EG as $2r$. (See Figure 4.37.)

$\triangle EGH$ is a right triangle, and so we may apply the Pythagorean Theorem to get

Figure 4.37

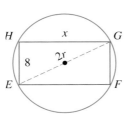

$$x^2 + 8^2 = (2r)^2 \qquad \text{We solve for } x.$$
$$x = \pm\sqrt{4r^2 - 64} \qquad \text{Since } x \text{ is a length, we take the positive square root.}$$
$$x = \sqrt{4r^2 - 64} = \sqrt{4(r^2 - 16)} = 2\sqrt{r^2 - 16}$$

Now we substitute for x in the area formula to get $A = 8x = 8\left(2\sqrt{r^2 - 16}\right)$. Since the diameter of the circle must be greater than 8, r must be greater than 4 and so the domain of $A(r)$ is the set of real numbers greater than 4.

$$A(r) = 16\sqrt{r^2 - 16} \qquad \text{for } r > 4$$
∎

This process of extracting a function from specific information describing a real-life situation is often called **mathematical modeling**. Based on these examples, we can suggest the following outline.

Outline for Obtaining a Mathematical Model

1. Read the problem over until you clearly understand what information is given and what is being sought. Often a diagram is extremely valuable in organizing and relating the given information.

2. Write an expression (formula) for the quantity being sought. This is what we are calling the *generic formula*. For example, if a volume is being sought, write a formula for the volume; if a total cost is being sought, write a formula for the cost.

3. If necessary, apply the generic formula to the particular situation under consideration. That is, replace items such as base, height, radius, or price by the letters or values given in the problem or that appear in the diagram.

4. Use the given information to express all variables in terms of the specified variable.

5. Using the results obtained in steps 3 and 4, write the required quantity as a function of the one specified variable.

6. Use the given information or the physical limitations of the problem to determine the domain of the function.

Note that some useful geometric formulas are summarized inside the back cover of the textbook.

Example 8 An oil tank is in the shape of a right circular cone with an altitude (height) of 15 feet and a radius of 4 feet. Suppose the tank is filled to a depth of h feet. Let x be the radius of the circle on the surface of the oil. Express the volume of oil in the tank as a function of x.

Solution We follow the given outline.

1. Figure 4.38 illustrates the situation described in the example.

2. The formula for the volume, V, of a right circular cone is $V = \frac{1}{3}\pi r^2 h$. *(This is the generic formula.)*

3. Thus the volume of oil in the tank is $V = \frac{1}{3}\pi x^2 h$. *(This is the generic formula applied to this particular example.)* To express the volume as a function of x, we need to find a relationship between x and h.

4. Looking carefully at Figure 4.38, we can see that $\triangle ABC$ is similar to $\triangle ADE$, and so we may write the following proportion:

$$\frac{h}{x} = \frac{15}{4} \qquad \text{We solve for } h.$$

$$h = \frac{15}{4}x \qquad \text{Now we substitute for } h \text{ in the volume formula.}$$

$$V(x) = \frac{1}{3}\pi x^2\left(\frac{15}{4}x\right) = \frac{5}{4}\pi x^3$$

5. $V(x) = \frac{5}{4}\pi x^3$

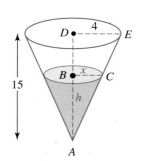

Figure 4.38

6. Since x is the radius of the circle on the surface of the oil, x can be any value greater than or equal to 0 (if the tank is empty) and less than or equal to the radius of the tank (if the tank if full). Therefore, the domain for $V(x)$ is

$$\{x \mid 0 \le x \le 4\}$$

As we proceed through the text, we will continue to apply these ideas about extracting functions to the wider variety of situations we encounter. In the next section we examine some optimization problems that can be solved without any methods from calculus.

4.3 Exercises

In the following exercises, extract the required function and determine its domain.

1. Use the accompanying figure.
   (a) Express y as a function of x.
   (b) Express the area of $\triangle ABC$ as a function of x.

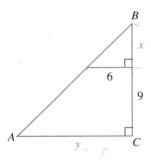

2. In the figure, the point (x, y) is on the line passing through the points $(0, 2)$ and $(5, 0)$.
   (a) Express y as a function of x.
   (b) Express the area of $\triangle ABC$ as a function of x.

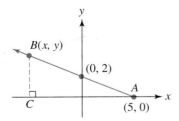

3. The following figure illustrates $\triangle ABC$ inscribed in a semicircle of radius r. Express the area of the shaded portion of the semicircle as a function of r.

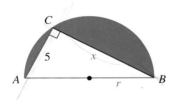

4. The figure below illustrates isosceles trapezoid $ABCD$. Express the area of the trapezoid as a function of x.

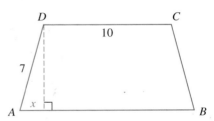

5. A closed box has a square base of side x and a height of h. If the volume of the box is 80 cu cm, express the surface area of the box as a function of x.

6. An open box is to be made from a 30-cm by 40-cm rectangular sheet of metal by cutting out identical squares of side x from each of the corners of the sheet and folding up the sides to form a box, as shown in the figure. Express the volume of the resulting box as a function of x.

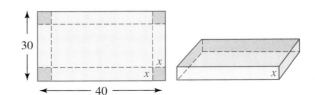

7. Express the area of a circle as a function of its circumference.

8. The volume of a closed right circular cylinder is 20 cu in. Express the surface area of the cylinder as a function of its radius r.

9. A 6-foot man is standing x feet away from a 20-foot pole with a light at the top, as shown in the figure. Express the length of the shadow s cast by the man as a function of x.

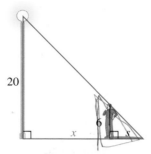

10. A water tank is in the shape of a right circular cone of height 15 feet and radius 5 feet. See the following figure. If the tank is filled so that the radius of the circle at the surface of the water is r, express the volume of the water in the tank as a function of the depth, d.

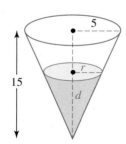

11. The point $(4, 1)$ lies on the line passing through the points $(a, 0)$ and $(0, b)$. Use the figure to express the area of $\triangle ABC$ as a function of a.

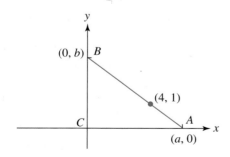

12. The point $D(x, y)$ is chosen on the line joining the origin to the point $(4, 7)$. Use the figure to express the area of rectangle $ABCD$ as a function of x.

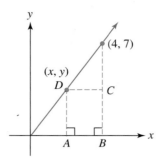

13. A running track is in the shape of a rectangle with a semicircle of radius r at each end, as shown. If the total distance around the track is 200 meters, express the area enclosed by the track as a function of r.

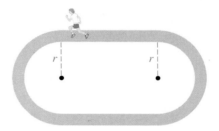

14. A window is constructed in the shape of a rectangle with a semicircle of radius r on top, as shown. If the area of the window is 10 sq ft, express the perimeter of the window as a function of r.

15. At 6:00 A.M. a weather balloon is released from ground level; it rises at a constant rate of 5 ft/sec. An observer is situated at a point 120 feet from the point

of release. If t represents the number of seconds that have elapsed after the release of the balloon, express the distance from the observer to the balloon as a function of t. See the figure.

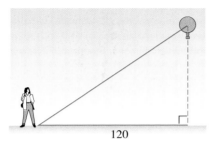

16. The following figure illustrates a telephone line that is to be attached from a point P, x feet above the ground on a telephone pole, to a point 8 ft above ground on a house located 30 feet from the pole.
    (a) Express the length of the wire as a function of x.
    (b) Where should the wire be attached to the pole so that the length of the wire is 50 feet?

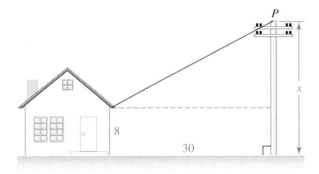

17. A piece of wire 30 cm long is to be cut into two pieces. One piece is to be formed into a square and the other piece formed into a circle.
    (a) Express the total area enclosed by the square and circle as a function of the side of the square s.
    (b) Express the total area enclosed by the square and circle as a function of the radius of the circle r.

18. Suppose that in Exercise 17 the piece of wire is of length L. Express the total area enclosed by the circle and square as a function of L and s.

19. A shoe manufacturer can sell 3500 pairs of shoes at $24 each. It is determined that for each 50¢ the price is reduced, 80 more pairs of shoes can be sold. Let n

be the number of 50¢ reductions. Express the total revenue as a function of n.

20. Given the scenario in Exercise 19, we can also analyze the problem by expressing the revenue as a function of the price per pair of shoes as follows.
    (a) If we let p be the reduced price per pair of shoes, determine how many 50¢ reductions in the original $24 price were made. For example, if the reduced price is $22.50, then there were three 50¢ reductions in the original $24 price.
    (b) Using the answer to part (a), determine how many pairs of shoes will be sold at this price p.
    (c) Since revenue is the price per pair times the number of pairs sold, use the answers to parts (a) and (b) to express the revenue R as a function of the price p.

21. An orange grove has 800 trees, each of which yields 750 oranges annually. It is determined that for each additional tree planted in the same grove, the annual yield for each tree drops by 30 oranges. Let n be the number of additional trees planted. Express the total annual yield of oranges as a function of n.

22. Given the scenario in Exercise 21, we can also analyze the problem by expressing the annual yield as a function of the total number of trees as follows:
    (a) If we let t be the total number of trees planted, determine who many additional trees were added to the original 800 trees. For example, if there are 815 trees, then 15 trees were added to the original 800 trees.
    (b) Using the answer to part (a), determine how many oranges each tree will yield if t trees are planted.
    (c) Since the total yield of oranges is the number of trees planted times the yield per tree, use the answers to parts (a) and (b) to express the total yield as function of the number of trees planted t.

23. A professional basketball team determines that an average of 8500 people per game will purchase general-admission tickets at a price of $9 per ticket, and that for each 50¢ added to the price, 225 fewer tickets will be sold. If n represents the number of times the ticket price is increased by 50¢, express the revenue taken in for general-admission tickets as a function of n.

24. Given the scenario in Exercise 23, we can also analyze the problem by expressing the revenue as a function of the price per ticket as follows.
    (a) If we let p be the increased price per ticket, determine how many 50¢ increases in the original

$9 price were made. For example, if the increased price is $11.50, then there were five 50¢ increases in the original $9 price.

(b) Using the answer to part (a), determine how many tickets will be sold at this price p.

(c) Since revenue is the price per ticket times the number of tickets sold, use the answers to parts (a) and (b) to express the revenue R as a function of the price p.

25. A small computer manufacturer finds that it is selling 750 computers per month at a price of $1250 each and that for each $75 that the price is reduced, 35 more computers are sold each month. If r represents the number of times the price is reduced by $75, express the revenue generated each month on computer sales as a function of r.

26. Given the scenario in Exercise 25, we can also analyze the problem by expressing the revenue as a function of the price per computer as follows.

(a) If we let p be the decreased price per computer, determine how many $75 reductions in the original $1250 price were made. For example, if the reduced price is $1100, then there were two $75 reductions in the original $1250 price.

(b) Using the answer to part (a), determine how many computers will be sold at th is price p.

(c) Since revenue is the price per computer times the number of computers sold, use the answers to parts (a) and (b) to express the revenue R as a function of the price p.

27. A homeowner wishes to enclose a rectangular garden next to the house. One side of the garden will be bounded by the house itself and the other three sides will be bounded by 80 feet of fencing. Express the area of the garden as a function of one variable.

28. A factory needs to fence in a rectangular parking lot next to a building. One side of the parking lot will be bounded by the building itself, so that fencing is needed only for the other three sides. If the parking lot is to have an area of 1000 sq meters, express the length of fence needed as a function of one variable.

29. A farmer wishes to enclose a rectangular plot of land and subdivide, as indicated in the figure. Heavy-duty fencing for the perimeter costs $5 per foot, and regular fencing for the interior costs $3 per foot. Express the total area that can be enclosed for $240 as a function of one variable.

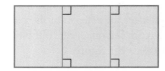

30. If the farmer in Exercise 29 wishes to use the same configuration to enclose an area of 480 sq ft, express the total cost of the fencing as a function of one variable.

31. A partition, as illustrated in the accompanying diagram, is to be set up so that it encloses an office area of 400 sq ft.

(a) Express the length of the partition as a function of x.

(b) If the partition costs $18 per foot, express the total cost of the partition as a function of x.

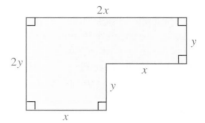

32. A hot air balloon is released from point A on the ground and rises vertically at a rate of 3 meters/sec. An observer is situated on the ground 200 meters from the release point A. If t represents the number of seconds after the balloon is released, express the distance from the observer to the balloon as a function of t.

33. A telephone company needs to lay a telephone cable from point A on one side of a river that is 1 mile wide to point D on the opposite side and 6 miles downriver, as indicated in the following diagram. The company will lay the cable underwater from A to some point C on the opposite shore, x miles from B, and then the cable will be above ground from C to D. Express the total length of the cable as a function of x.

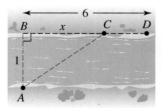

34. A graphics designer is creating a poster whose dimensions are to be 18 inches by 30 inches. The poster will have a uniform border x inches wide all around with an illustration in the center.
   (a) Express the area of the border as a function of x.
   (b) Express the central illustration area as a function of x.

35. A credit card company charges interest on an account's unpaid balance. The monthly interest rates are 1.65% on the first $2000 of the unpaid balance and 1.25% on any amount over $2000. Express the monthly interest, I, as a function of the unpaid balance b.

36. A salesperson is paid a commission of 6% on the first $20,000 in sales made each month and 10% on any amount over $20,000 in sales. Express the salesperson's monthly commission, C, as a function of the monthly sales, s.

37. A rectangular sheet of aluminum is 18 feet long by 1 foot wide. It is to be formed into a rain gutter by bending up a strip x feet wide along each edge of the sheet, as illustrated in the accompanying figure. Express the volume of rain the gutter can handle as a function of x.

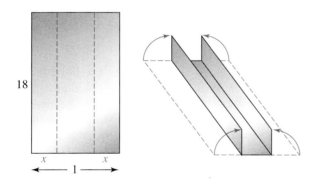

38. A manufacturer produces a product at a cost of $17.50 per unit. There are also fixed daily costs of $1900 regardless of how many units are produced. Each unit is sold for $29.75. Let x represent the number of units produced during a 10-day period.
   (a) Express the total cost, C, for producing x items during a 10-day period.
   (b) Express the revenue, R, earned over the 10-day period.
   (c) Express the profit, P, earned over the 10-day period as a function of x. (Recall that $P = R - C$.)
   (d) How many items must be manufactured and sold during the 10-day period in order to break even (that is, for R to be equal to C)?

39. Express the volume of a sphere as a function of its surface area.

40. Express the surface area of a sphere as a function of its volume.

41. Express the surface area of a closed right circular cylinder of height 10 as a function of its volume V.

42. Express the distance, D, between the point $(1, 4)$ and a point (x, y) on the parabola $y = x^2$ as a function of x.

43. Express the distance, D, between the point $(-1, 2)$ and a point (x, y) on the parabola $y = 9 - x^2$ as a function of x.

44. Express the distance, D, between the point $(-2, 4)$ and a point (x, y) on the parabola $y = (x + 2)^2$ as a function of y.

45. Express the distance, D, between the point $(5, 3)$ and a point (x, y) on the parabola $y = -(x - 5)^2 + 3$ as a function of y.

46. Express the distance, D, between the point $(6, 0)$ and a point (x, y) on the curve whose equation is $y = \sqrt{x}$ as a function of x.

47. Express the distance, D, between the point $(0, 2)$ and a point (x, y) on the curve whose equation is $y = 2 - \sqrt{x}$ as a function of x.

4.4   Quadratic Functions

A quadratic function is also called a second-degree function.

In Section 4.3 we developed the fact that the graph of $f(x) = a(x - h)^2 + k$ (called a parabola) can be obtained by applying the graphing principles to the graph of the basic parabola $f(x) = x^2$. In this section we extend these ideas to the general **quadratic function** of the form $f(x) = Ax^2 + Bx + C$, where $A \neq 0$. Let's begin with the following example.

Example 1 | Sketch a graph of the function $y = f(x) = -(x - 3)^2 + 4$.

Solution | See Figure 4.39.

Figure 4.39

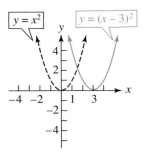

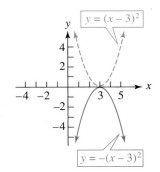

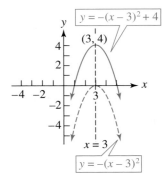

Starting with the basic parabola $f(x) = x^2$, the $(x - 3)^2$ tells us to shift the graph horizontally 3 units to the right.

(a)

The first negative sign in $-(x - 3)^2$ tells us to reflect the graph of $f(x) = (x - 3)^2$ about the x-axis.

(b)

The $+4$ tells us to shift the graph vertically up 4 units, for the graph of $f(x) = -(x - 3)^2 + 4$.

(c)

Notice that the vertex is shifted from $(0, 0)$ three units right and four units up to $(3, 4)$, and the axis of symmetry is now the vertical line $x = 3$. As usual, the x-intercepts are found by solving $f(x) = 0$. ∎

In general, the graph of $f(x) = a(x - h)^2 + k$, called the **standard form of the quadratic function**, can be found by shifting and stretching (and possibly reflecting about the x-axis) the graph of $f(x) = x^2$. Observe that in shifting the graph of $f(x) = x^2$ to obtain the graph of $f(x) = a(x - h)^2 + k$, the vertex is shifted from $(0, 0)$ to (h, k), and the axis of symmetry is shifted from $x = 0$ (the y-axis) to the vertical line $x = h$. Hence,

---

The graph of $f(x) = a(x - h)^2 + k$ is a parabola with vertex (h, k) and axis of symmetry $x = h$.

It may seem strange that h is preceded by a minus sign, and k is preceded by a plus sign. This form is constructed this way to maintain functional form (that is, keep $f(x)$ isolated).

As with the circle, a parabola is actually defined geometrically. Using this geometric definition (which we do in Chapter 8), we can derive the fact that the equation of a parabola can be expressed in the form $f(x) = Ax^2 + Bx + C$, where A, B, and C are constants with $A \neq 0$. This is called the **general form of the quadratic function.**

Notice that if we multiply out $f(x) = a(x - h)^2 + k$, we then get $f(x) = ax^2 - 2ahx + ah^2 + k$, which is of the form $f(x) = Ax^2 + Bx + C$, a quadratic function (remember that a, h, and k are constants).

On the other hand, we can take any quadratic function $f(x) = Ax^2 + Bx + C$ and put it into the form $f(x) = a(x - h)^2 + k$ by completing the square as discussed in Chapter 2. For example, to put the function $f(x) = x^2 + 6x + 5$ into the standard form $f(x) = a(x - h)^2 + k$, we proceed as follows:

$f(x) = x^2 + 6x + 5$ First group the terms involving powers of x together.

$\quad = (x^2 + 6x \quad) + 5$ Next complete the square for the expression within the parentheses. We take one-half the coefficient of x and square it:

$$\left[\frac{1}{2}(6)\right]^2 = (3)^2 = 9$$

In order to complete the square for $x^2 + 6x$, we want to have $x^2 + 6x + 9$.

Keep in mind that we do not want to add 9 to both sides of the equation because we want $f(x)$ to remain isolated. Thus to compensate for the 9 being added inside the parentheses, we must also *subtract* 9 inside the parentheses if we are to ensure the transformed function is equivalent to the original function. We have

$f(x) = (x^2 + 6x + 9 - 9) + 5$ Regroup and put the parabola in standard form.

$\quad = (x^2 + 6x + 9) - 9 + 5$

$\quad = (x + 3)^2 - 4$ This is the standard form of the parabola with $a = 1$, $h = -3$, and $k = -4$.

We can now easily identify the vertex of the parabola as $(-3, -4)$.

Example 2 Identify the vertex and axis of symmetry of the graph of the function $f(x) = 2x^2 - 6x + 7$.

Solution In cases where the leading coefficient A is not 1, we proceed as follows:

$f(x) = 2x^2 - 6x + 7$ Factor the coefficient 2 from $2x^2 - 6x$.

$\quad = 2(x^2 - 3x \quad) + 7$ Complete the square for $x^2 - 3x$:
$\left[\frac{1}{2}(-3)\right]^2 = \frac{9}{4}$. We add and subtract $\frac{9}{4}$ inside the parentheses.

$\quad = 2\left(x^2 - 3x + \frac{9}{4} - \frac{9}{4}\right) + 7$ Now distribute the 2 to the $-\frac{9}{4}$, leaving a perfect square within the parentheses.

$\quad = 2\left(x^2 - 3x + \frac{9}{4}\right) - 2\left(\frac{9}{4}\right) + 7$ Put the parabola in standard form.

$f(x) = 2\left(x - \frac{3}{2}\right)^2 + \frac{5}{2}$ This is in standard form with $a = 2$, $h = \frac{3}{2}$, and $k = \frac{5}{2}$.

Hence, the vertex is $\left(\frac{3}{2}, \frac{5}{2}\right)$; the axis of symmetry is $x = \frac{3}{2}$.   ∎

Since the graph of $f(x) = Ax^2 + Bx + C$, for $A > 0$, is a parabola opening upward, the vertex is the lowest point on the graph of $f(x)$; the y-coordinate of the vertex is called the ***minimum of*** $f(x)$. If $A < 0$, the parabola opens downward and the vertex is the highest point on the graph; its y-coordinate is called the ***maximum of*** $f(x)$. In Example 2, we found that the vertex of $f(x) = 2x^2 - 6x + 7$ is $(\frac{3}{2}, \frac{5}{2})$. Therefore the minimum value of $f(x) = 2x^2 - 6x + 7$ is $\frac{5}{2}$ and this minimum occurs when $x = \frac{3}{2}$.

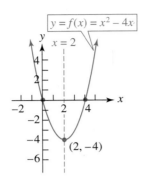

Figure 4.40

Note that the axis of symmetry is the vertical line midway between the x-intercepts.

A Formula for the Vertex of a Parabola

Let's think about the vertex of the general parabola $y = f(x) = Ax^2 + Bx + C$ from a slightly different point of view. Recall that the axis of symmetry divides the parabola into two identical mirror-image parts. In particular, if there are x-intercepts, *the axis of symmetry must cut through the midpoint between the x-intercepts.*

For example, the x-intercepts of $y = f(x) = x^2 - 4x$ can easily be found by factoring. $0 = x^2 - 4x = x(x - 4) \Rightarrow x = 0$ or $x = 4$, and so the axis of symmetry must be the vertical line $x = 2$. See Figure 4.40. Since the vertex falls on the axis of symmetry, we know that the x-coordinate of the vertex is $x = 2$, and we can find the y-coordinate by substituting $x = 2$ into $y = x^2 - 4x$, obtaining $y = -4$. Thus the vertex is $(2, -4)$ as indicated in Figure 4.40.

We can use this same approach to find the x-coordinate of the vertex of any parabola of the form $y = f(x) = Ax^2 + Bx$. We find the x-intercepts by setting $y = 0$ and solving for x:

$$0 = Ax^2 + Bx = x(Ax + B) \quad \Rightarrow \quad x = 0 \text{ or } x = -\frac{B}{A}$$

Since the x-intercepts are $x = 0$ and $x = -\frac{B}{A}$, the axis of symmetry will again be the vertical line that passes through the midpoint between them. Using the midpoint formula, this midpoint has x-coordinate

$$x = \frac{0 + \left(-\dfrac{B}{A}\right)}{2} = -\frac{B}{2A}$$

Thus we have found that the x-coordinate of the vertex of any parabola of the form $y = Ax^2 + Bx$ is $x = -\frac{B}{2A}$.

Knowing that the axis of symmetry is vertical and passes through the point $\left(-\dfrac{B}{2A}, 0\right)$, what is its equation?

The vertical shift principle tells us that to obtain the graph of the general parabola $y = f(x) = Ax^2 + Bx + C$ we shift the parabola $y = Ax^2 + Bx$ up or down C units. This shifts the y-coordinate of the vertex vertically up or down *but has no effect on the x-coordinate of the vertex,* and so the x-coordinate of the vertex of $y = Ax^2 + Bx + C$ is still given by $x = -\frac{B}{2A}$. Knowing the x-coordinate of the

How are the graphs of $y = Ax^2 + Bx$ and $y = Ax^2 + Bx + C$ related?

vertex $-\frac{B}{2A}$ allows us to find the vertex without the necessity of competing the square. We summarize this discussion as follows:

> An equation of the form $y = f(x) = Ax^2 + Bx + C$, $A \neq 0$, is a parabola with axis of symmetry $x = -\dfrac{B}{2A}$. The x-coordinate of the vertex is $-\dfrac{B}{2A}$.

Example 3 Graph the function $y = f(x) = -x^2 + 6x - 7$. Label the intercepts, vertex, and the axis of symmetry.

Solution This is the quadratic function $f(x) = Ax^2 + Bx + C$ with $A = -1$, $B = 6$, and $C = -7$. Hence, the x-coordinate of the vertex is

$$x = -\frac{B}{2A} = -\frac{6}{2(-1)} = 3$$

and the axis of symmetry is $x = 3$. Since the x-coordinate of the vertex is 3, the y-coordinate of the vertex is

$$f(3) = -(3)^2 + 6(3) - 7 = 2$$

Hence, the coordinates of the vertex are (3, 2).

The y-intercept is -7 since $f(0) = -(0)^2 + 6(0) - 7 = -7$.

The x-intercepts are found by using the quadratic formula:

$$x = \frac{-6 \pm \sqrt{6^2 - 4(-1)(-7)}}{2(-1)} = \frac{-6 \pm \sqrt{8}}{-2} = \frac{-6 \pm 2\sqrt{2}}{-2} = 3 \pm \sqrt{2}$$

Since $\sqrt{2} \approx 1.41$, the x-intercepts are

$$x = 3 + \sqrt{2} \approx 4.41 \quad \text{and} \quad x = 3 - \sqrt{2} \approx 1.59$$

The graph is in Figure 4.41.

Figure 4.41

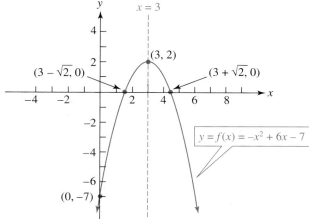

Note that we did not bother to develop a formula to find the y-coordinate of the vertex, since it is much easier to substitute the x-value in the equation to find the y-coordinate.

Let's look at Figure 4.42 and note a few things about the quadratic formula and its relationship to the axis of symmetry, vertex, and x-intercepts of the graph of $f(x) = Ax^2 + Bx + C$. The two solutions to the equation $Ax^2 + Bx + C = 0$ are the x-intercepts of the graph of $f(x) = Ax^2 + Bx + C$. The two solutions are found by applying the quadratic formula, which tells us to add and subtract the same quantity, $\dfrac{\sqrt{B^2 - 4AC}}{2A}$, to and from $\dfrac{-B}{2A}$, which is the x-coordinate of the vertex.

Figure 4.42

If a parabola has x-intercepts, they are equidistant from the axis of symmetry.

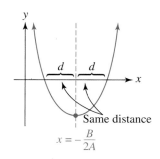

When solving a quadratic equation, the quadratic formula tells us that if the expression under the radical, $B^2 - 4AC$, is 0, then there is only one solution to $Ax^2 + Bx + C = 0$. This means the parabola $f(x) = Ax^2 + Bx + C$ has only

What does the fact that the roots of $Ax^2 + Bx + C = 0$ are imaginary say about the x-intercepts of the graph of $f(x) = Ax^2 + Bx + C$?

one x-intercept and therefore its vertex is *on* the x-axis. If $B^2 - 4AC < 0$, then there are two solutions, but the solutions are imaginary; hence, the graph of $f(x) = Ax^2 + Bx + C$ has no x-intercepts. On the other hand, if $B^2 - 4AC > 0$, then there are two real solutions and, therefore, two x-intercepts for the graph of $f(x) = Ax^2 + Bx + C$. See Figure 4.43. $B^2 - 4AC$ is called the **discriminant** of the equation $Ax^2 + Bx + C = 0$. In Figure 4.43 we consider the three possibilities for the discriminant and illustrate some possible graphs of $y = Ax^2 + Bx + C$ for each of these cases.

Figure 4.43

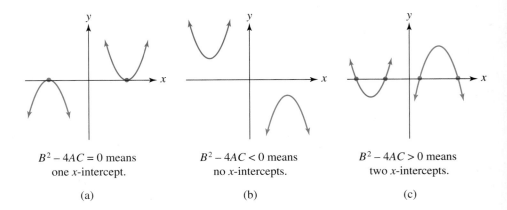

$B^2 - 4AC = 0$ means one x-intercept.

(a)

$B^2 - 4AC < 0$ means no x-intercepts.

(b)

$B^2 - 4AC > 0$ means two x-intercepts.

(c)

Sometimes relationships between variables can be described by quadratic functions, and we find that the vertex is an important point to identify. We can apply what we have learned about quadratic functions in precalculus to find this important point.

Example 4

The weekly profit earned by Ajax Inc. is related to the number of items produced each week in the following way:

$$P(x) = -2x^2 + 96x - 52$$

where $P(x)$ is the weekly profit earned in hundreds of dollars and x is the number of items produced each week. How many items must be produced weekly for the maximum weekly profit? What is the maximum weekly profit?

Solution

We note that the relationship is quadratic; therefore, if we let the horizontal axis be the number of items produced weekly, x, and let the vertical axis be the weekly profit in hundreds of dollars, $P(x)$, the graph of $P(x)$ is a parabola that opens down. (Why?) The ordered pairs satisfying the equation are of the form $(x, P(x))$. Since the parabola opens down, the vertex is the highest point. This means that the vertex is the point that yields the highest, or maximum, profit, $P(x)$. Thus, finding the number of items that will yield the maximum profit is equivalent to finding the x-coordinate of the vertex of the parabola $P(x) = -2x^2 + 96x - 52$. To find the x-coordinate of the vertex, we use $-\dfrac{B}{2A}$, with $A = -2$ and $B = 96$:

$$x = -\frac{B}{2A} = -\frac{96}{2(-2)} = 24$$

Therefore, Ajax Inc. must produce 24 items weekly to get the maximum profit. To find what this profit is, we simply find $P(24)$, the value of $P(x)$ when $x = 24$:

$$P(24) = -2(24)^2 + 96(24) - 52 = 1100$$

Since $P(x)$ is in hundreds of dollars, the profit is $110,000$ weekly. The graph of this function appears in Figure 4.44.

Figure 4.44

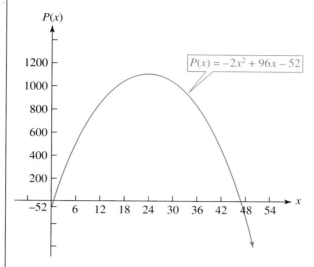

$$P(x) = -2x^2 + 96x - 52$$

Since one cannot produce a negative number of items, we have restricted the domain of $P(x)$ to nonnegative values. (In fact, the number of items must be a whole number, but for the sake of simplicity we draw the graph for all real $x \geq 0$.) However, it is possible for range values, $P(x)$ to include negative numbers (as losses), but note that the maximum value of $P(x)$ is 110,000. Hence the range is all real numbers less than or equal to 110,000. ∎

Example 5

Sarah wants to fence in a rectangular dog pen against her house. Since the house is to serve as one of the sides, she needs to fence in only three sides. If she uses 80 linear feet of fencing, what are the dimensions of the pen that would give the maximum area possible? See Figure 4.45.

Solution

We adapt the outline given in the last section for obtaining a mathematical model.

1. The diagram is given in the statement of the problem.

2. We write a formula for the area of a rectangle.

$$A = bh$$

3. We apply this formula to the rectangular pen in Figure 4.45:

$$A = xy$$

This is the quantity we are trying to maximize. We would like to express the area as a function of one variable.

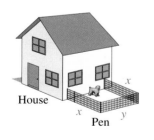

House   Pen

Figure 4.45

4. Since the length of fence to be used is 80 feet, we have

$$2x + y = 80 \quad \Rightarrow \quad y = 80 - 2x$$

5. We now substitute into the area equation:

$$A = xy \qquad \text{Substitute } y = 80 - 2x.$$
$$A = A(x) = x(80 - 2x) = -2x^2 + 80x$$

which we recognize as a quadratic function.

6. If we graph this function with x as the horizontal axis and $A(x)$ as the vertical axis, we get a parabola opening downward. The vertex of this parabola yields the maximum $A(x)$, or area. We will use the method of completing the square to find the vertex:

$$A(x) = -2x^2 + 80x \qquad \text{Factor out } -2.$$
$$= -2(x^2 - 40x \qquad) \qquad \text{Complete the square: } \left[\frac{1}{2}(-40)\right]^2 = 400.$$
$$\text{Add and subtract 400 inside the parentheses.}$$
$$= -2(x^2 - 40x + 400 - 400)$$
$$= -2(x^2 - 40x + 400) + 800 \qquad \text{Put in } a(x - h)^2 + k \text{ form.}$$
$$A(x) = -2(x - 20)^2 + 800 \qquad \text{Hence } h = 20 \text{ and } k = 800, \text{ and the vertex is } (20, 800).$$

This means that the maximum value of $A(x)$, the area of the rectangle, is 800 sq ft, and this occurs when $x = 20$ feet. Since x is the length of one side, the other side labeled y is $80 - 2x = 80 - 2(20) = 40$ feet. Hence the dimensions to produce the maximum area are 20 ft × 40 ft .

Observe that in completing the square for this problem, we simultaneously find both the x and the $A(x)$ values of the vertex. It is instructive to examine the graph of the area function $A(x) = -2x^2 + 80x$, which appears in Figure 4.46. Since the area of the pen cannot be zero or negative, we can see that the domain for x is the interval $(0, 40)$.

Figure 4.46

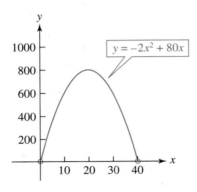

As we indicated in our discussion of straight lines, when we describe a graph we always describe it as we move from left to right. Thus we can also see that the area increases as x increases on the interval $(0, 20]$ until it reaches a maximum when $x = 20$, and then the area decreases on the interval $[20, 40)$.

4.4 Exercises

In Exercises 1–30, sketch the graph of the given equation. Label the axis of symmetry, vertex, and intercepts where appropriate.

1. $f(x) = 3x^2$

2. $f(x) = -3x^2$

3. $f(x) = (x - 3)^2$

4. $f(x) = (x + 3)^2$

5. $f(x) = x^2 + 8$

6. $f(x) = x^2 - 9$

7. $f(x) = (x - 3)^2 + 1$

8. $f(x) = -(x + 1)^2 - 1$

9. $f(x) = -3(x - 5)^2 + 4$

10. $f(x) = 3(x - 5)^2 + 4$

11. $f(x) = 2(x - 4)^2 + 3$

12. $f(x) = -2(x - 4)^2 + 3$

13. $y = x^2 - 4$

14. $y = x^2 - 4x$

15. $y = x - 4$

16. $y = x^2 - 4x + 4$

17. $y = 10 - 2x^2$

18. $y = 10 - 2x$

19. $f(x) = 3x^2 - 9x$

20. $y = 4x^2 + 8x$

21. $y = x^2 - 2x - 8$

22. $f(x) = x^2 - 4x - 12$

23. $y = f(x) = x^2 - 10x + 21$

24. $y = f(x) = x^2 - 5x + 5$

25. $y = 2x^2 + 12x + 16$

26. $y = -3x^2 + 6x + 24$

27. $f(x) = -2x^2 + 6x - 4$

28. $f(x) = 3x^2 + 3x + 6$

29. $y = 25x^2 - 1$

30. $f(x) = -3x^2 + 3x - 3$

In Exercises 31–42, find the maximum or minimum value of $f(x)$ if it exists.

31. $y = x^2 - 9$

32. $y = 1 - x^2$

33. $y = 6x - x^2$

34. $y = x^2 + 4x + 3$

35. $f(x) = -x^2 + 4x + 32$

36. $f(x) = -x^2 - 6x - 5$

37. $f(x) = -2x^2 + 6x - 18$

38. $f(x) = -3x^2 + 18x - 24$

39. $f(x) = x^2 - 4x - 32$

40. $f(x) = x^2 - 5x + 6$

41. $f(x) = 2x^2 - 6x + 18$

42. $f(x) = 3x^2 - 18x + 24$

43. A factory finds that its profit, $P(x)$, is related to the number of items it produces, x, in the following way:

$$P(x) = -x^2 + 80x$$

where $P(x)$ is the daily profit in dollars and x is the number of items produced daily.
 (a) Sketch the graph of $P(x)$.
 (b) How many items must be produced daily in order to maximize the profit?
 (c) What is the maximum profit?

44. The profit, $P(x)$, made on a concert is related to the price of a ticket, x, in the following way:

$$P(x) = 10{,}000(-x^2 + 10x - 24)$$

 (a) Sketch the graph of $P(x)$.
 (b) What ticket price would produce the maximum profit?

45. The daily cost for a table manufacturer is

$$C(x) = 0.02x^2 - 0.48x + 428.8$$

where $C(x)$ is the daily production cost in dollars, and x is the number of tables produced daily. How many tables should the manufacturer produce in order to minimize the daily production cost? What is this minimal production cost?

46. The daily profit earned by the Weldon factory is related to the number of cases of candy canes produced in the following way:

$$P(x) = -x^2 + 160x - 3400$$

where $P(x)$ is the daily profit in dollars, and x is the number of cases of candy canes produced daily. Find the number of cases of candy canes to be made daily in order to maximize the daily profit. What is the maximum profit?

47. Robin fires a rocket upward and the rocket travels according to the equation

$$s(t) = -16t^2 + 864t$$

where $s(t)$ is the height (in feet) of the rocket above the ground t seconds after it is fired. How many seconds does it take for the rocket to reach maximum height? What is the maximum height of the rocket?

48. Stacey stands on the roof of a building and throws a ball upward. The ball travels according to the equation

$$s(t) = -16t^2 + 64t + 60$$

where $s(t)$ is the height (in feet) of the ball above the *ground* t seconds after it is thrown. See the accompanying figure.
 (a) How high does the ball travel?
 (b) How many seconds does it take for the ball to hit the ground?
 (c) How high above the ground is the ball when it is thrown?

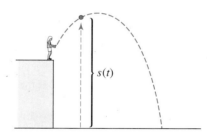

49. For a fixed perimeter of 100 feet, what dimensions will yield a rectangle with the maximum area? HINT: Draw a picture, label one side x, find the other sides in terms of x, and find the equation of the area in terms of x.

50. Repeat Exercise 49 if the fixed perimeter is P.

51. What two numbers whose sum is 104 will yield a maximum product?

52. What two numbers whose difference is 104 will yield a minimum product? Is there a maximum product? Explain.

53. A farmer wants to fence in two rectangular pens as illustrated in the following figure. If the farmer has

800 feet of fencing, what dimensions will yield the maximum area for the pens? HINT: Describe the total area in terms of one variable.

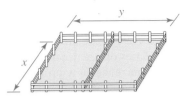

54. Suppose a farmer wants to set up a rectangular fencing scheme as illustrated in the figure. In addition, suppose that due to the strength of the prevailing winds, the farmer must use fence that costs $6 per linear foot for the fence in the north–south direction and $12 per linear foot for the fence in the east–west direction. What is the maximum area that can be enclosed if the farmer has a total of $1440 to spend on the fence?

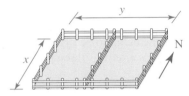

55. In economics, the demand function for a given item indicates how the price per unit p (in dollars) is related to the number of units x that are sold. Suppose a company finds that the demand function for one of the items it produces is

$$p = p(x) = 10 - \frac{x}{5} \quad \text{where } 0 \le x \le 50$$

(a) How many items would be sold if the price were $7 per unit?
(b) What should the price per unit be if 30 units are to be sold?
(c) Sketch the graph of this demand function.
(d) The revenue function, $R(x)$, is found by multiplying the price per unit p and the number of items sold x ($R = xp$). Find the revenue function corresponding to this demand function.
(e) Sketch the graph of this revenue function.
(f) How many items should be sold to maximize the revenue? What is the corresponding unit price?

56. Suppose that the unit price p (in dollars) and the quantity x sold of a particular item satisfy the demand equation

$$p = p(x) = 200 - \frac{x}{4} \quad \text{where } 0 \le x \le 800$$

(a) Find the price if 150 items are sold.
(b) If the price is $75, how many items are sold?
(c) Sketch the graph of this demand function.
(d) Find the revenue function $R(x)$; sketch its graph.
(e) How many items should be sold to maximize the revenue, and what is the corresponding unit price?

57. Suppose that the unit price p (in dollars) and the quantity x sold of a particular item satisfy the demand equation

$$p = p(x) = -\frac{x}{9} + 240 \quad \text{where } 0 \le x \le 2160$$

(a) If the price is $175, how many items are sold?
(b) Find the price if 1800 items are sold.
(c) Sketch the graph of this demand function.
(d) Find the revenue function $R(x)$; sketch its graph.
(e) How many items should be sold to maximize the revenue, and what is the corresponding unit price?

58. An athletic field is to be constructed in the shape of a rectangle with a semicircle at each end, as indicated in the accompanying figure. The perimeter of the field is to be a $\frac{1}{4}$-mile running track. Find the dimensions of r and x that yield the athletic field with the greatest possible area.

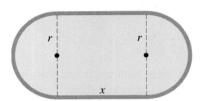

59. A Norman window is in the shape of a rectangle surmounted by a semicircle, as shown in the accompanying figure. If the perimeter of the window is 12 feet, show that the window will have a maximum area when both r and x are equal to $\frac{12}{\pi + 4}$, and find this maximum area.

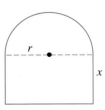

60. A manufacturer finds that the revenue, R, earned on the production and sale of n items is given by the function

$$R = R(n) = 1.56n - 0.0002n^2$$

Keep in mind that a negative number of items cannot be produced; assume that the manufacturer will stop production when the revenue becomes negative.

(a) Sketch the graph of the function $R(n)$.

(b) How many items can the manufacturer produce and not have a negative revenue?

(c) Compute $R(7000)$.

(d) What is the maximum possible revenue and how many items must be manufactured and sold to achieve this revenue?

In Exercises 61–68, we revisit some of the functions we obtained in the exercises of Section 4.3.

61. In Exercise 19 of Section 4.3, we obtained the total revenue as a function of n, the number of price increases. Find the value of n that maximizes this revenue function.

62. In Exercise 20 of Section 4.3, we obtained the total revenue as a function of p, the price per pair of shoes. Find the value of p that maximizes this revenue function, and show that this price is equivalent to the answer obtained in Exercise 61.

63. In Exercise 21 of Section 4.3, we obtained the annual yield of oranges as a function of n, the number of additional trees planted. Find the value of n that maximizes the total yield.

64. In Exercise 22 of Section 4.3, we obtained the annual yield of oranges as a function of the number of trees t. Find the value of t that maximizes the total yield, and show that this number of trees is equivalent to the answer obtained in Exercise 63.

65. In Exercise 23 of Section 4.3, we obtained the total revenue as a function of n, the number of price increases. Find the value of n that maximizes this revenue function.

66. In Exercise 24 of Section 4.3, we obtained the total revenue as a function of p, the price per ticket. find the value of p that maximizes this revenue function, and show that this price is equivalent to the answer obtained in Exercise 65.

67. In Exercise 25 of Section 4.3, we obtained the total revenue as a function of r, the number of price reductions. Find the value of r that maximizes this revenue function.

68. In Exercise 26 of Section 4.3, we obtained the total revenue as a function of p, the price per computer. Find the value of p that maximizes this revenue function, and show that this price is equivalent to the answer obtained in Exercise 67.

69. If a baseball is hit at a point 4 ft above home plate at a speed of 115 ft/sec and at an upward 45° angle, then (neglecting air resistance) its path is approximately the parabola with equation

$$y = -0.0024x^2 + x + 4$$

where x is the number of feet the ball is from home plate.

Determine whether the ball will clear an 18-foot-high fence 400 feet from home plate.

70. If a baseball is hit at a point 3.5 ft above home plate at a speed of 120 ft/sec and at an upward 42° angle, then (neglecting air resistance) its path is approximately the parabola with equation

$$y = -0.0022x^2 + 0.90x + 3.5$$

where x is the number of feet the ball is from home plate.

Determine whether the ball will clear a 15-foot-high fence 400 feet from home plate.

71. A real estate management firm is handling an apartment complex of 200 units. At a monthly rent of $800 all the units are occupied. The managers estimate that for each $30 the monthly rent per unit is increased, five additional units will become vacant. For each occupied unit the management must pay $320 dollars per month for taxes and maintenance, whereas for each empty unit the management must pay $170 in taxes. What rent should be charged in order to maximize the monthly profit?

72. Suppose that a motor vehicle consumer ratings service finds that the distance, d, in miles, that a certain car can travel on one tank of gasoline depends on its speed, v, according to the function

$$d = d(v) = 45v - \left(\frac{v}{1.46}\right)^2 \qquad \text{for } 10 \le v \le 90$$

What speed maximizes the distance d and consequently minimizes the fuel consumption?

73. To encourage the use of mass transit, a local transit authority projects that 36,000 people will use buses if the fare is $4 and that for each $0.25 decrease in the fare, 3000 additional passengers will decide to take the bus rather than drive.

(a) What fare should be charged if the transit authority wants to maximize the revenue from the bus fare?

(b) What fare will maximize the revenue if the buses can handle no more than 40,000 passengers?

74. A company finds that if it spends x dollars on advertising, it earns a profit, P, given by the function

$$P = P(x) = 250 + 30x - 0.02x^2$$

Find the advertising expenditure that generates the greatest profit.

75. For infants, the growth rate g, in pounds per month, can be approximated by the function

$$g = g(w) = kw(21 - w)$$

where w is their present weight and k is a positive constant. At what weight is an infant's rate of growth a maximum?

76. Suppose that a piece of wire 20 cm long is bent into the shape of a rectangle with length x and width y.
  (a) Express the width y as a function of x.
  (b) Express the area, A, of the rectangle as a function of x.
  (c) Use the function obtained in part (b) to show that the area A is a maximum when the rectangle is a square.

77. Given the x-coordinate of the vertex for the parabola $f(x) = Ax^2 + Bx + C$ is $-\dfrac{B}{2A}$, find the y-coordinate of the vertex in terms of A, B, and C.

78. Show that the coordinates of the vertex of the parabola $f(x) = Ax^2 + Bx + C$ are $\left(-\dfrac{B}{2A}, \dfrac{-B^2 + 4AC}{4A}\right)$ by putting $f(x) = Ax^2 + Bx + C$ in standard form, and identifying h and k.   HINT: Complete the square for $f(x) = Ax^2 + Bx + C$.

79. Sketch a graph of $f(x) = Ax^2 + Bx + C$ satisfying the following conditions.
  (a) $A > 0$, and $B^2 - 4AC < 0$
  (b) $A < 0$, and $B^2 - 4AC < 0$
  (c) Is it possible for the vertex of the graph described in parts (a) to be below the x-axis? Why or why not?

80. Sketch the graph of $f(x) = Ax^2 + Bx + C$ satisfying the following conditions:
  (a) $A > 0$, and $B^2 - 4AC > 0$
  (b) $A < 0$, and $B^2 - 4AC > 0$
  (c) Is it possible for the vertex of the graph described in part (a) to be above the x-axis? Why or why not?

81. Sketch the graph of $f(x) = 2x^2 - 6x - 8$ and answer the following questions using the graph.
  (a) When is $f(x) = 0$?
  (b) On what interval(s) of x is $f(x) > 0$?
  (c) On what interval(s) of x is $f(x) < 0$?
  (d) How does your answer to part (b) relate to solving the inequality $2x^2 - 6x - 8 > 0$?

82. Sketch the graph of $f(x) = -2x^2 + 6x + 8$ and answer the following questions using the graph.
  (a) When is $f(x) = 0$?
  (b) On what interval(s) of x is $f(x) > 0$?
  (c) On what interval(s) of x is $f(x) < 0$?
  (d) How does your answer to part (b) relate to solving the inequality $-2x^2 + 6x + 8 > 0$?

83. (a) What is the maximum value of the function $f(x) = -x^2 + 4$?
  (b) Use the information in part (a) to find the maximum value of the function $F(x) = \sqrt{-x^2 + 4}$.

84. (a) What is the maximum value of the function $h(x) = -x^2 + 6x - 8$?
  (b) What is the maximum value of the function $H(x) = \sqrt{-x^2 + 6x - 8}$?

85. (a) What is the minimum value of the function $f(x) = x^2 - 4x + 7$?
  (b) Use the information in (a) to find the maximum value of the function $F(x) = \dfrac{1}{x^2 - 4x + 7}$.

86. (a) What is the maximum value of the function $h(x) = -x^2 + 6x - 10$?
  (b) What is the minimum value of the function $H(x) = \dfrac{1}{-x^2 + 6x - 10}$?

87. We have noted that if a parabola has x-intercepts, they are equidistant from the axis of symmetry. Suppose the parabola $y = Ax^2 + Bx + C$ has x-intercepts. Verify that the equation of the axis of symmetry is $x = -\dfrac{B}{2A}$ by using the quadratic formula to find the x-intercepts and then using the midpoint formula to find the midpoint between them.

88. What happens to the reasoning in Exercise 87 if there is only one x-intercept?

Question for Thought

89. Show that the x-intercepts of the parabola $y = a(x - h)^2 + k$ are $x = h \pm \sqrt{\dfrac{-k}{a}}$. What must the signs of a and k be in order for the parabola to have two x-intercepts? Describe what these signs of a and k mean in terms of the graph of the parabola.

4.5 Operations on Functions

To analyze the behavior of a function such as $h(x) = \dfrac{x^2 - 1}{x^3}$, it will be advantageous to view $h(x)$ as the quotient of the two functions $f(x) = x^2 - 1$ and $g(x) = x^3$; that is, we will view $h(x)$ as being "built" from simpler functions via the operations of arithmetic.

Just as we can add, subtract, multiply, and divide two real numbers to create another real number, we now define what it means to perform the arithmetic operations on functions. This is often called the **algebra of functions**.

Definition of the Sum, Difference, Product, and Quotient of Two Functions

Let $f(x)$ and $g(x)$ be two functions with domains D_f and D_g, respectively. We define the following four functions:

1. $(f + g)(x) = f(x) + g(x)$ 　　　　The **sum** of the two functions

2. $(f - g)(x) = f(x) - g(x)$ 　　　　The **difference** of the two functions

3. $(f \cdot g)(x) = f(x) \cdot g(x)$ 　　　　The **product** of the two functions

4. $\left(\dfrac{f}{g}\right)(x) = \dfrac{f(x)}{g(x)}$ 　　　　The **quotient** of the two functions (provided $g(x) \neq 0$)

Since an x-value must be an input into both f and g, the domain of $(f + g)(x)$ is the set of all x common to the domains of f and g. This is usually written as $D_{f+g} = D_f \cap D_g$. Similar statements hold for the domains of the difference and product of two functions. In the case of the quotient, we must impose the additional restriction that all elements in the domain of g for which $g(x) = 0$ are excluded.

Example 1 　Let $f(x) = 3x^2 + 2$ and $g(x) = 5x - 4$. Find each of the following and its domain.

(a) $(f + g)(x)$ 　　(b) $(f - g)(x)$ 　　(c) $(f \cdot g)(x)$ 　　(d) $\left(\dfrac{f}{g}\right)(x)$

Solution 　(a) $(f + g)(x) = f(x) + g(x) = (3x^2 + 2) + (5x - 4) = 3x^2 + 5x - 2$

(b) $(f - g)(x) = f(x) - g(x) = (3x^2 + 2) - (5x - 4)$
$$= 3x^2 + 2 - 5x + 4 = 3x^2 - 5x + 6$$

(c) $(f \cdot g)(x) = f(x) \cdot g(x) = (3x^2 + 2)(5x - 4) = 15x^3 - 12x^2 + 10x - 8$
Since D_f and D_g are each the set of all real numbers, so D_{f+g}, D_{f-g}, and $D_{f\cdot g}$ are each the set of all real numbers.

(d) $\left(\dfrac{f}{g}\right)(x) = \dfrac{f(x)}{g(x)} = \dfrac{3x^2 + 2}{5x - 4}$

For x to be in the domain of $\dfrac{f}{g}$, we require that $5x - 4 \neq 0$. Thus we have

$$D_{f/g} = \left\{ x \mid x \neq \dfrac{4}{5} \right\}$$

Example 2 　Given $f(x) = \sqrt{x}$ and $g(x) = \dfrac{1}{x^2 - 5x + 6}$. Find $(f \cdot g)(x)$ and its domain.

Solution

$$(f \cdot g)(x) = f(x) \cdot g(x) = \sqrt{x} \cdot \frac{1}{x^2 - 5x + 6} = \boxed{\frac{\sqrt{x}}{x^2 - 5x + 6}}$$

For x to be in the domain of $f(x)$, we require that x be nonnegative, whereas for x to be in the domain of $g(x)$, we require that the denominator not be zero, that is

$$x^2 - 5x + 6 = (x - 2)(x - 3) \neq 0$$

We have $D_f = \{x | x \geq 0\}$ and $D_g = \{x | x \neq 2, 3\}$. Therefore,

$$\boxed{D_{f \cdot g} = \{x | x \geq 0, x \neq 2, 3\}}$$

Composing Functions

There is yet another way of producing a new function from two given functions.

Definition of the Composition of Functions

Given two functions $f(x)$ and $g(x)$, the **composition** of the two functions is denoted by $f \circ g$ and is defined by

$$(f \circ g)(x) = f[g(x)]$$

$(f \circ g)(x)$ is read "f composed with g of x."

The domain of $f \circ g$ consists of those x's in the domain of g whose range values are in the domain of f, that is, those x's for which $g(x)$ is in the domain of f.

When we compose two functions f and g, we begin with an input value x in the domain of g and get a unique output value $g(x)$ in the range of g. This output value is then used as the input value for $f(x)$, giving the unique output value $f[g(x)]$. Thus $g(x)$ must be in the domain of f.

For example, suppose we have the function $F = \{(2, z), (3, q)\}$ and the function $G = \{(a, 2), (b, 3), (c, 5)\}$. The function $(F \circ G)(x) = F[G(x)]$ is found by taking elements in the domain of G and evaluating as follows:

$$(F \circ G)(a) = F[G(a)] = F(2) = z \qquad (F \circ G)(b) = F[G(b)] = F(3) = q$$

If we attempt to find $F(G(c))$ we get $F(5)$, but 5 is not in the domain of $F(x)$ and so we cannot find $(F \circ G)(c)$. Hence $F \circ G = \{(a, z), (b, q)\}$. Figure 4.47 illustrates this situation.

Figure 4.47

Diagram illustrating $F \circ G$

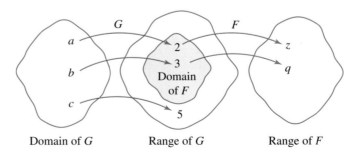

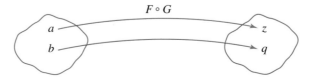

Notice that to evaluate $(F \circ G)(x)$ we first apply $G(x)$.

Figure 4.48 illustrates the situation for the general function $F \circ G$.

Figure 4.48

Diagram illustrating $(F \circ G)(x)$

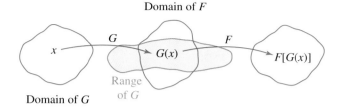

Example 3 | Given $f(x) = 5x^2 - 3x + 2$ and $g(x) = 4x + 3$, find

(a) $(f \circ g)(-2)$ **(b)** $(g \circ f)(2)$ **(c)** $(f \circ g)(x)$ **(d)** $(g \circ f)(x)$

Solution | **(a)** $(f \circ g)(-2) = f[g(-2)]$ First evaluate $g(-2) = 4(-2) + 3 = -5$.

$$= f(-5)$$

$$= 5(-5)^2 - 3(-5) + 2 = \boxed{142}$$

(b) $(g \circ f)(2) = g[f(2)]$ First evaluate $f(2) = 5(2)^2 - 3(2) + 2 = 16$.

$$= g(16)$$

$$= 4(16) + 3 = \boxed{67}$$

(c) $(f \circ g)(x) = f[g(x)]$ But $g(x) = 4x + 3$.

$$= f(4x + 3)$$

$$= 5(4x + 3)^2 - 3(4x + 3) + 2$$

$$= 5(16x^2 + 24x + 9) - 12x - 9 + 2 = \boxed{80x^2 + 108x + 38}$$

(d) $(g \circ f)(x) = g[f(x)]$ But $f(x) = 5x^2 - 3x + 2$.

$$= g(5x^2 - 3x + 2)$$

$$= 4(5x^2 - 3x + 2) + 3 = \boxed{20x^2 - 12x + 11}$$

Comparing the results of parts (c) and (d), we see that $(f \circ g)(x)$ is not the same as $(g \circ f)(x)$. ∎

Example 4 | Given $f(x) = \dfrac{x}{x + 1}$ and $g(x) = \dfrac{2}{x - 1}$, find

(a) $(f \circ g)(x)$ and its domain   **(b)** $(g \circ f)(x)$ and its domain.

Solution | **(a)** $(f \circ g)(x) = f[g(x)] = f\left(\dfrac{2}{x - 1}\right)$ In $f(x)$, we substitute $\dfrac{2}{x - 1}$ for x.

$$= \dfrac{\dfrac{2}{x - 1}}{\dfrac{2}{x - 1} + 1}$$ We simplify the complex fraction by multiplying the numerator and denominator by $x - 1$.

$$= \dfrac{\dfrac{2}{x - 1} \cdot (x - 1)}{\left(\dfrac{2}{x - 1} + 1\right) \cdot (x - 1)}$$

$$= \dfrac{2}{2 + x - 1} = \dfrac{2}{x + 1}$$ Keep in mind that a function is not completely defined until we specify its domain.

The Technology Corner at the end of this section discusses using a graphing calculator to compose functions.

Looking at the final answer, we can easily see that x cannot be equal to -1. However, it is not sufficient to look only at the final form of $(f \circ g)(x)$. As stated before, x must first be an input into $g(x)$ and so must be in the domain of g. Since 1 is not in the domain of g (Why?), 1 is not in the domain of $f \circ g$. Therefore, x cannot be equal to 1. Thus our final answer for the domain of $f \circ g$ is

$$D_{f \circ g} = \{x \mid x \neq \pm 1\}$$

(b) $(g \circ f)(x) = g[f(x)] = g\left(\dfrac{x}{x+1}\right)$ In $g(x)$, we substitute $\dfrac{x}{x+1}$ for x.

$$= \dfrac{2}{\dfrac{x}{x+1} - 1}$$ We simplify the complex fraction by multiplying the numerator and denominator by $x+1$.

$$= \dfrac{2x+2}{x - (x+1)}$$

$$= \dfrac{2x+2}{-1} = \boxed{-2x - 2}$$

Looking at the final form of $f \circ g$, we might mistakenly think that its domain is the set of all real numbers. Again, if we keep in mind that x must first be an input into $f(x)$ and so must be in the domain of f, we see that the domain of $g \circ f$ is

Why is -1 not in the domain of $f(x)$?

$$D_{g \circ f} = \{x \mid x \neq -1\}.$$ ∎

Decomposing Functions

Perhaps the most important aspect of the composition of functions is that it allows us to express a given function in terms of simpler functions.

In the composition $(f \circ g)(x) = f[g(x)]$, it is very useful to view g as the "inner function" and f as the "outer function." It is often useful to represent a given function $h(x)$ as the composition of two functions $f(x)$ and $g(x)$. The process of identifying possible functions $f(x)$ and $g(x)$ is called *decomposing* the function $h(x)$. We illustrate the process of decomposing a function in the next example.

Example 5   Let $h(x) = \sqrt{x^2 + 1}$. Find two functions $f(x)$ and $g(x)$ so that $h(x) = (f \circ g)(x)$.

Solution   Looking at $h(x) = \sqrt{x^2 + 1}$, we can view $x^2 + 1$ as the inner function and the square root function as the outer function. This suggests that we can let $g(x) = x^2 + 1$ and $f(x) = \sqrt{x}$; then

$$(f \circ g)(x) = f[g(x)] = f(x^2 + 1) = \sqrt{x^2 + 1} = h(x) \quad \text{as required}$$

It is worthwhile noting that this is not the only solution. Instead we may let $g(x) = x^2$ and $f(x) = \sqrt{x + 1}$; then

$$(f \circ g)(x) = f[g(x)] = f(x^2) = \sqrt{x^2 + 1} = h(x) \quad \text{as required}$$

Although both solutions are correct, the first has the advantage of adhering most closely to our description of g being the inner function and f being the outer function. Keep this in mind when you use the answer key to check the answers to the exercises at the end of this section.   ∎

If you look back at a number of the examples in Section 4.1 (especially Example 10) and Example 1 in Section 4.4, you will see that we graphed the various functions by mentally decomposing them.

We conclude this section with an application of composite functions.

Example 6

Suppose that the radius, r, of a circle is increasing so that its length in centimeters after t seconds is given by the function

$$r = r(t) = 5t + 3 \qquad \text{where } 0 \le t \le 8$$

(a) Use a composite function to express the area of the circle as a function of t.
(b) Find the area of the circle after 6 seconds.
(c) To the nearest second, find how many seconds it will take for the area to be 3500 sq cm.

Solution

(a) The area is a function of r; r is, in turn, a function of t, so that if we compose the two functions we will get the area as a function of t. We know that the formula for the area of a circle is $A = \pi r^2$. Thus the area of a circle is a function of its radius, and we may write

$$A = A(r) = \pi r^2$$

We are also given that the length of the radius is a function of t, $r(t) = 5t + 3$, and so we may write the area as a function of t as

$$A = A[r(t)] = A(5t + 3) = \pi(5t + 3)^2$$

and so the area of the circle as a function of t is $\boxed{A = A_1(t) = \pi(5t + 3)^2}$

We cannot denote the area A as a function of t as $A(t)$ because that would mean substitute t for r in $A(r)$.

(b) Once we have expressed the area as a function of t we simply substitute $t = 6$ to find the area after 6 seconds:

$$A = A_1(t) = \pi(5t + 3)^2 \qquad \text{Substitute } t = 6.$$
$$= \pi(5(6) + 3)^2 = \pi(33)^2$$
$$= \boxed{1089\pi} \text{ sq cm}$$

(c) We want to find the value of t that makes the area 3500. We can use the result of part (a), which expresses the area as a function of t.

$$A_1(t) = \pi(5t + 3)^2 \qquad \text{We substitute 3500 for the area.}$$
$$3500 = \pi(5t + 3)^2$$
$$\frac{3500}{\pi} = (5t + 3)^2 \qquad \text{Take square roots.}$$
$$\pm\sqrt{\frac{3500}{\pi}} = 5t + 3 \qquad \text{We solve for } t.$$
$$t = \frac{\pm\sqrt{\dfrac{3500}{\pi}} - 3}{5} \qquad \text{Since } t \text{ must be positive we have}$$
$$t = \frac{\sqrt{\dfrac{3500}{\pi}} - 3}{5} \approx \quad 6 \text{ seconds}$$

Technology Corner

We can construct composite functions on a T1-83 Plus graphing calculator. For example, if we want to construct the function $(f \circ g)(x)$ of Example 4, we would type in $f(x) = \dfrac{x}{x+1}$ as Y_1 and $g(x) = \dfrac{2}{x-1}$ as Y_2 in the $\boxed{Y=}$ menu.

Next, in the $\boxed{Y=}$ menu we construct the composition of the two functions in Y_3 as follows: go into the $\boxed{VARS}$ Y-VARS menu, choose 1:Function, and then select $1:Y_1$, which pastes Y_1 into Y_3. Finally, we open a parentheses and again go into the $\boxed{VARS}$ Y-VARS menu, choose 1:Function, and then select $2:Y_2$, which pastes Y_2 into the Y_1. (Don't forget to close your parentheses.) See the screen TC.8.

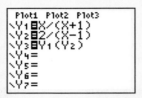

Figure TC.8

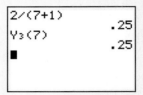

Figure TC.9

We can check the answer to example 4(a) by choosing an x-value, say, $x = 7$, and substitute it into $\dfrac{2}{x+1}$, the answer we got in Example 4, and into the composite function we just constructed. See TC.9. Although this check is not conclusive, it does give us a great deal of confidence that the answer we obtained in Example 4(a) is correct.

4.5 Exercises

In Exercises 1–32, use the following functions f, g, h, r, s, and t.

$$f(x) = 2x^2 - x - 3 \quad g(x) = 3x - 2 \quad h(x) = 5$$

$$s(x) = \frac{1}{x} \quad r(x) = x^3 - 1 \quad t(x) = \frac{4}{x+2}$$

In Exercises 1–14, find the required value.

1. $(g + r)(2)$
2. $(f - t)(-3)$
3. $(h \cdot s)(-6)$
4. $(r \cdot t)(3)$
5. $\left(\dfrac{g}{f}\right)(4)$
6. $\left(\dfrac{s}{t}\right)(x)$
7. $\left(\dfrac{h}{r}\right)(-1)$
8. $(t - s)(8)$
9. $(f \circ g)(4)$
10. $(g \circ f)(4)$
11. $(h \circ s)(-6)$
12. $(s \circ h)(8)$
13. $(f + g)(x)$
14. $(s \cdot r)(x)$

In Exercises 15–32, use the functions f, g, h, r, s, and t to find the required function and its domain.

15. $(f - g)(x)$
16. $(r + s)(x)$
17. $\left(\dfrac{s}{h}\right)(x)$
18. $\left(\dfrac{h}{s}\right)(x)$
19. $(r - f)(x)$
20. $(s + t)(x)$
21. $(g \cdot t)(x)$
22. $\left(\dfrac{g}{t}\right)(x)$
23. $(f \circ g)(x)$
24. $(g \circ f)(x)$
25. $(r \circ s)(x)$
26. $(s \circ r)(x)$
27. $(h \circ t)(x)$
28. $(t \circ h)(x)$
29. $(r \circ g)(x)$
30. $(g \circ r)(x)$
31. $(s \circ t)(x)$
32. $(t \circ s)(x)$
33. Let $f(x) = \sqrt{x+1}$ and $g(x) = 2x - 7$.
    (a) Find $(f \circ g)(x)$ and its domain.
    (b) Find $(g \circ f)(x)$ and its domain.

34. Let $F(x) = \dfrac{x-1}{x+1}$ and $G(x) = \dfrac{2}{x}$. Find

(a) $(F \circ G)(x)$ (b) $(G \circ F)(x)$

35. Let $f(t) = t^2 + t$ and $g(t) = \dfrac{6}{t-3}$. Find

(a) $(f \circ g)(t)$ (b) $(g \circ f)(t)$
(c) $(f \circ f)(t)$ (d) $(g \circ g)(t)$

36. Let $f(x) = 3x - 5$ and $g(x) = \dfrac{1}{3}(x+5)$. Find

(a) $(f \circ g)(x)$ (b) $(g \circ f)(x)$

37. We may define the composition of three functions as follows:

$$(f \circ g \circ h)(x) = f\{g[h(x)]\}$$

Let $f(x) = 2x + 3$, $g(x) = \sqrt{x}$, and $h(x) = \dfrac{1}{x}$. Find

(a) $(h \circ g \circ f)(x)$ (b) $(f \circ h \circ g)(x)$
(c) $(g \circ f \circ h)(x)$

38. Let $f(x) = x^2$ and $g(x) = \sqrt{x}$.
(a) Find $(f \circ g)(x)$ and its domain.
(b) Find $(g \circ f)(x)$ and its domain.
(c) Are $(f \circ g)(x)$ and $(g \circ f)(x)$ the same function? Explain.

In Exercises 39–44, find two functions $f(x)$ and $g(x)$ so that the given function $h(x) = (f \circ g)(x)$.

39. $h(x) = (x+3)^3$ **40.** $h(x) = \sqrt{5x-3}$

41. $h(x) = \left(\dfrac{x+4}{x-1}\right)^2$ **42.** $h(x) = \sqrt[3]{x^2 - 4x + 5}$

43. $h(x) = \dfrac{1}{x} + 6$ **44.** $h(x) = \dfrac{1}{x+6}$

45. Let $f(x) = 5x - 3$. Find $g(x)$ so that

$(f \circ g)(x) = 2x + 7$

46. Let $f(x) = 2x + 1$. Find $g(x)$ so that

$(f \circ g)(x) = 3x - 1$

47. Let $f(x) = 8 - 5x$. Find $g(x)$ so that $(f \circ g)(x) = x$.

48. Let $f(x) = \dfrac{2x+3}{5}$. Find $g(x)$ so that $(f \circ g)(x) = x$.

49. Suppose a laboratory technician is growing a bacteria culture in which the number of bacteria present, N, depends on the Celsius temperature, C, of the surrounding air and is given by the function

$N = N(C) = 3C^2 + 250C + 10{,}200$ for $15 \le C \le 40$

The Celsius temperature, C is, in turn, dependent on the number of hours, h, after the culture begins growing and is given by the function

$C(h) = 5h + 15$ for $0 \le h \le 5$

(a) Express the number of bacteria, N, as a function of h.
(b) How many bacteria are present after 4 hours?
(c) After how many hours are there 30,000 bacteria?

50. Suppose that the base of a rectangular box is a square of side x inches and its height is 6 inches. The length of x is dependent on the number of minutes, t, that have elapsed after $t = 0$ minutes according to the function

$x(t) = 25 - t^2$ for $0 \le t \le 4$

(a) Express the volume of the box as a function of x.
(b) Express the volume of the box as a function of t.
(c) What is the volume of the box after 3 minutes?

51. A calculator manufacturer sets the price, P, of a calculator at 30% above the cost to manufacture it. The cost for manufacturing n calculators is given by the function

$c = c(n) = 32n + 370$

(a) If 1 calculator is produced, what will its price be?
(b) If 10 calculators are produced, what will the price of each be?
(c) If n calculators are produced, express the price per calculator, P, as a function of n.

52. The volume, V, of a sphere of radius r is given by the formula

$$V = \dfrac{4}{3}\pi r^3$$

The radius is decreasing with time, t, according to the formula

$$r = \dfrac{1}{\sqrt{t+1}} \qquad \text{for } t \ge 0$$

Express the volume as a function of t.

Questions for Thought

53. We have seen through a variety of examples and exercises that, in general, $(f \circ g)(x) \ne (g \circ f)(x)$. Can you find a specific example of two functions f and g for which $(f \circ g)(x) = (g \circ f)(x)$? Can you describe some general examples or categories of functions for which $(f \circ g)(x) = (g \circ f)(x)$?

54. Write the domain of Example 2 in this section using interval notation. Which form of the answer is easier to understand?

4.6   Inverse Functions

In our initial discussions we stressed the critical aspect of a function—that to each x there corresponds *exactly one y*. We noted that it is quite possible for a function to assign the same y-value to two different x-values.

In addition, we have been very specific in our designations of the variables we use in defining functions; x is the independent variable and y is the dependent variable, or, as we have been saying, "Pick x, and get y." We have agreed to always list the independent variable (the value we pick) as the first coordinate and the dependent variable (the value we get) as the second coordinate. However, this arrangement is quite arbitrary.

Let's examine the following rather simple function F defined by the following set of ordered pairs:

$$F: \{(-2, 2), (0, 4), (1, 5), (3, 7)\}$$

with domain $D_F = \{-2, 0, 1, 3\}$ and range $R_F = \{2, 4, 5, 7\}$

Looking carefully at these ordered pairs, we note that not only do these ordered pairs define y as a function of x, but they can equally well be viewed as defining x as a function of y. That is, since each y-value has exactly one x-value associated with it, these ordered pairs allow us to "pick y and get x." Let us call this function G; that is, G is the result of using these ordered pairs to define x as a function of y. We would record this function G as

$$G: \{(2, -2), (4, 0), (5, 1), (7, 3)\}$$

with domain $D_G = \{2, 4, 5, 7\}$ and range $R_G = \{-2, 0, 1, 3\}$

Remember that the independent variable is recorded as the first coordinate, so that G is obtained by *interchanging* the x and y coordinates of F.

Let's examine what happens when we compose the functions F and G in either order.

$(F \circ G)(x)$	$(G \circ F)(x)$
$(F \circ G)(2) = F[G(2)] = F(-2) = 2$	$(G \circ F)(-2) = G[F(-2)] = G(2) = -2$
$(F \circ G)(4) = F[G(4)] = F(0) = 4$	$(G \circ F)(0) = G[F(0)] = G(4) = 0$
$(F \circ G)(5) = F[G(5)] = F(1) = 5$	$(G \circ F)(1) = G[F(1)] = G(5) = 1$
$(F \circ G)(7) = F[G(7)] = F(3) = 7$	$(G \circ F)(3) = G[F(3)] = G(7) = 3$

Figure 4.49 illustrates $(F \circ G)(x)$ and $(G \circ F)(x)$ with a diagram. We see that $(F \circ G)(x) = x$ for all x's in the domain of G and $(G \circ F)(x) = x$ for all x's in the domain of F. Functions that have this property are called **inverse functions**.

Figure 4.49

Diagram illustrating $(F \circ G)(x)$ and $(G \circ F)(x)$

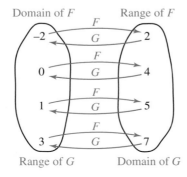

Definition of Inverse Functions

Two functions f and g are said to be **inverse functions** if and only if

$$f[g(x)] = x \qquad \text{for all } x \text{ in the domain of } g$$

and

$$g[f(x)] = x \qquad \text{for all } x \text{ in the domain of } f$$

Example 1 Verify that the functions $f(x) = \dfrac{1}{2}x - 5$ and $g(x) = 2x + 10$ are inverse functions.

Solution According to the definition, to verify that f and g are inverse functions, we must check that $f[g(x)] = x$ and $g[f(x)] = x$.

$$f[g(x)] = f(2x + 10) \qquad\qquad g[f(x)] = g\left(\frac{1}{2}x - 5\right)$$

$$= \frac{1}{2}(2x + 10) - 5 \qquad\qquad = 2\left(\frac{1}{2}x - 5\right) + 10$$

$$= x + 5 - 5 \qquad\qquad\qquad = x - 10 + 10$$

$$= x \quad \text{As required} \qquad\qquad = x \quad \text{As required}$$

Thus we have verified that f and g are inverse functions. ∎

Note that, based on this definition, the functions F and G described previously, which have their x- and y-coordinates interchanged, are inverse functions.

It is customary to denote the inverse function of $f(x)$ as $f^{-1}(x)$, which is read "f inverse of x." Using this notation, we can express the result of Example 1 by writing

$$f(x) = \frac{1}{2}x - 5 \qquad \text{and} \qquad f^{-1}(x) = 2x + 10$$

or we can write

$$g(x) = 2x + 10 \qquad \text{and} \qquad g^{-1}(x) = \frac{1}{2}x - 5$$

Important Even though in general we use an exponent of -1 to indicate a reciprocal, inverse function notation is an exception to this rule. Please be aware that $f^{-1}(x)$ is *not* the reciprocal of f.

$$f^{-1}(x) \neq \frac{1}{f(x)}$$

If we want to write the reciprocal of the function of $f(x)$ by using a negative exponent, we must write

$$\frac{1}{f(x)} = [f(x)]^{-1}$$

Since f^{-1} denotes the inverse function of f, the definition of inverse functions just given yields

$$f^{-1}[f(x)] = x \qquad \text{for all } x \text{ in the domain of } f$$

and

$$f[f^{-1}(x)] = x \qquad \text{for all } x \text{ in the domain of } f^{-1}$$

As we have noted before, a function may be viewed as a machine that takes an input x and does something to it. Inverse functions are in some sense "opposites" in that they have the property that what one function "does," the other function reverses, or "undoes." Thus when we compose two inverse functions, we get the original input value x back again. See Figure 4.50.

Figure 4.50

The interaction of inverse functions

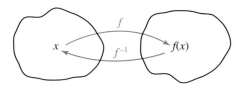

Let us now examine this idea from a more general standpoint. Suppose we have a function, which we know means that to each x there corresponds exactly one y. We may ask if this assignment also defines x as a function of y: If we pick y, does the equation determine a unique value of x?

For example, $F = \{(2, 4), (3, 6), (5, 4)\}$ is a function that has a unique y assigned to each x. Note, however, that it does not define x as a function of y. That is, *each y is not assigned a unique x.* Alternatively, we can say that if we reversed the x- and y-coordinates, the resulting set of ordered pairs, $\{(4, 2), (6, 3), (4, 5)\}$, is not a function.

Why is the set $\{(4, 2), (6, 3), (4, 5)\}$ not a function?

If we consider the function defined by the equation $y = f(x) = x^2$ and we choose $y = 4$, we get $x^2 = 4$, and so $x = \pm 2$. Thus in general we can see that an equation that defines y as a function of x will *not necessarily define x as a function of y.*

Why doesn't $y = x^2$ define x as a function of y?

A natural question to ask is, Can we specify conditions under which a function $y = f(x)$ also specifies x as a function of y? For y to be a function of x, each x is assigned a unique y, and for x to be a function of y, each y is assigned a unique x. Hence there must be a one-to-one correspondence between the elements of the domain x and elements of the range y. This means that we cannot have two x's assigned the same y-value as illustrated previously. In the language of functions we can formulate these ideas as follows:

Definition of a One-to-One Function

A function $f(x)$ is said to be **one-to-one** if and only if $x_1 \neq x_2$ implies that $f(x_1) \neq f(x_2)$. In words, this condition says that different x-values necessarily give different y-values.

Let's examine this definition graphically. Suppose we have a function $y = f(x)$ and we know its graph. If we want the equation $y = f(x)$ also to define x as a function of y, then to each y-value there must correspond exactly one x-value. Recalling that the vertical line test graphically expresses the idea that to each x-value there corresponds exactly one y-value, we recognize that an analogous **horizontal line test** (defined here) will ensure that to each y-value there must correspond exactly one x-value.

The Horizontal Line Test

A function $y = f(x)$ is one-to-one provided that any horizontal line intersects the graph in at most one point.

Figures 4.51(a) and (b) illustrate one function that does satisfy the horizontal line test and one that does not.

Note that although only one of the graphs in Figure 4.51 satisfies the horizontal line test, both are the graphs of functions, since they both satisfy the vertical line test.

Figure 4.51

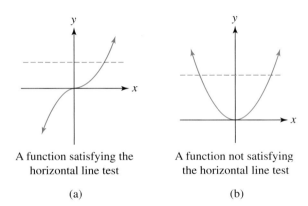

A function satisfying the horizontal line test

(a)

A function not satisfying the horizontal line test

(b)

Basically, the horizontal line test says that the graph of a function does not repeat any y-values and so the following two statements are equivalent:

1. The graph of $y = f(x)$ satisfies the horizontal line test.

2. The function $y = f(x)$ is one-to-one.

In summary, then, based on our discussion thus far we have seen that a one-to-one function $y = f(x)$ not only defines y as a function of x, but also defines x as a function of y. Also, the function obtained by viewing x as a function of y gives us the inverse function of $f(x)$. In fact, we can state the following theorem.

Theorem 4.1

$f(x)$ has an inverse, $f^{-1}(x)$, if and only if $f(x)$ is a one-to-one function.

Different Perspectives: **One-to-One Functions**

Graphical Interpretation

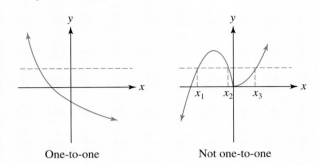

One-to-one   Not one-to-one

For a function to be one-to-one, any horizontal line can intersect the graph in at most one point.

Algebraic Interpretation

A one-to-one function necessarily assigns different y-values to different x-values.

In other words, a function *is* one-to-one if and only if $x_1 \neq x_2 \Rightarrow f(x_1) \neq f(x_2)$.

Equivalently, a function is one-to-one if and only if $f(x_1) = f(x_2) \Rightarrow x_1 = x_2$.

Theorem 4.1 tells us when a function will have an inverse, but how do we go about actually finding the inverse function of a one-to-one function? Consider the function $y = f(x) = 2x - 1$, whose graph is a straight line of slope 2. This function is certainly one-to-one, and so we can view the equation $y = 2x - 1$ as we usually do, defining y as a function of x *or* defining x as a function of y.

The ordered pairs $(0, -1)$, $(-1, -3)$, $(2, 3)$, and $(5, 9)$ all satisfy the function $y = 2x - 1$, and based on our earlier ground rules, we understand this to mean that if x is chosen to be $0, -1, 2$, and 5, then the corresponding y-values are, respectively, -1, -3, 3, and 9.

Keep in mind that the various graphing techniques we have developed are predicated upon the independent variable being represented along the horizontal axis (that is, being the first coordinate) and the dependent variable being represented along the vertical axis (that is, being the second coordinate). Because we do not want to lose all the graphing machinery we have built, we will insist that this continue to be the case.

Therefore, if we now return to the ordered pairs for $y = 2x - 1$ and we want to view them as "pick y and get x," we record them as

$$(-1, 0), (-3, -1), (3, 2), \text{ and } (9, 5)$$

What we are in effect doing is *interchanging the roles of x and y*. As we saw at the beginning of this section, if we interchange the x- and y-coordinates of the points of a one-to-one function, we obtain its inverse function. Since most of the functions we will encounter are defined by an equation rather than a set of ordered pairs, instead of interchanging the coordinates of points, we will interchange the roles of x and y in the equation. However, keeping in mind that we are generally used to having functions solved explicitly for the dependent variable ($y = f(x)$), after we interchange x and y, we want this new inverse function to be solved explicitly for the dependent variable y.

Example 2   Given $y = f(x) = \dfrac{1}{4}x + 3$, find $f^{-1}(x)$.

Solution   We first note that the graph of $y = f(x) = \dfrac{1}{4}x + 3$ is a nonhorizontal line.

Thus $f(x)$ satisfies the horizontal line test and therefore has an inverse. Based on the preceding discussion, we interchange x and y and then solve for y.

$$y = \frac{1}{4}x + 3 \qquad \text{We interchange } x \text{ and } y.$$

$$x = \frac{1}{4}y + 3 \qquad \text{Now we solve for } y.$$

$$4x = y + 12$$

$$4x - 12 = y \qquad \text{This is the inverse function.}$$

Thus $\boxed{f^{-1}(x) = 4x - 12}$.

It is left to the reader to verify that when we compose f and f^{-1} in either order we get x. (See Example 1.)   ∎

Thus we have seen that an equation of a one-to-one function can be viewed as defining y as a function of x or vice versa and that if we interchange the roles of x and y, we obtain a function that is the inverse of the original function.

Keep in mind that since we are interchanging the roles of x and y, we also interchange the domain and range. That is, the domain of the inverse function is the range of the original function and vice versa.

The following box summarizes what we have discussed thus far.

Inverse Functions

To find the inverse of a one-to-one function $y = f(x)$,

1. Interchange x and y in the equation $y = f(x)$.

2. Solve the resulting equation for y, obtaining the inverse function.

3. The domain of the inverse function is the range of the original function and the range of the inverse function is the domain of the original function.

Step 2 of this outline is not always possible as we will see later in the text.

Example 3 | Given $y = f(x) = x^3$, find $f^{-1}(x)$ and its domain.

Solution | Again we begin by interchanging x and y, and then we solve for y.

$$y = x^3 \qquad \text{Interchange } x \text{ and } y.$$
$$x = y^3 \qquad \text{Take the cube root of both sides.}$$
$$\sqrt[3]{x} = y \qquad \text{This is the inverse function. Thus } f^{-1}(x) = \sqrt[3]{x}.$$

The domain of the inverse function is the set of all real numbers.

The functions $y = f(x) = x^3$ and $y = f^{-1}(x) = \sqrt[3]{x}$ clearly illustrate the idea of a function and its inverse undoing each other. What the cubing function does, the cube root function "undoes." ∎

> ## Calculator Exploration
>
> Use a graphing calculator to graph the following in the standard $[-10, 10]$ by $[-10, 10]$ graphing window.
>
> 1. Graph $y = \frac{1}{4}x + 3$ (from Example 2), its inverse $y = 4x - 12$, and $y = x$ on the same set of coordinate axes. Can you make a conjecture about the relationship between the graphs of the first two equations with respect to the graph of $y = x$?
>
> 2. Graph $y = x^3$ (from Example 3), its inverse $y = \sqrt[3]{x}$, and $y = x$ on the same set of coordinate axes. You may find it helpful to change the viewing window. Can you make a conjecture about the relationship between the graphs of the first two equations with respect to the graph of $y = x$?

Example 4 | Let $y = f(x) = \dfrac{x}{x + 2}$. Find $f^{-1}(x)$ and verify that is satisfies the definition of an inverse function.

Solution |

$$y = \frac{x}{x + 2} \qquad \text{Interchange } x \text{ and } y.$$
$$x = \frac{y}{y + 2} \qquad \text{Now solve for } y.$$
$$x(y + 2) = y$$
$$xy + 2x = y \qquad \text{Which becomes}$$
$$2x = y - xy$$
$$2x = y(1 - x) \qquad \text{And so we have}$$
$$\frac{2x}{1 - x} = y$$

Hence $f^{-1}(x) = \dfrac{2x}{1 - x}$.

According to the definition, we need to verify that $f^{-1}[f(x)] = x$ and $f[f^{-1}(x)] = x$.

$$f^{-1}[f(x)] = f^{-1}\left(\frac{x}{x+2}\right) = \frac{2\left(\dfrac{x}{x+2}\right)}{1 - \dfrac{x}{x+2}} = \frac{2\left(\dfrac{x}{x+2}\right)\cdot(x+2)}{\left(1 - \dfrac{x}{x+2}\right)\cdot(x+2)} = \frac{2x}{x+2-x} = \frac{2x}{2} = x$$

and

$$f[f^{-1}(x)] = f\left(\frac{2x}{1-x}\right) = \frac{\dfrac{2x}{1-x}}{\dfrac{2x}{1-x}+2} = \frac{\left(\dfrac{2x}{1-x}\right)\cdot(1-x)}{\left(\dfrac{2x}{1-x}+2\right)\cdot(1-x)} = \frac{2x}{2x+2(1-x)} = \frac{2x}{2} = x$$

as required. ∎

In Example 1 we found that $y = f(x) = \frac{1}{2}x - 5$ and $y = g(x) = 2x + 10$ are inverse functions. Let's graph these two functions on the same coordinate system. The graphs appear in Figure 4.52.

Figure 4.52

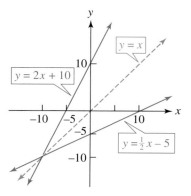

Note that we have also drawn in the line $y = x$ and that the graphs of the inverse functions $y = f(x) = \frac{1}{2}x - 5$ and $y = g(x) = 2x + 10$ appear to be symmetric with respect to the line $y = x$. In other words, if we placed a mirror along the line $y = x$, the graphs of $f(x)$ and $g(x)$ would be reflections of each other. (Keep in mind that we have already seen symmetry with respect to two other particular lines—the x- and y-axes.)

The graphs of inverse functions always exhibit this type of symmetry.

> The graph of $y = f^{-1}(x)$ can be obtained by reflecting the graph of $y = f(x)$ about the line $y = x$ (and vice versa).

A more detailed discussion of symmetry with respect to a line appears in Exercise 72 at the end of this section.

Example 5 | Use the graph in Figure 4.53 to answer the following.
(a) Is this the graph of a function?
(b) Does this function have an inverse?
(c) Let $f(x)$ be the function whose graph is given. Sketch the graph of $f^{-1}(x)$.

Figure 4.53

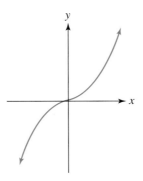

Solution | (a) Since this graph satisfies the vertical line test, it is the graph of a function.
(b) Since this graph satisfies the horizontal line test, this function does have an inverse.
(c) We draw the original graph and the line $y = x$. The graph of $f^{-1}(x)$ is obtained by reflecting the graph of $f(x)$ about the line $y = x$. See Figure 4.54.

Figure 4.54

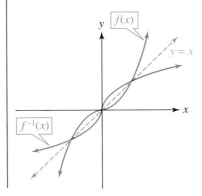

Suppose we are given the function $f(x)$ explicitly as a set of ordered pairs, formula, or graph, and that $f(x)$ has an inverse $f^{-1}(x)$. To evaluate $f^{-1}(a)$, the value of the function $f^{-1}(x)$ at some $x = a$, we do not need to construct $f^{-1}(x)$. We can get this information directly from $f(x)$ by noting how the function of f^{-1} is defined: f assigns x to y, if and only if f^{-1} assigns y to x, or

If the inverse of $f(x)$ exists, then $f(x) = y$ if and only if $f^{-1}(y) = x$.

If we are given $y = f(x)$, then finding $f(a)$ means finding y given $x = a$. On the other hand, finding $f^{-1}(a)$ means finding x given $y = a$; this requires us to solve the equation $f(x) = a$.

Example 6 (a) For $f(x) = \dfrac{x + 3}{x}$, find $f(3)$ and $f^{-1}(3)$.

(b) For $f(x)$ given by the accompanying graph, find $f(3)$ and $f^{-1}(3)$.

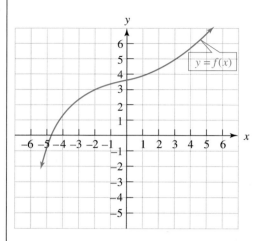

Solution (a) $f(x) = \dfrac{x + 3}{x}$. To find $f(3)$, we substitute 3 for x in f:

$$f(3) = \frac{3 + 3}{3} = 2 \qquad \text{Hence } \boxed{f(3) = 2}.$$

On the other hand, finding $f^{-1}(3)$ means finding the value of x such that $f(x) = 3$:

$$f(x) = \frac{x + 3}{x} \qquad \text{Let } f(x) = 3.$$

$$3 = \frac{x + 3}{x} \qquad \text{Now solve for } x.$$

$$3x = x + 3$$

$$x = \frac{3}{2}$$

Hence $\boxed{f^{-1}(3) = \dfrac{3}{2}}$. (This also means $f(\tfrac{3}{2}) = 3$. Verify this.)

(b) We use the graph to find the values $f(3)$ and $f^{-1}(3)$ as follows. To find $f(3)$, note that we are looking for the y at $x = 3$. Figure 4.55(a) shows that to find $f(3)$ we start on the x-axis at $x = 3$, project vertically up to the graph, then project horizontally to the y-axis and read off 5. Hence $f(3) = 5$. On the other hand, to find $f^{-1}(3)$, we are now given a y-value of 3 and asked to find the x-value. We start on the y-axis at $y = 3$, project horizontally to the graph, then project vertically to the x-axis to get $x = -2$. Hence $f^{-1}(3) = -2$. See Figure 4.55(b).

Figure 4.55

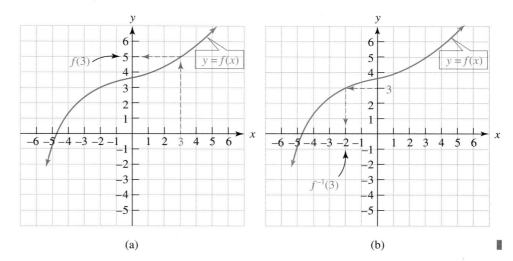

(a)   (b)

Example 7 | Given $f(x) = x^3 + 2x + 2$, use a graphing calculator to find $f^{-1}(4)$ to three decimal places.

Solution | It is important that you first understand that you are being given the y-value of $f(x)$ and you are being asked for the x-value. That is, $f^{-1}(4) = x$ means $f(x) = 4$, so you want to solve the equation $x^3 + 2x + 2 = 4$. There are several ways to arrive at the answer using a graphing calculator. As you can see in the screen in Figure 4.56, we chose to find the intersection of the equations $Y1=X^3+2X+2$ and $Y2=4$. Since $x = 0.771$ (rounded to three places) when $y = 4$, we have $f^{-1}(4) = 0.771$. ∎

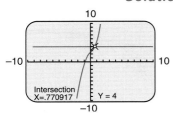

Figure 4.56

We will return to this idea of an inverse function in conjunction with a number of special functions we will encounter later in this text.

4.6 Exercises

In Exercises 1–6, use the given graph to determine whether the function has an inverse.

1.

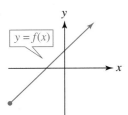

2.

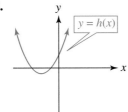

3.

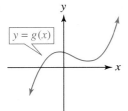

4.

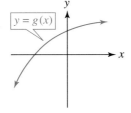

5.

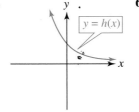

6.

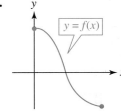

In Exercises 7–14, verify that $f(x)$ and $g(x)$ are inverse functions.

7. $f(x) = x - 1$; $g(x) = x + 1$

8. $f(x) = 6x$; $g(x) = \dfrac{x}{6}$

9. $f(x) = x^3$; $g(x) = \sqrt[3]{x}$

10. $f(x) = 5x + 4;$ $g(x) = \dfrac{x - 4}{5}$

11. $f(x) = \dfrac{1}{3}x + \dfrac{2}{5};$ $g(x) = 3x - \dfrac{6}{5}$

12. $f(x) = \sqrt[5]{x} + 6;$ $g(x) = (x - 6)^5$

13. $f(x) = \dfrac{4x}{x + 4};$ $g(x) = \dfrac{4x}{4 - x}$

14. $f(x) = \dfrac{x + 5}{2x + 1};$ $g(x) = \dfrac{5 - x}{2x - 1}$

15. Show that $f(x) = \dfrac{1}{x}$ is its own inverse function.

16. Verify that $f(x) = \dfrac{x + 1}{x - 1}$ is its own inverse function.

In Exercises 17–32, find the inverse of the given function.

17. $y = f(x) = 5x - 9$ **18.** $y = g(x) = \dfrac{1}{3}x + 6$

19. $y = F(x) = 2x^3 + 1$ **20.** $y = G(x) = 8 - x^5$

21. $y = h(x) = \dfrac{1}{x + 4}$ **22.** $y = H(x) = \dfrac{1}{x} + 4$

23. $y = f(x) = 2\sqrt{x} - 7$ **24.** $y = F(x) = \sqrt{2x - 7}$

25. $y = h(x) = x^2 - 1$ for $x \geq 0$

26. $y = h(x) = (x - 1)^2$ for $x \geq 1$

27. $y = g(x) = \dfrac{2x + 5}{3x - 2}$ **28.** $y = G(x) = \dfrac{6x}{x + 3}$

29. $y = f(x) = x^{3/5}$ **30.** $y = f(x) = -x^{-5/3}$

31. $y = f(x) = x^{-5/7} + 1$

32. $y = g(x) = \sqrt[3]{x^3 - 5}$

In Exercises 33–38, find $f^{-1}(x)$ and verify that $f^{-1}[f(x)] = f[f^{-1}(x)] = x$.

33. $f(x) = 7x - 6$ **34.** $f(x) = \dfrac{2x - 9}{4}$

35. $f(x) = 1 - \dfrac{3}{x}$ **36.** $f(x) = \dfrac{4 - x}{3x}$

37. $f(x) = \dfrac{5x + 3}{1 - 2x}$ **38.** $f(x) = \sqrt[3]{x + 1}$

39. Find the inverse of $y = f(x) = x^2$ for $x \geq 0$. Explain why the restriction $x \geq 0$ is needed. Find the domain and range of both f and f^{-1}.

40. Find the inverse of $y = f(x) = (x + 1)^2$ for $x \geq -1$. Explain why the restriction $x \geq -1$ is needed. Find the domain and range of both f and f^{-1}.

In Exercises 41–48, determine whether the given graph of a function has an inverse. If it does, sketch the graph of the inverse function.

41.
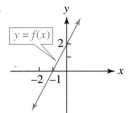

42.
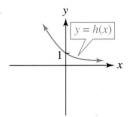

43.
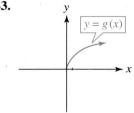

44.
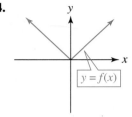

45.
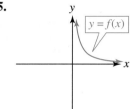

46.
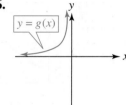

47.
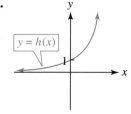

48.
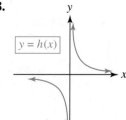

In Exercises 49–54, find $f(2)$ and $f^{-1}(2)$.

49. $f(x) = -2x + 7$

50. $f(x) = \dfrac{2x - 3}{x}$

51. $f(x) = x^2 - 2x - 6$ for $x \geq 1$

52. $f(x) = \dfrac{x - 1}{3x + 1}$

53.

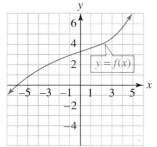

54.

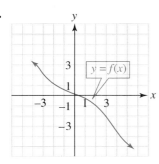

55. The Fahrenheit temperature, F, can be expressed as a function of the Celsius temperature, C, by the equation

$$F = F(C) = \frac{9}{5}C + 32$$

(a) Explain why this function has an inverse and find it.
(b) Find $F(100)$ and then find $F^{-1}(212)$ without any additional computation. Explain your reasoning.
(c) If you want to find the Celsius temperature that corresponds to a Fahrenheit temperature of $80°$, is it easier to use F or F^{-1}? Explain your reasoning.
(d) Compute $F(100)$ and $F^{-1}(100)$, and explain the difference in meaning between them.

56. In Exercise 55 of Section 4.4, we discussed the demand function

$$p = p(x) = 10 - \frac{x}{5} \quad \text{where } 0 \le x \le 50$$

which expresses the price per unit as a function of the number of units x that are sold.
(a) Explain why this demand function has an inverse and find it.
(b) Describe what this inverse function represents.
(c) Find $p(40)$ and then find $p^{-1}(2)$ without doing any additional computation. Explain.

57. There are 453.6 grams in one pound.
(a) Write a function of $g = f(p)$ that expresses the number of grams g in p pounds.
(b) Find $f^{-1}(p)$ and describe what relationship is expressed by the inverse function.

(c) Find $f(50)$ and then find $f^{-1}(50)$ and explain what each represents.

58. There are 1.609 km in one mile.
(a) Write a function of $k = f(m)$ that expresses the number of kilometers k in m miles.
(b) Find $f^{-1}(m)$ and describe what relationship is expressed by the inverse function.
(c) Find $f(100)$ and $f^{-1}(100)$ and explain what each represents.

59. Consider a simple "code" in which the letter A corresponds to 1, B corresponds to 2, C corresponds to 3, . . . , and Z corresponds to 26.
(a) Use this function to encode the word MOTHER.
(b) Can you decode the "word" 6 1 20 8 5 18?
(c) Explain the connection between being able to decode a "word" and the idea of having an inverse function.

60. Consider the letter–number correspondence of Exercise 59. Suppose we encode a word by adding the values of letters in the word. Thus, since C = 3, A = 1, and T = 20, the word CAT would be encoded as $3 + 1 + 20 = 24$.
(a) How would the word EASY be encoded?
(b) How would the words STAIR and MOST be encoded?
(c) Can an encoded word be decoded in this system? Explain the connection between being able to decode a "word" and the idea of having an inverse function.

In Exercises 61–64, find $f(2)$ and $f^{-1}(2)$ using a graphing calculator. Round your answers to two decimal places.

61. $f(x) = x^3 + 3x + 3$
62. $f(x) = x^2 + 2x + 2 \quad \text{for } x \ge -1$
63. $f(x) = 2x^4 + x - 3 \quad \text{for } x \ge 0$
64. $y = 2x^4 + x^2 + 1 \quad \text{for } x \le 0$

In Exercises 65–68, find the inverse of the given function, then use a graphing calculator to graph the given function, its inverse, and $y = x$ on the same coordinate system.

65. $y = 3x - 6$
66. $y = x^3 - 1$
67. $y = \sqrt[3]{x + 4}$
68. $y = \frac{5}{x} - 1$

Questions for Thought

69. Suppose $y = f(x)$ has an inverse. Discuss the meaning of the following set of implications.

$$y = f(x) \quad \Rightarrow \quad f^{-1}(y) = f^{-1}[f(x)] \quad \Rightarrow \quad f^{-1}(y) = x$$

70. Once we understand that $f^{-1}(x)$ is the function that has the property that $(f \circ f^{-1})(x) = x$, we may use this property as an alternative approach to finding f^{-1} as follows: Let $f(x) = 5x - 3$; then f^{-1} must satisfy

$$f[f^{-1}(x)] = x \qquad \text{Use the definition of } f(x).$$
$$5[f^{-1}(x)] - 3 = x \qquad \text{Solve for } f^{-1}(x).$$
$$f^{-1}(x) = \frac{x + 3}{5}$$

Use this approach to find $f^{-1}(x)$ for $f(x) = \sqrt[3]{x + 7}$.

71. In this section we outline a procedure for finding the inverse of a one-to-one function. Describe what happens when you try to apply this procedure that is *not* one-to-one.

72. We have stated that the graphs of inverse functions are symmetric with respect to the line $y = x$. This exercise expands on the idea of symmetry with respect to a line.

Definition Two points P and Q are said to be **symmetric with respect to line L** if and only if L is the perpendicular bisector of the line segment joining P and Q. The accompanying figure illustrates this definition.

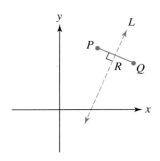

The points P and Q are symmetric with respect to the line L. Note that $\overline{PR} = \overline{QR}$. In Exercise 70 we saw that $y = 5x - 3$ and $y = \dfrac{x + 3}{5}$ are inverse functions. The point $P(2, 7)$ is on the graph of $y = 5x - 3$, and, of course, $Q(7, 2)$ is on the graph of $y = \dfrac{x + 3}{5}$. Verify that these two points are symmetric with respect to the line $y = x$ by using this definition. HINT: First, show that the line joining the points $P(2, 7)$ and $Q(7, 2)$ is perpendicular to the line $y = x$. Second, let $R(a, a)$ be the point where the line segment $\overline{PQ}$ intersects the line $y = x$ and show $|PR| = |QR|$. Note that it was not necessary to determine the value of a.

Chapter 4   *Summary*

After completing this chapter you should:

1. Be able to use the basic graphing principles. (Sections 4.1 and 4.2)
Specific algebraic changes made to the equation of a function have a predictable effect on its graph.
For example:
Use the graph of $y = f(x)$ in Figure 4.57 to sketch the graph of

Figure 4.57

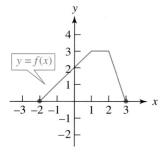

(a) $y = 2f(x)$ **(b)** $y = \dfrac{1}{2}f(x)$

(c) $y = f(x - 2)$ **(d)** $y = f(x) - 2$

(e) $y = f(-x)$ **(f)** $y = -f(x)$

Solution:

(a) The graph of $2f(x)$ is obtained by deflecting the graph of $f(x)$ away from the x-axis. The graph appears in Figure 4.58(a).

Figure 4.58(a)

(b) The graph of $y = \dfrac{1}{2}f(x)$ is obtained by pulling the graph of $f(x)$ toward the x-axis. The graph appears in Figure 4.58(b).

Figure 4.58(b)

(c) The graph of $y = f(x - 2)$ is obtained by shifting the original graph 2 units to the right. The graph appears in Figure 4.58(c).

Figure 4.58(c)

(d) The graph of $f(x) - 2$ is obtained by shifting the original graph 2 units down. The graph appears in Figure 4.58(d).

Figure 4.58(d)

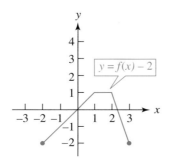

(e) The graph of $y = f(-x)$ is obtained by reflecting the graph of $y = f(x)$ about the y-axis. The graph appears in Figure 4.58(e).

Figure 4.58(e)

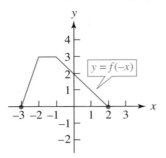

(f) The graph of $y = -f(x)$ is obtained by reflecting the graph of $y = f(x)$ about the x-axis. The graph appears in Figure 4.58(f).

Figure 4.58(f)

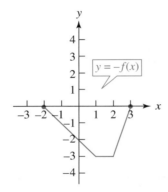

2. Be able to extract functions from *real-life* situations. (Section 4.3).
For example:
Express the hypotenuse y of $\triangle ADE$, in Figure 4.59, as a function of its height, h.

Figure 4.59

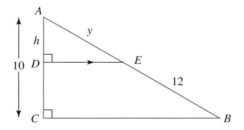

Solution:
We can see that $\triangle ADE$ is similar to $\triangle ACB$. Therefore,

$$\frac{y}{h} = \frac{y + 12}{10}$$
$$10y = hy + 12h$$
$$10y - hy = 12h$$
$$y(10 - h) = 12h \qquad \text{Solving for } y, \text{ we get } y \text{ as a function of } h.$$
$$y = y(h) = \frac{12h}{10 - h}$$

Since h is the height of $\triangle ADE$, h must be greater than or equal to 0 and less than or equal to the height of $\triangle ACB$, which is 10. Therefore the domain of $y(h)$ is $\{h \mid 0 \leq h \leq 10\}$.

3. Be able to sketch the graphs of quadratic functions. (Section 4.4)

The graph of a quadratic function is a parabola.

For example:

Sketch the graph of $y = f(x) = x^2 + 4x + 3$.

Solution:

Since we know that the graph of a quadratic function $y = Ax^2 + Bx + C$ is a parabola, we begin by finding the vertex. The x-coordinate of the vertex is

$$x = -\frac{B}{2A} = -\frac{4}{2(1)} = -2.$$ The y-coordinate of the vertex is found by computing $f(-2)$. We get $y = (-2)^2 + 4(-2) + 3 = -1$. Thus the vertex is $(-2, -1)$.

We find the y-intercept, $f(0)$, to be 3, and the x-intercepts (found by setting $y = 0$) to be $x = -1$ and $x = -3$. The graph appears in Figure 4.60.

Figure 4.60

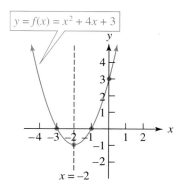

$y = f(x) = x^2 + 4x + 3$

$x = -2$

4. Recognize that the maximum or minimum values of a quadratic function can be found by identifying the vertex of the parabola. (Section 4.4)

5. Understand the algebra of functions. (Section 4.5)

You can perform the arithmetic operations on functions. The composition of functions is another way of producing a new function from two given functions.

For example:

Given $f(x) = 3x^2 - 5x + 2$ and $g(x) = x + 4$, find **(a)** $(f \cdot g)(x)$ and **(b)** $(f \circ g)(x)$.

Solution:

(a) $(f \cdot g)(x) = f(x) \cdot g(x)$

$$= (3x^2 - 5x + 2)(x + 4)$$

$$= 3x^3 + 7x^2 - 18x + 8$$

(b) $(f \circ g)(x) = f[g(x)]$

$$= f(x + 4)$$

$$= 3(x + 4)^2 - 5(x + 4) + 2$$

$$= 3(x^2 + 8x + 16) - 5x - 20 + 2$$

$$= 3x^2 + 19x + 30$$

6. Understand inverse functions. (Section 4.6)

Two functions f and g are inverse functions if and only if $(f \circ g)(x) = (g \circ f)(x) = x$. The graphs of inverse functions are symmetric with respect to the line $y = x$.

For example:

Given $y = f(x) = 5x - 2$, find its inverse function and sketch the graphs of f and f^{-1} on the same coordinate system.

Solution:
To find the inverse function of $y = f(x)$, interchange x and y and then solve for y.

$y = 5x - 2$ Interchange x and y.

$x = 5y - 2$ Solve for y.

$$y = f^{-1}(x) = \frac{x + 2}{5}$$

Note that the graphs of f and f^{-1} in Figure 4.61 are symmetric with respect to the line $y = x$.

Figure 4.61

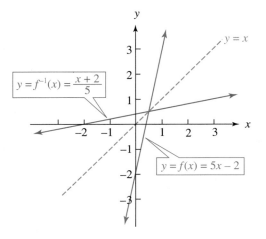

Chapter 4   *Review Exercises*

In Exercises 1–16, use the graphs of $y = f(x)$ and $y = g(x)$ given in the figures to graph each of the following functions.

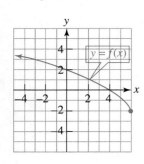 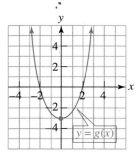

1. $f(-x)$

2. $-f(x)$

3. $-g(x)$

4. $g(-x)$

5. $f(x + 2)$

6. $f(x) + 2$

7. $-f(x) + 2$

8. $g(x) - 3$

9. $g(x) + 3$

10. $-f(x - 2)$

11. $g(3 - x)$

12. $f(x) - 2$

13. $-f(x) - 1$

14. $g(-x) + 3$

15. $-f(x - 4)$

16. $-g(x) - 1$

In Exercises 17–20, use the graph of $y = f(x)$ given in the figure to determine the equation for each of the following graphs in terms of $f(x)$.

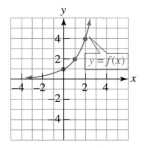

17.

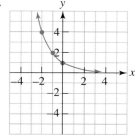

18.

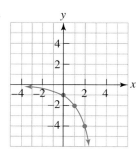

19.

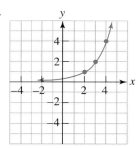

20.

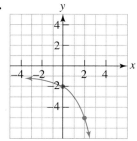

21. Find the vertex and the axis of symmetry of the quadratic function $f(x) = -(x - 4)^2 + 6$.

22. Find the vertex and the axis of symmetry of the parabola whose equation is $f(x) = x^2 - 6x + 5$.

23. What is the maximum value of the function $y = -2x^2 - 9x$?

24. What is the minimum value of the function $f(x) = 3x^2 - 4x + 2$?

In Exercises 25–36, sketch the graph of each of the following functions. Label the intercepts where appropriate.

25. $y = x^2 - 5$

26. $y = -(x - 3)^2 - 2$

27. $y = |2x - 6|$

28. $f(x) = |x + 1| - 3$

29. $f(x) = 2x^2 - 3x - 2$

30. $y = -x^2 + 4x - 5$

31. $f(x) = 36 - 9x^2$

32. $y = 2(x - 1)^2 - 8$

33. $y = 2x^2 - 10x$

34. $y = x^2 - 2x - 3$

35. $y = 3x^2 - 6x + 2$

36. $y = -3x^2 - 4$

Use the functions f, g, h, r, s, and t as defined here for Exercises 37–70.

$$f(x) = 3x^2 - 4x + 1 \quad g(x) = 5 - 2x \quad h(x) = -3$$

$$r(x) = x^3 + 8 \qquad s(x) = \frac{-2}{x} \qquad t(x) = \frac{3}{x - 4}$$

In Exercises 37–50, find the required value.

37. $(g + h)(2)$

38. $(r - t)(-3)$

39. $(f \cdot s)(-6)$

40. $(t \cdot f)(3)$

41. $\left(\dfrac{g}{f}\right)(4)$

42. $\left(\dfrac{s}{t}\right)(4)$

43. $\left(\dfrac{r}{h}\right)(-1)$

44. $(s - t)(8)$

45. $(f \circ g)(4)$

46. $(g \circ f)(4)$

47. $(h \circ s)(-6)$

48. $(s \circ h)(8)$

49. $(r \circ g)(a)$

50. $(s \circ h)(c)$

In Exercises 51–70, use the functions f, g, h, r, s, and t as defined previously to find the required function and its domain.

51. $(f - g)(x)$

52. $(r \cdot s)(x)$

53. $\left(\dfrac{s}{h}\right)(x)$

54. $\left(\dfrac{h}{s}\right)(x)$

55. $(r - f)(x)$

56. $(s + t)(x)$

57. $(g \cdot t)(x)$

58. $\left(\dfrac{g}{t}\right)(x)$

59. $(g \circ f)(x)$

60. $(f \circ g)(x)$

61. $(s \circ r)(x)$

62. $(r \circ s)(x)$

63. $(t \circ h)(x)$

64. $(h \circ t)(x)$

65. $(g \circ g)(x)$

66. $(t \circ t)(x)$

67. $(t \circ s)(x)$

68. $(s \circ t)(x)$

69. $(s \circ g \circ t)(x)$

70. $(t \circ s \circ g)(x)$

In Exercises 71–74, find functions $f(x)$ and $g(x)$ so that the given function $h(x) = (f \circ g)(x)$.

71. $h(x) = \dfrac{1}{\sqrt{x + 2}}$

72. $h(x) = (3x + 2)^4$

73. $h(x) = (5x - 7)^{-3}$

74. $h(x) = \sqrt[3]{\dfrac{x}{x + 1}}$

75. Suppose a laboratory technician is growing a bacteria culture in which the number of bacteria present, N, depends on the Celsius temperature, C, of the surrounding air, as given by the function

$$N(C) = C^2 + 125C + 1000 \qquad \text{for } 20 \le C \le 32$$

The Celsius temperature, C, depends in turn on the number of hours, h, after the culture begins growing, as given by the function

$$C(h) = 4h + 10 \qquad \text{for } 0 \le h \le 6$$

(a) Express the number of bacteria, N, as a function of h.

(b) How many bacteria are present after 4 hours?

(c) After approximately how many hours are there 10,000 bacteria?

76. Suppose that the length of the radius of a circle is dependent on the number of minutes, t, that have elapsed after $t = 0$ minutes according to the function

$$r(t) = 64 - t^2 \qquad \text{for } 0 \le t \le 8$$

(a) Express the area of the circle as a function of t.

(b) What is the area of the circle after 4 minutes?

77. Use the accompanying diagram to express the area of $\triangle ABC$ as a function of x. The point (x, y) is chosen in the first quadrant on the line through the points $(0, 0)$ and $(3, 4)$.

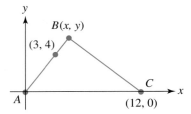

78. A closed right circular cylinder has a surface area of 20π sq cm. Express the volume of the cylinder as a function of r.

79. A closed rectangular box has a base that is x units wide and twice that long. If we let h be the height of the box and the total surface area of the box is 180 sq cm, express the volume of the box as a function of x.

80. A farmer wished to enclose a rectangular garden and subdivide it with a fence as indicated in the figure. Fencing costs $6 per foot for the outside and $4 per foot for the interior dividers.

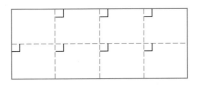

(a) Express the area that can be enclosed for a total cost of $240 as a function of one variable.

(b) What is the maximum area that can be enclosed for $240?

In Exercises 81–82, determine whether the function whose graph is given has in inverse.

81.

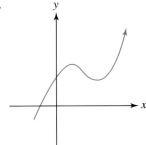

82.

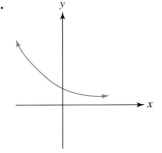

In Exercises 83–86, sketch the graph of the inverse of the given function f. Give the domain and range for both f and f^{-1}.

83.

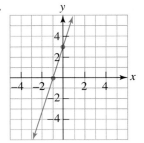

84.

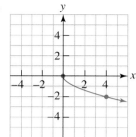

85.

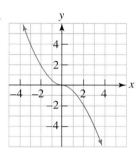

86.

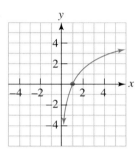

In Exercises 87–92, find the inverse of the given function and verify that

$$(f^{-1} \circ f)(x) = (f \circ f^{-1})(x) = x$$

87. $f(x) = \dfrac{1}{4}x + 5$ **88.** $f(x) = 8 - 5x$

89. $f(x) = \dfrac{x + 4}{x - 3}$ **90.** $f(x) = \sqrt{x - 5}$

91. $f(x) = \dfrac{3}{x + 6}$ **92.** $f(x) = \dfrac{1}{x} - 2$

In Exercises 93–95, find $f(3)$ and $f^{-1}(3)$.

93. $f(x) = \dfrac{x + 7}{2x}$ **94.** $f(x) = x^3 + x + 4$
(Use a graphing calculator.)

95.

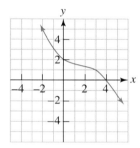

Chapter 4   *Practice Test*

1. Use the following graphs for $y = f(x)$ and $y = g(x)$ to sketch the graph of each function in parts (a) through (g).

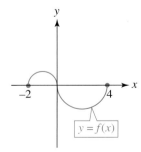
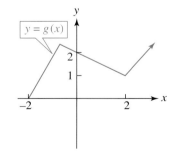

(a) $f(-x)$ **(b)** $-f(x)$ **(c)** $|f(x)|$
(d) $f(x) - 2$ **(e)** $g(x) + 2$
(f) $g(x + 2)$ **(g)** $-f(x - 2)$

2. Find the vertex and axis of symmetry of the following quadratic functions.
(a) $y = 3(x + 5)^2 - 3$ **(b)** $y = -2x^2 - 5x + 1$

3. Sketch the graphs of each of the following functions. Label the intercepts where appropriate.
(a) $y = 25 - x^2$
(b) $f(x) = |4x + 8|$
(c) $y = -2(x + 3)^2 + 8$
(d) $f(x) = x^2 + 4x - 5$
(e) $y = 3x^2 - 48x$
(f) $f(x) = -x^2 + 3x - 4$
(g) $y = 10 - |x|$
(h) $f(x) = 2x^2 - 5x - 1$

4. Given $f(x) = 7 - x^2$, $g(x) = \dfrac{3}{x + 2}$, and $h(x) = \sqrt{x - 1}$, find each of the following.
(a) $(f \cdot h)(3)$ **(b)** $(g + h)(10)$

(c) $\left(\dfrac{f}{g}\right)(-3)$

(d) $(f \circ g)(x)$

(e) $(g \circ f)(x)$

(f) $(g \circ g)(x)$

(g) $(g \circ f \circ h)(x)$

5. Let $f(x) = \dfrac{2x}{x+1}$ and $g(x) = \dfrac{x}{x-1}$.

  (a) Find the domain of $(f \circ g)(x)$.

  (b) Find the domain of $(g \circ f)(x)$.

6. A rectangle whose length is 12 and width is x is inscribed in a circle of radius r. Use the following figure to express the area of the shaded portion as a function of r.

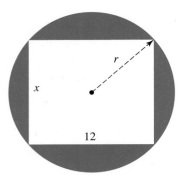

7. Explain why the function whose graph appears in the accompanying figure has an inverse, and sketch the graph of the inverse function.

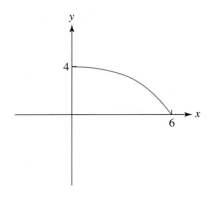

8. Find the inverse of the following functions and verify that $(f^{-1} \circ f)(x) = (f \circ f^{-1})(x) = x$.

  (a) $f(x) = \dfrac{5x-4}{3}$

  (b) $f(x) = \sqrt[3]{2x+9}$

  (c) $f(x) = \dfrac{2}{x} - 5$

9. Given $f(x) = \dfrac{5x}{x-1}$, find $f^{-1}(3)$.

10. On November 3, 2002, a British pound was worth 1.56 U.S. dollars.

  (a) Write a function $d = f(p)$ that expresses the dollar value of p pounds.

  (b) Find $f^{-1}(p)$ and describe what relationship is expressed by the inverse function.

  (c) Find $f(1000)$ and $f^{-1}(1000)$ and explain what each represents.

5

Polynomial, Rational, and Radical Functions

In this chapter we continue our discussion of the different types of functions mentioned in Chapter 4. Many of the ideas introduced in this chapter have wide applications throughout mathematics.

In the first four sections we investigate polynomial functions. We will examine rational functions in Section 5.5, and as we do we compare and contrast their behavior to that of polynomial functions. The chapter concludes with a discussion of radical functions and variation.

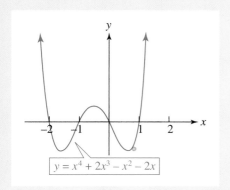

$$y = x^4 + 2x^3 - x^2 - 2x$$

The functions described in this chapter frequently occur as mathematical models of real-life situations. For example:

In business, the *demand function* gives the price per item, p, in terms of the number of items sold, x. Suppose a company finds that the price p (in dollars) for its model GC-5 calculator is related to the number of calculators sold, x (in millions), and is given by the demand function

This example will be discussed further in Example 4 in Section 5.2.

$$p = 80 - 4x^2$$

The manufacturer's revenue is determined by multiplying the number of items sold (x) by the price per item (p). Thus the *revenue function R* for this calculator manufacturer is

$$R = xp = x(80 - 4x^2) = 80x - 4x^3$$

where R is calculated in millions of dollars.

These demand and revenue functions are examples of *polynomial functions*. We will see additional applications that give rise to some of the other functions discussed in this chapter.

A major aim of this chapter is to better understand the significance of applied functions (such as this demand function). In order to do this, we must analyze the domain, range, and behavior of such functions.

5.1   Polynomial Functions

Recall that in Section 4.2 we gave the following definition of a polynomial function.

Definition of a Polynomial Function

A **polynomial function** is a function of the form

$$y = p(x) = a_n x^n + a_{n-1} x^{n-1} + a_{n-2} x^{n-2} + \cdots + a_2 x^2 + a_1 x + a_0 \quad a_n \neq 0$$

Each a_i is assumed to be a real number, and n is a nonnegative integer. a_n is called the **leading coefficient**. Such a polynomial function is said to be of **degree n**.

From its definition we can see that the domain of a polynomial function is always the set of all real numbers. Recall that a polynomial of degree 1 is called a *linear function* and a polynomial of degree 2 is called a *quadratic function*.

Calculator Exploration

1. Graph the following on the same set of coordinate axes:

$$y = x^2 \qquad y = x^4 \qquad y = x^6$$

Use the $[-2, 2]$ by $[-1, 4]$ viewing window.

Which graph is highest and which graph is lowest on the interval $(-1, 1)$? Outside this interval? Describe the similarities and differences among the graphs. For even values of n, conjecture how changing n affects the graph of $y = x^n$. You may need to zoom in at various locations to get a better look at how the graphs compare.

2. Graph the following on the same set of coordinate axes:

$$y = x^3 \qquad y = x^5 \qquad y = x^7$$

Use the $[-2, 2]$ by $[-2, 2]$ viewing window.

Which graph is highest and which graph is lowest on the interval $(-1, 1)$? Outside this interval? Describe the similarities and differences among the graphs. For odd values of n, conjecture how changing n affects the graph of $y = x^n$. You may need to zoom in at various locations to get a better look at how the graphs compare.

We begin by examining the simplest polynomial functions of the form $y = f(x) = ax^n$. These are often called **pure power functions**. If $n = 0$ or 1, then we are dealing with a function of the form $y = a$ or $y = ax$, which we recognize as a function whose graph is a straight line.

We first consider positive even integers n. If $n = 2$, then we are dealing with a quadratic function whose graph we know to be a parabola. To get a sense of how these functions behave, we consider the graphs of $y = x^2$ and $y = x^4$. Table 5.1 contains some values for these functions and Figure 5.1 illustrates their graphs.

Table 5.1

x	$y = x^2$	$y = x^4$
-2	$y = (-2)^2 = 4$	$y = (-2)^4 = 16$
-1.5	$y = (-1.5)^2 = 2.25$	$y = (-1.5)^4 = 5.0625$
-1	$y = (-1)^2 = 1$	$y = (-1)^4 = 1$
-0.5	$y = (-0.5)^2 = 0.25$	$y = (-0.5)^4 = 0.0625$
0	$y = (0)^2 = 0$	$y = (0)^4 = 0$
0.5	$y = (0.5)^2 = 0.25$	$y = (0.5)^4 = 0.0625$
1	$y = (1)^2 = 1$	$y = (1)^4 = 1$
1.5	$y = (1.5)^2 = 2.25$	$y = (1.5)^4 = 5.0625$
2	$y = (2)^2 = 4$	$y = (2)^4 = 16$

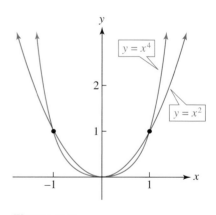

Figure 5.1
The graphs of $y = x^2$ and $y = x^4$

Looking at the graphs, we can see that on the interval $(-1, 1)$, the graph of $y = x^4$ is flatter than the graph of $y = x^2$, whereas outside this interval (that is, for $x < -1$ or $x > 1$), the graph of $y = x^4$ is steeper than the graph of $y = x^2$. In other words, for $|x| < 1$, the graph of $y = x^4$ is below the graph of $y = x^2$, but for $|x| > 1$, the positions of the graphs are reversed.

A moment's reflection should convince us that for all *even* integers $n > 1$, the graph of a function of the form $y = f(x) = x^n$ will have a shape similar to the graph of $y = x^2$. We call this the "x^2 look."*

---

*It is very important to realize that although the graphs of $y = x^n$ for n a positive even integer are *similar* to a parabola, they *are not* parabolas. The only function of the form $y = x^n$ that is a parabola is $y = x^2$.

It is important to recognize that although the graphs of $y = x^n$ for n even are similar, they are not identical. As n increases (but still remains even), the graphs of $y = x^n$ get flatter on the interval $(-1, 1)$ and steeper outside this interval (just as we saw in Figure 5.1).

Now let's turn our attention to the graphs of $y = x^n$ for odd integers $n \geq 3$. To get a sense of how these functions behave, we consider the graphs of $y = x^3$ and $y = x^5$. Table 5.2 contains some values for these functions and Figure 5.2 contains their graphs.

Table 5.2

x	$y = x^3$	$y = x^5$
-3	$y = (-3)^3 = -27$	$y = (-3)^5 = -243$
-2	$y = (-2)^3 = -8$	$y = (-2)^5 = -32$
-1	$y = (-1)^3 = -1$	$y = (-1)^5 = -1$
-0.5	$y = (-0.5)^3 = -0.125$	$y = (-0.5)^5 = -0.0325$
0	$y = (0)^3 = 0$	$y = (0)^5 = 0$
0.5	$y = (0.5)^3 = 0.125$	$y = (0.5)^5 = 0.0325$
1	$y = (1)^3 = 1$	$y = (1)^5 = 1$
2	$y = (2)^3 = 8$	$y = (2)^5 = 32$
3	$y = (3)^3 = 27$	$y = (3)^5 = 243$

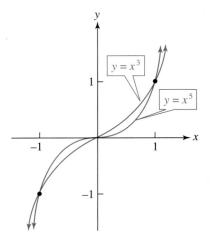

Figure 5.2
The graphs of $y = x^3$ and $y = x^5$

Again, it is important to recognize that although the graphs of $y = x^3$ and $y = x^5$ are similar, they are not identical. As n increases (but still remains odd), the graphs of $y = x^n$ have a similar shape but get flatter on the interval $(-1, 1)$ and steeper outside this interval (just as we saw in Figure 5.2). We call this the "x^3 look."

We note the similarities and differences between the graphs of $y = x^2$ and $y = x^3$: For positive x-values, both $y = x^2$ and $y = x^3$ are positive, whereas for negative x-values, $y = x^2$ is positive and $y = x^3$ is negative. In particular, for both functions we see that "out at the ends" (that is, way out to the right and left), the graphs are going way up or way down. For both $y = x^2$ and $y = x^3$ we express this observation by writing that as $x \to \infty$ then $y \to \infty$. This means that as x gets larger and larger (we often say that x increases without bound), the accompanying y-values are also getting larger and larger. For $y = x^2$, we can write that as $x \to -\infty$ then $y \to \infty$. This means that as x gets smaller and smaller (that is, as x is farther and farther to the left or, equivalently, that the x-values are decreasing without bound), the accompanying y-values are getting larger and larger. However, for $y = x^3$ we would write that as $x \to -\infty$ then $y \to -\infty$.

Different Perspectives: | **End Behavior**

$x \to \infty$ means x increases without bound (right end behavior), and $x \to -\infty$ means x decreases without bound (left end behavior).

Algebraic Notation

Consider $y = x^2$:
As x gets larger and larger ($x \to \infty$), y gets larger and larger ($y \to \infty$). As x gets smaller and smaller ($x \to -\infty$), y gets larger and larger ($y \to \infty$).

Graphical Interpretation

Note the behavior of the graph of $y = x^2$ "at the ends."

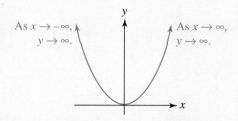

Consider $y = x^3$:
As x gets larger and larger ($x \to \infty$), y gets larger and larger ($y \to \infty$). As x gets smaller and smaller ($x \to -\infty$), y gets smaller and smaller ($y \to -\infty$).

Note the behavior of the graph of $y = x^3$ "at the ends."

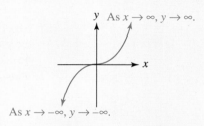

The graphs in the following box summarize these observations.

The Graphs of $y = x^n$
(for positive integers $n > 1$)

These graphs should be added to our catalog of basic functions whose graphs we recognize immediately.

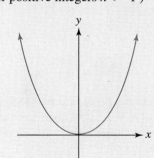

The graph of $y = x^n$ for n even

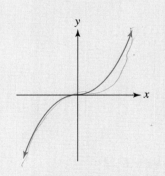

The graph of $y = x^n$ for n odd

Based on the stretching principle, we recognize that the description in the box applies equally well to functions of the form $y = ax^n$. The coefficient a merely stretches the graph of $y = x^n$ if a is positive and also reflects the graph about the x-axis if a is negative.

Example 1

Compare the graphs of $y = x^4$, $y = x^4 + 2$, and $y = (x + 2)^4$.

Sketch the graphs of

(a) $y = f(x) = (x + 2)^4$ **(b)** $y = g(x) = 1 - x^5$ **(c)** $y = h(x) = 5x^3$

Solution

(a) The graph of $y = (x + 2)^4$ is similar to the graph of $y = x^4$ (it will have the x^2 look) but is shifted 2 units to the left. By substituting $x = 0$, we find the y-intercept to be $y = 16$. The graph appears in Figure 5.3.

Figure 5.3

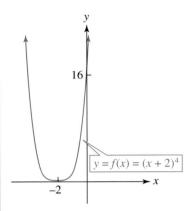

(b) To get the graph of $y = 1 - x^5$, it is helpful to view the equation as $y = -x^5 + 1$. The graph of $y = -x^5 + 1$ can be obtained from the graph of $y = x^5$ (which has the x^3 look) by reflecting it about the x-axis and shifting it 1 unit up. By substituting $x = 0$, we find the y-intercept to be $y = 1$. The x-intercept(s) is (are) found by setting $y = 0$. These steps are shown in Figure 5.4.

Figure 5.4

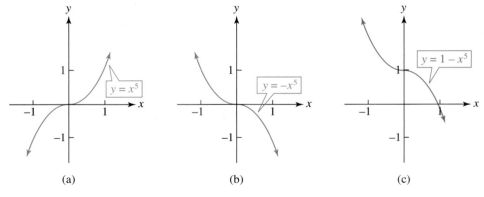

(a)   (b)   (c)

(c) Based on the preceding discussion, the graph of $y = 5x^3$ will be similar to the graph of $y = x^3$. The coefficient of 5 stretches the graph of $y = x^3$. The graph appears in Figure 5.5.

Figure 5.5

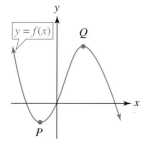

Figure 5.6

P and *Q* are turning points of the graph of $y = f(x)$.

Before looking at more polynomials, let's review some useful terminology. By a **turning point** of a graph, we mean a point on the graph where the function changes from increasing (graph rising) to decreasing (graph falling) or vice versa. In Figure 5.6, the points *P* and *Q* are turning points.

One example of a turning point that we encountered previously is the vertex of a parabola.

To sketch the graphs of more general polynomials, we will need to accept some basic properties of the graphs of polynomials. (These properties can be proven using the ideas of calculus.)

Properties of Polynomial Functions

1. The graph of a polynomial is a smooth unbroken curve. The word *smooth* means that the graph does not have any sharp corners as turning points. See the following figure.

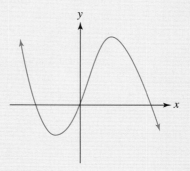

2. If $p(x)$ is a polynomial of degree n, then the polynomial equation $p(x) = 0$ has *at most* n distinct solutions; that is, $p(x)$ has at most n zeros. This is equivalent to saying that the graph of $y = p(x)$ crosses the x-axis at most n times. Thus a polynomial of degree 5 can have *at most* 5 x-intercepts.

3. The graph of a polynomial function of degree n can have *at most* $n - 1$ turning points. For example, the graph of a polynomial of degree 5 can have *at most* 4 turning points. In particular, the graph of a quadratic polynomial (degree 2) always has exactly one turning point—its vertex.

4. The graph of a polynomial always exhibits the characteristic that as $|x|$ gets very large, $|y|$ gets very large.

Figure 5.7 illustrates two graphs that *could not* be the graphs of a polynomial function.

Figure 5.7

These graphs cannot be the graphs of polynomial functions.

The points *P* and *Q* in Figure 5.7(a) are turning points even though this is not the graph of a polynomial.

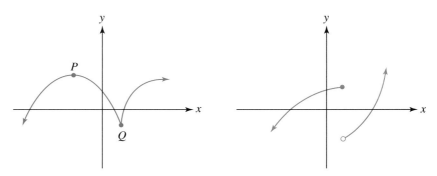

This graph has a sharp corner.

(a)

This graph has a break.

(b)

Calculator Exploration

Use a graphing calculator to graph the polynomial $y = x^4 - 2x^2 + 1$ and the functions $y = \sqrt{x^4 - 2x^2 + 1}$ and $y = \dfrac{1}{x^4 - 2x^2 + 1}$, which are not polynomials. Does the graph of the polynomial exhibit all the properties we have listed? Do the nonpolynomial graphs?

Our discussion throughout this and the next few sections elaborates on these properties. What is important to recognize is that these properties of polynomials highlight the intimate relationship between the algebraic properties of polynomials (e.g., the number of solutions to the equation $p(x) = 0$) and graphical properties (e.g., the number of x-intercepts of the graph). In fact, we often use information about a graph to find the zeros of a function, and in other cases we use information about the zeros of a function to help us sketch the graph.

Example 2 Suppose Figure 5.8 shows the graph of a polynomial function. What is the minimum possible degree of the polynomial?

Figure 5.8

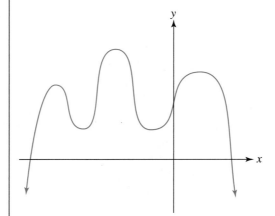

Solution

What do I need to find?	The minimum possible degree of the polynomial whose graph is given
What information do we get from the graph?	The graph is a polynomial function with 2 zeros (x-intercepts) and 5 turning points.
What conclusion can be drawn from the number of zeros?	Property 2 for polynomials tells us that if a polynomial has 2 zeros, then it must have degree at least 2.
What conclusion can be drawn from the number of turning points?	Property 3 for polynomials tell us that if a polynomial has degree n then it can have *at most* $n - 1$ turning points. In other words, the degree of a polynomial must be at least one more than the number of turning points. Since this graph has 5 turning points, the degree of the polynomial must be at least 6.

In summary then, based on the number of zeros, the polynomial must be of degree at least 2, whereas based on the number of turning points, the polynomial must be of degree at least 6. Therefore, we conclude that the minimum possible degree of the polynomial is 6.

Keep in mind that although a sixth-degree polynomial *may* have as many as six real zeros, it need not have that many.   ▪

End Behavior of Functions

Another property of the graph of a polynomial that we can determine by inspection of its equation is the behavior of the graph for extreme values of x: that is, how it behaves at its "ends." We examined end behavior for pure power functions previously in this section.

In general, for a polynomial function, the behavior of its graph for extreme values of x is primarily dictated by its highest-power term. Consider the polynomial $f(x) = 2x^5 - 8x^4 + x^3 + x^2 + 5$. Let's examine a table of values comparing the highest-power term, $2x^5$, with the value of $f(x) = 2x^5 - 8x^4 - x^3 + x^2 + 5$, for a few very large values of x.

Table 5.3

x	$2x^5$	$f(x)$
10,000	2×10^{20}	1.9992×10^{20}
50,000	6.25×10^{23}	6.2495×10^{23}
100,000	2×10^{25}	1.99992×10^{25}
500,000	6.25×10^{28}	6.24995×10^{28}
1,000,000	2×10^{30}	1.999992×10^{30}

Table 5.3 illustrates that as x gets extremely large, the values of $f(x)$ are relatively close to the values of $2x^5$, its highest-power term. In short, the highest-power term dictates the end behavior of the polynomial. The same is true for extremely small values of x (the left end of the graph of $f(x)$).

Hence, the end behavior for any polynomial is the same as the end behavior of its highest-power term, as discussed at the beginning of this section. For example, the polynomial $f(x) = 2x^5 - 8x^4 + x^3 - x^2 + 5$ must have the end behavior shown in Figure 5.9(a) (p. 282) because that is the end behavior of $y = 2x^5$. The polynomial $g(x) = 5x^6 - 3x^2 + 2x - 5$ must have the end behavior shown in Figure 5.9(b) because that is the end behavior of $y = 5x^6$. In contrast, the polynomial $h(x) = -5x^6 - 3x^2 + 2x - 5$ will have the end behavior shown in Figure 5.9(c) because that is the end behavior of $y = -5x^6$. Note that since the leading coefficient is negative, the graph of $y = -5x^6$ is the reflection of the graph of $y = 5x^6$ about the x-axis. Thus, the effect of a negative coefficient is to cause $h(x)$ to have the "opposite" end behavior of $g(x)$.

Because the degree of the polynomial is the degree of the highest-power term, the degree and the leading coefficient dictate the end behavior of the polynomial.

In Example 2, we were given the graph shown in Figure 5.8 and asked to find the minimum possible degree of the polynomial. We determine this to be 6 because we knew that there were 5 turning points. Now we can say a bit more about the polynomial by seeing the end behavior of the graph: The degree of the polynomial must be even and the leading coefficient must be negative.

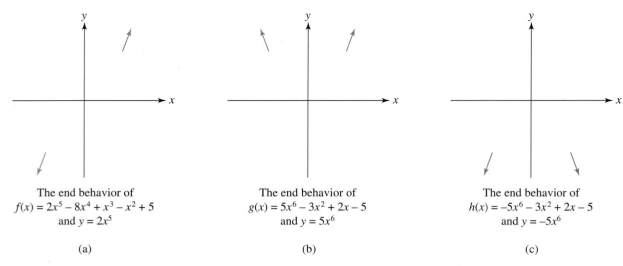

The end behavior of
$f(x) = 2x^5 - 8x^4 + x^3 - x^2 + 5$
and $y = 2x^5$

(a)

The end behavior of
$g(x) = 5x^6 - 3x^2 + 2x - 5$
and $y = 5x^6$

(b)

The end behavior of
$h(x) = -5x^6 - 3x^2 + 2x - 5$
and $y = -5x^6$

(c)

Figure 5.9

End behavior information can also help us determine whether we have a "complete" graph of a polynomial function on the graphing calculator. By a "complete" graph we mean that the window shows all the turning points and its end behavior. For example, using the standard $[-10, 10]$ by $[-10, 10]$ window, we present the graph of $f(x) = 0.01x^4 - 0.27x^3 + 2.07x^2 - 3.65x - 6$ in Figure 5.10.

Figure 5.10

The graph of $f(x) =$
$0.01x^4 - 0.27x^3 + 2.07x^2$
$\qquad - 3.65x - 6$
in the $[-10, 10]$ by
$[-10,10]$ window

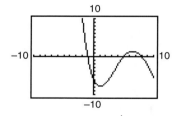

Because the function $f(x) = 0.01x^4 - 0.27x^3 + 2.07x^2 - 3.65x - 6$ is a polynomial of degree 4 (even) and because the leading coefficient is positive, we would expect both "ends" to point upward. Hence, we know by its end behavior that Figure 5.10 cannot be a "complete" graph of this function. After trial and error, we use the window shown in Figure 5.11 for the complete graph of $y = f(x)$.

Figure 5.11

The graph of $f(x) =$
$0.01x^4 - 0.27x^3 + 2.07x^2$
$\qquad - 3.65x - 6$
in the $[-20, 20]$
by $[-20,20]$ window

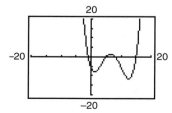

Given a polynomial function such as $y = f(x) = x^3 + x^2 - 2x$, we could get a rough sketch of its graph by plotting a sufficient number of points (probably a large number of points would be needed to get a good idea of what the graph actually looks like). In fact, this is basically the way a graphing calculator plots the graph of a function. To get a detailed description of the graph of a polynomial and understand *why* it looks the way it does, it is necessary to employ the ideas of calculus. Nevertheless, for certain polynomials we can use some of the ideas about solving quadratic inequalities developed in Chapter 2. We illustrate this with the following example.

Example 3   Sketch the graph of $y = f(x) = x^3 + x^2 - 2x$. Label the intercepts.

Solution   In Section 3.6 we discussed how we can use the graph of a function to determine the intervals where the function is positive and negative. In this example we are going to do the opposite. We find x-intercepts and the intervals where $f(x)$ is positive and negative to help us sketch the graph. We do this by performing a sign analysis on $f(x)$. We begin by finding the zeros of $f(x)$:

$$y = f(x) = x^3 + x^2 - 2x = 0$$
$$x(x^2 + x - 2) = 0$$
$$x(x + 2)(x - 1) = 0 \qquad \text{Therefore, the cut points are } x = 0,$$
$$\qquad\qquad\qquad\qquad\qquad\qquad x = -2, \text{ and } x = 1.$$

We draw a number line and complete the sign analysis by picking a test point on each interval. (We have chosen -10, -1, $\frac{1}{2}$, and 5 as the test points.) See Figure 5.12.

Figure 5.12

The sign of $f(x) = x(x + 2)(x - 1)$

$$- - - - - - - - \;|\; + + + + + + + + + + + + \;|\; - - - - - - \;|\; + + + + +$$
$$\qquad\qquad\quad -2 \qquad\qquad\qquad\qquad\qquad 0 \qquad\quad 1$$

Test point $x = -10$ $x = -1$ $x = \dfrac{1}{2}$ $x = 5$

Wherever $f(x)$ is positive, its graph will be above the x-axis, and wherever $f(x)$ is negative, the graph will be below the x-axis. From this sign analysis we know the graph of $f(x)$ will fall below the x-axis for x less than -2 and between 0 and 1, whereas the graph of $f(x)$ will rise above the x-axis for x between -2 and 0 and for x greater than 1. Keep in mind that the cut points are the x-intercepts of the graph of $f(x)$. In other words, the graph will fall in the shaded regions in Figure 5.13.

Remember that the graph of a polynomial cannot have any breaks.

Figure 5.13

The graph of $f(x)$ must lie in the shaded regions.

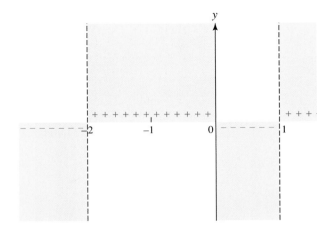

Clearly, then, because polynomial functions are smooth *unbroken* curves, the graph must have a turning point for $-2 < x < 0$ and another turning point for $0 < x < 1$. Because $f(x)$ is a polynomial of degree 3, by property 3 we know that it has at most 2 turning points, and so we have *all* the turning points. Because the precise x-values at which these turning points occur is a matter for calculus, we will content ourselves with finding several more values of $f(x)$ to help us sketch the graph. We include values to help us see the behavior of $f(x)$ to the extreme left and to the extreme right.

$$f(x) = x^3 + x^2 - 2x \quad \text{We choose several values for } x.$$
$$f(-1) = (-1)^3 + (-1)^2 - 2(-1) = -1 + 1 + 2 = 2$$
$$f\left(\frac{1}{2}\right) = \left(\frac{1}{2}\right)^3 + \left(\frac{1}{2}\right)^2 - 2\left(\frac{1}{2}\right) = \frac{1}{8} + \frac{1}{4} - 1 = -\frac{5}{8}$$
$$f(-100) = (-100)^3 + (-100)^2 - 2(-100) = -989,800$$
$$f(100) = (100)^3 + (100)^2 - 2(100) = 1,009,800$$

These values agree with the polynomial property that $|f(x)|$ gets very large as $|x|$ gets large. We make a reasonable guess as to the turning points; the graph appears in Figure 5.14. The end behavior shown is consistent with an odd-degree polynomial with a positive leading coefficient.

Figure 5.14

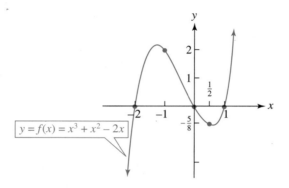

$$y = f(x) = x^3 + x^2 - 2x$$

Calculator Exploration

Use a graphing calculator to graph $y = f(x) = x^3 + x^2 - 2x$ verifying our analysis in Example 3. Also estimate the turning points of the graph and see whether they agree with the graph drawn in Figure 5.14.

Example 4 Use a graphing calculator to sketch the graph of the function $y = 2x^4 - 3x^2 + x - 1$ and describe the graph with respect to its turning points, x-intercepts, and "end" behavior. Round the values to the nearest hundredth.

Solution The graph of $y = 2x^4 - 3x^2 + x - 1$ appears in Figure 5.15.

Figure 5.15

The graph of
$y = 2x^4 - 3x^2 + x - 1$

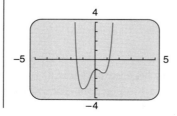

We can see that (as we move from left to right) the function is decreasing to a turning point. Using a graphing calculator, we find the coordinates of this turning point to be $(-0.94, -3.03)$. Recall that this type of turning point is called a relative minimum.

Then as we continue to move to the right, the function increases to a relative maximum. Again using a graphing calculator, we find the coordinates of this turning point to be $(0.17, -0.91)$.

The graph then decreases to another relative minimum, which we find to have coordinates $(0.77, -1.31)$, after which the function increases as we move to the right.

The graph shows us that there are two x-intercepts: one between -2 and -1 and one between 1 and 2. The calculator allows us to approximate these zeros to be -1.44 and 1.20.

As for the end behavior, the y-values are increasing out at both ends. This is consistent with the function being a polynomial of even degree with a positive leading coefficient. We note that since this function is a fourth-degree polynomial, it can have at most three turning points. Since this graph does indeed have three turning points, there can be no others and so we have a complete description of the behavior of this function.

See Exercise 67 for a further discussion of some other issues that also need to be addressed when using a graphing calculator to analyze the graph of a function.

We will continue our discussion of polynomial functions in the next section.

5.1 Exercises

In Exercises 1–8, if the given graph can be the graph of a polynomial, what is its minimum degree? Is the degree odd or even? Is the leading coefficient positive or negative? If it cannot be the graph of a polynomial, explain why.

1.
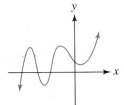

2.
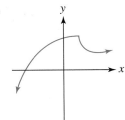

3.
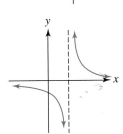

4.
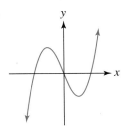

5.
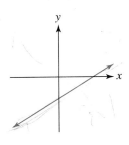

6.
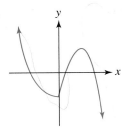

7.
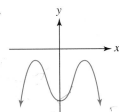

8.
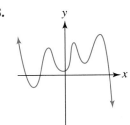

In Exercises 9–22, sketch the graph of the given function. Label the intercepts where appropriate.

9. $y = x^3 + 1$

10. $y = (x + 1)^3$

11. $y = (x - 1)^4$

12. $y = x^4 - 1$

13. $y = 8 - x^3$

14. $y = 9 - x^2$

15. $y = x^2 - 4$

16. $y = (x - 4)^2$

17. $y = -(x + 2)^5$

18. $y = (x - 2)^4 - 16$

19. $y = 2x^3 - 16$

20. $y = -4x^6 - 2$

21. $y = -(x - 1)^5 + 1$

22. $y = (x - 2)^3 - 1$

23. On the same set of coordinate axes, sketch and label the graphs of $y = x^3$, $y = 2x^3$, $y = 3x^3$, and $y = \frac{1}{2}x^3$.

24. On the same set of coordinate axes, sketch and label the graphs of $y = -x^4$, $y = -2x^4$, $y = -3x^4$, and $y = -\frac{1}{2}x^4$.

25. On the same set of coordinate axes, sketch and label the graphs of $y = x^2$, $y = x^4$, $y = x^6$, and $y = x^8$ for $x \geq 0$.

26. On the same set of coordinate axes, sketch and label the graphs of $y = -x^3$, $y = -x^5$, $y = -x^7$, and $y = -x^9$ for $x \geq 0$.

In Exercises 27–30, sketch the graph of a polynomial function that satisfies the given conditions. If no such polynomial function is possible, explain why not.

27. The polynomial $y = p(x)$ is of degree 5 with five real zeros and four turning points. Also, as $x \to -\infty$ then $y \to \infty$, and as $x \to \infty$ then $y \to -\infty$.

28. The polynomial $y = p(x)$ is of degree 5 with four real zeros and five turning points.

29. The polynomial $y = p(x)$ is of degree 4 with two real zeros and three turning points.

30. The polynomial $y = p(x)$ is of degree 6 with four real zeros and five turning points. Also, as $x \to \infty$ then $y \to -\infty$, and as $x \to -\infty$ then $y \to -\infty$.

In Exercises 31–44, use a sign analysis and your knowledge of end behavior to sketch the graph of the given polynomial function.

31. $y = f(x) = x^3 - x^2 - 2x$

32. $y = f(x) = -x^3 + x^2 + 6x$

33. $y = g(x) = x^2 - 6x + 8$

34. $y = h(x) = 3x^2 + 2x - 8$

35. $y = p(x) = x^3 - 9x^2$

36. $y = h(x) = x^3 + 6x^2 + 8x$

37. $y = f(x) = x^4 + x^2$

38. $y = g(x) = 5x - 8$

39. $y = h(x) = x(x - 1)(x + 1)(x + 2)$

40. $y = r(x) = x^4 - 5x^2 + 4$

41. $y = f(x) = (x + 1)(x - 2)(x + 3)(x - 4)$

42. $y = g(x) = (x - 1)(x + 2)(3 - x)(x + 4)$

43. $y = f(x) = x^3 + x^2 - 2x - 2$

44. $y = g(x) = x^3 + 2x^2 - 4x - 8$

Graphing Calculator Exercises

In Exercises 45–50, use a graphing calculator to graph the given polynomial and describe its behavior with respect to the number of zeros, the number and types of turning points, and its behavior at the ends.

45. $y = x^3 - x - 2$

46. $y = x^3 + x - 2$

47. $y = x^4 - 7x^3 + 12x^2 + 4x - 16$

48. $y = x^4 - 10x^2 + 9$

49. $y = 4x^2 - x^5$

50. $y = x^6 + 2x^3$

In Exercises 51–56, use a graphing calculator to graph the given polynomial. Find its x-intercepts and the coordinates of its turning points rounded to the nearest hundredth.

51. $y = x^3 + 4x^2 + x - 4$

52. $y = x^3 - 10x + 8$

53. $y = x^4 + 4x^3 - 9x - 1$

54. $y = x^4 - 5x^2 + x + 2$

55. $y = x^6 - x^5 + x^4 - x^2 + x - 4$

56. $y = 2x^5 - 5x^3 + 3x^2 + 6x + 5$

57. Use a graphing calculator to sketch the graphs of $y = x^3 + kx$ for $k = \pm 1, \pm 2, \pm 3$, and describe how the value of k affects the graph.

58. Repeat Exercise 57 for $y = x^3 + kx^2$.

59. Repeat Exercise 57 for $y = x^4 + kx$.

60. Repeat Exercise 57 for $y = x^4 + kx^2$.

Questions for Thought

61. Can a polynomial have two distinct real zeros and no turning points? Explain.

62. We have stated that a polynomial of degree n can have at most n zeros and $n - 1$ turning points. Show that if a polynomial of degree n has n distinct zeros, then it *must* have $n - 1$ turning points. Also show that the converse of this statement is false.

63. For the polynomial $p(x) = 2x^3 - 10x^2 - 3x - 100$, compute $p(10)$, $p(100)$, $p(1000)$, $p(10,000)$, and $p(100,000)$. What is happening to the values of $p(x)$ as x gets larger and larger? Compute $p(-100)$, $p(-1000)$, $p(-10,000)$, and $p(-100,000)$. What is happening to the values of $p(x)$ as x gets smaller and smaller? We may write

$$p(x) = 2x^3 - 10x^2 - 3x - 100$$

$$= x^3\left(2 - \frac{10}{x} - \frac{3}{x^2} - \frac{100}{x^3}\right)$$

Analyze this factored form of $p(x)$ to support the claim that $|p(x)|$ gets very large as $|x|$ gets very large.

64. Consider the general third-degree polynomial

$$p(x) = ax^3 + bx^2 + cx + d$$

Suppose $x \to +\infty$; what determines whether $p(x) \to +\infty$ or $p(x) \to -\infty$? Suppose $x \to -\infty$; what determines whether $p(x) \to +\infty$ or $p(x) \to -\infty$? [HINT: Use the factorization technique illustrated in Exercise 63.]

65. Repeat Exercise 64 for the general fourth-degree polynomial $p(x) = ax^4 + bx^3 + cx^2 + dx + e$.

66. Generalize the results of Exercises 63, 64, and 65 for a polynomial of degree n. Explain how the leading coefficient and the degree of the polynomial determine the end behavior of a polynomial.

67. If you have worked on Exercises 63–66, perhaps you have come to the conclusion that if a polynomial is of even degree with a positive leading coefficient, then its graph will go up at both ends.

    (a) Use a graphing calculator to graph

$$y = p(x) = 0.00001(x^8 - 298x^6$$
$$+ 23{,}625x^4 - 209{,}952x^2 + 186{,}624)$$

    in the standard $[-10, 10]$ by $[-10, 10]$ window.

    (b) How many zeros and turning points does $p(x)$ appear to have?

    (c) Could this be a complete picture of the behavior of the graph? Is it?

    (d) Graph this function in the $[-20, 20]$ by $[-50, 50]$ window. How many zeros and turning points does $p(x)$ appear to have now?

5.2   More on Polynomial Functions and Mathematical Models

We next turn our attention to the exact shape of the graphs we have been sketching.

Concavity

Before we look at another example, let's take a moment to discuss the shape of a curve. Throughout our discussions we have always described a graph as we move from left to right. Figure 5.16 illustrates four graphs. The first two graphs are rising between points A and B and yet have different shapes, and the second two graphs are falling between points A and B and also have different shapes.

Figure 5.16

Rising and falling graphs with different shapes

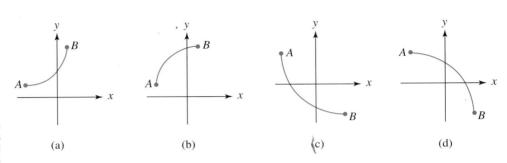

(a)   (b)   (c)   (d)

The functions illustrated by the first two graphs in Figure 5.16 are both increasing. How would you describe the difference between the two graphs?

As the first graph rises, it bends or curves upward; as the second graph rises, it bends or curves downward. Similarly, the third is falling and curved upward, whereas the fourth is falling and curved downward. A graph that curves upward as in Figure 5.16(a) and (c) is called *concave up;* a graph that curves downward as in Figure 5.16(b) and (d) is called *concave down.*

Figure 5.17(a) illustrates two graphs that are concave up, and Figure 5.17(b) illustrates two graphs that are concave down.

Figure 5.17

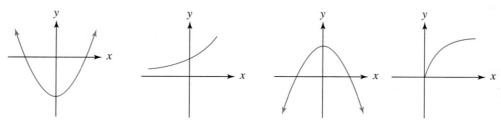

These graphs are concave up.

These graphs are concave down.

(a)

(b)

Figure 5.18 illustrates a graph that is rising from A to B and is concave up part of the way (until point C) and then becomes concave down. This point at which the concavity changes is called a **point of inflection**.

Looking at the graph of $y = x^3 + x^2 - 2x$ of Example 3 of the previous section, which we repeat in Figure 5.19, we can see that the graph appears to have an inflection point a bit to the left of the y-axis.

How is a point of inflection related to the concavity of a graph?

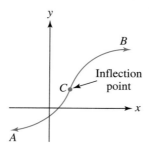

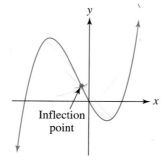

Figure 5.18

C is a point of inflection.

Figure 5.19

The graph of $y = f(x) = x^3 + x^2 - 2x$ has one inflection point.

Example 1 Use the graph in Figure 5.20.
(a) Find the intervals where $f(x)$ is increasing.
(b) Find the intervals where $f(x)$ is decreasing.
(c) Find the intervals where the graph of $f(x)$ is concave up.
(d) Find the intervals where the graph of $f(x)$ is concave down.
(e) Find the number of inflection points for $f(x)$.

Figure 5.20

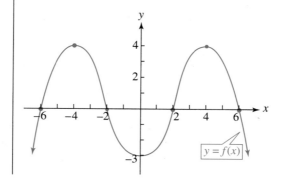

Solution

(a) The intervals where $f(x)$ is increasing are the x-intervals on which the graph is rising, which are $(-\infty, -4] \cup [0, 4]$.

(b) The intervals where $f(x)$ is decreasing are the x-intervals on which the graph is falling, which are $[-4, 0] \cup [4, \infty)$.

(c) It appears that the graph of $f(x)$ is concave up on the interval $(-2, 2)$.

(d) It appears that the graph of $f(x)$ is concave down on the interval $(-\infty, -2)$ and on the interval $(2, \infty)$.

(e) As we move from left to right, the graph changes from concave down to concave up and then back to concave down. Each change occurs at an inflection point, so $f(x)$ has two inflection points. ∎

Example 2

Suppose $p(x)$ is a polynomial function that satisfies the following conditions: $p(x)$ has exactly three turning points at $(-3, -4)$, $(0, 2)$, $(3, -4)$ and exactly two inflection points at $(-1, 0)$ and $(1, 0)$.

(a) Sketch a graph of $p(x)$ based on this information.

(b) How many real zeros does $p(x)$ have?

Figure 5.21

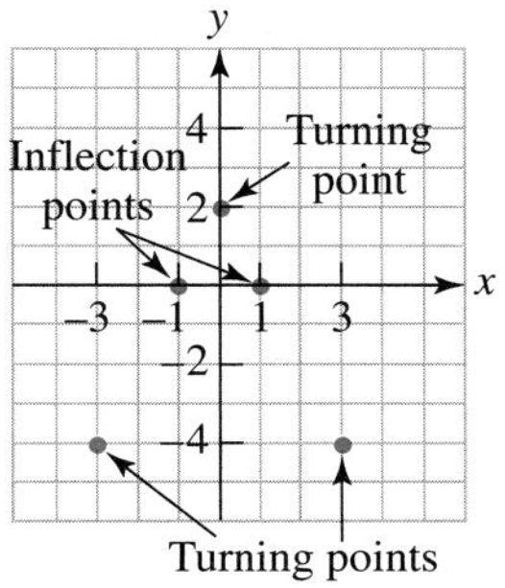

Solution

(a) We begin by plotting the given points and labeling them as turning points or inflection points. See Figure 5.21. Let's analyze the given information logically. Suppose we imagine the given turning points of $p(x)$ look like the graph in Figure 5.22(a). We would then connect the points with the dashed graph in the same figure. However, this would force the graph to have more than the three turning points we are given. Similarly, if we try to connect the points as in Figure 5.22(b), then the graph would have *four* inflection points instead of two.

Figure 5.22

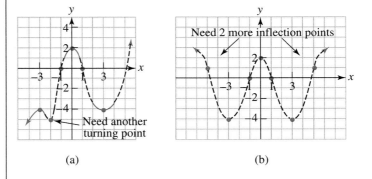

(a)   (b)

Because we know that $p(x)$ is a polynomial, its graph will be a smooth curve with no breaks or corners, and because there are no turning points other than the ones we are given, the graph in Figure 5.23 (page 290) is a fair representation of $p(x)$.

Figure 5.23

Graph of $p(x)$

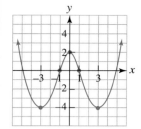

What polynomial properties have we
used in this example?

Note that the inflection points help us sketch the shape of the graph. Also note that the graph shows that the y-values on the graph are getting very large as $|x|$ gets large, as expected for a polynomial function.

(b) Based on the given information and our analysis of the graph, we can see that the graph must have four x-intercepts. Therefore, $p(x)$ has four real zeros.

Knowing the turning points and inflection points for a function is important, but we still need to develop the ability to synthesize all this information in order to obtain the graph of the function. That is exactly what we are attempting to do throughout this chapter.

One of the important properties of polynomials is that their graphs have no breaks. (The technical terminology is that a polynomial function is *continuous*.) Thus, if a polynomial has -3 and 4 as range values, then all the values between -3 and 4 must also be range values. See Figure 5.24.

This result can be formally stated as the *Intermediate Value Theorem for Polynomials*.

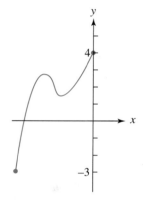

Figure 5.24

If -3 and 4 are range values of a polynomial, then all the values between -3 and 4 must also be in the range.

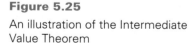

Theorem 5.1   Intermediate Value Theorem for Polynomials

Consider a polynomial $p(x)$ and a closed interval $[a, b]$. For any value c between $p(a)$ and $p(b)$, there is at least one x-value in $[a, b]$ such that $p(x) = c$.

Figure 5.25 illustrates the content of the Intermediate Value Theorem.

Figure 5.25

An illustration of the Intermediate Value Theorem

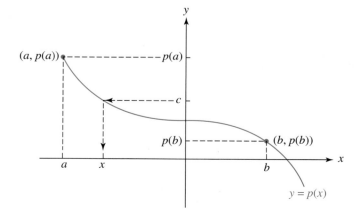

In words, this theorem says that a polynomial function will take on all the values between any two range values. In particular, this theorem tell us that *if we know that p(a) is negative and p(b) is positive, then there must be an x-value between a and b for which p(x) = 0.* Let's see how we can apply this idea to approximate the zero of a polynomial.

Calculator Exploration

Use a graphing calculator to graph the following polynomials:

$$p(x) = x^4 - x^2 - 2 \qquad q(x) = x^5 - 2x^2 + 1$$

1. Use the graphs to observe the location of the zeros of these functions between consecutive integers.
2. Use $\boxed{\text{TRACE}}$ and $\boxed{\text{ZOOM}}$ to locate the zeros between consecutive tenths.
3. Discuss the zeros of these functions as they relate to the symmetry of these graphs.

Example 3 Approximate the real zeros of $p(x) = x^3 - x^2 - 2$ to the nearest tenth.

Solution Since we cannot factor $p(x)$, we will begin with a graph of $p(x)$ to get a general idea of how many real zeros there are and where they are. Figure 5.26 illustrates the graph of $p(x)$ in the $[-5, 5]$ by $[-3, 3]$ viewing window.

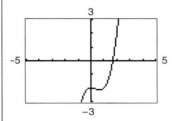

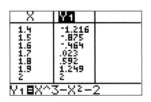

Figure 5.26
The graph of
$p(x) = x^3 - x^2 - 2$

Table 5.4

As we can see from the graph, it appears that $p(x)$ has one real zero and it falls between 1 and 2. Rather than use $\boxed{\text{TRACE}}$ to approximate the zero, we construct a table of values, which appears as Table 5.4. Based on this table, we can see that $p(1.6) = -0.464$ and $p(1.7) = 0.023$. Thus the polynomial is equal to zero between $x = 1.6$ and $x = 1.7$. It appears that, to the nearest tenth, the zero is 1.7.

Modify the table settings to verify that the zero is 1.7 to the nearest tenth.

Let's now return to the situation described in the introduction to this chapter.

Example 4 In business, the *demand function* gives the price per item, p, in terms of the number of items sold, x. Suppose that a calculator manufacturer finds that for its model GC-5 calculator, the price p (in dollars) is related to the number of items produced, x (in millions), by the demand function

$$p = 80 - 4x^2$$

The manufacturer's revenue function is determined by multiplying the number of items sold (x) by the price per item (p). Thus the *revenue function R* for this calculator manufacturer is

Revenue $=$ $\underbrace{\text{price per item}}_{x}$ $\cdot$ $\underbrace{\text{\# of items sold}}_{(80 - 4x^2)}$ Thus we have

$$R = R(x) = x(80 - 4x^2) = 80x - 4x^3$$

where R is measured in millions of dollars. Use a graphing calculator to estimate the price per item that maximizes the revenue.

Solution   This example gives us an opportunity to take advantage of the technology available to us. Rather than go through the detailed analysis necessary to obtain the graph of the revenue function (as was done in the previous section), we graph the revenue function $R = 80x - 4x^3$ on a graphing calculator.

After a bit of trial and error, we decide to use the viewing window of $[-10, 10]$ by $[-200, 200]$ to get a reasonable view of the graph of the revenue function. We now view the graph, which appears in Figure 5.27.

Figure 5.27

The graph of $R = 80x - 4x^3$

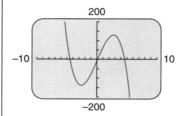

Given that $R(x)$ is a third-degree polynomial, we know that the graph of $R(x)$ can have at most two turning points. Since we see that the graph in Figure 5.27 has two turning points, we recognize that we have a complete picture of the behavior of the revenue function.

However, since x represents the number of calculators sold (in millions), x must be a nonnegative number. On the other hand, the revenue *can* be negative. We interpret negative revenue as a loss. Consequently (again after a bit of trial and error), we change the viewing window to $[0, 10]$ by $[-200, 200]$ and obtain the graph in Figure 5.28.

Figure 5.28

The graph of $R = 80x - 4x^3$

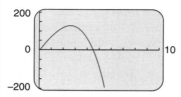

We can see that as the price increases, the revenue increases to some maximum value and then decreases.

We have previously discussed using TRACE and ZOOM to find the maximum and/or minimum value of a function by observing where a function changes from increasing to decreasing or vice versa. This approach allows us to actually "feel" the

Graphing calculators provide the capability of finding maximum and minimum values directly. On the TI-83 Plus, for example, choose **maximum** from the **CALC** menu.

maximum or minimum value. Following that procedure here, we approximate that the maximum revenue will occur when $x = 2.58$, which means that 2.58 million calculators are sold. See Figure 5.29.

Figure 5.29

Finding the maximum value of
$R = 80x - 4x^3$

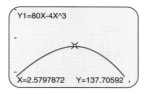

Y1=80X-4X^3

X=2.5797872   Y=137.70592

Keep in mind that the *Y*-value shown on the screen in Figure 5.25 is the **revenue**. This example is asking us for the **price** that produces the maximum revenue.

Keep in mind that we were asked to estimate the *price* that maximizes the revenue. Since $p = 80 - 4x^2$, when $x = 2.58$ we have $p = 80 - 4(2.58)^2 = 53.37$. This means that the company will earn its maximum revenue if it sets the calculator price to be $53.37. ∎

Example 5   If $g(x) = -x^4 - 2x^3 + 4$ and $f(x) = x^2 + 1$, use a graphing calculator to solve the equation $g(x) = f(x)$ accurate to two decimal places.

Solution   First, we should understand that we are looking for values of x that make $g(x) = f(x)$, or, equivalently, we are solving the equation $-x^4 - 2x^3 + 4 = x^2 + 1$ for x. That being said, we graph $g(x)$ and $f(x)$ together on the graphing calculator. The "complete" graphs of both functions using a standard $[-10, 10]$ by $[-10, 10]$ window are in Figure 5.30(a). We are being asked, When is the function, $g(x)$, equal to the function, $f(x)$? Or, more precisely, for which values of x are the y values equal?

Figure 5.30

The graphs of $f(x)$ and $g(x)$

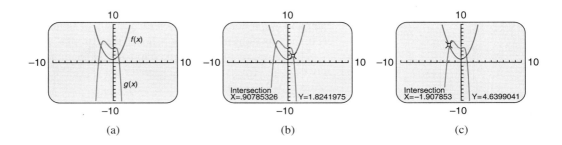

(a)   (b)   (c)

We want to find where the graph of $y = g(x)$ intersects the graph of $y = f(x)$. Using INTERSECT in the [2nd] [CALC] menu, we find that the points of intersection (accurate to two places) are $(0.91, 1.82)$ and $(-1.91, 4.64)$. See Figures 5.30(b) and (c). We must be careful here as the question asks for the values of x for which $g(x) = f(x)$, not the values of y. The points of intersection are *ordered pairs* (x, y); however, the question asks for values of x. Therefore, the answer is $x = -1.91$ and $x = 0.91$.

An alternative way to solve this equation is to start with $-x^4 - 2x^3 + 4 = x^2 + 1$, put this equation into the equivalent form $-x^4 - 2x^3 - x^2 + 3 = 0$, and then graph the function $h(x) = -x^4 - 2x^3 - x^2 + 3$ on a graphing calculator. See

Figure 5.31(a). Again we have a "complete" graph of $h(x)$ (ZOOM IN on the "flat top" of $h(x)$ and see that there are actually three turning points.) Notice we are now asking, When is $h(x) = 0$?, or, equivalently, What are the zeros (x-intercepts) of $h(x)$?

Figure 5.31

The graph of $h(x)$

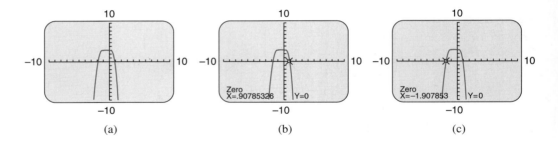

(a)   (b)   (c)

Using ZERO in the $\boxed{\text{2nd}}$ $\boxed{\text{CALC}}$ menu, we find the zeros (accurate to two places) to be 0.91 and -1.91. See Figures 5.31(b) and (c). With this alternative method, we arrived at the same x values as we did using the original method. ∎

In Example 5 we led you to the approach by predefining the functions and asking you to solve the functional equation $g(x) = f(x)$. The same problem could have been simply stated as, Solve $-x^4 - 2x^3 + 4 = x^2 + 1$. It would have been up to you to define two functions (or one) and use your graphing calculator to graph those functions to find the solutions. This is illustrated in Example 6.

Example 6   Using a graphing calculator, solve the inequality $-x^4 - 2x^3 + 4 > x^2 + 1$ accurate to two decimal places.

Solution   As with Example 5, we can approach this by defining the expression on each side of the inequality as a function: Let $g(x) = -x^4 - 2x^3 + 4$ and $f(x) = x^2 + 1$. Next we enter and graph both functions on the graphing calculator. These functions are identical to the functions in Example 5, hence their graphs in Figure 5.30 are the same. This time, however, we are being asked to solve an inequality. The original inequality can now be restated as, When is $g(x)$ greater than $f(x)$? More precisely, we ask, For which values of x is $g(x) > f(x)$?

Figure 5.32

The graphs of $f(x)$ and $g(x)$

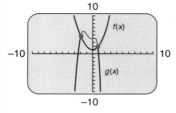

We see in Figure 5.32 that what we really want to find is when the graph of the fourth-degree polynomial $g(x)$ is *higher* than the graph of the quadratic polynomial $f(x)$. we see that the graph of $g(x)$ is higher than the graph of $f(x)$ between the two points of intersection of the two graphs. In Example 5 we found that the points of intersection occur at $x = -1.91$ and $x = 0.91$ accurate to two places. Hence, $g(x) > f(x)$ or $-x^4 - 2x^3 + x^2 + 2 > 0.5x^2 - 2$ for $-1.91 < x < 0.91$, or when x is in the *interval* $(-1.91, 0.91)$.

As with Example 5, we could have found the solution alternatively by putting $-x^4 - 2x^3 + 4 > x^2 + 1$ into its equivalent form $-x^4 - 2x^3 - x^2 + 3 > 0$, graphing the function $h(x) = -x^4 - 2x^3 - x^2 + 3$ and asking these questions: When is $h(x) > 0$? When is $h(x)$ positive? When is the graph of $h(x)$ above the x-axis? See Figure 5.33.

Figure 5.33

The graph of $h(x)$

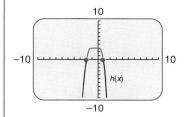

We see that the graph of $h(x)$ is above the x-axis between the two x-intercepts or between the two zeros of $h(x)$. Hence, to find the interval(s) where $h(x) > 0$, we need to identify the zeros (where $h(x) = 0$). In Example 5, we found the zeros (accurate to two places) to be -1.91 or 0.91, and again, we find that the solution is all x in the interval $(-1.91, 0.91)$.   ∎

Much of our work with polynomials has been based on finding the zeros of polynomial functions, which we have been able to do by factoring. In the next section we digress a bit to discuss polynomial division, which will allow us to develop some additional methods for finding the zeros of polynomial functions.

5.2 Exercises

In Exercises 1–6, the given information pertains to a polynomial function $p(x)$. Sketch a graph of $p(x)$ consistent with this information.

1. $p(x)$ has exactly one turning point at $(2, 0)$ and is always concave down.

2. $p(x)$ has exactly three turning points at $(-3, 2)$, $(0, -3)$, and $(3, 2)$ and exactly two inflection points at $(-1, -1)$ and $(1, -1)$.

3. $p(x)$ has exactly one turning point at $(0, 0)$ and exactly two inflection points at $(1, 2)$ and $(3, 5)$.

4. $p(x)$ has exactly one turning point at $(2, -6)$, two inflection points, one of which is at $(6, 0)$, and one additional x-intercept at $(-4, 0)$.

5. $p(x)$ has exactly two turning points at $(-1, 4)$ and $(2, -1)$ and exactly one inflection point at $(1, 0)$.

6. $p(x)$ has exactly three turning points at $(-1, 0)$, $(0, 3)$, and $(2, -10)$ and two inflection points at $(-0.5, 2)$ and $(1.5, -6)$.

In Exercises 7–12, place the real zeros of $p(x)$ within consecutive tenths of a unit; that is, the zero is between 3.6 and 3.7. Each exercise indicates how many real zeros there are.

7. $p(x) = x^3 - 2$;   1 real zero

8. $p(x) = 4 - x^5$;   1 real zero

9. $p(x) = x^3 + x^2 - 8x + 8$;   1 real zero

10. $p(x) = \frac{1}{3}x^3 - x^2 - 3x - 6$;   1 real zero

11. $p(x) = x^4 + 2x^3 - 1$;   2 real zeros

12. $p(x) = x^4 - 3x^3 + 3x^2 - x - 1$;   2 real zeros

13. Find the points of intersection of the graph of $y = x^3$ with the graph of $y = 9x$.

14. Find the points of intersection of the graph of $y = 3x^2$ with the graph of $y = x^3 - 10x$.

15. Find the points of intersection of the graph of $y = x^5$ with the graph of $y = 2x^4 + 3x^3$.

16. Find the points of intersection of the graph of $y = x^2$ with the graph of $y = 49x^2 - 4x^3$.

17. Let $f(x) = x^3$; find and simplify the difference quotient $\dfrac{f(x) - f(2)}{x - 2}$.

18. Let $f(x) = x^4$; find and simplify the difference quotient $\dfrac{f(x) - f(3)}{x - 3}$.

19. Let $f(x) = x^4$; find and simplify the difference quotient $\dfrac{f(x + h) - f(x)}{h}$.

20. Let $f(x) = x^3 + 1$; find and simplify the difference quotient $\dfrac{f(x + h) - f(x)}{h}$.

21. Suppose n is a positive integer and $|x| < \dfrac{1}{2}$. What is the smallest value of n for which $|x^n| < 0.005$?

22. What is the range of x-values for which $|x^4| < 10^{-4}$?

23. What is the range of x-values for which $\left|\dfrac{x^3}{6}\right| < 0.001$?

24. Suppose n is a positive integer and $|x| < \dfrac{1}{3}$. What is the smallest value of n for which $\left|\dfrac{x^n}{n}\right| < 10^{-5}$?

Use a graphing calculator in Exercises 25–36. Round your answers to the nearest hundredth.

25. A candy manufacturer finds that when he charges d dollars per pound, he can sell $430 - d^2$ pounds of candy per week.
 (a) Express the revenue R as a function of d.
 (b) Use a graphing calculator to sketch the graph of the revenue function, and use the graph to approximate the price per pound that maximizes the revenue.
 (c) What is the maximum revenue to the nearest dollar?

26. An electronics distributor finds that when she charges x dollars per unit, she can sell $108 - x^2$ units.
 (a) Express the revenue R as a function of x.
 (b) Use a graphing calculator to sketch the graph of the revenue function, and use the graph to approximate the price that maximizes the revenue.
 (c) What is the maximum revenue to the nearest dollar?

27. A hardware manufacturer finds that he can sell n items at a price of $2 + 0.45n - 0.001n^2$ dollars per item.
 (a) Express the revenue R as a function of n.

 (b) Use a graphing calculator to sketch the graph of the revenue function, and use the graph to approximate the *price* that maximizes the revenue.
 (c) What is the maximum revenue to the nearest dollar?

28. A cellular phone company finds that it can lease n phones per month at a price of $24 + 0.58n - 0.003n^2$ dollars per phone per month.
 (a) Express the monthly revenue R as a function of n.
 (b) Use a graphing calculator to sketch the graph of the revenue function, and use the graph to approximate the *price* that maximizes the revenue.
 (c) What is the maximum revenue to the nearest dollar?

29. The base of a closed rectangular box has length equal to three times its width, which is w inches, and its height is 6 inches less than its width. Express the surface area S of the box as a function of w, and use this function to compute the surface area when the width of the box is 20 in., 26 in., and 42 in.

30. The base of a rectangular box has length equal to twice its width, which is w cm, and its height is 3 cm less than its width. Express the volume V of the box as a function of w, and use this function to compute the volume when the width of the box is 12 cm, 18 cm, and 36 cm.

31. A pharmaceutical company finds that the amount A (in milligrams) of a new medication present in the bloodstream m minutes after the medication is taken can be approximated by the function

$$A = A(m) = 0.28m + 0.39m^2 - 0.01m^3$$

 (a) Use this function to approximate the number of milligrams of the medication present in the bloodstream at 5-minute intervals for the first 30 minutes after it is taken.
 (b) Use the information obtained in part (a) to estimate, to the nearest minute, when the amount of medication in the bloodstream is at a maximum.

32. If an object is thrown straight up from ground level with an initial velocity of 45 feet per second, then its height h above the ground t seconds after the object is thrown is given by the function

$$h = h(t) = 45t - 16t^2$$

 (a) Use this function to find the height of the object at intervals of one-half second.
 (b) Use the information obtained in part (a) to estimate, to the nearest tenth of a second, when the object will hit the ground.

33. An open box with a square base is to be made from a square sheet of material 20 inches on a side by cutting a square corner of side x from each corner of the sheet and folding up the edges, as indicated in the accompanying diagram.

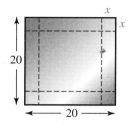

(a) Express the volume V of the box as a function of x. What is the domain?

(b) Find the dimensions of the box that maximize the volume of the box. Round to 2 decimal places.

34. A cardboard carton has a rectangular bottom whose length is twice its width x. See the accompanying figure. The total length of all 12 edges of the box is 120 inches.

(a) Express the volume of the box as a function of x.

(b) Find the dimensions of the box that maximize the volume of the box.

35. The following table lists the amount in mortgage loans (in millions of dollars) a bank has lent in each year from 1990 through 1998.

Year	Amount
1990	9.9
1991	10.3
1992	11.2
1993	11.9
1994	12.7
1995	14.3
1996	15.5
1997	17.9
1998	20.5

Which of the following functions is the best mathematical model for the amount of mortgage money lent in year x where $x = 0$ is 1990, $x = 1$ is 1991, etc.? Explain.

(a) $f(x) = 1.07x + 8.3$

(b) $f(x) = 0.15x^2 + 10.4$

(c) $f(x) = 1.4\sqrt{x} + 9.1$

36. The following table lists the number of visitors each year (in millions) to a national chain of amusement parks in each year from 1990 through 1998.

Year	Number
1990	8.5
1991	8.59
1992	8.62
1993	8.67
1994	8.75
1995	9.42
1996	9.67
1997	10.13
1998	12.46

Which of the following functions do you think is the best mathematical model for the number of visitors to the amusement parks in year x where $x = 0$ is 1990 and $x = 1$ is 1991, etc.? Explain.

(a) $f(x) = 0.003x^4 - 0.005x^3 + 0.084x + 8.5$

(b) $f(x) = 0.002x^4 - 0.007x^3 + 0.062x + 8.5$

(c) $f(x) = 0.004x^4 - 0.009x^3 + 0.078x + 8.5$

In Exercises 37–50, use a graphing calculator to solve the following accurate to two decimal places.

37. If $f(x) = x - 1$ and $g(x) = x^2 - 6x + 1$, find x such that $f(x) = g(x)$.

38. If $g(x) = 3x^2 - x - 2$ and $h(x) = 2x + 1$, find x such that $g(x) = h(x)$.

39. If $f(x) = 2x^2 - 1$ and $g(x) = -x^2 + 1$, solve $f(x) = g(x)$ for x.

40. If $h(x) = -x^2 + x + 2$ and $k(x) = 2x^2 + 1$, solve $h(x) = k(x)$ for x.

41. If $f(x) = x^3 - 1$ and $g(x) = 2x + 1$, solve $f(x) < g(x)$ for x.

42. If $h(x) = -x^2 + x + 2$ and $k(x) = x^3 - 1$, solve $h(x) > k(x)$ for x.

43. $2x - 5 = -x^3 - 2x + 4$

44. $x^2 - 3 = -x^3 + 2x^2 + 1$

45. $-x^2 - 5x + 2 > x - 1$

46. $-x^2 + 3x + 4 < 2x + 1$

47. $x - 5 < -x^3 - 2x + 4$

48. $-x^2 + 5x + 6 < x^2 + 1$

49. $\frac{1}{2}x^2 - 5 > -x^3 - 5x^2 + 2$

50. $x^4 - 3x^3 + 5x + 1 > -x^2 + 8$

Questions for Thought

51. Use a graphing calculator to examine the graph of the polynomial $p(x) = (x - 2)^3(x + 3)^2$. What are the

zeros? Notice that at $x = 2$ the sign of $p(x)$ changes, whereas at $x = -3$ it does not. Explain why.

52. Graph $y = x^3 - 4x^2 + 2x + 2$ on a graphing calculator. How many inflection points does this graph appear to have? Does the graphing calculator allow you to estimate the inflection point very well? (The inflection point is actually at $x = 1\frac{1}{3}$.)

5.3   Polynomial Division, Roots of Polynomial Equations: The Remainder and Factor Theorems

In the previous section we saw that the real zeros of a polynomial function provide valuable information that can be helpful in sketching its graph. We briefly discussed how to approximate the real zeros of a polynomial, but for the most part we focused on finding the zeros by factoring the polynomial. However, we have no general method for factoring polynomials of degree greater than 2. In this section and the next we turn our attention to methods that will allow us to find zeros of higher-degree polynomials. To do this, we first need to discuss the division process for polynomials.

The same four-step process of long division (divide, multiply, subtract, and bring down) used in dividing numbers can be applied to long division of polynomials.

Let's begin by illustrating the division $\dfrac{2x^2 - 11x + 5}{x - 3}$. We set up the long division as we would for numbers and we carry out each of the four steps. Note that in setting up the long division process, we make sure that both dividend and divisor are in standard form (written from highest power to lowest).

1. First, we divide $2x^2$ by x: $\dfrac{2x^2}{x} = 2x.$

This becomes the first term in the quotient.

2. Multiply the divisor $x - 3$ by the term $2x$ obtained in step 1. This produces $2x^2 - 6x$, which is written below the dividend, as shown.

$$
\begin{array}{r}
2x - 5 \\
x - 3 \overline{)\, 2x^2 - 11x + 5} \\
-(2x^2 - 6x) \\
\hline
-5x + 5 \\
-(-5x + 15) \\
\hline
-10
\end{array}
$$

This is the long division **tableau**.

3. Next we subtract $2x^2 - 6x$ from $2x^2 - 11x$ in the dividend, obtaining $-5x$. Note that we indicate subtraction by enclosing the $2x^2 - 6x$ in parentheses with a subtraction sign in front of it.

4. Finally, we bring down the next term or terms in the dividend and continue by repeating steps 1 through 3.

Thus, when $2x^2 - 11x + 5$ is divided by $x - 3$, the quotient is $2x - 5$ and the remainder is -10. We may write this result in two ways, either as

$$\frac{2x^2 - 11x + 5}{x - 3} = 2x - 5 + \frac{-10}{x - 3} \tag{1}$$

or, if we multiply both sides of the equation by $x - 3$,

$$\underbrace{2x^2 - 11x + 5}_{\text{Dividend}} = \underbrace{(x - 3)}_{\text{divisor}} \cdot \underbrace{(2x - 5)}_{\text{quotient}} + \underbrace{(-10)}_{\text{remainder}} \tag{2}$$

It is important to note that the division process ends when the degree of the remainder is less than the degree of the divisor, which is quite similar to division of natural numbers, where we stop when the remainder is less than the divisor. Also note that equation (1) is valid for all $x \neq 3$, whereas equation (2) is valid for all values of x.

We illustrate this process again in the next example.

Example 1 Divide: $\dfrac{6x^3 + 19x^2 - 25}{3x + 5}$

Solution As mentioned before, when we set up the long division, we want the dividend and divisor to be in standard form. In addition, if there are any "missing" terms in the *dividend,* we insert these missing terms with zero coefficients, as indicated here. These zero-coefficient terms serve as placeholders to make the "bookkeeping" a bit easier.

$$
\begin{array}{r}
2x^2 + 3x - 5 \\
3x + 5 \overline{)\; 6x^3 + 19x^2 + 0x - 25} \\
\underline{-(6x^3 + 10x^2)} \\
9x^2 + 0x \\
\underline{-(9x^2 + 15x)} \\
-15x - 25 \\
\underline{-(-15x - 25)} \\
0
\end{array}
$$

The fact that the remainder is 0 means that the division is exact. Therefore, we have
$$6x^3 + 19x^2 - 25 = (3x + 5)(2x^2 + 3x - 5)$$
and so $3x + 5$ is a factor of $6x^3 + 19x^2 - 25$. ∎

Example 2 Divide: $\dfrac{x^4 - 1}{x^2 + 2x}$

Solution

$$
\begin{array}{r}
x^2 - 2x + 4 \\
x^2 + 2x \overline{)\; x^4 + 0x^3 + 0x^2 + 0x - 1} \\
\underline{-(x^4 + 2x^3)} \\
-2x^3 + 0x^2 \\
\underline{-(-2x^3 - 4x^2)} \\
4x^2 + 0x \\
\underline{-(4x^2 + 8x)} \\
-8x - 1
\end{array}
$$

Again this long division means
$$\underbrace{x^4 - 1}_{\text{Dividend}} = \underbrace{(x^2 + 2x)}_{\text{divisor}} \cdot \underbrace{(x^2 - 2x + 4)}_{\text{quotient}} + \underbrace{(-8x - 1)}_{\text{remainder}}$$
It is left to the reader to verify this last equation. ∎

An algorithm is a precise sequence of steps that can be followed to solve a specific problem.

The result of this division for polynomials can be summarized in the following theorem, called the **Division Algorithm**.

Theorem 5.2   The Division Algorithm

Let $p(x)$ and $d(x)$ be polynomials with $d(x) \neq 0$ and with the degree of $d(x)$ less than or equal to the degree of $p(x)$. Then there are polynomials $q(x)$ and $r(x)$ such that

$$\underbrace{p(x)}_{\text{Dividend}} = \underbrace{d(x)}_{\text{divisor}} \cdot \underbrace{q(x)}_{\text{quotient}} + \underbrace{r(x)}_{\text{remainder}}$$

where either $r(x) = 0$ or the degree of $r(x)$ is less than the degree of $d(x)$.

Later on we will see that this theorem is instrumental in deriving important results about polynomials.

Synthetic Division

When we divide a polynomial by a divisor of the form $x - c$ (where c is a constant), there is quite of bit of repetitious writing. We can shorten this process by recognizing that all the essential information in the division process is carried by the various coefficients and their specific location in the tableau.

Let's twice rewrite the division problem with which we began this section, once as before and a second time with just the coefficients.

$$
\begin{array}{r}
2x \; - \; 5 \\
x-3\overline{)2x^2 - 11x + \; 5} \\
\underline{2x^2 - \; 6x} \\
-5x + \; 5 \\
\underline{-5x + 15} \\
10
\end{array}
\qquad
\begin{array}{r}
2 \; - \; 5 \\
-3\overline{)2 \; - \; 11 \; + \; 5} \\
\underline{②- \; 6} \\
\;⊖+ \;⑤ \\
\underline{⊖+ \; 15} \\
-10
\end{array}
$$

The circled coefficients are merely duplicates of those in the quotient or dividend, so we delete these and consolidate the tableau vertically as follows:

$$
\begin{array}{r}
2 \; - \; 5 \\
-3\overline{)2 \; - \; 11 \; + \; 5} \\
\underline{- \; 6} \\
+ \; 15 \\
\overline{-10}
\end{array}
\qquad
\begin{array}{r}
2 \quad -5 \\
-3\overline{)2 \quad -11 \quad 5} \\
\underline{-6 \quad 15} \\
-10
\end{array}
$$

Next we move the coefficients of the quotient to the bottom row.

$$
\begin{array}{r}
-3\overline{)2 \quad -11 \quad 5} \\
\underline{-6 \quad 15} \\
2 \quad -5 \quad -10
\end{array}
$$

Finally, to avoid subtraction (which sometimes introduces careless errors), we change the sign of the divisor and the signs in the second row, which allows us to *add* in each column instead of subtract.

The final format that follows is called **synthetic division** for dividing $2x^2 - 11x + 5$ by $x - 3$. We include an explanation of how to construct this table.

1. Since the divisor is $x - 3$, we represent the divisor by 3, written at the left of the top row. We write the coefficients of the dividend in the first row of the table. Remember to hold a place for any missing terms by including a coefficient of 0.

> When we divide $x - r$ we represent the divisor by r.

$$
\begin{array}{r}
3\overline{)2 \quad -11 \quad 5} \\
\underline{6 \quad -15} \\
2 \quad -5 \quad \boxed{-10}
\end{array} \quad \longleftarrow \textit{Remainder}
$$

2. Bring down the 2.

3. Multiply 3×2; enter 6 in second column and add to get -5.

4. Multiply $3 \times (-5)$; enter -15 in third column and add to get -10.

5. Recognizing the last number in the bottom row as the remainder, we read the coefficients of the quotient as 2 and -5. Therefore, the quotient is $2x - 5$ and the remainder is -10.

We can summarize the synthetic division procedure as follows:

Synthetic Division

To divide $a_4x^4 + a_3x^3 + a_2x^2 + a_1x + a_0$ by $x - c$ using synthetic division, we set up the following table.

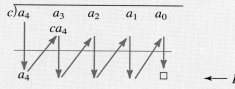

Note that the diagonal arrows indicate multiplication by c; the vertical arrows indicate addition in each column.

It is most important to remember that the synthetic division process works *only* when the divisor is of the form $x - c$.

Example 3   Use synthetic division to find the quotient $\dfrac{x^5 + x^4 - 2x^2 + 24}{x + 2}$.

Solution   Since the divisor $x + 2$ corresponds to $x - c$, we have $c = -2$ for the synthetic division process. (That is, $x - (-2) = x + 2$.) Also, we must remember to insert 0 for the coefficients of the missing terms. In other words, we view the dividend as $x^5 + x^4 + 0x^3 - 2x^2 + 0x + 24$.

$$
\begin{array}{r|rrrrrr}
-2) & 1 & 1 & 0 & -2 & 0 & 24 \\
    &   & -2 & 2 & -4 & 12 & -24 \\
\hline
    & 1 & -1 & 2 & -6 & 12 & \boxed{0}
\end{array}
$$
   ← *Remainder*

Quotient: $x^4 - x^3 + 2x^2 - 6x + 12$ ∎

Roots of Polynomial Equations: The Factor and Remainder Theorems

As we saw in Section 5.1, finding the zeros of a polynomial function can help in sketching its graph. With the aid of the Division Algorithm, we can derive two important theorems that will allow us to recognize the zeros of polynomials.

The Division Algorithm states that if we divide a polynomial $p(x)$ by a polynomial $d(x)$, where the degree of $d(x)$ is less than or equal to the degree of $p(x)$, then there are polynomials $q(x)$ and $r(x)$ such that

$$p(x) = d(x) \cdot q(x) + r(x)$$

where either $r(x) = 0$ or the degree of $r(x)$ is less than the degree of $d(x)$.

If we apply the Division Algorithm where the divisor, $d(x)$, is linear (that is, of the form $x - c$), we get

Why must r be a constant?

$$p(x) = (x - c)q(x) + r$$

Note that since the divisor is of the first degree, the remainder, r, must be a constant. If we now substitute $x = c$ into this equation, we get

$$p(c) = (c - c)q(c) + r = 0 \cdot q(c) + r$$ Therefore,
$$p(c) = r$$

The result we just proved is called the **Remainder Theorem**.

Theorem 5.3   The Remainder Theorem

When a polynomial $p(x)$ of degree at least 1 is divided by $x - c$, the remainder is $p(c)$.

Example 4   Use synthetic division to divide $p(x) = x^3 - x^2 + 3x - 1$ by $x - 2$ and verify the Remainder Theorem.

Solution   We set up the synthetic division:

$$
\begin{array}{r|rrrr}
2) & 1 & -1 & 3 & -1 \\
   &   & 2 & 2 & 10 \\
\hline
   & 1 & 1 & 5 & \boxed{9}
\end{array}
\quad \longleftarrow \textit{Remainder}
$$

In this example $x - c = x - 2$ and so $c = 2$; therefore, according to the Remainder Theorem, the remainder should be equal to $p(2)$.

$$p(x) = x^3 - x^2 + 3x - 1 \qquad \text{Substitute } x = 2.$$

$$p(2) = 2^3 - 2^2 + 3(2) - 1 = 9 \qquad \text{Thus}$$

$$p(2) = 9 \qquad \text{Which is the remainder, as required} \qquad \blacksquare$$

> According to the Remainder Theorem, what does it mean that $p(2) = 9$?

As an immediate corollary to the Remainder Theorem, we recognize that if $x - c$ is a factor of $p(x)$, then the remainder $p(c)$ must be 0. Conversely, if the remainder $p(c)$ is 0, then $x - c$ is a factor. This is known as the **Factor Theorem**.

Theorem 5.4   The Factor Theorem

$x - c$ is a factor of $p(x)$ if and only if $p(c) = 0$.

Recall that when we solve a quadratic equation $p(x) = x^2 - 3x - 10 = 0$, we get

$$p(x) = x^2 - 3x - 10 = 0$$

$$(x - 5)(x + 2) = 0 \qquad \text{The factors of } p(x) \text{ are } x - 5 \text{ and } x + 2.$$

$$x = 5 \quad \text{or} \quad x = -2 \qquad \text{The roots are 5 and } -2.$$

Thus the Factor Theorem is merely a generalization of what we have already seen for quadratic polynomials: Roots produce factors and factors produce roots.

Example 5   Given that one root of the equation $p(x) = x^3 - 23x + 10 = 0$ is $x = -5$, find the other roots.

Solution   Since $x = -5$ is given as a root of $p(x) = 0$, the Factor Theorem tells us that $x + 5$ is a factor. If we divide $p(x)$ by $x + 5$, we know the remainder will be 0 and the quotient will be of degree 2, to which we can apply the techniques for finding the roots of quadratic equations. We may use synthetic division to divide the polynomial $x^3 - 23x + 10$ by $x + 5$.

> If $x + 5$ is a factor of $p(x)$, what will the remainder be when $p(x)$ is divided by $x + 5$?

Note that if -5 is the root, then -5 appears as the divisor in the synthetic division tableau.

$$
\begin{array}{r|rrrr}
-5) & 1 & 0 & -23 & 10 \\
    &   & -5 & 25 & -10 \\
\hline
    & 1 & -5 & 2   & \boxed{0}
\end{array}
$$

The quotient polynomial is $q(x) = x^2 - 5x + 2$, and the original equation becomes

$$x^3 - 23x + 10 = (x + 5)(x^2 - 5x + 2) = 0$$

Thus any additional roots must be solutions to $x^2 - 5x + 2 = 0$. We can find the roots of $x^2 - 5x + 2 = 0$ by using the quadratic formula:

$$x = \frac{5 \pm \sqrt{17}}{2}$$

Recall that we stated in Section 5.1 that for a polynomial $p(x)$ of degree n, the equation $p(x) = 0$ has at most n solutions. Since we have three roots of the third-degree equation $x^3 - 23x + 10 = 0$, we have found *all* the roots of this equation. The roots are $x = -5$, $x = \dfrac{5 \pm \sqrt{17}}{2}$.

Example 6 Factor $p(x) = 2x^4 + x^3 - 14x^2 - 19x - 6$ as completely as possible, given that $x = 3$ and $x = -\dfrac{1}{2}$ are roots of $p(x) = 0$.

Solution Since $x = 3$ is a root, we know that $x - 3$ is a factor and so we begin by dividing $p(x)$ by $x - 3$. (Note that we could also begin by dividing $p(x)$ by $x + \dfrac{1}{2}$.) As before, we use synthetic division:

$$
\begin{array}{r|rrrr}
3) & 2 & 1 & -14 & -19 & -6 \\
   &   & 6 & 21 & 21 & 6 \\
\hline
   & 2 & 7 & 7   & 2   & \boxed{0}
\end{array}
$$

Thus we have

$$p(x) = 2x^4 + x^3 - 14x^2 - 19x - 6 = (x - 3)(2x^3 + 7x^2 + 7x + 2)$$

Since we are given that $x = -\dfrac{1}{2}$ is also a root of $p(x) = 0$, it follows that $x + \dfrac{1}{2}$ must be a factor of $2x^3 + 7x^2 + 7x + 2$.

We again use synthetic division to divide $2x^3 + 7x^2 + 7x + 2$ by $x + \dfrac{1}{2}$. Remember that the divisor for the synthetic division is $-\dfrac{1}{2}$.

$$
\begin{array}{r|rrrr}
-\frac{1}{2}) & 2 & 7 & 7 & 2 \\
   &   & -1 & -3 & -2 \\
\hline
   & 2 & 6 & 4 & \boxed{0}
\end{array}
$$

Thus we have

$$p(x) = 2x^4 + x^3 - 14x^2 - 19x - 6 = (x - 3)(2x^3 + 7x^2 + 7x + 2)$$
$$= (x - 3)\left(x + \frac{1}{2}\right)(2x^2 + 6x + 4)$$

We now factor $2x^2 + 6x + 4$ as $2(x^2 + 3x + 2) = 2(x + 2)(x + 1)$, and so the complete factorization of $p(x)$ is

$$p(x) = 2x^4 + x^3 - 14x^2 - 19x - 6 = 2(x + 2)(x + 1)(x - 3)\left(x + \frac{1}{2}\right)$$

We may incorporate the factor of 2 into $x + \dfrac{1}{2}$ and rewrite the factorization as

$$p(x) = 2x^4 + x^3 - 14x^2 - 19x - 6 = (x + 2)(x + 1)(x - 3)(2x + 1)$$

Zeros of Polynomials

The Factor and Remainder Theorems establish the intimate relationship between the factors of a polynomial $p(x)$ and the roots of the equation $p(x) = 0$. In Section 5.1 we mentioned several properties of polynomial functions, one of which was that a polynomial of degree n can have at most n zeros. We now elaborate on this idea.

To make the counting of the zeros of polynomials uniform, we introduce the notion of *multiplicity*. For example, the polynomial equation $x^2 - 2x + 1 = (x - 1)(x - 1) = 0$ has as its only root $x = 1$; however, since the factor $x - 1$ appears twice, we call 1 a **double root** or a repeated **root of multiplicity 2**. In general, if $(x - a)$ appears as a factor of a polynomial k times, we say that a is a repeated **root of multiplicity k**.

Does every polynomial equation (of degree at least 1) have a root? Our answer depends on the number system in which we are working. If we restrict ourselves to the real number system, then we are already familiar with the fact that an equation such as $x^2 + 1 = 0$ has no *real* solutions. However, this equation does have two roots in the complex number system. (The roots are i and $-i$.)

Carl Friedrich Gauss (1777–1855) is generally considered to be one of the greatest mathematicians that the world has known.* In 1799 Gauss, in his doctoral dissertation, proved the rather remarkable fact that within the complex number system every polynomial equation of degree greater than or equal to 1 has at least one root. This fact has such far-reaching significance that it is usually referred to as the **Fundamental Theorem of Algebra**.

Theorem 5.5   The Fundamental Theorem of Algebra

If $p(x)$ is a polynomial of degree $n > 0$ whose coefficients are complex numbers, then the equation $p(x) = 0$ has at least one root in the complex number system.

Keep in mind that since all real numbers are also complex numbers, a polynomial with real coefficients also satisfies the Fundamental Theorem of Algebra.

As an immediate consequence of the Fundamental Theorem, we can prove the following.

Theorem 5.6   The Linear Factorization Theorem

If $p(x) = a_n x^n + a_{n-1} x^{n-1} + \cdots + a_1 x + a_0$, where $n \geq 1$ and $a_n \neq 0$, then

$$p(x) = a_n(x - c_1)(x - c_2) \cdots (x - c_n)$$

where the c_i are complex numbers (possibly real and not necessarily distinct).

In words, this theorem says that any polynomial of degree at least 1 with complex number coefficients can be expressed as a product of linear factors.

Proof   By virtue of the Fundamental Theorem we know that the equation $p(x) = 0$ has at least one complex root; call it c_1. By the Factor Theorem we know that $x - c_1$ is a factor of $p(x)$. Therefore, we can write

$$p(x) = (x - c_1)q_1(x)$$

where the degree of $q_1(x)$ is $n - 1$ and again has leading coefficient a_n. (This is true because we are dividing by $x - c_1$, which has a leading coefficient of 1. See Exercise 88.)

---

*Although there have been many very great mathematicians, three—Archimedes, Newton, and Gauss—are generally considered to be in a class of their own.

If the degree of $q_1(x)$ is zero, then we are done. (Why?) If not, then the Fundamental Theorem applied to $q_1(x)$ says that $q_1(x)$ has at least one zero, call it c_2. Applying the Factor Theorem to $q_1(x)$, we get

$$q_1(x) = (x - c_2)q_2(x)$$

where the degree of $q_2(x)$ is $n - 2$ and it again has leading coefficient a_n. Therefore, we have

$$p(x) = (x - c_1)q_1(x) = (x - c_1)(x - c_2)q_2(x)$$

We can continue this process until the last quotient $q_n(x)$ is a constant; as explained before, it must be a_n. Thus $q_n(x) = a_n$, which gives us

$$p(x) = (x - c_1)(x - c_2) \cdots (x - c_n)a_n$$

as required.   ∎

An immediate consequence of the Linear Factorization Theorem is the fact, quoted earlier, that a polynomial of degree n can have at most n distinct zeros. In fact, we can state even more.

Theorem 5.7

Every polynomial equation of degree $n \geq 1$ has exactly n roots in the complex number system, where a root of multiplicity k is counted k times.

Example 7   In each of the following equations, express the polynomial in the form described by the Linear Factorization Theorem. List each root and its multiplicity.
(a) $p(x) = x^3 - 6x^2 - 16x = 0$ (b) $q(x) = 3x^2 - 10x + 8 = 0$
(c) $f(x) = 2x^4 + 8x^3 + 10x^2 = 0$

Solution   (a) We may factor $p(x)$ as follows:

$$p(x) = x^3 - 6x^2 - 16x = x(x^2 - 6x - 16)$$
$$= x(x - 8)(x + 2)$$

To fit the Linear Factorization Theorem, we write $x + 2 = x - (-2)$.

The factor $(x - 0)$ is usually just written as x.

$$= x(x - 8)(x - (-2))$$

The roots of $p(x) = 0$ are 0, 8, and -2, each of multiplicity one.
(b) We may factor $q(x)$ as follows:

$$q(x) = 3x^2 - 10x + 8 = (3x - 4)(x - 2)$$

To fit the Linear Factorization Theorem, we write $3x - 4$ as $3\left(x - \dfrac{4}{3}\right)$.

$$= 3\left(x - \frac{4}{3}\right)(x - 2)$$

The roots of $p(x) = 0$ are $\dfrac{4}{3}$ and 2, each of multiplicity one.
(c) We may factor $f(x)$ as follows:

$$f(x) = 2x^4 + 8x^3 + 10x^2 = 2x^2(x^2 + 4x + 5)$$

We can find the roots of $x^2 + 4x + 5$ by using the quadratic formula. The roots are $-2 \pm i$. Therefore, we have

$$= 2x^2[x - (-2 + i)][x - (-2 - i)]$$

The roots of $f(x) = 0$ are: 0 with multiplicity two and $-2 + i$ and $-2 - i$, each with multiplicity one.   ∎

Example 8

See Exercises 83–85 for a further investigation of the behavior of polynomial functions at zeros of even/odd multiplicity.

(a) Find a polynomial equation $p(x) = 0$ with exactly the following roots and multiplicities:

Root	Multiplicity
-1	3
2	4
5	2

Are there any other polynomials that give these same roots and multiplicities?

(b) Find a polynomial $f(x)$ having the zeros described in part (a) such that $f(1) = 32$.

Solution

(a) Based on the Factor Theorem, we may write the polynomial

$$p(x) = (x - (-1))^3(x - 2)^4(x - 5)^2 = (x + 1)^3(x - 2)^4(x - 5)^2$$

which gives the required roots and multiplicities.

Because these are the only roots and hence the only linear factors, any other polynomial must be a constant multiple of $p(x)$. Therefore, any polynomial of the form $kp(x)$, where k is a nonzero constant, will give the same roots and multiplicities.

(b) Based on part (a), we know that $f(x) = k(x + 1)^3(x - 2)^4(x - 5)^2$. Since we want $f(1) = 32$, we have

$$f(1) = k(1 + 1)^3(1 - 2)^4(1 - 5)^2$$

$$32 = k(8)(1)(16) \implies 32 = 128k \implies k = \frac{1}{4}$$

Thus $f(x) = \dfrac{1}{4}(x + 1)^3(x - 2)^4(x - 5)^2$. ∎

Our experience in using the quadratic formula on quadratic equations with real coefficients has shown us that complex roots always appear in conjugate pairs. For example, the roots of $x^2 - 2x + 5 = 0$ are $1 + 2i$ and $1 - 2i$. In fact, this property extends to all polynomial equations with real coefficients.

Theorem 5.8   Conjugate Roots Theorem

Let $p(x)$ be a polynomial with real coefficients. If complex number $a + bi$ (where a and b are real numbers) is a root of $p(x) = 0$, then so is its conjugate $a - bi$.

The proof of this theorem is a direct consequence of the properties of conjugates and is outlined in Exercise 89 at the end of this section.

Example 9

Let $r(x) = x^4 + 2x^3 - 9x^2 + 26x - 20$. Given that $1 - i\sqrt{3}$ is a root, find the other roots of $r(x) = 0$.

Solution

According to the Conjugate Roots Theorem, if $1 - i\sqrt{3}$ is a root, then its conjugate, $1 + i\sqrt{3}$, must also be a root. Therefore, $x - (1 - i\sqrt{3})$ and $x - (1 + i\sqrt{3})$ are both factors of $r(x)$, and so their product must be a factor of $r(x)$. The reader should verify that

$$[x - (1 - i\sqrt{3})][x - (1 + i\sqrt{3})] = x^2 - 2x + 4$$

We divide $r(x)$ by $x^2 - 2x + 4$:

$$
\begin{array}{r}
x^2 + 4x - 5 \\
x^2 - 2x + 4 \overline{)\ x^4 + 2x^3 - 9x^2 + 26x - 20} \\
\underline{-(x^4 - 2x^3 + 4x^2)} \\
4x^3 - 13x^2 + 26x \\
\underline{-(4x^3 - 8x^2 + 16x)} \\
-5x^2 + 10x - 20 \\
\underline{-(-5x^2 + 10x - 20)} \\
0
\end{array}
$$

Therefore, we have

$$r(x) = (x^2 - 2x + 4)(x^2 + 4x - 5) = (x^2 - 2x + 4)(x + 5)(x - 1)$$

and so the roots of $r(x) = 0$ are $1 - i\sqrt{3}, 1 + i\sqrt{3}, -5$, and 1.   ∎

The theorems we have discussed in this section are called *existence theorems* because they ensure the existence of zeros and linear factors of polynomials. However, it is important to recognize that these theorems do not tell us *how* to find the zeros or the linear factors. For example, the Linear Factorization Theorem guarantees that we can factor a polynomial of degree at least one into linear factors, but it doesn't tell us how! We will discuss some techniques for actually finding the zeros of certain polynomials in the next section.

5.3 Exercises

In Exercises 1–20, perform the required long division.

1. $\dfrac{x^2 - 5x - 14}{x - 4}$

2. $\dfrac{x^2 + 6x - 16}{x + 8}$

3. $\dfrac{x^2 + 7x - 18}{x + 9}$

4. $\dfrac{x^2 - 3x - 10}{x - 4}$

5. $\dfrac{6t^2 - 23t + 15}{3t - 4}$

6. $\dfrac{10a^2 + 7a - 10}{2a + 3}$

7. $\dfrac{y^2 + 2}{y^2 + 1}$

8. $\dfrac{x^3 - 4}{x^2 - 1}$

9. $\dfrac{2x^3 + x^2 - 5x + 2}{x + 2}$

10. $\dfrac{x^4 - 4x^3 - x^2 + x - 8}{x - 3}$

11. $\dfrac{3x^4 - x^3 + 6}{x - 1}$

12. $\dfrac{5x^3 + 3x^2 + 15}{x + 3}$

13. $\dfrac{8c^3 + 1}{2c + 1}$

14. $\dfrac{27z^3 - 1}{3z - 1}$

15. $\dfrac{2w^4 + w^3 - 3w^2 - 5w + 4}{w^2 + w + 1}$

16. $\dfrac{x^6 + 3x^5 - 2x^4 - 6x^3 + x^2 + 3x + 2}{x^2 + 3x}$

17. $\dfrac{x^4 - 16}{x - 2}$

18. $\dfrac{125 - x^3}{x - 5}$

19. $\dfrac{3x^2 + 6 - x + x^3}{x - 2}$

20. $\dfrac{8 - 2x^3 + 6x^2}{3 - x}$

In Exercises 21–40, use synthetic division to find the quotient and remainder.

21. $\dfrac{x^2 - 8x + 7}{x - 2}$

22. $\dfrac{5x^2 + 6x - 9}{x + 3}$

23. $\dfrac{2x^3 + x^2 - 3x + 15}{x + 2}$

24. $\dfrac{x^4 - 5x^3 + 3x^2 - 3x - 2}{x - 3}$

25. $\dfrac{2x^4 + x^3 - 6x^2 - 10x + 13}{x - 1}$

26. $\dfrac{6x^3 + 4x^2 - 3x - 5}{x + 1}$

27. $\dfrac{5x^2 - 23}{x + 4}$

28. $\dfrac{5x^3 - 4x - 12}{x - 2}$

29. $\dfrac{2x^3 + 4x^2 + 6}{x + 3}$

30. $\dfrac{x^4 - 7x^2 + 8x - 42}{x - 3}$

31. $\dfrac{x^4 - 64}{x - 4}$ **32.** $\dfrac{x^5 + 1}{x + 1}$

33. $\dfrac{6x^3 - 3x^2 - 8x + 4}{x - \dfrac{1}{2}}$

34. $\dfrac{12x^3 + 4x^2 - 15x - 5}{x + \dfrac{1}{3}}$

35. $\dfrac{2x^3 - 3x^2 + 4x - 6}{x - \dfrac{3}{2}}$

36. $\dfrac{3x^3 - 2x^2 + 12x - 8}{x - \dfrac{2}{3}}$

37. $\dfrac{x^3 + a^3}{x + a}$ **38.** $\dfrac{x^3 - a^3}{x - a}$

39. $\dfrac{x^4 - a^4}{x - a}$

40. $\dfrac{x^5 - a^5}{x - a}$

41. Given $p(x) = x^3 + x^2 + 4x - 5$ and $d(x) = x - 5$, use the Division Algorithm to find $q(x)$ and $r(x)$.

42. Given $p(x) = x^4 - 16$ and $d(x) = x - 2$, use the Division Algorithm to find $q(x)$ and $r(x)$.

43. When $x^3 - kx + 5$ is divided by $x - 2$, the remainder is 1. Find k.

44. When $2x^3 + kx^2 - 6$ is divided by $x + 1$, the remainder is -3. Find k.

45. When $p(x) = x^3 + kx + 4$ is divided by $x + 2$, the remainder is the same as when $q(x) = 2x^3 + kx^2 - 2$ is divided by $x - 2$. Find k.

46. When $p(x) = 3x^4 + kx^2 + 7$ is divided by $x - 1$, the remainder is the same as when $q(x) = x^4 + kx - 6$ is divided by $x - 2$. Find k.

In Exercises 47–54, perform the division by $x - a$. Find the quotient and remainder and verify the Remainder Theorem by computing $p(a)$.

47. Divide $p(x) = x^2 - 5x + 8$ by $x + 4$.

48. Divide $p(x) = 2x^2 - 7x + 3$ by $x - 3$.

49. Divide $p(x) = 2x^3 - 7x^2 + x + 4$ by $x - 4$.

50. Divide $p(x) = 4x^3 - x^2 + 1$ by $x + 2$.

51. Divide $p(x) = 1 - x^4$ by $x - 1$.

52. Divide $p(x) = x^3 + 27$ by $x + 3$.

53. Divide $p(x) = x^5 - 2x^2 - 3$ by $x + 1$.

54. Divide $p(x) = x^6 - 16x^3 + 64$ by $x - 2$.

In Exercises 55–64, use the Remainder Theorem to find $p(c)$.

55. $p(x) = x^2 - 5x + 3, \quad c = 4$

56. $p(x) = 3x^2 + 7x - 2, \quad c = -3$

57. $p(x) = x^3 - 2x^2 + 3x - 5, \quad c = -2$

58. $p(x) = -2x^3 + 7x - 4, \quad c = 3$

59. $p(x) = 3x^4 - 5x^2 - 6x - 16, \quad c = 2$

60. $p(x) = x^5 + 2x^4 - 5x^3 + 6x^2 - 15x + 36, \quad c = -4$

61. $p(x) = -x^6 + 6x^4 + 23, \quad c = -3$

62. $p(x) = x^7 - 3x^5 - x^3 - x^2 - x + 18, \quad c = -2$

63. $p(x) = 8x^4 + 4x^3 - 6x + 3, \quad c = \dfrac{1}{2}$

64. $p(x) = 9x^3 + 3x^2 - 12x + 7, \quad c = -\dfrac{1}{3}$

65. Given that $p(4) = 0$, factor
$$p(x) = 2x^3 - 11x^2 + 10x + 8$$
as completely as possible.

66. Given that $q(x) = 6x^3 - 5x^2 - 7x + 4$ and $q(-1) = 0$, find the remaining zeros of $q(x)$.

67. Given that $r(x) = 4x^3 - x^2 - 36x + 9$ and $r\left(\dfrac{1}{4}\right) = 0$, find the remaining zeros of $r(x)$.

68. Given that $s\left(-\dfrac{1}{5}\right) = 0$, factor
$$s(x) = 30x^3 - 19x^2 - 35x - 6$$
as completely as possible.

69. Given that 3 is a double root of
$$p(x) = x^4 - 3x^3 - 19x^2 + 87x - 90 = 0$$
find all the roots of $p(x) = 0$.

70. Given that -2 is a double root of
$$q(x) = 3x^4 + 10x^3 - x^2 - 28x - 20 = 0$$
find all the roots of $q(x) = 0$.

71. Let $p(x) = x^3 - 3x^2 + x - 3$. Verify that $p(3) = 0$ and find the other roots of $p(x) = 0$.

72. Let $q(x) = 3x^3 + x^2 - 4x - 10$. Verify that $q\left(\dfrac{5}{3}\right) = 0$ and find the other roots of $q(x) = 0$.

73. Express $p(x) = x^3 - x^2 - 12x$ in the form described in the Linear Factorization Theorem. List each zero and its multiplicity.

74. Express $q(x) = 6x^5 - 33x^4 - 63x^3$ in the form described in the Linear Factorization Theorem. List each zero and its multiplicity.

75. (a) Write the general polynomial $p(x)$ whose only zeros are 1, 2, and 3, with multiplicities 3, 2, and 1, respectively. What is its degree?
(b) Find the $p(x)$ described in part (a) if $p(0) = 6$.

76. (a) Write the general polynomial $q(x)$ whose only zeros are -4 and -3 with multiplicities 4 and 6, respectively. What is its degree?
(b) Find the $q(x)$ described in part (a) if $q(1) = 48$.

77. Write a polynomial with zeros 0, -2, and 1 with multiplicities 2, 2, and 1, respectively. What is its degree?

78. Write a polynomial whose only zero is 6 with multiplicity 7. What is its degree?

In Exercises 79–82, use the given information to find the remaining roots.

79. $2 - 3i$ is a root of $2x^3 - 5x^2 + 14x + 39 = 0$

80. $1 + 2i$ is a root of $x^4 - x^3 - 9x^2 + 29x - 60 = 0$

81. $3 + 4i$ is a root of
$$4x^4 - 28x^3 + 129x^2 - 130x + 125 = 0$$

82. $i\sqrt{2}$ and $3i$ are roots of
$$x^6 - 2x^5 + 12x^4 - 22x^3 + 29x^2 - 36x + 18 = 0$$

Graphing Calculator Exercises

83. Graph $y = p(x) = (x - 1)^2(x + 2)^4$. What do you notice about the graph at zeros of even multiplicity?

84. Sketch the graph of $y = p(x) = (x - 2)^3(x + 1)^5$. What do you notice about the graph at zeros of odd multiplicity?

85. Graph $y = p(x) = (2x - 3)^2(3x + 1)^3$. What difference do you notice between how the polynomial behaves at a zero of even multiplicity as compared to how it behaves at a zero of odd multiplicity?

86. Sketch the graph of $y = p(x) = x^4 + x^2 + 1$. Does $p(x)$ have any real zeros? Does this violate the Fundamental Theorem of Algebra? Explain.

Questions for Thought

87. Let $p(x) = x^3 - 5x^2 + 3x - 2$. Compute $p\left(\dfrac{3}{4}\right)$ directly and by using synthetic division. Which method do you think is easier in this case?

88. Suppose a polynomial has a leading coefficient of $a_n \neq 0$; that is, suppose
$$p(x) = a_n x^n + a_{n-1}x^{n-1} + \cdots + a_1 x + a_0$$
Show that if $p(x)$ is divided by $x - a$, the leading coefficient of the quotient will also be a_n.

89. The conjugate of a complex number z is frequently denoted as $\bar{z}$. Thus if $z = a + bi$, then $\bar{z} = a - bi$. Use the following properties of conjugates to prove the Conjugate Roots Theorem.
(a) $\bar{r} = r$ for all real numbers r
(b) $\overline{u + v} = \bar{u} + \bar{v}$
(c) $\overline{uv} = \bar{u}\,\bar{v}$
(d) $\overline{u^m} = \bar{u}^m$ HINT: Show that for a polynomial, $p(\bar{z}) = \overline{p(z)}$ and so if $p(z) = 0$, then $p(\bar{z}) = 0$.

5.4   Roots of Polynomial Equations (continued): The Rational Root Theorem and Descartes' Rule of Signs

The Linear Factorization Theorem we discussed in the last section guarantees that we can factor a polynomial of degree at least one into linear factors, but it doesn't tell us how!

We know from experience that if $p(x)$ happens to be a quadratic polynomial, then we may solve $p(x) = Ax^2 + Bx + C = 0$ by using the quadratic formula to obtain the roots

$$x = \frac{-B \pm \sqrt{B^2 - 4AC}}{2A}$$

A natural question one might then ask is whether there is some algebraic "formula" involving the coefficients, the four basic arithmetic operations, and various radicals with which to compute the roots of a polynomial equation, $p(x) = 0$, of degree greater than 2.

It turns out that this question interested mathematicians for many years. In the early sixteenth century significant progress was made, and formulas for the solutions to the general cubic and quartic (fourth-degree) equations were derived. For example, given the general cubic equation $p(x) = x^3 + a_2x^2 + a_1x + a_0 = 0$,* there is an explicit, albeit quite messy, formula for its three roots. It is known as **Cardan's Formula**, and part of what it says is the following:

Let

$$p = a_1 - \frac{(a_2)^2}{3} \quad \text{and} \quad q = \frac{2(a_2)^3}{27} - \frac{a_1a_2}{3} + a_0$$

and let

$$P = \sqrt[3]{-\frac{q}{2} + \sqrt{\frac{p^3}{27} + \frac{q^2}{4}}} \quad \text{and} \quad Q = \sqrt[3]{-\frac{q}{2} - \sqrt{\frac{p^3}{27} + \frac{q^2}{4}}}$$

Then one of the three roots of $p(x) = 0$ is given by $x = P + Q - \frac{a_2}{3}$. The other two roots are given by similar formulas.

If $p(x) = x^3 - 2x - 21$, we have $a_2 = 0$, $a_1 = -2$, and $a_0 = -21$ (the fact that $a_2 = 0$ simplifies the formula significantly) and yet it is still quite a chore to use the formula to obtain $x = 3$ as a root. (Try it!)

Mathematicians continued to search for a formula for the roots of the general fifth-degree (and higher) polynomial equation. In 1828, the Norwegian mathematician Niels Abel proved the remarkable fact that the search was futile. He proved that the problem in finding such a formula was not that mathematicians were not clever enough but rather that there can be no such formula for solving the general polynomial equation of degree 5 or higher. Keep in mind this does not mean that these equations don't have solutions (in fact, our previous theorems ensure that they do); rather, it just says that, *in general,* we cannot express the solutions in terms of the coefficients and radicals. Finally, in 1830, the French mathematician Evariste Galois (at the age of 18) settled the question by describing exactly which polynomials can be solved in terms of their coefficients and radicals.

The rest of this section is devoted to developing some special methods for finding the roots of polynomial equations.

As we have seen, even though we have no general techniques for factoring polynomials of degree greater than 2, if we happen to know a root, say, c, we can use synthetic division to divide $p(x)$ by $x - c$ and obtain a quotient polynomial of *lower* degree. If we can get the quotient polynomial down to a quadratic, then we are able to determine all the roots. But how do we find a root to start this process? The following theorem can be most helpful.

Theorem 5.9   The Rational Root Theorem

Suppose that

$$f(x) = a_nx^n + a_{n-1}x^{n-1} + \cdots + a_2x^2 + a_1x + a_0$$

where $n \geq 1$, $a_n \neq 0$

is an nth-degree polynomial with *integer* coefficients. If $\frac{p}{q}$ is a rational root of $f(x) = 0$, where p and q have no common factors other than ± 1, then p is a factor of a_0 and q is a factor of a_n.

---

*Without loss of generality, we may assume that $a_3 = 1$; otherwise, we can divide both sides of the equation $a_3x^3 + a_2x^2 + a_1x + a_0 = 0$ by a_3.

A general proof of this theorem is outlined in Exercise 53 at the end of this section; however, by looking at an example, we can get a feeling as to why this theorem is true. Suppose $\frac{3}{2}$ is a root of the third-degree polynomial equation

$$a_3 x^3 + a_2 x^2 + a_1 x + a_0 = 0$$

Then

$$a_3\left(\frac{3}{2}\right)^3 + a_2\left(\frac{3}{2}\right)^2 + a_1\left(\frac{3}{2}\right) + a_0 = 0$$

$$\frac{27a_3}{8} + \frac{9a_2}{4} + \frac{3a_1}{2} = -a_0 \qquad \text{Multiply both sides of the equation by 8.}$$

$$27a_3 + 18a_2 + 12a_1 = -8a_0 \qquad (1)$$

Which can also be written as

$$27a_3 = -18a_2 - 12a_1 - 8a_0 \qquad (2)$$

If we look carefully at equation (1) we can see that the left-hand side is divisible by 3, and therefore the right-hand side must also be divisible by 3. But if $-8a_0$ is divisible by 3, than a_0 must be divisible by 3 (because 8 is not divisible by 3). Similarly, by looking at the right-hand side of equation (2) we can see that it is divisible by 2, and since 27 is not divisible by 2, a_3 must be divisible by 2. This is exactly what the Rational Root Theorem asserts.

It is very worthwhile to note that if a polynomial has a leading coefficient of 1, then the Rational Root Theorem tells us that any rational roots must be integers and divisors of the constant term a_0.

Why does the Rational Root Theorem tell us that if the leading coefficient of a polynomial is 1, then its rational roots must be factors of the constant term?

Example 1   Find all the roots of the equation $p(x) = 2x^3 + 3x^2 - 23x - 12 = 0$.

Solution   We begin by searching for any possible rational roots to the given equation. According to the Rational Root Theorem, if $\frac{p}{q}$ is a rational root of the given equation, then p must be a factor of -12 and q must be a factor of 2. Thus we have

$$\text{Possible values of } p: \quad \pm 1, \pm 2, \pm 3, \pm 4, \pm 6, \pm 12$$
$$\text{Possible values of } q: \quad \pm 1, \pm 2$$
$$\text{Possible rational roots } \frac{p}{q}: \quad \pm 1, \pm\frac{1}{2}, \pm 2, \pm 3, \pm\frac{3}{2}, \pm 4, \pm 6, \pm 12$$

Note that the possible quotients $\frac{p}{q}$ repeat certain roots, which we list only once. For example, $\frac{6}{2} = \frac{3}{1}$.

We may check these possible roots by substituting the values into $p(x)$; however, it is more efficient to check the values by using synthetic division, since in case we do find a root, we will already have the quotient available.

$$
\begin{array}{r|rrrr}
1 & 2 & 3 & -23 & -12 \\
  &   & 2 & 5 & -18 \\
\hline
  & 2 & 5 & -18 & \boxed{-30}
\end{array}
\qquad
\begin{array}{r|rrrr}
-1 & 2 & 3 & -23 & -12 \\
   &   & -2 & -1 & 24 \\
\hline
   & 2 & 1 & -24 & \boxed{12}
\end{array}
$$

How do we know that $p(1) = -30$ and $p(-1) = 12$?

Therefore, we know that $p(1) = -30$ and $p(-1) = 12$. Since $p(1)$ is negative and $p(-1)$ is positive, the Intermediate Value Theorem guarantees that $p(x)$ has a zero between -1 and 1. Thus it seems reasonable next to check the rational roots on our list between -1 and 1.

$$\begin{array}{r|rrrr} \frac{1}{2} & 2 & 3 & -23 & -12 \\ & & 1 & 2 & -\frac{21}{2} \\ \hline & 2 & 4 & -21 & \boxed{-\frac{45}{2}} \end{array} \qquad \begin{array}{r|rrrr} -\frac{1}{2} & 2 & 3 & -23 & -12 \\ & & -1 & -1 & +12 \\ \hline & 2 & 2 & -24 & \boxed{0} \end{array}$$

Therefore, $-\dfrac{1}{2}$ is a root, and we can read the quotient from the bottom line of the synthetic division:

$$p(x) = 2x^3 + 3x^2 - 23x - 12 = \left(x + \frac{1}{2}\right)(2x^2 + 2x - 24)$$

Now we may try to find another root by using the Rational Root Theorem on $2x^2 + 2x - 24$. However, it is much easier to simply try to factor the quadratic directly:

$$2x^2 + 2x - 24 = 2(x^2 + x - 12) = 2(x + 4)(x - 3)$$

Therefore, the original equation becomes

$$\begin{aligned} p(x) = 2x^3 + 3x^2 - 23x - 12 &= \left(x + \frac{1}{2}\right)(2x^2 + 2x - 24) \\ &= 2\left(x + \frac{1}{2}\right)(x + 4)(x - 3) = 0 \end{aligned}$$

and all the roots are $-\dfrac{1}{2}, -4, 3$.

Note that since $p(x)$ is a third-degree polynomial, we know that it has *at most* three zeros. Having found three zeros, we know that we have found them all. ▌

Example 2 Use a graphing calculator as an aid in finding the rational zeros of

$$p(x) = 12x^4 - 76x^3 + 57x^2 + 169x - 60$$

Solution We can easily recognize that the list of possible rational roots generated by the Rational Root Theorem is quite long. However, a glance at the graph of $p(x)$ can help us enormously in the trial-and-error process of finding any rational roots. After a bit of trial and error in choosing an appropriate viewing window, we obtain the graph of $p(x)$ as illustrated in Figure 5.34.

Be aware: Sometimes when you use a graphing calculator to find a zero, the answer may not be exact even when the zero is an integer. For instance, if the zero is at $x = 3$, the calculator may return the value 2.999998.

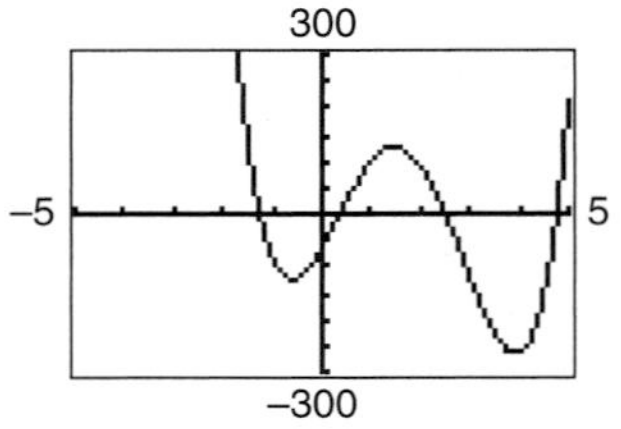

Figure 5.34
The graph of $y = 12x^4 - 76x^3 + 57x^2 + 169x - 60$

According to the Rational Root Theorem, there is an extensive list of possible rational roots (any factor of 60 divided by any factor of 12). Examining Figure 5.34, we notice that the graph has four zeros and so there can be no more. (Why?) It is clear from the graph that there is a zero between -2 and -1, between 0 and 1, between 2 and 3, and between 4 and 5. However, we do not know whether these zeros are ra-

Write out a list of the possible rational roots.

tional. To check for rational zeros, we need only try those possible rational zeros that fall in one of these four intervals.

For example, in the entire list of possible rational roots, the only one that falls between 2 and 3 is $\frac{5}{2}$. Let's try $\frac{5}{2}$:

$$
\begin{array}{r|rrrrr}
\frac{5}{2}) & 12 & -76 & 57 & 169 & -60 \\
 & & 30 & -115 & -145 & 60 \\
\hline
 & 12 & -46 & -58 & 24 & \boxed{0}
\end{array}
$$

Thus, $x = \frac{5}{2}$ is a rational zero and we have $p(x) = (x - \frac{5}{2})(12x^3 - 46x^2 - 58x + 24)$.

Keep in mind that if we now want to check any other possible zeros, we can work with $12x^3 - 46x^2 - 58x + 24$ rather than the original polynomial $p(x)$.

We leave it to the student to verify that none of the possible rational zeros fall between 4 and 5. Let's try some of the possible rational zeros between 0 and 1. If we try $\frac{1}{2}$ we find that it is not a zero. If we try $\frac{1}{3}$ we get

$$
\begin{array}{r|rrrr}
\frac{1}{3}) & 12 & -46 & -58 & 24 \\
 & & 4 & -14 & -24 \\
\hline
 & 12 & -42 & -72 & \boxed{0}
\end{array}
$$

Thus, $x = \frac{1}{3}$ is a rational zero and we have $p(x) = (x - \frac{5}{2})(x - \frac{1}{3})(12x^2 - 42x - 72)$.

We can now use the quadratic formula on $12x^2 - 42x - 72 = 0$ to see that this has two real (irrational) zeros. Therefore the only rational zeros of $p(x)$ are $\frac{1}{3}$ and $\frac{5}{2}$. ∎

We have seen that an nth-degree polynomial will have exactly n zeros (counting multiplicity) in the complex number system. However, if we are sketching the graph of a polynomial, it is only the *real* zeros in which we are interested. (Keep in mind that even if we use a graphing calculator or a computer to sketch the graph of a function, it will show us only the *real* zeros of the function.) The following two theorems are useful in determining the number of real zeros a polynomial has.

Because there is no difficulty determining whether 0 is a zero of a polynomial (we simply see whether x is a factor of $p(x)$), for the purposes of this discussion, we may as well assume that $p(x) = a_n x^n + a_{n-1}x^{n-1} + \cdots + a_1 x + a_0$ where $a_0 \neq 0$. This assures us that 0 is not a zero of $p(x)$. Again, for the purposes of this discussion we will assume that all polynomials are written in standard form—that is, from the highest to the lowest power.

To state the first theorem, we must introduce the following idea. By a *variation in sign* of the real coefficients of a polynomial, we mean that two consecutive coefficients have opposite signs. For example, the coefficients of the polynomial $p(x) = 5x^3 - 7x^2 - 4x + 6$ change from 5 to -7 (one sign variation) and then from -4 to 6 (a second variation in sign). So $p(x) = 5x^3 - 7x^2 - 4x + 6$ has two variations in sign.

We can now state **Descartes' Rule of Signs**, which deals with the number of positive and negative real zeros* a polynomial can have.

---

*The terminology *positive real* number is actually redundant, since complex numbers cannot be positive or negative. Only real numbers can be positive or negative. Nevertheless, we feel the redundancy is useful to emphasize the fact that we are talking about the real zeros of $p(x)$.

> **Theorem 5.10   Descartes' Rule of Signs**
>
> Let $p(x)$ be a polynomial with real coefficients such that $p(0) \neq 0$. Then
>
> **1.** The number of *positive real zeros* of $p(x)$ is either equal to the number of variations in sign of $p(x)$ or is less than that number by an even integer.
>
> **2.** The number of *negative real zeros* of $p(x)$ is either equal to the number of variations in sign of $p(-x)$ or is less than that number by an even integer.

Keep in mind that this theorem tells us about the *real* zeros of $p(x)$, and to apply this theorem $p(x)$ must be in standard form and have a nonzero constant term. We will not prove this theorem.

When applying Descartes' Rule of Signs we count roots with their multiplicity. For example, in the equation $p(x) = x^2 - 6x + 9 = 0$, there are two variations in sign and so the equation has either two positive real roots or an even number fewer than 2, which would mean no roots. The factored form of this equation is $p(x) = (x - 3)^2 = 0$, so we see that 3 is a root of multiplicity 2, which agrees with the theorem.

Example 3   Apply Descartes' Rule of Signs to find the real zeros of

$$p(x) = x^5 + 2x^4 + x^3 + 2x^2 + 3x + 6$$

Solution   First we examine $p(x)$ and see that there are 0 variations in the signs of its coefficients. Therefore, according to Descartes' Rule of Signs, $p(x)$ must have 0 positive real zeros.

Next we examine

$$p(-x) = (-x)^5 + 2(-x)^4 + (-x)^3 + 2(-x)^2 + 3(-x) + 6$$
$$= -x^5 + 2x^4 - x^3 + 2x^2 - 3x + 6$$

Why don't we need to test the possible positive rational roots?

and sees that $p(-x)$ has 5 variations in the signs of its coefficients. So by the second part of Descartes' Rule of Signs, $p(x)$ must have 5, 3, or 1 negative real zeros.

Applying the Rational Root Theorem to $p(x)$, we see that the possible rational roots are $\pm 1, \pm 2, \pm 3, \pm 6$. Testing the possible negative rational roots by synthetic division, we find that -2 is a root and that

$$p(x) = x^5 + 2x^4 + x^3 + 2x^2 + 3x + 6 = (x + 2)(x^4 + x^2 + 3)$$

Examining $q(x) = x^4 + x^2 + 3$, we find that both $q(x)$ and $q(-x)$ have 0 sign variations and so $q(x)$ has no real zeros. Therefore, the only real zero of $p(x)$ is -2.   ∎

As we have seen, the Rational Root Theorem can involve a fair number of possible rational roots. The following theorem can be helpful in reducing the number of possible rational roots that have to be considered.

We say that a real number U is an **upper bound** for the real zeros of a function f if all the real zeros of f are less than or equal to U. Similarly, a real number L is a lower bound for the real zeros of f if all the real zeros of f are greater than or equal to L. In other words, if a function has an upper bound U and a lower bound L for its real zeros, then all its real zeros must fall in the interval $[L, U]$ on the number line. See Figure 5.35.

Figure 5.35

All the real zeros of f will fall in the closed interval $[L, U]$.

> **Theorem 5.11   The Upper and Lower Bound Theorem**
>
> Let $p(x)$ be a polynomial with real coefficients whose leading coefficient is positive. Suppose we divide $p(x)$ by $x - c$ using synthetic division.
>
> 1. If $c > 0$ and all the numbers in the bottom row of the synthetic division table are positive or zero, then c is an upper bound for the real roots of $p(x) = 0$.
> 2. If $c < 0$ and the numbers in the bottom row of the synthetic division table are alternately positive and negative (zero entries count as either positive or negative), then c is a lower bound for the real roots of $p(x) = 0$.

A justification of part 2 of this theorem is outlined in Exercise 54.

Example 4   Find the upper and lower bounds for the real roots of $p(x) = x^3 - 3x^2 - 2x + 10 = 0$.

Solution   We test 1, 2, 3, 4, . . . as possible upper bounds by synthetically dividing $p(x)$ by $x - 1, x - 2, \ldots$.

$$
\begin{array}{r|rrrr}
1) & 1 & -3 & -2 & 10 \\
   &   & 1 & -2 & -4 \\
\hline
   & 1 & -2 & -4 & \boxed{6}
\end{array}
\qquad
\begin{array}{r|rrrr}
2) & 1 & -3 & -2 & 10 \\
   &   & 2 & -2 & -8 \\
\hline
   & 1 & -1 & -4 & \boxed{2}
\end{array}
$$

$$
\begin{array}{r|rrrr}
3) & 1 & -3 & -2 & 10 \\
   &   & 3 & 0 & -6 \\
\hline
   & 1 & 0 & -2 & \boxed{4}
\end{array}
\qquad
\begin{array}{r|rrrr}
4) & 1 & -3 & -2 & 10 \\
   &   & 4 & 4 & 8 \\
\hline
   & 1 & 1 & 2 & \boxed{18}
\end{array}
$$

Since the entries of the last row of the synthetic division of $p(x)$ by $x - 4$ are all positive, according to the Upper and Lower Bound Theorem, 4 is an upper bound on the real zeros of $p(x)$. [Let's take a moment to see what part 1 of this theorem is saying. Based on this last division, we have

$$p(x) = x^3 - 3x^2 - 2x + 10 = (x - 4)(x^2 + x + 2) + 18$$

Looking at the right-hand side of this equation, we can see that if $x > 4$, then $x - 4$ is positive, as is $x^2 + x + 2$. Therefore, for $x > 4$, the right-hand side of the equation is positive. Hence $p(x)$ must be positive and so cannot be equal to zero for $x > 4$.]

To find a lower bound, we divide $p(x)$ synthetically by $x - (-1), x - (-2), \ldots$.

$$
\begin{array}{r|rrrr}
-1) & 1 & -3 & -2 & 10 \\
    &   & -1 & 4 & -2 \\
\hline
    & 1 & -4 & 2 & \boxed{8}
\end{array}
\qquad
\begin{array}{r|rrrr}
-2) & 1 & -3 & -2 & 10 \\
    &   & -2 & 10 & -16 \\
\hline
    & 1 & -5 & 8 & \boxed{-6}
\end{array}
$$

Using an analysis similar to the preceding one, which showed that 4 is an upper bound, show that -2 is a lower bound on the zeros of $p(x)$.

Because the entries of the last row of the synthetic division of $p(x)$ by $x - (-2)$ are alternately positive and negative, according to the Upper and Lower Bound Theorem, -2 is a lower bound on the real zeros of $p(x)$. All the zeros of $p(x)$ lie in the interval $[-2, 4]$. ∎

As we saw in Section 5.1, knowing the location of the zeros of a function can be very helpful in sketching its graph. We conclude this with an example in which we use the three theorems in this section dealing with finding the zeros of polynomials.

Example 5   Find all the real zeros of $y = f(x) = x^4 - x^3 + x^2 - 3x - 6$ and sketch its graph.

Solution   Starting with Descartes' Rule of Signs, we see that $f(x)$ has 3 variations in sign, so that $f(x)$ has either 3 or 1 positive real zeros. Because

$$f(-x) = (-x)^4 - (-x)^3 + (-x)^2 - 3(-x) - 6 = x^4 + x^3 + x^2 + 3x - 6$$

has one variation in sign, $f(x)$ has exactly one negative real zero.

According to the Rational Root Theorem we consider zeros of the form

$$\frac{\text{factors of 6}}{\text{factors of 1}} \quad \begin{array}{c} \rightarrow \\ \rightarrow \end{array} \quad \frac{\pm 1, \pm 2, \pm 3, \pm 6}{\pm 1}$$

Therefore the possible rational zeros of $f(x)$ are $\pm 1, \pm 2, \pm 3, \pm 6$. Trying $x = 1$ and $x = 2$ as roots for synthetic division, we get

> *If $x - 2$ is a factor of $p(x)$, what will the remainder be when $p(x)$ is divided by $x - 2$?*

$$
\begin{array}{r|rrrrr}
1) & 1 & -1 & 1 & -3 & -6 \\
   &   & 1 & 0 & 1 & -2 \\
\hline
   & 1 & 0 & 1 & -2 & \boxed{-8}
\end{array}
\qquad
\begin{array}{r|rrrrr}
2) & 1 & -1 & 1 & -3 & -6 \\
   &   & 2 & 2 & 6 & 6 \\
\hline
   & 1 & 1 & 3 & 3 & \boxed{0}
\end{array}
$$

Not only do we see that 2 is a zero, but since all the numbers in the bottom row of the division by 2 are positive, the theorem on upper and lower bounds tells us that 2 is an upper bound on the positive zeros of $f(x)$. Therefore, we may ignore 3 and 6 as possible zeros, and we have $x = 2$ as the only positive rational zero.

As noted before, we know that $f(x)$ has exactly one negative zero, and so we now test the possible negative rational zeros in the qotient we obtained in the last synthetic division.

$$
\begin{array}{r|rrr}
-1) & 1 & 1 & 3 & 3 \\
    &   & -1 & 0 & -3 \\
\hline
    & 1 & 0 & 3 & \boxed{0}
\end{array}
$$

Therefore, $x = -1$ is also a root and we have 2 and -1 as the only rational roots of $f(x)$. Knowing that $x - 2$ and $x + 1$ are factors of $f(x)$, we can factor $f(x)$:

$$f(x) = (x - 2)(x + 1)(x^2 + 3)$$ Since $x^2 + 3$ has no real zeros, 2 and -1 are the only real zeros of $f(x)$.

> *Why doesn't the factor $(x^2 + 3)$ contribute a cut point?*

Doing a sign analysis on $f(x) = (x - 2)(x + 1)(x^2 + 3)$ and recognizing that $x^2 + 3$ is always positive, we get Figure 5.36.

Sign of $f(x)$ $+ + + + + + + +$ $- - - - - - - -$ $+ + + + + + + +$
                                          -1 2

Figure 5.36

The sign analysis for $f(x) = (x - 2)(x + 1)(x^2 + 3)$

We have $f(1) = -8$ (from the Remainder Theorem and the synthetic division) and the y-intercept is $f(0) = -6$. Checking some values, we see that y becomes positive as $|x|$ gets large. A rough sketch of the graph of $f(x)$ appears in Figure 5.37. ∎

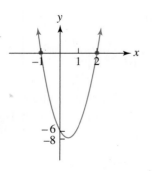

Figure 5.37

A rough sketch of the graph of
$y = x^4 - x^3 + x^2 - 3x - 6$

If after employing the Rational Root Theorem we determine that a polynomial has no rational zeros, a graphing calculator can be very helpful in estimating its zeros. Keep in mind that we can still use Descartes' Rule of Signs and the Upper and Lower Bound Theorem to obtain information that helps us pick the proper viewing rectangle and use the $\boxed{\text{TRACE}}$ and $\boxed{\text{ZOOM}}$ functions more effectively.

5.4 Exercises

In Exercises 1–18, determine the rational roots of the given equation.

1. $x^3 - 4x^2 - 7x + 10 = 0$
2. $x^4 + 7x^3 + 17x^2 + 17x + 6 = 0$
3. $2x^3 - 5x^2 + 15 = 28x$
4. $3x^3 + 13x^2 + 2x = 8$
5. $x^3 - 4x^2 + 5x - 2 = 0$
6. $2x^3 + 3x^2 = 1$
7. $4x^3 - 3x = 1$
8. $4y^3 + 3y^2 + 8y + 6 = 0$
9. $t^5 - 4t^3 + t^2 - 4 = 0$
10. $30z^3 - 31z^2 - 15z + 4 = 0$
11. $6u^3 + u^2 - 4u + 1 = 0$
12. $w^5 + 9w^3 - w^2 - 9 = 0$
13. $-x^4 + 8x^2 - 15 = 0$
14. $-x^6 + 4x^4 + x^2 - 4 = 0$
15. $8x^3 + 18x^2 + 45x + 27 = 0$
16. $9x^4 + 15x^3 - 20x^2 - 20x + 16 = 0$
17. $9x^6 - 18x^5 - 28x^4 + 38x^3 + 39x^2 - 4x - 4 = 0$
18. $4x^6 - 21x^4 + 11x^2 - 4 = 0$

In Exercises 19–26, use Descartes' Rule of Signs to determine the possible number of positive and negative real zeros of $p(x)$.

19. $p(x) = 8x^3 + 1$
20. $p(x) = x^3 - 5x^2 - 6x + 3$
21. $p(x) = 3x^4 + 5x^2 + 2$
22. $p(x) = x^4 - x^3 - 6$
23. $p(x) = x^4 - 3x^3 + 7x$
24. $p(x) = -x^4 + x^3 - 3x^2 - 2x + 1$
25. $p(x) = x^5 - x^4 + x^3 - x^2 + x - 8$
26. $p(x) = 4x^3 - 3x^2 + 7x - 4$

In Exercises 27–40, use the Upper and Lower Bound Theorem to find upper and lower bounds for the roots of the given equation.

27. $x^2 - 5x + 3 = 0$
28. $3x^2 + 9x - 2 = 0$
29. $x^3 - 3x^2 - 18x + 4 = 0$
30. $x^3 - 4x^2 - 5x + 8 = 0$
31. $x^4 - x^2 + 3x + 2 = 0$
32. $x^4 - 2x^3 + 4x - 3 = 0$

33. $2x^3 - 5x + 1 = 0$
34. $3x^3 - 6x^2 - 14 = 0$
35. $-x^4 - 6x^3 + 3x + 7 = 0$
36. $-2x^3 + 6x^2 - 4x + 3 = 0$
37. $x^3 + 3x^2 + 5 = 0$
38. $2x^3 - 3x^2 + 12x + 9 = 0$
39. $\frac{1}{3}x^3 - x^2 - 3x + 4 = 0$
40. $\frac{1}{2}x^3 - \frac{1}{3}x^2 - 3x + 4 = 0$

In Exercises 41–48, use the various theorems in this section to find all the roots of the given equation.

41. $3x^3 - 7x^2 + 5x - 1 = 0$
42. $2x^3 + 13x^2 - 2x - 4 = 0$
43. $2x^4 + 7x^3 - 8x + 3 = 0$
44. $x^4 + 4x^3 - x^2 - 20x - 20 = 0$
45. $6x^4 - 19x^3 + 21x^2 - 19x + 15 = 0$
46. $x^3 - 5x^2 + 6x - 8 = 0$
47. $3x^5 - 9x^4 - 28x^3 + 84x^2 + 9x - 27 = 0$
48. $2x^4 + 13x^3 + 4x^2 - 13x - 6 = 0$

In Exercises 49–52, use a sign analysis to determine the intervals on which $f(x)$ is positive and on which $f(x)$ is negative.

49. $f(x) = x^4 + 2x^3 - 13x^2 - 14x + 24$
50. $f(x) = x^4 - 2x^3 + 5x^2 - 8x + 4$
51. $f(x) = x^5 - 5x^3 - x^2 + 5$
52. $f(x) = x^5 + x^4 - 2x^3 + 8x^2 + 8x - 16$

Graphing Calculator Exercises

53. **a.** Sketch the graph of
$$y = p(x) = 6x^3 - 31x^2 + 25x + 12$$
and estimate its zeros.
   (b) Use the Rational Root Theorem on $p(x)$ to find its zeros. Which approach was easier? More accurate?

54. Graph $y = p(x) = 12x^3 - 40x^2 + 13x + 30$. Use the graph to estimate the zeros of $p(x)$, and use this information to apply the Rational Root Theorem more efficiently and find the exact zeros of $p(x)$.

55. Repeat the process used in Exercise 54 for

$$y = p(x) = 48x^4 + 4x^3 - 128x^2 - 29x + 15$$

Questions for Thought

56. Look back at the discussion following the statement of the Rational Root Theorem. Then use the following outline to prove the theorem. Suppose

$$f(x) = a_n x^n + a_{n-1}x^{n-1} + \cdots + a_1 x + a_0$$

and suppose that $\dfrac{p}{q}$ is a solution to $f(x) = 0$, where $\dfrac{p}{q}$ is reduced to lowest terms.

(a) Show that if $f\left(\dfrac{p}{q}\right) = 0$, then

$$a_n p^n + a_{n-1}p^{n-1}q + \cdots + a_1 pq^{n-1} + a_0 q^n = 0$$

(b) Show that this implies that p must divide a_0 and that q must divide a_n.

57. Use the following outline to explain part 2 of the Upper and Lower Bound Theorem.
(a) Use synthetic division to divide

$$p(x) = 2x^3 + 5x^2 - x + 3 \text{ by } x + 3$$

(b) Examine the bottom line of the synthetic division tableau and see that it satisfies the condition of part 2 of the Upper and Lower Bound Theorem.
(c) Translate the division into the form
$$p(x) = (x + 3)q(x) + r.$$
(d) Verify that for $x < -3$, we must have $p(x) < 0$, so $p(x)$ cannot have any zeros less than -3, which is what the Upper and Lower Bound Theorem asserts.

58. Is it possible for the graph of a polynomial of even degree to have no x-intercepts? Explain. Is it possible for the graph of a polynomial of odd degree to have no x-intercepts? Explain.

59. Prove that every polynomial of odd degree must have at least one real zero.

60. As mentioned at the beginning of this section, show that 3 can be obtained as a root of $x^3 - 2x - 21 = 0$ by using Cardan's Formula. Now use the Rational Root Theorem to list the possible rational roots and then check that 3 is a root. Which approach do you think is easier?

5.5　Rational Functions

An environmental consultant for a large factory finds that the cost C (in thousands of dollars) to remove p percent of the pollutants released by the factory can be approximated by the function

$$C(p) = \frac{70p}{100 - p}$$

Understanding how this function behaves would be very useful in determining the monetary investment necessary to make specific improvements in the amount of pollution removed. This cost function $C(p)$ is called a *rational function*, and will be investigated further later in this section.

Definition of a Rational Function

A **rational function** is a function of the form

$$y = f(x) = \frac{p(x)}{q(x)} \qquad \text{where } p(x) \text{ and } q(x) \text{ are polynomials and } q(x) \neq 0$$

Some examples of rational functions are

$$f(x) = \frac{3}{x + 5} \qquad g(x) = \frac{x - 1}{x^2 - 9} \qquad h(x) = \frac{1}{x^2 + 4}$$

Unlike a polynomial function, whose domain is always the set of all real numbers, a rational function may have a restricted domain, since we must exclude any values that make the denominator equal to zero. Consider the functions f, g, and h defined previously. The domain of $f(x)$ excludes $x = -5$, and the domain of $g(x)$ excludes the values $x = \pm 3$. On the other hand, since $x^2 + 4$ cannot be equal to zero (for real values of x), the domain of $h(x)$ is the set of all real numbers.

Example 1   Find the domain and zeros of the function $f(x) = \dfrac{3x - 5}{x^2 - x - 12}$.

Solution   Those values of x for which $x^2 - x - 12 = 0$ are excluded from the domain of $f(x)$. Since $x^2 - x - 12 = (x - 4)(x + 3)$, we have $D_f = \{x \,|\, x \neq -3, 4\}$.

To find the zeros of $f(x)$, we recognize that the only way for a fraction to be equal to zero is if the numerator is zero. In general,

$$\frac{p(x)}{q(x)} = 0 \qquad \text{if and only if} \qquad p(x) = 0 \quad \text{and} \quad q(x) \neq 0$$

Therefore, to find the zeros of $f(x)$ we solve $3x - 5 = 0$, giving $x = \frac{5}{3}$. Since $\frac{5}{3}$ does not make the denominator equal to zero, it is the only zero of $f(x)$. ∎

To analyze the behavior of a rational function, we will need to determine what happens near the value(s) that make the denominator equal to zero (if any). To avoid unnecessary complications (such as that described in Exercise 61), we will assume that each rational function under discussion is reduced to lowest terms. In other words, throughout the discussion of rational functions that follows, we assume that the numerator and denominator have no common factors.

See Exercise 61 for an illustration of why we insist that rational functions be reduced to lowest terms.

In Section 5.1 we used a sign analysis on polynomial functions to sketch their graphs. Knowing where a polynomial is positive, negative, and zero, along with the behavior of the polynomial when $|x|$ gets large, allows us to get a fairly accurate sketch of its graph. To sketch the graph of rational functions, we also need to investigate the behavior of the function near those values of x that make the denominator equal to zero.

Let us begin our discussion with the following example.

Example 2   Sketch the graph of $y = f(x) = \dfrac{1}{x}$.

Solution   As usual, rather than resorting to plotting points, we will try to analyze this function to obtain information that will help us sketch the graph.

The domain of $f(x)$ is all $x \neq 0$, and so the graph of $f(x)$ has no y-intercept. (Why?) Similarly, since there are no values of x for which $\dfrac{1}{x} = 0$ (remember that for a fraction to be zero, the numerator must be zero), the graph has no x-intercepts.

Try solving $\dfrac{1}{x} = 0$.

Looking at the equation $y = \dfrac{1}{x}$, we can see that if x is positive, then y is positive and if x is negative, then y is negative. In terms of the graph, this means that the graph will appear only in quadrants I and III. However, since we have already established that there are no x- or y-intercepts, the graph must consist of two separate pieces, one in quadrant I and one in quadrant III. See Figure 5.38 on page 320.

In other words, this graph has a break in it, unlike the graph of a polynomial, which cannot have any breaks.

We might also make note of the fact that for $f(x) = \dfrac{1}{x}$,

$$f(-x) = \frac{1}{-x} = -\frac{1}{x} = -f(x)$$

which means that the graph of this function exhibits origin symmetry. This agrees with our observation that the graph is in quadrants I and III.

Figure 5.38

The graph of $y = \dfrac{1}{x}$ lies in

quadrants I and III

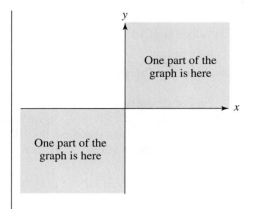

Next let's examine what happens as $|x|$ gets very large. As we move along the x-axis farther and farther to the right, say, for $x = 10, 100, 1000, 10{,}000, \ldots$, the values of $y = \dfrac{1}{x}$ get closer and closer to 0 (the y-values would be 0.1, 0.01, 0.001, 0.0001, ...). Similarly, as we move along the x-axis farther and farther to the left, say for $x = -10, -100, -1000, -10{,}000, \ldots$, the values of $y = \dfrac{1}{x}$ again become closer and closer to 0 (the y-values would be $-0.1, -0.01, -0.001, -0.0001, \ldots$).

Thus "out at the ends," the graph gets closer and closer to the x-axis, but it never touches the x-axis, because—as we have already established—there are no x-intercepts. In such a situation the x-axis (which has equation $y = 0$) is called a **horizontal asymptote**. See Figure 5.39.

Figure 5.39

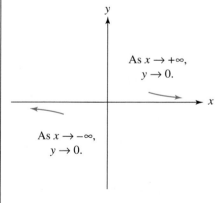

In general, we say that a line is an **asymptote** for a graph if the distance between the graph and the line approaches 0 as we move farther and farther out along the graph.*

It remains for us to determine the behavior of y for values of x near 0, where y is undefined. We want to determine what happens to y-values as the x-values get closer and closer to zero. Let's examine some values of y for values of x close to 0 (both positive and negative).

---

*Note that this description allows a graph to cross a horizontal asymptote *before* we get out to the ends.

To indicate that x is approaching 0 *from the right,* we write $x \to 0^+$.

x	1	0.5	0.1	0.01	0.001	0.0001	$x \to 0^+$
$y = \dfrac{1}{x}$	1	2	10	100	1000	10,000	$y \to \infty$

$y \to \infty$ means that y increases without bound. In other words, as x gets closer and closer to zero from the right side of the y-axis, the values of y get larger and larger. See Figure 5.40.

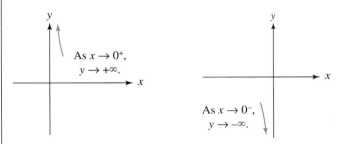

Figure 5.40   **Figure 5.41**

To indicate that x is approaching 0 *from the left,* we write $x \to 0^-$.

x	-1	-0.5	-0.1	-0.01	-0.001	-0.0001	$x \to 0^-$
$y = \dfrac{1}{x}$	-1	-2	-10	-100	-1000	$-10,000$	$y \to -\infty$

$y \to -\infty$ means that y decreases without bound; that is, as x gets closer and closer to zero from the left side of the y-axis, the values of y are negative but getting larger in absolute value. See Figure 5.41.

To summarize, as x gets close to zero from the right-hand side, the graph is getting higher and higher but never touches the y-axis (why not?), whereas as x gets closer to zero from the left-hand side, the graph is getting lower and lower but never touches the y-axis. Thus the y-axis (which has equation $x = 0$) is a **vertical asymptote**.

Putting together all the information we have gathered about the function and its graph, we obtain Figure 5.42.

Figure 5.42

Notice that the graph approaches but never touches the x- and y-axes.

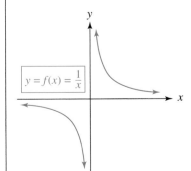

Keep in mind that all our information does not tell us the exact shape of the curve. As mentioned previously, determining the concavity of a graph requires some techniques from calculus. It turns out that Figure 5.42 describes the shape of $y = \dfrac{1}{x}$ accurately. ∎

Calculator Exploration

Use a graphing calculator to graph $y = \dfrac{1}{x}$, $y = \dfrac{2}{x}$, $y = \dfrac{3}{x}$ simultaneously. Explain how the constant c affects the graph of $y = \dfrac{c}{x}$. Does your description agree with the stretching principle discussed in Chapter 4?

Having gone through this detailed analysis for $y = \dfrac{1}{x}$, we can now go through a similar analysis much more quickly.

Example 3   Sketch the graph of $y = f(x) = \dfrac{1}{x^2}$.

Solution   Our analysis of this function will lead us to many of the same conclusions we came to for $y = \dfrac{1}{x}$. The graph will have no x- or y-intercepts and will have the x-axis as a horizontal asymptote and the y-axis as a vertical asymptote.

However, for $y = \dfrac{1}{x^2}$, whether x is positive or negative, y will be positive, and so the graph will be in quadrants I and II only. We also note that

$$f(-x) = \frac{1}{(-x)^2} = \frac{1}{x^2} = f(x)$$

and so the graph of $f(x)$ exhibits y-axis symmetry, which agrees with our observation that the graph is in quadrants I and II.

Notice that when x gets close to zero, whether from the right or from the left, x^2 will be a small positive number getting closer and closer to zero, and so $y = \dfrac{1}{x^2}$ will be increasing without bound. See Figure 5.43(a). The graph is in Figure 5.43(b).

Figure 5.43

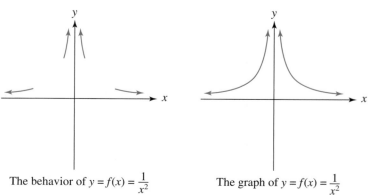

The behavior of $y = f(x) = \dfrac{1}{x^2}$

(a)

The graph of $y = f(x) = \dfrac{1}{x^2}$

(b)

∎

Calculator Exploration

1. Graph the following in a $[-4, 4]$ by $[-3, 3]$ viewing window.

$$y = \frac{1}{x} \qquad y = \frac{1}{x^3} \qquad y = \frac{1}{x^5}$$

For odd values of n, what can you conclude about how changing n affects the graph of $y = \frac{1}{x^n}$?

2. Graph the following in a $[-3, 3]$ by $[-1, 5]$ viewing window.

$$y = \frac{1}{x^2} \qquad y = \frac{1}{x^4} \qquad y = \frac{1}{x^6}$$

For even values of n, what can you conclude about how changing n affects the graph of $y = \frac{1}{x^n}$?

Example 4 | Sketch the graphs of **(a)** $y = f(x) = \frac{1}{x^3}$ **(b)** $y = g(x) = \frac{1}{x^4}$

Solution | **(a)** It is left to the student to analyze the function $y = \frac{1}{x^3}$ and recognize that it has all the essential characteristics of $y = \frac{1}{x}$, including the same domain and range. This is directly due to the fact that the exponent 3 is odd, and hence when x is negative, y is negative, putting the graph in quadrants I and III. The graph appears in Figure 5.44. The graph of $y = \frac{1}{x}$ is also shown. Note that the higher exponent (3) pulls the graph closer to the x-axis.

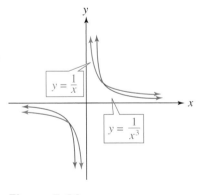

Figure 5.44
The graph of $y = \frac{1}{x^3}$

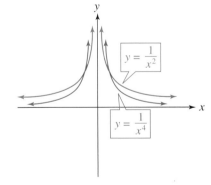

Figure 5.45
The graph of $y = \frac{1}{x^4}$

(b) Again it is left to the student to analyze the function $y = \frac{1}{x^4}$ and recognize that it has all the essential characteristics of $y = \frac{1}{x^2}$, including the same domain and range. This is directly due to the fact that the exponent 4 is even, and hence when x is negative, y is positive, putting the graph in quadrants I and II. The graph appears in Figure 5.45. The graph of $y = \frac{1}{x^2}$ is also shown. Note that the higher exponent (4) pulls the graph closer to the x-axis.

Based on these examples, we can generalize as follows:

The Graphs of $y = \dfrac{1}{x^n}$

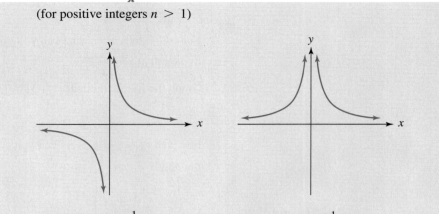

(for positive integers $n > 1$)

The graph of $y = \dfrac{1}{x^n}$ for n odd   The graph of $y = \dfrac{1}{x^n}$ for n even

The vertical asymptote is $x = 0$ and the horizontal asymptote is $y = 0$.

These graphs should be added to our catalog of basic functions whose graphs we recognize immediately.

Example 5 Sketch the graphs of

(a) $y = \dfrac{5}{x + 2}$ (b) $y = \dfrac{1}{x} + 2$ (c) $y = -\dfrac{1}{(x - 3)^2}$

Solution (a) Knowing the graph of $y = \dfrac{1}{x}$, our basic graphing principles tell us that

$y = \dfrac{5}{x} = 5\left(\dfrac{1}{x}\right)$ will have the same basic shape and behavior as that of $y = \dfrac{1}{x}$.

As we saw with the power functions, multiplying a function by a constant simply stretches the graph. The graph of $y = \dfrac{5}{x + 2}$ is obtained by shifting the graph of $y = \dfrac{5}{x}$ two units to the left. The graph appears in Figure 5.46. Note that the hor-

Figure 5.46

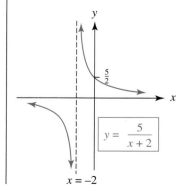

$$y = \frac{5}{x + 2}$$

$x = -2$

izontal asymptote remains the x-axis ($y = 0$) but the vertical asymptote is shifted from the y-axis ($x = 0$) to the line $x = -2$. As usual, we find the y-intercept by setting $x = 0$.

(b) Again applying our basic graphing principles, we obtain the graph of $y = \dfrac{1}{x} + 2$ by shifting the graph of $y = \dfrac{1}{x}$ up 2 units. The y-axis will still be the vertical asymptote, but now the horizontal asymptote is shifted up 2 units from the x-axis ($y = 0$) to the line $y = 2$. It would be appropriate to find the x-intercept (we set $y = 0$):

$$0 = \frac{1}{x} + 2 \qquad \text{Solving for } x \text{ gives} \qquad x = -\frac{1}{2}$$

The graph is shown in Figure 5.47.

Why don't we bother to look for a y-intercept?

Figure 5.47

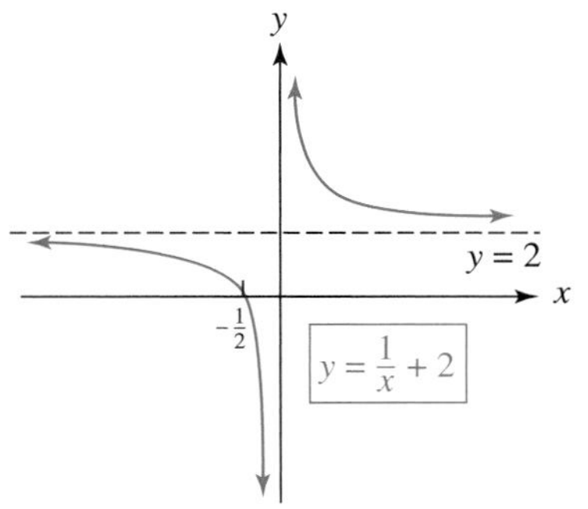

(c) The graph of $y = -\dfrac{1}{(x-3)^2}$ is obtained by shifting the graph of $y = \dfrac{1}{x^2}$ three units to the right and reflecting it about the x-axis. The horizontal asymptote remains the x-axis and the vertical asymptote is shifted to $x = 3$. It is appropriate to find the y-intercept. Setting $x = 0$, we get $y = -\dfrac{1}{9}$. The graph appears in Figure 5.48.

Figure 5.48

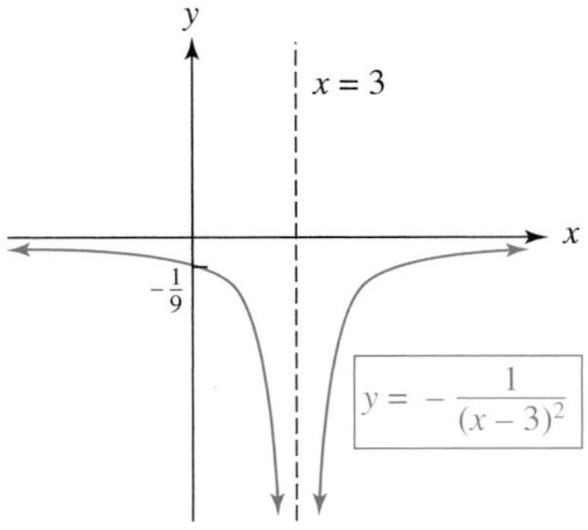

Example 6   Sketch the graph of $y = \dfrac{1 - x}{x - 3}$.

Solution   At first glance, $y = \dfrac{1 - x}{x - 3}$ does not look like any of the rational functions we have graphed thus far. We can use this as an opportunity to see how the technology we have available can work hand in hand with the knowledge we have already accumulated.

Use a graphing calculator to explore the end behavior of the graph.

Let's take a look at the graph of $y = \dfrac{1-x}{x-3}$ on a graphing calculator and see whether the graph looks at all familiar. See Figure 5.49(a).

Figure 5.49

The graph of $y = \dfrac{1-x}{x-3}$

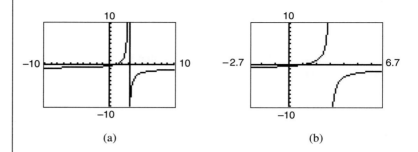

(a)   (b)

Depending on the viewing window, a graphing calculator may not give an accurate representation of the graph of a rational function. The "vertical" segment appearing in Figure 5.49(a) is *not* a representation of the vertical asymptote $x = 3$. It is the result of the calculator evaluating the function to the left and right of $x = 3$ but not at $x = 3$, and then mistakenly connecting these points with a nearly vertical line segment. To avoid this, we could reset the calculator from connect mode to dot mode (try it), or we can make sure that the viewing window is one in which the function is evaluated at $x = 3$, where it is undefined, so that this value is skipped and the points on either side are not joined. See Figure 5.49(b).

On a TI-83 Plus calculator, whenever $X_{max} - X_{min}$ is a convenient multiple or fraction of 94, we get a window that evaluates the function at "nice" values of x. The window settings we have chosen are such that $6.7 - (-2.7) = 9.4$, which ensures that the calculator will evaluate the function at $x = 3$. Try it.

The graph certainly looks like a reflection and shift of the $y = \dfrac{1}{x}$ graph, but the given equation does not exhibit the usual appearance we would expect from a reflection and shift of $y = \dfrac{1}{x}$. This suggests that perhaps we can use some algebraic procedure to expose the behavior we see in the graph. Looking at the given equation $y = \dfrac{1-x}{x-3}$, we can see that the degrees of the numerator and denominator are equal, so we can use the long-division process (or synthetic division) to divide $x - 3$ into $1 - x$:

$$\frac{1-x}{x-3} = x - 3 \overline{\smash{\big)}\begin{array}{r} -1 \\ -x+1 \\ \underline{-(-x+3)} \\ -2 \end{array}}$$

Therefore, we have

$$\frac{1-x}{x-3} = -1 - \frac{2}{x-3} = -\frac{2}{x-3} - 1$$

We can now see that the graph can be obtained by shifting the graph of $\dfrac{2}{x}$ three units to the right, reflecting it about the x-axis, and then shifting it down 1 unit.

Graph $y = -1$ together with $y = \dfrac{1-x}{x-3}$ on a graphing calculator. Does the picture agree with your understanding of what a horizontal asymptote should look like?

We leave it to the student to verify that the y-intercept is $-\dfrac{1}{3}$ and the x-intercept is $x = 1$. The graph is in Figure 5.50.

Figure 5.50

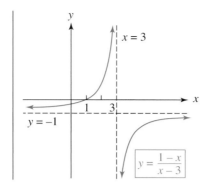

The same type of approach used in Example 6 will work for any rational function of the form

$$f(x) = \frac{ax + b}{cx + d}$$

Before we continue, let's summarize some of the observations we have made thus far:

Vertical Asymptotes

The line $x = a$ is a vertical asymptote if as $x \to a^+$ or $x \to a^-$, then $f(x) \to \infty$ or $f(x) \to -\infty$. See Figure 5.51.

Figure 5.51

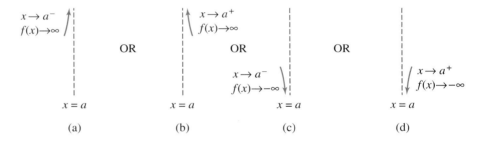

Horizontal Asymptotes

The line $y = b$ is a horizontal asymptote if as $x \to \infty$ or $x \to -\infty$, then $f(x) \to b$. See Figure 5.52.

Figure 5.52

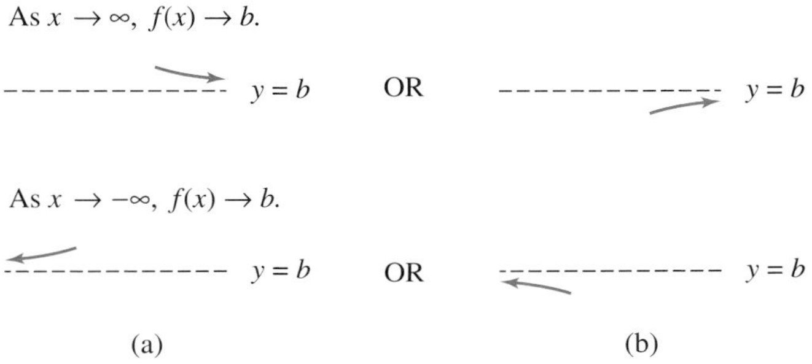

As $x \to \infty$, $f(x) \to b$.

---------------→ $y = b$ OR   ------------- $y = b$
                                        ------→

As $x \to -\infty$, $f(x) \to b$.

←------------- $y = b$ OR   ------------- $y = b$
                                    ←-------

(a)   (b)

A rational function will have a vertical asymptote at every x-value that makes the denominator (but not the numerator) zero and only at those values, so a rational function need not have any vertical asymptotes. See Exercises 62–64 and 67 for more about asymptotes.

We can find the horizontal asymptotes (if they exist) by the algebraic manipulation demonstrated in the next example.

Example 7   Find the vertical and horizontal asymptotes for the graph of

$$y = \frac{x^2 - 2x - 8}{2x^2 - 5x - 3}$$

Solution   To find the vertical asymptotes, we set the denominator equal to zero:

Recall that at the beginning of this section we insisted that all rational functions be reduced to lowest terms. This ensures that the graph will have a vertical asymptote at each zero of the denominator. See Exercise 61.

$$2x^2 - 5x - 3 = 0$$

$$(2x + 1)(x - 3) = 0$$

$$2x + 1 = 0 \quad \text{or} \quad x - 3 = 0$$

$$x = -\frac{1}{2} \quad \text{or} \quad x = 3$$

Hence the *equations* of the vertical asymptotes are $x = -\dfrac{1}{2}$ and $x = 3$.

To find the horizontal asymptotes, we divide the numerator and the denominator by the highest power of x appearing in the denominator, in this case, x^2:

$$y = \frac{x^2 - 2x - 8}{2x^2 - 5x - 3} = \frac{\dfrac{x^2 - 2x - 8}{x^2}}{\dfrac{2x^2 - 5x - 3}{x^2}} = \frac{\dfrac{x^2}{x^2} - \dfrac{2x}{x^2} - \dfrac{8}{x^2}}{\dfrac{2x^2}{x^2} - \dfrac{5x}{x^2} - \dfrac{3}{x^2}} = \frac{1 - \dfrac{2}{x} - \dfrac{8}{x^2}}{2 - \dfrac{5}{x} - \dfrac{3}{x^2}}$$

Looking at the rightmost expression for y, we observe that as x gets really large (as $x \to \infty$), the expressions $\frac{2}{x}$, $\frac{8}{x^2}$, $\frac{5}{x}$, and $\frac{3}{x^2}$ all approach 0. Hence, the entire expression, y, approaches $\frac{1}{2}$. The same is true as $x \to -\infty$. Therefore, the horizontal asymptote is $y = \frac{1}{2}$. ∎

In general we follow this order when graphing a rational function:

1. Find and plot the y-intercept (if any).
2. Find and plot any x-intercept(s).
3. Find and graph any vertical asymptote(s).
4. Find and graph any horizontal asymptote(s).
5. Check the behavior of the function near its asymptotes and x-intercepts.

Example 8   Sketch the graph of the function $f(x) = \dfrac{x - 1}{x^2 - 4}$.

Solution   **1.** We first find the y-intercept:

$$y = \frac{x - 1}{x^2 - 4} \qquad \text{Let } x = 0 \text{ and find } y.$$

$$= \frac{0 - 1}{0^2 - 4} = \frac{1}{4} \qquad \text{The } y\text{-intercept is } \tfrac{1}{4}.$$

2. Next find the x-intercept:

$$y = \frac{x - 1}{x^2 - 4}$$

Let $y = 0$ and find x.

$$0 = \frac{x - 1}{x^2 - 4}$$

For a fraction to be 0, the numerator must be 0.

$$0 = x - 1 \rightarrow x = 1$$

The x-intercept is 1.

3. Find the vertical asymptote(s):

$$x^2 - 4 = 0$$

Set the denominator equal to 0.

$$(x - 2)(x + 2) = 0$$

$$x = 2 \quad \text{or} \quad x = -2$$

The vertical asymptotes are $x = 2$ and $x = -2$

4. Find the horizontal asymptote(s): Divide the numerator and the denominator by the highest power of x in the denominator, x^2:

$$f(x) = \frac{x - 1}{x^2 - 4} = \frac{\dfrac{x - 1}{x^2}}{\dfrac{x^2 - 4}{x^2}} = \frac{\dfrac{x}{x^2} - \dfrac{1}{x^2}}{\dfrac{x^2}{x^2} - \dfrac{4}{x^2}} = \frac{\dfrac{1}{x} - \dfrac{1}{x^2}}{1 - \dfrac{4}{x^2}}$$

As x gets large, the expressions $\dfrac{1}{x}, \dfrac{1}{x^2}$, and $\dfrac{4}{x^2}$ all approach 0, so $f(x)$ approaches $\dfrac{0 - 0}{1 - 0} = 0$. Thus, the horizontal asymptote is $y = 0$ (the x-axis). We plot the intercepts and sketch the asymptotes as shown in Figure 5.53.

Figure 5.53

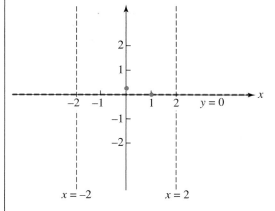

We now check the values between the intercepts and vertical asymptotes much as we did in graphing polynomials or solving quadratic inequalities. The test values we used are shown in Table 5.5 and plotted in Figure 5.54(a) on page 330.

Table 5.5

x	-3	$-2\frac{1}{2}$	$-1\frac{1}{2}$	0	$1\frac{1}{2}$	$2\frac{1}{4}$	4
y	$-\frac{4}{5}$	$-1\frac{5}{9}$	$1\frac{3}{7}$	$\frac{1}{4}$	$-\frac{2}{7}$	$1\frac{3}{17}$	$\frac{1}{4}$

Next we check values near and about the asymptotes. Let's carefully analyze what happens to the y-values as x approaches 2. First, we examine what happens to y as x approaches 2 *from the right*. Recall that we write this as $x \rightarrow 2^+$. As $x \rightarrow 2^+$, x is a little bit bigger than 2, hence x^2 is a little bit bigger than 4, and therefore the denominator, $x^2 - 4$, is a positive number close to zero. The numerator, $x - 1$, is a little bit bigger than 1. Therefore,

$$\text{As } x \to 2^+, \quad y = \frac{\text{a number close to 1}}{\text{a positive number close to 0}}$$

so as x gets closer and closer to 2 from the right, y increases without bound ($x \to 2^+$, $y \to +\infty$).

Similarly, as $x \to 2^-$, x is a little smaller than 2, hence x^2 is a little bit smaller than 4, and therefore the denominator, $x^2 - 4$, is a negative number close to zero. The numerator, $x - 1$, is a little bit smaller than 1. Therefore,

$$\text{As } x \to 2^-, \quad y = \frac{\text{a number close to 1}}{\text{a negative number close to 0}}$$

so as x gets closer and closer to 2 from the left, y decreases without bound ($x \to 2^-$, $y \to -\infty$).

We leave it to the reader to carry out the same type of analysis as x approaches -2 from the right and left. The results are summarized in Table 5.6.

Table 5.6

As $x \to 2^+$	$x^2 - 4$ is a positive number near 0	$y = \dfrac{x - 1}{x^2 - 4} \to +\infty$
As $x \to 2^-$	$x^2 - 4$ is a negative number near 0	$y = \dfrac{x - 1}{x^2 - 4} \to -\infty$
As $x \to -2^+$	$x^2 - 4$ is a negative number near 0	$y = \dfrac{x - 1}{x^2 - 4} \to +\infty$
As $x \to -2^-$	$x^2 - 4$ is a positive number near 0	$y = \dfrac{x - 1}{x^2 - 4} \to -\infty$

This information allows us to visualize those portions of the graph close to the vertical asymptotes, as shown in Figure 5.54.

To check the behavior of $f(x)$ "out at the ends," we can compute $f(x)$ for some large values of x:

$$f(100) \approx 0.01 \qquad f(1000) \approx 0.001 \qquad f(10,000) \approx 0.0001$$

Similar numerical results apply for $x = -100, -1000$, and $-10,000$. Thus it appears that as $x \to +\infty$ and as $x \to -\infty$, $y \to 0$.

Figure 5.54

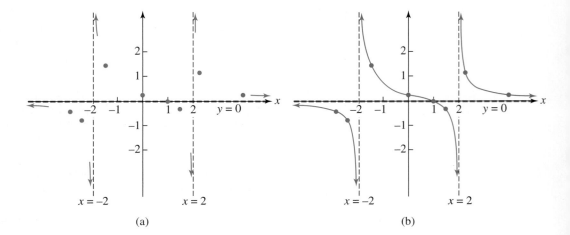

(a)   (b)

Do a sign analysis for $\dfrac{x-1}{x^2-4}$. Do your results agree with the graph in Figure 5.54(b)?

When we put this information together with what we know about the asymptotes, we get the graph shown in Figure 5.54(b).

Figure 5.55 illustrates the graph of $y = \dfrac{x-1}{x^2-4}$ as displayed by a graphing calculator. Observe that this graph supports our analysis about the vertical and horizontal asymptotes of the function.

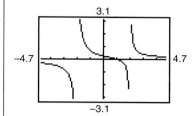

Figure 5.55

The graph of $y = \dfrac{x-1}{x^2-4}$ using the ZDecimal window on the TI-83 Plus calculator

Many students find that they can obtain the various pieces of information about a function but have trouble sketching a graph consistent with all this information. It is important that you go back over this problem and make sure you see how all the information we gathered about $f(x)$—its intercepts and its behavior near the vertical asymptotes and the horizontal asymptote—has been integrated to produce the graph in Figure 5.55. Although we have not been able directly to address the question of concavity, the information we have about the horizontal and vertical asymptotes does suggest the concavity pictured in the graph. ∎

Example 9   Graph $f(x) = \dfrac{2x^2}{x^2-1}$ using a graphing calculator. Analyze the end behavior of the function both graphically and numerically.

Solution   We should immediately recognize that the graph will have vertical asymptotes where the denominator is equal to 0. Thus the lines $x = -1$ and $x = 1$ are vertical asymptotes. Figure 5.56(a) illustrates the graph of $f(x)$.

Figure 5.56

The graph of $y = \dfrac{2x^2}{x^2-1}$

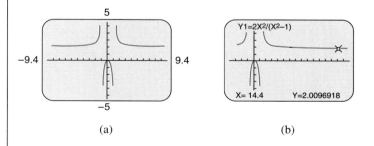

(a)   (b)

Do a symmetry analysis and sign analysis for $f(x) = \dfrac{2x^2}{x^2-1}$. Do your conclusions agree with the graph in Figure 5.56?

The graph appears to have a horizontal asymptote of $y = 2$. By using $\boxed{\text{TRACE}}$, we can analyze numerically the end behavior of the graph. As x moves out to the right, the y-values do appear to get closer and closer to 2. See Figure 5.56(b). It is left to the reader to verify that as x moves out to the left, the y-values also get closer and closer to 2. ∎

We must point out that the techniques we have developed are somewhat limited in that they do not tell the entire story for many rational functions, as illustrated in Exercises 62–67.

Let's return to the situation described at the beginning of this section.

Example 10 An environmental consultant for a large factory finds that the cost C (in thousands of dollars) to remove p percent of the pollutants released by the factory can be approximated by the function

$$C(p) = \frac{70p}{100 - p}$$

Use a graphing calculator to examine the graph of this function, describe its behavior, and interpret this behavior as it relates to the cost function.

Solution Since p represents the percent of pollutants removed, we recognize that p must be a number between 0 and 100. After a bit of trial and error, we choose the window settings appearing in Figure 5.57, which illustrates the graph of $y = \dfrac{70p}{100 - p}$.

Figure 5.57

The graph of $y = C(p) = \dfrac{70p}{100 - p}$

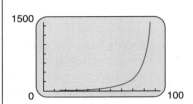

We can see that as the percent of pollutants removed (p) increases, the cost to remove this level of pollutants also increases. However, it is interesting to note that as p increases, the cost increases more and more rapidly. In particular, as p approaches 100, the cost increases enormously. For example, $C(75) = 210$, which means that it costs $210,000 to remove 75% of the pollutants, and $C(80) = 280$, which means that it costs $280,000 to remove 80% of the pollutants. Thus, if the factory is already removing 75% of the pollutants, removing an additional 5% of pollutants would cost the factory an additional $70,000.

However, $C(90) = 630$ and $C(95) = 1330$, which means that if the factory is already removing 90% of the pollutants, in order to remove an additional 5% of pollutants, the factory would require an additional $700,000! The graph shows us that as p approaches 100%, the cost of each additional percentage removed is increasing enormously. ∎

Note that as the percentage of pollutants removed approaches 100%, the cost becomes prohibitive.

5.5 Exercises

In Exercises 1–6, find the domain and the real zeros of the given function.

1. $f(x) = \dfrac{3}{x^2 - 25}$

2. $g(x) = \dfrac{x - 3}{x^2 - 4x - 12}$

3. $h(x) = \dfrac{x^2 - 4x}{2x^2 + 3x - 20}$

4. $f(x) = \dfrac{x^2 - 16}{x^2 + 4}$

5. $g(x) = \dfrac{x^2 - 2x + 5}{x^2}$

6. $g(x) = \dfrac{(x - 3)^2}{x^3 - 3x^2 + 2x}$

7. Sketch the graphs.

(a) $y = \dfrac{2}{x}$.   (b) $y = \dfrac{2}{x - 3}$ (c) $y = \dfrac{2}{x} - 3$

8. Sketch the graphs.

(a) $y = \dfrac{3}{x^2}$ (b) $y = \dfrac{3}{x^2} + 2$ (c) $y = \dfrac{3}{(x + 2)^2}$

9. Sketch the graphs.

(a) $y = -\dfrac{1}{x^4}$ (b) $y = -\dfrac{1}{(x - 1)^4}$

(c) $y = -\dfrac{1}{x^4} - 1$

10. Sketch the graphs.

(a) $y = \dfrac{2}{x^3}$ (b) $y = \dfrac{2}{x^3} - 4$

(c) $y = \dfrac{2}{(x + 4)^3}$

11. On the same set of coordinate axes, sketch the graphs of $y = \dfrac{1}{x}$, $y = \dfrac{6}{x}$, and $y = \dfrac{10}{x}$.

12. On the same set of coordinate axes, sketch the graphs of $y = -\dfrac{1}{x^2}$, $y = -\dfrac{4}{x^2}$, and $y = -\dfrac{9}{x^2}$.

In Exercises 13–40, sketch the graph of each function. Be sure to identify the intercepts and the horizontal and vertical asymptotes wherever appropriate.

13. $y = \dfrac{4}{x^6}$ **14.** $y = \dfrac{3}{x^5}$

15. $y = \dfrac{1}{x} - 5$ **16.** $y = \dfrac{1}{x - 5}$

17. $y = \dfrac{1}{(x - 2)^2}$ **18.** $y = \dfrac{1}{x^2} - 2$

19. $y = \dfrac{1}{x^3} - 8$ **20.** $y = \dfrac{1}{(x - 1)^4}$

21. $y = 9 - \dfrac{1}{x^2}$ **22.** $y = 4 - \dfrac{1}{x}$

23. $y = \dfrac{x}{x + 5}$ **24.** $y = \dfrac{-2x}{x - 2}$

25. $y = \dfrac{3x - 5}{x}$ **26.** $y = \dfrac{5x + 4}{2x}$

27. $y = \dfrac{x + 2}{x - 2}$ **28.** $y = \dfrac{x - 3}{x + 6}$

29. $y = \dfrac{5 - x}{x + 4}$ **30.** $y = \dfrac{6x - 3}{2x - 3}$

31. $y = \dfrac{2x}{x^2 - 1}$ **32.** $y = \dfrac{-x}{x^2 - 9}$

33. $y = \dfrac{x^2}{x^2 - 4}$ **34.** $y = \dfrac{3x^2}{1 - x^2}$

35. $y = \dfrac{x - 1}{x^2 - 2x - 3}$ **36.** $y = \dfrac{x + 1}{x^2 - x - 6}$

37. $y = \dfrac{2 - x}{2x^2 - x - 3}$ **38.** $y = \dfrac{x^2 - 4}{x^2 + 2x - 3}$

39. $y = \dfrac{x^2 - 4x + 3}{x^2 - 2x}$ **40.** $y = \dfrac{x^2 - 4x + 3}{x^2 - 4x}$

41. Let $f(x) = \dfrac{1}{x}$; compute and simplify the difference quotient $\dfrac{f(x) - f(5)}{x - 5}$.

42. Let $f(x) = \dfrac{1}{x + 1}$; compute and simplify the difference quotient $\dfrac{f(x) - f(2)}{x - 2}$.

43. Let $f(x) = \dfrac{1}{x^2}$; compute and simplify the difference quotient $\dfrac{f(x + h) - f(x)}{h}$.

44. Let $f(x) = \dfrac{1}{x^2 - 4}$; compute and simplify the difference quotient $\dfrac{f(x) - f(a)}{x - a}$.

45. Determine the behavior of $\dfrac{x^3 - 8x - 3}{x - 3}$ when x is near 3.

46. Determine the behavior of $\dfrac{2x^4 - 3x^2 + 1}{x + 1}$ when x is near -1.

47. A fence is to be set up to enclose a rectangular field and divide it down the middle, as indicated in the diagram. If the area of the field is 60,000 sq ft, express the length, L, of fence needed as a function of x.

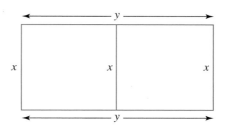

48. Use the fact that the volume of a closed right circular cylinder is 200 cu cm to express the surface area, S, of the cylinder as a function of r, the radius of the cylinder.

Certain procedures in calculus applied to rational functions give rise to the types of problems demonstrated in Problems 49–52.

49. Show that $\dfrac{(x^2 + 1)(2x) - (x^2 - 1)(2x)}{(x^2 + 1)^2} = \dfrac{4x}{(x^2 + 1)^2}$.

50. Show that
$$\dfrac{x^3(2x - 2) - (x^2 - 2x - 3)3x^2}{x^6} = \dfrac{-x^2 + 4x + 9}{x^4}$$

51. Show that
$$\dfrac{(x + 2)^3(2x) - (x^2 - 1)3(x + 2)^2}{(x + 2)^6} = \dfrac{3 + 4x - x^2}{(x + 2)^4}$$

52. Show that
$$\dfrac{2(x^2 - 4)^2 - (2x - 1)(x^2 - 4)(4x)}{(x^2 - 4)^4} = -\dfrac{2(3x^2 - 2x + 4)}{(x^2 - 4)^3}$$

53. A manufacturer finds that the cost C (in dollars) to produce x items is given by $C = 0.4x + 8000$. It then follows that the *average* cost per unit to produce x items is given by

$$A(x) = \frac{0.4x + 8000}{x}$$

Use a graphing calculator to examine the graph of this function, describe its behavior, and interpret this behavior as it relates to the average cost per item. In particular, what is the horizontal asymptote of this function? What is its significance?

54. A state environmental control department finds that the function

$$C(p) = \frac{180p}{100 - p}$$

serves as a good model for the cost C (in millions of dollars) to remove $p\%$ of the water pollutants in the state's lakes and rivers. Use a graphing calculator to examine the graph of this function, describe its behavior, and interpret this behavior as it relates to the cost of removing $p\%$ of the pollutants.

   (a) What is the vertical asymptote of this function? What is its significance?
   (b) The state legislature receives a report that the state spent 300 million dollars to increase the percentage of pollutants removed from the state's lakes and rivers from 85% to 88%. Consequently, the legislature budgets an additional 600 million dollars to raise the level of pollutants removed to 94%. Comment on the reasonableness of the legislature's plan.

55. An entomologist uses the function

$$n(d) = \frac{500(d + 4)}{0.4d + 2}$$

to model the number n of insects present in a colony d days after the initial formation of the colony. Use a graphing calculator to examine the graph of this function, describe its behavior, and interpret this behavior as it relates to the number of insects in the colony as time goes by.

   (a) How many insects are present after 30 days? 60 days? 90 days?
   (b) What appears to be the horizontal asymptote of this function? What is its significance?

56. An ecological study group suggests that under certain conditions the following function relates the number (n) of deer than can be realistically supported on a acres of foraging land:

$$n(a) = \frac{40a}{0.4a + 8} \qquad \text{for } 0 \le a \le 200$$

   (a) Compute $n(1)$, $n(10)$, and $n(100)$ and explain what each of these values means.
   (b) Use a graphing calculator to examine the graph of this function, describe its behavior, and interpret this behavior as it relates to the number of deer that can be sustained on a acres of foraging land.

57. When two resistors, one of 6 ohms and one of x ohms, are connected in parallel the resulting circuit has a combined resistance R given by

$$R(x) = \frac{6x}{6 + x}$$

   (a) Compute $R(1)$, $R(5)$, and $R(10)$ and explain what each of these values means.
   (b) Use a graphing calculator to examine the graph of this function, describe its behavior, and interpret this behavior as it relates to the total resistance of the circuit as the resistance x is increased.

58. A pharmaceutical company finds that, when a certain drug is administered to a patient, the concentration c of the drug in the bloodstream (in milligrams per liter) h hours after the drug was administered is given by

$$c(h) = \frac{50h}{h^2 + 1}$$

   (a) Compute $c(1)$, $c(2)$, $c(6)$, and $c(24)$ and explain what each of these values means.
   (b) Use a graphing calculator to examine the graph of this function, describe its behavior, and interpret this behavior as it relates to the concentration of the drug in the bloodstream as time passes by.

59. Suppose that a manufacturer finds that the cost C of managing and storing g gallons of a certain chemical is given by the function

$$C = C(g) = 2g + \frac{5000}{g}$$

   (a) What should the domain of this function be?
   (b) Use a graphing calculator to graph this function in an appropriate viewing window.
   (c) Describe the behavior of this cost function.
   (d) Approximate (to the nearest gallon) the number of gallons for which the cost is a minimum.

60. Suppose that a company finds that the cost C *per unit* of producing x units is given by the function

$$C = C(x) = 325 + \frac{750}{x^2}$$

   (a) Use a graphing calculator to find an appropriate viewing window for this function and examine its graph.
   (b) Describe the behavior of this cost function.
   (c) What is happening to the cost per item as the company manufactures more and more items?

Questions for Thought

61. In this section we mentioned that we are assuming that rational functions are reduced to lowest terms. The reason is illustrated in the following. Consider the function

$$y = f(x) = \frac{x^2 - x - 6}{x - 3}$$

(a) What is the domain of this function?

(b) Reduce this function to lowest terms. Does this affect the domain of the function?

(c) Sketch the graphs of $y = \dfrac{x^2 - x - 6}{x - 3}$ and $y = x + 2$. Are the graphs identical? Should they be?

(d) Explain why the graph of $f(x)$ should be a straight line with a "hole" in it at $x = 3$.

(e) Graph both functions in part (c) on a graphing calculator. Does the graph of $f(x)$ have a "hole" in it at $x = 3$? If not, it is due to the window you are using. On the TI-83 Plus use the ZDecimal window and you will see the "hole."

62. As we mentioned in this section, a rational function may have no horizontal or vertical asymptotes.

Consider the function $y = f(x) = \dfrac{x^4 + 2}{x^2 + 1}$.

(a) Verify that $f(x)$ has no vertical asymptotes.

(b) Compute $f(x)$ for $x = 10, 100,$ and 1000 and for $x = -10, -100,$ and -1000. What do you think is happening to $f(x)$ as $x \to +\infty$? As $x \to -\infty$?

(c) Verify your conjecture is part (b). HINT: Divide the numerator and denominator of $f(x)$ by x^2 and then determine what happens to $f(x)$ as $|x| \to \infty$.

(d) Why does this imply that $f(x)$ does not have any horizontal asymptotes?

63. Although a rational function need not have any horizontal or vertical asymptotes, it may have what is known as an *oblique*, or *slant asymptote*. The accompanying figure shows the graph $y = f(x) = \dfrac{x^2 + 1}{x}$.

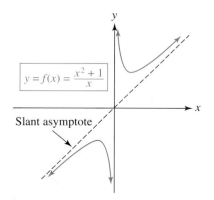

(a) Show that $y = f(x) = x + \dfrac{1}{x}$ and that as $|x| \to \infty$, y gets closer and closer to x. Thus the graph of $f(x)$ gets closer and closer to the graph of $y = x$.

(b) Also show that as $x \to 0$, $y \approx \dfrac{1}{x}$, so that near the y-axis, the graph of $f(x)$ looks like the graph of $y = \dfrac{1}{x}$.

64. The graph of any rational function in which the degree of the numerator is exactly one more than the degree of the denominator will have a slant asymptote. (See Exercise 63.)

(a) Use long division to show that

$$y = f(x) = \frac{x^2 - x + 6}{x - 2} = x + 1 + \frac{8}{x - 2}$$

(b) Show that this means that the line $y = x + 1$ is a slant asymptote for the graph and sketch the graph of $y = f(x)$.

65. (a) Try to sketch the graph of $y = \dfrac{x^2 - 4}{x^2 + 1}$ using the techniques developed in this section.

(b) Does our approach develop enough information to get an accurate graph?

(c) Use a graphing calculator to sketch the graph of this function and see whether you can justify the various features of the graph from its equation.

(d) Does the graph as exhibited on the calculator agree with your attempt at sketching the graph in part (a)?

66. (a) Try to sketch the graph of $y = \dfrac{x^2 - 1}{x + 2}$ using the techniques developed in this section.

(b) Does our approach develop enough information to get an accurate graph?

(c) Use a graphing calculator to sketch the graph of this function and see whether you can justify the various features of the graph from its equation.

(d) Sketch the graph of $y = x - 2$ with a calculator on the same set of coordinate axes. What conclusion would you draw?

67. Use a graphing calculator to investigate the end behavior of the rational function

$$f(x) = \frac{ax^2 + 5x + 2}{bx^2 + 2x + 3}$$

for a variety of values of a and b. Can you predict what the equation of the horizontal asymptote of the graph of such a function will be?

5.6   Radical Functions

When analyzing the scene of an automobile accident, the police can sometimes estimate the speed v of a vehicle (before the brakes were applied) by measuring the length s of the skid marks and using the function

$$v(s) = 2\sqrt{5s}$$

where v is the speed in miles per hour, and s is the length of the skid marks in feet. This function $v(s)$ is called a *radical function*, and will be discussed further later in this section.

Definition of a Radical Function

A **radical function** is a function that contains roots of variables. The following are examples of radical functions.

$$f(x) = \sqrt{x} \qquad g(x) = \frac{\sqrt[3]{x+1}}{x^2+2} \qquad h(x) = \frac{(x-4)^{1/6}}{3}$$

Keep in mind that $h(x)$ is a radical function because $(x-4)^{1/6} = \sqrt[6]{x-4}$.

Given a radical function, the first order of business is, as usual, to determine its domain.

Example 1   Find the domain of each function.

(a) $f(x) = \dfrac{1}{\sqrt{12 + 4x - x^2}}$ (b) $g(x) = \dfrac{\sqrt{x+5}}{x-3}$ (c) $h(x) = \sqrt[3]{x-3}$

Solution   (a) The domain of $f(x)$ consists of those real numbers for which $\sqrt{12 + 4x - x^2}$ is a real number and not equal to zero. (Why?) Thus we need $12 + 4x - x^2 > 0$, which we can solve by doing a sign analysis.

$$12 + 4x - x^2 > 0 \qquad \text{Multiply both sides by } -1.$$
$$x^2 - 4x - 12 < 0$$
$$(x + 2)(x - 6) < 0 \qquad \text{The cut points are } -2 \text{ and } 6.$$

Figure 5.58

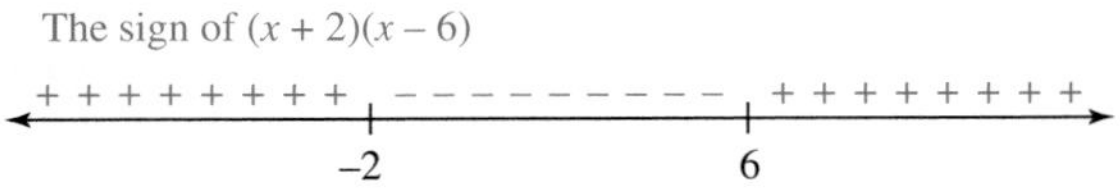

The sign of $(x + 2)(x - 6)$

$$\begin{array}{ccc} + + + + + + + + & - - - - - - - - - - & + + + + + + + + \\ \big| & & \big| \\ -2 & & 6 \end{array}$$

Figure 5.58 shows that $(x + 2)(x - 6)$ is negative for $-2 < x < 6$ and therefore the domain of f is $D_f = \{x \mid -2 < x < 6\}$.

(b) For x to be in the domain of $g(x)$, we require

$$x + 5 \geq 0 \quad \text{and} \quad x - 3 \neq 0$$

and so $D_g = \{x \mid x \geq -5 \text{ and } x \neq 3\}$.

(c) Since the cube root is defined for all real numbers, there are no restrictions on the domain of $h(x)$ and so $D_h = $ all real numbers.   ∎

Calculator Exploration

Use the graph of $y = \dfrac{\sqrt{x+3}}{x-1}$ to determine its domain and range.

Radical functions can exhibit rather complicated behavior, and it can be difficult to sketch their graphs even with the tools of calculus. However, using some of the ideas we have developed, we can describe the graphs of a fairly large class of radical functions.

Suppose we want to find the graph of the radical function $y = f(x) = \sqrt{x}$, whose domain is $\{x \mid x \geq 0\}$. At first glance, this function does not appear to be related to any of the functions we have studied thus far and so its graph is not immediately apparent. Consider the following strategy for obtaining the graph of $y = \sqrt{x}$. Looking at the equation $y = f(x) = \sqrt{x}$, we can see that $f(x)$ is a one-to-one function. (Each y-value comes from a unique x-value.) Therefore, we know that $f(x)$ has an inverse function. Perhaps the inverse function has a graph with which we *are* familiar and that we can reflect through the line $y = x$ to get the graph of $y = \sqrt{x}$.

So let's find the inverse of $y = \sqrt{x}$, keeping in mind that y is defined only for $x \geq 0$.

Are the equations $y^2 = x$ and $y = \sqrt{x}$ equivalent?

$$y = \sqrt{x} \quad \textit{for } x \geq 0 \qquad \text{Interchange } x \text{ and } y.$$
$$x = \sqrt{y} \quad \textit{for } x \geq 0 \qquad \text{Now solve for } y.$$
$$x^2 = y \quad \textit{for } x \geq 0$$

Thus we see that the inverse function of $y = f(x) = \sqrt{x}$ for $x \geq 0$ is the function $y = f^{-1}(x) = x^2$ for $x \geq 0$. We can get the graph of $y = \sqrt{x}$ by reflecting the graph of $y = x^2$ for $x \geq 0$ about the line $y = x$. See Figure 5.59(a) and (b).

Figure 5.59

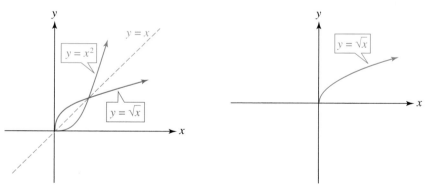

The graph of $y = f(x) = x^2$ for $x \geq 0$

(a)

The graph of $y = f(x) = \sqrt{x}$

(b)

The graph of $y = \sqrt{x}$ is an important graph and should be added to your catalog of basic graphs.

Example 2 | Sketch the graph of $y = f(x) = \sqrt[3]{x}$.

Solution | Using the same approach outlined above, we note that $y = f(x) = \sqrt[3]{x}$ is a one-to-one function. We will find the inverse function, sketch its graph, and use the reflection principle for inverse functions to obtain the graph we are interested in.

$$y = \sqrt[3]{x} \qquad \text{Interchange } x \text{ and } y.$$
$$x = \sqrt[3]{y} \qquad \text{Now solve for } y.$$
$$x^3 = y$$

Thus the inverse function of $y = f(x) = \sqrt[3]{x}$ is the function $y = f^{-1}(x) = x^3$. We can get the graph of $y = \sqrt[3]{x}$ by reflecting the graph of $y = x^3$ about the line $y = x$. See Figure 5.60(a) and (b).

Figure 5.60

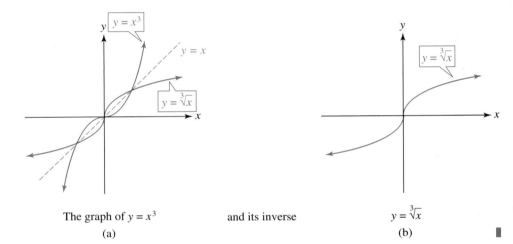

The graph of $y = x^3$ and its inverse
(a)

$y = \sqrt[3]{x}$
(b)

Calculator Exploration

In the following, it is suggested that you try a variety of viewing windows to aid in your analysis.

1. Graph the following on the same set of coordinate axes:

$$y = \sqrt{x} = x^{1/2} \qquad y = \sqrt[4]{x} = x^{1/4} \qquad y = \sqrt[6]{x} = x^{1/6}$$

For even values of n, what can you conclude about how changing n affects the graph $y = x^{1/n}$?

2. Graph the following on the same set of coordinate axes:

$$y = \sqrt[3]{x} = x^{1/3} \qquad y = \sqrt[5]{x} = x^{1/5} \qquad y = \sqrt[7]{x} = x^{1/7}$$

For odd values of n, what can you conclude about how changing n affects the graph $y = x^{1/n}$?

We know that the graphs of $y = x^n$ for n a positive even integer are similar to the graph of $y = x^2$, and that the graphs of $y = x^n$ for n an odd integer greater than 1 are similar to the graph of $y = x^3$. Therefore, based on the previous analysis in Example 1, we conclude that the graph of $y = \sqrt[n]{x}$ will be similar to the graph of $y = \sqrt{x}$

when the index n is *even,* and similar to the graph of $y = \sqrt[3]{x}$ when the index n is *odd.* These results are summarized in the following box.

The Graphs of $y = f(x) = \sqrt[n]{x}$

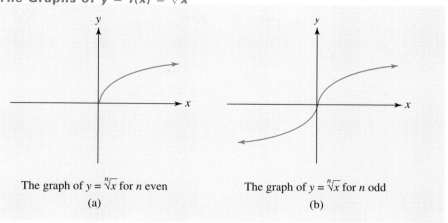

The graph of $y = \sqrt[n]{x}$ for n even

(a)

The graph of $y = \sqrt[n]{x}$ for n odd

(b)

These graphs should be added to your catalog of basic graphs.

Example 3 Sketch the graphs of **(a)** $y = \sqrt{x - 4}$ **(b)** $y = x^{1/3} - 2$

Solution **(a)** The graph of $y = \sqrt{x - 4}$ can be obtained by shifting the graph of $y = \sqrt{x}$ horizontally 4 units to the right. See Figure 5.61.

(b) We recognize that $y = x^{1/3} - 2$ is the same as $y = \sqrt[3]{x} - 2$. The graph of $y = \sqrt[3]{x} - 2$ can be obtained by shifting the graph of $y = \sqrt[3]{x}$ down 2 units. See Figure 5.62. We find the x-intercept of $y = \sqrt[3]{x} - 2$ by setting $y = 0$:

$$0 = \sqrt[3]{x} - 2$$
$$2 = \sqrt[3]{x} \qquad \text{Cube both sides of the equation.}$$
$$8 = x$$

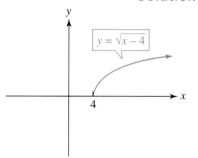

Figure 5.61

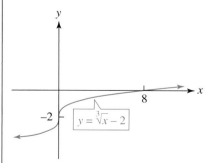

Figure 5.62

Often, we can use algebraic techniques to manipulate an equation into a more familiar form.

Example 4 | Sketch the graph of $y = \sqrt{9 - x^2}$.

Solution

Do you think the graphs of $y = \sqrt{9 - x^2}$ and $y = \sqrt{9 - x}$ will be similar?

This equation does not fit the form of any of the radical functions we have seen thus far. The expression under the radical symbol is not linear. Nevertheless, if we manipulate and analyze this equation a bit, we can determine its graph.

Let's see what happens when we square both sides of this equation:

$$y = \sqrt{9 - x^2} \qquad \text{Square both sides.}$$
$$y^2 = 9 - x^2$$
$$x^2 + y^2 = 9$$

We recognize $x^2 + y^2 = 9$ as the equation of a circle with center $(0, 0)$ and radius 3, whose graph appears in Figure 5.63(a). However, every point that satisfies $y = \sqrt{9 - x^2}$ satisfies $x^2 + y^2 = 9$ but *not* vice versa. For example, $(0, -3)$ satisfies $x^2 + y^2 = 9$ but does not satisfy $y = \sqrt{9 - x^2}$. In fact, in the original equation $y = \sqrt{9 - x^2}$, y must be nonnegative, since the square root is, by definition, nonnegative. Therefore, to get the graph of $y = \sqrt{9 - x^2}$, we delete that portion of the graph $x^2 + y^2 = 9$ for which y is negative; that is, we delete that portion of the graph below the x-axis. The graph of $y = \sqrt{9 - x^2}$ appears in Figure 5.63(b).

Figure 5.63

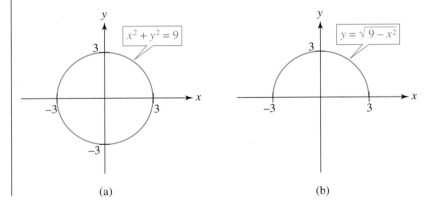

(a)   (b)

Example 5 | Sketch the graph of $y = \sqrt{x + 2}$ and $y = x$.
(a) Estimate the points of intersection of the two graphs.
(b) Find the points of intersection of the two graphs algebraically.

Solution | We sketch the graphs of $y = \sqrt{x + 2}$ and $y = x$ on the same coordinate system. The graph of $y = \sqrt{x + 2}$ is obtained by shifting the graph of $y = \sqrt{x}$ to the left 2 units. The graph of $y = x$ is the familiar line of slope 1 passing through the origin. See Figure 5.64.

Figure 5.64

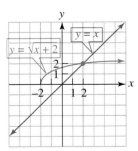

(a) By examining the graphs, we can see that the graphs intersect in one point. We estimate the point of intersection of the two graphs to be (2, 2).

(b) We can find the points of intersection algebraically by recognizing that at any point of intersection, the y-values on the graphs must be equal. If we equate the y-values, we get

$$\sqrt{x + 2} = x \qquad \text{Square both sides.}$$
$$x + 2 = x^2$$
$$x^2 - x - 2 = 0$$
$$(x - 2)(x + 1) = 0$$
$$x = 2 \quad \text{or} \quad x = -1$$

Substituting these x-values into the equation $y = x$, we get $x = 2$, $y = 2$ and $x = -1$, $y = -1$. However, if we check these points in the equation $y = \sqrt{x + 2}$, we see that (2, 2) satisfies it, but $(-1, -1)$ does not. (Remember that $\sqrt{x + 2}$ means the nonnegative square root.) This agrees with the result we found in part (a), where we saw that there is only one point of intersection. ∎

We discuss finding points of intersection and solving systems of nonlinear equations in greater detail in Chapter 8.

Real-life situations can also give rise to radical functions.

Example 6 Points A and B are on opposite sides of a straight river that is 1 mi wide. Point C is 3 mi down the river on the same side as B. A person swims from A to some point P between B and C and then runs from P to C.

(a) Express the total distance covered, D, as a function of one variable.

(b) If the person can swim at 1.5 mi/hr and run at 7 mi/hr, also express the total time, T, to get from A to C as a function of one variable.

Solution (a) Figure 5.65 illustrates the given situation. Note that we have labeled the distance from B to P by x.

Figure 5.65

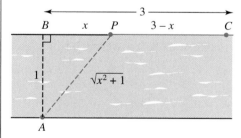

By using the Pythagorean Theorem, we find the distance between A and P to be $\sqrt{x^2 + 1}$, and the distance between P and C is $3 - x$. Therefore, the total distance is $D = D(x) = \sqrt{x^2 + 1} + (3 - x)$. From the diagram we can clearly see that the domain of this function is $\{x \mid 0 \le x \le 3\}$.

(b) Since each distance is being covered at a uniform rate, the time to cover each part of the trip is computed by dividing each distance by its rate. Therefore, the total time needed to cover the distance D is

$$T = T(x) = \frac{\sqrt{x^2 + 1}}{1.5} + \frac{(3 - x)}{7} \qquad \text{for } 0 \le x \le 3$$

Calculus techniques can be applied to this function to determine where the point P should be chosen to *minimize* the total time needed to get from A to C. ∎

Let's return to the situation described in the introduction.

Example 7 When analyzing the scene of an automobile accident, the police can sometimes estimate the speed v of a vehicle (before the brakes were applied) by measuring the length s of the skid marks and using the function

$$v(s) = 2\sqrt{5s}$$

where v is the speed in miles per hour, and s is the length of the skid marks in feet. A motorist involved in an automobile accident claims that he was driving within the posted speed limit of 45 mph. A police officer called to the scene of the accident measures the skid marks left by the motorist's vehicle to be 155.4 ft. Using this mathematical model for the velocity v, is the motorist's claim plausible?

Solution To evaluate the motorist's claim, we compute the speed of the car as predicted by the function and compare the result to the claim.

$$v(s) = 2\sqrt{5s} \qquad \text{Substitute } s = 155.4.$$
$$= 2\sqrt{5(155.4)} = 55.75$$

Thus the function predicts that skid marks of length 155.4 feet would be produced when the vehicle is traveling at approximately 56 mph.

Thus, based on the given mathematical model, the motorist's claim of producing skid marks of 155.4 ft while driving at 45 mph seems implausible. ∎

The types of functions we have discussed in this chapter, polynomial functions, rational functions, and radical functions, are part of a general category called *algebraic functions*.

Definition of Algebraic Function

An **algebraic function** is a function obtained by applying a finite number of additions, subtractions, multiplications, divisions and by taking roots of constants and/or variables.

In the next chapter we will begin to examine functions that are not algebraic.

5.6 Exercises

In Exercises 1–24, sketch the graph of the given equation. Label the intercepts where appropriate.

1. $y = \sqrt{x} + 3$

2. $y = \sqrt{x + 3}$

3. $y = \sqrt{x} - 4$

4. $y = \sqrt{x - 4}$

5. $y = \sqrt{2x - 6}$

6. $y = \sqrt{2x} - 6$

7. $y = 2\sqrt{x} - 6$

8. $y = -\sqrt{x}$

9. $y = \sqrt{-x}$

10. $y = \sqrt{|x|}$

11. $y = \sqrt[3]{x + 2}$

12. $y = \sqrt[3]{x} + 2$

13. $y = \sqrt[3]{x} - 3$

14. $y = 2 - \sqrt[3]{x}$

15. $y = 1 - \sqrt[4]{x}$

16. $y = -\sqrt{4 - x}$

17. $y = \sqrt[5]{x} - 1$

18. $y = \sqrt[5]{x - 1}$

19. $y = \sqrt[6]{x} - 1$

20. $y = \sqrt[6]{x - 1}$

21. $y = \sqrt{16 - x^2}$

22. $x = \sqrt{16 - y^2}$

23. $y = -\sqrt{16 - x^2}$

24. $x = -\sqrt{16 - y^2}$

Exercises 25–28 illustrate how radical expressions, particularly those obtained from difference quotients, can be simplified.

25. Let $f(x) = \sqrt{x}$; compute the difference quotient and show that $\dfrac{f(x) - f(4)}{x - 4} = \dfrac{1}{\sqrt{x} + 2}$.

26. Let $f(x) = \sqrt{x}$; compute the difference quotient and show that $\dfrac{f(9+h) - f(9)}{h} = \dfrac{1}{\sqrt{9+h}+3}$.

27. Let $f(x) = \dfrac{1}{\sqrt{x}}$; compute the difference quotient and show that $\dfrac{f(t) - f(3)}{t-3} = \dfrac{-1}{\sqrt{3t}(\sqrt{3}+\sqrt{t})}$.

28. Let $f(x) = 2\sqrt{x}$; compute the difference quotient and show that $\dfrac{f(u) - f(5)}{u-5} = \dfrac{2}{\sqrt{u}+\sqrt{5}}$.

In Exercises 29–34, algebraically find the points of intersection of the graphs of the two given functions.

29. $y = \sqrt{x}, \quad y = x - 2$

30. $y = \sqrt{x-4}, \quad y = 4 - x$

31. $y = \sqrt{x} - 2, \quad y = -x$

32. $y = \sqrt{x} + 5, \quad y = x - 1$

33. $y = \sqrt{9-x}, \quad y = x + 3$

34. $y = \sqrt{x} - 2, \quad y = \frac{1}{6}(x-4)$

35. Why is it difficult to determine the behavior of the expression $\dfrac{\sqrt{x}-2}{x-4}$ when x is near 4? Rationalize the numerator of this expression. Is it now easier to analyze the behavior of this expression when x is near 4? What is happening to the expression when x is near 4?

36. Using the procedure outlined in Exercise 35, analyze the behavior of $\dfrac{\sqrt{x+3}-3}{x-6}$ when x is near 6.

37. (a) Write an equation that identifies those points equidistant from the points $(2, 3)$ and $(5, 1)$.
(b) Verify that the equation obtained in part (a) is the equation of the perpendicular bisector of the line segment joining the two points.

38. Write an equation that identifies the points equidistant from the point $(0, 4)$ and the x-axis.

39. Points A and B are directly opposite each other along the banks of a straight river that is 3 miles wide. Point C is 8 miles down the river on the same side as B. An oil company wishes to lay a pipeline so that oil can be pumped from point A to point C. However, it costs twice as much to lay each mile of pipe under water as it does on dry land. Therefore, the company considers laying the pipe from point A to some point P between B and C, say, x miles from B.
(a) Express the total length of pipeline needed, L, as a function of x.

(b) If it costs D dollars per mile to lay pipe on land and $2D$ dollars per mile to lay pipe under water, express the total cost of the pipeline, C, as a function of x.
(c) If the company chooses point P to be 5 mi down the shoreline from B, approximate the total cost of the pipeline in terms of D.
(The methods of calculus allow us actually to determine where the point P should be chosen to minimize the cost of the pipeline.)

40. Two poles, 20 feet and 30 feet high, are to be anchored to the ground with guy wires from the top of each to a point directly between the two poles, which are 50 feet apart. See the figure. Express the total length of wire, L, needed as a function of x.

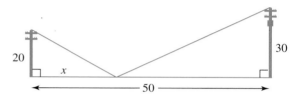

41. The slant height, s, of a right circular cone is as indicated in the following figure.

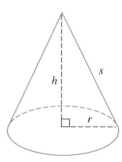

If the height of a cone is 8 cm, express the volume of the cone as a function of s.

42. An open right circular cylinder of height h is inscribed in a sphere of radius 12 cm. See the following figure.

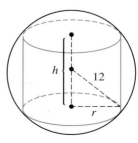

(a) Express the surface area of the cylinder as a function of r.
(b) Express the volume of the cylinder as a function of r.

43. A rectangle of base b and height h is inscribed in a circle of radius 18 meters. Express the area of the rectangle as a function of h.

44. Use the formula for the surface area, S, of a sphere to express the radius of the sphere as a function of S.

45. The *period* T of a pendulum (in seconds) is the time it takes to complete one full cycle (swinging out and returning back), and it is given by the formula

$$T = 2\pi\sqrt{\frac{L}{980}}$$

where L is measured in centimeters.
- **(a)** To the nearest tenth of a second, determine the period of a pendulum of length 12 cm.
- **(b)** To the nearest tenth, determine the length of a pendulum whose period is 1 second.

46. If you want to double the period of a pendulum, what must you do to its length? Use the formula of Exercise 45.

Certain procedures in calculus applied to radical functions give rise to the following types of problems.

47. Show that $x^{1/2} + (x + 8)\frac{1}{2}x^{-1/2} = \dfrac{3x + 8}{2\sqrt{x}}$.

48. Show that $\dfrac{x^{1/2} - (x - 1)\frac{1}{2}x^{-1/2}}{x} = \dfrac{x + 1}{2\sqrt{x^3}}$.

49. Show that

$$x\left(\frac{1}{2}\right)(x^2 + 4)^{-1/2}(2x) + (x^2 + 4)^{1/2} = \frac{2x^2 + 4}{\sqrt{x^2 + 4}}$$

50. Show that

$$\frac{\sqrt{x^2 + 1} - x[\frac{1}{2}(x^2 + 1)^{-1/2}(2x)]}{(\sqrt{x^2 + 1})^2} = \frac{1}{(x^2 + 1)^{3/2}}$$

51. Show that $\sqrt[3]{x} + (x - 3)x^{-2/3} = \dfrac{2x - 3}{\sqrt[3]{x^2}}$.

52. Show that

$$\frac{x^2\frac{1}{3}(x - 1)^{-2/3} - (x - 1)^{1/3}(2x)}{x^4} = \frac{6 - 5x}{3x^3\sqrt[3]{(x - 1)^2}}$$

5.7   Variation

In Section 4.3 we discussed how functions can be used to describe relationships between various physical quantities. We called the equation describing such a relationship a mathematical model of the real-life situation. In this section we extend these ideas to particular types of relationships using the polynomial, rational, and radical functions studied in this chapter.

We are familiar with the fact that if we travel at a constant rate of 50 mi/hr for a period of t hours, then the distance traveled is $d = 50t$. Here d is a function of t, and we often say "The distance varies directly as the time."

Suppose someone drops a heavy metal ball from a high bridge and is able to determine the distance the object has fallen (in feet) after t seconds, the results of which are recorded in the following table.

t	1	2	3	4
d	16.3	64.7	144.5	256.8

Based on these observed data, we might conjecture that a mathematical model for the distance fallen is $d = 16t^2$, which would give distances of 16, 64, 144, and 256 ft after 1, 2, 3, and 4 seconds, respectively. In fact, it turns out that if the ball were falling in a vacuum with no air resistance, the distance the ball falls is given exactly by $d = 16t^2$. This gives the distance fallen as a function of t.

In both of these cases, $d = 50t$ and $d = 16t^2$, the distance is a function of the time t. However, there is a special vocabulary used to describe these functions and indicate the way in which the distance is a function of time. In the case of $d = 50t$, we say that the distance varies *directly as the time,* whereas in the case $d = 16t^2$, we say that the distance varies *directly as the square of the time.*

A formula from physics called Boyle's law says that all other factors remaining the same, the volume, V, occupied by a specific amount of a gas is a function of the pressure, P, according to the formula

$$V = \frac{k}{P} \quad \text{where } k \text{ is a nonzero constant}$$

Another formula from physics says that the intensity of illumination, I, from a source of light is a function of the distance, d, from the source according to the formula

$$I = \frac{k}{d^2} \quad \text{where } k \text{ is a nonzero constant}$$

Again, although V is a function of P and I is a function of d, the language used to describe these functions is not the same. We often say "The volume of the gas varies *inversely as its pressure*" and "The intensity of illumination varies *inversely as the square of the distance* from the source of light."

The following makes this terminology precise.

Direct Variation

The following three statements are equivalent:

1. y **varies directly** as x.

2. y is **directly proportional** to x.

3. $y = kx$ for some constant k not equal to zero.

Inverse Variation

The following three statements are equivalent:

1. y **varies inversely** as x.

2. y is **inversely proportional** to x.

3. $y = \dfrac{k}{x}$ for some constant k not equal to zero.

k is called the **constant of variation** or the **constant of proportionality**.

For example, from the formula for the circumference of a circle, $C = 2\pi r$, we see that the circumference of a circle varies directly as its radius, and the constant of variation for this relationship is 2π. On the other hand, from the formula for the area of a circle, $A = \pi r^2$, we see that the area of a circle is directly proportional to the *square* of its radius, and the constant of proportionality is π.

Example 1   Suppose that s varies directly as t and that $s = 12$ when $t = 5$. Find
(a) s when $t = 9$
(b) t when $s = \dfrac{3}{4}$

Solution   (a) Since s varies directly as t, we use the definition of direct variation to write

If y varies directly as x, does it follow that x varies directly as y?

$$s = kt \qquad\qquad \text{Now we subtitute } s = 12 \text{ and } t = 5 \text{ to find } k.$$

$$12 = k(5) \quad \Rightarrow \quad k = \frac{12}{5} \qquad\qquad \text{Thus the variation equation becomes}$$

$$s = \frac{12}{5}t \qquad\qquad \text{Now we substitute } t = 9.$$

$$s = \frac{12}{5}(9) \quad \text{or} \quad s = \frac{108}{5}$$

An alternative approach is to recognize that if s varies directly as t, then we can write

$$s = kt \quad \text{or} \quad \frac{s}{t} = k$$

Thus the ratio $\dfrac{s}{t}$ is constant and we can set up the following proportion:

$$\frac{12}{5} = \frac{s}{9} \quad \Rightarrow \quad \boxed{s = \frac{108}{5}}$$

The first approach offers the advantage that we obtain the variation equation $s = \dfrac{12}{5}t$, which can be used to find additional values of s or t, as we illustrate in part (b).

(b) Using the variation equation $s = \dfrac{12}{5}t$ obtained in part (a), we can substitute $s = \dfrac{3}{4}$ to find t:

$$s = \frac{12}{5}t \qquad \text{Substitute } s = \frac{3}{4}.$$

$$\frac{3}{4} = \frac{12}{5}t \quad \Rightarrow \quad \boxed{t = \frac{5}{16}}$$

$\blacksquare$

Example 2   As mentioned before, the intensity of illumination, I, of a light source is inversely proportional to the square of the distance from the source. If a light source has an intensity of 1000 lumens (lm) at a distance of 1.5 feet, find the intensity at a distance of 10 feet.

Solution   The variation equation for this example is $I = \dfrac{k}{d^2}$. We use the given information to find the constant of proportionality:

If y varies inversely as x, does it follow that x varies inversely as y?

$$I = \frac{k}{d^2} \qquad\qquad \text{Substitute } I = 1000 \text{ and } d = 1.5.$$

$$1000 = \frac{k}{1.5^2} \quad \Rightarrow \quad k = 2250 \qquad \text{The variation equation becomes}$$

$$I = \frac{2250}{d^2} \qquad\qquad \text{Now we substitute } d = 10.$$

$$I = \frac{2250}{10^2} \quad \Rightarrow \quad \boxed{I = 22.5 \text{ lumens}}$$

At a distance of 10 feet, the light source has an intensity of 22.5 lumens. $\blacksquare$

Having the variation model for a particular situation allows us to analyze the relationship among the variables, as illustrated in the following example.

Example 3   The volume of a sphere varies directly as the cube of its radius, according to the formula

$$V = \frac{4}{3}\pi r^3$$

What happens to the volume of a sphere if the radius is tripled?

Solution   Tripling the radius means changing the radius from r to $3r$. When the radius is r, the volume is

See what happens to the volume of a sphere if the radius is doubled from 3 inches to 6 inches.

$$V = \frac{4}{3}\pi r^3$$

When the radius is $3r$, the volume is

$$V = \frac{4}{3}\pi(3r)^3 = \frac{4}{3}\pi(27)r^3 = 27\left(\frac{4}{3}\pi r^3\right)$$

What do you get when you divide the volume of the sphere when the radius is $3r$ by the volume of the sphere when the radius is r?

When the radius is $3r$, we can see that the volume is 27 times the volume when the radius is r. Therefore, we conclude that tripling the radius causes the volume to increase by a factor of 27. ∎

There are many other types of variation. For example, if z varies directly as the product of x and y, then $z = kxy$ for some nonzero constant k and we say that z **varies jointly as x and y.** If z varies directly as x and inversely as y, we write

$$z = k\left(\frac{x}{y}\right) = \frac{kx}{y}$$

If z varies directly as the square of x and inversely as the square root of y, we write

$$z = \frac{kx^2}{\sqrt{y}}$$

Example 4 The resistance, R, of a wire to an electrical current (measured in ohms) is directly proportional to the length, L, of the wire and inversely proportional to the square of the diameter, d, of the wire. If the resistance of the 50-meter-long wire of diameter 0.1 cm is 8 ohms, find the resistance of 100 meters of the same type of wire with a diameter of 0.05 cm.

Solution The given relationship can be translated into the following variation equation:

$$R = \frac{kL}{d^2} \qquad \text{We substitute } R = 8, L = 50, \text{ and } d = 0.1 \text{ and solve for } k.$$

$$8 = \frac{k(50)}{(0.1)^2} \;\Rightarrow\; k = 0.0016 \qquad \text{The variation equation becomes}$$

$$R = \frac{0.0016L}{d^2} \qquad \text{Now we substitute } L = 100 \text{ and } d = 0.05.$$

$$R = \frac{0.0016(100)}{(0.05)^2} = 64 \text{ ohms}$$

Note that doubling the length of the wire (from 50 to 100 meters) and halving the diameter (from 0.1 cm to 0.05 cm), increases the resistance by a factor of 8 (from 8 ohms to 64 ohms). ∎

5.7 Exercises

In the following exercises, be sure to identify the constant of variation. Round answers to the nearest hundredth where necessary.

1. If y varies directly as x and $y = 15$ when $x = 8$, find y when $x = 25$.

2. If y varies inversely as x and $y = 15$ when $x = 8$, find y when $x = 25$.

3. If u varies inversely as the square of t and $u = 4$ when $t = 8$, find u when $t = 10$.

4. If a varies directly as the square root of b and $a = 5$ when $b = 12$, find b when $a = 6$.

5. If z varies jointly as m and p, and $z = 20$ when $m = \frac{1}{2}$ and $p = 7$, find z when $m = -3$ and $p = \frac{2}{9}$.

6. If v varies directly as r and inversely as s and $v = -12$ when $r = 3$ and $s = 4$, find s when $v = 2$ and $r = -6$.

7. If z varies directly as the square of x and inversely with the cube of y, and $z = 1$ when $x = 2$ and $y = 3$, find z when $x = 3$ and $y = 2$.

8. If z varies jointly as the cube of x and the fourth power of y, and $z = 2$ when $x = 0.25$ and $y = 0.5$, find x when $z = 0.2$ and $y = 1.2$.

9. Suppose y varies inversely as x. What happens to y if x is multiplied by a factor of 4?

10. Suppose y varies directly as x. What happens to y if x is multiplied by a factor of 4?

11. Suppose y varies directly as the square root of x. What happens to y if x is multiplied by a factor of 9?

12. Suppose y varies inversely as the cube root of x. What happens to y if x is multiplied by a factor of 8?

13. Suppose s varies jointly as t and u. What happens to s if t is doubled and u is tripled?

14. Suppose r varies directly as c and inversely as d. What happens to r if c is tripled and d is halved?

15. Hooke's law for springs states that the force or weight required to stretch a spring x units beyond its natural length is directly proportional to x. If a weight of 5 pounds is necessary to stretch a spring from its natural length of 8 cm to 8.4 cm, find the weight necessary to stretch the spring to a length of 9.2 cm.

16. Hooke's law also applies to the force or weight necessary to compress a spring x units within its natural length. If a force of 25 pounds is needed to compress a spring 4 inches shorter than its natural length, find the force necessary to compress the spring 6 additional inches.

17. If a force of 30 pounds stretches a spring 6.5 inches, how far will a force of 42 pounds stretch the spring? See Exercise 15.

18. A force of 45 pounds stretches a spring 6 cm. If the spring has a natural length of 30 cm and can be stretched to at most twice its natural length, what is the maximum weight that the spring can support? See Exercise 15.

19. If a searchlight has an intensity of 50,000 lumens at a distance of 100 feet, what will the intensity be at a distance of 200 feet? (See Example 2.)

20. If a light source has an intensity of 1600 lumens at a distance of 10 feet, at what distance will the intensity be 2 lumens? (See Example 2.)

21. According to Boyle's law, if all other factors remain the same, the volume, V, occupied by a specific amount of a gas varies inversely as the pressure, P. If 100 cu in. of a gas exerts a pressure of 40 pounds per square inch (psi), what pressure will be exerted if the same amount of gas is compressed to 30 cu in.?

22. To what volume must the gas in Exercise 21 be allowed to expand if it is to exert a pressure of 15 psi?

23. The electrical resistance of a wire varies directly with the length of the wire and inversely with the square of its diameter. If a wire 100 meters long with a diameter of 0.36 cm has a resistance of 80 ohms, how much resistance will there be if only 40 meters of the wire is used?

24. If you want to use a wire made of the same material as in Exercise 23 (which means it has the same variation constant) and you want 100 feet of this wire to have a resistance of 40 ohms, how thick must the wire be?

25. When analyzing an accident scene, the police can sometimes estimate the speed of a vehicle (before the brakes were applied) by measuring the length of the skid marks and using the fact that the speed is directly proportional to the square root of the length of the skid marks. If skid marks of 50 feet are made at 35 mi/hr, estimate the speed of a car that makes skid marks of 125 feet.

26. The driver of a car involved in an accident claims that he was driving at a legal speed of 55 mi/hr. His car is tested and leaves a skid mark of 20 feet at 30 mi/hr. If his vehicle left skid marks of 160 feet at the accident site, is his claim plausible? See Exercise 25.

27. The destructive force, F, of a car in an automobile accident can be described approximately by saying that it varies jointly with the weight of the car, w, and the square of the speed of the car, v. How would F be affected if
    (a) The speed of a car is doubled?
    (b) The weight of a car is doubled?
    (c) The speed and weight of a car are doubled?

28. The range of a projectile such as an artillery shell shot out of a cannon or a motorcycle stunt rider flying off a ramp is directly proportional to the square of the velocity at takeoff. If a motorcycle stunt rider has jumped a distance of 168 ft with a takeoff speed of 70 mi/hr, how far can she expect to jump with a takeoff speed of 85 mi/hr?

29. The load, L, that can be safely supported by a beam with a rectangular cross section varies directly with the product of the width, w, and the square of the depth, d, of the cross section and inversely with the length of the beam, ℓ.

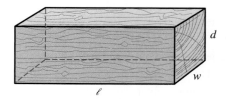

For the purposes of this discussion we assume that the width is the shorter dimension of the cross section. If a 3-inch by 5-inch beam that is 10 feet long can safely carry a load of 750 pounds, what is the maximum load that can be safely supported by a beam of the same material if it is 2.5 in. by 4 in. by 15 ft long?

30. Using the information in Exercise 29, how is the safe load affected by doubling each of the dimensions of the beam individually? All together?

31. Newton's law of gravitation states that the force, F, of attraction between two objects of mass m_1 and m_2 is

$$F = G \frac{m_1 m_2}{d^2}$$

where G is a constant and d is the distance between the two masses. (In using this formula, we perform the computations as if the entire mass were concentrated at the center of each object.)

(a) Use the given formula to describe how the force of attraction between two objects varies in relation to their masses and the distance between them.

(b) Suppose it is known that if two 1-kg masses are placed 1 meter apart they will exert a force of attraction of 6.67×10^{-11} newtons (N). Find the variation constant G.

(c) The earth has a mass of approximately 5.98×10^{24} kg. Find the gravitational force exerted by the earth on a 1000-kg satellite orbiting the earth at an altitude of 400 km. (The radius of the earth is approximately 6400 km.)

32. The moon has a mass of 6.7×10^{22} kg and is a distance of approximately 385,000 km from the earth. Use the information obtained in part (b) of Exercise 31 to determine the gravitational attraction between the earth and the moon.

In Exercises 33–38, use the given equation to describe in words how the variable on the left-hand side of the equation varies with respect to the *variables* on the right-hand side. In each case, k is constant.

33. $S = 2\pi r h$

34. $V = \pi r^2 h$

35. $V = \frac{4}{3}\pi r^3$

36. $y = \frac{kx^4}{\sqrt{z}}$

37. $z = \frac{kx^2 y^3}{4w}$

38. $s = \frac{\sqrt[3]{rt}}{kv^2}$

Questions for Thought

39. In the formula $d = rt$, relating distance, rate, and time, describe which two of the three quantities vary directly and which vary inversely.

40. The electrical resistance of a wire varies directly with its length and inversely with its cross-sectional area. Does this statement agree or disagree with the statement made in Example 4?

Chapter 5   Summary

After completing this chapter you should:

1. Recognize the graphs of the basic power functions. (Section 5.1)
   For $n > 1$, the graph of $y = x^n$ looks like the graph of $y = x^2$ when n is even and like the graph of $y = x^3$ when n is odd.
   For example:
   Sketch the graphs of
   (a) $y = (x - 1)^3$ (b) $y = (x + 2)^4 - 16$

   Solution:
   (a) The graph of $y = (x - 1)^3$ is obtained by shifting the graph of $y = x^3$ one unit to the right.
   (b) The graph of $y = (x + 2)^4 - 16$ is obtained by shifting the graph of $y = x^4$ two units to the left and 16 units down. The intercepts are found as usual. The graphs are in Figure 5.66 (a) and (b) on page 350.

Figure 5.66

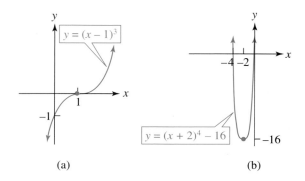

(a)   (b)

2. Be able to sketch the graph of a polynomial function when its zeros can be found. (Sections 5.1, 5.2)

We can use the properties of polynomial functions to sketch the graphs of more general polynomials. The zeros of f are the x-intercepts of the graph of $f(x)$. A sign analysis tell us where the graph is above the x-axis and where the graph is below the x-axis.

For example:

Sketch the graph of $y = f(x) = x^3 + 4x^2 + 3x$.

Solution:

We begin by finding the zeros of f and doing a sign analysis on $f(x)$.

$$f(x) = x^3 + 4x^2 + 3x = x(x^2 + 4x + 3) = 0$$
$$= x(x + 1)(x + 3) = 0$$

The cut points are $0, -1, -3$. See Figure 5.67.

Figure 5.67

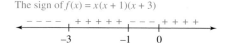

Keeping in mind that $|f(x)| \to \infty$ as $|x| \to \infty$, we sketch the graph, which appears in Figure 5.68.

Figure 5.68

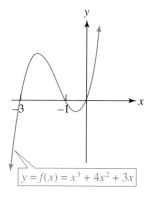

3. Be able to determine the end behavior of a polynomial function. (Section 5.1)

For example:

Determine the end behavior of the function $p(x) = -3x^7 + 2x^3 - 6$.

Solution:

The polynomial function $p(x)$ has the same end behavior as its highest-degree term, $-3x^7$. Figure 5.69(a) shows the end behavior of $y = 3x^7$ (which has the same end behavior as $y = x^3$). Since the leading coefficient is negative, the end behavior of $y = -3x^7$ is reflected about the x-axis, resulting in the graph shown in Figure 5.69(b).

Figure 5.69

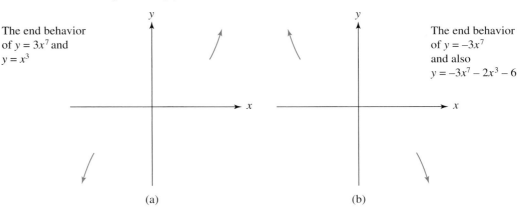

The end behavior of $y = 3x^7$ and $y = x^3$

The end behavior of $y = -3x^7$ and also $y = -3x^7 - 2x^3 - 6$

(a)   (b)

4. Be able to use long division to divide polynomials and synthetic division to divide a polynomial by $x - a$. (Section 5.3)

5. Understand the Factor and Remainder Theorems and use them to determine whether $x - a$ is a factor of $p(x)$. (Section 5.3)
   According to the Factor Theorem, $x - a$ is a factor of $p(x)$ if and only if $p(a) = 0$.
   For example:
   Determine whether $x + 2$ is a factor of $p(x) = 2x^4 + x^3 - 3x^2 - 12$.
   Solution:
   We look at $x + 2$ as $x - a$, which implies $a = -2$. Because
   $$p(-2) = 2(-2)^4 + (-2)^3 - 3(-2)^2 - 12$$
   $$= 32 - 8 - 12 - 12 = 0$$
   we know that when $p(x)$ is divided by $x - (-2) = x + 2$ the remainder is 0, it follows that $x + 2$ is a factor of $p(x)$.

6. Understand the Rational Root Theorem, Descartes' Rule of Signs, and the Upper and Lower Bound Theorem, and use them to help factor a polynomial. (Section 5.4) By determining the factors of a_0 and a_n in the polynomial equation
   $$f(x) = a_n x^n + a_{n-1} x^{n-1} + \cdots + a_1 x + a_0 = 0$$
   where $n \geq 1$, $a_n \neq 0$, we can determine the possible rational roots of $f(x) = 0$. We can then use synthetic division to check these possible roots.
   For example:
   Find all the real zeros of $f(x) = x^3 - 3x^2 + 2x - 6$.
   Solution:
   According to the Rational Root Theorem, if $\dfrac{p}{q}$ is a rational root of $f(x) = 0$, then p must be a factor of -6 and q must be a factor of 1. Thus we have

   Possible values of p: $\pm 1, \pm 2, \pm 3, \pm 6$
   Possible values of q: ± 1
   Possible rational roots $\dfrac{p}{q}$: $\pm 1, \pm 2, \pm 3, \pm 6$

   By using synthetic division, we can divide $f(x)$ by $x \pm 1$, $x \pm 2$, $x \pm 3$, and $x \pm 6$. We find that the remainder when $f(x)$ is divided by $x - 3$ is zero, and so by the Factor Theorem, $x - 3$ is a factor and we have
   $$f(x) = x^3 - 3x^2 + 2x - 6 = (x - 3)(x^2 + 2)$$
   Because $x^2 + 2$ has no real zeros, 3 is the only real zero of $f(x)$.

7. Be familiar with the graphs of basic rational functions. (Section 5.5)

For $n \geq 1$, the graph of $y = \dfrac{1}{x^n}$ looks like the graph of $y = \dfrac{1}{x}$ when n is odd and like the graph of $y = \dfrac{1}{x^2}$ when n is even.

For example:

Sketch the graphs of

(a) $y = \dfrac{1}{x} + 1$　　　　**(b)** $y = \dfrac{1}{(x-3)^2}$

Solution:

(a) The graph of $y = \dfrac{1}{x} + 1$ can be obtained by shifting the graph of $y = \dfrac{1}{x}$ up 1 unit. The graph appears in Figure 5.70.

Figure 5.70

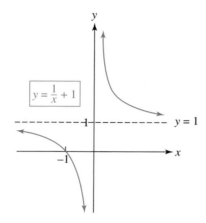

(b) The graph of $y = \dfrac{1}{(x-3)^2}$ can be obtained by shifting the graph of $y = \dfrac{1}{x^2}$ to the right 3 units. The graph appears in Figure 5.71.

Figure 5.71

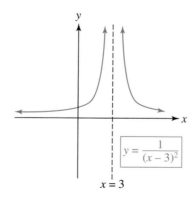

8. Be familiar with the graphs of basic radical functions. (Section 5.6)

The graph of $y = \sqrt[n]{x}$ will look like the graph of $y = \sqrt{x}$ when n is even and like the graph of $y = \sqrt[3]{x}$ when n is odd.

For example:

Sketch the graphs of

(a) $y = \sqrt{x} + 3$　　　　**(b)** $y = \sqrt[5]{x} - 2$

Solution:

(a) The graph of $y = \sqrt{x + 3}$ is obtained by shifting the graph of $y = \sqrt{x}$ to the left 3 units. The graph appears in Figure 5.72.

Figure 5.72

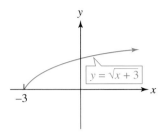

(b) The graph of $y = \sqrt[5]{x} - 2$ is obtained by shifting the graph of $y = \sqrt[5]{x}$ down 2 units. The graph appears in Figure 5.73.

Figure 5.73

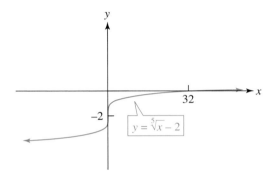

9. Be able to translate statements of variation and solve variation problems. (Section 5.7)

The relationship between some physical quantities can be expressed as a variation. The mathematical model may be a polynomial, rational, or radical function that can be used to describe the real-life problem.

For example:

Suppose that z varies directly with the square root of x and inversely with the square of y. If $z = 20$ when $x = 4$ and $y = 3$, find z when $x = 10$ and $y = 2$.

Solution:

The given variation relationship is translated as $z = \dfrac{k\sqrt{x}}{y^2}$. We can find the constant of variation, k, by substituting the given values.

$$z = \frac{k\sqrt{x}}{y^2} \qquad\qquad \text{Substitute } z = 20, \ x = 4, \text{ and } y = 3.$$

$$20 = \frac{k\sqrt{4}}{3^2} \ \Rightarrow\ k = 90 \qquad \text{Therefore, we have}$$

$$z = \frac{90\sqrt{x}}{y^2} \qquad\qquad \text{Substitute } x = 10 \text{ and } y = 2.$$

$$z = \frac{90\sqrt{10}}{2^2}$$

$$= \frac{45\sqrt{10}}{2}$$

In Exercises 1–22, sketch the graph of the given function. Be sure to indicate the intercepts and asymptotes where appropriate.

1. $y = x^3 + 8$

2. $y = (x - 2)^6$

3. $y = (x - 1)^4 - 16$

4. $y = (x + 2)^5 - 1$

5. $y = x^3 - x^2 - 6x$

6. $y = x^4 - 4x^2$

7. $y = x^2 - 6x + 9$

8. $y = 1 - x^4$

9. $y = \dfrac{1}{(x - 1)^2}$

10. $y = \dfrac{-1}{x} + 2$

11. $y = \dfrac{1}{x + 2} - 3$

12. $y = \dfrac{1}{(x - 3)^4} + 2$

13. $y = \dfrac{x + 3}{x + 2}$

14. $y = \dfrac{x}{x + 1}$

15. $y = \dfrac{x}{x^2 - 1}$

16. $y = \dfrac{x^2 + 2}{4 - x^2}$

17. $y = \sqrt{2x - 5}$

18. $y = \sqrt[3]{4 - x} + 1$

19. $y = \sqrt[5]{x} - 2$

20. $y = \sqrt[4]{3x + 6}$

21. $y = 8 - x^3$

22. $y = (x - 2)^3 + 8$

In Exercises 23–26, use long division to find the quotient and remainder.

23. $\dfrac{2x^3 - 3x^2 + 4x - 7}{2x - 3}$

24. $\dfrac{x^4 - x + 1}{x^2 + 1}$

25. $\dfrac{2x^4 + x^3 + 2x + 1}{2x + 1}$

26. $\dfrac{x^5 + 3x^4 - 4x^2 + 2}{x^2 + x - 1}$

In Exercises 27–30, use synthetic division to find the quotient and remainder.

27. $\dfrac{2x^3 - 3x^2 - 4x - 15}{x - 3}$

28. $\dfrac{x^4 - 5x + 3}{x + 2}$

29. $\dfrac{x^5 - 2x^4 + x^3 - 3x^2 + 5x - 4}{x - 2}$

30. $\dfrac{x^6 - x - 3}{x + 1}$

In Exercises 31–36, find all the zeros of the given polynomial.

31. $p(x) = x^3 - 13x - 12$

32. $p(x) = x^3 + 2x^2 - 3x - 6$

33. $p(x) = x^3 - 6x^2 + 7x - 10$

34. $p(x) = 3x^3 - 2x^2 - 27x + 18$

35. $p(x) = 2x^4 + 7x^3 - 2x^2 + 7x - 4$

36. $p(x) = x^4 - 2x^3 - 3x^2 + 4x + 4$

37. If -2 and $5 - 2i$ are two zeros of a third-degree polynomial, how many more zeros does $p(x)$ have? What are they? What can $p(x)$ be?

38. If $2 + 3i$ and $1 - i$ are two zeros of a fourth-degree polynomial $p(x)$, how many more zeros does $p(x)$ have? What are they? What can $p(x)$ be?

39. How many positive real zeros can the polynomial $p(x) = 2x^3 + 4x^2 + 5x + 3$ have?

40. How many negative real zeros can the polynomial $p(x) = x^4 + x^3 - 2x^2 + 3x + 4$ have?

41. How many positive real zeros can the polynomial $p(x) = 3x^5 - 4x^2 + 3x - 5$ have?

42. How many negative real zeros can the polynomial $p(x) = x^6 - x^5 + x^4 - x^3 + x^2 - x + 1$ have?

In Exercises 43–46, find an upper and lower bound for the real zeros of $p(x)$.

43. $p(x) = x^4 - 4x^3 + 15$

44. $p(x) = 2x^3 - 3x^2 - 12x + 9$

45. $p(x) = x^3 - \dfrac{2}{3}x^2 + \dfrac{1}{2}x - \dfrac{1}{3}$

46. $p(x) = x^4 - 4x^3 + 16x - 17$

47. The perimeter of the square $ABCD$ is 20 cm. Express the area of the triangle as a function of x. See the accompanying figure.

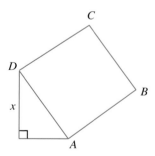

48. The base of a closed rectangular box is a square of side x. If the volume of the box is 100 cm^3, express the surface area, S, of the box as a function of x.

In Exercises 49–52, find the points of intersection of the graphs of the two given functions algebraically.

49. $y = \sqrt{4 - x}$, $y = x + 2$

50. $y = \sqrt{x} - 3$, $y = x - 9$

51. $y = \sqrt{2x + 5} - 2,$ $y = x - 7$

52. $y = 4 - \sqrt{x},$ $y = 2x + 1$

53. If x varies inversely as the cube root of y and $x = 27$ when $y = 27$, find x when $y = 8$.

54. If z varies jointly as x and the square of y, and $z = 30$ when $x = 5$ and $y = 2$, find x when $z = 10$ and $y = 3$.

55. Suppose that on a certain planet the distance that an object falls is directly proportional to the length of time it falls raised to the $\frac{3}{2}$ power. If an object on this planet falls 350 feet in 4 seconds, how long will it take for an object to fall 500 feet?

56. The volume of a cone varies jointly as its height and the square of its radius. If a cone with a height of 3 cm and a radius of 2 cm has a volume of 4π cu cm, find the volume of a cone whose height is 2 cm and radius is 3 cm.

Chapter 5   *Practice Test*

1. Sketch the graph of the given function. Be sure to indicate the intercepts and asymptotes where appropriate.

(a) $y = \dfrac{1}{x - 2} + 3$

(b) $y = \dfrac{-3}{(x + 1)^2}$

(c) $y = 2 - \sqrt{x}$

(d) $y = \dfrac{x + 3}{x - 2}$

(e) $y = x^3 - x^2 - 12x$

(f) $y = \dfrac{2x}{9 - x^2}$

(g) $y = \sqrt[3]{x - 2} + 1$

(h) $y = (x + 1)^2(x - 2)^2$

2. Divide: $\dfrac{3x^4 - 10x^3 - 7x^2 + 17x + 3}{3x - 4}$

3. Which of the following are factors of
$$p(x) = 2x^4 - x^3 - 11x^2 + 4x + 12?$$
(a) $x + 3$ **(b)** $x - 2$ **(c)** $x - 1$ **(d)** $x + 2$

4. Find all the zeros of the following polynomials.
(a) $p(x) = 3x^3 - 20x^2 + 29x + 12$
(b) $p(x) = 2x^4 - 9x^3 + 19x^2 - 15x$

5. According to Poiseuille's law, the blood pressure, P, in a blood vessel varies directly as the length of the vessel and inversely as the fourth power of its radius. What happens to the blood pressure in a particular blood vessel if, due to a buildup in the blood vessel, the radius decreases from 3 mm to 2 mm?

6

Exponential and Logarithmic Functions

In Chapter 5 we examined polynomial, rational, and radical functions, all of which are types of *algebraic functions*. In this chapter we introduce two new *nonalgebraic* functions, the exponential function and its inverse, the logarithmic function. As we will see, these new functions have a wide variety of applications.

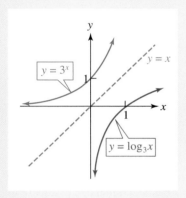

6.1   Exponential Functions

Let's consider the following example.

Example 1   A student decides she wants to save money to buy a used car, which costs $2600. She comes up with what she thinks is a very modest savings plan. She decides to save 2¢ the first day and double the amount she saves each day thereafter. On the second day she plans to save 4¢, on the third day, 8¢, and so on. Write an expression that represents the amount saved on day n and determine how long it will take her to save enough money to buy the car. (The answer may surprise you.)

Solution   We construct a table that records the amount of money earned on day n.

Day	A = Amount saved *on* that day	T = Total amount saved *by* that day
1	$2¢ = 2^1¢$	2¢
2	$4¢ = 2^2¢$	$2¢ + 4¢ = 6¢$
3	$8¢ = 2^3¢$	$2¢ + 4¢ + 8¢ = 14¢$
4	$16¢ = 2^4¢$	$2¢ + 4¢ + 8¢ + 16¢ = 30¢$
$\vdots$	$\vdots$	
n	$2^n¢$	$\vdots$
$\vdots$	$\vdots$	
16	$65{,}536¢ = 2^{16}¢$	$131{,}070¢ = \$1310.70$
17	$131{,}072¢ = 2^{17}¢$	$262{,}142¢ = \$2621.42$

The expression $A = 2^n$ represents the amount saved on day n. As the table illustrates, the expression 2^n grows quite rapidly. If the student can manage to adhere to this rather aggressive savings plan, she would be able to purchase the car in just 17 days!

(In Chapter 9 we develop certain formulas that will allow us to directly compute the values of T, the total amount saved by day n, without having to actually add the amounts from the previous days.)   ∎

In Example 1 we should recognize that the amount, A, saved on day n is a *function* of n. This is our first exposure to a function where the variable appears in the exponent and our first example of a nonalgebraic function, usually called a **transcendental function**. The function $A = 2^n$ is a particular type of transcendental function called an **exponential function**.

In Section 5.1 we examined functions of the form $f(x) = x^n$, where n is constant. How is this different from $f(x) = n^x$?

Definition of an Exponential Function

A function of the form $y = f(x) = b^x$, where $b > 0$ and $b \neq 1$, is called an **exponential function**. b is called the **base** of the exponential function.

Exercise 77 discusses why we have restricted b to be nonnegative and not equal to 1.

As we have seen repeatedly, the first order of business when we encounter a new function is to determine its domain. Since rational exponents are well defined, we know that any rational number will be in the domain of an exponential function.

For example, let $f(x) = 3^x$. Then as x takes on the rational values $x = 4, -2, \frac{1}{2}$, and $\frac{4}{5}$ we have

$$f(4) = 3^4 = 3 \cdot 3 \cdot 3 \cdot 3 = 81$$

$$f(-2) = 3^{-2} = \frac{1}{3^2} = \frac{1}{9}$$

$$f\left(\frac{1}{2}\right) = 3^{1/2} = \sqrt{3}$$

$$f\left(\frac{4}{5}\right) = 3^{4/5} = \sqrt[5]{3^4} = \sqrt[5]{81}$$

Note that even though we do not know the *exact* value of $\sqrt{3}$ or $\sqrt[5]{81}$, we do know exactly what they mean.

However, what about $f(x)$ for irrational values of x?

$$f(\sqrt{2}) = 3^{\sqrt{2}} = ?$$

We have not defined the meaning of irrational exponents.

In fact, a precise formal definition of b^x where x is irrational requires the ideas of calculus. However, we can get an idea of what $3^{\sqrt{2}}$ should be by using successive *rational* approximations to $\sqrt{2}$. For example, we have

$$1.414 < \sqrt{2} < 1.415$$

Thus it would seem reasonable to expect that

$$3^{1.414} < 3^{\sqrt{2}} < 3^{1.415}$$

Since 1.414 and 1.415 are rational numbers, $3^{1.414}$ and $3^{1.415}$ are well defined, even though we cannot compute their values by hand. Using a calculator, we get

$$4.7276950 < 3^{\sqrt{2}} < 4.7328918$$

If we use better approximations to $\sqrt{2}$, we get

$$3^{1.4142} < 3^{\sqrt{2}} < 3^{1.4143}$$

Using a calculator again, we get

$$4.7287339 < 3^{\sqrt{2}} < 4.7292535$$

Computing $3^{\sqrt{2}}$ directly on a calculator gives $3^{\sqrt{2}} \approx 4.7288044$.

This numerical evidence suggests that as x approaches $\sqrt{2}$, the values of 3^x approach a unique real number that we designate as $3^{\sqrt{2}}$, and so we will accept, without proof, the fact that the domain of the exponential function is the set of all real numbers.

In fact, we state even more.

> The exponential function $y = b^x$, where $b > 0$ and $b \neq 1$, is defined for all real values of x. In addition, all the rules for rational exponents hold for real number exponents as well.

Before we state some general facts about exponential functions, let's see if we can determine what the graph of an exponential function will look like.

Calculator Exploration

1. Graph $y = 2^x$, $y = 4^x$, and $y = 6^x$ on the same set of axes. Discuss the similarities and differences among the three graphs. What would you conjecture about how the curve changes as the base b varies for $b > 1$?

2. Graph $y = \left(\dfrac{1}{2}\right)^x$, $y = \left(\dfrac{1}{3}\right)^x$, and $y = \left(\dfrac{1}{5}\right)^x$ on the same set of axes. Discuss the similarities and differences among the three graphs. What would you conjecture about how the curve changes as the base b varies for $0 < b < 1$?

Example 2   Sketch the graph of the function $y = 2^x$ and identify its domain and range.

Solution   To aid in our analysis, we set up a short table of values to give us a frame of reference (Table 6.1).

Table 6.1

x	$y = f(x) = 2^x$	y
-3	$y = 2^{-3}$	$\frac{1}{8}$
-2	$y = 2^{-2}$	$\frac{1}{4}$
-1	$y = 2^{-1}$	$\frac{1}{2}$
0	$y = 2^0$	1
1	$y = 2^1$	2
2	$y = 2^2$	4
3	$y = 2^3$	8

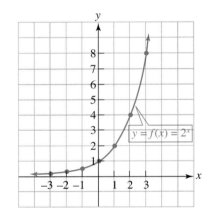

Figure 6.1

With these points in hand, we draw a smooth curve through the points, obtaining the graph in Figure 6.1. It is important to recognize that by joining these points, we are assuming that the exponential function is defined for *all real numbers*.

We note several important aspects of this graph. First, as $x \to +\infty$ the y-values are increasing very rapidly, whereas as $x \to -\infty$ the y-values are getting closer and closer to 0. Thus the x-axis is a horizontal asymptote.

Second, there is no x-intercept. In fact, the graph of an exponential function $y = b^x$ will never have an x-intercept, since $b^x \neq 0$ for any value of x.

Third, the y-intercept is 1. The graph of every exponential function of the form $y = f(x) = b^x$ will have a y-intercept of 1, since $b^0 = 1$ (remember $b \neq 0$).

Fourth, from the graph we can see that the range of $y = f(x) = 2^x$ is the *set of positive real numbers*. ∎

> Recall that $x \to +\infty$ means that x gets extremely large (x takes on values to the extreme right), and $x \to -\infty$ means that x gets extremely small (x takes on values to the extreme left).

> Are there any values of x for which $3^x = 0$? Are there any values of x for which $3^x < 0$?

Example 3   Sketch the graph of $y = f(x) = \left(\dfrac{1}{2}\right)^x$.

Solution   It would be instructive to compute a table of values as we did in Example 2 (you are urged to do so). However, we will take a different approach. We note that

$$y = f(x) = \left(\frac{1}{2}\right)^x = \frac{1}{2^x} = 2^{-x}$$

The graphing principle for $f(-x)$ is discussed in Section 4.2.

If $f(x) = 2^x$, then $f(-x) = 2^{-x}$. Thus by the graphing principle for $f(-x)$, we can obtain the graph of $y = 2^{-x}$ by reflecting the graph of $y = 2^x$, which we found in Example 2, about the y-axis. The graph appears in Figure 6.2.

Here again the x-axis is a horizontal asymptote, there is no x-intercept, 1 is the y-intercept, and the range is the set of positive real numbers. However, the graph is now decreasing rather than increasing.

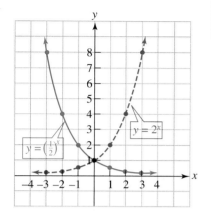

Figure 6.2

If we construct a table of values similar to Table 6.1 for $b = 3$ or 5 or $\dfrac{7}{2}$, in fact, for all $b > 1$, we can see that the graph of the exponential function $y = b^x$ will look very much like the graph in Example 2. This function is said to exhibit **exponential growth**. On the other hand, a similar table for $b = \dfrac{1}{3}$ or $\dfrac{4}{5}$, in fact, for all $0 < b < 1$, would reveal that the graph of the exponential function $y = b^x$ will look very much like the graph of $y = \left(\dfrac{1}{2}\right)^x$ in Example 3. This function is said to exhibit **exponential decay**. We also note that the graphs in Examples 2 and 3 satisfy the horizontal line test and so the functions $y = 2^x$ and $y = \left(\dfrac{1}{2}\right)^x$ are one-to-one functions.

The following box summarizes the important facts about exponential functions and their graphs.

The Exponential Function $y = f(x) = b^x$

1. The domain of the exponential function is the set of all real numbers.
2. The range of the exponential function is the set of all *positive* real numbers.
3. The graph of $y = b^x$ exhibits *exponential growth* if $b > 1$ or *exponential decay* if $0 < b < 1$.

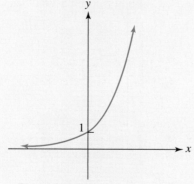

The graph of $y = b^x$ for $b > 1$
Exponential growth

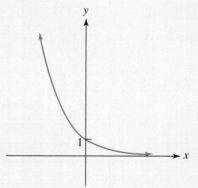

The graph of $y = b^x$ for $0 < b < 1$
Exponential decay

Example 5 on page 363 illustrates the algebraic meaning of the fact that exponential functions are one-to-one.

4. The y-intercept is 1. There are no x-intercepts.
5. The x-axis is a horizontal asymptote.
6. Since the graphs satisfy the horizontal line test, the exponential function is one-to-one. *Algebraically, this means if $b^{x_1} = b^{x_2}$, then $x_1 = x_2$.*

These graphs of exponential growth and decay should be added to our catalog of basic graphs.

Example 4 | Sketch the graph of each of the following. Find the domain, range, intercepts, and asymptotes.

(a) $y = f(x) = 3^x + 1$ (b) $y = g(x) = 3^{x+1}$ (c) $y = h(x) = \left(\dfrac{2}{3}\right)^x$

Solution | (a) To get the graph of $y = 3^x + 1$, we start the graph of $y = 3^x$, which is the basic exponential growth graph, and shift it up 1 unit. The graph is shown in Figure 6.3. From the graph we can see that the domain is the set of all real numbers; the range is the set of real numbers greater than 1; the y-intercept is 2; there are no x-intercepts; and the line $y = 1$ is a horizontal asymptote.

Figure 6.3

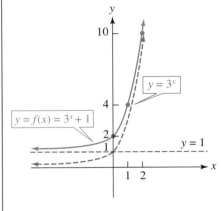

It is necessary to include the horizontal asymptote for an accurate sketch of the graph.

(b) To get the graph of $y = 3^{x+1}$, we start with the graph of $y = 3^x$, and shift it 1 unit to the left. The graph appears in Figure 6.4. From the graph we can see that the domain is the set of all real numbers; the range is the set of real numbers greater than 0; there is no x-intercept; by setting $x = 0$, we find that the y-intercept is 3; the x-axis is a horizontal asymptote.

Figure 6.4

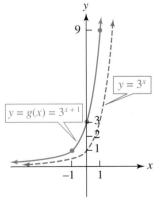

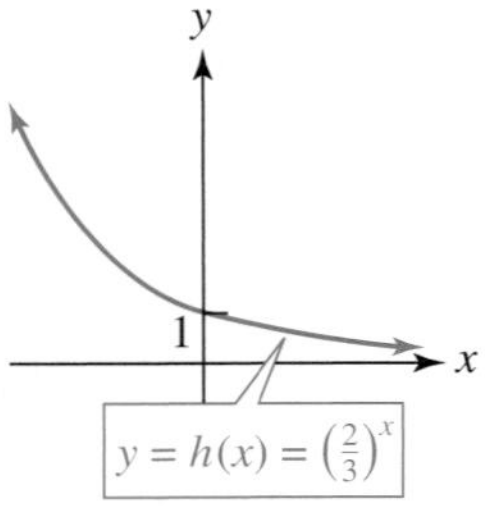

Figure 6.5

(c) Since $\dfrac{2}{3}$ is less than 1, the graph of $y = \left(\dfrac{2}{3}\right)^x$ is the basic exponential decay curve with the same domain, range, intercepts, and asymptote as described in the box on page 361. The graph appears in Figure 6.5. ∎

The fact that all exponential functions are one-to-one allows us to solve equations in which the variable appears in the exponent, as illustrated in the next example. Such equations are called **exponential equations**.

Example 5 | Solve for x: $8^x = 16$

Solution | As we just noted, an exponential function is one-to-one. Recall that if a function is one-to-one, it means that we cannot have the same y-value for two different x-values. Symbolically, we may write that, for a one-to-one function, $f(x_1) = f(x_2)$ implies that $x_1 = x_2$. In the case of the exponential function, this means, for example, that $2^a = 2^b \Rightarrow a = b$. Thus if we can rewrite the given equation so that both sides are expressed in terms of the same base, we should be able to solve the equation. We proceed as follows.

$$8^x = 16 \qquad \text{We can rewrite both sides using base 2.}$$
$$(2^3)^x = 2^4$$
$$2^{3x} = 2^4 \qquad \text{Since the exponential function is one-to-one, this implies}$$
$$3x = 4 \qquad \text{and so } x = \frac{4}{3}. \text{ The check is left to the student.} \qquad \blacksquare$$

We have seen that we can solve an equation such as $9^x = 27$ by expressing both sides of the equation as powers of the same base. However, an equation such as $2^x = x + 4$ does not yield to this method of solution. We will discuss more general exponential equations in Section 6.4

Example 6 | Sketch the graph of $y = -9^{-x} + 3$. Find the domain, range, intercepts, and asymptotes.

Solution | To find the graph of $y = -9^{-x} + 3$, we start with the basic exponential decay graph of $y = 9^{-x}$. See Figure 6.6(a). We then reflect this graph about the x-axis, which gives the graph of $y = -9^{-x}$. See Figure 6.6(b). Finally, we shift this graph up 3 units to get the required graph of $y = -9^{-x} + 3$. See Figure 6.6(c).

Keep in mind that $9^{-x} = \left(\frac{1}{9}\right)^x$, and that -9^x is not the same as $(-9)^x$. In -9^x the base is 9, whereas in $(-9)^x$ the base is -9, which is not allowed for exponential functions.

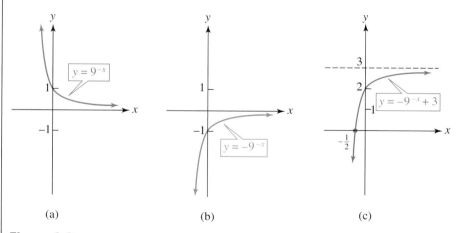

Figure 6.6
The steps leading to the graph of $y = -9^{-x} + 3$.

From this graph we can see the following for $y = -9^{-x} + 3$:

1. The domain is the set of all real numbers.
2. The range is the set of real numbers less than 3.
3. The line $y = 3$ is a horizontal asymptote.
4. By setting $x = 0$, we find that the y-intercept is 2.

The *x*-intercept is not immediately apparent from the graph, so we substitute $y = 0$ and try to solve the resulting equation:

$$y = -9^{-x} + 3 \qquad \text{Substitute } y = 0.$$
$$0 = -9^{-x} + 3$$

Again we solve this exponential equation by expressing both sides of the equation in terms of the *same* base:

$$9^{-x} = 3 \qquad \text{We express both sides in terms of base 3.}$$
$$(3^2)^{-x} = 3$$
$$3^{-2x} = 3^1 \qquad \begin{array}{l}\text{Because the exponential function is one-to-one,} \\ 3^a = 3^b \;\Rightarrow\; a = b; \text{ therefore}\end{array}$$

$$-2x = 1$$

$$x = -\frac{1}{2} \qquad \text{Thus the } x\text{-intercept is } -\frac{1}{2}.$$

We have already described how the base determines exponential growth if $b > 1$ and exponential decay if $0 < b < 1$. However, the size of b will also have a bearing on the relative shape of the graph of $y = b^x$. For $b > 1$, as b increases, the exponential growth graph increases more rapidly, whereas for $0 < b < 1$, as b gets closer to 0, the exponential decay graph decreases more rapidly. Figure 6.7 shows a number of graphs of exponential functions for a variety of values of b.

Figure 6.7

The graphs of $y = b^x$ for a variety of values of b

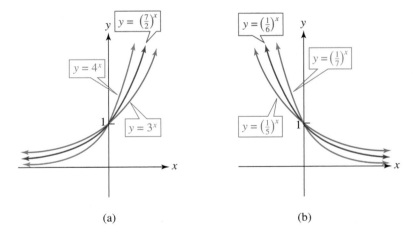

(a)   (b)

The Base *e* and the Natural Exponential Function

You recall from basic geometry that the circumference of a circle is given by the formula $C = \pi d$, which is another way of saying that the ratio of the circumference (C) of *any* circle to its diameter (d) is always the number π. This is the definition of π. We know that π has an approximate value of 3.14159, but because it is an irrational number its decimal representation is an infinite decimal that never stops and never repeats. It would have been nice if the ratio of the circumference of a circle to its diameter had turned out to be 3—but it did not! Thus π is a number that is just built into the natural world.

As we will see in Section 6.5, there are many real-life applications in which another particular irrational number naturally arises. This number is denoted as e and its decimal expansion begins as

$$e = 2.718281828459045\ldots$$

```
e
           2.718281828
e^(3)
           20.08553692
e^(5)
           148.4131591
```

We will have more to say about the significance of the number *e* later in this chapter, but for computational purposes it is important to remember that is *e* approximately 2.72.

Let's examine the number *e* as a base for exponential functions.

Definition of the Natural Exponential Function

The exponential function with base *e*, $f(x) = e^x$, is called the **natural exponential function**.*

Example 7 Sketch the graph of $y = e^x$.

Solution Based on our previous discussion and the fact that $e \approx 2.72$, we know that the graph of $y = e^x$ will exhibit exponential growth and its graph will fall between the graphs of $y = 2^x$ and $y = 3^x$. The graph appears in Figure 6.8.

Figure 6.8

The graph of $y = e^x$

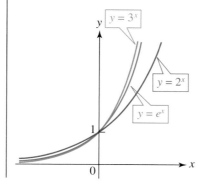

Example 8 The number *N* of bacteria in a certain colony is growing according to the equation $N = 1000e^{0.25t}$, where *t* is the number of hours that have elapsed since the colony began growing.
(a) Use an approximation for *e* to estimate the number of bacteria present after 8 hours.
(b) Use the given growth equation to compute the number of bacteria present after 8 hours to the nearest hundred.

Solution (a) Since $e \approx 2.72$, we could replace *e* by 3 in the given formula to approximate the number of bacteria present after 8 hours:

$N = 1000e^{0.25t}$ We substitute *t* = 8, and use 3 as an approximation for *e*.
$N \approx 1000(3^{(0.25)8}) = 1000(3^2) = 9000$

Thus there are approximately 9000 bacteria present after 8 hours. Since *e* is actually less than 3, we know that the number of bacteria computed from the given growth equation is less than 9000.
(b) According to the given equation, after 8 hours we have

$N = 1000e^{0.25t}$ We substitute *t* = 8.
$= 1000(e^{(0.25)8}) = 1000(e^2) = 7389.056099$

Thus, according to the given growth equation, there will be approximately 7400 bacteria present after 8 hours.

*The letter e is used to denote the base of the natural exponential function in honor of the brilliant Swiss mathematician Leonhard Euler (1707–1783), who is credited with first using it.

Example 9   Use the graph of $y = e^x$ to:

(a) Evaluate e^5. (Compare your answer to that which appears in the margin comment on page 365.)

(b) Find the value of x for which $e^x = 100$. Round your answer to four decimal places.

Solution   (a) We can evaluate e^5 by graphing $y = e^x$ and identifying the y-coordinate corresponding to $x = 5$. In Figure 6.9 we have done exactly that by having the calculator evaluate $y = e^x$ when $x = 5$. We thus have $e^5 = 148.41316$, which agrees with the value given previously rounded to five decimal places.

Figure 6.9

The graph of $y = e^x$

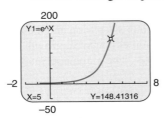

(b) As we have discussed previously, to find the value of x for which $e^x = 100$, we can find the point where the graph of $y = e^x$ intersects the graph of $y = 100$. This is illustrated in Figure 6.10, and we have $x = 4.6052$ rounded to four decimal places.

Check this solution by evaluating $e^{4.6052}$.

Figure 6.10

The graphs of $y = e^x$ and $y = 100$ in the $[-2, 8]$ by $[-50, 200]$ window

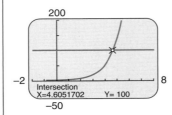

This equation is solved explicitly in Example 5(c) in Section 6.4.

Note that in both parts of this example we used the graph of $y = e^x$. However, in part (a) we are given the x-value and need to find the y-value (this is fairly direct), whereas in part (b) we are given the y-value and need to find the x-value (this is somewhat indirect). In Example 5(c) in Section 6.4, we will talk about how to find the answer to part (b) more directly.   ∎

Knowing the behavior of exponential functions—in particular, the fact that the exponential function is never equal to 0—allows us to determine the zeros of related functions, as in the next example.

Example 10   Find the zeros of $f(x) = 2(x^3)5^x - (8x)5^x$.

Solution

What does a zero of $f(x)$ mean?

We solve $f(x) = 0$ by factoring $f(x)$:

$$2(x^3)5^x - (8x)5^x = 0 \qquad \text{There is a common factor of } 2x5^x.$$
$$2x5^x(x^2 - 4) = 0$$

Is there a value of x for which $5^x = 0$?

We can ignore the factor of 5^x since it is never equal to 0. Alternatively, we may divide both sides of the equation by 5^x without losing any solutions. Therefore, we have

$$x = 0 \quad \text{or} \quad x^2 - 4 = 0$$
$$x = 0 \quad \text{or} \qquad x = \pm 2$$

Therefore, the zeros of $f(x)$ are 0, 2, and -2.   ∎

Example 11   A population of bacteria type DT3 doubles every 3 hours in a petri dish. If the initial population is 4000 bacteria, how many bacteria will be in the petri dish **(a)** after 6 hours? **(b)** After 12 hours? **(c)** After t hours? **(d)** After 17 hours?

Solution   Since the bacteria double every 3 hours, we let A = the number of bacteria in the petri dish and chart its growth as follows:

at 0 hours, $A = 4000$ This is the initial number of bacteria.

after 3 hours, $A = 4000(2)$ $= 8000$ The number of bacteria has doubled once.

after 6 hours, $A = [4000(2)](2) = 4000(2)^2 = 16{,}000$ The number of bacteria has doubled twice.

after 9 hours, $A = [4000(2)^2](2) = 4000(2)^3 = 32{,}000$ The number of bacteria has doubled three times.

after 12 hours, $A = [4000(2)^3](2) = 4000(2)^4 = 64{,}000$ The number of bacteria has doubled four times.

(a) By the chart we can see that the number of bacteria at the end of 6 hours is $16{,}000$.

(b) At the end of 12 hours it is $64{,}000$.

(c) Notice the pattern: After 15 hours the 4000 bacteria have doubled 5 times ($15 \div 3 = 5$), and so the number of bacteria present is $4000(2)^5$.

    Because the number of bacteria doubles every 3 hours, in t hours the number will double $\frac{t}{3}$ times. Hence, if we continue the pattern, we see that the number of bacteria at the end of time t is

$$A = 4000(2^{t/3}) \qquad \text{where } t \text{ is the time in hours}$$

(d) To find the number of bacteria in the petri dish at the end of 17 hours, we need to use the formula found in part (c):

$A = 4000(2^{t/3})$ Since $t = 17$, we get

$\quad = 4000(2^{17/3})$ We first evaluate $2^{17/3}$ on a calculator to get

$\quad = 4000(50.796834) = 203{,}187.33$ Which we round to

$\quad = 203{,}187 \text{ bacteria}$

Example 12   An ecological study finds that if a new state park is built today and stocked with 400 deer, then the number N of deer in t years will be given by the function

$$N(t) = \frac{1200e^{t/4}}{e^{t/4} + 2} \qquad \text{for } t \geq 0$$

Graph this function, describe its behavior, and explain what it says about the deer population in the future.

Solution   Even though this function involves exponential functions, it is certainly more complicated than anything we have examined thus far. We use a graphing calculator to examine its behavior.

    We could use trial and error to choose a reasonable viewing window; however, it is more efficient just to compute some values. We know that we are starting with 400 deer so that $N(0) = 400$. Using a calculator, we compute $N(10) = 1030.8$,

Evaluate $N(0)$ from the formula and verify that the result is 400.

$N(20) = 1184$, and $N(30) = 1198.7$. This means that, according to this mathematical model, after 10 years the deer population will be about 1030, after 20 years it will be about 1184, and after 30 years it will be about 1200. Thus it seems reasonable to choose a [0, 50] by [0, 1500] viewing window. The graph is in Figure 6.11.

Figure 6.11

The graph of $N = \dfrac{1200e^{t/4}}{e^{t/4} + 2}$

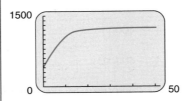

Looking at this graph, we can see that, according to this model, the deer population seems to increase rapidly over the first 10 years or so, and then levels off. By tracing the values on the graph, we see that the values of the function are leveling off at 1200. Thus this model predicts that, although the deer population increases, it will not exceed about 1200. This number is called the *carrying capacity* of the particular environment and is usually the result of limitations of area, food, weather conditions, predators, etc.

We noted that the exponential function is one-to-one, and we know that a one-to-one function has an inverse function. The inverse of the exponential function is the subject of the next section.

6.1 Exercises

In Exercises 1–10, solve the given exponential equation.

1. $2^{x-1} = 8$
2. $2^{x^2-1} = 8$
3. $3^{2x} = 243$
4. $25^{a+3} = 5$
5. $9^t = 27$
6. $4^x = \dfrac{1}{2}$
7. $\left(\dfrac{1}{2}\right)^{x+2} = 16$
8. $\left(\dfrac{1}{3}\right)^{s+1} = 9$
9. $8^x = \sqrt{2}$
10. $16^{3a-2} = \dfrac{1}{4}$

In Exercises 11–16, find the domain of the given function.

11. $f(x) = \dfrac{1}{6^x}$
12. $g(x) = \sqrt{3^x + 1}$
13. $h(x) = \sqrt{2^x - 8}$
14. $f(x) = \dfrac{5}{4^x - 1}$
15. $g(x) = \dfrac{1}{2^{3x} - 2}$
16. $h(x) = \dfrac{1}{2^{3x-2}}$

In Exercises 17–30, sketch the graph of the given function. Identify the domain, range, intercept(s), and asymptote. Round your answer to two decimal places where necessary.

17. $y = 5^{-x}$
18. $y = -5^x$
19. $y = 4^x - 1$
20. $y = 4^{x-1}$
21. $y = 9 - 3^x$
22. $y = 3^{-x} + 2$
23. $y = 2^{x-3} - 8$
24. $y = 3^{x+2} - 9$
25. $y = -3^{-x} + 1$
26. $y = \left(\dfrac{1}{2}\right)^x - 4$
27. $y = e^{x-2}$
28. $y = e^x - 2$
29. $y = 1 - e^x$
30. $y = 1 - e^{-x}$

In Exercises 31–36, let $f(x) = 5^x$, $g(x) = x^2 - 6x + 2$, $h(x) = \dfrac{1}{x+1}$, and $t(x) = \sqrt{x}$. Find each of the following.

31. $(f \circ g)(x)$
32. $(g \circ f)(x)$
33. $(f \circ t)(x)$
34. $(t \circ f)(x)$
35. $(h \circ f)(x)$
36. $(f \circ h)(x)$

In Exercises 37–42, find all the real zeros of the given function.

37. $f(x) = x4^x$

38. $g(x) = \dfrac{2^x - 8}{x}$

39. $h(x) = 9(2^x) - x^2 2^x$

40. $F(x) = 6(3^x) - 5x3^x + x^2 3^x$

41. $f(x) = \dfrac{1 - e^x}{x^2}$

42. $g(x) = 5(2^x) - 10$

43. Solve the equation $3^x + 3^{-x} = 2$. HINT: Let $u = 3^x$, making this a quadratic equation in u.

44. Solve the equation $2^x - 4(2^{-x}) = 3$. HINT: Let $u = 2^x$ making this a quadratic equation in u.

45. Let $f(x) = 2^x$. Show that $f(x + 3) = 8f(x)$.

46. Let $g(x) = 5^x$. Show that $g(x - 2) = \dfrac{1}{25}g(x)$.

47. Let $f(x) = 3^x$. Show that $\dfrac{f(x + 2) - f(x)}{2} = 4(3^x)$.

48. Let $f(x) = 4^x$. Show that
$$\frac{f(x + h) - f(x)}{h} = 4^x\left(\frac{4^h - 1}{h}\right)$$

49. On the same coordinate system, sketch the graphs of
$$y = 2^x, \quad y = \left(\frac{5}{2}\right)^x, \quad \text{and } y = 3^x$$

50. On the same coordinate system, sketch the graphs of
$$y = \left(\frac{1}{2}\right)^x, \quad y = \left(\frac{2}{5}\right)^x, \quad \text{and } y = \left(\frac{1}{3}\right)^x$$

The following exercises deal with applications of exponential growth and decay. You may find a calculator helpful for many of the exercises. Round your answers to two decimal places.

51. The population of Rabbitville doubles every 2 months. If the initial population is 4, how many people will be in Rabbitville at the end of 6 months? At the end of t months? At the end of 2 years?

52. Since 1980, the population of Lewistown has doubled every 10 years. If this trend continues and if there were 800 people living in Lewistown in 1992, how many people would you expect to be living there in 2015?

53. The book value, B, of a used car t years after it was new is often computed using the formula
$$B = N\left(\frac{S}{N}\right)^{t/20} \quad \text{for} \quad 0 \le t \le 20$$

where N is the value of the car when it was new and S is the salvage value of the car after 20 years. Suppose that a new car costs \$20,000 and will have a salvage value of \$600 after 20 years.
(a) Sketch the graph of B as a function of t.
(b) What will the book value of the car be when it is 3 years old?

54. Radioactive argon-39 has a half-life of 4 minutes. This means that every 4 minutes one-half of the amount of argon-39 present changes into another substance as a result of radioactive decay.
(a) If we start with A_0 milligrams of argon-39, explain why the amount A remaining after t minutes is given by the formula
$$A = A_0(2^{-t/4})$$
(b) If $A_0 = 50$ mg, sketch the graph of
$$A = 50(2^{-t/4})$$
Note the values of A for $t = 0, 1, 2, 4, 8, 16$, and 20 minutes.
(c) If $A_0 = 50$ mg, how much argon-39 will be left after 1 hour?

55. The demand function for a certain product is given by the function
$$d = 600 - 0.4(e^{0.005p})$$
where d is the number of items in demand when the price per unit is p. Find the demand d when the price is (a) $p = 800$ and (b) $p = 1200$.

56. A company estimates that the annual profit, P (in dollars), from the sales of a particular item x years after the product is first introduced can be computed as
$$P = P(x) = 120{,}000 - 80{,}000e^{-x}$$
(a) What will the annual profit be after 3 years? After 6 years?
(b) Sketch the graph of $y = P(x)$ for $x \ge 0$.
(c) What is the maximum profit the company can expect from this item? Is this maximum profit ever actually achieved?

57. The atmospheric pressure, P, at a height x feet above sea level can be approximated by the function
$$P = P(x) = 3^{-0.000035x}$$
where the pressure is given in atmospheres. [One atmosphere (1atm) is approximately equal to 14.69 pounds of pressure per square inch (psi)].
(a) What is the atmospheric pressure at sea level?
(b) What is the atmospheric pressure at the top of Mount Everest, which is 29,028 feet above sea level?

(c) What is the atmospheric pressure at the Dead Sea, which is 1290 feet *below* sea level?

58. An alternate formula for computing the atmospheric pressure m miles above sea level is

$$P = P(x) = 14.7e^{-0.21m}$$

where the pressure is given in (psi). Use this formula to answer the questions in Exercise 57. How do the answers compare with those obtained using the formula in Exercise 57?

59. The population, P, of a town is growing according to the formula

$$P = P(t) = P_0 2^{0.05t}$$

where P_0 is the population in a certain year and P is the population after t years have elapsed. Suppose the town's population in 1990 is 12,000.
   (a) Find the population 4 years later and 5 years later.
   (b) Find the population 19 years later and 20 years later.
   (c) Find the ratio of the two populations found in part (a) and the two populations in part (b).
   (d) What is the relationship between the population t years later and $t + 1$ years later?

60. The *E. coli* bacteria grow according to the formula

$$N = N_0 2^{t/25}$$

where N_0 is the number of bacteria present initially and N is the number of bacteria present after t minutes.
   (a) If an *E. coli* colony begins with a population of 1000, how long will it take for the population to double to 2000?
   (b) How long will it take the population to double from 2000 to 4000?
   (c) Based on your answers to parts (a) and (b), what might you conclude about the time it takes for the population of an *E. coli* bacterial colony to double?
   (d) Show that the doubling time for *E. coli* bacteria is 25 minutes.

61. The fruit fly *Drosophila* is often used for genetic studies. A typical fruit fly population will double in 2.5 days.
   (a) If we start with an initial population of N_0 fruit flies, explain why the number of fruit flies, N, after t days is given by the formula

$$N = N_0 2^{t/2.5}$$

   (b) If we start with an initial population of 10 male and 10 female flies, how many flies will there be after 1 week?

62. The use of certain insecticides has been discontinued because of their long-lasting activity in the environment. Suppose that a certain insecticide has a half-life of 16 years; that is, it takes 16 years for one-half of a given amount of the insecticide to become inactive and harmless.
   (a) If we start with an initial amount of insecticide, A_0, explain why the amount of insecticide, A, still active after t years is given by the formula

$$A = A_0 2^{-t/16}$$

   (b) If a farmer uses 100 pounds of this insecticide on his crop, how much will still be active after 10 years? After 100 years?

63. Radioactive isotopes that have relatively short half-lives are often used in medical imaging procedures. Suppose that a certain radioisotope has a half-life of 4 hours.
   (a) Write an equation that gives the amount, A, of this isotope still radioactive t hours after an initial amount A_0 is injected into the body.
   (b) If 10 mg are injected into the body, how much would still be radioactive after 2 hours? After 24 hours?

64. Suppose that the charge remaining in a battery is decreasing exponentially according to the formula

$$C = C(T) = C_0 e^{-T}$$

where C is the charge in coulombs remaining T days after the battery receives an initial charge of C_0. If a battery has a charge of 3.5×10^{-5} coulombs remaining after 14 days, find the initial charge.

65. According to Newton's law of cooling, the rate at which an object cools is directly proportional to the difference in temperature between the object and the surrounding environment. A metal rod whose initial temperature is 100°C cools as it is placed in surrounding air, which is kept at a temperature of 15°C. Suppose that for a certain metal rod the temperature, T, of the rod will decrease exponentially according to the equation

$$T = 15 + 85e^{-0.4m}$$

where m is the number of minutes the rod has been exposed to the cooler air. How long will it take for the rod to cool down to a temperature of 40°C?

66. The biological half-life of a medication is the amount of time it takes an initial amount of medication to lose half its effectiveness. Suppose that a certain medication has a half-life of 9 hours in the body of a nonsmoker but a half-life of only 5 hours in the body of a smoker. If equal doses of this drug are adminis-

tered to a nonsmoker and a smoker, compare the amounts of drug remaining in each person after 24 hours.

67. The rate at which certain electrical items fail can often be described using exponential functions. For example, suppose that a large corporate office has 2400 lightbulbs and that the maintenance department replaces all the old bulbs at the same time, with new bulbs that have an average life of 800 hours. In such a situation, the number of bulbs, N, that should be expected to burn out after t hours can be given by the equation

$$N = N(t) = 2400(1 - 0.998^t)$$

(a) Using appropriate units along the axes, sketch the graph of $y = N(t)$.
(b) How many lightbulbs would be expected to have burned out after 400 hours? 600 hours? 800 hours?
(c) Interpret the results of part (b) in terms of the *percentage* of bulbs that have burned out after 400, 600, and 800 hours.
(d) Explain why the equation for N suggests that it might be more economical to replace all the lightbulbs at once rather than as they burn out.

68. Suppose that 18 wolves (9 male and 9 female) are introduced into an uninhabited area. Further suppose that the wolf population increases by 20% per year for the next 15 years, at which time the wolf population becomes so large that it overwhelms the environment by destroying important elements of the food chain. As a result, the wolf population decreases by 15% per year for the next 25 years.
(a) Using a split function, express the wolf population, P, as a function of the time, t in years over this 40-year period.
(b) Estimate the largest and smallest wolf populations and in which years they occur.

69. Medical technicians often give intravenous infusions of a particular nutrient or medication. The concentration, C, of the nutrient or medication in the blood t minutes after the infusion begins might be given by a formula such as

$$C = C(t) = F + (I - F)3^{-0.01t}$$

(a) Explain the significance of I. HINT: Try finding the concentration when $t = 0$.
(b) Explain the significance of F. HINT: Try finding the concentration when $t = 100, 200, \ldots$.

70. We have seen previously that if a heavy object (such as a rock) is dropped from a high place and we neglect air resistance, then its velocity, v, after t seconds is given by $v = -32t$ feet per second. On the other

hand, if a sky diver jumps out of a plane with arms and legs spread out, then air resistance has a significant effect on the velocity. For example, after jumping from a plane at an altitude of 10,000 feet, the sky diver's velocity, v, in feet per second, would be given by an equation such as

$$v = v(t) = -220(1 - 0.9^t)$$

where t is the number of seconds after the jump but before the parachute is opened.
(a) Sketch the graph of $y = v(t)$.
(b) Find the sky diver's velocity after 4 seconds, and compare it to the velocity of a rock 4 seconds after it is dropped from an altitude of 10,000 feet.
(c) As t increases, what happens to the sky diver's velocity? Find v after 10 sec, 20 sec, 30 sec, ... (This "limiting" velocity is often called the *terminal velocity*.)
(d) What is this terminal velocity? [You may be more impressed if you convert to miles per hour (88 ft/s = 60 mi/hr).]

71. Oceanographers know that the amount of light that penetrates to x meters below the surface decays exponentially as a function of x. For example, at a certain location if the light intensity at the surface is 12 lumens (lm), then the intensity, I, of the light x meters below the surface can be given by

$$I = 12e^{-0.8x}$$

Find the intensity of the light at a depth of 5 meters.

72. Suppose that if 20 grams of sugar are added to a quantity of water, then the amount of sugar, S, that remains undissolved after t minutes is given by

$$S = S(t) = 20\left(\frac{5}{7}\right)^t$$

Find the amount of sugar that remains undissolved after 5 minutes.

73. An epidemiologist is investigating the rate at which a particularly virulent infection spreads through a population if it goes unchecked. She proposes the following model:

$$p(d) = 1 - e^{-0.0235d}$$

where p is the percentage of the population infected d days after the first outbreak of the infection.
(a) According to this model, what percentage of the population would be infected after two weeks?
(b) To the nearest day, approximately how many days will it take for 50% of the population to be infected? (HINT: Use the graph.)

74. An animal control unit estimates that the stray cat population in a certain town is 150, and that the future population can be approximated by the function

$$n(t) = \frac{600}{1 + 3e^{-0.52t}}$$

where n is the number of stray cats t years in the future.
 (a) According to this model, approximately how many stray cats will there be in 5 years?
 (b) To the nearest tenth, approximately how many years will it take for there to be 400 stray cats? (HINT: Use the graph.)

Questions for Thought

75. Earlier in this section we noted that the exponential function is one-to-one, and so we know that the exponential function has an inverse function. In fact, we have used this idea to solve some exponential equations.
 (a) Solve for x: $2^x = 16$. Explain how the idea of the one-to-oneness of the exponential function is used in the solution.
 (b) Is there a solution to $2^x = 15$? How would you go about finding it?

76. Try to find the inverse function of $y = 2^x$. Are there any difficulties in following our familiar outline for finding an inverse function: Interchange x and y and then solve for y?

77. In the definition of the exponential function, we restricted the base b to $b > 0$ and $b \neq 1$.
 (a) If we allow $b = 1$, what happens to the function $y = b^x$?
 (b) If we allow b to be negative, what happens to the domain of $y = b^x$? Think about what happens when we try to find the y-values corresponding to $x = \frac{1}{2}, \frac{3}{4}$, etc. . . . .

78. Let $y = f(x) = 2^x$.
 (a) Show that $\frac{1}{2}f(x) = f(x - 1)$.
 (b) How do we get the graph of $y = \frac{1}{2}f(x)$ from the graph of $y = f(x)$?
 (c) How do we get the graph of $y = f(x - 1)$ from the graph of $y = f(x)$?
 (d) Using the result of part (a), what can we say about the relationship between a stretch and a shift of the graph of $y = 2^x$?
 (e) What stretching factor is equivalent to shifting the graph of $y = 2^x$ three units to the left?

79. Even though $f(x) = e^x$ is a transcendental function, it can be approximated by polynomial functions. Use a graphing calculator to do the following.
 (a) Sketch the graphs of $y = f(x) = e^x$ and $y = g(x) = 1 + x + \frac{x^2}{2}$. What do you notice?
 (b) Sketch the graphs of $y = f(x) = e^x$ and $y = h(x) = 1 + x + \frac{x^2}{2} + \frac{x^3}{6}$. What do you notice?
 (c) Sketch the graphs of $y = f(x) = e^x$ and $y = r(x) = 1 + x + \frac{x^2}{2} + \frac{x^3}{6} + \frac{x^4}{24}$. What do you notice?
 (d) Compute $f(0.5)$, $g(0.5)$, $h(0.5)$, and $r(0.5)$. Do these values agree with your conclusion?

80. A function such as $y = ke^{-(x-a)^2}$ is commonly used in probability and statistics to represent populations that have a *normal distribution*.
 (a) Use a graphing calculator to sketch the graph of this function for $k = 3$ and $a = 2$. Describe the graph.
 (b) Try some other values of k and a and determine how these affect the graph. Relate these results to the graphing principles we have been using.
 (c) Can you see why the graph of a normal distribution is called a *bell-shaped curve*?

81. In a restricted environment, populations often grow rapidly at first and then more slowly as time passes. A function of the form $y = \dfrac{a}{1 + be^{-r(x-h)}}$ is commonly used to describe such a situation and is called a *logistic growth model* where x is the time.
 (a) Use a graphing calculator to sketch the graph of this function for $a = 3$, $b = 1$, $r = 2$, and $h = 2$.
 (b) Try different values of b and r and determine how they affect the curve.
 (c) Do these curves exhibit the characteristic that they grow rapidly at first and then more slowly as x increases?
 (c) By changing the viewing window, describe what happens to the curve as time continues to increase. As mentioned in Example 12, the value at which the population levels off is called the carrying capacity of the environment.

6.2 Logarithmic Functions

In the last section we noted that the exponential function $y = f(x) = b^x$ (where $b > 0$ and $b \neq 1$) is one-to-one. From our previous work with functions, we know this means that the exponential function has an inverse function.

Let's review the process for finding an inverse function by comparing the process for the polynomial function $y = x^3$ and the exponential function $y = 3^x$. Keep in mind that throughout this text we let x be the *independent* variable and y be the *dependent* variable, and so whenever possible we want a function solved explicitly for y.

To find the inverse of $y = x^3$:

$y = x^3$	Interchange x and y.
$x = y^3$	Solve for y.
$y = \sqrt[3]{x}$	

To find the inverse of $y = 3^x$:

$y = 3^x$	Interchange x and y.
$x = 3^y$	Solve for y.
$y = ?$	

There is no algebraic procedure we can use to solve $x = 3^y$ for y. This comparison brings into sharp focus the difference between an algebraic function, such as $y = x^3$, and a transcendental function, such as $y = 3^x$.

In fact, had we studied functions and their inverses before learning about radicals, we would have had to invent radical notation so that we could express the inverse of $y = x^3$ explicitly in the form $y = \sqrt[3]{x}$. In other words, $y^3 = x$ and $y = \sqrt[3]{x}$ both mean exactly the same thing: y is the number whose cube is x. The only difference is that $y = \sqrt[3]{x}$ is solved explicitly for y.

Similarly, if we want to express $x = 3^y$ explicitly as a function of y, we need to invent a special notation for this. The key idea is to take the equation $x = 3^y$ and express it verbally:

$x = 3^y$ means y is the exponent to which 3 must be raised to yield x

We introduce the following notation, which expresses this same idea in a much more compact form.

Definition of $\log_b x$

For $b > 0$ and $b \neq 1$, we write $y = \log_b x$ to mean y is the exponent to which b must be raised to yield x. In other words,

$$x = b^y \iff y = \log_b x$$

The word *log* is short for logarithm.

We read $y = \log_b x$ as "y equals log base b of x" or "y equals the logarithm of x to the base b."

Remember: $y = \log_b x$ is just an alternative way of writing $x = b^y$.

When an expression is written in the form $x = b^y$, it is said to be in **exponential form**. When an expression is written in the form $y = \log_b x$, it is said to be in **logarithmic form**. All our work with logarithms will be made easier if we keep in mind the fact that, according to the definition, *a logarithm is simply an exponent.*

Table 6.2 (page 374) illustrates the equivalence of the exponential and logarithmic forms of a number of statements. Be sure you understand the equivalence of the two forms on each line in Table 6.2.

Table 6.2

Exponential Form	Logarithmic Form
$4^2 = 16$	$\log_4 16 = 2$
$2^4 = 16$	$\log_2 16 = 4$
$5^{-3} = \dfrac{1}{125}$	$\log_5 \dfrac{1}{125} = -3$
$6^{1/2} = \sqrt{6}$	$\log_6 \sqrt{6} = \dfrac{1}{2}$
$7^0 = 1$	$\log_7 1 = 0$
$b^m = u$	$\log_b u = m$

Example 1   Write each of the following in exponential form.

(a) $\log_3 \dfrac{1}{9} = -2$ (b) $\log_{16} 2 = \dfrac{1}{4}$

Solution   We use the fact that $y = \log_b x$ is equivalent to $b^y = x$.

(a) $\log_3 \dfrac{1}{9} = -2$ means $3^{-2} = \dfrac{1}{9}$.

(b) $\log_{16} 2 = \dfrac{1}{4}$ means $16^{1/4} = 2$.

∎

Example 2   Write each of the following in logarithmic form.
(a) $10^{-3} = 0.001$ (b) $27^{2/3} = 9$

Solution   We use the fact that $b^y = x$ is equivalent to $y = \log_b x$.
(a) $10^{-3} = 0.001$ means $\log_{10} 0.001 = -3$.

(b) $27^{2/3} = 9$ means $\log_{27} 9 = \dfrac{2}{3}$.

∎

Example 3   Evaluate each of the following.

(a) $\log_3 81$ (b) $\log_8 \dfrac{1}{64}$ (c) $\log_{16} 8$ (d) $\log_2(-2)$

Solution   (a) When we want to find $\sqrt[3]{243}$, we restate the question using the inverse operation: To find $\sqrt[3]{243}$ we ask, What number raised to the third power yields 243? In the same way, when we want to find $\log_3 81$, we may restate this as, What exponent of 3 will yield an answer of 81? Once we have restated $\log_3 81$ in this way, we recognize that the answer is 4, because $3^4 = 81$. However, the answer is not always quite so obvious, so we offer a more formal approach, which hinges on an idea we discussed in the last section. We let $t = \log_3 81$ and then rewrite this equation in exponential form, obtaining $3^t = 81$. Now if we can express both sides in terms of the same base, we can solve the resulting exponential equation, as follows:

Let $t = \log_3 81$ — Translate into exponential form.

$$3^t = 81$$ — Express both sides in terms of the same base.

$$3^t = 3^4$$ — Since the exponential function is one-to-one

$$t = 4$$

Therefore, $\log_3 81 = 4$.

(b) We apply the same procedure as in part (a). Let

$$t = \log_8 \frac{1}{64}$$ — Rewrite in exponential form.

$$8^t = \frac{1}{64}$$ — Express both sides in terms of the same base.

$$8^t = \frac{1}{8^2} = 8^{-2}$$ — Therefore,

$$t = -2$$

Hence, $\log_8 \dfrac{1}{64} = -2$.

(c) Let $t = \log_{16} 8$

$$16^t = 8$$ — We express both sides in terms of the base 2.

$$(2^4)^t = 2^3$$

$$2^{4t} = 2^3$$ — Therefore,

$$4t = 3 \quad \text{and so} \quad t = \frac{3}{4}$$

Therefore, $\log_{16} 8 = \dfrac{3}{4}$.

Remember that because an exponential function and its corresponding logarithmic function are inverses, the domain of the logarithmic function is the same as the range of the exponential function.

(d) Let $t = \log_2(-2)$, which means

$$2^t = -2$$

But no value of t can make 2^t negative; $2^t > 0$ for all values of t. Hence, $\log_2(-2)$ is undefined. ∎

The result of Example 3(d) illustrates that, because we are restricting the base of the exponential and logarithmic functions to positive numbers, *the log of a nonpositive number is undefined.*

Example 4 | Find $\log_5 5^7$.

Solution | Again, let

$$t = \log_5 5^7$$ — We rewrite this in exponential form.

$$5^t = 5^7 \Rightarrow t = 7$$

Therefore, $\log_5 5^7 = 7$. Alternatively, if we simply understand that $\log_5 5^7$ means "the exponent to which 5 must be raised to yield 5^7," then the answer of 7 should be immediately apparent. ∎

As was pointed out at the beginning of this section, logarithm notation was invented to express the inverse of the exponential function. Thus $\log_b x$ is a function of x. We usually write $f(x) = \log_b x$ rather than writing $f(x) = \log_b (x)$ and use parentheses only when needed to clarify the input to the log function. In this context, the word *argument* is often used in place of the word *input*. For example, for the function $f(x) = \log_5 (4 - x)$, the parentheses indicate that $4 - x$ is the argument of the logarithmic function. Thus we may say that the argument of a logarithmic function

If $f(x) = \log_5(4 - x)$, then $f(a) = \log_5(4 - a)$ and a is the input to $f(x)$; $4 - a$ is the argument of the logarithmic function.

must be positive.

For example,

If $f(x) = \log_5(4 - x)$, then $f(-1) = \log_5[4 - (-1)] = \log_5 5 = 1$, whereas if $f(x) = 4 - \log_5 x$, then $f(-1) = 4 - \log_5(-1)$, which is undefined.

Example 5   Given $f(x) = \log_5 x$, find

(a) $f(25)$ (b) $f\left(\dfrac{1}{25}\right)$ (c) $f(0)$ (d) $f(-125)$

Solution   Because $f(x) = \log_5 x$,

(a) $f(25) = \log_5 25 = \boxed{2}$ (since $5^2 = 25$)

(b) $f\left(\dfrac{1}{25}\right) = \log_5\left(\dfrac{1}{25}\right) = \boxed{-2}$ $\left(\text{since } 5^{-2} = \dfrac{1}{25}\right)$

(c) $f(0) = \log_5 0$ is not defined. (What power of 5 will yield 0?) We say that 0 is not in the domain of $f(x)$.

(d) $f(-125) = \log_5(-125)$ is not defined. (What power of 5 will yield -125?) We say that -125 is not in the domain of $f(x)$. ∎

Explain in words why $\log_b b^n = n$.

The following example illustrates several *logarithmic equations*. As in the last two examples, we will often find it helpful to rewrite a logarithmic equation in exponential form.

Example 6   Solve each of the following equations for t.

(a) $\log_5 t = 4$ (b) $\log_8 \dfrac{1}{2} = t$ (c) $\log_t 216 = 3$

Solution   (a) We rewrite $\log_5 t = 4$ in exponential form to obtain $t = 5^4 = \boxed{625}$.

(b) Again we rewrite the given logarithmic equation in exponential form:

$$\log_8 \frac{1}{2} = t \;\Rightarrow\; 8^t = \frac{1}{2}$$
$$(2^3)^t = 2^{-1}$$
$$3t = -1 \;\Rightarrow\; \boxed{t = -\frac{1}{3}}$$

(c) $\log_t 216 = 3 \;\Rightarrow\; t^3 = 216 \;\Rightarrow\; t = \sqrt[3]{216} \;\Rightarrow\; \boxed{t = 6}$. ∎

Since $b^1 = b$ and $b^0 = 1$, we make note of the following useful logarithmic statements (which are true for all $b > 0$).

$$\log_b b = 1 \quad \text{and} \quad \log_b 1 = 0$$

Acknowledging that the logarithmic and exponential functions are inverses, we can derive a great deal of information about the logarithmic function and its graph from the exponential function and its graph.

Example 7   Sketch the graph of each of the following equations. Find the domain and range of each.   **(a)** $y = \log_3 x$ **(b)** $y = \log_{1/2} x$

Solution   **(a)** Because $y = \log_3 x$ is the inverse of $y = 3^x$, we can obtain the graph of $y = \log_3 x$ by reflecting the graph of $y = 3^x$ about the line $y = x$. The graph appears in Figure 6.12(a).

Figure 6.12

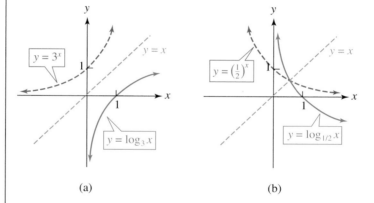

(a)   (b)

(b) To get the graph of $y = \log_{1/2} x$, we relect the graph of $y = \left(\frac{1}{2}\right)^x$ about the line $y = x$. The graph appears in Figure 6.12(b).

We take note of the features of these two graphs, which are the result of the logarithmic function being the inverse of the exponential function.

1. Since the graphs of $y = b^x$ all pass through the point $(0, 1)$, the graphs of $y = \log_b x$ all pass through $(1, 0)$.

2. When the graph of an exponential function has the negative x-axis as a horizontal asymptote, the graph of its inverse logarithmic function has the negative y-axis as a vertical asymptote. See Figure 6.12(a).
   When the graph of an exponential function has the positive x-axis as a horizontal asymptote, the graph of its inverse logarithmic function has the positive y-axis as a vertical asymptote. See Figure 6.12(b).

3. From its graph we can see that the domain of $y = \log_3 x$ is the set of positive real numbers and its range is the set of all real numbers (which are, respectively, the range and domain of the exponential function). A similar statement is true for $y = \log_{1/2} x$.

These statements agree with the fact that inverse functions interchange their domain and range. ∎

The following box summarizes some important information about the logarithmic function.

The Logarithmic Function $y = \log_b x$

Notice that the domain of $f(x) = \log_b x$ is $(0, \infty)$ which agrees with our earlier comment that the logarithm of a nonpositive number is undefined.

1. The domain of the logarithmic function is the set of *positive* real numbers.
2. The range of the logarithmic function is the set of all real numbers.
3. The graph of $y = \log_b x$ exhibits *logarithmic growth* if $b > 1$ or *logarithmic decay* if $0 < b < 1$.
4. The x-intercept is 1. There is no y-intercept.
5. The y-axis is a vertical asymptote.

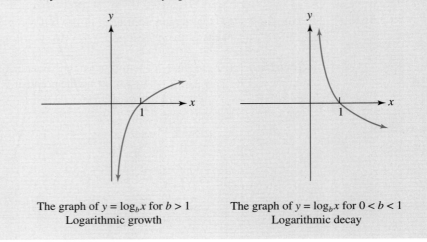

The graph of $y = \log_b x$ for $b > 1$
Logarithmic growth

The graph of $y = \log_b x$ for $0 < b < 1$
Logarithmic decay

These logarithmic graphs should be added to our catalog of basic graphs. Figure 6.13 illustrates the graphs of the logarithmic functions with bases 2, 3, 5, and 10. As usual we can apply the various graphing principles to the basic logarithm graph.

Figure 6.13

The graphs of various logarithmic functions

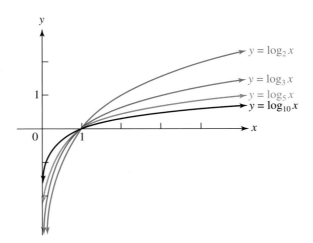

$y = \log_2 x$
$y = \log_3 x$
$y = \log_5 x$
$y = \log_{10} x$

Example 8 | Sketch the graph of $f(x) = 1 + \log_3(x-2)$. Find the domain, range, asymptote, and intercepts.

Solution We can obtain the graph of $y = f(x) = 1 + \log_3(x-2)$ by applying the graphing principles to shift the basic logarithmic growth graph 2 units to the right and 1 unit up. The graph appears in Figure 6.14.

Figure 6.14

The graph of
$f(x) = 1 + \log_3(x - 2)$

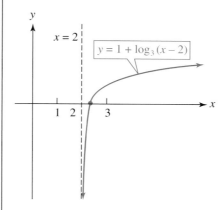

Alternatively, we know that the argument of the logarithm function must be positive. So we can find the domain by solving the inequality $x - 2 > 0$

We can see that the domain of the function is the set of all real numbers greater than 2, the range is the set of all real numbers, and the graph has the line $x = 2$ as a vertical asymptote.

To find the x-intercept, we set $y = 0$ and solve for x:

$$y = 1 + \log_3(x - 2) \qquad \text{Set } y = 0.$$
$$0 = 1 + \log_3(x - 2)$$
$$-1 = \log_3(x - 2) \qquad \text{Translate into exponential form.}$$
$$3^{-1} = x - 2 \quad \Rightarrow \quad \frac{1}{3} = x - 2 \quad \Rightarrow \quad x = 2\frac{1}{3}$$

Thus the x-intercept is at $2\frac{1}{3}$. ∎

We have been using the terminology of *exponential growth* and *logarithmic growth*. Let's take a moment to get a feeling for what these types of growth are like. While we are at it, we might as well include an example of polynomial growth as well.

One of the characteristics of exponential growth is that small changes in x can cause large changes in y. For example, for the function $y = f(x) = 2^x$, we have $f(9) = 512$ and $f(10) = 1024$. Thus a 1-unit change in x causes a 512-unit change in y. On the other hand, if we consider the function $y = \log_2 x$, we have $g(512) = 9$ and $g(1024) = 10$, so that a 512-unit change in x causes only a 1-unit change in y.

It is instructive to choose some values of x and compute the values of $y = 2^x$, $y = \log_2 x$, and $y = x^2$ for a comparison of how these three functions grow. Consider the following table:

x	$y = 2^x$	$y = x^2$	$y = \log_2 x$
1	2	1	0
2	4	4	1
4	16	16	2
8	256	64	3
16	65,536	256	4
32	4,294,967,296	1024	5

Based upon this table of values, not only can we see numerically that the values of the exponential function dwarf the values of the polynomial function, which are in turn much greater than the values of the logarithm function, but we can also see how quickly the exponential function grows and how slowly the logarithm function grows.

See Exercise 67 for a further discussion of the relative growth of these three functions.

For example, as x changes from 16 to 32 the exponential function changes by over 4 billion while the polynomial function changes by 768 and the logarithm function changes by only 1 unit!

Example 9 | Use the following graph of $y = a^x$ to determine **(a)** a^3 and   **(b)** $\log_a 3$.

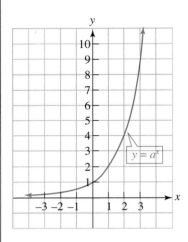

Solution | **(a)** We can exhibit a^3 by recognizing that it is the y-coordinate on the graph of $y = a^x$ corresponding to $x = 3$. See the point labeled P in Figure 6.15.

Figure 6.15

The graph of $y = a^x$

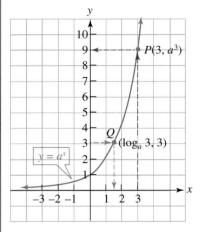

From the graph we can see that the y-coordinate of P is 9 and therefore $a^3 = 9$.

See Example 6 in Section 4.6 for a review of the idea of using the graph of $f(x)$ to find a value of $f^{-1}(x)$.

(b) Similarly, recognizing that $y = a^x$ means the same thing as $x = \log_a y$, then $\log_a 3$ is the x-coordinate on the graph of $y = a^x$ corresponding to $y = 3$. See the point labeled Q in Figure 6.15. From the graph we can see that the x-coordinate of Q appears to be 1.5 and therefore $\log_a 3 = 1.5$. ∎

Example 10 Find the inverse function for
(a) $y = f(x) = 3^x + 4$ **(b)** $y = g(x) = \log_3(x - 2)$

Solution Following the procedure for finding an inverse function, we interchange the x and y variables, and then solve explicitly for y by first isolating 3^y.

(a)

$$y = 3^x + 4 \qquad \text{Interchange } x \text{ and } y.$$

$$x = 3^y + 4 \qquad \text{Solve explicitly for } y.$$

$$x - 4 = 3^y \qquad \text{Write in logarithmic form.}$$

$$\log_3(x - 4) = y \qquad \text{Thus } f^{-1}(x) = \log_3(x - 4).$$

(b)

$$y = \log_3(x - 2) \qquad \text{Interchange } x \text{ and } y.$$

$$x = \log_3(y - 2) \qquad \text{Solve explicitly for } y; \text{ write in exponential form.}$$

$$y - 2 = 3^x$$

$$y = 3^x + 2 \qquad \text{Thus } g^{-1}(x) = 3^x + 2.$$

It is left to the reader to check these answers by composing each function with its inverse and verifying that the result is x. ∎

Once we clearly understand that a logarithm is simply an exponent and that the logarithmic function and exponential functions are inverses of each other, we can derive some very useful relationships.

Let $f(x) = \log_b x$ and $g(x) = b^x$. Then f and g are inverse functions, and by the definition of inverse functions $f[g(x)] = x$ and $g[f(x)] = x$. In other words,

$$f[g(x)] = f(b^x) = \log_b b^x = x$$

$$g[f(x)] = g(\log_b x) = b^{\log_b x} = x$$

These basic logarithmic relationships are very important for much of our future work and so we highlight them in the following box.

Basic Logarithmic Relationships
1. $\log_b b^x = x$ **2.** $b^{\log_b x} = x$

Although these two relationships may seem a bit obscure, they become more understandable if we analyze verbally what they are saying.

For the first, $\log_b b^x$ means the exponent to which b must be raised to yield b^x and so the answer is x. In other words, $\log_b b^x$ is asking $b^? = b^x$ and therefore, the answer is x.

For the second, $\log_b x$ means the exponent to which b must be raised to yield x. Therefore, if b is raised to this power the answer must be x. In other words, $b^{\log_b x} = x$.

Example 11 Simplify: **(a)** $3^{\log_3 7}$ **(b)** $\log_4(\log_2 16)$

Solution **(a)** Even though we do not know the value of $\log_3 7$, we can use the second basic logarithmic relationship to get $3^{\log_3 7} = 7$

Think about why $3^{\log_3 9} = 9$ and why $3^{\log_3 7} = 7$.

(b) $\log_4(\log_2 16) = \log_4 4 \qquad \text{Because } 2^4 = 16, \text{ we have } \log_2 16 = 4.$

$$= 1 \qquad \text{Because } 4^1 = 4$$

Thus $\log_4(\log_2 16) = \boxed{1}$. ∎

In the next section we develop additional properties of the logarithmic function, which will allow us to apply logarithms to a variety of situations.

6.2 Exercises

In Exercises 1–20, translate the given logarithmic statements into exponential form and the given exponential statements into logarithmic form.

1. $\log_7 49 = 2$

2. $10^4 = 10{,}000$

3. $3^{-4} = \dfrac{1}{81}$

4. $\log_9 3 = \dfrac{1}{2}$

5. $\log_{1/4} 64 = -3$

6. $\log_8 4 = \dfrac{2}{3}$

7. $27^{-1/3} = \dfrac{1}{3}$

8. $\left(\dfrac{1}{5}\right)^{-2} = 25$

9. $\log_8 \dfrac{1}{2} = -\dfrac{1}{3}$

10. $7^0 = 1$

11. $5^{1/4} = \sqrt[4]{5}$

12. $\log_7 7 = 1$

13. $\log_2 \dfrac{1}{2} = -1$

14. $11^{1/2} = \sqrt{11}$

15. $\log_{2/3} \dfrac{27}{8} = -3$

16. $\log_{16} 8 = \dfrac{3}{4}$

17. $4^5 = 1024$

18. $\log_8 1 = 0$

19. $\left(\dfrac{1}{9}\right)^{-1/2} = 3$

20. $\log_4 \dfrac{1}{2} = -\dfrac{1}{2}$

In Exercises 21–46, evaluate the given logarithmic expression (where it is defined).

21. $\log_2 32$

22. $\log_{10} 1000$

23. $\log_6 \dfrac{1}{6}$

24. $\log_{1/3} 9$

25. $\log_8 16$

26. $\log_{16} 8$

27. $\log_{16} \dfrac{1}{8}$

28. $\log_8 \dfrac{1}{16}$

29. $\log_3(-9)$

30. $\log_9 \left(\dfrac{1}{27}\right)$

31. $\log_4 \dfrac{1}{8}$

32. $\log_5(-25)$

33. $\log_{10} 0.0001$

34. $\log_{1/2} 8$

35. $\log_5 \sqrt[3]{25}$

36. $\log_4 \sqrt{128}$

37. $\log_5(\log_3 243)$

38. $\log_6(\log_7 7)$

39. $\log_2(\log_9 3)$

40. $\log_3(\log_8 2)$

41. $4^{\log_4 7}$

42. $\log_8 8^{0.3}$

43. $\log_6 \dfrac{1}{\sqrt{6}}$

44. $2^{\log_2 \sqrt{5}}$

45. $\log_b b^6$

46. $\log_b b^{2/3}$

In Exercises 47–52, solve the given equation for t.

47. $\log_3 t = 4$

48. $\log_{36} \dfrac{1}{6} = t$

49. $\log_t \dfrac{1}{8} = -3$

50. $\log_5 t = 0$

51. $\log_{32} 16 = t$

52. $\log_t 1 = 0$

53. Given $F(x) = \log_2(x^2 - 4)$, find $F(6)$ and the domain of $F(x)$.

54. Given $g(x) = \log_3(x^2 - 4x + 3)$, find $g(4)$ and the domain of $g(x)$.

In Exercises 55–62, sketch the graph of the given function and identify the domain, range, intercepts, and asymptotes.

55. $y = f(x) = \log_2(x - 3)$

56. $y = f(x) = -3 + \log_2 x$

57. $y = f(x) = \log_5(x + 4)$

58. $y = f(x) = 4 + \log_5 x$

59. $y = f(x) = -\log_3(-x)$

60. $y = f(x) = 3\log_5 x$

61. $y = f(x) = \log_3 |x|$

62. $y = f(x) = 2 + \log_3(x - 1)$

In Exercises 63–66, find the function of the form $y = \log_b x$ whose graph is given.

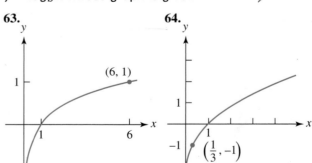

63.

64.

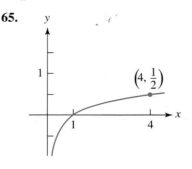

65.

66.

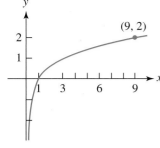

Questions for Thought

67. In this section we discussed the fact that exponential growth is much more rapid than polynomial growth. Use a graphing calculator to graph $y = 2^x$ and $y = x^2$ on the same screen. Is $y = 2^x$ always greater than $y = x^2$? If not, for what values is $y = x^2$ greater than $y = 2^x$? HINT: Determine whether $x^2 = 2^x$ has any solutions.

68. Use the meaning of $\log_b x$ to answer each question.
 (a) Which is larger, $\log_2 35$ or $\log_2 41$? Explain.
 (b) Which is larger, $\log_2 35$ or $\log_3 35$? Explain.
 (c) Which is larger, $\log_4 60$ or $\log_3 40$? Explain.

69. Sketch the graphs of $y = f(x) = -\log_4 x$ and $y = g(x) = \log_{1/4} x$ on the same coordinate system. What do you notice?

70. Show that $\log_{1/6} x = -\log_6 x$. HINT: Write both sides in exponential form.

71. Show that $\log_{1/6} x = \log_6 \dfrac{1}{x}$. HINT: Write both sides in exponential form.

72. State and prove a generalization of the result of Exercise 69.

73. Suppose $\log_b u = m$ and $\log_b v = n$. Express
 (a) $\log_b(uv)$ in terms of m and n.
 (b) $\log_b \dfrac{u}{v}$ in terms of m and n.
 (c) $\log_b(u^r)$ in terms of m and n, where r is a real number.

6.3 Properties of Logarithms; Logarithmic Equations

Historically, logarithms were first developed to simplify long and complex numerical computations. Inexpensive and powerful hand-held calculators have made the use of logarithms for computing virtually obsolete. Nevertheless, the logarithmic function, by virtue of the logarithmic properties we develop in this section, retains an important role in mathematics.

We have noted on many occasions that, in general, $f(s + t) \neq f(s) + f(t)$. This fact applies to the logarithmic functions as well: $\log_b(s + t) \neq \log_b s + \log_b t$. If we encounter a function for which $f(s + t) = f(s) + f(t)$, we would consider this a "special property" exhibited by the function, which other functions, in general, do not have. In the last section we repeatedly emphasized that a logarithm is merely an exponent, and so it should come as no surprise that the logarithmic function does have some special properites that it "inherits" from the properties of exponents. The proofs of the properties listed here highlight the intimate connection between the properties of logarithms and the basic rules for exponents.

The following box contains the basic properties of logarithms. Properties 4 and 5 were derived in the last section as direct consequences of the definitions. We include them here for ease of reference.

Properties of Logarithms

The following properties assume b, u, and v are positive ($b \neq 1$).

1. $\log_b(uv) = \log_b u + \log_b v$

In words, this says the log of a product is equal to the sum of the logs of the factors.

2. $\log_b\left(\dfrac{u}{v}\right) = \log_b u - \log_b v$

In words, this says the log of a quotient is the log of the numerator minus the log of the denominator.

3. $\log_b(u^r) = r \log_b u$

In words, this says the log of a power is the exponent times the log.

4. $\log_b b^x = x$

5. $b^{\log_b x} = x$

We prove these properties by using the fact that

$$x = b^y \iff \log_b x = y$$

Proof of Property 1   Let $u = b^m$ and $v = b^n$, and write these exponential statements in logarithmic form.

$$u = b^m \iff \log_b u = m \qquad (1)$$
$$v = b^n \iff \log_b v = n \qquad (2)$$

How is property 1 of logarithms related to the property of exponents that states $b^m b^n = b^{m+n}$?

Thus

$$uv = b^m b^n = b^{m+n} \qquad \text{Rewriting this in logarithmic form, we get}$$

$$uv = b^{m+n} \iff \log_b(uv) = m + n \qquad \text{Using equations (1) and (2), we substitute for } m \text{ and } n \text{ to get}$$

$$\log_b(uv) = \log_b u + \log_b v \qquad \text{As required} \quad \blacksquare$$

How is property 2 of logarithms related to the property of exponents that states $\dfrac{b^m}{b^n} = b^{m-n}$?

Property 1 for logarithms is, in fact, nothing more than a logarithmic statement of a property of exponents. The proof of property 2 is quite similar and is left as an exercise for the student.

How is property 3 of logarithms related to the property of exponents that states $(b^m)^n = b^{nm}$?

Proof of Property 3   Again we let $u = b^m$ and write this exponential statement in logarithmic form.

$$u = b^m \iff \log_b u = m \qquad (1)$$

Thus

$$u^r = (b^m)^r = b^{rm} \qquad \text{Rewriting this in logarithmic form, we get}$$

$$u^r = b^{rm} \iff \log_b(u^r) = rm \qquad \text{Using equation (1), we substitute for } m \text{ to get}$$

$$\log_b(u^r) = r \log_b u \qquad \text{As required} \quad \blacksquare$$

Again property 3 for logarithms is merely the logarithmic form of the property of exponents, which says that $(b^m)^n = b^{n \cdot m}$.

As the following examples illustrate, these logarithmic properties allow us to rewrite fairly complex logarithmic expressions in terms of simpler logarithmic expressions. This process is especially useful in applications (Section 6.5) and in calculus.

Example 1   Express in terms of simpler logarithms:
   (a) $\log_b(x^3 y)$ **(b)** $\log_b(x^3 + y)$

Solution   **(a)** $\log_b(x^3 y)$ Using log property 1, we get

$$= \log_b x^3 + \log_b y \qquad \text{Using log property 3, we get}$$

$$= \boxed{3 \log_b x + \log_b y}$$

Note the difference between parts (a) and (b) of Example 1.

(b) Examining the three properties of logarithms, we see that they deal with the log of a product, quotient, and power. Thus $\log_b(x^3 + y)$, which is the log of a *sum*, *cannot* be simplified using the log properties. ∎

Example 2   Express in terms of simpler logarithms: $\log_b\left(\dfrac{\sqrt{xy}}{z^3}\right)$

Solution   Since a logarithm is an exponent, it is usually easier to simplify a logarithmic expression if we rewrite radical expressions in exponential form.

$$\log_b\left(\frac{\sqrt{xy}}{z^3}\right) = \log_b\left(\frac{(xy)^{1/2}}{z^3}\right) \qquad \text{Use property 2 on the quotient.}$$

$$= \log_b(xy)^{1/2} - \log_b z^3 \qquad \text{Use property 3 on the powers.}$$

$$= \frac{1}{2}\log_b(xy) - 3\log_b z \qquad \text{Use property 1 on the product.}$$

$$= \boxed{\frac{1}{2}\left(\log_b x + \log_b y\right) - 3\log_b z}$$

∎

Example 3   Show that $\log_b \dfrac{1}{2} = -\log_b 2$.

Solution   We offer two approaches.

Appoach 1:   **Approach 2:**

$\log_b \dfrac{1}{2}$ Use log property 2 to get $\log_b \dfrac{1}{2} = \log_b 2^{-1}$ Now use log property 3.

$= \log_b 1 - \log_b 2$ But $\log_b 1 = 0$

$= \boxed{-\log_b 2}$ $= \boxed{-\log_b 2}$ ∎

The previous example can be generalized to the following useful result.

$$\boxed{\log_b \frac{1}{x} = -\log_b x}$$

It is also important to recognize what the log properties *do not* say.

Correct	Incorrect
$\log_b(uv) = \log_b u + \log_b v$	$\log_b(u + v) = \log_b u + \log_b v$
$\log_b\left(\dfrac{u}{v}\right) = \log_b u - \log_b v$	$\dfrac{\log_b u}{\log_b v} = \log_b u - \log_b v$
$\log_b(u^r) = r\log_b u$	$(\log_b u)^r = r\log_b u$

Thus

$$\log_b(x^2 + y^3) \neq 2 \log_b x + 3 \log_b y$$

and

$$\frac{\log_b x^3}{\log_b y^4} \neq 3 \log_b x - 4 \log_b y$$

The properties of logarithms can also be used to rewrite an expression involving several logarithmic expressions as a single logarithm. This process is useful in solving logarithmic equations.

Example 4   Express as a single logarithm: $\log_b x + 3 \log_b y - \dfrac{1}{2} \log_b z$

Solution   We use the log properties in reverse.

$$\log_b x + \underbrace{3 \log_b y} \quad - \quad \underbrace{\tfrac{1}{2} \log_b z} \qquad \text{Use log property 3.}$$

$$= \quad \underbrace{\log_b x + \log_b y^3} \quad - \log_b z^{1/2} \qquad \text{Use log property 1.}$$

$$= \qquad \log_b(xy^3) \qquad\quad - \log_b z^{1/2} \qquad \text{Use log property 2.}$$

$$= \quad \log_b\!\left(\frac{xy^3}{z^{1/2}}\right)$$

The logarithm properties can be applied to solving logarithmic equations, as illustrated in the next three examples.

Example 5   Solve for x: $\log_6 x - \log_6 3 = 2$

Solution   In the last section we solved a simple logarithmic equation involving a single logarithm by translating it into exponential form. Thus our first step is to use the log properties to rewrite the left-hand side of the given equation as a single logarithm.

$$\log_6 x - \log_6 3 = 2 \qquad \text{Use log property 2.}$$

$$\log_6 \frac{x}{3} = 2 \qquad \text{Rewrite in exponential form.}$$

$$\frac{x}{3} = 6^2 = 36$$

$$x = 108$$

Check $x = 108$:

$$\log_6 108 - \log_6 3 = \log_6\!\left(\frac{108}{3}\right) = \log_6 36 \stackrel{\checkmark}{=} 2$$

Note that even though we do not know the exact values of $\log_6 108$ and $\log_6 3$, the log properties allow us to compute their difference exactly.

Example 6 | Solve for x: $\log_2 x + \log_2(x + 2) = 3$

Solution | Again we begin by rewriting the equation so that it involves a single logarithm:

$$\log_2 x + \log_2(x + 2) = 3 \qquad \text{Use log property 1.}$$
$$\log_2[x(x + 2)] = 3 \qquad \text{Rewrite in exponential form.}$$
$$x(x + 2) = 2^3$$
$$x^2 + 2x - 8 = 0$$
$$(x + 4)(x - 2) = 0$$
$$x = -4 \quad \text{or} \quad x = 2$$

Be careful! $x = -4$ does satisfy $\log_2[x(x + 2)] = 3$ (it makes the argument $-4(-2) = 8$), but it does not satisfy the original equation because $\log_2 x$ and $\log_2(x + 2)$ are undefined for $x = -4$. *Remember that the argument of a logarithmic function cannot be negative.* Thus the only solution to the given equation is $x = 2$.

Check $x = 2$:

$$\log_2 x + \log_2(x + 2) = 3 \qquad \text{Substitute } x = 2.$$
$$\log_2 2 + \log_2 4 \overset{?}{=} 3$$
$$1 + 2 \overset{\checkmark}{=} 3$$

We have already made use of the fact that the exponential function is one-to-one. Of course, since the logarithmic function is the inverse of the exponential function, it too is one-to-one. We can also use this fact in solving logarithmic equations.

Example 7 | Solve for x: $\log_b(8 - x) - \log_b(2 - x) = \log_b 3$

Solution | $$\log_b(8 - x) - \log_b(2 - x) = \log_b 3 \qquad \text{Use log property 2.}$$

The log function being one-to-one means that $\log_b U = \log_b V \implies U = V$.

$$\log_b\left(\frac{8 - x}{2 - x}\right) = \log_b 3 \qquad \begin{array}{l}\text{Since the log function}\\\text{is one-to-one, we have}\end{array}$$
$$\frac{8 - x}{2 - x} = 3$$
$$8 - x = 6 - 3x \implies 2x = -2 \implies x = -1$$

Note that even though $x = -1$ is a negative number, it is a perfectly valid solution to the original equation. When $x = -1$, both

$$\log_b(8 - x) = \log_b(8 - (-1)) \qquad \text{and} \qquad \log_b(2 - x) = \log_b(2 - (-1))$$
$$= \log_b 9 \qquad\qquad\qquad\qquad\qquad\qquad = \log_b 3$$

are well defined. It is left to the student to complete the check.

Example 8 | Evaluate $2^{3 \log_2 4}$.

Solution | We know that $b^{\log_b x} = x$. Unfortunately, the expression $2^{3 \log_2 4}$ is not quite in that form. However, we can use log property 3 to put the expression into the form $b^{\log_b x}$ as follows:

$$2^{3 \log_2 4} \qquad \text{Use property 3 to rewrite } 3 \log_2 4 \text{ as } \log_2 4^3.$$
$$= 2^{\log_2 4^3} \qquad \text{Since } b^{\log_b x} = x \text{ we get}$$
$$= 4^3 = \boxed{64}$$

As a final comment, we want to point out that the log properties apply only when both sides make sense. That is, $\log_b x^2$ is defined for all $x \neq 0$, whereas $2 \log_b x$ is defined for $x > 0$. Therefore, $\log_b x^2 = 2 \log_b x$ is true only for those values for which both sides are defined—that is, for $x > 0$. Another way of saying the same thing is that the domain of $f(x) = \log_b x^2$ is all real numbers not equal to zero, whereas the domain of $g(x) = 2 \log_b x$ is all positive real numbers.

6.3 Exercises

In Exercises 1–14, use the properties of logarithms to write the given expression in terms of simpler logarithms. Express the answer so that it does not contain the logarithm of products, quotients, or powers. If the expression cannot be simplified, say so.

1. $\log_4(x^2 y z^3)$

2. $\log_2\left(\dfrac{5a}{b^2}\right)$

3. $\log_b\left(\dfrac{x^3}{yz^4}\right)$

4. $\log_b b^5$

5. $\log_5 \sqrt{5x^3}$

6. $\log_t \sqrt[3]{\dfrac{4}{t}}$

7. $\log_b(x^3 + y^2 - z^5)$

8. $\log_2\left(\dfrac{4x}{\sqrt{y^3}}\right)$

9. $\log_b(x^2 - 4)$

10. $\log_b \sqrt{x^2 - x - 6}$

11. $\log_4 \sqrt[3]{\dfrac{x-3}{2x^2}}$

12. $\log_9 \sqrt{\dfrac{x}{3}}$

13. $\log_b \sqrt{\dfrac{x^2 - 16}{x^2 - 2x - 8}}$

14. $\log_3 \sqrt{\dfrac{27r^5 s}{t^3}}$

In Exercises 15–18, determine whether the given statement is true or false.

15. $\log_b(x^3 - y^4) = 3 \log_b x - 4 \log_b y$

16. $\dfrac{\log_b x^3}{\log_b y^4} = 3 \log_b x - 4 \log_b y$

17. $\log_b\left(\dfrac{x^3}{y^4}\right) = 3 \log_b x - 4 \log_b y$

18. $(\log_b x)^3 = 3 \log_b x$

In Exercises 19–30, write the given expression as a single logarithm with a coefficient of 1 and simplify as completely as possible.

19. $4 \log_b 2 + \log_b 3$

20. $2 \log_b x + \log_b y - 3 \log_b z$

21. $\dfrac{1}{2} \log_b s - \dfrac{3}{2} \log_b t$

22. $5 \log_b 3 - 2 \log_b 4$

23. $\log_{10} 50 - \log_{10} 5$

24. $\log_8 80 - \log_8 5$

25. $\log_6 9 + \log_6 24$

26. $\log_{10} 200 + \log_{10} 5$

27. $3 \log_b x - 4 \log_b y - 2 \log_b z$

28. $\dfrac{1}{3} \log_b (x + 1) - \dfrac{1}{3} \log_b(x + 2)$

29. $\dfrac{1}{4} \log_b x + \dfrac{1}{3} \log_b y - \dfrac{1}{2} \log_b z$

30. $\log_b \dfrac{x}{4} + \log_b \dfrac{y}{3} - \log_b \dfrac{z}{2}$

In Exercises 31–36, let $\log_{10} 2 = A$, $\log_{10} 3 = B$, and $\log_{10} 5 = C$. Verify each of the following statements.

31. $\log_{10} 16 = 4A$

32. $\log_{10} 18 = A + 2B$

33. $\log_{10} \sqrt{6} = \dfrac{1}{2}(A + B)$

34. $\log_{10} \dfrac{1}{20} = -A - 1$

35. $\log_{10} 300 = B + 2$

36. $\log_{10} \sqrt[3]{24} = A + \dfrac{1}{3} B$

In Exercises 37–40, use $\log_b 2 = 0.69$, $\log_b 3 = 1.10$, and $\log_b 5 = 1.61$ to evaluate the given logarithm.

37. $\log_b 12$

38. $\log_b 25$

39. $\log_b \dfrac{1}{18}$

40. $\log_b 60$

In Exercises 41–54, solve each logarithmic equation.

41. $\log_3 4 + \log_3 x = 2$

42. $\log_2 x = 4 + \log_2 3$

43. $2 \log_5 x = \log_5 49$

44. $\dfrac{1}{2} \log_4 x = \log_4 9$

45. $\log_2 t - \log_2(t - 2) = 3$

46. $\log_6 2 - \log_6 x = 1$

47. $\log_{10} 8 + \log_{10} x + \log_{10} x^2 = 3$

48. $\log_4 x + \log_4 (x - 6) = 2$

49. $\log_9 (2x + 7) - \log_9 (x - 1) = \log_9 (x - 7)$

50. $\log_2 6 + \log_2 x - \log_2 (x + 2) = 2$

51. $\log_4 x - \log_4 (x - 4) = \log_4 (x - 6)$

52. $\log_3 7 - \log_3 x - \log_3 (x - 2) = 2$

53. $\dfrac{1}{2} \log_3 x = \log_3 (x - 6)$

54. $\dfrac{1}{2} \log_2 (x + 1) = 2 + \dfrac{1}{2} \log_2 5$

55. Evaluate $3^{2 \log_3 5}$.

56. Evaluate $5^{3 \log_5 2}$.

Question for Thought

57. At the end of this section we explained why, despite the properties of logarithms, the functions $f(x) = \log_b x^2$ and $g(x) = 2 \log_b x$ are not the same. Sketch the graphs of these functions.

6.4 Common and Natural Logarithms; Exponential Equations and Change of Base

In Sections 6.1 and 6.2, we introduced the exponential and logarithmic functions without stressing the particular base chosen. However, there are two bases of special importance in science and mathematics.

Common Logarithms

We have already mentioned that, historically, logarithms were developed to simplify complex numerical computations. Since we do arithmetic in a base 10 number system, logarithms to the base 10, called **common logarithms**, play a significant role. In fact, $\log_{10} x$ is usually just written $\log x$. In other words, if a logarithmic function appears without a base, the base is assumed to be 10.

> **Common Logarithms**
> $f(x) = \log_{10} x$ is called the **common logarithmic function**. We write
> $$\log_{10} x = \log x$$

Example 1 Evaluate: **(a)** $\log 1000$ **(b)** $\log 0.01$

Solution **(a)** Let $a = \log 1000$ Write this in exponential form. Remember the base is 10.
$$10^a = 1000 = 10^3 \implies a = 3$$
(b) Let $t = \log 0.01$ Write this in exponential form.
$$10^t = 0.01 = 10^{-2} \implies t = -2 \quad\blacksquare$$

Natural Logarithms

The inverse of the natural exponential function is called the **natural logarithmic function** and has its own special notation.

> **Natural Logarithms**
> $f(x) = \log_e x$ is called the **natural logarithmic function**. We write
> $$\log_e x = \ln x$$

Example 2 | Find: **(a)** $\ln 1$ **(b)** $\ln e$

Solution | **(a)** Since $\ln 1 = \log_e 1$, we are asking for the exponent of e that gives 1, so the answer is 0. Therefore, $\ln 1 = 0$.

(b) Since $\ln e = \log_e e$, we are asking for the exponent of e that gives e; clearly the answer is 1. Therefore $\ln e = 1$.   ∎

Calculator Exploration

Graph the functions $y = e^x + 1$, $y = \ln(x - 1)$, and $y = x$ on the same set of coordinate axes. Keeping in mind that $\boxed{\text{LN}}$ on your calculator means $\log_e$, what do the three graphs suggest?

Example 3 | Find the inverse function of $y = f(x) = e^x + 1$, and sketch the graphs of $f(x)$ and $f^{-1}(x)$ on the same set of coordinate axes.

Solution | To find $f^{-1}(x)$, we start with $y = e^x + 1$, interchange x and y, and then solve for y:

$$y = e^x + 1 \qquad \text{Interchange } x \text{ and } y.$$
$$x = e^y + 1 \qquad \text{Solve for } y.$$
$$x - 1 = e^y \qquad \text{Rewrite in logarithmic form.}$$
$$\ln(x - 1) = y \qquad \text{Thus } f^{-1}(x) = \ln(x - 1).$$

The graph of $y = e^x + 1$ is obtained by shifting the graph of $y = e^x$ up 1 unit. Since e is greater than 1, the graph of $y = \ln x$ exhibits logarithmic growth; the graph of $y = \ln(x - 1)$ is obtained by shifting the graph of $y = \ln x$ to the right 1 unit. The graphs of $f(x)$ and $f^{-1}(x)$ are in Figure 6.16.

Figure 6.16

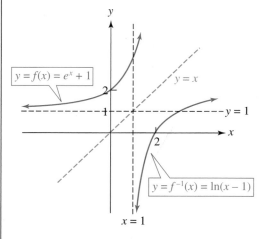

Since these are the graphs of inverse functions, they are symmetric about the line $y = x$. In fact, we could have gotten the graph of $y = \ln(x - 1)$ by reflecting the graph of $y = e^x + 1$ about the line $y = x$.   ∎

Exponential Equations

We are able to solve certain exponential equations such as $3^x = \dfrac{1}{9}$ by expressing each side as a power of the same base ($3^x = 3^{-2}$), thus obtaining the solution

$x = -2$. However, this approach does not work when solving an equation such as $3^{x+1} = 7$. How do we find, with some accuracy, the exponent x such that $3^{x+1} = 7$?

The definition of a function requires that each input x produces a unique output $f(x)$. Symbolically we say: If $a = b$, then $f(a) = f(b)$. Because the logarithm is a function, it follows that: If $U = V$, then $\log_b U = \log_b V$.

Thus if two expressions are equal, their logarithms are equal. We say that when we have an equation, we can "take the log of both sides." Once we have the equation in log form, we can often apply the log properties, which then allows us to solve the equation. In particular, log property 3 applied to an exponential equation allows us to *get the variable out of the exponent* and solve the equation more easily.

Example 4 Solve $3^{x+1} = 7$ for x. Round to four decimal places.

Solution Since we cannot easily express both sides in terms of the same base, we take the logarithm of both sides of the equation. Although we may use any base we choose, most calculators come equipped with a log key and a ln key, and so we will generally use either base 10 logarithms (that is, use log) or base e logarithms (that is, use ln) in solving problems.

$$3^{x+1} = 7 \qquad \text{Take ln of both sides.}$$

$$\ln 3^{x+1} = \ln 7 \qquad \text{Use log property 3.}$$

$$(x+1)\ln 3 = \ln 7 \qquad \text{Solve for } x.$$

$$x + 1 = \frac{\ln 7}{\ln 3}$$

$$x = \frac{\ln 7}{\ln 3} - 1 \qquad \text{This is the exact answer.}$$

Check this solution by using your calculator to substitute $x = 0.7712$ into the original equation.

Using a calculator, we can approximate this answer to be $x = 0.7712$. ∎

Example 5 Solve for x.
(a) $5^{x+2} = 3^{2x+1}$ (b) $5^{2x} = 25^{3x+1}$ (c) $e^x = 100$

Solution (a) We proceed as we did before.

$$5^{x+2} = 3^{2x+1} \qquad \text{Take the log of each side.}$$

$$\log(5^{x+2}) = \log(3^{2x+1}) \qquad \text{Apply log property 3.}$$

$$(x+2)\log 5 = (2x+1)\log 3 \qquad \text{Remember: log 5 and log 3 are numbers. Now we have a linear equation similar to } (x+2)7 = (2x+1)4. \text{ Multiply out both sides.}$$

$$x \log 5 + 2 \log 5 = x(2 \log 3) + \log 3 \qquad \text{Collect the } x \text{ terms on one side and factor out } x.$$

$$x \log 5 - x(2 \log 3) = \log 3 - 2 \log 5$$

$$x(\log 5 - 2 \log 3) = \log 3 - 2 \log 5$$

$$x = \frac{\log 3 - 2 \log 5}{\log 5 - 2 \log 3} \qquad \text{This is the exact answer.}$$

Using a calculator, this answer is approximately 3.6071991. Substituting this value into the original equation, the left-hand side becomes $5^{5.6071991}$, which a calculator computes as 8303.5533, whereas the right-hand side becomes $3^{8.2143982}$, which the calculator also computes as 8303.5533.

(b) $5^{2x} = 25^{3x+1}$

We could start by taking the log of each side of the equation, but in this equation we can get an exact rational answer by expressing each side in terms of the same base.

Parts (a) and (b) of Example 5 illustrate two approaches to solving exponential equations. How do you know which approach to take?

$$5^{2x} = (5^2)^{3x+1}$$
$$5^{2x} = 5^{6x+2}$$

Since the exponential function is one-to-one we have

$$2x = 6x + 2 \quad \Rightarrow \quad \boxed{x = -\frac{1}{2}}$$

(c) In Example 9 of Section 6.1 we solved the equation $e^x = 100$ using the graph of $y = e^x$. We can now solve this equation explicitly as follows:

$$e^x = 100$$ Rewrite this equation in logarithmic form.

$$x = \ln 100$$ Using a calculator we get

Rewriting the equation in logarithmic form is equivalent to taking the ln of both sides.

$$\boxed{x = 4.6052}$$ rounded to 4 decimal places.   ∎

Example 6   Solve for x: $e^{2x} - e^x = 6$

Solution   If we recognize that $e^{2x} = (e^x)^2$, then this is an equation in quadratic form. In other words, if we substitute $u = e^x$, then the given equation becomes

$$e^{2x} - e^x = 6$$
$$(e^x)^2 - e^x = 6$$ Substitute $u = e^x$.
$$u^2 - u = 6$$ This is a quadratic equation that we can solve by factoring.
$$u^2 - u - 6 = 0$$
$$(u - 3)(u + 2) = 0$$

This gives two solutions: $u = 3$ and $u = -2$. However, because $u = e^x$,

$$e^x = 3 \quad \text{or} \quad e^x = -2$$

The equation $e^x = -2$ has no solutions. Why? Solving $e^x = 3$ gives

$$\boxed{x = \ln 3}$$

The check is left to the student.   ∎

A few things should be kept in mind when solving exponential equations. First, if we need or are asked to find the *exact* answer to an exponential equation, we may need to leave the answer in terms of logs rather than using the calculator to approximate the answer (just as we would give $\sqrt{2}$ as an exact answer rather than $1.414214\ldots$).

Second, as we remarked in Section 6.1, exponential equations may not lend themselves to an *algebraic* solution. In Example 5, part (b) yielded to an algebraic solution whereas parts (a) and (c) did not (we had to use logarithms). Sometimes, however, the only method available to us is to find approximate solution(s) by using a graphing calculator or some other graphing utility, as illustrated in the next example.

Example 7   Find all positive solutions to the following equation (round the answer to three decimal places): $2^x = x + 4$

Solution   This is an exponential equation; however, the methods we illustrated in the previous examples will not work here. (Try them!) In such a case, we may use a graphing calculator to find approximate solutions to this equation.

As we discussed in Chapter 3, we may view a graphical solution to the equation $2^x = x + 4$ in two ways. We may graph the two equations $y = 2^x$ and $y = x + 4$ and find their point(s) of intersection, or we may graph the equation $y = 2^x - (x + 4)$ and find its zeros.

The fact that we know what the graphs of $y = 2^x$ and $y = x + 4$ look like allows us to conclude that there are exactly two points of intersection. See Figure 6.17(a). Since we are being asked to find the *positive* solutions(s) to the equation, we are looking for the first quadrant point of intersection. Figure 6.17(a) illustrates the point of intersection as found on a TI-83 Plus graphing calculator. Thus the solution is $x = 2.756$ rounded to three decimal places.

See Exercise 54 for a discussion of the care that must be employed when using a graphing calculator to solve equations by finding the point(s) of intersection of two graphs.

Figure 6.17(a)

The points of intersection of $y = 2^x$ and $y = x + 4$

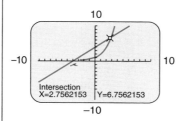

Figure 6.17(b) illustrates the graph of $y = 2^x - x - 4$ and the positive zero as found on a TI-83 Plus graphing calculator.

Figure 6.17(b)

The positive x-intercept of the graph of $y = 2^x - x - 4$

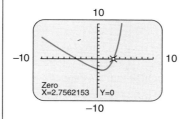

The calculator indicates that the x-intercept and hence the solution is again approximately 2.756.

We can check this solution by substituting $x = 2.756$ into both sides of the original equation. The calculator returns the following results, which confirm our approximate solution.

```
2^2.756
       6.755207099
2.756+4
          6.756
```

The method of taking the logarithm of both sides of an equation can be used to solve a broader range of exponential equations.

We are able to evaluate certain logarithmic expressions, such as $\log_5 25$, by recognizing that we are actually solving the exponential equation $5^x = 25$. However, if we want to evaluate $\log_5 2$, we run into difficulty. In the next example we utilize the technique we have developed thus far to deal with this problem, and in so doing we obtain a general formula for changing logarithms from one base to another.

Example 8   (a) Express $\log_5 2$ in terms of base 10 logarithms.
(b) Express $\log_5 2$ in terms of natural logarithms.
(c) Use these results to approximate $\log_5 2$ to three decimal places.

Solution

Can you find $\log_5 2$ directly on your calculator?

Let $x = \log_5 2$, which is equivalent to the exponential equation $5^x = 2$. We can now take the base 10 (or base e) logarithm of both sides of this equation.

(a)

$$5^x = 2 \qquad \text{Take the base 10 log of both sides.}$$

$$\log 5^x = \log 2 \qquad \text{Use log property 3.}$$

$$x \log 5 = \log 2$$

$$x = \frac{\log 2}{\log 5} \qquad \text{Thus}$$

$$\log_5 2 = \frac{\log 2}{\log 5}$$

(b) We now repeat this process with base e logarithms.

$$5^x = 2 \qquad \text{Take ln of both sides.}$$

$$\ln 5^x = \ln 2 \qquad \text{Use log property 3.}$$

$$x \ln 5 = \ln 2$$

$$x = \frac{\ln 2}{\ln 5} \qquad \text{Thus}$$

$$\log_5 2 = \frac{\ln 2}{\ln 5}$$

What we have done in parts (a) and (b) is convert an expression from base 5 logarithms to base 10 and base e logarithms.

(c) We first recognize that we cannot use a calculator to approximate $\log_5 2$ directly, because there is no $\log_5$ key on a calculator. To approximate $\log_5 2$, we use a calculator to compute the answers obtained in parts (a) and (b):

$$\log_5 2 = \frac{\log 2}{\log 5} \qquad\qquad \log_5 2 = \frac{\ln 2}{\ln 5}$$

$$= 0.4306765581 \qquad\qquad\qquad = 0.4306765581$$

$$= 0.431 \quad \text{Rounded to three places} \qquad = 0.431 \quad \text{Rounded to three places}$$

∎

The method illustrated in the last example to convert from base 5 logarithms to base 10 logarithms or base e logarithms can easily be generalized to the following **change-of-base formula**. You are asked to prove this formula in Exercise 57.

Change-of-Base Formula

$$\log_b c = \frac{\log_a c}{\log_a b}$$

Example 9   Graph $y = \log_2 x$ on a graphing calculator.

Solution   To the best of our knowledge, all graphing calculators contain the $\boxed{\text{LOG}}$ key, which gives the function $y = \log_{10} x$, and the $\boxed{\text{LN}}$ key, which gives $y = \log_e x$. The function $y = \log_2 x$ is not directly available.

However, the change-of-base formula gives us

We could just as well have used the natural logarithm in the change-of-base formula.

$$y = \log_2 x = \frac{\log_{10} x}{\log_{10} 2}$$

which *can* be created on the calculator. The graph of $y = \log_2 x$ appears in Figure 6.18. We have chosen a convenient x-value as a check that we actually have the correct graph.

Figure 6.18
The graph of $y = \log_2 x = \dfrac{\log_{10} x}{\log_{10} 2}$

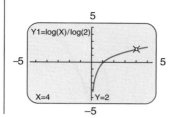

In the next section we look at a variety of applications of exponential and logarithmic functions.

6.4 Exercises

In Exercises 1–6, simplify the given expression.

1. $\ln e^5$

2. $\log 0.001$

3. $\log \sqrt{10}$

4. $\ln \dfrac{1}{e}$

5. $e^{\ln(x+1)}$

6. $\log 10^{(2t+3)}$

In Exercises 7–14, sketch the graph of the given function. Indicate the domain, range, intercepts, and asymptotes, where appropriate.

7. $y = 10^{x+1}$

8. $y = \log(x + 1)$

9. $y = 2 - \ln x$

10. $y = e^{x-2}$

11. $y = \log_{10}(x - 3)$

12. $y = \log_{10}(2x + 5)$

13. $y = \log x^2$

14. $y = \ln(ex)$

15. Find the inverse function for $y = f(x) = e^{(3x-1)}$.

16. Find the inverse function for $y = f(x) = 10^{(5-x)}$.

17. Let $f(x) = e^{\sqrt{x}}$. Find a function $g(x)$ so that $(f \circ g)(x) = (g \circ f)(x) = x$.

18. Let $g(x) = 10^{(x-3)/2}$. Find a function $h(x)$ so that $(g \circ h)(x) = (h \circ g)(x) = x$.

In Exercises 19–24, evaluate the given logarithm using the change-of-base formula, once with base 10 logarithms and once with natural logarithms. Round the answer to two decimal places.

19. $\log_7 4$

20. $\log_4 7$

21. $\log_{1/2} 5$

22. $\log_8 30$

23. $\log_9 \dfrac{2}{3}$

24. $\log_6 14$

25. Express $\log_3 x$ in terms of $\log_9 x$.

26. Express $\log_8 x$ in terms of $\log_4 x$.

27. Express $\log_{10} x$ in terms of $\ln x$.

28. Express $\ln x$ in terms of $\log_{10} x$.

In Exercises 29–42, solve the given equation. You may leave your answer in terms of natural logarithms.

29. $e^{1-2x} = 5$

30. $3e^{x-3} = 10$

31. $\dfrac{e^{3x+2}}{4} = 5$

32. $6e^{x/5} = 2$

33. $3^{4-x} = 7$

34. $5^{2x+3} = 4^{x-2}$

35. $2^{2x-2} = 6^{-x}$

36. $3^{-x^2} = 4$

37. $e^{2x} - 4e^x = -4$

38. $5^{6x} - 6 = 5^{3x}$

39. $2^{8x} + 3(2^{4x}) - 18 = 0$

40. $e^{-2x} - 3e^{-x} = 10$

41. $(\log_2 x)^2 - 5\log_2 x = 6$

42. $(\log_3 x)^2 - 6\log_3 x + 8 = 0$

In Exercises 43–50, solve the given equation. Round your answers to three decimal places where necessary.

43. $3^x = 2 - 3x$

44. $2^x = 3x + 2$

45. $5^x = x^5$ [HINT: There are two solutions.]

46. $3^{-x} = x^3$

47. $e^x = x^2 - 4$

48. $e^{-x} = 2x + 3$

49. $\ln(x + 2) = x^2 - 1$

50. $\log(x - 1) = 4 - x^2$

51. Use the change-of-base formula to show that
$$\log_b a = \frac{1}{\log_a b}.$$

52. Graph $y = \log_{1/2} x$ on a graphing calculator.

53. Graph $y = \log_4 x$ on a graphing calculator.

Questions for Thought

54. When using a graphing calculator to find the solutions to an equation, it is important to recognize that the approach we take can influence what we "see" on the calculator screen.

(a) Solve the equation $2^x = x^2$ by graphing $y = 2^x$ and $y = x^2$ and looking for the points of intersection. When you graph the two equations on the same screen, is it clear how many points of intersection there are? How many are there?

(b) Now solve the equation $2^x = x^2$ by graphing $y = 2^x - x^2$ and find the zeros. Is it clearer now how many zeros there are?

55. Keeping in mind the observations made in Exercise 54, use a graphing calculator to find all the solutions to the following equations:

(a) $3^x = x^3$

(b) $x^4 - x^3 = 4^x - 3^x$

(c) $3^x - 9 = 2x^3$

(d) $x^2 - x = 2^{x-1}$

56. Use a calculator to evaluate the expression $\left(1 + \dfrac{1}{n}\right)^n$

for $n = 10; 20; 50; 100; 1000;$ and $10,000$. Note that the values seem to be tending toward the value of e. In fact, this is another way of defining the number e.

57. Use the method outlined in Example 8 to prove the change-of-base formula $\log_b c = \dfrac{\log_a c}{\log_a b}.$

58. Use the change-of-base formula to express $\log_4 x$ in terms of $\log_{16} x$ and explain why the answer makes sense.

59. If $\ln A = 1.5$ and $\ln B = 2.4$, which is larger, A^3 or B^2?

6.5   Applications

There are many real-world quantities that increase or decrease with time in proportion to the amount of the quantity present. Some examples are human population, bacteria in a culture, radioactivity, the concentration of drugs in the bloodstream, and the value of certain types of investments. It can be shown that this type of increase or decrease can be described by an exponential function.

Let's discuss some ideas from the world of finance, where it may come as a bit of a surprise that the number e makes a dramatic appearance. We begin by introducing some terminology. Suppose that $1000 is invested in an account at an interest rate of 6% per year. The $1000 is called the **principal** and is often designated as P. At the end of one year the amount in the account would be

new principal = original principal + interest

$$= \qquad 1000 \qquad + 0.06(1000) \qquad \text{Factor out 1000.}$$
$$= 1000(1 + 0.06) = 1000(1.06) = \$1060 \qquad (1)$$

If the interest earned each year remains in the account so that in subsequent years the interest itself earns interest, the account is said to be earning **compound interest**.

Let's start with P dollars in an account paying an interest rate of r (expressed as a decimal) compounded annually (meaning that the interest is computed at the end of each year) and attempt to develop a formula for the amount, A, present in an account after t years.

Our computation will be simplified if we recognize that as in equation (1), *the amount in the account at the end of any year is* $(1 + r)$ *times the previous principal* (where, again, r is the interest rate expressed as a decimal). We therefore have the following:

at the end of 1 year: $A = P(1 + r)$ This is now the new principal, as in equation (1).

at the end of 2 years: $A = \underbrace{P(1 + r)}_{\text{Previous principal}} [1 + r] = P(1 + r)^2$

at the end of 3 years: $A = \underbrace{P(1 + r)^2}_{\text{Previous principal}} [1 + r] = P(1 + r)^3$

$$\vdots$$

at the end of t years: $A = P(1 + r)^t$ (2)

Example 1

Suppose $1000 is placed into a savings account paying 6% per year.
(a) How much money will be in the account after 10 years?
(b) How long will it take the initial principal of $1000 to double?
(c) How long will it take an initial principal of $1000 to double if the annual interest rate is 8%?

Solution

(a) Using the formula $A = P(1 + r)^t$, we substitute $P = 1000$, $r = 0.06$, and $t = 10$ to get

$$A = 1000(1 + 0.06)^{10} = 1000(1.06)^{10}$$ Using a calculator, we get

$$= \$1790.85 \text{ in the account at the end of 10 years}$$

(b) Since we are looking for the number of years it takes the money to double, we want the amount A to be $2000. We substitute $A = 2000$, $P = 1000$, and $r = 0.06$ in the formula $A = P(1 + r)^t$ and solve for t.

$$2000 = 1000(1.06)^t$$ To isolate t, first divide both sides of the equation by 1000.

$$2 = 1.06^t$$ Next, take ln (or $\log_{10}$) of both sides.

$$\ln 2 = \ln(1.06^t)$$ Use log property 3.

$$\ln 2 = t \ln 1.06$$

$$t = \frac{\ln 2}{\ln 1.06} \approx 11.9 \text{ years}$$

Since interest is paid at the end of each year, it will take 12 years for the money to double.

(c) If the annual interest rate is 8%, then $1 + r = 1 + 0.08 = 1.08$, and the same procedure used in part (b) gives

$$t = \frac{\ln 2}{\ln 1.08} \approx 9$$

Thus increasing the interest rate to 8% reduces the doubling time from almost 12 years to about 9 years.

Suppose we change the way the interest is compounded. Most banks do not compute the interest on an account once yearly; rather, they compound the interest a number of times per year. For example, if a bank compounds the interest quarterly, it computes and credits the interest 4 times per year (every 3 months) using one-quarter of the annual interest rate each time. Similarly, banks may compound monthly by computing and adding the interest to the account 12 times per year using one-twelfth the annual interest rate each time, or daily by computing $\frac{1}{365}$ of the interest rate each day.

Essentially what is happening is that we are making shorter and shorter interest periods with smaller and smaller interest rates. If we are compounding $1000 quarterly at an annual interest rate of 6%, we are computing a new principal 4 times per year by using an interest rate of $\frac{0.06}{4} = 0.015$. (This situation will yield the same amount of money as if $1000 were invested at 1.5% per year for 4 years.)

For illustrative purposes we again begin with $1000 invested at an annual interest rate of 6%. Let's see what happens if we compound this annual interest rate quarterly, monthly, and daily for one year.

compounding quarterly: $A = 1000\left(1 + \frac{0.06}{4}\right)^4 = \$1061.36$

compounding monthly: $A = 1000\left(1 + \frac{0.06}{12}\right)^{12} = \$1061.68$

compounding daily: $A = 1000\left(1 + \frac{0.06}{365}\right)^{365} = \$1061.83$

Note that increasing the number of interest periods from 12 per year to 365 per year increases the annual interest by only 15¢.

Also note that if we were compounding the annual interest monthly for a period of 5 years, there would be $5(12) = 60$ interest periods and so

$$A = 1000\left(1 + \frac{0.06}{12}\right)^{5(12)} = 1000(1.005)^{60} = \$1348.85$$

As a result of this discussion, we state the following formula.

Compound Interest Formula

If P dollars is invested at an annual interest rate r, compounded n times per year, then A, the amount of money present after t years, is given by

$$A = P\left(1 + \frac{r}{n}\right)^{nt}$$

Example 2   In Example 1 we saw that it takes 11.9 years for $1000 invested at 6% per year to double. If the $1000 is invested at 6% compounded daily, how long will it take to double?

Solution   Using the compound interest formula, substitute $A = 2000$, $P = 1000$, $r = 0.06$, and $n = 365$ solve for t:

$2000 = 1000\left(1 + \frac{0.06}{365}\right)^{365t}$ Divide both sides by 1000.

$2 = \left(1 + \frac{0.06}{365}\right)^{365t}$ Take ln of both sides (or log10 of both sides), and apply log property 3.

$\ln 2 = \ln\left(1 + \frac{0.06}{365}\right)^{365t}$ Apply log property 3.

$\ln 2 = 365t \ln\left(1 + \frac{0.06}{365}\right)$ Isolate t.

Compare this reduction in doubling time to the reduction in doubling time if the interest rate is increased from 6% to 8% (See Example 1(c)).

$$t = \frac{\ln 2}{365 \ln(1 + \frac{0.06}{365})} \approx 11.55$$

Thus compounding daily reduces the doubling time from 11.9 years to a little more than $11\frac{1}{2}$ years. ∎

Example 3 What simple interest rate would be equivalent to an interest rate of 6% compounded monthly? (Banks often call such a simple annual interest rate the *effective annual yield*.)

Solution Using the compound interest formula, we have

$$A = P\left(1 + \frac{0.06}{12}\right)^{12} \approx P(1.0617)$$

which means that at the end of 1 year we have 1.0617 times the original principal. This is equivalent to a simple interest rate of 6.17% per year. Thus an account paying 6% per year compounded monthly has an effective annual yield of 6.17%. ∎

A natural question one might ask is, What happens if we start with P dollars and compound an interest rate of r every minute? Every second? Will the amount A keep growing larger? Is there any limit to how much the amount A will grow as we compound more and more frequently? As we saw before, increasing the number of interest periods does increase the amount A, but not by as much as we might have expected. It turns out there is a limit to how much A can grow regardless of how many times interest is compounded per year.

Let's suppose that $1 is invested at an annual rate of 100% per year compounded n times per year. Using the formula for compound interest with $r = 1$, the amount A at the end of 1 year is given by

100% written as a decimal is 1.00.

$$A = \left(1 + \frac{1}{n}\right)^n$$

if we compound it annually, then $n = 1$ $\quad A = \left(1 + \frac{1}{1}\right)^1 = 2$

semiannually, then $n = 2$ $\quad A = \left(1 + \frac{1}{2}\right)^2 = 2.25$

quarterly, then $n = 4$ $\quad A = \left(1 + \frac{1}{4}\right)^4 \approx 2.44$

monthly, then $n = 12$ $\quad A = \left(1 + \frac{1}{12}\right)^{12} \approx 2.61$

daily, then $n = 365$ $\quad A = \left(1 + \frac{1}{365}\right)^{365} \approx 2.7146$

each minute, then $n = 525,600$ $\quad A = \left(1 + \frac{1}{525,600}\right)^{525,600} \approx 2.71828$

each second, then $n = 31,536,000$ $\quad A = \left(1 + \frac{1}{31,536,000}\right)^{31,536,000} \approx 2.71828$

As soon as n gets larger than 365, the amount remains steady at $2.72, assuming that the bank will round to the nearest cent. If we allow the number of interest periods to get larger and larger, we approach what we might describe as **continuous compounding.**

Note that as n increases, the value of $\left(1 + \dfrac{1}{n}\right)^n$ seems to be approaching e, the base for the natural exponential and logarithmic function. In fact, this is an alternative way of defining e.

The following formula is derived in calculus.

See Exercise 61 to see how the two compound interest formulas are related.

Compound Interest Formula (Continuous Compounding)

If P dollars is invested at an annual interest rate r compounded continuously, then A, the amount of money present after t years, is given by

$$A = Pe^{rt}$$

Example 4   At what annual rate, compounded continuously, must \$10,000 be invested to grow to \$25,000 in 8 years?

Solution   We use the formula $A = Pe^{rt}$ with $A = 25{,}000$, $P = 10{,}000$, and $t = 8$ and solve for r:

$$25{,}000 = 10{,}000e^{8r} \qquad \text{Divide both sides by 10,000.}$$

$$2.5 = e^{8r} \qquad\qquad \text{Take ln of both sides and use log property 3.}$$

$$\ln 2.5 = 8r \Rightarrow r = \frac{\ln 2.5}{8} \approx 0.1145$$

Thus the annual rate must be 11.45%.

As we mentioned previously, when an equation describes a real-life situation (as the formula $A = Pe^{rt}$ describes the amount in an investment), we often say that the equation is a **model** of the situation. We can generalize this model to other situations where a quantity increases or decreases at a continuous rate r over time t.

If we start with a population or amount of a substance, A_0, which is growing at a continuous rate of r per unit of time (per year, per day, per minute, etc.), then after t units of time (t years, t days, t minutes, etc.) the number or amount present will have grown to A, where A is given by the exponential equation that appears in the following box. Populations or substances whose growth is described by such an equation are said to conform to an **exponential growth model**.

The Exponential Growth/Decay Model

Let A_0 represent the initial number or amount present, and let A represent the amount present after t units of time. If a population or substance is changing at a continuous rate r per unit of time, then A is given by

$$A = A_0 e^{rt}$$

Keep in mind that if the growth rate r is given as a percent, it must be converted to decimal form in order to use the growth and decay formula.

If the amount or number present is *increasing* as time passes, then r is positive and vice versa. In such a case, we say that the population or substance conforms to the **exponential growth model**.

If the amount or number present is *decreasing* as time passes, then r is negative and vice versa. In such a case, we say that the population or substance conforms to the **exponential decay model**.

Example 5 According to a world almanac, the population of the world in 1986 was estimated to be 4.7 billion people. Assuming that the world's population is growing at the continuous rate of 1.8% per year,
(a) Estimate the population of the world in the year 2008.
(b) In what year will the earth's population be 10 billion?

Solution We use the exponential growth model $A = A_0 e^{rt}$ with $A_0 = 4.7$ and $r = 0.018$ (the decimal equivalent of 1.8%). A represents the population (in billions) t years after 1986, and A_0 represents the population (in billions) of the earth in 1986 (this is the initial population). Thus we have

$$A = 4.7e^{0.018t}$$

(a) Since we want to estimate the earth's population in the year 2008, 22 years will have elapsed since the initial year of 1986. We substitute $t = 22$ into the growth equation:

$$A = 4.7e^{0.018t} \qquad \text{Substitute } t = 22.$$
$$= 4.7e^{.018(22)} \qquad \text{Using a calculator, we get}$$
$$A \approx 6.98$$

Thus the earth's population in the year 2008 will be approximately 6.98 billion people according to this model.

(b) To determine in what year the earth's population will be 10 billion, we again use the equation $A = 4.7e^{0.018t}$, but this time we want to know what value of t will make $A = 10$.

$$A = 4.7e^{0.018t} \qquad \text{We let } A = 10 \text{ and solve for } t.$$
$$10 = 4.7e^{0.018t} \qquad \text{Divide both sides by 4.7.}$$
$$\frac{10}{4.7} = e^{0.018t} \qquad \text{Take ln of both sides.}$$
$$\ln \frac{10}{4.7} = \ln e^{0.018t} = 0.018t$$
$$\frac{\ln \frac{10}{4.7}}{0.018} = t \qquad \text{Using a calculator, we get}$$
$$t \approx 41.95$$

It would take almost 42 years, or until the year 2028 (42 years after 1986), for the earth's population to reach 10 billion. ∎

Example 6 Using the exponential growth model, how long, in terms of r, will it take a population to double if it increases at a rate of r per year?

Solution We begin with the population model $A = A_0 e^{rt}$ and note that we are looking for the time t it takes for the amount A to become $2A_0$, twice the initial amount.

$$A = A_0 e^{rt} \qquad \text{Substitute } 2A_0 \text{ for } A.$$
$$2A_0 = A_0 e^{rt} \qquad \text{Divide both sides of the equation by } A_0.$$
$$2 = e^{rt} \qquad \text{Take ln of both sides (or rewrite in ln form).}$$
$$\ln 2 = rt \qquad \text{Solve for } t.$$
$$t = \frac{\ln 2}{r}$$

Hence, if a population were growing at a rate of 5% a year, then $r = 0.05$ and it would take $\dfrac{\ln 2}{0.05} \approx \dfrac{0.6931}{0.05} \approx 13.86$ years to double, whereas a population growing at a rate of 15% would take $\dfrac{\ln 2}{0.15} \approx 4.6$ years to double. ∎

Example 7 Suppose we have 100 grams of a radioactive substance that decays at the continuous rate of 4% per hour.

(a) How long will it take until only 50 grams of radioactive substance remain?
(b) How long will it take until only 25 grams of radioactive substance remain?
(c) Find the time it takes for an initial amount A_0 of this radioactive substance to decay to half this amount. This time is called the **half-life** of the substance.

Solution The exponential decay equation for this substance is

$$A = e^{-0.04t} \quad \text{Note that } r = -0.04 \text{ because the substance is decaying.}$$

(a) To determine how long it will take the 100 grams to decay to 50 grams, we set $A = 50$ and solve for t:

$$50 = 100e^{-0.04t} \qquad\qquad \text{Divide both sides by 100.}$$

$$0.5 = e^{-0.04t} \qquad\qquad \text{Take ln of both sides.}$$

$$\ln 0.5 = -0.04t$$

$$t = \dfrac{\ln 0.5}{-0.04} \approx 17.33$$

Thus it takes approximately 17.33 hours for the 100 grams to decay to 50 grams.

(b) We can determine the time it takes for 25 grams of radioactive substance to remain either by using exactly the same approach as in part (a) or by finding the time it takes for 50 grams to decay to 25 grams and then adding this time to the answer obtained in part (a). We illustrate both approaches.

$$25 = 100e^{-0.04t} \qquad\qquad\qquad 25 = 50e^{-0.04t}$$

$$0.25 = e^{-0.04t} \qquad\qquad\qquad 0.5 = e^{-0.04t}$$

$$\ln 0.25 = -0.04t \qquad\qquad\qquad \ln 0.5 = -0.04t$$

$$t = \dfrac{\ln 0.25}{-0.04} \approx 34.66 \qquad\qquad t = \dfrac{\ln 0.5}{-0.04} \approx 17.33$$

Thus it takes 34.66 hours for 100 grams to decay to 25 grams. | So it takes another 17.33 hours for 50 grams to decay to 25 grams. Thus it takes a total of 34.66 hours for 100 grams to decay to 25 grams.

(c) To determine the half-life of the substance, we want to know how long it takes an initial amount A_0 to decay to an amount $\dfrac{1}{2}A_0$.

$$\dfrac{1}{2}A_0 = A_0 e^{-0.04t} \qquad \text{Divide both sides by } A_0.$$

$$0.5 = e^{-0.04t} \quad \Rightarrow \quad \ln 0.5 = -0.04t$$

$$t = \dfrac{\ln 0.5}{-0.04t} \approx 17.33$$

How does the answer to part (c) relate to the answer to Example 6?

Thus the half-life of this substance is approximately 17.33 hours. Note that (as parts (a) and (b) suggested) this answer is independent of the initial amount. This fact is crucial for scientists' ability to use the radioactive decay of carbon-14 to determine the age of fossils and archeological artifacts. (See Exercise 47.) ∎

Example 8   Suppose a colony of bacteria starts with 600 and grows to 4500 in 12 hours. How long will it take for the colony to have 15,000 bacteria?

Solution   We start with the exponential growth model $A = A_0e^{rt}$ and note that we are given that $A = 4500$, $A_0 = 600$, and $t = 12$, and we need to solve for r.

$$4500 = 600e^{12r} \qquad \text{Divide both sides by 600.}$$

$$7.5 = e^{12r} \qquad \text{Take ln of both sides.}$$

$$\ln 7.5 = 12r \;\Rightarrow\; r = \frac{\ln 7.5}{12} \approx 0.17$$

Note that $\dfrac{\ln 7.5}{12}$ is positive as it should be if the population is growing.

Thus an exponential growth model for these bacteria is $A = A_0 \, e^{(t \ln 7.5)/12}$.

Now that we have the growth model, finding how long it takes the population to grow to 15,000 is simply a matter of finding t when $A = 15,000$.

$$A = A_0e^{(t \ln 7.5)/12} \qquad \text{We substitute } A = 15{,}000 \text{ and } A_0 = 600.$$

$$15000 = 600e^{(t \ln 7.5)/12} \qquad \text{Divide both sides by 600.}$$

$$25 = e^{(t \ln 7.5)/12} \qquad \text{Take the ln of both sides.}$$

$$\ln 25 = \frac{t \ln 7.5}{12} \qquad \text{Now solve for t.}$$

$$t = \frac{12 \ln 25}{\ln 7.5} \approx 19.2$$

Thus it takes approximately 19.2 hours for the number of bacteria to reach 15,000. ∎

Example 9   The half-life of a radioactive substance is 8 minutes. How long does it take 80 grams of this substance to decay to 7 grams?

Solution

What do we need to do?	Find how long it takes for 80 grams of a substance to decay to 7 grams, given its half-life.
How do we start?	First, write the model for exponential decay: $A = A_0e^{rt}$. Then determine what variables are given and what needs to be found.
We have $A_0 = 7$ and $A = 80$. We need to find t, but we are not given r.	Is there information given in the problem that can help us to find r?
We are given the half-life of the substance. How can we use that information to find r?	Remember that the half-life of a substance is the amount of time it takes for half the initial amount to decay. If we let $A = \frac{1}{2}A_0$ and $t = 8$ minutes, we can substitute these values into the exponential decay model to solve for r.

$$A = A_0 e^{rt}$$

The half-life, the time it takes for half the substance to decay, is 8 minutes, which means it takes 8 minutes for A to become $\frac{1}{2} A_0$, so we substitute $A = \frac{1}{2} A_0$ and $t = 8$ to get

$$\frac{1}{2} A_0 = A_0 e^{8r}$$

Remember, we first need to solve for r. Divide both sides by A_0.

$$\frac{1}{2} = e^{8r}$$

Now translate this into ln form.

$$\ln 0.5 = 8r$$

Solve for r.

$$r = \frac{\ln 0.5}{8} \approx -0.086643$$

Rounded to six decimal places. Note that r is negative, as it should be in a decay model.

Now that we have r, what do we need next?

Given r, we now have the decay model for the substance: $A = A_0 e^{\frac{\ln 0.5}{8} t}$, where t is measured in minutes. Now we can substitute $A_0 = 80$ and $A = 7$ in the model to find t, the time it takes for 80 grams to decay to 7 grams:

$$A = A_0 e^{\frac{\ln 0.5}{8} t}$$

Substitute $A_0 = 80$ and $A = 7$.

$$7 = 80 e^{\frac{\ln 0.5}{8} t}$$

Divide both sides by 80.

$$0.0875 = e^{\frac{\ln 0.5}{8} t}$$

Rewrite in ln form.

$$\ln 0.0875 = \frac{\ln 0.5}{8} t$$

Solve for t.

See Exercise 59 for a slightly different approach to this example.

$$t = \frac{8 \ln 0.0875}{\ln 0.5} \approx 28.12 \text{ min}$$

∎

Logarithms also arise in numerous fields. The American geologist Charles Richter conceived of a scale that is used to measure the magnitude of earthquakes. He defined the magnitude M of an earthquake to be

$$M = \log\left(\frac{I}{S}\right)$$

where I is the intensity of the measured earthquake (as measured by a seismograph) and S is the intensity of a "standard" earthquake. Thus the magnitude of a standard earthquake is

A standard earthquake is often called a zero-level earthquake.

$$M = \log\left(\frac{S}{S}\right) = \log 1 = 0$$

Example 10   How many times more intense is an earthquake of magnitude 5 than one of magnitude 3?

Solution   This example is asking us to compare the intensities of two earthquakes with known magnitudes. Let I_1 be the intensity of the earthquake of magnitude 5 and let I_2 be the intensity of the earthquake of magnitude 3. We need to compute the ratio $\frac{I_1}{I_2}$.

$$\frac{I_1}{I_2} = \frac{\dfrac{I_1}{S}}{\dfrac{I_2}{S}}$$

We divided numerator and denominator by S. Now take log of both sides.

$$\log \frac{I_1}{I_2} = \log\left(\frac{\dfrac{I_1}{S}}{\dfrac{I_2}{S}}\right)$$

Use log property 3.

$$\log \frac{I_1}{I_2} = \log \frac{I_1}{S} - \log \frac{I_2}{S}$$

Using the definition of magnitude and the given magnitudes, we get

$$\log \frac{I_1}{I_2} = M_1 - M_2$$

$$\log \frac{I_1}{I_2} = 5 - 3 = 2$$

Translating into exponential form, we get

$$\frac{I_1}{I_2} = 10^2 = 100$$

Exercise 60 asks you to generalize the result of this example.

This means that I_1 is 100 times I_2. Thus on the Richter scale, an earthquake of magnitude 5 is 100 times more intense than an earthquake of magnitude 3. ∎

The exercises illustrate many areas in which exponential and logarithmic functions arise.

6.5 Exercises

1. If a store raises its prices by 20% and then reduces its prices by 20%, are the items back to their original prices? Explain.

2. Are successive price reductions of 25% and 15% equivalent to a price reduction of 40%? Explain.

In the following exercises, assume that the situations and populations described are governed by an exponential growth or decay model.

3. Plutonium-239 (^{239}Pu), one of the radioactive waste by-products resulting from the production of nuclear energy, has a half-life of approximately 25,000 years. Find the exponential decay model for ^{239}Pu.

4. The formula for the radioactive decay of radium is $A = A_0 e^{-0.0004279t}$. Find the half-life of radium.

5. The half-life of carbon-14 (^{14}C) is approximately 5730 years. Of 100 grams of ^{14}C, how much will be left after 1000 years?

6. How much of 50 grams of ^{14}C will have decayed in 100 years?

7. Neptunium-239 (^{239}Np) has a short half-life of approximately 2.24 days. Find the decay model for neptunium and determine how much of an initial amount of 40 grams will be radioactive after 30 days.

8. Calculate how long it will take for 10 grams of neptunium to decay to less than 0.1 gram.

9. If it takes 100 years for 80 grams of radioactive substance to decay to 60 grams, how long will it take for an amount of this substance to decay to one-fifth that amount?

10. Strontium-90, another waste by-product of nuclear fission reactors, has a half-life of approximately 28 years. How long will it take a sample of strontium-90 to be reduced by a factor of 100?

11. A population of a colony of bacteria increases according to the growth model $A = A_0 3^{t/30}$ (where t is measured in hours). How long will it take for the population to grow from 100 to 200? From 100 to 300? (You should be able to answer this second question without using a calculator.)

12. A certain type of bacteria grows according to the growth model $A = A_0 e^{0.357t}$, where the time t is measured in hours. If a colony of bacteria starts with approximately 400 bacteria, how many will there be after 6 hours? After 2 days?

13. According to 1998 *World Almanac* census data, the population of the United States increased from approximately 226.5 million in 1980 to 248.7 million in 1990. Find the annual growth rate during this 10-year period.

14. According to 1998 *World Almanac* census data, the population of the United States increased from approximately 38.6 million in 1870 to 50.2 million in 1880. Find the annual growth rate during this 10-year period. Compare this rate with the result of Exercise 13. Does the result surprise you? What do you think caused the significantly higher growth rate for the period between 1870 and 1880?

15. It is estimated that the world's population was 5.8 billion in 1996. Assuming a growth rate of 1.8% for the foreseeable future, what will the earth's population be in the year 2025?

16. It is estimated that in 1996, the population of the "third world" was 4.6 billion, or about 79% of the world's population. If the third world's population is growing at a rate of 2.1%, use the result of Exercise 15 to determine what part of the world's population the third world population will be in the year 2025.

17. According to United Nations' estimates, the population of the Republic of China was 686.4 *million* in 1960 and 1.032 *billion* in 1984. Find the growth rate of the Republic of China's population. Is this growth rate higher or lower than the world's current growth rate of 1.8%?

18. Bacteria B is known to satisfy the exponential growth model $A = A_0 e^{0.37t}$, where t is the number of hours the initial number of bacteria are placed in a particular growth medium at a certain temperature. Suppose a lab technician places 800 bacteria of an unknown type in the same growth medium at the same temperature and observed that after 18 hours there are approximately 12,800 bacteria present. Would you conclude that these bacteria are of type B? Explain.

19. If $10,000 is deposited in a bank account paying an interest rate of 7.6% per year compounded quarterly, how long will it take for there to be $15,000 in the account?

20. If $8500 is invested at 6.5% per year compounded daily, how long will it take for the investment to double?

21. What is the effective annual yield if $3000 is invested at 6.7% per year compounded monthly? Does your answer depend on the amount of money in the account?

22. How many years will it take $500 invested at 8.3% compounded weekly to grow to $1500?

23. What rate of continuous compounding gives an effective annual yield of 6.7%?

24. Find the doubling time for an investment at 6% compounded continuously. What must the interest rate be if the doubling time is to be cut in half?

25. How much money must be invested at 5.8% compounded continuously to yield $6,000 after 5 years?

26. How much money must be invested at 8% compounded continuously to yield A dollars after 10 years?

27. Find the number of years it takes $1000 invested at 6% and compounded continuously to double. Compare this to the doubling time if the interest rate is compounded daily, which was computed in Example 2. How does this reflect on the statement made in the text that increasing the number of compoundings does not significantly increase the amount of interest accumulated?

28. A common credit card company charges an annual interest rate of 18.5% compounded monthly. If you make a credit card purchase of $600 and don't make any payments for a year, how much do you owe at the end of the year?

In Exercises 29–32, use the information given in Exercise 29.

29. The repayment schedule for most long-term loans, such as mortgages and car loans, is calculated so that the amount borrowed plus the interest is paid back in equal monthly payments. The general formula for the month payment, M, required on an amount A borrowed at an annual rate r for t years is

$$M = \frac{Ar}{12\left[1 - \left(1 + \frac{r}{12}\right)^{-12t}\right]}$$

(a) What are the monthly payments required on a car loan of $8000 at a rate of 9% per year for a period of 4 years?

(b) Over this 4-year period, how much has been paid back to borrow the $8000?

30. If someone can afford a repayment schedule of $200 per month for 3 years, how large a car loan can they afford at a rate of 10%?

31. (a) What would the monthly payments be on a 30-year mortgage of $80,000 at 8.2%?

(b) Over the course of this 30-year mortgage, how much money has been paid back in total?

32. If a couple wishes to budget $700 per month for their mortgage payments, how large a 30-year mortgage can they afford at 7.8%?

The Richter Scale

33. Seismologists measure the magnitude of earthquakes using the Richter scale. This scale defines the magnitude, M, of an earthquake as

$$M = \log\left(\frac{I}{S}\right) \quad \text{(remember that log means } \log_{10})$$

where I is the intensity of the earthquake being measured and S is the intensity of a *zero-level* earthquake.
(a) Find the Richter scale measure of an earthquake that is 1000 times as intense as a zero-level earthquake—that is, $I = 1000S$.
(b) The great 1906 earthquake in San Francisco had a Richter scale measure of approximately 8.3. Compare the intensity of this earthquake to that of a zero-level earthquake.

34. The Loma Prieta earthquake, which interrupted the 1989 World Series, registered 7.1 on the Richter scale. Compare the intensities of the Loma Prieta quake and the great 1906 quake, which registered 8.3 on the Richter scale.

35. Suppose that three earthquakes E_1, E_2, and E_3 register $M_1 = 2$, $M_2 = 6$, and $M_3 = 10$, respectively, on the Richter scale. Note that the differences between M_1 and M_2, and between M_2 and M_3 are both 4. Compare the intensities of the three earthquakes.

36. Show that if earthquake E_2 has a Richter scale magnitude of M_2 and twice the intensity of earthquake E_1, which has a Richter scale magnitude of M_1, then $M_2 - M_1 = \log 2$.

pH Levels

37. Chemists have defined the pH (which stands for *hydrogen potential*) of a solution to be

$$pH = -\log[H_3O^+]$$

where $[H_3O^+]$ stands for the concentration of the hydronium ion in the solution (measured in moles per liter). The pH is a measure of the acidity or alkalinity of the solution. Water, which is neutral, has a pH of 7. Solutions with a pH below 7 are acidic, whereas those with a pH above 7 are alkaline. What is the pH of a glass of orange juice if its hydronium ion concentration is 6.82×10^{-5}?

38. What is the hydronium ion concentration of a solution with a pH of 9.1?

39. Environmentalists are constantly monitoring the pH levels of rain and snow because of the destructive effects of "acid rain," which is caused primarily by sulfur dioxide emissions. Rain and snow have a natural concentration of $[H_3O^+] = 2.5 \times 10^{-6}$ due to dissolved carbon dioxide that is normally in the atmosphere. Find the natural pH of rain and snow.

40. If a rain sample shows that the concentration of $[H_3O^+]$ has increased by a factor of 100, what will its pH level be? Use the information in Exercise 39.

Psychophysics

41. The *Weber–Fechner law* in psychophysics relates the intensity S of a sensation (a psychological reaction) to the intensity P of the physical stimulus. This law was published in 1930 by Gustav Fechner and was based on experiments performed by Ernst Weber a year earlier. It states that the change in the intensity of a sensation caused by a small change in the intensity of a stimulus is proportional not to the amount of change in P, as we might expect, but rather to the percentage change in P. Using calculus, this relationship gives the Weber–Fechner law, which says that

$$S = K \ln\left(\frac{P}{P_0}\right)$$

where P_0 is the minimum intensity that can be perceived, and K is some constant that determines the units in which S is measured.

The brightness of stars, as seen by the naked eye, is measured in units called *magnitudes*. The Greek astronomer Ptolemy set up six categories; the dimmest stars are of magnitude 6 and the brightest stars are of magnitude 1. If we let I be the brightness of a star of magnitude M and I_0 be the minimal brightness a star must have to be visible, then using the Weber–Fechner law, we can derive the following formula for the magnitude M:

$$M = 6 - 2.5 \log\left(\frac{I}{I_0}\right)$$

(a) Calculate the ratio of the light intensity of a star of magnitude 1 to that of a star of magnitude 2.
(b) Calculate the ratio of the light intensity of a star of magnitude 4 to that of a star of magnitude 5.
(c) Based on the results of parts (a) and (b), how much less intense is the light of a star of any magnitude than a star of magnitude one less?
(d) If a star's light intensity increases by a factor of 40 before it goes nova, how many magnitudes does it decrease?

42. Suppose that the smallest electrical current you can feel is 1 milliampere and that the minimum *difference* in current you can detect is 25 milliamperes. Use the Weber–Fechner law to develop a scale of perceived electrical current. Call one unit on this scale a *jolt*, and write a formula for the number of jolts corresponding to a current of A milliamperes.

Sound Intensity

43. The unit of measurement frequently used to measure sound levels is the *decibel* (dB). The number of decibels, N, of a sound with intensity I (usually measured in watts per square centimeter, or W/sq cm) is defined to be

$$N = 160 + 10 \log I$$

The faintest sound audible to the human ear is a sound with an intensity of approximately 10^{-16} W/sq cm. What is the decibel level of this sound?

44. A sound at the threshold of pain to the human ear is a sound with an intensity of approximately 10^{-4} W/sq cm. What is the decibel level of this sound?

Natural Resources

45. Many of the world's natural resources are being consumed at a rate that is increasing exponentially with respect to time. Assume an exponential growth model for the consumption of petroleum products.
  (a) If the world consumed approximately 1.5 billion barrels of petroleum products in 1940 and 3.6 billion barrels in 1960, estimate the rate at which the consumption was increasing over this 20-year period.
  (b) If the world consumed approximately 12.1 billion barrels of petroleum products in 1965 and 20.4 billion barrels in 1975, estimate the rate at which the consumption was increasing over this 10-year period.

46. It can be shown (using calculus) that if an amount A_0 is consumed in a certain year, and assuming an annual growth rate of r, then the amount A of oil consumed in the next T years is given by the formula

$$A = \frac{A_0}{r}(e^{rT} - 1)$$

  (a) Show that solving this formula for T gives

$$T = \frac{\ln\left[\dfrac{rA}{A_0} + 1\right]}{r}$$

  (b) In 1990 it was estimated that confirmed available world oil reserves were 983.4 billion barrels of oil and that 21.3 billion barrels of oil were consumed that year. Assuming a rate of growth in oil consumption of 2.5%, use the formula derived in part (a) to determine how many years the 1990 oil reserves will last.

Carbon Dating

47. All living plant and animal tissue contains both carbon-12, which is not radioactive, and carbon-14, which is radioactive with a half-life of approximately 5730 years. While the organism is alive, the ratio of carbon-14 to carbon-12 remains constant. After the organism dies (we will call this $t = 0$), the carbon-14 begins to decay. Thus the smaller the ratio of carbon-14 to carbon-12, the older the remains.
  (a) Verify that the decay model for carbon-14 is $A = A_0 e^{-0.000121t}$.
  (b) Given that the half-life of carbon-14 is 5730 years, explain why an alternative decay model is $A = A_0\left(\dfrac{1}{2}\right)^{t/5730}$. HINT: Examine what happens when $t = 5730$, $t = 11{,}460 = 2(5730)$, etc.
  (c) Verify that both equations give the same result for the amount of carbon-14 left from an initial amount of 60 grams after 1000 years.
  (d) Suppose that a fossil now contains 65% of the carbon-14 it contained originally. Estimate the age of the fossil.

48. Skeletal remains are found to contain 35% of their original carbon-14. Estimate the age of the skeleton.

49. In 1947 an Arab Bedouin herdsman climbed into a cave near Qumran on the shores of the Dead Sea in search of a stray goat. He found some earthenware jars containing what we now know as the *Dead Sea Scrolls*. The wrappings of the scrolls were analyzed and found to contain 76% of their original carbon-14. Estimate the age of the Dead Sea Scrolls.

50. If a fossil is estimated to be 10,000 years old, what percentage of its original carbon-14 remains?

Miscellaneous Problems

51. Suppose that water and chlorine are continuously being added to a swimming pool in such a way that the number of grams of chlorine in the pool at time t hours is given by

$$c(t) = 100 - 30e^{-t/10}$$

  (a) How much chlorine is in the pool initially (at time $t = 0$)?

(b) How much chlorine is in the pool after 5 hours?
(c) How much chlorine is in the pool after 10 hours?
(d) How much chlorine is in the pool after 100 hours?

52. A basic cooling principle in physics (called Newton's law of cooling) states that if an object at temperature T_0 is placed into a surrounding environment, which is at a constant temperature C, then the temperature, T, of the object after t minutes is given by

$$T = C + (T_0 - C)e^{-kt}$$

where k is a constant that depends on the particular object.
(a) Determine the constant k (to the nearest hundredth) for a bottle of orange juice that takes 10 minutes to cool from 70°F to 55°F after being placed in a refrigerator that is maintaining a constant temperature of 45°F.
(b) What will the temperature of the juice be $\frac{1}{2}$ hour after it is placed in the refrigerator?
(c) According to the given formula for T, is it possible for the temperature of the juice to reach 45°F? Explain.

53. If inflation is running at a steady rate of r per year over an extended period of time, then the value of A_0 dollars after t years is given by

$$A = A_0(1 - r)^t$$

Assume an inflation rate of 4.8%.
(a) How much will $100 be worth after 5 years?
(b) How long will it take until $100 is worth only $50?

54. An altimeter is an instrument that measures altitude. The altimeters used in most aircraft measure the altitude by measuring the outside barometric pressure, P, and then displaying the altitude by means of a scale that is calibrated by using the following *barometric equation:*

$$a = (30T + 8000)\ln\left(\frac{P_0}{P}\right)$$

which relates the altitude a in meters above sea level, the air temperature T in degrees Celsius, the atmospheric pressure P_0 at sea level, and the atmospheric pressure P at altitude a. (Atmospheric pressure is measured in centimeters of mercury.) Suppose the atmospheric pressure at a certain altitude is 24.9 cm of mercury and the temperature is -3°C. If the atmospheric pressure at sea level is 76 cm of mercury, use the barometric equation to find the altitude to the nearest foot (1 meter is approximately 3.3 ft).

55. The number $n!$ (read n *factorial*) is defined as

$$n! = n(n - 1)(n - 2) \cdots 3 \cdot 2 \cdot 1$$

for all positive integers n. For example,

$$5! = 5 \cdot 4 \cdot 3 \cdot 2 \cdot 1 = 120$$

Stirling's formula, which states

$$n! \approx \left(\frac{n}{e}\right)^n \sqrt{2\pi n}$$

can be used to approximate very large factorials. Use Stirling's formula to approximate 20! and compare it to the actual value.

Questions for Thought

56. Develop a formula for the half-life of a substance that is decaying at the rate of r per year.

57. If we are exposed to two sounds with intensities I_1 and I_2 and decibel levels N_1 and N_2, respectively, then we are experiencing sound at an intensity level of $I_1 + I_2$. However, the decibel level of the combination of sounds is *not* $N_1 + N_2$. In other words, combing two sounds of 80 decibels and 90 decibels does not produce a sound of 170 decibels. Suppose that union regulations prohibit employers from constructing a work environment in which a worker is exposed to a sound level greater than 100 decibels. If an employee works at a machine that produces 90 decibels and then a new machine is moved nearby that produces 80 decibels, has the union regulation been violated?

58. Explain how the decibel scale discussed in Exercise 43 is an application of the Weber–Fechner law discussed in Exercise 41.

59. In Example 9 we used the exponential decay model to find how long it would take 80 grams of a substance with a half-life of 8 minutes to decay to 7 grams.
(a) Explain why an alternate exponential decay model for this substance is $A = A_0\left(\frac{1}{2}\right)^{t/8}$.
(b) Use this alternate decay model to answer the question of Example 9. Which model was easier to use?

60. In Example 10 we showed that an earthquake of magnitude 5 is 100 times more intense that an earthquake of magnitude 3.
(a) Show that regardless of the magnitudes, if one earthquake has a magnitude 2 greater than a second earthquake, then it is 100 times more intense.
(b) Show that if one earthquake has magnitude M_1 and a second earthquake has magnitude M_2, with $M_2 - M_1 = d$, then the intensity of the second earthquake is 10^d times that of the first.

61. In this section we derived the compound interest formula $A = P\left(1 + \dfrac{r}{n}\right)^{nt}$ and we also recognized that the expression $\left(1 + \dfrac{1}{n}\right)^{n}$ gets closer to e as n gets larger and larger. Let's see how the compound interest formula and the formula for continuous compounding are related.

(a) Explain why the compound interest formula $A = P\left(1 + \dfrac{r}{n}\right)^{nt}$ is equivalent to

$$A = P\left[\left(1 + \dfrac{r}{n}\right)^{\frac{n}{r}}\right]^{rt}$$

(b) Make the substitution $u = \dfrac{n}{r}$ and rewrite the equivalent form of the compound interest formula obtained in part (a) as

$$A = P\left[\left(1 + \dfrac{1}{u}\right)^{u}\right]^{rt}$$

(c) Since r is a constant, what happens to $u = \dfrac{n}{r}$ as n gets very large?

(d) In the expression $\left(1 + \dfrac{1}{n}\right)^{n}$, note that the exponent n and $\dfrac{1}{n}$ are reciprocals and that as n gets larger, $\dfrac{1}{n}$ gets closer to zero. Explain why the same situation exists for the expression $\left(1 + \dfrac{1}{u}\right)^{u}$.

(e) Conclude that as n gets larger, $\left(1 + \dfrac{1}{u}\right)^{u}$ is also getting closer to e.

(f) Therefore, as n gets larger (which is the situation approaching continuous compounding) we have

$$P\left[\left(1 + \dfrac{r}{n}\right)^{\frac{n}{r}}\right]^{rt} = P\left[\left(1 + \dfrac{1}{u}\right)^{u}\right]^{rt} = Pe^{rt}$$

which is the formula for continuous compound interest.

Chapter 6   ***Summary***

After completing this chapter you should:

1. Recognize the graphs of the basic exponential and logarithmic functions. See Figure 6.19. (Sections 6.1 and 6.2)

Figure 6.19

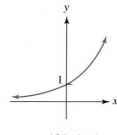

$y = b^x$ for $b > 1$

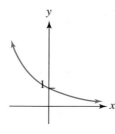

$y = b^x$ for $0 < b < 1$

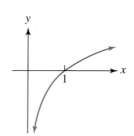

$y = \log_b x$ for $b > 1$

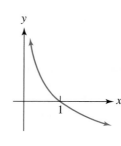

$y = \log_b x$ for $0 < b < 1$

2. Be able to translate statements from exponential form to logarithmic form, and vice versa. (Section 6.2)

A logarithm is simply an exponent:

$$x = b^y \iff y = \log_b x$$

That is, y is the exponent to which b must be raised to yield x.
For example:

The exponential statement $4^{-3} = \dfrac{1}{64}$ is equivalent to the logarithmic statement

$$\log_4 \dfrac{1}{64} = -3.$$

3. Be able to use the definition of logarithms to evaluate logarithmic expressions. (Section 6.2)

For example:

Find $\log_8 \dfrac{1}{4}$.

Solution:

We let $\log_8 \dfrac{1}{4} = a$, translate this into exponential form, and solve the resulting exponential equation.

$\log_8 \dfrac{1}{4} = a$ is equivalent to

$8^a = \dfrac{1}{4}$ Express both sides in terms of base 2.

$(2^3)^a = \dfrac{1}{2^2}$

$2^{3a} = 2^{-2} \implies 3a = -2 \implies a = -\dfrac{2}{3}$

Thus $\log_8 \dfrac{1}{4} = -\dfrac{2}{3}$.

4. Be able to use the properties of logarithms to rewrite logarithmic expressions. (Section 6.3)

The logarithm properties allow us to express the logarithm of products, quotients, and powers as the sum and difference of simpler logarithms.

For example:

Write $\log_b\left(\dfrac{\sqrt[3]{xy}}{z^4}\right)$ in terms of simpler logarithmic expressions:

Solution:

$$\log_b\left(\dfrac{\sqrt[3]{xy}}{z^4}\right) = \log_b\left(\dfrac{(xy)^{1/3}}{z^4}\right) \qquad \text{Use property 2.}$$

$$= \log_b(xy)^{1/3} - \log_b z^4 \qquad \text{Use property 3.}$$

$$= \dfrac{1}{3}\log_b(xy) - 4\log_b z \qquad \text{Use property 1.}$$

$$= \dfrac{1}{3}\left(\log_b x + \log_b y\right) - 4\log_b z$$

5. Be able to use the logarithm properties to solve logarithmic equations. (Section 6.3)

For example:

Solve for x: $\log_2(3x - 1) + \log_2(x + 1) = 5$

Solution:

$$\log_2(3x - 1) + \log_2(x + 1) = 5 \qquad \text{Write as a single logarithm.}$$

$$\log_2[(3x - 1)(x + 1)] = 5 \qquad \text{Write in exponential form.}$$

$$(3x - 1)(x + 1) = 2^5 = 32$$

$$3x^2 + 2x - 33 = 0$$

$$(3x + 11)(x - 3) = 0$$

$$x = -\dfrac{11}{3} \quad \text{or} \quad x = 3$$

We reject $x = -\frac{11}{3}$ because it makes the arguments negative; the solution to the equation is $x = 3$.

6. Be able to use logarithms to solve exponential equations. (Section 6.4)

If two expressions are equal, their logarithms are equal and we can take the log of both sides. We then apply the log properties to solve the equation.
For example:
Solve for t: $7^{2t-1} = 5$
Solution:
Since we cannot easily express both sides in terms of the same base, we can use logarithms to solve the equation.

$$7^{2t-1} = 5 \qquad \text{Take the natural logarithm of both sides.}$$
$$\ln 7^{2t-1} = \ln 5 \qquad \text{Use the log properties.}$$
$$(2t - 1)\ln 7 = \ln 5 \implies t = \frac{\ln 5 + \ln 7}{2 \ln 7} \approx 0.914$$

7. Solve applied problems involving exponential or logarithmic functions. (Section 6.5)

Many real-life situations, such as continuous compounding, population growth, and radioactive decay, can be solved using the exponential growth model or the exponential decay model.
For example:
What sum of money must be invested at 7.3% compounded continuously to yield $10,000 in 12 years?
Solution:
We use the formula $A = A_0 e^{rt}$ with $A = 10,000$, $r = 0.073$, and $t = 12$ and solve for A_0:

$$10,000 = A_0 e^{0.073(12)}$$
$$10,000 = A_0 e^{0.876}$$
$$A_0 = \frac{10,000}{e^{0.876}} \approx \$4164 \quad \text{to the nearest dollar}$$

Chapter 6   *Review Exercises*

In Exercises 1–8, sketch the graph of the given function. Be sure to indicate any asymptotes and label the intercepts.

1. $y = 2^{x-1}$

2. $f(x) = \left(\frac{1}{4}\right)^x - 1$

3. $f(x) = \log_{1/3}(x + 3)$

4. $y = 5 + \log_5 x$

5. $y = e^{x+2} - 1$

6. $f(x) = 2 + \ln x^2$

7. $y = 8 - 2^{x+3}$

8. $y = -\ln x$

In Exercises 9–14, translate the logarithmic statements into exponential form and the exponential statements into logarithmic form.

9. $\log_6 \frac{1}{6} = -1$

10. $9^{1/2} = 3$

11. $8^{-2/3} = \frac{1}{4}$

12. $\log_3 81 = 4$

13. $\log_b b^6 = 6$

14. $b^{\log_b t} = t$

In Exercises 15–26, evaluate the given expression without using a calculator.

15. $\log 1000$

16. $\log 0.000001$

17. $\log_3 \frac{1}{9}$

18. $\log_{1/2} 8$

19. $\log_8 4$

20. $\log_{32} 16$

21. $3^{\log_3 7}$

22. $\ln e^5$

23. $\log_2 16^5$

24. $10^{\log 17}$

25. $\log_b \sqrt{b^3}$

26. $\log_b 1$

In Exercises 27–32, express the given logarithm in terms of simpler logarithms where possible.

27. $\log_b(x^3 y^4 z^2)$

28. $\log_2\left(\frac{8}{\sqrt{xy}}\right)$

29. $\log_5 \sqrt[3]{\dfrac{8x}{5y^4}}$

30. $\log_b(x^2 + y^5)$

31. $\dfrac{\sqrt{\log_b x}}{\sqrt[3]{\log_b y}}$

32. $\log_b \dfrac{\sqrt{x}}{\sqrt[3]{y}}$

In Exercises 33–50, solve the given exponential or logarithmic equation.

33. $8^x = \dfrac{1}{64}$

34. $\left(\dfrac{1}{9}\right)^x = 27$

35. $\left(\dfrac{1}{9}\right)^x = 29$

36. $4^{x^2 - x} = 16$

37. $\log_b(3x) + \log_b(x + 2) = \log_b 9$

38. $\log_2 x + \log_2(x + 1) = 1$

39. $7^{x-1} = 3$

40. $\log_2(t + 1) + \log_2(t - 1) = 3$

41. $\log_5(6x) - \log_5(x + 2) = 1$

42. $3^x = 5^{x+2}$

43. $3^x = 5(2^x)$

44. $\log_4 x - \log_4(x - 4) = \log_4(x - 6)$

45. $\dfrac{1}{2}\log_3 x = \log_3(x - 6)$

46. $2 \log_b x = \log_b(6x - 5)$

47. $8^{3x-2} = 9^{x+2}$

48. $\log x = 2 + \log(x - 1)$

49. $\log_b 125 = 3$

50. $\log_8 128 = x$

51. Verify that $f(x) = e^{2x-3}$ and $g(x) = \dfrac{3 + \ln x}{2}$ are inverse functions.

52. Find the inverse function of $y = f(x) = 2^{x+3}$.

53. Find the inverse function of
$$y = f(x) = 5 + \log_3(x - 1)$$

54. Express $\log_6 x$ in terms of base 3 logarithms.

55. Estimate $\log_7 11$. Round your answer to two decimal places.

56. If $2000 is invested in an account paying 6.5% annual interest compounded daily, how much money will be in the account after 6 years?

57. How much money must be invested at 7.2% annual interest compounded continuously so that the investment will be worth $10,000 in 8 years?

58. What annual interest rate compounded continuously will make an investment of $1000 grow to $2000 in 6 years?

59. A colony of bacteria is growing according to the exponential growth model. Suppose 800 bacteria grow to 2000 in 16 hours.
  (a) How many bacteria will be present after 10 hours?
  (b) How long will it take for there to be 3000 bacteria?

60. Find the doubling time of a colony of bacteria with growth equation $A = A_0 2^{0.06t/4}$, where t is measured in hours.

61. A radioactive substance has a half-life of 53 days. Assuming an exponential decay model, find the decay equation for this substance.

62. Certain manufactured radioactive substances have extremely short half-lives. Find the exponential decay model for a substance with a half-live of 5 minutes.

63. In the exponential decay model $A = A_0 e^{-rt}$, describe how the half-life is affected if r doubles from 0.1 to 0.2.

64. The magnitude of the great San Francisco earthquake of 1906 measured 8.3 on the Richter scale. If an earthquake is 5 times as intense, what would its magnitude be on the Richter scale? See Example 10 in Section 6.5.

65. Find the intensity of a sound that registers 100 decibels. See Exercise 43 in Section 6.5.

Chapter 6   *Practice Test*

1. Sketch the graphs of the following functions. Label the intercepts.
  (a) $y = \left(\dfrac{1}{2}\right)^{x+1} - 4$
  (b) $y = 1 + \log_3 x$

2. Given $y = f(x) = 2^{x+1}$, find its inverse function and sketch the graphs of both $f(x)$ and its inverse on the same coordinate system.

3. Evaluate each of the following. Round to the nearest hundredth where necessary.
  (a) $\log_9 \dfrac{1}{3}$
  (b) $\log_7 4$
  (c) $\log_8 16$
  (d) $\log_{1/2} \dfrac{1}{4}$

4. Solve each of the following equations. Round your answer to the nearest hundredth where necessary.
  (a) $3^{2x-1} = 9$
  (b) $\log_3(x - 9) + \log_3(x + 1) = 2$
  (c) $\dfrac{1}{2}\log_2 x - \log_2(x - 3) = 1$
  (d) $5^{x+1} = 10$

5. A colony of bacteria is growing according to the exponential growth model. Suppose 600 bacteria grow to 1500 in 20 hours.
  (a) How many bacteria will be present after 10 hours?
  (b) How long will it take for there to be 6000 bacteria?

6. A radioactive substance has a half-life of 750 years. Assuming an exponential decay model, find the decay equation for this substance. How much of 100 grams of this substance will be left in 200 years?

7

Systems of Linear Equations and Inequalities

In this chapter, we will discuss systematic procedures for finding solutions to problems that involve several conditions or constraints. These conditions may be represented by systems of equations or systems of inequalities.

In the first section, we will discuss solving systems of two linear equations in two variables by manipulating the equations. Although much of this material will be a review, it is important that you thoroughly understand these fundamental concepts, as they will be generalized to larger systems in Section 7.2, and will form the basis for developing other systematic methods to be discussed in the remaining sections.

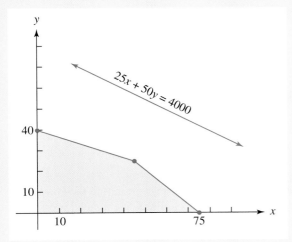

415

7.1   2 × 2 Linear Systems: Elimination and Substitution

Consider the following situation: As an employee in a sales department, you are offered the option of receiving your salary two possible ways: You can either be paid a straight commission of 9% of your gross sales, or you can receive a base pay of $240 per week plus a commission of 6% of your gross sales. Which salary plan should you choose?

To get a better idea of how the two salary plans differ, let's examine the weekly salary earned for different amounts of weekly sales. We can write an equation for each salary plan expressing the weekly salary s in terms of the gross sales g, as follows:

$$s = 0.09g \qquad \text{The weekly salary under the straight commission plan}$$

$$s = 240 + 0.06g \qquad \text{The weekly salary under the base pay + 6\% commission plan}$$

Thus, for example, if you expect to sell $5000 worth of merchandise in a week, you would earn $0.09 \times 5000 = \$450$ under the straight 9% commission plan, but $240 + 0.06 \times 5000 = \$540$ under the base pay plus 6% commission plan. On the other hand, if you expect to sell $10,000 worth of merchandise in a week, you would earn $0.09 \times 10,000 = \$900$ under the straight 9% commission plan, but $240 + 0.06 \times 10,000 = \$840$ under the base pay plus 6% commission plan.

It seems that if you expect to sell "a lot" of merchandise, you earn more under the straight 9% commission plan, whereas if you do not expect to sell "a lot," then the base pay plus 6% commission plan is better. To make an intelligent choice, it would be helpful to know the level of sales at which each plan is advantageous.

Note that each of the salary equations is a first-degree (linear) equation in two variables and hence has a straight line as its graph. Let's graph the two equations on a graphing calculator where the horizontal axis represents g and the vertical axis represents s.

Figure 7.1

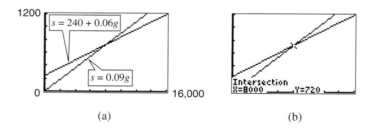

(a)   (b)

By determining the point of intersection of the lines, we can find the gross sales amount that yields the same income under both plans. For this example, gross sales of $8000 will yield the same income, $720, for both plans. The point of intersection of the two lines, (8000, 720), is important for our purposes because it tells us when one plan is more advantageous than the other. You can see from the graph in Figure 7.1 that if you sell less than $8000 worth of merchandise per week, the base pay plus commission plan yields a higher salary; if you sell more than $8000 worth of merchandise per week, the straight commission plan yields a higher salary. Your decision as to which plan to take reduces to whether you think you can average above $8000 in gross sales per week.

For this problem, finding the point of intersection helped to clarify the information we are given in order to make a decision as to which plan of income to select. We are often interested in finding points of intersection, but determining such points from hand-drawn graphs can be imprecise. In this chapter we discuss more systematic procedures for finding solutions to problems like these, and then expand these methods to more general cases.

Linear Systems: Two Variables

First we review some terminology. Two or more equations considered together are called a **system of equations**. In particular, if the equations are of the first degree, it is called a **linear system**. The system

$$\begin{cases} x - 5y = 6 \\ 2x + 3y = 4 \end{cases}$$

is an example of a linear system in two variables called a **2 × 2 system**, since there are two equations and two variables.

Solving a system of equations means finding all ordered pairs that satisfy all the equations in the system.

Since the graph of an equation is a picture of its solutions and the graph of a first-degree equation in two variables is a straight line, we have the following three possibilities for a linear system:

1. The lines intersect at exactly one point. The coordinates of this point are the solution to this system. In this case we say that the system is **consistent and independent**. See Figure 7.2(a).

2. The lines are parallel; that is, they never intersect. In this case we say that the system is **inconsistent** and that there is no solution. See Figure 7.2(b).

3. The lines coincide; that is, all solutions to one of the equations are also solutions to the other (the two equations are equivalent). There are an infinite number of solutions. In this case, we call the system **dependent**. See Figure 7.2(c).

Figure 7.2

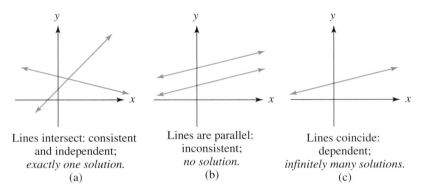

| Lines intersect: consistent and independent; *exactly one solution.* (a) | Lines are parallel: inconsistent; *no solution.* (b) | Lines coincide: dependent; *infinitely many solutions.* (c) |

In much of this chapter we will focus on algebraic techniques for solving these systems. In this section we will discuss the **elimination method** and the **substitution method** for solving systems of equations.

The idea behind the **elimination method**, as its name suggests, is to eliminate variables by adding equations until we have a single equation in one variable. We illustrate this method with some examples.

The elimination method is also called the addition method.

Example 1 Solve the following systems of equations:

(a) $\begin{cases} 2x + y = 5 \\ 3x - y = 10 \end{cases}$ (b) $\begin{cases} 2x - 5y = -1.5 \\ 3x - 9y = -3 \end{cases}$ (c) $\begin{cases} 6x - 3y = 9 \\ 8x - 4y = 5 \end{cases}$

Solution (a) The addition property of equality allows us to add equal quantities to both sides of an equation. From the second equation we know that $3x - y$ and 10 are equal quantities. Hence we can add $3x - y$ to the left-hand side of $2x + y = 5$, and 10 to the right-hand side of $2x + y = 5$ as follows:

$$\begin{array}{rl} 2x + y = & 5 \\ \underline{3x - y = } & \underline{10} \quad \text{Add.} \\ 5x \quad\;\; = & 15 \end{array}$$

We *eliminated* one of the variables; we can now solve this first-degree equation in *one* variable, $5x = 15$, to get $x = 3$.

A solution to a system of equations in two variables consists of two numbers: an x-value and a y-value. We have found x. To find its corresponding y-value, we substitute the value found for x into either of the original equations and solve for y:

$$2x + y = 5 \qquad \text{Substitute } x = 3 \text{ in the first equation and solve for } y.$$
$$2(3) + y = 5 \qquad \text{Which yields } y = -1$$

The solution is $(3, -1)$. We check this solution by substituting $x = 3$ and $y = -1$ in both of the original equations. We leave this check to the student.

(b) $\begin{cases} 2x - 5y = -1.5 \\ 3x - 9y = -3 \end{cases}$

Adding these two equations at this point will not eliminate either of the variables. However, we can transform one or both of the equations by using the multiplication property of equality to change the coefficients of either x or y so that they are exact opposites. Then we can add the two equations and eliminate the variable with opposite coefficients. We choose to eliminate the x variable in this problem.

$$\begin{array}{lll} 2x - 5y = -1.5 & \xrightarrow{\;\text{Multiply by 3}\;} & 6x - 15y = -4.5 \\ 3x - 9y = -3 & \xrightarrow{\;\text{Multiply by } -2\;} & \underline{-6x + 18y = 6} \quad \text{Add.} \\ & & 3y = 1.5 \quad \text{Which yields } y = 0.5 \end{array}$$

Now we find x by substituting $y = 0.5$ into the first equation:

$$2x - 5y = -1.5 \qquad \text{Substitute } y = 0.5 \text{ to get}$$
$$2x - 5(0.5) = -1.5 \qquad \text{Which yields } x = 0.5$$

The solution is $(0.5, 0.5)$. Again, the check is left to the student.

(c) $\begin{cases} 6x - 3y = 9 \\ 8x - 4y = 5 \end{cases}$

In this system we choose to eliminate y by multiplying the first equation by 4 and the second by -3.

$$\begin{array}{lll} 6x - 3y = 9 & \xrightarrow{\;\text{Multiply by 4}\;} & 24x - 12y = 36 \\ 8x - 4y = 5 & \xrightarrow{\;\text{Multiply by } -3\;} & \underline{-24x + 12y = -15} \quad \text{Add.} \\ & & 0 = 21 \quad \text{This is a contradiction.} \end{array}$$

This means that it is impossible for both equations to have a common solution; therefore, there is no solution.

If you write the two equations in part (c) in slope–intercept form, you will see that both equations have the same slope but different y-intercepts. This means that the two lines are parallel, and therefore the system has no solution. ∎

For a system of two equations in two variables, what does it mean graphically when we arrive at no solution?

Recall that we discussed substitution in Section 2.6.

We can also solve systems of equations using the **substitution method**. Our goal remains the same, to obtain a single equation in one variable, but the approach is slightly different than the elimination method, as illustrated next.

Example 2 Solve the following systems of equations:

(a) $\begin{cases} 5x + 3y = 5 \\ \quad\quad x = 2y - 10 \end{cases}$ (b) $\begin{cases} 6x - y = 9 \\ 8x - \dfrac{4}{3}y = 12 \end{cases}$

Solution (a) The second equation, $x = 2y - 10$, is solved explicitly for x; hence we can substitute $2y - 10$ in place of x into the first equation as follows:

$$5x + 3y = 5 \qquad \text{Replace } x \text{ by } 2y - 10.$$
$$5(2y - 10) + 3y = 5 \qquad \text{Now solve for } y.$$
$$10y - 50 + 3y = 5 \qquad \text{We get } y = \dfrac{55}{13}$$

Now solve for x by substituting $\dfrac{55}{13}$ for y in the *second* equation, since it is already explicitly solved for x.

$$x = 2y - 10$$
$$x = 2\left(\frac{55}{13}\right) - 10 = -\frac{20}{13}$$

Check the solution by using a graphing calculator to find the intersection of the graphs of the equations of Example 2(a). Compare the fractional answers with the decimal values given by the calculator.

The solution is $\left(-\dfrac{20}{13}, \dfrac{55}{13} \right)$. The check is left to the student.

(b) $\begin{cases} 6x - y = 9 \\ 8x - \dfrac{4}{3}y = 12 \end{cases}$

We can easily solve the first equation explicitly for y:

$$6x - y = 9 \implies y = 6x - 9$$

Now substitute $6x - 9$ for y in the second equation:

$$8x - \frac{4}{3}y = 12 \qquad \text{Replace } y \text{ with } 6x - 9.$$
$$8x - \frac{4}{3}(6x - 9) = 12$$
$$8x - 8x + 12 = 12$$
$$12 = 12 \qquad \text{This is an identity.}$$

If we examine the two equations closely, we find that they both have the same slope and y-intercept. If we put both equations into slope–intercept form, we get $y = 6x - 9$. Hence, both equations represent the same line. The equations are dependent. The solution set is all points lying on the line; that is, $\{(x, y) \,|\, y = 6x - 9\}$.

This notation for the final answer means the solutions are all ordered pairs (x, y) such that $y = 6x - 9$.

There are infinitely many solutions to this system: For any value we choose for x, we can come up with a corresponding y-value, $6x - 9$, such that the ordered pair (x, y) satisfies the system. A more convenient way to express this type of solution is by letting x be a constant; for example, let $x = a$, where a is any real number. Then $y = 6x - 9 \implies y = 6a - 9$. This allows us to write the solution as $(a, 6a - 9)$, where a is any real number .

For a system of two equations in two variables, what does a dependent system mean graphically?

Calculator Exploration

Using a graphing calculator to find the solution to the following systems by finding the intersection of the graphs of the two lines.

(a) $\begin{cases} 2x - 7y = 5 \\ \quad x + \ y = 1 \end{cases}$ **(b)** $\begin{cases} x - 5y = 8 \\ \qquad x = 10y + 3 \end{cases}$

Solve the same problems algebraically and compare the solutions.

To summarize, we solve a 2×2 system by reducing it to a 1×1 system—that is, one equation in one variable. Now that we are able to solve systems of equations with two variables, we have more flexibility in solving verbal problems.

Example 3 How much of each of a 30% alcohol solution and a 15% alcohol solution should be mixed together to get 50 liters of a 20% alcohol solution?

Solution We approach this problem using two variables. If we let x be the number of liters of the 30% alcohol solution and y be the number of liters of the 15% alcohol solution, then we need to write *two* equations expressing the relationships between x and y.

We know that $x + y = 50$, since we want the final mixture to be 50 liters. This is the first equation. The second equation is written in terms of the amounts of pure alcohol. The 30% solution contains $0.30x$ liters of pure alcohol, and the 15% solution contains $0.15y$ liters of pure alcohol. Because we want to end up with 50 liters of a 20% alcohol solution, this means we want to have $0.20(50) = 10$ liters of pure alcohol in the mixture when we are done. Hence our second equation is:

$$0.30x \qquad + \qquad 0.15y \qquad = \qquad (0.20)(50)$$

| Amount of pure alcohol in 30% solution | + | Amount of pure alcohol in the 15% solution | = | Amount of pure alcohol in the final mixture |

So the system is

$$\begin{cases} \quad x + \qquad y = 50 \\ 0.30x + 0.15y = 10 \end{cases}$$

We will use the substitution method and begin by solving for y in the first equation:

$$x + y = 50 \implies y = 50 - x$$

Now substitute $50 - x$ for y into the second equation:

$$0.30x + 0.15y = 10 \qquad \text{Replace } y \text{ with } 50 - x.$$
$$0.30x + 0.15(50 - x) = 10 \qquad$$

This is the same equation we would arrive at if we used the single-variable approach. We clear the decimals by first multiplying each side by 100.

$$30x + 15(50 - x) = 1000$$
$$30x + 750 - 15x = 1000$$
$$x = 16.67$$

Hence we have $\boxed{16.67 \text{ liters of the 30\% solution}}$. Since $y = 50 - x$, we have $y = 50 - 16.67 = \boxed{33.33 \text{ liters of the 15\% solution}}$. The check is left to the student.

Example 4

An object is thrown in the air. If v_0 is the initial velocity with which it is thrown upward and s_0 is the height of the object above the ground when it is first thrown (called its initial height), then the equation

$$s = -16t^2 + v_0t + s_0$$

gives the height (s) in feet the object is above the ground t seconds after it is thrown. See Figure 7.3. If the object is 64 feet above the ground after 1 second, and 46 feet above the ground after 2 seconds, find its initial velocity (v_0) and initial height (s_0).

Solution

In this example we are asked to find v_0 and s_0 for the equation $s = -16t^2 + v_0t + s_0$. It is important that you understand that this equation has a real meaning: It tells us how high above the ground, s, the object is (in feet) as t (the time in seconds) changes. We are given that the object is 64 feet above the ground after 1 second. This translates into $s = 64$ when $t = 1$. If we substitute these values into the equation, we get

$$s = -16t^2 + v_0t + s_0 \qquad \text{Substitute } s = 64 \text{ and } t = 1 \text{ into the equation.}$$
$$64 = -16(1)^2 + v_0(1) + s_0$$
$$64 = -16 + v_0 + s_0 \qquad \text{Which simplifies to } 80 = v_0 + s_0$$

Next we are given that after 2 seconds, the object is 46 feet above the ground. We translate this as $s = 46$ when $t = 2$. We substitute these values into the same equation to generate another equation in the variables v_0 and s_0:

$$s = -16t^2 + v_0t + s_0 \qquad \text{Substitute } s = 46 \text{ and } t = 2 \text{ into the equation.}$$
$$46 = -16(2)^2 + v_0(2) + s_0$$
$$46 = -64 + 2v_0 + s_0 \qquad \text{Which simplifies to } 110 = 2v_0 + s_0$$

Hence, we have two equations in two unknowns:

$$\begin{cases} v_0 + s_0 = 80 \\ 2v_0 + s_0 = 110 \end{cases}$$

We solve this system as follows:

$$\begin{array}{l} v_0 + s_0 = 80 \\ 2v_0 + s_0 = 110 \end{array} \quad \begin{array}{l} \xrightarrow{\text{Multiply by } -1} \\ \xrightarrow{\text{As is}} \end{array} \quad \begin{array}{l} -v_0 - s_0 = -80 \\ 2v_0 + s_0 = 110 \qquad \text{Add.} \\ \hline v_0 = 30 \qquad \text{Now we find } s_0. \end{array}$$

$$v_0 + s_0 = 80 \qquad \text{Substitute } v_0 = 30 \text{ into the first equation to get}$$
$$30 + s_0 = 80 \qquad \text{Which yields } s_0 = 50$$

Hence, the upward velocity of the object is $v_0 = 30$ ft/sec and the initial height of the object is $s_0 = 50$ feet . That is, the object starts out at 50 feet above the ground and is thrown with an initial upward velocity of 30 ft/sec. The equation for s is $s = -16t^2 + 30t + 50$. ∎

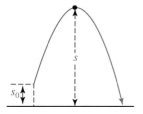

Figure 7.3

Given that factors such as air resistance and friction are negligible, the equation $s = -16t^2 + v_0t + s_0$ is a mathematical model of a real physical relationship.

Partial Fractions

In algebra, we are used to finding the sum of two simple fractions such as $\dfrac{3}{x-4} + \dfrac{1}{x+1}$ and arriving at the answer $\dfrac{4x-1}{(x-4)(x+1)}$.

In calculus, it is occasionally necessary to do the opposite—that is, to take a rational expression and express it as a sum of *simpler* rational expressions. For example, we may need to rewrite the expression $\dfrac{x-5}{(x-2)(x+1)}$ as the sum $\dfrac{A}{x-2} + \dfrac{B}{x+1}$, where A and B are constants. Rewriting a fraction as a sum of simpler fractions in this way is called **partial fraction decomposition**. It will be our goal to find the constants A and B, as illustrated in the following example.

Example 5 | Find A and B such that $\dfrac{x - 5}{(x - 2)(x + 1)} = \dfrac{A}{x - 2} + \dfrac{B}{x + 1}$.

Solution

$$\frac{x - 5}{(x - 2)(x + 1)} = \frac{A}{x - 2} + \frac{B}{x + 1}$$

Rewrite the right-hand side as a single fraction.

$$\frac{x - 5}{(x - 2)(x + 1)} = \frac{A(x + 1) + B(x - 2)}{(x - 2)(x + 1)}$$

Since the denominators are identical, the two fractions are equivalent if and only if the two numerators are equal.

$$x - 5 = A(x + 1) + B(x - 2)$$

Multiply out the right-hand side.

$$x - 5 = Ax + A + Bx - 2B$$

Rewrite this equation so that the right-hand side is in polynomial form.

$$x - 5 = Ax + Bx + A - 2B$$
$$x - 5 = (A + B)x + (A - 2B)$$

This last equation must be true for *all x*. In effect, we are saying that the *polynomials* $x - 5$ and $(A + B)x + (A - 2B)$ are equal. But two polynomials of the same degree are equal if and only if *the coefficients of each power of x are identical*. Hence, for these two polynomials to be equal, the coefficients of x for both polynomials must be equal and the numerical terms for both polynomials must be equal.

$$x - 5$$

The coefficient of x is 1: the numerical term is -5.

$$(A + B)x + (A - 2B)$$

The coefficient of x is $A + B$; the numerical term is $A - 2B$.

Hence we have $\begin{cases} 1 = A + B \\ -5 = A - 2B \end{cases}$ which is a 2 × 2 system solved as follows:

$$
\begin{array}{lll}
A + B = 1 & \xrightarrow{\text{As is}} & A + B = 1 \\
A - 2B = -5 & \xrightarrow{\text{Multiply by } -1} & \underline{-A + 2B = 5} \quad \text{Add.}\\
& & 3B = 6 \quad \text{Which yields } B = 2
\end{array}
$$

Now we find A by substituting $B = 2$ into the first equation:

$$A + B = 1 \implies A + (2) = 1 \quad \text{which yields} \quad A = -1.$$

Because $A = -1$ and $B = 2$, we have

$$\frac{x - 5}{(x - 2)(x + 1)} = \frac{A}{x - 2} + \frac{B}{x + 1} \implies \frac{x - 5}{(x - 2)(x + 1)} = \frac{-1}{x - 2} + \frac{2}{x + 1}$$

The approach we took in this partial fraction decomposition problem is to use methods of solving systems of linear equations. Another way to approach this problem is to realize that if the equation $x - 5 = A(x + 1) + B(x - 2)$ is true for all values of x, then, in particular, it must be true for $x = -1$. Therefore, if we substitute $x = -1$ into this equation we get

$$x - 5 = A(\ x + 1) + B(\ x - 2)$$

Let $x = -1$. Then we have

$$-1 - 5 = A(-1 + 1) + B(-1 - 2)$$

Which becomes

$$-6 = -3B$$

Which yields $B = 2$

If we substitute $x = 2$ into this equation we get

$$x - 5 = A(x + 1) + B(x - 2)$$ Let $x = 2$. Then we have

$$2 - 5 = A(2 + 1) + B(2 - 2)$$ Which becomes

$$-3 = 3A$$ Which yields $A = -1$

Note that each value we chose for x eliminated either A or B.   ∎

7.1 Exercises

In Exercises 1–4, solve the system of equations by graphing both equations and finding where they intersect. Round your answers to the nearest hundredth where necessary.

1. $\begin{cases} 4x + 6y = 4 \\ 6x - 3y = 2 \end{cases}$

2. $\begin{cases} 8x - 5y = 13 \\ 12x - 10y = 20 \end{cases}$

3. $\begin{cases} 3x - y = 6 \\ x + 4y = 8 \end{cases}$

4. $\begin{cases} x - 5y = 10 \\ 4x - y = 6 \end{cases}$

In Exercises 5–26, solve the system of equations.

5. $\begin{cases} 2x + y = 12 \\ 3x - y = 8 \end{cases}$

6. $\begin{cases} -x + 2y = -4 \\ x - y = 3 \end{cases}$

7. $\begin{cases} 3x - 2y = 15 \\ 2x + y = 10 \end{cases}$

8. $\begin{cases} 5x + 2y = -8 \\ 3x + y = -4 \end{cases}$

9. $\begin{cases} 5x - 2y = 1 \\ x - 5y = -32 \end{cases}$

10. $\begin{cases} x = 3y - 1 \\ 2x - 5y = -3 \end{cases}$

11. $\begin{cases} y = 3x - 15 \\ 8x - 2y = 10 \end{cases}$

12. $\begin{cases} x + 7y = -1 \\ 3x + 21y = -3 \end{cases}$

13. $\begin{cases} 3x + 4y = 12 \\ 4x + 5y = 15 \end{cases}$

14. $\begin{cases} 2x - 7y = 8 \\ 5x - 3y = 20 \end{cases}$

15. $\begin{cases} 5x - 6y = 13 \\ 3x + 8y = 2 \end{cases}$

16. $\begin{cases} 6x - 7y = 9 \\ 9x - 2y = 5 \end{cases}$

17. $\begin{cases} 0.1x + 0.4y = 21 \\ 0.2x + 0.3y = 17 \end{cases}$

18. $\begin{cases} 0.35x - 0.36y = 540 \\ 0.7x - 0.18y = 1215 \end{cases}$

19. $\begin{cases} \dfrac{2}{3}y + \dfrac{3}{2}x = 2 \\[2mm] \dfrac{3}{4}x + \dfrac{1}{12}y = 1 \end{cases}$

20. $\begin{cases} \dfrac{5x}{4} + \dfrac{2y}{3} = -4 \\[2mm] \dfrac{3x}{5} + \dfrac{5y}{2} = -15 \end{cases}$

21. $\begin{cases} \dfrac{a}{6} + \dfrac{b}{8} = \dfrac{3}{4} \\[2mm] \dfrac{a}{4} + \dfrac{b}{3} = \dfrac{17}{12} \end{cases}$

22. $\begin{cases} \dfrac{s}{5} + \dfrac{t}{3} = 1 \\[2mm] \dfrac{s}{4} + \dfrac{t}{3} = 2 \end{cases}$

23. $\begin{cases} \dfrac{x + 3}{2} + \dfrac{y - 4}{3} = \dfrac{5}{3} \\[2mm] \dfrac{x - 2}{3} + \dfrac{y - 2}{2} = 2 \end{cases}$

24. $\begin{cases} 0.01x + 0.003y = 6 \\ 0.2x = 0.05y - 2 \end{cases}$

25. $\begin{cases} \dfrac{x}{2} - 0.03y = 0.6 \\[2mm] 0.0004y + \dfrac{x}{2} = 0.2 \end{cases}$

26. $\begin{cases} 0.3x - 0.002y = 0.6 \\ 12x = 0.08y - 2 \end{cases}$

In Exercises 27–42, write an equation or system of equations to solve the problem, then solve the problem. Round your answer to the nearest hundredth where necessary.

27. How much of each of a 30% alcohol solution and a 45% alcohol solution must be mixed together to get 30 liters of a 40% solution of alcohol?

28. How much of each of a 20% alcohol solution and a 45% alcohol solution must be mixed together to get 30 liters of a 30% solution of alcohol?

29. Carol wants to invest a total of $20,000 so that her yearly interest is $1690. If she invests part in a certificate yielding 8% and the other part in stock at 9.5%, how much should she invest at each rate?

30. Rob invests money at 10% and 8%, earning a yearly interest of $640. Had the amounts invested been reversed, he would have received $610. How much is invested altogether?

31. A car rental agency charges a flat fee plus a mileage rate for a 1-day rental. If the charge for a 1-day rental with 90 miles is $58.30 and the charge for a 1-day rental with 140 miles is $74.30, find the flat fee and the charge per mile.

32. A photocopy machine company charges a flat monthly fee plus a usage rate for its XS-8600 duplicating machine. The charge for the first month, when 3000 copies were made, was $220 and the charge for the second month, when 2600 copies were made, was $210. Find the flat monthly fee and charge per copy for the XS-8600.

33. A plane can cover a distance of 2520 miles in 4 hours with a tailwind (with the wind) and a distance of 2280 miles in the same time with a headwind (against the wind). Find the speed of the plane and the speed of the wind.

34. One train travels 30 km/hr faster than another. After 2 hours they have traveled a total of 430 km. Find the speed of each train.

35. A manufacturer produces two types of telephones. The more expensive model requires 1 hour to manufacture and 30 minutes to assemble. The less expensive model requires 45 minutes to manufacture and 15 minutes to assemble. If the company can allocate 150 hours for manufacturing and 60 hours for assembly, how many of each type can be produced?

36. The physics department hires graders and tutors. For January the department budgets $900 for 60 hours of grading and 35 hours of tutoring. The next month the department budgets $700 for 40 hours of grading and 30 hours of tutoring. How much does the department pay for each hour of tutoring and for each hour of grading?

37. The Speedy car rental agency charges a daily flat fee of $18, plus a mileage rate of $0.28 per mile. The Hirsch car rental agency charges a daily flat fee of $14 plus a mileage rate of $0.32 per mile for the same car. If you need the car for only a day, explain under what conditions one company may be better than the other.

38. The Frugal car rental agency charges a daily flat fee of $18 plus a mileage rate of $0.28 per mile. The Hirsch car rental agency charges a daily flat fee of $14 plus a mileage rate of $0.32 per mile for the same car. If you need the car for 3 days, explain under what conditions one company may be cheaper than the other.

39. A car-telephone company has two plans: plan A and plan B. In plan A, the company charges air time of $0.90 per minute; for plan B, the company charges a flat fee of $20 a month, but the air charge is reduced to $0.70 per minute. Explain under what conditions one plan may be cheaper than the other.

40. Ace Loan Company charges 18% simple annual interest on loans plus a $100 loan processing fee. Deuce Loan Company charges 20% simple annual interest but does not charge a processing fee. If you intend to borrow money and pay the entire principal and interest back at the end of 1 year, explain under what conditions one company might be giving a cheaper deal than the other.

41. An object is thrown straight up in the air. If v_0 is its initial velocity and s_0 is its initial height, then the position equation

$$s = -16t^2 + v_0t + s_0$$

gives the distance s in feet the object is above the ground t seconds after it is thrown. If the object is 64 feet above the ground after 1 second and 96 feet above the ground after 2 seconds, find its initial velocity and initial height.

42. An object is thrown straight up in the air. If v_0 is its initial velocity and s_0 is its initial height, then the position equation

$$s = -16t^2 + v_0t + s_0$$

gives the distance s in feet the object is above the ground t seconds after it is thrown. If the object is 32 feet above the ground after $\frac{1}{2}$ second, and 36 feet above the ground after 1 second, find its initial velocity and initial height.

43. (a) Find the area of the triangle bounded by the x-axis and the lines $2x - y = 10$ and $2y + x = 10$.
(b) Find the area of the triangle bounded by the y-axis and the lines $2x - y = 10$ and $2y + x = 10$.

44. (a) Find the area of the triangle bounded by the x-axis and the lines $3x - 4y = 12$ and $4x + 3y = 12$.
(b) Find the area of the triangle bounded by the y-axis and the lines $3x - 4y = 12$ and $4x + 3y = 12$.

In Exercises 45–48, find A and B such that:

45. $\dfrac{10}{(x - 2)(2x + 1)} = \dfrac{A}{x - 2} + \dfrac{B}{2x + 1}$

46. $\dfrac{7x + 1}{(x + 3)(x - 1)} = \dfrac{A}{x + 3} + \dfrac{B}{x - 1}$

47. $\dfrac{2x - 1}{(x + 1)^2} = \dfrac{A}{x + 1} + \dfrac{B}{(x + 1)^2}$

48. $\dfrac{4x - 15}{(x - 3)^2} = \dfrac{A}{x - 3} + \dfrac{B}{(x - 3)^2}$

49. Find the point (x, y) so that the line passing through (x, y) and $(-1, 2)$ has a slope of 3 and intersects the line passing through (x, y) and $(2, -1)$ with a slope of 1.

50. Find the values of A and B so that the line whose equation is $Ax + By = 4$ passes through the points $(1, 3)$ and $(-2, 4)$.

51. Suppose we arrive at the following solution for a 2 × 2 system: $(a, 2a - 5)$ for all real a. How many solutions to this system are there? Give examples of some numerical solutions, and explain in words the conditions under which a pair of numbers can be a solution to the system.

52. Suppose we arrive at the following solution for a 2 × 2 system: $\left(\dfrac{a + 5}{2}, a\right)$ for all real a. How does this dependent solution compare to the dependent solution in Exercise 51?

Questions for Thought

53. Explain what it means geometrically when we find that there is no solution to a system, or that the system is inconsistent.

54. Explain what it means geometrically when we find that there are an infinite number of solutions to a 2 × 2 system—that is, if the system is dependent.

55. Explain how you solve the following system of equations:

$$\begin{cases} 5x - 3y = 8 \\ \qquad 2y = -10 \end{cases}$$

Do you need to eliminate a variable?

56. Explain how you solve the following system of equations:

$$\begin{cases} 2x - 3y + z = 1 \\ \qquad 2y - z = 9 \\ \qquad\qquad 3z = -3 \end{cases}$$

Do you need to eliminate a variable?

7.2 3 × 3 Linear Systems: Elimination and Gaussian Elimination

Consider the following situation: A computer manufacturing company produces thre models: A, B, and C. The company knows how much time it takes for the production, assembly, and testing of each model. This information is found in the accompanying table.

Model	Production hours	Assembly hours	Testing hours
A	1.7	3.0	0.5
B	2.2	3.2	0.8
C	3.0	3.5	1.0

This problem is solved in Example 3.

To minimize costs, the company decides that 385 hours should be allocated for production, 560 hours for assembly, and 128 hours for testing. How many of each model can the company manufacture if it wants to use all the allotted time for each phase of production?

Translating this problem mathematically gives rise to a system of three equations in three variables. Solving such a system is the topic of this section.

For a first-degree equation in three variables, a solution is an ordered triple (x, y, z). Thus, for example, a few solutions to the equation $3x - 2y + z = 11$ are $(3, -1, 0)$, $(2, 1, 7)$, and $(1, 1, 10)$.

Geometrically, we know that we can represent the solutions to a first-degree equation in two variables as a straight line in a two-dimensional coordinate system.

A first-degree equation in three variables can be represented geometrically as a plane in a three-dimensional coordinate system, as shown in Figure 7.4.

Figure 7.4

On a three-dimensional coordinate system, the graph of $x + y + z = 6$ is a plane.

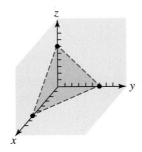

A 3×3 system is a system of three equations in three variables. For example,

$$\begin{cases} 2x + 3y - z = 13 \\ 5x - y + z = 0 \\ x - 3y - z = -6 \end{cases}$$

is a system satisfied by the ordered triple $(1, 3, -2)$. It is natural to ask, Is this the only solution? As with a 2×2 system, a 3×3 system has three possible types of solutions, illustrated by the pictures of intersections of planes in Figure 7.5.

Figure 7.5

Three possible intersections of three planes

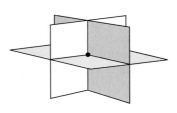

The three planes intersect at a point;
one unique solution.
(a)

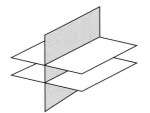

The three planes do not intersect;
no solutions.
(b)

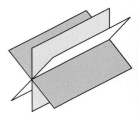

The three planes intersect in a line;
infinitely many solutions.
(c)

In general, we take the same approach solving larger systems of linear equations as we did in solving 2×2 systems; that is, we reduce a larger system to a smaller system that we already know how to solve. In this case, we can use elimination or substitution to reduce a 3×3 system (three equations in three variables) to a 2×2 system and solve the 2×2 system by the methods discussed in the previous section.

Example 1 Solve the following system of equations:

$$\begin{cases} 2x + 3y - z = 13 & \text{(1)} \\ 5x - y + z = 0 & \text{(2)} \\ x - 3y - z = -6 & \text{(3)} \end{cases}$$

We number the equations for ease of reference.

Solution | Our goal is to eliminate one of the variables and arrive at two equations in two variables. As with 2 × 2 systems, choosing the "right" variable to eliminate can simplify the procedure. Looking at this system, it appears that z could be eliminated most easily.

(1) $2x + 3y - z = 13$ Add equations (1) and (2).
(2) $\underline{5x -\ \ y + z =\ \ 0}$
         $7x + 2y\ \ \ \ \ \ = 13$ We call this equation (4).

Now we have to find another equation in the *same* two variables, x and y. We must use equation (3). (We explain why after completing this example.) Hence

(2) $5x -\ \ y + z =\ \ \ 0$ Add equations (2) and (3).
(3) $\underline{\ \ x - 3y - z = -6}$
         $6x - 4y\ \ \ \ \ \ = -6$ We call this equation (5).

Equations (4) and (5) form a 2 × 2 system, which we can solve:

(4) $7x + 2y = 13$ $\xrightarrow{\text{Multiply by 2}}$ $14x + 4y =\ \ 26$
(5) $6x - 4y = -6$ $\xrightarrow{\text{As is}}$ $\underline{\ \ 6x - 4y = -6}$ Add.
                                         $20x\ \ \ \ \ \ = 20$ Which yields $x = 1$

We have not completed solving the 2 × 2 system until we find y. We substitute $x = 1$ in equation (4) and solve for y:

(4) $7x + 2y = 13$ Substitute $x = 1$ and solve for y.
             $7(1) + 2y = 13$ We get $y = 3$.

To find z, we substitute $x = 1$ and $y = 3$ into any of the three original equations. We will use equation (1):

(1) $2x + 3y - z = 13$ Substitute $x = 1$ and $y = 3$ and solve for z.
             $2(1) + 3(3) - z = 13$ We get $z = -2$.

Hence our solution is $(1, 3, -2)$. The student should check this solution in all three original equations.   ∎

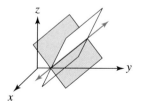

Figure 7.6

The intersection of two distinct planes is a line.

What kind of solution would you expect to get if we had one first-degree equation in two variables?

We already noted that a *single* equation in *two* variables, such as $x - 3y = 5$, does not have a unique solution. The same is true of a simultaneous system of two equations in three variables: If the system *has* a solution, the solution will not be unique. If we represent the two equations in three variables as planes in a three-dimensional coordinate system, we can see that if the planes intersect (if there is a solution), their intersection will be a line. (See Figure 7.6.)

In general, if the number of variables is greater than the number of equations in a system, the system will not have a unique solution.

In Example 1, we started out by combining equations (1) and (2) to arrive at an equation in two variables. Then, to create a second equation in the same two variables, we noted that we *must* somehow use equation (3). If we did not use equation (3), we would be changing the problem, and rather than looking at the intersection of the *three* planes described by the three equations, we would be looking only at the intersection of the two planes described by equations (1) and (2), which would produce a different solution.

In general, when solving 3 × 3 linear systems, we choose the most convenient variable to eliminate and produce a 2 × 2 system using all three equations. We use two equations to get one equation in two variables and then use a different pair of equations to get another equation in the *same* two variables. We solve the 2 × 2 system and then substitute these values into one of the original equations to find the third variable.

Example 2   Solve the following system of equations:

$$\begin{cases} x - 3y + z = -1 & \text{(1)} \\ 2x - 6y + 2z = -2 & \text{(2)} \\ x - y - z = 1 & \text{(3)} \end{cases}$$

Solution   We choose to eliminate x. We combine equations (1) and (2) as follows:

(1) $x - 3y + z = -1$ $\xrightarrow{\text{Multiply by } -2}$ $-2x + 6y - 2z = 2$

(2) $2x - 6y + 2z = -2$ $\xrightarrow{\text{As is}}$ $\underline{2x - 6y + 2z = -2}$ Add.

$ 0 = 0$ This is always true.

Equations (1) and (2) are equivalent. This means our system is really a system of two equations in three unknowns; therefore, if a solution to the system does exist, there must be an infinite number of them. Let's set aside equation (2) for a moment, since it is equivalent to equation (1), and examine the system consisting of equations (1) and (3):

$$\begin{cases} x - 3y + z = -1 & \text{(1)} \\ x - y - z = 1 & \text{(3)} \end{cases}$$

If, in the process of eliminating one variable from this system, we end up with an identity such as $0 = 0$, we would know that equations (1) and (3) are equivalent. Then all three equations in the 3×3 system would be equivalent, and we would choose one of the equations to represent the system and express the dependent solution as we did when we found two equations in two variables equivalent (see Example 2(b) in Section 7.1).

If all three equations were equivalent, how would we express the solution?

On the other hand, equations (1) and (3) may represent parallel planes, and there would be no solution to the system; if we tried to eliminate a variable, we would end up with a contradiction, such as in Example 1(c) in Section 7.1.

Let's see what happens when we try to eliminate z in this 2×3 system:

(1) $x - 3y + z = -1$ Add equations (1) and (3).

(3) $\underline{x - y - z = 1}$

$ 2x - 4y = 0$ Which we can simplify and rewrite as

$ x = 2y$ We call this equation (4).

Because this equation is not a contradiction, there must be solutions (in fact, infinitely many solutions, since we have two equations in three unknowns).

How do we represent the solution? We know that whatever the solutions are, equation (4) tells us that x must be twice y. To find the restrictions on z and conveniently express the solutions, we proceed as follows: We let $y = a$, where a is a constant, and we substitute $y = a$ into equation (4) and solve for x, so that we may express x in terms of a:

(4) $x = 2y$ To express x in terms of a, let $y = a$.

$ x = 2a$

So far we have expressed x and y in terms of a. Next, we want to express z in terms of a. We can choose any one of the original equations, so we select equation (3); we substitute $y = a$ and $x = 2a$ into this equation and solve for z:

(3) $x - y - z = 1$ Substitute $x = 2a$ and $y = a$.

$ 2a - a - z = 1$ Solve for z to get

$ z = a - 1$

Hence if $y = a$, then $x = 2a$ and $z = a - 1$, which can be written as the ordered triple $(2a, a, a - 1)$.

This means that for every real number a, we get a different solution for the system: Any ordered triple where the x-coordinate is twice the y-coordinate and the z-coordinate is 1 less than the y-coordinate will be a solution to the system. For example, if we let $a = 3$, then $(6, 3, 2)$ is a solution. As with 2×2 dependent solutions, there are an infinite number of solutions; what is being expressed in the preceding solution is how x, y, and z should be related if the ordered triple is to be one of the solutions. Had we decided to let $x = a$ and then solved for y and z in terms of a, the solution would have had a different appearance, but the relationships among the variables would have been the same: x would be twice y and z would be 1 less than y. ∎

Example 3 A computer manufacturing company produces three models: model A, model B, and model C. The company knows how much time it takes for the production, assembly, and testing of each model. This information is found in the accompanying table. To minimize costs, the company decides that 385 hours should be allocated for production, 560 hours for assembly, and 128 hours for testing. How many of each model can the company produce if it wants to use all the allotted time for each phase of the process?

Model	Production hours	Assembly hours	Testing hours
A	1.7	3.0	0.5
B	2.2	3.2	0.8
C	3.0	3.5	1.0

Solution If we let a, b, and c represent the number of computers of models A, B, and C, respectively, then we can create three equations in three variables; each equation represents *the number of hours allotted for each of the three phases:*

$$\begin{cases} 1.7a + 2.2b + 3c = 385 & (1) & \textit{No. hours for production} \\ 3a + 3.2b + 3.5c = 560 & (2) & \textit{No. hours for assembly} \\ 0.5a + 0.8b + c = 128 & (3) & \textit{No. hours for testing} \end{cases}$$

We choose to eliminate c. Combine equations (1) and (3) as follows:

(1) $1.7a + 2.2b + 3c = 385$ $\xrightarrow{\textit{As is}}$ $1.7a + 2.2b + 3c = 385$
(3) $0.5a + 0.8b + c = 128$ $\xrightarrow{\textit{Multiply by} -3}$ $\underline{-1.5a - 2.4b - 3c = -384}$ Add.
$$0.2a - 0.2b = 1$$
Call this equation (4).

Combine equations (2) and (3) as follows:

(2) $3a + 3.2b + 3.5c = 560$ $\xrightarrow{\textit{As is}}$ $3a + 3.2b + 3.5c = 560$
(3) $0.5a + 0.8b + c = 128$ $\xrightarrow{\textit{Multiply by} -3.5}$ $\underline{-1.75a - 2.8b - 3.5c = -448}$ Add.
$$1.25a + 0.4b = 112$$
Call this equation (5).

Now we solve the 2×2 system consisting of equations (4) and (5):

(4) $\begin{cases} 0.2a - 0.2b = 1 \end{cases}$ $\xrightarrow{\textit{Multiply by 2}}$ $0.4a - 0.4b = 2$
(5) $\begin{cases} 1.25a + 0.4b = 112 \end{cases}$ $\xrightarrow{\textit{As is}}$ $\underline{1.25a + 0.4b = 112}$ Add.
$$1.65a = 114$$
Which yields $a = 69.09$

If we do not keep the necessary degree of accuracy in a preliminary answer, we may not get the correct final answer. It is usually sufficient to keep two decimal places more than you need in the final answer. In this example, the answers must be whole numbers, so we rounded the preliminary answers for a, b, and c to the nearest hundredth.

Now find b by substituting $a = 69.09$ in equation (4):

(4) $0.2a - 0.2b = 1$ Substitute $a = 69.09$.

           $0.2(69.09) - 0.2b = 1$ Which yields $b = 64.09$

Finally, find c by substituting $a = 69.09$ and $b = 64.09$ in equation (3):

(3) $0.5a + 0.8b + c = 128$ Substitute $a = 69.09$ and $b = 64.09$.

      $0.5(69.09) + 0.8(64.09) + c = 128$ Which yields $c = 42.18$ rounded to the nearest hundredth

Keep in mind that a, b, and c represent the number of model A, B, and C computers the company will manufacture, and so fractional values of a, b, and c are not acceptable. Hence, using the largest possible whole number values for a, b, and c, the company should produce

> 69 of model A,   64 of model B,   and   42 of model C

It is left to the student to check that there are enough hours available for production, assembly, and testing to produce this number of computers of each type. Also notice that, although there are some hours left over, there aren't enough hours to manufacture any additional computers. For example, this solution uses 558.8 hours of assembly time, leaving 1.2 hours of unused assembly time, which is not enough time to assemble any additional computers. ∎

The procedures we have discussed in this section can be generalized to any size system of linear equations. If we had to solve a 5×5 system, we would start by eliminating a variable and an equation, reducing it to a 4×4 system. Then we would reduce the 4×4 system to a 3×3 system, and so on.

Gaussian Elimination

Gaussian elimination is a variation of the elimination method that uses the same algebraic procedures as elimination, but rather than reducing the number of equations and variables, as we did previously in reducing a 3×3 system to a 2×2 system, the goal is to reduce only the number of variables in each successive equation in the original system.

Consider the following two system of equations:

$$\text{I.} \begin{cases} x + 3y - z = -4 \\ 2x + 8y + z = -4 \\ x + 5y + 6z = 8 \end{cases} \qquad \text{II.} \begin{cases} x + 3y - z = -4 \\ 2y + 3z = 4 \\ 4z = 8 \end{cases}$$

You may verify for yourself that these two systems are equivalent, since they each have the same unique solution: $(1, -1, 2)$. System II, however, is much easier to solve. Notice that the last equation in system II is an equation in one variable, z, which can be solved easily:

$$4z = 8 \implies z = 2$$

Now that we have found z, we can find y by substituting the value for z in the second equation of system II:

       $2y + 3z = 4$ Substitute $z = 2$.

       $2y + 3(2) = 4$ Which yields $y = -1$

Finally, we can find x by substituting values for z and y in the first equation of system II:

$$x + 3y - z = -4 \qquad \text{Substitute } z = 2 \text{ and } y = -1.$$
$$x + 3(-1) - (2) = -4 \qquad \text{Which yields } x = 1; \text{ the solution is } (1, -1, 2).$$

This process is called **back substitution**.

To solve a system by using substitution in this manner, we have to put the system into a form like that of system II, called (upper) **triangular form**—in which each successive equation in the system is an equation containing at least one less variable than the preceding equation. The process of putting a system into triangular form is called **Gaussian elimination**.

To put a system into triangular form, we use the same procedures as we did with the elimination method. Let's pause to summarize some of these procedures, because we will refer to them in subsequent sections.

We remind you that *two systems are equivalent if their solutions are identical.*

A system of equations can be changed into an *equivalent* system by the following transformations:

1. Multiplying both sides of any equation by a nonzero constant
2. Interchanging any two equations
3. Adding a multiple of one equation to another equation

Example 4 | Solve the following system by Gaussian elimination:

$$\begin{cases} x - y + 3z = 11 & \textbf{(1)} \\ 3x - y + z = 5 & \textbf{(2)} \\ x + 3y - z = -9 & \textbf{(3)} \end{cases}$$

Solution | Our approach is to put the system into triangular form and then use back substitution to find the solution. We start by transforming the system into an equivalent system in triangular form; that is, we first want to eliminate x from the last two equations and then eliminate y from the last equation. We start by eliminating x from equation (2) by multiplying each side of equation (1) by -3 and adding this equation to equation (2) to get equation (4):

(1) $\quad x - y + 3z = 11 \quad \xrightarrow{\text{Multiply by } -3} \quad -3x + 3y - 9z = -33$
(2) $\quad 3x - y + z = 5 \quad \xrightarrow{\text{As is}} \quad \underline{\quad 3x - y + z = 5 \quad}$ Add.
$$2y - 8z = -28$$

This is equation (4).

Next, we eliminate x using equations (1) and (3): Multiply both sides of equation (1) by -1 and add the resultant equation to equation (3) to get equation (5):

(1) $\quad x - y + 3z = 11 \quad \xrightarrow{\text{Multiply by } -1} \quad -x + y - 3z = -11$
(3) $\quad x + 3y - z = -9 \quad \xrightarrow{\text{As is}} \quad \underline{\quad x + 3y - z = -9 \quad}$ Add.
$$4y - 4z = -20$$

This is equation (5).

This is almost in triangular form. What has to be done next?

So now we have the following system:

$$\begin{cases} x - y + 3z = 11 & \textbf{(1)} \\ 2y - 8z = -28 & \textbf{(4)} \\ 4y - 4z = -20 & \textbf{(5)} \end{cases}$$

Finally, to get the new system into triangular form, we still need to eliminate y from equation (5). We do this by multiplying both sides of equation (4) by -2 and adding the resultant equation to equation (5):

(4) $2y - 8z = -28$ $\xrightarrow{\text{Multiply by } -2}$ $-4y + 16z = 56$
(5) $4y - 4z = -20$ $\xrightarrow{\text{As is}}$ $4y - 4z = -20$ Add.
$12z = 36$ This is equation (6).

We now have the following equivalent system:

$$\begin{cases} x - y + 3z = 11 & \textbf{(1)} \\ 2y - 8z = -28 & \textbf{(4)} \\ 12z = 36 & \textbf{(6)} \end{cases}$$

This system is now in triangular form. We can now solve the system using back substitution. Equation (6) in this system yields $z = 3$. We use equation (4) to find y:

(4) $2y - 8z = -28$ Substitute $z = 3$ and solve for y.
$\quad 2y - 8(3) = -28$ To get $y = -2$

We use equation (1) to find x:

(1) $x - y + 3z = 11$ Substitute $z = 3$ and $y = -2$ and solve for x.
$\quad x - (-2) + 3(3) = 11$ To get $x = 0$

Hence the solution is $x = 0$, $y = -2$, and $z = 3$, or $(0, -2, 3)$. ∎

When a system is in triangular form and the coefficient of the leftmost variable in each equation is 1, we say that the system is in **echelon form**. For example, the following system is in echelon form:

What is the difference between a system in triangular form and one in echelon form?

$$\begin{cases} x - 5y + z = 0 \\ y - 2z = -8 \\ z = 12 \end{cases}$$

The process of putting a system into echelon form lends itself well to programming. However, for all practical purposes, if you are transforming the system by hand, putting a system into triangular form usually requires less computation.

In the next sections we discuss systematic procedures for solving linear systems that are essentially variations on the elimination methods.

7.2 Exercises

In Exercises 1–10, solve the system of equations by elimination.

1. $\begin{cases} x + y + z = 6 \\ 3x + y - z = 6 \\ 2x + y - z = 4 \end{cases}$

2. $\begin{cases} x - 2y + z = -5 \\ 3x - y + z = 2 \\ x + 2y - z = 9 \end{cases}$

3. $\begin{cases} x + y - z = 1 \\ 2x + 2y + 2z = 0 \\ x - y + z = 3 \end{cases}$

4. $\begin{cases} 3x + y + 2z = 0 \\ x - 2y - z = -3 \\ x + y + z = 1 \end{cases}$

5. $\begin{cases} x + y + z = 6 \\ 3x + y - z = 4 \\ 2x + 2y + 2z = 4 \end{cases}$

6. $\begin{cases} x + y - z = 3 \\ 2x + 2y - 2z = 6 \\ x - y + z = 5 \end{cases}$

7. $\begin{cases} 0.01a + 0.2b + c = 0.45 \\ 0.3a - b + 0.2c = 1.06 \\ a + b + c = 5.8 \end{cases}$

8. $\begin{cases} x + y = 2x + z \\ x - 2y = z + 3 \\ 2x - y = 3z - 9 \end{cases}$

9. $\begin{cases} \dfrac{x}{4} + \dfrac{y}{6} - \dfrac{z}{3} = 1 \\[2mm] \dfrac{x}{2} + \dfrac{y}{3} + z = -3 \\[2mm] \dfrac{x}{8} + \dfrac{y}{4} - z = 5 \end{cases}$ **10.** $\begin{cases} x + y = 5 \\ x + 2z = 5 \\ y - z = 1 \end{cases}$

In Exercises 11–16, solve the system of equations by Gaussian elimination.

11. $\begin{cases} x + y - z = 7 \\ x + 2y + 2z = 4 \\ 2x - y + z = -1 \end{cases}$ **12.** $\begin{cases} x - y + 3z = 14 \\ x + 2y - z = -8 \\ 2x + y + z = 2 \end{cases}$

13. $\begin{cases} x + y - 2z = 3 \\ 2x + y + z = -4 \\ x + 2y - 3z = 5 \end{cases}$ **14.** $\begin{cases} x + z = 4 \\ x - y = 1 \\ y - z = 1 \end{cases}$

15. $\begin{cases} 2x + y + z = 4 \\ x - y - z = 5 \\ 2x - 2y - 2z = 1 \end{cases}$ **16.** $\begin{cases} x - y - z = 5 \\ 2x - 2y - 2z = 10 \\ 2x + y + z = 4 \end{cases}$

17. Elaine has a total of $19,750 in three investments. She has a bank account paying 6%, a certificate of deposit paying 8%, and a stock yielding 10.5%. Her annual interest from the three investments is $1680. If the interest from her stock is equal to the sum of the interest of the other two investments, how much was invested at each rate?

18. A theater group plans to sell 520 tickets for a play. They are charging $12 for orchestra seats, $10 for mezzanine seats, and $6 for balcony seats, and they plan to collect $5080. If there are 100 more mezzanine tickets than orchestra and balcony tickets combined, how many of each type are there?

19. A compact disc manufacturer produces three models of compact disc players, model A, model B, and model C. The company knows how much production, assembly, and testing time is needed for each model. This information is found in the accompanying table. If the company has allocated 586 hours for production, 227 hours for assembly, and 147 hours for testing, how many of each model can the company produce if it wants to use all the time allocated for each phase of the process?

Model	Production hours	Assembly hours	Testing hours
A	1.2	0.4	0.2
B	1.4	0.5	0.3
C	1.8	0.8	0.6

20. A nutritionist wants to create a food supplement out of three substances; X, Y, and Z. He wants the food supplement to have the following characteristics: 10 grams of the supplement should supply 3.5 grams of iron and cost 77¢. The iron content and cost of the substances X, Y, and Z are entered in the accompanying table. How many grams of each substance should be used to make such a food supplement?

Substance	Iron content per gram	Cost per gram
X	0.5	10¢
Y	0.2	6¢
Z	0.4	8¢

21. A nutritionist wants to create a supplement that contains 46 mg of zinc, 53 mg of iron, and 148 mg of calcium. She is choosing from three foods A, B, and C whose content of these minerals per ounce is given in the accompanying table. How many ounces of each food should be combined to satisfy the given requirements?

	Food A	Food B	Food C
No. of mg of Zinc	5	6	2
No. of mg of Iron	6	2	6
No. of mg of Calcium	16	10	14

22. A store sells VCR tapes in packages of 3, 5 and 10 tapes for $2, $4, and $8, respectively. If a store sold 20 packages containing a total of 127 video tapes and collected a total of $100, how many of packages of each type were sold?

23. A chemist needs to make a mixture that contains 12 liters of substance A, 16 liters of substance B, and 26 liters of substance C. Tank X contains 20% substance A, and 40% each of substances B and C. Tank Y contains only substance C. Tank Z contains equal amounts of substances A and B. How many liters must be taken from each tank to create the required mixture?

24. A store wants to receive 67 units of item A, 102 units of item B, and 105 units of item C. The store can order these items in any of three combination packages whose content is given in the table at the top of page 600. How many of each combination package should be ordered to result in the required number of each item?

Combination Package	No. of item A	No. of item B	No. of item C
Economy	3	2	5
Super	4	7	6
Ultra	5	9	8

25. A chemist needs to mix three solutions containing 10%, 25%, and 40% adipic acid, respectively, in order to get 100 liters of a solution that is 30% acid. If she needs to use all three solutions and there are 20 liters more of the 25% solution than of the 10% solution in the mixture, how much of each solution will be needed?

26. A horticulturist wants to mix three types of fertilizers, which contain 20%, 30%, and 40% nitrogen, respectively. If the final mixture should be 1000 pounds of 32% nitrogen, all three type are used, and there is twice as much of the 40% type as of the 20% type, how much of each is in the final mixture?

27. A function of the form $f(x) = ax^2 + bx + c$, where a, b, and c are constants, is called a quadratic function. Find the quadratic function $f(x)$ such that $f(1) = 2, f(-1) = 6$, and $f(2) = 9$.

28. A function of the form $f(x) = ax^2 + bx + c$, where a, b, and c are constants, is called a quadratic function. Find the quadratic function $f(x)$ such that $f(1) = 0, f(-1) = 8$, and $f(2) = 2$.

In Exercises 29 and 30, find A, B, and C such that:

29. $\dfrac{2x + 5}{x(x - 1)^2} = \dfrac{A}{x} + \dfrac{B}{x - 1} + \dfrac{C}{(x - 1)^2}$

30. $\dfrac{5x + 6}{x(x - 2)(x + 3)} = \dfrac{A}{x} + \dfrac{B}{x - 2} + \dfrac{C}{x + 3}$

Question for Thought

31. Find A, B, C, and D such that

$$\frac{x^3 + 2x^2 + 3x + 5}{(x^2 + 1)^2} = \frac{Ax + B}{x^2 + 1} + \frac{Cx + D}{(x^2 + 1)^2}$$

7.3 Solving Linear Systems Using Augmented Matrices

When we solve a system of linear equations using the elimination method, we first make sure that the equations are written in such a way that the same variables are aligned vertically. Once we have the system in this form, we concentrate on the coefficients of the variables, transforming equations so that the coefficients of the variables to be eliminated are "matched up." This process suggests that if we can develop some systematic procedure to keep track of the variables, we can ignore the variables and focus on the coefficients in order to solve a system of linear equations. The matrix methods we develop in this section provide such a procedure.

A rectangular array of numbers is called a **matrix**. The numbers in a matrix are called the **elements** or **entries** of the matrix. The horizontal arrays of entries are called the **rows**, and the vertical arrays of entries are called the **columns**. The following are two examples of matrices:

$$\begin{bmatrix} 1 & -2 & 3 & 4 \\ 0 & 5 & 0 & -3 \end{bmatrix}$$

Row 1 →
Row 2 →
Row 3 →

$$\begin{bmatrix} 2 & 3 & -1 \\ 0 & -5 & 2 \\ 1 & 3 & 5 \end{bmatrix}$$

Column 1 ↓ *Column 2* ↓ *Column 3* ↓

We usually specify the size of a matrix by first giving the number of rows and then the number of columns. The first example is a 2 × 4 (2 by 4) matrix; the second is a 3 × 3 matrix. A **square matrix** is a matrix that contains an equal number of rows and columns; the 3 × 3 matrix above is a square matrix.

We will now describe how we can use matrices to solve systems of equations. We begin by taking a system and rewriting it as an **augmented matrix** as follows:

$$\begin{cases} 2x - 3y = 4 \\ 5x + y = -1 \end{cases} \text{ is written as the augmented matrix } \begin{bmatrix} 2 & -3 & | & 4 \\ 5 & 1 & | & -1 \end{bmatrix}.$$

The matrix $\begin{bmatrix} 2 & -3 \\ 5 & 1 \end{bmatrix}$, made up of the coefficients of the variables of the system, is called the **coefficient matrix**. The **augmented matrix** of a system of equations consists of the coefficient matrix of the system with the constants of the system adjoined to the right. Consider another example:

$$\begin{cases} 3x + 5y - z = 7 \\ 2x + 3z = -2 \\ x - 2y - 2z = 4 \end{cases} \text{ is written as } \begin{matrix} x & y & z \\ \begin{bmatrix} 3 & 5 & -1 & | & 7 \\ 2 & 0 & 3 & | & -2 \\ 1 & -2 & -2 & | & 4 \end{bmatrix} \end{matrix}$$

The relative positions of the column entries in the coefficient matrix indicate the variable: The first column consists of the coefficients of the x variable; the second column, the coefficients of the y variable; and the third column, the coefficients of the z variable. The constants are on the augmented side. Thus it is important that the variables be lined up columnwise before we represent a system by its augmented matrix.

Notice that a 0 is entered in the second row of the last augmented matrix for the missing y variable in the second equation.

Recall that the goal of Gaussian elimination (discussed in Section 7.2) is to begin with any system and, using certain transformations, change it to a system in triangular form. Once we have the system in triangular form, we can solve it using back substitution. The following system is in triangular form. Beside this system, we write its associated augmented matrix:

Note the "triangle" of zeros below the main diagonal of the matrix.

$$\begin{cases} x + 2y + 2z = 12 \\ y - 3z = -5 \\ 2z = 6 \end{cases} \quad \begin{bmatrix} 1 & 2 & 2 & | & 12 \\ 0 & 1 & -3 & | & -5 \\ 0 & 0 & 2 & | & 6 \end{bmatrix}$$

We refer to this system's augmented matrix as being in (upper) *triangular* form. The elements 1, 1, and 2 outlined in the augmented matrix of the system are called the **main diagonal**. The main diagonal of an augmented matrix actually consists of the diagonal elements of the coefficient matrix. A matrix is said to be in (upper) **triangular form** if it has all zero entries below the main diagonal.

We saw in Section 7.2 that we can use certain operations on a system to transform it into a simpler equivalent system, one in triangular (or echelon) form. In this section, we will do exactly the same thing, but rather than working with the equations in the system, we will work instead with its augmented matrix of coefficients and constants. We begin with the following definition.

Definition of Row Equivalence of Two Matrices

Two matrices are **row-equivalent** if and only if their associated systems of equations are equivalent.

In the previous section, we used the properties of equality to transform a given system of equations into a "simpler" *equivalent* system of equations. Now that we have defined row-equivalence for matrices, the procedures we illustrated in the previous section translate into the following list of transformations that we can apply to a matrix to produce a row-equivalent matrix. We call these transformations the **elementary row operations**.

Elementary Row Operations

1. Multiply each entry in a given row by a nonzero constant.

2. Interchange any two rows.

3. Add a multiple of one row to another row.

The following are three illustrations of the elementary row operations.

$$\begin{bmatrix} 3 & 2 & 2 \\ 2 & -6 & 0 \\ 1 & 2 & -1 \end{bmatrix}$$

The following notation means multiply each entry in row 2 by -3 to get the new row 2.

$$-3R_2 \rightarrow R_2$$

$$\begin{bmatrix} 3 & 2 & 2 \\ -6 & 18 & 0 \\ 1 & 2 & -1 \end{bmatrix}$$

$$\begin{bmatrix} 3 & 2 & 2 \\ 2 & -6 & 0 \\ 1 & 2 & -1 \end{bmatrix}$$

The following notation means interchange row 1 and row 3 (row 2 remains unchanged).

$$R_1 \leftrightarrow R_3$$

$$\begin{bmatrix} 1 & 2 & -1 \\ 2 & -6 & 0 \\ 3 & 2 & 2 \end{bmatrix}$$

$$\begin{bmatrix} 3 & 2 & 2 \\ 2 & -6 & 0 \\ 1 & 2 & -1 \end{bmatrix}$$

The following notation means multiply each entry in row 1 by 2 and add the resulting entries to each entry in row 3 to get a new row 3 (rows 1 and 2 remain unchanged).

$$2R_1 + R_3 \rightarrow R_3$$

$$\begin{bmatrix} 3 & 2 & 2 \\ 2 & -6 & 0 \\ 7 & 6 & 3 \end{bmatrix}$$

Since this last illustration is a bit more complex than the first two, we will demonstrate how we found the new row 3.

Multiply each entry in row 1 by 2:

$$3 \quad 2 \quad 2 \quad \longrightarrow \quad 6 \quad 4 \quad 4 \qquad \text{This is } 2R_1.$$

Then add each entry in this multiple of row 1 to each entry in row 3:

$$+ \quad \underline{1 \quad 2 \quad -1} \qquad \text{This is } R_3.$$

This gives the new row 3:

$$7 \quad 6 \quad 3 \qquad \text{This is the new } R_3.$$

The elementary row operations are equivalent to the transformations we performed on systems of equations. For example,

$$\begin{array}{l} 5x + y = 7 \\ x - 3y = -5 \end{array} \quad \begin{array}{c} \xrightarrow{\text{Multiply by 3}} \\ \xrightarrow{\text{As is}} \end{array} \quad \begin{array}{l} 15x + 3y = 21 \\ x - 3y = -5 \end{array}$$

is the same as using elementary row operation 1:

$$\begin{bmatrix} 5 & 1 & | & 7 \\ 1 & -3 & | & -5 \end{bmatrix} \qquad 3R_1 \rightarrow R_1 \qquad \begin{bmatrix} 15 & 3 & | & 21 \\ 1 & -3 & | & -5 \end{bmatrix}$$

Hence elementary row operations produce matrices whose associated systems are equivalent to each other.

> Performing the elementary row operations on a matrix produces a row-equivalent matrix.

Now we can use matrices to solve systems of equations by **Gaussian elimination**, a process in which we use the elementary row operations to convert the augmented matrix of a system into a row-equivalent matrix in triangular form. Once we have the augmented matrix in triangular form, we write out its associated system and solve the system by back substitution. This process is outlined in the following box and illustrated in the next few examples.

Method for Solving Linear Systems Using Augmented Matrices (Gaussian Elimination)

1. Set up the augmented matrix of the system.
2. Use the elementary row operations to transform the augmented matrix into a row-equivalent matrix in triangular form.
3. For the augmented matrix in triangular form, write the corresponding system of equations.
4. Solve this system by back substitution.
5. Check your solution in the original equations.

Example 1 Solve the following systems of equations by Gaussian elimination.

(a) $\begin{cases} x - 5y = 4 \\ 2x - 3y = 7 \end{cases}$

(b) $\begin{cases} x + 2y + 3z = -5 \\ 2x + 6y + 7z = -10 \\ x + 8y + 2z = 3 \end{cases}$

Solution (a) First we write the associated augmented matrix of the system:

$$\begin{cases} x - 5y = 4 \\ 2x - 3y = 7 \end{cases} \Rightarrow \begin{bmatrix} 1 & -5 & | & 4 \\ 2 & -3 & | & 7 \end{bmatrix} \qquad \text{Next we get it into triangular form.}$$

Triangular form for a 2×2 matrix means the bottom left entry of the matrix (row 2, column 1) is 0.

$$\begin{bmatrix} 1 & -5 & | & 4 \\ 2 & -3 & | & 7 \end{bmatrix} \quad \underset{-2R_1 + R_2 \to R_2}{} \quad \begin{bmatrix} 1 & -5 & | & 4 \\ 0 & 7 & | & -1 \end{bmatrix}$$

REMEMBER: $-2R_1 + R_2 \to R_2$ means multiply the entries in row 1 by -2, add the result to row 2, and replace row 2 with this result.

This matrix is now in triangular form. If we write its associated system we get

$$\begin{bmatrix} 1 & -5 & | & 4 \\ 0 & 7 & | & -1 \end{bmatrix} \Rightarrow \begin{cases} x - 5y = 4 \\ 7y = -1 \end{cases}$$

The last equation in this new (equivalent) system yields $y = -\dfrac{1}{7}$. We use the first equation in this new system to find x:

$$x - 5y = 4 \qquad \text{Substitute } y = -\dfrac{1}{7} \text{ and solve for } x.$$

$$x - 5\left(-\dfrac{1}{7}\right) = 4 \qquad \text{To get } x = \dfrac{23}{7}$$

The student can verify that the solution is $\left(\dfrac{23}{7}, -\dfrac{1}{7}\right)$.

(b) First, we write the associated augmented matrix of the system.

$$\begin{cases} x + 2y + 3z = -5 \\ 2x + 6y + 7z = -10 \\ x + 8y + 2z = 3 \end{cases} \Rightarrow \left[\begin{array}{ccc|c} 1 & 2 & 3 & -5 \\ 2 & 6 & 7 & -10 \\ 1 & 8 & 2 & 3 \end{array}\right]$$

Now use the elementary row operations to transform the matrix into an equivalent matrix in triangular form; that is, we want 0 to be the first entry in row 2 and the first two entries in row 3. First, we get 0 to be the first entry in row 2:

$$\left[\begin{array}{ccc|c} 1 & 2 & 3 & -5 \\ 2 & 6 & 7 & -10 \\ 1 & 8 & 2 & 3 \end{array}\right]$$

$-2R_1 + R_2 \rightarrow R_2$
Multiply entries in row 1 by -2, add the result to row 2, and replace row 2 with this result.

$$\left[\begin{array}{ccc|c} 1 & 2 & 3 & -5 \\ 0 & 2 & 1 & 0 \\ 1 & 8 & 2 & 3 \end{array}\right]$$

Next, we get 0 as the first entry in row 3:

$$\left[\begin{array}{ccc|c} 1 & 2 & 3 & -5 \\ 0 & 2 & 1 & 0 \\ 1 & 8 & 2 & 3 \end{array}\right]$$

Multiply row 1 by -1, add the result to row 3 to get the new row 3.

$-R_1 + R_3 \rightarrow R_3$

$$\left[\begin{array}{ccc|c} 1 & 2 & 3 & -5 \\ 0 & 2 & 1 & 0 \\ 0 & 6 & -1 & 8 \end{array}\right]$$

To get 0 as the second entry in row 3, why not multiply the entries in row 1 by -3, add the result to row 3, and replace row 3 with this result?

Finally, we get 0 as the second entry in row 3:

$$\left[\begin{array}{ccc|c} 1 & 2 & 3 & -5 \\ 0 & 2 & 1 & 0 \\ 0 & 6 & -1 & 8 \end{array}\right]$$

Multiply row 2 by -3, add the result to row 3 to get the new row 3.

$-3R_2 + R_3 \rightarrow R_3$

$$\left[\begin{array}{ccc|c} 1 & 2 & 3 & -5 \\ 0 & 2 & 1 & 0 \\ 0 & 0 & -4 & 8 \end{array}\right]$$

Remember that equivalent matrices produce equivalent associated systems.

This matrix is now in triangular form (zeros below the main diagonal). Now we write the associated system for the transformed matrix:

$$\left[\begin{array}{ccc|c} 1 & 2 & 3 & -5 \\ 0 & 2 & 1 & 0 \\ 0 & 0 & -4 & 8 \end{array}\right] \Rightarrow \begin{cases} x + 2y + 3z = -5 \\ 2y + z = 0 \\ -4z = 8 \end{cases}$$

The third equation in this equivalent system yields $z = -2$. Now we find the rest of the variables by back substitution. We use the second equation to find y:

$2y + z = 0$ Substitute $z = -2$ and solve for y.
$2y + (-2) = 0$ To get $y = 1$

We use the first equation to find x:

$x + 2y + 3z = -5$ Substitute $y = 1$ and $z = -2$, and solve for x.
$x + 2(1) + 3(-2) = -5$ To get $x = -1$

Hence the solution is $x = -1, y = 1,$ and $z = -2$, or $(-1, 1, -2)$. ∎

Example 2 | Solve the following systems of equations using Gaussian elimination. ,

(a) $\begin{cases} 2x - 6y = 8 \\ 3x - 9y = 12 \end{cases}$ **(b)** $\begin{cases} x + 2y + 3z = 2 \\ 2x + 4y + 6z = 8 \\ 2y - 5z = -7 \end{cases}$

Solution   **(a)** We first write its augmented matrix:

$$\begin{cases} 2x - 6y = 8 \\ 3x - 9y = 12 \end{cases} \Rightarrow \left[\begin{array}{cc|c} 2 & -6 & 8 \\ 3 & -9 & 12 \end{array}\right]$$ Get it into triangular form.

Getting 1 in the upper left-hand corner usually makes later computations easier.

$$\left[\begin{array}{cc|c} 2 & -6 & 8 \\ 3 & -9 & 12 \end{array}\right] \qquad \begin{array}{c} \frac{1}{2}R_1 \rightarrow R_1 \end{array} \qquad \left[\begin{array}{cc|c} 1 & -3 & 4 \\ 3 & -9 & 12 \end{array}\right]$$

Multiply row 1 by $\frac{1}{2}$ to get a new row 1.

> We could also have arrived at triangular form in one step by applying the row operation $-\frac{3}{2}R_1 + R_2 \rightarrow R_2$.

Next we get 0 to be the first entry in row 2:

$$\left[\begin{array}{cc|c} 1 & -3 & 4 \\ 3 & -9 & 12 \end{array}\right] \qquad \begin{array}{c} -3R_1 + R_2 \rightarrow R_2 \end{array} \qquad \left[\begin{array}{cc|c} 1 & -3 & 4 \\ 0 & 0 & 0 \end{array}\right]$$ This is in triangular form.

Multiply row 1 by -3, add the result to row 2 to get the new row 2.

This matrix is now in triangular form. If we write its associated system we get

$$\left[\begin{array}{cc|c} 1 & -3 & 4 \\ 0 & 0 & 0 \end{array}\right] \Rightarrow \begin{cases} x - 3y = 4 \\ \quad\quad 0 = 0 \end{cases}$$

> Equivalently, we could have let $x = a$, and the solution would have the form $\left(a, \dfrac{a-4}{3}\right)$.

The second equation in this system is always true, so our system consists of only one equation in two variables. Hence the system is dependent. Recall from Section 9.1 that we can put the dependent solution into ordered pair form as follows: If we let $y = a$ and solve for x, we get $x = 4 + 3a$. Hence, the solutions are all ordered pairs of the form $(4 + 3a, a)$, where a is any real number .

(b) First we write the system's augmented matrix:

$$\begin{cases} x + 2y + 3z = 2 \\ 2x + 4y + 6z = 8 \\ \quad\quad 2y - 5z = -7 \end{cases} \Rightarrow \left[\begin{array}{ccc|c} 1 & 2 & 3 & 2 \\ 2 & 4 & 6 & 8 \\ 0 & 2 & -5 & -7 \end{array}\right]$$

Now use the elementary row operations to transform the matrix into an equivalent matrix in triangular form. Get 0 as the first entry in row 2:

$$\left[\begin{array}{ccc|c} 1 & 2 & 3 & 2 \\ 2 & 4 & 6 & 8 \\ 0 & 2 & -5 & -7 \end{array}\right] \qquad \begin{array}{c} R_2 \leftrightarrow R_3 \end{array} \qquad \left[\begin{array}{ccc|c} 1 & 2 & 3 & 2 \\ 0 & 2 & -5 & -7 \\ 2 & 4 & 6 & 8 \end{array}\right]$$

Switch rows 2 and 3.

Next we need to get 0 as the first entry in row 3:

$$\left[\begin{array}{ccc|c} 1 & 2 & 3 & 2 \\ 0 & 2 & -5 & -7 \\ 2 & 4 & 6 & 8 \end{array}\right] \qquad \begin{array}{c} -2R_1 + R_3 \rightarrow R_3 \end{array} \qquad \left[\begin{array}{ccc|c} 1 & 2 & 3 & 2 \\ 0 & 2 & -5 & -7 \\ 0 & 0 & 0 & 4 \end{array}\right]$$

Multiply row 1 by -2, add the result to row 3 to get the new row 3.

We write the associated system of the matrix in triangular form:

What does the row 0 0 0 4 in the augmented matrix mean for the associated system of equations?

$$\begin{bmatrix} 1 & 2 & 3 & | & 2 \\ 0 & 2 & -5 & | & -7 \\ 0 & 0 & 0 & | & 4 \end{bmatrix} \Rightarrow \begin{cases} x + 2y + 3z = 2 \\ 2y - 5z = -7 \\ 0 = 4 \end{cases}$$

The last equation in this system is a contradiction and therefore has no solution. Hence the *system* has no solution . ∎

Another method of solving systems of linear equations using matrices is **Gauss–Jordan elimination**. As with Gaussian elimination, we use elementary row operations to transform a matrix into a row-equivalent matrix, except rather than having triangular form as the final goal, Gauss–Jordan elimination requires that we put a system's associated matrix into **reduced echelon form**—a form where each entry on the main diagonal is 1 and all other entries in the coefficient matrix are 0:

$$\begin{bmatrix} 1 & 0 & 0 & | & a \\ 0 & 1 & 0 & | & b \\ 0 & 0 & 1 & | & c \end{bmatrix} \quad \text{where } a, b, \text{ and } c \text{ are constants}$$

A matrix in this form has the associated system

$$\begin{cases} x & & = & a \\ & y & & = & b \\ & & z & = & c \end{cases}$$

If we have a system's augmented matrix in this form, we can simply read the solution: $x = a$, $y = b$, and $z = c$. The following is a formal definition of reduced echelon form.

We mentioned echelon form for a system in the previous section. What would echelon form for a matrix look like, and how would it be different from reduced echelon form?

A matrix is in **reduced echelon form** if

1. The first nonzero entry in each row, called the *pivot element*, is 1.
2. The pivot element of each nonzero row occurs farther to the right than the pivot element in the preceding row.
3. In each column containing a pivot element, the remaining entries are 0.
4. Rows consisting of all zeros are at the bottom of the matrix.

The next example illustrates Gauss–Jordan elimination—that is, putting a system's augmented matrix into reduced echelon form.

Example 3   Solve the following system using Gauss–Jordan elimination:

$$\begin{cases} 2x + 2y + 4z = 8 \\ 2x + 4y + 6z = 10 \\ x + 2y + 5z = 1 \end{cases}$$

Solution   First we write the system's augmented matrix: $\begin{bmatrix} 2 & 2 & 4 & | & 8 \\ 2 & 4 & 6 & | & 10 \\ 1 & 2 & 5 & | & 1 \end{bmatrix}$.

Now we use the elementary row operations to transform the matrix into an equivalent matrix in reduced echelon form.

1. First we want to get a 1 in the upper left-hand corner. A quick way to do this is to interchange rows 1 and 3:

$$\begin{bmatrix} 2 & 2 & 4 & | & 8 \\ 2 & 4 & 6 & | & 10 \\ 1 & 2 & 5 & | & 1 \end{bmatrix} \quad R_1 \leftrightarrow R_3 \quad \begin{bmatrix} 1 & 2 & 5 & | & 1 \\ 2 & 4 & 6 & | & 10 \\ 2 & 2 & 4 & | & 8 \end{bmatrix}$$

2. Now we want 0's as the remaining entries of the first column. (We call this sweeping out the column.) We will take two steps at once:

$$\begin{bmatrix} 1 & 2 & 5 & | & 1 \\ 2 & 4 & 6 & | & 10 \\ 2 & 2 & 4 & | & 8 \end{bmatrix} \quad \begin{matrix} -2R_1 + R_2 \to R_2 \\ -2R_1 + R_3 \to R_3 \end{matrix} \quad \begin{bmatrix} 1 & 2 & 5 & | & 1 \\ 0 & 0 & -4 & | & 8 \\ 0 & -2 & -6 & | & 6 \end{bmatrix}$$

3. Next we want to get 1 in the middle of the main diagonal (in row 2, column 2). First, we interchange rows 2 and 3, then we divide row 2 by -2:

$$\begin{bmatrix} 1 & 2 & 5 & | & 1 \\ 0 & 0 & -4 & | & 8 \\ 0 & -2 & -6 & | & 6 \end{bmatrix} \quad R_2 \leftrightarrow R_3 \quad \begin{bmatrix} 1 & 2 & 5 & | & 1 \\ 0 & -2 & -6 & | & 6 \\ 0 & 0 & -4 & | & 8 \end{bmatrix}$$

$$\begin{bmatrix} 1 & 2 & 5 & | & 1 \\ 0 & -2 & -6 & | & 6 \\ 0 & 0 & -4 & | & 8 \end{bmatrix} \quad -\tfrac{1}{2}R_2 \to R_2 \quad \begin{bmatrix} 1 & 2 & 5 & | & 1 \\ 0 & 1 & 3 & | & -3 \\ 0 & 0 & -4 & | & 8 \end{bmatrix}$$

Notice that in using the elementary row operations to transform the matrix, we are taking particular advantage of the rows (or columns) with zero entries.

4. Now we want 0's in the remaining entries of the second column. (We sweep out the rest of the second column.)

$$\begin{bmatrix} 1 & 2 & 5 & | & 1 \\ 0 & 1 & 3 & | & -3 \\ 0 & 0 & -4 & | & 8 \end{bmatrix} \quad -2R_2 + R_1 \to R_1 \quad \begin{bmatrix} 1 & 0 & -1 & | & 7 \\ 0 & 1 & 3 & | & -3 \\ 0 & 0 & -4 & | & 8 \end{bmatrix}$$

5. Next, we want to get 1 in the lower right entry of the main diagonal:

$$\begin{bmatrix} 1 & 0 & -1 & | & 7 \\ 0 & 1 & 3 & | & -3 \\ 0 & 0 & -4 & | & 8 \end{bmatrix} \quad -\tfrac{1}{4}R_3 \to R_3 \quad \begin{bmatrix} 1 & 0 & -1 & | & 7 \\ 0 & 1 & 3 & | & -3 \\ 0 & 0 & 1 & | & -2 \end{bmatrix}$$

6. Finally, sweep out the third column:

$$\begin{bmatrix} 1 & 0 & -1 & | & 7 \\ 0 & 1 & 3 & | & -3 \\ 0 & 0 & 1 & | & -2 \end{bmatrix} \quad \begin{matrix} R_3 + R_1 \to R_1 \\ -3R_3 + R_2 \to R_2 \end{matrix} \quad \begin{bmatrix} 1 & 0 & 0 & | & 5 \\ 0 & 1 & 0 & | & 3 \\ 0 & 0 & 1 & | & -2 \end{bmatrix}$$

This matrix is in reduced echelon form. We can simply read the solution (or write its associated system) to get $x = 5, y = 3, z = -2$. ∎

A matrix can have an infinite number of equivalent triangular forms, but its reduced echelon form is unique. This is one advantage of Gauss–Jordan elimination. Another advantage is that Gauss–Jordan elimination lends itself well to programming. However, putting a matrix into reduced echelon form can require tedious computations. If you are not required to use one particular approach, you may start out trying to put a matrix into reduced echelon form, but you may want to stop if you find that the matrix is in triangular form and you can use back substitution to find your solution. Actually, you can stop anytime you see that you can arrive at a solution fairly easily from the transformed matrix.

7.3 Exercises

In Exercises 1–4, set up the augmented matrix of the system of equations.

1. $\begin{cases} 7x + y = 11 \\ x - 3y = 4 \end{cases}$

2. $\begin{cases} 6x + 2y = -1 \\ 4x - y = 4 \end{cases}$

3. $\begin{cases} 3x - 2y + z = 15 \\ 2x + y - 3z = 10 \\ 5x - 3y + z = -2 \end{cases}$

4. $\begin{cases} 5x + 2y - z = -8 \\ 3x + y = -4 \\ 3y - 4z = 7 \end{cases}$

In Exercises 5–8, put the matrix into triangular form by following the given elementary row operations.

5. $\begin{bmatrix} 1 & 3 & | & 2 \\ 5 & 4 & | & 0 \end{bmatrix}$ Perform the row operation $-5R_1 + R_2 \to R_2$

6. $\begin{bmatrix} 5 & 4 & | & 0 \\ 1 & 3 & | & 2 \end{bmatrix}$ First perform row operation $R_1 \leftrightarrow R_2$; then row operation $-5R_1 + R_2 \to R_2$ on the resultant matrix.

7. $\begin{bmatrix} 1 & 2 & 1 & | & 0 \\ 2 & 6 & 3 & | & 1 \\ 3 & 8 & 7 & | & -8 \end{bmatrix}$ Perform the following row operations: $-2R_1 + R_2 \to R_2$ and $-3R_1 + R_3 \to R_3$; then perform row operation $-R_2 + R_3 \to R_3$ on the resultant matrix.

8. $\begin{bmatrix} 1 & 2 & 3 & | & 5 \\ 3 & 1 & 0 & | & 4 \\ -1 & 3 & 0 & | & 2 \end{bmatrix}$ Perform the following row operations: First $-3R_1 + R_2 \to R_2$ and $R_1 + R_3 \to R_3$; then perform row operation $R_2 + R_3 \to R_3$ on the resultant matrix.

In Exercises 9–28, solve the system of equations by Gaussian elimination.

9. $\begin{cases} x - y = 4 \\ 2x + y = 11 \end{cases}$

10. $\begin{cases} x + 2y = -2 \\ x - y = 4 \end{cases}$

11. $\begin{cases} x + y = -1 \\ 3x - 2y = 12 \end{cases}$

12. $\begin{cases} 5x - 2y = 0 \\ x - 2y = 8 \end{cases}$

13. $\begin{cases} 5x - 2y = 1 \\ x - 5y = -32 \end{cases}$

14. $\begin{cases} x = 3y - 1 \\ 2x - 5y = -3 \end{cases}$

15. $\begin{cases} 8x - 2y = 10 \\ y = 3x - 15 \end{cases}$

16. $\begin{cases} x + 2y = -1 \\ 3x + 6y = 3 \end{cases}$

17. $\begin{cases} 2y + 3x = 2 \\ 3x + y = 1 \end{cases}$

18. $\begin{cases} 5x + 2y = -8 \\ 3x + 2y = -4 \end{cases}$

19. $\begin{cases} x + y + z = 6 \\ 2x + y - z = 4 \\ 3x + y - z = 6 \end{cases}$

20. $\begin{cases} x + y - z = 1 \\ 2x + 2y + 2z = 0 \\ x - y + z = 3 \end{cases}$

21. $\begin{cases} x + y - 2z = -5 \\ x - y + 2z = 9 \\ 3x + y - z = 2 \end{cases}$

22. $\begin{cases} x - 2y + z = 4 \\ x - y - 2z = -3 \\ 3x + 2y + z = 0 \end{cases}$

23. $\begin{cases} x + y + z = 6 \\ 3x + y - z = 4 \\ 2x + 2y + 2z = 4 \end{cases}$

24. $\begin{cases} x + y - z = 3 \\ 2x + 2y - 2z = 6 \\ x - y + z = 5 \end{cases}$

25. $\begin{cases} 2a - 2b + c = 0 \\ 4a - b + 2c = 6 \\ 2a + 3b - c = 4 \end{cases}$

26. $\begin{cases} x + 3y - z = 0 \\ 2x + y - z = 4 \\ 3x - y + 2z = 5 \end{cases}$

27. $\begin{cases} w + x + y - z = 4 \\ w + 2x + 2y + z = 3 \\ 2w + x + 3y - 2z = 7 \\ 3w + x - y + 2z = 5 \end{cases}$

28. $\begin{cases} w - x + y - z = 4 \\ w - 3y - z = 0 \\ w + 2x - 2z = 4 \\ -x + 2y - z = 3 \end{cases}$

In Exercises 29–30, put the given matrix into reduced echelon form by successively applying the given elementary row operations to the resultant matrix.

29. $\begin{bmatrix} 3 & 6 & | & 0 \\ 3 & 1 & | & 5 \end{bmatrix}$

(a) $\frac{1}{3}R_1 \to R_1$ (b) $-3R_1 + R_2 \to R_2$

(c) $-\frac{1}{5}R_2 \to R_2$ (d) $-2R_2 + R_1 \to R_1$

30. $\begin{bmatrix} 2 & 0 & 2 & | & -2 \\ 2 & 3 & 5 & | & 4 \\ 4 & 3 & 9 & | & 6 \end{bmatrix}$

(a) $\frac{1}{2}R_1 \to R_1$

(b) $-2R_1 + R_2 \to R_2$ and $-4R_1 + R_3 \to R_3$

(c) $\frac{1}{3}R_2 \to R_2$

(d) $-3R_2 + R_3 \to R_3$

(e) $\frac{1}{2}R_3 \to R_3$

(f) $-R_3 + R_2 \to R_2$ and $-R_3 + R_1 \to R_1$

In Exercises 31–34, put the augmented matrix in reduced echelon form.

31. $\begin{bmatrix} 1 & 4 & | & 0 \\ 2 & 3 & | & -5 \end{bmatrix}$ **32.** $\begin{bmatrix} 5 & 2 & | & 0 \\ 2 & 1 & | & 1 \end{bmatrix}$

33. $\begin{bmatrix} 1 & 1 & 2 & | & 0 \\ 2 & 4 & 2 & | & 6 \\ 3 & 1 & 2 & | & 6 \end{bmatrix}$ **34.** $\begin{bmatrix} 2 & 1 & 3 & | & 4 \\ 1 & 2 & 1 & | & 1 \\ 2 & 3 & 2 & | & 6 \end{bmatrix}$

In Exercises 35–40, solve by the Gauss–Jordan method.

35. $\begin{cases} x - 2y = 4 \\ 3x + 5y = 1 \end{cases}$ **36.** $\begin{cases} x + 2y = 3 \\ 6x - y = 31 \end{cases}$

37. $\begin{cases} x + 2y + z = 8 \\ x + 4y - z = 12 \\ x - 2y + z = -4 \end{cases}$ **38.** $\begin{cases} x + z = 2 \\ 2x - 2y = -6 \\ 2y - z = 4 \end{cases}$

39. $\begin{cases} x + y + 2z = 1 \\ 2x + 4y + 2z = 6 \\ 3x + y + 2z = -4 \end{cases}$

40. $\begin{cases} w - x + y = -2 \\ x - z = -3 \\ -2x + 2y - z = -7 \\ w + 2y + z = -1. \end{cases}$

41. Find A, B, C, and D such that

$$\frac{5x^3 + 3x^2 + 2}{(x^2 + 5)(x^2 + 4)} = \frac{Ax + B}{x^2 + 5} + \frac{Cx + D}{x^2 + 4}$$

HINT: See Example 5 on page 422.

42. Find A, B, C, and D such that

$$\frac{3x^2 + 2x + 4}{(x^2 + 3)^2} = \frac{Ax + B}{x^2 + 3} + \frac{Cx + D}{(x^2 + 3)^2}$$

HINT: See Example 5 on page 422.

Questions for Thought

43. In our approach using Gaussian elimination, our first goal was to put the system's augmented matrix into (upper) triangular form, a form where there are all zeros below the main diagonal. We could just as easily have solved the systems by putting the matrix into *lower* triangular form, a form where there are all zeros *above* the main diagonal, and using forward substitution. Solve Exercises 21 and 22 by putting the matrix into lower triangular form.

44. Discuss the differences among the following matrix forms: triangular form, echelon form, and reduced echelon form.

7.4 The Algebra of Matrices

In the previous section we defined and used matrices in developing methods for solving linear systems of equations. In this section we expand our discussion on matrices and discuss matrix operations.

Recall that a matrix is a rectangular array of numbers; the numbers in the matrix are called the *elements*, or *entries*, of the matrix. When we refer to a matrix, we usually refer to its size by specifying the number of rows by the number of columns. The following matrix is called an $m \times n$ matrix (*m rows* by *n columns*).

$$\begin{bmatrix} a_{11} & a_{12} & a_{13} & \cdots & a_{1n} \\ a_{21} & a_{22} & a_{23} & \cdots & a_{2n} \\ a_{31} & a_{32} & a_{33} & \cdots & a_{3n} \\ \vdots & \vdots & \vdots & & \vdots \\ a_{m1} & a_{m2} & a_{m3} & \cdots & a_{mn} \end{bmatrix}$$

Take a careful look at the subscripts of a, the entries of the above matrix. The subscripts indicate the position of the entry in the matrix. The first number indicates the row; the second indicates the column. For example a_{34} is the entry in the third row and fourth column. In general, a_{ij} is the entry in the ith row and jth column.

To designate matrices we will either use uppercase letters such as A, B, X, or Y, or use the notation $[a_{ij}]$, whichever is more convenient. (*Remember:* Although a_{ij} refers to a single entry in a matrix, $[a_{ij}]$ refers to the whole matrix.)

When we introduced number systems, such as the real or complex numbers, we started out defining the elements of the system (what a real number or complex number looks like), we next defined operations on these elements, and then we examined the properties of the elements and their operations. We approach matrices in the same way. We have already defined what a matrix is; next we make sure these elements are well defined.

Equivalence of Matrices

If $A = [a_{ij}]$ and $B = [b_{ij}]$ are $m \times n$ matrices, then

$$A = B \quad \text{if and only if} \quad a_{ij} = b_{ij}$$

for i and j such that $1 \le i \le m$ and $1 \le j \le n$.

In words, two matrices are equal if and only if corresponding elements in each matrix are equal.

This requires, of course, that the two matrices have the same number of rows and columns. For example, $\begin{bmatrix} 2 & 5 \\ -3 & 8 \end{bmatrix}$ is equivalent to $\begin{bmatrix} 2 & \sqrt{25} \\ 2-5 & |-8| \end{bmatrix}$, whereas the

two matrices $\begin{bmatrix} 2 & 4 \\ 6 & 8 \\ 1 & 5 \end{bmatrix}$ and $\begin{bmatrix} 2 & 6 & 1 \\ 4 & 8 & 5 \end{bmatrix}$ *cannot* be equal because they do not have

the same number of rows and columns.

Now we can define addition of matrices.

Definition of Addition of Matrices

If $A = [a_{ij}]$ and $B = [b_{ij}]$ are *both* $m \times n$ matrices, then

$$A + B = [a_{ij}] + [b_{ij}] = [a_{ij} + b_{ij}]$$

for i and j such that $1 \le i \le m$ and $1 \le j \le n$. This definition states that the sum of two matrices is found by adding their corresponding entries.

Example 1 | Find the sum: $\begin{bmatrix} 3 & -2 & 0 & 1 \\ 2 & 8 & -1 & 4 \end{bmatrix} + \begin{bmatrix} 5 & 0 & -1 & 3 \\ -2 & 1 & 9 & 10 \end{bmatrix}$

Solution | First we observe that both matrices are the same size, 2×4; therefore, we can add them. (If they are not the same size, then addition is not defined.) To find their sum, all we do is add their corresponding elements:

$$\begin{bmatrix} 3 & -2 & 0 & 1 \\ 2 & 8 & -1 & 4 \end{bmatrix} + \begin{bmatrix} 5 & 0 & -1 & 3 \\ -2 & 1 & 9 & 10 \end{bmatrix}$$

$$= \begin{bmatrix} 3+5 & -2+0 & 0+(-1) & 1+3 \\ 2+(-2) & 8+1 & -1+9 & 4+10 \end{bmatrix}$$

$$= \begin{bmatrix} 8 & -2 & -1 & 4 \\ 0 & 9 & 8 & 14 \end{bmatrix}$$

Using the definition of matrix addition, we can easily show that matrix addition, where it makes sense, is both commutative and associative.

As far as additive identities for matrices, keep in mind that matrices can be added only if they are the same size. So we must describe an additive identity for each class of $m \times n$ matrices. Let $\mathbf{O}$ be the $m \times n$ matrix consisting of all zero entries. Then the following can be proven.

Theorem 7.1

The $m \times n$ matrix $\mathbf{O}$, consisting of all zero entries, is the additive identity for all $m \times n$ matrices. That is, if A is any $m \times n$ matrix, then $A + \mathbf{O} = \mathbf{O} + A = A$.

For example, $\begin{bmatrix} 0 & 0 \\ 0 & 0 \end{bmatrix}$ is the $\mathbf{O}$ matrix for *all* 2×2 matrices. Hence

$$A + \mathbf{O} = \begin{bmatrix} -1 & 3 \\ 5 & -8 \end{bmatrix} + \begin{bmatrix} 0 & 0 \\ 0 & 0 \end{bmatrix} = \begin{bmatrix} -1+0 & 3+0 \\ 5+0 & -8+0 \end{bmatrix} = \begin{bmatrix} -1 & 3 \\ 5 & -8 \end{bmatrix} = A$$

The additive inverse of an $m \times n$ matrix A is the $m \times n$ matrix designated $-A$, consisting of entries that are opposite in sign to the entries of A. We summarize these properties as follows.

Matrix Addition Properties

If A, B, and C, are $m \times n$ matrices, then

1. $A + B = B + A$

2. $A + (B + C) = (A + B) + C$

3. There exists a unique $m \times n$ matrix, called $\mathbf{O}$, such that

$$A + \mathbf{O} = \mathbf{O} + A = A$$

4. For each A, there exists a unique matrix, called $-A$, such that

$$A + (-A) = (-A) + A = \mathbf{O}$$

We define multiplying a matrix by a constant, which is called **scalar multiplication**, as follows.

Definition of Scalar Multiplication

If $A = [a_{ij}]$ is an $m \times n$ matrix and c is a real number, then

$$cA = c[a_{ij}] = [ca_{ij}]$$

for i and j such that $1 \le i \le m$ and $1 \le j \le n$.

In words, multiplying a matrix by a constant yields a matrix in which *each* entry is mutiplied by the constant.

For example,

Notice that the matrix $-A$ can be viewed as $(-1)A$, which is the matrix consisting of entries that are opposite in sign to the entries of A.

$$2 \begin{bmatrix} 5 & 2 & 0 \\ -2 & 1 & -1 \\ 2 & -3 & 7 \end{bmatrix} = \begin{bmatrix} 2(5) & 2(2) & 2(0) \\ 2(-2) & 2(1) & 2(-1) \\ 2(2) & 2(-3) & 2(7) \end{bmatrix} = \begin{bmatrix} 10 & 4 & 0 \\ -4 & 2 & -2 \\ 4 & -6 & 14 \end{bmatrix}$$

We can similarly define Ac, where A is a matrix and c is a real number. We can easily show that scalar multiplication and addition have the following properties.

Scalar Multiplication Properties

If A and B are $m \times n$ matrices, and c and d are real numbers, then

1. $cA = Ac$ **2.** $(cd)A = c(dA)$

3. $(c + d)A = cA + dA$ **4.** $c(A + B) = cA + cB$

Example 2   If $A = \begin{bmatrix} 3 & -2 \\ 0 & 1 \\ 2 & 8 \end{bmatrix}$ and $B = \begin{bmatrix} 7 & 0 \\ -1 & 3 \\ -2 & 1 \end{bmatrix}$, find $5A - B$.

Solution   We make sure addition makes sense by noting that since A is a 3×2 matrix, $5A$ will also be 3×2, which *can* be added to B, also a 3×2 matrix.

$$5A - B = 5 \cdot A \qquad\qquad + (-1) \cdot B$$

$$= 5\begin{bmatrix} 3 & -2 \\ 0 & 1 \\ 2 & 8 \end{bmatrix} + (-1)\begin{bmatrix} 7 & 0 \\ -1 & 3 \\ -2 & 1 \end{bmatrix} \qquad \text{Find } 5A \text{ and } (-1)\,B.$$

$$= \begin{bmatrix} 15 & -10 \\ 0 & 5 \\ 10 & 40 \end{bmatrix} + \begin{bmatrix} -7 & 0 \\ 1 & -3 \\ 2 & -1 \end{bmatrix} = \begin{bmatrix} 8 & -10 \\ 1 & 2 \\ 12 & 39 \end{bmatrix}$$

Example 3   If $A = \begin{bmatrix} -4 & 0 \\ 1 & 2 \end{bmatrix}$ and $B = \begin{bmatrix} 2 & -12 \\ 0 & 5 \end{bmatrix}$, find the matrix X such that $3X + A = B$.

Solution   We begin by solving the equation $3X + A = B$ for X as we would any linear equation:

$$3X + A = B \qquad \text{Solve for } X \text{ to get}$$

$$X = \frac{1}{3}(B - A)$$

Since both A and B are 2×2 matrices, then their sum (and difference) is a 2×2 matrix and hence X is a 2×2 matrix.

$$X = \frac{1}{3}\left(\begin{bmatrix} 2 & -12 \\ 0 & 5 \end{bmatrix} - \begin{bmatrix} -4 & 0 \\ 1 & 2 \end{bmatrix}\right) = \frac{1}{3}\begin{bmatrix} 6 & -12 \\ -1 & 3 \end{bmatrix} \qquad \text{Hence}$$

$$X = \begin{bmatrix} 2 & -4 \\ -\frac{1}{3} & 1 \end{bmatrix}$$

The student should check that the matrix X satisfies the equation $3X + A = B$.

Matrix Multiplication

The product of two matrices is not as easy to define as addition or scalar multiplication. To more easily describe the process of matrix multiplication, we first introduce a special operation called the *inner product,* defined for matrices with one row or matrices with one column. We call a matrix with one row a **row vector**, and a matrix with one column, a **column vector**.

Definition of Inner Product

The **inner product** of a $1 \times p$ row vector and a $p \times 1$ column vector is

$$\begin{bmatrix} a_{11} & a_{12} & a_{13} & \cdots & a_{1p} \end{bmatrix} \cdot \begin{bmatrix} b_{11} \\ b_{21} \\ b_{31} \\ \vdots \\ b_{p1} \end{bmatrix} = a_{11}b_{11} + a_{12}b_{21} + a_{13}b_{31} + \cdots + a_{1p}b_{p1}$$

The inner product is sometimes called the dot product.

The inner product of a row vector and column vector is the sum of the products of the entries in the row vector with their corresponding entries in the column vector. For example, we can find the inner product

$$[3 \quad -2 \quad 0 \quad 4] \cdot \begin{bmatrix} 7 \\ -1 \\ 2 \\ 5 \end{bmatrix} = 3(7) + (-2)(-1) + 0(2) + 4(5) = 43$$

It is important to note that the inner product is a *single number* and that we define inner products only for $1 \times p$ row vectors by $p \times 1$ column vectors, where the row vector is multiplied *left* of the column vector. Thus the *inner product*

$$\begin{bmatrix} 7 \\ -1 \\ 2 \\ 5 \end{bmatrix} \cdot [3 \quad -2 \quad 0 \quad 4]$$

is not defined. Keep in mind that the symbol • used here is special notation used to indicate the inner product of a row vector by a column vector; it is not the usual dot symbol used to indicate numerical multiplication. Now we can define matrix multiplication.

Definition of Matrix Multiplication

The product of A, an $m \times n$ matrix, and B, an $n \times p$ matrix, is the $m \times p$ matrix C, where the entry in the ith row and jth column is the inner product of the ith row vector of A with the jth column vector of B.

This definition requires that the number of columns of A, the matrix on the left, be equal to the number of rows of B, the matrix on the right. For example, if

Note that matrix A has 3 columns and matrix B has 3 rows.

$$A = \begin{bmatrix} 6 & -2 & 8 \\ 1 & 4 & 5 \end{bmatrix}, \text{ and } B = \begin{bmatrix} 1 & 3 \\ 5 & 0 \\ 2 & 7 \end{bmatrix}, \text{ the product } AB \text{ is found as follows:}$$

$$\begin{bmatrix} 6 & -2 & 8 \\ 1 & 4 & 5 \end{bmatrix} \begin{bmatrix} 1 & 3 \\ 5 & 0 \\ 2 & 7 \end{bmatrix} \quad 6(1) + (-2)5 + 8(2) = 12 \quad \begin{bmatrix} 12 & \\ & \end{bmatrix}$$

The first entry (row 1, column 1) of the product AB is the inner product of row 1 of A and column 1 of B:

$$\begin{bmatrix} 6 & -2 & 8 \\ 1 & 4 & 5 \end{bmatrix} \begin{bmatrix} 1 & 3 \\ 5 & 0 \\ 2 & 7 \end{bmatrix} \quad 6(3) + (-2)0 + 8(7) = 74 \quad \begin{bmatrix} 12 & 74 \\ & \end{bmatrix}$$

The entry for row 1, column 2 of the product AB is the inner product of row 1 of A and column 2 of B:

$$\begin{bmatrix} 6 & -2 & 8 \\ 1 & 4 & 5 \end{bmatrix} \begin{bmatrix} 1 & 3 \\ 5 & 0 \\ 2 & 7 \end{bmatrix} \quad \begin{bmatrix} 12 & 74 \\ 31 & \end{bmatrix}$$

The entry for row 2, column 1 of the product AB is the inner product of row 2 of A and column 1 of B:

$$1(1) + (4)5 + 5(2) = 31$$

$$\begin{bmatrix} 6 & -2 & 8 \\ 1 & 4 & 5 \end{bmatrix} \begin{bmatrix} 1 & 3 \\ 5 & 0 \\ 2 & 7 \end{bmatrix} \quad \begin{bmatrix} 12 & 74 \\ 31 & 38 \end{bmatrix}$$

Finally, the entry in row 2, column 2 of the product AB is the inner product of row 2 of A and column 2 of B:

$$1(3) + 4(0) + 5(7) = 38$$

The matrix product of a 2 × 3 matrix and a 3 × 2 matrix is a 2 × 2 matrix.

The product is $AB = \begin{bmatrix} 6 & -2 & 8 \\ 1 & 4 & 5 \end{bmatrix} \begin{bmatrix} 1 & 3 \\ 5 & 0 \\ 2 & 7 \end{bmatrix} = \begin{bmatrix} 12 & 74 \\ 31 & 38 \end{bmatrix}$. Notice that

we end up with a 2 × 2 matrix; the resulting matrix AB should have the same number of rows as A and the same number of columns as B.

On the other hand, we find a different result computing the matrix BA for the same A and B, as demonstrated in the next example.

Example 4 If $A = \begin{bmatrix} 6 & -2 & 8 \\ 1 & 4 & 5 \end{bmatrix}$ and $B = \begin{bmatrix} 1 & 3 \\ 5 & 0 \\ 2 & 7 \end{bmatrix}$, find BA.

Solution

$\begin{bmatrix} 1 & 3 \\ 5 & 0 \\ 2 & 7 \end{bmatrix} \begin{bmatrix} 6 & -2 & 8 \\ 1 & 4 & 5 \end{bmatrix}$ $1(6) + (3)1 = 9$

The entry for row 1, column 1 of the product BA is the inner product of row 1 of B and column 1 of A:

$\begin{bmatrix} 9 & & \\ & & \\ & & \end{bmatrix}$

$\begin{bmatrix} 1 & 3 \\ 5 & 0 \\ 2 & 7 \end{bmatrix} \begin{bmatrix} 6 & -2 & 8 \\ 1 & 4 & 5 \end{bmatrix}$

The entry for row 2, column 3 of the product BA is the inner product of row 2 of B and column 3 of A:

$\begin{bmatrix} 9 & & \\ & & 40 \\ & & \end{bmatrix}$

$5(8) + (0)5 = 40$

The remaining entries are as follows:

Many graphing calculators can do matrix multiplication. For instance, if we enter the matrix values of this example for matrices A and B, then the following display illustrates the matrix multiplication of AB and BA.

$BA = \begin{bmatrix} 1 & 3 \\ 5 & 0 \\ 2 & 7 \end{bmatrix} \begin{bmatrix} 6 & -2 & 8 \\ 1 & 4 & 5 \end{bmatrix} = \begin{bmatrix} 1(6) + 3(1) & 1(-2) + 3(4) & 1(8) + 3(5) \\ 5(6) + 0(1) & 5(-2) + 0(4) & 5(8) + 0(5) \\ 2(6) + 7(1) & 2(-2) + 7(4) & 2(8) + 7(5) \end{bmatrix}$

$= \begin{bmatrix} 9 & 10 & 23 \\ 30 & -10 & 40 \\ 19 & 24 & 51 \end{bmatrix}$

```
[A] [B]
       [[12 74]
        [31 38]]
[B] [A]
       [[9  10  23]
        [30 -10 40]
        [19 24  51]]
```

Notice that BA is a 3 × 3 matrix (the number of rows of B by the number of columns of A). ∎

As illustrated in the previous two examples, if A is a 2 × 3 matrix and B is a 3 × 2 matrix, the product AB is a 2 × 2 matrix, but the product BA is a 3 × 3 matrix. In general, we find that $AB \neq BA$; that is, matrix multiplication is not commutative. As a matter of fact, in many cases we may be able to compute a product of two matrices but not be able to compute the product of the same two matrices in reverse order. This is demonstrated in the next example.

Example 5 Given $A = \begin{bmatrix} 1 & -2 \\ 3 & 1 \\ 0 & -1 \end{bmatrix}$ and $B = \begin{bmatrix} 2 & 5 & 3 & 0 \\ 4 & -1 & 2 & 1 \end{bmatrix}$.

(a) Find AB.   **(b)** Find BA.

Solution **(a)** AB means we multiply the 3 × 2 matrix A by the 2 × 4 matrix B, where A is left of B. We first check to see whether the multiplication makes sense—that is, whether the number of columns in the left matrix (which is 2) is equal to the number of rows in the right matrix (which is 2). Note that we should end up with a 3 × 4 matrix.

For matrix multiplication, remember: multiply each left-row vector by right-column vector.

$$\begin{bmatrix} 1 & -2 \\ 3 & 1 \\ 0 & -1 \end{bmatrix} \begin{bmatrix} 2 & 5 & 3 & 0 \\ 4 & -1 & 2 & 1 \end{bmatrix} \xrightarrow{\;1(3)+(-2)2\;} = \begin{bmatrix} -6 & 7 & -1 & -2 \\ 10 & 14 & 11 & 1 \\ -4 & 1 & -2 & -1 \end{bmatrix} = AB$$

Matrix multiplicaton:
$$(A)(B) = C$$
$$(m \times n)(n \times p) = (m \times p)$$
same

(b) *BA* means we multiply the 2×4 matrix *B* by the 3×2 matrix *A*, where *B* is left of *A*. We first check to see whether the multiplication makes sense—that is, whether the number of columns in the left matrix (which is 4) is equal to the number of rows in the right matrix (which is 3). Since they are not equal, the product *BA* is not defined. ∎

Although matrix multiplication is not commutative, matrix multiplication is associative and does distribute over addition. The following box summarizes some useful properties:

Matrix Multiplication Properties

Suppose *A*, *B*, and *C* are matrices (where multiplication makes sense) and *k* is a real number. Then

1. $A(BC) = (AB)C$
2. $A(B + C) = AB + AC$
3. $(B + C)A = BA + CA$
4. $(kA)B = k(AB)$

Matrices can be used in applications where we have to perform repetitive operations on groups or tables of numbers, as illustrated next.

Example 6

Over a 6-month period, the mathematics and English departments use office supplies as illustrated by the accompanying table.

This type of problem is ideal for a spreadsheet program.

Supplies	English Department	Mathematics Department
Cases of legal pads	8	6
Cases of duplicating paper	12	18
Gross of pencils	6	3

Tom's Office Supply Company charges $230 for a case of legal pads, $65 for a case of duplicating paper, and $18 for a gross of pencils. Kuma's Office Supply Company charges $250 for a case of legal pads, $56 for a case of duplicating paper, and $16 for a gross of pencils. If supplies must be ordered from one company, which office supply company is less expensive for each department's 6-month order?

Solution

It is convenient to represent the department orders and costs as matrices. The matrix

There is some ambiguity left in the problem. Must we order both math and English supplies from one supplier? Or can separate departments order from different suppliers? We will explore all possibilities.

representing the order from each department can be written as $\begin{bmatrix} 8 & 6 \\ 12 & 18 \\ 6 & 3 \end{bmatrix}$. We can represent the cost matrix for Tom's Office Supplies as [$230 $65 $18] and for Kuma's Office Supplies as [$250 $56 $160].

To find the cost of the order for each department, we multiply the cost matrix by the order matrix:

The cost from Tom's Office Supplies is

	Legal Pads	Paper	Pencils		English	Math		Total For English	Total For Math

$$[\$230 \quad \$65 \quad \$18] \begin{bmatrix} 8 & 6 \\ 12 & 18 \\ 6 & 3 \end{bmatrix} = [\$2728 \quad \$2604]$$

The cost from Kuma's Office Supplies is

$$[\$250 \quad \$56 \quad \$16] \begin{bmatrix} 8 & 6 \\ 12 & 18 \\ 6 & 3 \end{bmatrix} = [\$2768 \quad \$2556]$$

If the departments order separately, then to lower the cost the English department should order from Tom, but the mathematics department should order from Kuma. By ordering separately from their less expensive vendor, they would spend all together:

$$\underbrace{\$2728}_{\text{English department from Tom}} + \underbrace{\$2556}_{\text{Math department from Kuma}} = \$5284$$

If both departments had to combine their orders, then the total cost from Tom's Office Supplies would be $\$2728 + \$2604 = \$5332$; the total cost from Kuma's Office Supplies would be $\$2768 + \$2556 = \$5324$. Hence, a combined order should be made from Kuma's Office Supplies. Notice that combining orders would cost more all together than each department ordering separately from its lower-priced supplier.   ∎

7.4 Exercises

In Exercises 1–8, perform the operations if possible.

1. $\begin{bmatrix} 1 & -3 \\ 5 & 2 \end{bmatrix} + \begin{bmatrix} 2 & 5 \\ 3 & 0 \end{bmatrix}$

2. $\begin{bmatrix} 2 & 9 \\ 15 & -23 \end{bmatrix} + \begin{bmatrix} -14 & 6 \\ 0 & -20 \end{bmatrix}$

3. $\begin{bmatrix} 1 & 3 \\ -5 & 0 \\ 2 & -7 \end{bmatrix} + \begin{bmatrix} -1 & -3 \\ 5 & 0 \\ -2 & 7 \end{bmatrix}$

4. $\begin{bmatrix} 2 & -3 & 8 \\ 5 & -1 & 0 \\ -2 & 7 & 18 \end{bmatrix} + \begin{bmatrix} -1 & -3 \\ 5 & 0 \\ -2 & 7 \end{bmatrix}$

5. $\begin{bmatrix} 0 & 3 & 6 \\ 80 & 0 & -19 \\ 2 & 12 & 4 \end{bmatrix} + \begin{bmatrix} 2 & 3 \\ 5 & 0 \end{bmatrix}$

6. $\begin{bmatrix} 2 & -3 & 8 \\ 5 & -1 & 0 \\ -2 & 7 & 18 \end{bmatrix} + \begin{bmatrix} 0 & 0 & 0 \\ 0 & 0 & 0 \\ 0 & 0 & 0 \end{bmatrix}$

7. $3 \begin{bmatrix} 0 & 3 \\ 17 & 0 \end{bmatrix}$

8. $-4 \begin{bmatrix} 2 & -3 & 8 \\ 5 & -1 & 0 \\ -2 & 7 & 18 \end{bmatrix}$

In Exercises 9–16, compute the matrix, if possible, given matrices A, B, C, and D.

$$A = \begin{bmatrix} 2 & -3 \\ -1 & 0 \end{bmatrix} \qquad B = \begin{bmatrix} 5 & 1 \\ -2 & 3 \end{bmatrix}$$

$$C = \begin{bmatrix} 1 & -1 & 3 \\ -2 & 3 & 0 \\ 5 & 8 & 2 \end{bmatrix} \qquad D = \begin{bmatrix} 3 & 1 & 0 \\ -2 & 4 & 6 \\ 3 & 8 & -10 \end{bmatrix}$$

9. $-A$ **10.** $2C$

11. $\frac{2}{3}B$ **12.** $-3D$

13. $A - B$ **14.** $B - A$

15. $2A - B$ **16.** $C - 3D$

In Exercises 17–20, solve for X, if possible, given the matrices A, B, C, and D defined above.

17. $2X = B$ **18.** $3X = B + C$

19. $2X - B = 2A$ **20.** $2(4D - X) = C$

In Exercises 21–36, compute the matrix products, if possible.

21. $[3 \quad -3 \quad 1]\begin{bmatrix} 5 \\ 1 \\ -2 \end{bmatrix}$

22. $[5 \quad 0 \quad -1]\begin{bmatrix} 3 \\ -9 \\ 0 \end{bmatrix}$

23. $[0 \quad -1 \quad 1 \quad 2]\begin{bmatrix} 1 \\ -1 \\ -2 \\ 2 \end{bmatrix}$

24. $\begin{bmatrix} 1 \\ -1 \\ -2 \\ 2 \end{bmatrix}\begin{bmatrix} 3 & 2 & 5 & 1 \\ 0 & -1 & 1 & 2 \end{bmatrix}$

25. $\begin{bmatrix} 0 & -1 \\ 3 & 2 \end{bmatrix}\begin{bmatrix} -1 & 3 \\ 0 & -2 \end{bmatrix}$

26. $\begin{bmatrix} 1 & 0 \\ 0 & 1 \end{bmatrix}\begin{bmatrix} 3 & 5 \\ -4 & 6 \end{bmatrix}$

27. $\begin{bmatrix} 3 & -1 & 4 \\ 3 & -2 & 5 \end{bmatrix}\begin{bmatrix} -1 & 3 & 5 \\ 0 & -2 & 1 \end{bmatrix}$

28. $\begin{bmatrix} 1 & 0 & -3 \\ 2 & 1 & -4 \end{bmatrix}\begin{bmatrix} 3 & 5 \\ -4 & 6 \\ 3 & -1 \end{bmatrix}$

29. $\begin{bmatrix} 0 & 1 & -5 \\ -3 & 1 & 6 \end{bmatrix}\begin{bmatrix} -1 & 3 \\ 0 & -2 \\ 6 & 0 \end{bmatrix}$

30. $\begin{bmatrix} -1 & 3 \\ 0 & -2 \\ 6 & 0 \end{bmatrix}\begin{bmatrix} 0 & 1 & -5 \\ -3 & 1 & 6 \end{bmatrix}$

31. $\begin{bmatrix} 1 & 0 & -2 \\ 3 & 1 & -1 \\ 2 & 0 & 1 \end{bmatrix}\begin{bmatrix} 1 & -3 & 0 \\ 0 & 2 & -2 \\ -6 & 0 & 5 \end{bmatrix}$

32. $\begin{bmatrix} 1 & -3 & 0 \\ 0 & 2 & -2 \\ -6 & 0 & 5 \end{bmatrix}\begin{bmatrix} 1 & 0 & -2 \\ 3 & 1 & -1 \\ 2 & 0 & 1 \end{bmatrix}$

33. $\begin{bmatrix} 2 & 0 & -2 & 1 \\ -2 & 3 & -1 & 1 \end{bmatrix}\begin{bmatrix} 1 & 3 \\ 1 & 0 \\ -1 & 5 \\ 1 & -3 \end{bmatrix}$

34. $\begin{bmatrix} 1 & 3 \\ 1 & 0 \\ -1 & 5 \\ 1 & -3 \end{bmatrix}\begin{bmatrix} 2 & 0 & -2 & 1 \\ -2 & 3 & -1 & 1 \end{bmatrix}$

35. $\begin{bmatrix} 1 & 0 & 0 \\ 0 & 1 & 0 \\ 0 & 0 & 1 \end{bmatrix}\begin{bmatrix} 3 & 1 & 0 \\ -2 & 4 & 6 \\ 3 & 8 & -10 \end{bmatrix}$

36. $\begin{bmatrix} 3 & 1 & 0 \\ -2 & 4 & 6 \\ 3 & 8 & -10 \end{bmatrix}\begin{bmatrix} 1 & 0 & 0 \\ 0 & 1 & 0 \\ 0 & 0 & 1 \end{bmatrix}$

In Exercises 37–40, given matrices A, B, C, and D, demonstrate whether the given statement is true or false.

$$A = \begin{bmatrix} 2 & 1 & 0 \\ 3 & 1 & -2 \end{bmatrix} \qquad B = \begin{bmatrix} 3 & 2 \\ -5 & 0 \end{bmatrix}$$

$$C = \begin{bmatrix} 2 & -2 \\ 4 & 1 \end{bmatrix} \qquad D = \begin{bmatrix} 5 & -1 \\ 0 & -2 \end{bmatrix}$$

37. $CB = BC$

38. $A(B + C) = AB + AC$

39. $(B + C)A = BA + CA$

40. $D(BC) = (DB)C$

41. Show that

$$\left(\frac{1}{2}\begin{bmatrix} 3 & 5 \\ 2 & 4 \end{bmatrix}\right)\begin{bmatrix} 0 & 2 \\ 1 & -1 \end{bmatrix} = \frac{1}{2}\left(\begin{bmatrix} 3 & 5 \\ 2 & 4 \end{bmatrix}\begin{bmatrix} 0 & 2 \\ 1 & -1 \end{bmatrix}\right)$$

42. If A is a 3×4 matrix and B is a 4×3 matrix, what size is the matrix AB? What size is the matrix BA?

43. If A is a 5×4 matrix and B is a 3×5 matrix, what size is the matrix AB? What size is the matrix BA?

44. When does it make sense to discuss A^2 if A is a matrix?

45. Prove that if A and B are both 3×3 matrices, then $A + B = B + A$.

46. Prove that if A, B, and C are all 2×2 matrices, then $A + (B + C) = (A + B) + C$.

47. If an additive inverse exists, does each $n \times n$ matrix A have an additive inverse? (In other words, for each matrix A, does there exist a matrix B such that $A + B = B + A = \mathbf{O}$?) If so, explain what it should look like.

48. Prove that if A is an $m \times n$ matrix, and c and d are real numbers, then $(c + d)A = cA + dA$.

49. Prove that if A and B are $m \times n$ matrices, and c is a real number, then $c(A + B) = cA + cB$.

50. Over a 6-month period, the political science and the sociology departments use office supplies as illustrated by the accompanying table.

Supplies	Sociology Department	Political Science Department
Cases of legal pads	7	4
Cases of duplicating paper	6	10
Gross of pencils	2	9

Tom's Office Supply Company charges $230 for a case of legal pads, $65 for a case of duplicating paper, and $18 for a gross of pencils. Kuma's Office Supply Company charges $250 for a case of legal pads, $56 for a case of duplicating paper, and $16 for a gross of pencils. If supplies are ordered for each department, which office supply company is less expensive for each 6-month order?

51. Three scientific supply companies each manufacture their own brand of pneumatic articulators: brand A, brand B, and brand C. Each brand contains a certain amount of plastic, rubber, and glass tubing, as illustrated by the accompanying table.

Type of tubing required	Brand A	Brand B	Brand C
Plastic tubing	12 inches	18 inches	14 inches
Rubber tubing	14 inches	6 inches	12 inches
Glass tubing	18 inches	12 inches	5 inches

Only two companies in the country manufacture tubing suitable for the pneumatic articulators: George's Tubing Company and Raju's Tubing Company. George's Tubing Company charges 1¢ per inch for plastic tubing, 1.5¢ per inch for rubber tubing, and 1.8¢ per inch for glass tubing. Raju's Tubing Company charges 1.2¢ per inch for plastic tubing, 1.4¢ per inch for rubber tubing, and 1.6¢ per inch for glass tubing. If tubing supplies are ordered for all three brands, which tubing company offers less expensive tubing for each brand?

52. A small chain of health food restaurants has three outlets in a certain city: on Main street, Oak Street, and Union Boulevard. The chain sells three different lunch meals: salad with pasta, salad with chicken, and salad with seafood. The following matrix gives the one-day sales at each of the outlets.

Number of lunches sold

	Main St.	Oak St.	Union Blvd.
Pasta Salad	452	328	596
Chicken Salad	394	287	478
Seafood Salad	184	198	209

$= A$

The price of each lunch is given by the following matrix.

Pasta Salad	Chicken Salad	Seafood Salad
[$4.25	$5.75	$6.45]

$= B$

(a) Compute the matrix product BA.
(b) Explain what the entries of the product matrix BA represent.

53. A knapsack manufacturer sells three different models: the Alpine, the Sierra, and the Outlander. These knapsacks are sold via three outlets: a company store, through a catalog, and over the Internet. The following matrix gives the average weekly sales of each model through each of the outlets during 2001 and 2002.

Number of knapsacks sold

	Alpine	Sierra	Outlander
Store	395	722	488
Catalog	576	1080	613
Internet	820	1245	763

$= A$

The price of knapsacks during 2001 and 2002 is given by the following matrix.

	2001	2002
Alpine	$42	$45
Sierra	$53	$57
Outlander	$67	$72

$= B$

(a) Compute the matrix product AB.
(b) Explain what the entries of the product matrix AB represent.
(c) What were the average total weekly sales (in dollars each year?

54. Evaluate the inner product $\begin{bmatrix} 2 & -3 & 1 \end{bmatrix} \cdot \begin{bmatrix} 1 \\ 0 \\ 4 \end{bmatrix}$ and

the matrix product $\begin{bmatrix} 2 & -3 & 1 \end{bmatrix} \begin{bmatrix} 1 \\ 0 \\ 4 \end{bmatrix}$. What are the

differences between the two?

55. Evaluate the inner product $\begin{bmatrix} 1 \\ 0 \\ 4 \end{bmatrix} \cdot \begin{bmatrix} 2 & -3 & 1 \end{bmatrix}$ and

the matrix product $\begin{bmatrix} 1 \\ 0 \\ 4 \end{bmatrix} \begin{bmatrix} 2 & -3 & 1 \end{bmatrix}$. What are the

differences between the two?

Question for Thought

56. If A is a matrix and c is a real number, how should Ac be defined such that $Ac = cA$?

7.5 Solving Linear Systems Using Matrix Inverses

In this section we examine another method for solving systems of linear equations. Using matrices, we approach and solve a linear system as though it were a simple linear equation. The process of solving simple linear equations requires the application of real number properties. To approach matrices in the same way, we need a bit more information regarding the algebra of matrices.

In Section 7.4 we mentioned that matrix addition, where it makes sense, is commutative and associative, and that each matrix has an additive inverse; we pointed out that matrix multiplication, where it makes sense, is generally not commutative, but it is associative and does distribute over addition. Now we will discuss multiplicative identities and inverses.

Multiplicative Identities and Inverses

For all real numbers, the multiplicative identity is 1; that is, if we multiply any real number by 1, it remains unchanged. Is there a matrix that will act as an identity for matrix multiplication? That is, can we find the matrix I such that $AI = IA = A$ for *all* matrices A? First, we note that multiplication must make sense; that is, AI may not be defined if the number of columns of A is not equal to the number of rows of I. For simplicity, we will restrict our discussion to *square* matrices and define the following:

Definition of the $n \times n$ Identity Matrix: I_n
I_n is the $n \times n$ matrix with 1's in the main diagonal and 0's elsewhere.

For example, $I_4 = \begin{bmatrix} 1 & 0 & 0 & 0 \\ 0 & 1 & 0 & 0 \\ 0 & 0 & 1 & 0 \\ 0 & 0 & 0 & 1 \end{bmatrix}$. The following theorem can be proven:

> **Theorem 7.2**
> I_n is the identity matrix for all $n \times n$ matrices. I_n is called the **multiplicative identity of order n**. Hence, if A is an $n \times n$ matrix, then $AI_n = I_nA = A$.

In particular, I_4 is the identity matrix for all 4×4 matrices.

Example 1

Prove that if A is a 2×2 matrix, then $AI_2 = I_2A = A$.

Solution

The identity matrix of order 2 is $I_2 = \begin{bmatrix} 1 & 0 \\ 0 & 1 \end{bmatrix}$. If we let $A = \begin{bmatrix} a & b \\ c & d \end{bmatrix}$, then

$$AI_2 = \begin{bmatrix} a & b \\ c & d \end{bmatrix}\begin{bmatrix} 1 & 0 \\ 0 & 1 \end{bmatrix} = \begin{bmatrix} a \cdot 1 + b \cdot 0 & a \cdot 0 + b \cdot 1 \\ c \cdot 1 + d \cdot 0 & c \cdot 0 + d \cdot 1 \end{bmatrix}$$

$$= \begin{bmatrix} a & b \\ c & d \end{bmatrix} = A$$

On the other hand,

$$I_2A = \begin{bmatrix} 1 & 0 \\ 0 & 1 \end{bmatrix}\begin{bmatrix} a & b \\ c & d \end{bmatrix} = \begin{bmatrix} 1 \cdot a + 0 \cdot c & 1 \cdot b + 0 \cdot d \\ 0 \cdot a + 1 \cdot c & 0 \cdot b + 1 \cdot d \end{bmatrix}$$

$$= \begin{bmatrix} a & b \\ c & d \end{bmatrix} = A$$ ∎

Now that we have defined and demonstrated the existence of multiplicative identities for square matrices, our next step is to determine what multiplicative inverses look like if they do exist. In other words, for each $n \times n$ matrix A, does there exist a matrix A^{-1} such that $AA^{-1} = A^{-1}A = I_n$? We call A^{-1} the inverse of the matrix A.

Throughout this section, whenever we refer to the inverse of a matrix A, we mean its multiplicative inverse A^{-1}.

It would be nice if we would find that A^{-1} is simply some sort of reciprocal of the elements of A; unfortunately, because of the complex way matrix multiplication is defined, finding the inverse of a matrix takes a bit more work. A discussion of the derivation of matrix inverses is beyond the scope of this text. The following outline demonstrates the procedure for finding the inverse of a square matrix.

To find the inverse of the matrix $A = \begin{bmatrix} 1 & 1 & 2 \\ 1 & 2 & 4 \\ 3 & 4 & 10 \end{bmatrix}$ we proceed as follows:

1. We are looking for the inverse, A^{-1}, of this 3×3 matrix. We first form a new matrix in the following way:

    We put the 3×3 matrix A together with the identity matrix of order 3, I_3, where the elements of I_3 are placed to the right of the elements of A to form a **grafted** 3×6 matrix $[A \mid I]$:

$$\left[\begin{array}{ccc|ccc} 1 & 1 & 2 & 1 & 0 & 0 \\ 1 & 2 & 4 & 0 & 1 & 0 \\ 3 & 4 & 10 & 0 & 0 & 1 \end{array}\right]$$

    This is known as **grafting** I_3 *onto a 3×3 matrix*.

2. Our goal is to try to transform the left-hand square matrix (the A portion of the grafted matrix) into the identity matrix I_3. We use the elementary row operations discussed in Section 7.3 as we did when we transformed a matrix into reduced echelon form. *All transformations performed on the A portion are applied to the entire 3×6 grafted matrix.*

We already have 1 in the upper left-hand corner of the main diagonal of A, so our first step is to sweep out (get zeros in the remaining entries of) the first column:

$$\left[\begin{array}{ccc|ccc} 1 & 1 & 2 & 1 & 0 & 0 \\ 1 & 2 & 4 & 0 & 1 & 0 \\ 3 & 4 & 10 & 0 & 0 & 1 \end{array}\right] \begin{array}{l} \\ -R_1 + R_2 \rightarrow R_2 \\ -3R_1 + R_3 \rightarrow R_3 \end{array} \left[\begin{array}{ccc|ccc} 1 & 1 & 2 & 1 & 0 & 0 \\ 0 & 1 & 2 & -1 & 1 & 0 \\ 0 & 1 & 4 & -3 & 0 & 1 \end{array}\right]$$

We already have 1 in the second row, second column; now we sweep out the second column:

$$\left[\begin{array}{ccc|ccc} 1 & 1 & 2 & 1 & 0 & 0 \\ 0 & 1 & 2 & -1 & 1 & 0 \\ 0 & 1 & 4 & -3 & 0 & 1 \end{array}\right] \begin{array}{l} -R_2 + R_1 \rightarrow R_1 \\ \\ -R_2 + R_3 \rightarrow R_3 \end{array} \left[\begin{array}{ccc|ccc} 1 & 0 & 0 & 2 & -1 & 0 \\ 0 & 1 & 2 & -1 & 1 & 0 \\ 0 & 0 & 2 & -2 & -1 & 1 \end{array}\right]$$

Next get 1 in row 3, column 3:

Remember that all transformations are applied to the entire grafted matrix.

$$\left[\begin{array}{ccc|ccc} 1 & 0 & 0 & 2 & -1 & 0 \\ 0 & 1 & 2 & -1 & 1 & 0 \\ 0 & 0 & 2 & -2 & -1 & 1 \end{array}\right] \begin{array}{l} \\ \\ \frac{1}{2}R_3 \rightarrow R_3 \end{array} \left[\begin{array}{ccc|ccc} 1 & 0 & 0 & 2 & -1 & 0 \\ 0 & 1 & 2 & -1 & 1 & 0 \\ 0 & 0 & 1 & -1 & -\frac{1}{2} & \frac{1}{2} \end{array}\right]$$

Finally, sweep out the third column:

$$\left[\begin{array}{ccc|ccc} 1 & 0 & 0 & 2 & -1 & 0 \\ 0 & 1 & 2 & -1 & 1 & 0 \\ 0 & 0 & 1 & -1 & -\frac{1}{2} & \frac{1}{2} \end{array}\right] \begin{array}{l} \\ -2R_3 + R_2 \rightarrow R_2 \\ \\ \end{array} \left[\begin{array}{ccc|ccc} 1 & 0 & 0 & 2 & -1 & 0 \\ 0 & 1 & 0 & 1 & 2 & -1 \\ 0 & 0 & 1 & -1 & -\frac{1}{2} & \frac{1}{2} \end{array}\right]$$

3. The right-hand 3×3 matrix in the grafted matrix is A^{-1}:

Notice that the entries in A^{-1} are not the reciprocals of the entries of A.

$$A^{-1} = \left[\begin{array}{ccc} 2 & -1 & 0 \\ 1 & 2 & -1 \\ -1 & -\frac{1}{2} & \frac{1}{2} \end{array}\right]$$

We will verify that $AA^{-1} = I_3$:

$$AA^{-1} = \left[\begin{array}{ccc} 1 & 1 & 2 \\ 1 & 2 & 4 \\ 3 & 4 & 10 \end{array}\right] \left[\begin{array}{ccc} 2 & -1 & 0 \\ 1 & 2 & -1 \\ -1 & -\frac{1}{2} & \frac{1}{2} \end{array}\right]$$

$$= \left[\begin{array}{ccc} (1)(2) + (1)(1) + (2)(-1) & (1)(-1) + (1)(2) + (2)(-\frac{1}{2}) & (1)(0) + (1)(-1) + (2)(\frac{1}{2}) \\ (1)(2) + (2)(1) + (4)(-1) & (1)(-1) + (2)(2) + (4)(-\frac{1}{2}) & (1)(0) + (2)(-1) + (4)(\frac{1}{2}) \\ (3)(2) + (4)(1) + (10)(-1) & (3)(-1) + (4)(2) + (10)(-\frac{1}{2}) & (3)(0) + (4)(-1) + (10)(\frac{1}{2}) \end{array}\right]$$

$$= \left[\begin{array}{ccc} 1 & 0 & 0 \\ 0 & 1 & 0 \\ 0 & 0 & 1 \end{array}\right] = I_3$$

We leave it to the student to verify $A^{-1}A$ is also equal to I_3.

Procedure for Finding A^{-1}

In general, if A is an $n \times n$ matrix, to find A^{-1}, we

1. Graft I_n onto A to get $[A \,|\, I_n]$.
2. Using the elementary row operations on the entire matrix, transform the A portion of the grafted matrix into I_n.
3. When the left-hand portion of the grafted matrix is I_n, the right-hand $n \times n$ matrix is A^{-1}; that is, if $[A \,|\, I_n]$ is row equivalent to $[I_n \,|\, B]$, then B is the inverse of A.

Example 2 If $A = \begin{bmatrix} 2 & 4 \\ 3 & 1 \end{bmatrix}$, find A^{-1}.

Solution **1.** Graft I_2 onto A: $\begin{bmatrix} 2 & 4 & | & 1 & 0 \\ 3 & 1 & | & 0 & 1 \end{bmatrix}$.

2. Try to transform the left-hand square matrix (the A portion of the grafted matrix) into the identity matrix, I_2, using the elementary row operations applied to the entire grafted matrix.

$\begin{bmatrix} 2 & 4 & | & 1 & 0 \\ 3 & 1 & | & 0 & 1 \end{bmatrix}$ $\quad \frac{1}{2}R_1 \to R_1 \quad$ $\begin{bmatrix} 1 & 2 & | & \frac{1}{2} & 0 \\ 3 & 1 & | & 0 & 1 \end{bmatrix}$

$\begin{bmatrix} 1 & 2 & | & \frac{1}{2} & 0 \\ 3 & 1 & | & 0 & 1 \end{bmatrix}$ $\quad -3R_1 + R_2 \to R_2 \quad$ $\begin{bmatrix} 1 & 2 & | & \frac{1}{2} & 0 \\ 0 & -5 & | & -\frac{3}{2} & 1 \end{bmatrix}$

$\begin{bmatrix} 1 & 2 & | & \frac{1}{2} & 0 \\ 0 & -5 & | & -\frac{3}{2} & 1 \end{bmatrix}$ $\quad -\frac{1}{5}R_2 \to R_2 \quad$ $\begin{bmatrix} 1 & 2 & | & \frac{1}{2} & 0 \\ 0 & 1 & | & \frac{3}{10} & -\frac{1}{5} \end{bmatrix}$

$\begin{bmatrix} 1 & 2 & | & \frac{1}{2} & 0 \\ 0 & 1 & | & \frac{3}{10} & -\frac{1}{5} \end{bmatrix}$ $\quad -2R_2 + R_1 \to R_1 \quad$ $\begin{bmatrix} 1 & 0 & | & -\frac{1}{10} & \frac{2}{5} \\ 0 & 1 & | & \frac{3}{10} & -\frac{1}{5} \end{bmatrix}$

3. Since the left-hand portion is now I_2, the right-hand portion is A^{-1}. Hence

$A^{-1} = \begin{bmatrix} -\frac{1}{10} & \frac{2}{5} \\ \frac{3}{10} & -\frac{1}{5} \end{bmatrix}$ which can be written as $\frac{1}{10}\begin{bmatrix} -1 & 4 \\ 3 & -2 \end{bmatrix}$

Many graphing calculators can find matrix inverses. For instance, if we enter the matrix values of this example as matrix A, the following display illustrates the matrix A^{-1}:

```
[A]⁻¹
      [[-.1  .4 ]
       [.3  -.2]]
```

We factor the scalar $\frac{1}{10}$ out in order to simplify computations performed with this matrix. For example, we next demonstrate that $A^{-1}A = I_2$.

$A^{-1}A = \left(\frac{1}{10}\begin{bmatrix} -1 & 4 \\ 3 & -2 \end{bmatrix}\right)\begin{bmatrix} 2 & 4 \\ 3 & 1 \end{bmatrix}$ $(cA)B = c(AB)$ where c is a scalar.

$= \frac{1}{10}\left(\begin{bmatrix} -1 & 4 \\ 3 & -2 \end{bmatrix}\begin{bmatrix} 2 & 4 \\ 3 & 1 \end{bmatrix}\right)$ It is easier to multiply the matrices first.

$= \frac{1}{10}\begin{bmatrix} (-1)(2) + (4)(3) & (-1)(4) + (4)(1) \\ (3)(2) + (-2)(3) & (3)(4) + (-2)(1) \end{bmatrix}$ Then we multiply the scalar.

$= \frac{1}{10}\begin{bmatrix} 10 & 0 \\ 0 & 10 \end{bmatrix} = \begin{bmatrix} 1 & 0 \\ 0 & 1 \end{bmatrix} = I_2$

We leave it to the student to verify $AA^{-1} = I_2$.   ∎

We have yet to answer the question, Does every square matrix have an inverse? (In the system of real numbers we know that a^{-1} does not exist for $a = 0$.) Let's consider the square matrix $A = \begin{bmatrix} 1 & 3 \\ -2 & -6 \end{bmatrix}$. To try to find A^{-1} we proceed as follows:

1. Graft I_2 onto the matrix: $\begin{bmatrix} 1 & 3 & | & 1 & 0 \\ -2 & -6 & | & 0 & 1 \end{bmatrix}$.

2. Then try to transform the left-hand square matrix into the identity matrix using the elementary row operations:

$\begin{bmatrix} 1 & 3 & | & 1 & 0 \\ -2 & -6 & | & 0 & 1 \end{bmatrix}$ $\quad 2R_1 + R_2 \to R_2 \quad$ $\begin{bmatrix} 1 & 3 & | & 1 & 0 \\ 0 & 0 & | & 2 & 1 \end{bmatrix}$

The entries of the bottom row of the left-hand A portion are all 0's. No matter what approach we take, we will find that we cannot transform A into I_2 using the elementary row operations. We assert the following:

The Existence of A^{-1}

If the $n \times n$ matrix A cannot be transformed into I_n using the elementary row operations, then A^{-1} does not exist.

If in the process of transforming A into I_n using elementary row operations, we arrive at either a row or column consisting entirely of 0's, then A cannot be transformed into I_n and, therefore, A^{-1} does not exist.

Now we have the tools to approach linear systems in the same way we approach simple linear equations.

Solving Linear Systems Using Matrix Inverses

We return to the linear system $\begin{cases} a_{11}x + a_{12}y = k_1 \\ a_{21}x + a_{22}y = k_2 \end{cases}$.

If we let $A = \begin{bmatrix} a_{11} & a_{12} \\ a_{21} & a_{22} \end{bmatrix}$, $X = \begin{bmatrix} x \\ y \end{bmatrix}$, and $K = \begin{bmatrix} k_1 \\ k_2 \end{bmatrix}$, then

$$AX = \begin{bmatrix} a_{11} & a_{12} \\ a_{21} & a_{22} \end{bmatrix} \begin{bmatrix} x \\ y \end{bmatrix}$$

$$= \begin{bmatrix} a_{11}x + a_{12}y \\ a_{21}x + a_{22}y \end{bmatrix} \qquad \text{Since } a_{11}x + a_{12}y = k_1 \text{ and } a_{21}x + a_{22}y = k_2,$$

$$= \begin{bmatrix} k_1 \\ k_2 \end{bmatrix} = K$$

Thus we can write this system in matrix terms as $AX = K$. We call A the **coefficient matrix**, X the **variable vector**, and K the **constant vector** of the system. Given $AX = K$, we can solve for X using the matrix properties as follows:

Notice the left-hand multiplication.

$$AX = K \qquad \text{Multiply both sides of this equation by } A^{-1} \text{ on the left.}$$
$$A^{-1}(AX) = A^{-1}K \qquad \text{Use associativity of matrix multiplication.}$$
$$(A^{-1}A)X = A^{-1}K \qquad \text{Since } A^{-1}A = I_2,$$
$$I_2X = A^{-1}K \qquad \text{Since } I_2 \text{ is the identity matrix for the matrix } X, \text{ we have}$$
$$X = A^{-1}K$$

Hence, if $AX = K$, then $X = A^{-1}K$ yields the matrix of solutions. In other words, the solution to $AX = K$ is the constant matrix *left-multiplied* by the inverse of the coefficient matrix. Although we approach a linear system as though we are solving a simple linear equation, unlike simple linear equations, matrix multiplication is not commutative; therefore, it is important that A^{-1} be multiplied to the *left* of K.

This example is actually easier to solve by elimination or Gaussian elimination. This example is simply an illustration of using matrix inverses to solve a system of equations. The next example will illustrate when matrix inverses can be very useful.

Example 3

Using matrix inverses, find the solution to the following system:

$$\begin{cases} 2x - 3y = 7 \\ 3x + 5y = 20 \end{cases}$$

Solution   We identify matrices A, X, and K:

$$A = \begin{bmatrix} 2 & -3 \\ 3 & 5 \end{bmatrix} \qquad X = \begin{bmatrix} x \\ y \end{bmatrix} \qquad K = \begin{bmatrix} 7 \\ 20 \end{bmatrix}$$

Hence $AX = K$, and our solution is found by $X = A^{-1}K$, which means we must find the inverse of the coefficient matrix and multiply it by the constant matrix to find the solution. First, we find A^{-1}:

1. Graft I_2 onto A:
$$\begin{bmatrix} 2 & -3 & | & 1 & 0 \\ 3 & 5 & | & 0 & 1 \end{bmatrix}$$

2. Transform the left 2×2 portion onto I_2:

$$\begin{bmatrix} 2 & -3 & | & 1 & 0 \\ 3 & 5 & | & 0 & 1 \end{bmatrix} \quad \tfrac{1}{2}R_1 \to R_1 \quad \begin{bmatrix} 1 & -\tfrac{3}{2} & | & \tfrac{1}{2} & 0 \\ 3 & 5 & | & 0 & 1 \end{bmatrix}$$

$$\begin{bmatrix} 1 & -\tfrac{3}{2} & | & \tfrac{1}{2} & 0 \\ 3 & 5 & | & 0 & 1 \end{bmatrix} \quad -3R_1 + R_2 \to R_2 \quad \begin{bmatrix} 1 & -\tfrac{3}{2} & | & \tfrac{1}{2} & 0 \\ 0 & \tfrac{19}{2} & | & -\tfrac{3}{2} & 1 \end{bmatrix}$$

$$\begin{bmatrix} 1 & -\tfrac{3}{2} & | & \tfrac{1}{2} & 0 \\ 0 & \tfrac{19}{2} & | & -\tfrac{3}{2} & 1 \end{bmatrix} \quad \tfrac{2}{19}R_2 \to R_2 \quad \begin{bmatrix} 1 & -\tfrac{3}{2} & | & \tfrac{1}{2} & 0 \\ 0 & 1 & | & -\tfrac{3}{19} & \tfrac{2}{19} \end{bmatrix}$$

$$\begin{bmatrix} 1 & -\tfrac{3}{2} & | & \tfrac{1}{2} & 0 \\ 0 & 1 & | & -\tfrac{3}{19} & \tfrac{2}{19} \end{bmatrix} \quad \tfrac{3}{2}R_2 + R_1 \to R_1 \quad \begin{bmatrix} 1 & 0 & | & \tfrac{5}{19} & \tfrac{3}{19} \\ 0 & 1 & | & -\tfrac{3}{19} & \tfrac{2}{19} \end{bmatrix}$$

Again, notice A^{-1} is multiplied on the left.

3. Hence, the right-hand portion is $A^{-1} = \begin{bmatrix} \tfrac{5}{19} & \tfrac{3}{19} \\ -\tfrac{3}{19} & \tfrac{2}{19} \end{bmatrix} = \dfrac{1}{19}\begin{bmatrix} 5 & 3 \\ -3 & 2 \end{bmatrix}$. Now we can solve for X:

Using a graphing calculator to solve this system, we can enter the coefficient matrix as A and the constant matrix as B, giving the following solution:

```
[A]-1[B]
         [[5]
          [1]]
```

$$X = A^{-1}K$$
$$= \dfrac{1}{19}\begin{bmatrix} 5 & 3 \\ -3 & 2 \end{bmatrix}\begin{bmatrix} 7 \\ 20 \end{bmatrix}$$
$$= \dfrac{1}{19}\begin{bmatrix} (5)(7) + (3)(20) \\ (-3)(7) + (2)(20) \end{bmatrix} = \dfrac{1}{19}\begin{bmatrix} 95 \\ 19 \end{bmatrix}$$

Perform matrix multiplication first. Then do scalar multiplication.

$$= \begin{bmatrix} 5 \\ 1 \end{bmatrix}$$

Hence $X = \begin{bmatrix} x \\ y \end{bmatrix} = \begin{bmatrix} 5 \\ 1 \end{bmatrix}$, which means $x = 5, y = 1$. Check the solution. ∎

If we were to use the elimination method or Gaussian elimination with back substitution on the last example, we would get to the solution much more quickly than by using matrix inverses. So a natural question to ask is, Why bother with this method? We answer this question by considering the following example.

Example 4 Danny invests $10,000, part at 6% per year in a conservative investment and part at 12% per year in a riskier investment. How much should he invest at each rate if he wants a return on his total investment of **(a)** 8%? **(b)** 10%? **(c)** 10.5%?

Solution If we let $x = $ the amount invested at 6% and $y = $ the amount invested at 12%, the first equation for all three parts of this problem is

$$x + y = 10{,}000 \qquad \text{The total amount invested}$$

The second equation involves the interest received on each investment. In general, if Danny invests x dollars at 6% and y dollars at 12%, then his total interest for the year is $0.06x + 0.12y$. The three parts of this example differ on what this amount should be:

For part **(a)**, interest is expressed in the equation

$$0.06x + 0.12y = 0.08(10{,}000) = 800$$

For part **(b)**, interest is expressed in the equation

$$0.06x + 0.12y = 0.10(10{,}000) = 1000$$

For part **(c)**, interest is expressed in the equation

$$0.06x + 0.12y = 0.105(10{,}000) = 1050$$

Hence we are being asked to solve three systems of equations:

(a) $\begin{cases} x + y = 10{,}000 \\ 0.06x + 0.12y = 800 \end{cases}$ **(b)** $\begin{cases} x + y = 10{,}000 \\ 0.06x + 0.12y = 1000 \end{cases}$

(c) $\begin{cases} x + y = 10{,}000 \\ 0.06x + 0.12y = 1050 \end{cases}$

Notice that all three parts of this example have the same matrix of coefficients. We could solve each system separately using Gaussian elimination; however, once we have the inverse of the coefficient matrix, we can solve all three quickly.

The coefficient matrix of all three systems above is the same:

$$A = \begin{bmatrix} 1 & 1 \\ 0.06 & 0.12 \end{bmatrix}$$

The student should verify that the inverse of this matrix is $A^{-1} = \begin{bmatrix} 2 & -\frac{50}{3} \\ -1 & \frac{50}{3} \end{bmatrix}$.

Given that we have the inverse of the coefficient matrix, we can find the solution to all three systems easily:

(a) $\begin{cases} x + y = 10{,}000 \\ 0.06x + 0.12y = 800 \end{cases}$

The constant matrix for this problem is $\begin{bmatrix} 10{,}000 \\ 800 \end{bmatrix}$. We have

$$X = A^{-1}K$$

$$\begin{bmatrix} x \\ y \end{bmatrix} = \begin{bmatrix} 2 & -\frac{50}{3} \\ -1 & \frac{50}{3} \end{bmatrix} \begin{bmatrix} 10{,}000 \\ 800 \end{bmatrix} = \begin{bmatrix} 6666.67 \\ 3333.33 \end{bmatrix}$$

Hence $x = 6666.67$ and $y = 3333.33$; therefore, Danny should invest $6666.67 at 6% and $3333.33 at 12% to gain a return of 8% on his total investment.

(b) $\begin{cases} x + y = 10{,}000 \\ 0.06x + 0.12y = 1000 \end{cases}$

The constant matrix for this problem is $\begin{bmatrix} 10{,}000 \\ 1000 \end{bmatrix}$. We have

$$X = A^{-1}K$$

$$\begin{bmatrix} x \\ y \end{bmatrix} = \begin{bmatrix} 2 & -\frac{50}{3} \\ -1 & \frac{50}{3} \end{bmatrix} \begin{bmatrix} 10{,}000 \\ 1000 \end{bmatrix} = \begin{bmatrix} 3333.33 \\ 6666.67 \end{bmatrix}$$

Therefore, Danny should invest $3333.33 at 6% and 6666.67 at 12% to gain a return of 10% on his total investment.

(c) $\begin{cases} x + \quad y = 10{,}000 \\ 0.06x + 0.12y = 1050 \end{cases}$

The constant matrix for this problem is $\begin{bmatrix} 10{,}000 \\ 1050 \end{bmatrix}$. We have

$$X = A^{-1}K$$

$$\begin{bmatrix} x \\ y \end{bmatrix} = \begin{bmatrix} 2 & -\frac{50}{3} \\ -1 & \frac{50}{3} \end{bmatrix}\begin{bmatrix} 10{,}000 \\ 1050 \end{bmatrix} = \begin{bmatrix} 2500 \\ 7500 \end{bmatrix}$$

Therefore, Danny should invest $2500 at 6% and $7500 at 12% to gain a return of 10.5% on his total investment. ∎

In the course of solving the system $AX = K$, we may find that A^{-1} does not exist; that is, we may arrive at a column or row consisting entirely of zeros and hence A cannot be transformed into I_n using the elementary row operations. If this is the case, then there are no unique solutions: the system may be either dependent or inconsistent.

7.5 Exercises

1. Compute $\begin{bmatrix} 3 & -1 \\ 3 & -2 \end{bmatrix}\begin{bmatrix} 1 & 0 \\ 0 & 1 \end{bmatrix}$.

2. Compute $\begin{bmatrix} 1 & 0 & 0 \\ 0 & 1 & 0 \\ 0 & 0 & 1 \end{bmatrix}\begin{bmatrix} 2 & -2 & 1 \\ 3 & 0 & -4 \\ 7 & -1 & 5 \end{bmatrix}$.

3. Show that if A is a 3×3 square matrix, then $AI_3 = I_3A = A$.

4. If A is a 3×2 matrix, then does $AI_3 = I_3A = A$?

In Exercises 5–20, find the inverse of the matrix if it exists.

5. $\begin{bmatrix} 1 & 2 \\ 2 & 0 \end{bmatrix}$

6. $\begin{bmatrix} 3 & -2 \\ 1 & 1 \end{bmatrix}$

7. $\begin{bmatrix} -1 & 5 \\ 3 & 1 \end{bmatrix}$

8. $\begin{bmatrix} 2 & -1 \\ 0 & 1 \end{bmatrix}$

9. $\begin{bmatrix} 4 & 2 \\ 2 & 1 \end{bmatrix}$

10. $\begin{bmatrix} -1 & 5 \\ 2 & -10 \end{bmatrix}$

11. $\begin{bmatrix} -1 & 5 \\ -2 & 10 \end{bmatrix}$

12. $\begin{bmatrix} 3 & 0 \\ -5 & 1 \end{bmatrix}$

13. $\begin{bmatrix} 1 & 1 & 2 \\ 2 & 0 & 1 \\ 3 & -1 & 1 \end{bmatrix}$

14. $\begin{bmatrix} 1 & -1 & 0 \\ -2 & 2 & 1 \\ 4 & -1 & 0 \end{bmatrix}$

15. $\begin{bmatrix} 1 & 0 & 1 \\ -2 & 3 & 2 \\ 2 & -2 & 0 \end{bmatrix}$

16. $\begin{bmatrix} 1 & 1 & 1 \\ -2 & 2 & 3 \\ 5 & 0 & 1 \end{bmatrix}$

17. $\begin{bmatrix} 2 & 1 & 3 \\ 1 & 0 & 1 \\ 4 & -2 & 2 \end{bmatrix}$

18. $\begin{bmatrix} 1 & -1 & 2 \\ -2 & 2 & -3 \\ 1 & 1 & 0 \end{bmatrix}$

19. $\begin{bmatrix} 1 & -1 & 2 \\ 2 & -2 & 1 \\ 3 & -1 & 0 \end{bmatrix}$

20. $\begin{bmatrix} 1 & 0 & 0 \\ -2 & 0 & 0 \\ 4 & -3 & 1 \end{bmatrix}$

In Exercises 21–38, solve the given system using matrix inverses.

21. $\begin{cases} x + 2y = 3 \\ x - \quad y = 6 \end{cases}$

22. $\begin{cases} x + y = 5 \\ 3x - y = 8 \end{cases}$

23. $\begin{cases} x - 2y = 4 \\ 2x + \quad y = 13 \end{cases}$

24. $\begin{cases} 3x + 2y = 3 \\ 3x + \quad y = 0 \end{cases}$

25. $\begin{cases} 5x - 2y = 11 \\ x - 5y = -7 \end{cases}$

26. $\begin{cases} x = 3y - 1 \\ 2x - 5y = -3 \end{cases}$

27. $\begin{cases} y = 3x - 15 \\ 8x - 2y = 10 \end{cases}$

28. $\begin{cases} x + \quad 7y = -1 \\ 3x + 21y = -3 \end{cases}$

29. $\begin{cases} \dfrac{3}{2}x + \dfrac{1}{3}y = 1 \\ x + \dfrac{1}{12}y = \dfrac{1}{4} \end{cases}$

30. $\begin{cases} 5x + 2y = -8 \\ 3x + \quad y = -4 \end{cases}$

31. $\begin{cases} x + y + z = 6 \\ 3x + y - z = 6 \\ 2x + y - z = 4 \end{cases}$

32. $\begin{cases} x + \quad y - \quad z = 1 \\ 2x + 2y + 2z = 0 \\ x - \quad y + \quad z = 3 \end{cases}$

33. $\begin{cases} x - 3y + z = 0 \\ 3x - y + z = 2 \\ x + 2y - z = 2 \end{cases}$

34. $\begin{cases} x + y + 2z = 7 \\ x - 2y - z = 1 \\ x + y + z = 4 \end{cases}$

35. $\begin{cases} x + y + z = 6 \\ 3x + y - z = 4 \\ 2x + 2y + 2z = 4 \end{cases}$

36. $\begin{cases} x + y - z = 3 \\ 2x + 2y - 2z = 6 \\ x - y + z = 5 \end{cases}$

37. $\begin{cases} x + 0.2y + z = 0.02 \\ 0.3x - y + 0.2z = 0 \\ x + y + z = 0.1 \end{cases}$

38. $\begin{cases} x + y = 2x + z \\ x - 2y = z - 1 \\ 4x - 2y = 6z - 5 \end{cases}$

39. Jenna has $20,000 to split between a low-risk certificate of deposit yielding 6% per year and a high-risk oil stock yielding 18% per year. How much should she invest at each rate if she wants a return on her total investment of (a) 8%? (b) 10%? (c) 12%?

40. A nutritionist wants to create a food supplement out of three substances: X, Y, and Z. The iron content and cost of the substances X, Y, and Z are entered in the accompanying table.

Substance	Iron content per gram	Cost per gram
X	0.5	10¢
Y	0.2	6¢
Z	0.4	8¢

How many grams of each substance should be used to make a food supplement if she wants 10 grams of the food supplement to supply 3.5 grams of iron and cost (a) 77¢? (b) 78¢? (c) 79¢?

41. A lab technician is responsible for feeding the mice and rats used for experimentation. She feeds them different mixtures of three types of food pellets: brands A, B, and C. She needs to create a mixture that supplies the exact daily requirement of protein, fat, and carbohydrates. Suppose that mice need

300 mg of protein, 280 mg of fat, and 400 mg of carbohydrates, and that rats need 450 mg of protein, 400 mg of fat, and 600 mg of carbohydrates. The amount of each nutrient provided by one gram of each brand of food pellet is given in the accompanying table. How many grams of each brand should be mixed together to feed the mice daily? the rats daily?

	Brand A	Brand B	Brand C
Protein (mg)	8	4	15
Fat (mg)	8	16	10
Carbohydrates (mg)	6	10	25

42. Matrix inverses can be used to provide a simple and efficient procedure for encoding and decoding messages. First, we assign the numbers 1 through 26 to the letters of the alphabet: A is 1, B is 2, and so on. We also assign the number 27 to a blank space so that we can put spaces between words. Thus the message PLEASE SEND MONEY corresponds to the sequence

16 12 5 1 19 5 27 19 5 14 4 27 13 15 14 5 25

Any square matrix of positive numbers that has an inverse can be used to encode this message. For example, we could use the 2×2 matrix

$$A = \begin{bmatrix} 3 & 4 \\ 2 & 1 \end{bmatrix}$$

as an *encoding matrix*. We do this by dividing the numbers in the sequence into groups of two and use these groups to form the columns of a matrix with two rows. If necessary, we add an extra blank at the end of the message to ensure that the columns are even. Then we multiply this matrix on the left by A:

$$\begin{bmatrix} 3 & 4 \\ 2 & 1 \end{bmatrix} \begin{bmatrix} 16 & 5 & 19 & 27 & 5 & 4 & 13 & 14 & 25 \\ 12 & 1 & 5 & 19 & 14 & 27 & 15 & 5 & 27 \end{bmatrix}$$

(a) Find this matrix product, which represents the encoded message.

(b) Verify that if we multiply this encoded message on the left by the *decoding matrix* A^{-1} we get the original matrix (message) back.

Since A^{-1} can be found fairly easily once A is known, the encoding matrix is the only key necessary to decode messages encoded in this way.

43. Find the inverse of the following matrices:

(a) $\begin{bmatrix} 2 & 0 & 0 \\ 0 & 4 & 0 \\ 0 & 0 & 10 \end{bmatrix}$ (b) $\begin{bmatrix} \frac{1}{5} & 0 & 0 \\ 0 & \frac{1}{3} & 0 \\ 0 & 0 & \frac{1}{6} \end{bmatrix}$

(c) Can you generalize the results of parts (a) and (b)

to any 3 × 3 matrix of the form $\begin{bmatrix} a & 0 & 0 \\ 0 & b & 0 \\ 0 & 0 & c \end{bmatrix}$

where a, b, and c are not 0?

44. Do the results of Exercise 43 generalize to 4 × 4 matrices? to $n \times n$ matrices?

7.6   Determinants and Cramer's Rule: 2 × 2 and 3 × 3 Systems

In Section 7.1 we used the elimination method to solve many systems of equations. You may have noticed that the same procedure was consistently repeated. This leads us to a natural question as to whether we can apply this procedure to a general case and devise a formula for solving systems. In this section we will discuss a "formula" for solving systems of equations called **Cramer's rule**. We will begin by solving a general 2 × 2 system.

Example 1   Solve the following system of equations for x and y:

$$\begin{cases} a_{11}x + a_{12}y = k_1 \\ a_{21}x + a_{22}y = k_2 \end{cases}$$

Solution   The subscripts of a indicate their position in the system. The first number indicates the equation number; the second number indicates the variable. (Usually the leftmost variable, in this case x, is considered the "first.") For example, a_{12} appears in the first equation and is the coefficient of the second variable, whereas a_{21} appears in the second equation, and is the coefficient of the first variable.

The advantages of this subscript notation will become apparent in the next few pages.

We will solve this general 2 × 2 system by elimination. First, we eliminate y:

$$a_{11}x + a_{12}y = k_1 \quad \xrightarrow{\text{Multiply by } a_{22}}$$
$$a_{21}x + a_{22}y = k_2 \quad \xrightarrow{\text{Multiply by } -a_{12}}$$

$$a_{11}a_{22}x + a_{12}a_{22}y = k_1 a_{22}$$
$$\underline{-a_{21}a_{12}x - a_{22}a_{12}y = -k_2 a_{12}} \quad \text{Add.}$$
$$a_{11}a_{22}x - a_{21}a_{12}x = k_1 a_{22} - k_2 a_{12}$$

Now solve for x.

$$(a_{11}a_{22} - a_{21}a_{12})x = k_1 a_{22} - k_2 a_{12}$$

$$x = \frac{k_1 a_{22} - k_2 a_{12}}{a_{11}a_{22} - a_{21}a_{12}}$$

Note that to have a solution we must have $a_{11}a_{22} - a_{21}a_{12} \neq 0$. We can solve for y using the same method as above—that is, eliminating x—to arrive at

$$y = \frac{a_{11}k_2 - a_{21}k_1}{a_{11}a_{22} - a_{21}a_{12}}$$

Again, this solution is valid only if $a_{11}a_{22} - a_{21}a_{12} \neq 0$. Hence the general solution for this 2 × 2 system is

$$x = \frac{k_1 a_{22} - k_2 a_{12}}{a_{11}a_{22} - a_{21}a_{12}}, \qquad y = \frac{a_{11}k_2 - a_{21}k_1}{a_{11}a_{22} - a_{21}a_{12}}$$

∎

This looks like a difficult formula to commit to memory. So we will digress a bit and define some notation that will help us to use this formula as well as provide us with a formula for larger system of equations. We begin by defining a **determinant**.

Definition of a 2 × 2 Determinant

The symbol $\begin{vmatrix} a & c \\ b & d \end{vmatrix}$ is called a **2 × 2 determinant**. It is a number whose value is defined to be

$$\begin{vmatrix} a & c \\ b & d \end{vmatrix} = ad - bc$$

Example 2 Evaluate the following determinants.

(a) $\begin{vmatrix} 3 & 2 \\ 8 & 1 \end{vmatrix}$ (b) $\begin{vmatrix} 5 & -1 \\ -2 & 6 \end{vmatrix}$

Solution (a) $\begin{vmatrix} 3 & 2 \\ 8 & 1 \end{vmatrix} = 3(1) - 8(2) = -13$

Note that a determinant is a single number.

(b) $\begin{vmatrix} 5 & -1 \\ -2 & 6 \end{vmatrix} = 5(6) - (-2)(-1) = 28$

∎

Returning to Example 1, we found the general solution to the system

$$\begin{cases} a_{11}x + a_{12}y = k_1 \\ a_{21}x + a_{22}y = k_2 \end{cases} \quad \text{to be} \quad x = \frac{k_1 a_{22} - k_2 a_{12}}{a_{11}a_{22} - a_{21}a_{12}}, \qquad y = \frac{a_{11}k_2 - a_{21}k_1}{a_{11}a_{22} - a_{21}a_{12}}$$

which, using determinants, could be written as $x = \dfrac{\begin{vmatrix} k_1 & a_{12} \\ k_2 & a_{22} \end{vmatrix}}{\begin{vmatrix} a_{11} & a_{12} \\ a_{21} & a_{22} \end{vmatrix}}, \quad y = \dfrac{\begin{vmatrix} a_{11} & k_1 \\ a_{21} & k_2 \end{vmatrix}}{\begin{vmatrix} a_{11} & a_{12} \\ a_{21} & a_{22} \end{vmatrix}}.$

Notice that each denominator of the general solution is the determinant of the coefficients of the variables. We usually denote this determinant by D:

$$D = \begin{vmatrix} a_{11} & a_{12} \\ a_{21} & a_{22} \end{vmatrix}$$

The numerators of the general solution are the determinants arrived at by taking D and replacing the coefficients of the variable for which we are solving by the constant terms. We denote these numerators by

$$D_x = \begin{vmatrix} k_1 & a_{12} \\ k_2 & a_{22} \end{vmatrix} \qquad \text{and} \qquad D_y = \begin{vmatrix} a_{11} & k_1 \\ a_{21} & k_2 \end{vmatrix}$$

Using this notation, we have derived the following, called **Cramer's rule**:

Cramer's Rule for 2 × 2 Systems of Equations

The solution to the general system

$$\begin{cases} a_{11}x + a_{12}y = k_1 \\ a_{21}x + a_{22}y = k_2 \end{cases} \quad \text{is} \quad x = \frac{D_x}{D} \text{ and } y = \frac{D_y}{D}$$

where $D = \begin{vmatrix} a_{11} & a_{12} \\ a_{21} & a_{22} \end{vmatrix} \neq 0,$ $D_x = \begin{vmatrix} k_1 & a_{12} \\ k_2 & a_{22} \end{vmatrix},$ $D_y = \begin{vmatrix} a_{11} & k_1 \\ a_{21} & k_2 \end{vmatrix}.$

Example 3 Solve the following system using Cramer's rule:

$$\begin{cases} 3x - 5y = 2 \\ 2x + 8y = 9 \end{cases}$$

Solution Using Cramer's rule, we first find D, D_x, and D_y:

$$D = \begin{vmatrix} 3 & -5 \\ 2 & 8 \end{vmatrix} = 3(8) - 2(-5) = 34$$

$$D_x = \begin{vmatrix} 2 & -5 \\ 9 & 8 \end{vmatrix} = 2(8) - 9(-5) = 61$$

$$D_y = \begin{vmatrix} 3 & 2 \\ 2 & 9 \end{vmatrix} = 3(9) - 2(2) = 23$$

By Cramer's rule we have

$$x = \frac{D_x}{D} = \frac{61}{34} \quad \text{and} \quad y = \frac{D_y}{D} = \frac{23}{34}$$

Hence the solution is $\left(\dfrac{61}{34}, \dfrac{23}{34} \right)$. The student should check this solution. ∎

If $D = 0$, then there is no unique solution: The system is either inconsistent and has no solutions (if $D_x \neq 0$ or $D_y \neq 0$) or it is dependent and has infinitely many solutions (if $D_x = 0$ and $D_y = 0$).

Cramer's rule can be extended to 3 × 3 linear systems as well. Before we do however, we must define a 3 × 3 determinant.

Definition of a 3 × 3 Determinant

A **3 × 3 determinant** is defined as

$$\begin{vmatrix} a_{11} & a_{12} & a_{13} \\ a_{21} & a_{22} & a_{23} \\ a_{31} & a_{32} & a_{33} \end{vmatrix} = a_{11} \begin{vmatrix} a_{22} & a_{23} \\ a_{32} & a_{33} \end{vmatrix} - a_{21} \begin{vmatrix} a_{12} & a_{13} \\ a_{32} & a_{33} \end{vmatrix} + a_{31} \begin{vmatrix} a_{12} & a_{13} \\ a_{22} & a_{23} \end{vmatrix}$$

The subscripts of a, the entries of the 3 × 3 determinant, indicate their positions. The first number indicates the row; the second indicates the column. For example, a_{32} is the entry of the third row in the second column. In general, a_{ij} is the entry of the ith row and jth column.

The 2 × 2 determinants in this definition are called the **minors** of their coefficients.

A minor is found by choosing an element of a determinant and then crossing off the row and column that contain it. Hence the minor of a_{11} is found as follows:

$$\begin{vmatrix} \cancel{a_{11}} & \cancel{a_{12}} & \cancel{a_{13}} \\ \cancel{a_{21}} & a_{22} & a_{23} \\ \cancel{a_{31}} & a_{32} & a_{33} \end{vmatrix} \quad \text{which gives} \quad \begin{vmatrix} a_{22} & a_{23} \\ a_{32} & a_{33} \end{vmatrix}$$

Example 4

Evaluate the determinant $\begin{vmatrix} 5 & 0 & -2 \\ 3 & 1 & -4 \\ 2 & -1 & 5 \end{vmatrix}$.

Solution

Most graphing calculators can evaluate determinants. On a TI-83 Plus, for example, we type in the determinant's entries as the elements of a matrix, and then the calculator evaluates the determinant of the matrix as illustrated below, where A is the matrix of the entries of the determinant of Example 4.

```
[A]
 [[5  0  -2]
  [3  1  -4]
  [2 -1  5 ]]
det([A])
            15
```

By definition we get

$$\begin{vmatrix} 5 & 0 & -2 \\ 3 & 1 & -4 \\ 2 & -1 & 5 \end{vmatrix} = 5\begin{vmatrix} \boxed{5} & 0 & -2 \\ 3 & 1 & -4 \\ 2 & -1 & 5 \end{vmatrix} - 3\begin{vmatrix} 5 & 0 & -2 \\ \boxed{3} & 1 & -4 \\ 2 & -1 & 5 \end{vmatrix} + 2\begin{vmatrix} 5 & 0 & -2 \\ 3 & 1 & -4 \\ \boxed{2} & -1 & 5 \end{vmatrix}$$

$$= 5\begin{vmatrix} 1 & -4 \\ -1 & 5 \end{vmatrix} - 3\begin{vmatrix} 0 & -2 \\ -1 & 5 \end{vmatrix} + 2\begin{vmatrix} 0 & -2 \\ 1 & -4 \end{vmatrix}$$

$$= 5[(1)(5) - (-1)(-4)] - 3[(0)(5) - (-1)(-2)] + 2[(0)(-4) - (1)(-2)]$$

$$= 5(5 - 4) - 3(0 - 2) + 2(0 + 2) = \boxed{15}$$

Actually we could have defined a 3 × 3 determinant by using its minors in several ways. The procedure given in the definition above is called *expanding down the first column,* since we use the coefficients of the first column along with their minors. Again, a minor of any entry is the determinant formed by deleting the row and column containing that entry. Hence in the determinant $\begin{vmatrix} a_{11} & a_{12} & a_{13} \\ a_{21} & a_{22} & a_{23} \\ a_{31} & a_{32} & a_{33} \end{vmatrix}$, the minor of a_{32} is $\begin{vmatrix} a_{11} & a_{12} & a_{13} \\ a_{21} & a_{22} & a_{23} \\ a_{31} & a_{32} & a_{33} \end{vmatrix} = \begin{vmatrix} a_{11} & a_{13} \\ a_{21} & a_{23} \end{vmatrix}$.

We can write a 3 × 3 determinant as an expansion of minors by expanding along any column or row provided we prefix each entry with the appropriate sign. The sign is determined by its position, which is represented by the following **sign array**:

$$\begin{vmatrix} + & - & + \\ - & + & - \\ + & - & + \end{vmatrix}$$

Notice that starting with $+$ at the top left, the signs alternate along each row and column.

For example, we can expand and evaluate the determinant of Example 4,

$\begin{vmatrix} 5 & 0 & -2 \\ 3 & 1 & -4 \\ 2 & -1 & 5 \end{vmatrix}$, *across the second row* as follows:

$$\begin{vmatrix} 5 & 0 & -2 \\ 3 & 1 & -4 \\ 2 & -1 & 5 \end{vmatrix} = -3\begin{vmatrix} 0 & -2 \\ -1 & 5 \end{vmatrix} + 1\begin{vmatrix} 5 & -2 \\ 2 & 5 \end{vmatrix} - (-4)\begin{vmatrix} 5 & 0 \\ 2 & -1 \end{vmatrix}$$

Note we use the signs of the second row of the sign array.

$$= -3(0 - 2) + 1(25 + 4) + 4(-5 - 0) = 15$$

In general, it is convenient to expand down columns or across rows that contain 0's and 1's.

Example 5

Evaluate the determinant $\begin{vmatrix} 3 & 2 & 0 \\ 6 & -2 & 0 \\ 2 & 3 & 4 \end{vmatrix}$.

Solution | Since we can expand the determinant along any row or column we choose, it would be convenient to expand this determinant along the third column, because there are two zeros in this column.

Why did we expand down the third column?

$$\begin{vmatrix} 3 & 2 & 0 \\ 6 & -2 & 0 \\ 2 & 3 & 4 \end{vmatrix} = +0 \begin{vmatrix} 6 & -2 \\ 2 & 3 \end{vmatrix} - 0 \begin{vmatrix} 3 & 2 \\ 2 & 3 \end{vmatrix} + 4 \begin{vmatrix} 3 & 2 \\ 6 & -2 \end{vmatrix}$$

Note we use the signs of the third column of the sign array.

$$= 0 - 0 + 4(-6 - 12) = \boxed{-72}$$

∎

Now we can generalize Cramer's rule for 3 × 3 systems of linear equations.

Cramer's Rule for 3 × 3 Systems of Equations

The solution to the general system

$$\begin{cases} a_{11}x + a_{12}y + a_{13}z = k_1 \\ a_{21}x + a_{22}y + a_{23}z = k_2 \\ a_{31}x + a_{32}y + a_{33}z = k_3 \end{cases}$$

is

$$x = \frac{D_x}{D}, \qquad y = \frac{D_y}{D}, \qquad z = \frac{D_z}{D}$$

where

$$D = \begin{vmatrix} a_{11} & a_{12} & a_{13} \\ a_{21} & a_{22} & a_{23} \\ a_{31} & a_{32} & a_{33} \end{vmatrix} \neq 0 \qquad D_x = \begin{vmatrix} k_1 & a_{12} & a_{13} \\ k_2 & a_{22} & a_{23} \\ k_3 & a_{32} & a_{33} \end{vmatrix}$$

$$D_y = \begin{vmatrix} a_{11} & k_1 & a_{13} \\ a_{21} & k_2 & a_{23} \\ a_{31} & k_3 & a_{33} \end{vmatrix} \qquad D_z = \begin{vmatrix} a_{11} & a_{12} & k_1 \\ a_{21} & a_{22} & k_2 \\ a_{31} & a_{32} & k_3 \end{vmatrix}$$

Example 6 | Solve the following system of equations using Cramer's rule:

$$\begin{cases} 2x + 2y + z = 7 \\ 4x - y + 3z = -1 \\ 6x - y - z = 0 \end{cases}$$

Solution | First we compute D, because if $D = 0$, then there is no unique solution. We compute D by expanding across the last row:

$$D = \begin{vmatrix} 2 & 2 & 1 \\ 4 & -1 & 3 \\ 6 & -1 & -1 \end{vmatrix} = 6 \begin{vmatrix} 2 & 1 \\ -1 & 3 \end{vmatrix} - (-1) \begin{vmatrix} 2 & 1 \\ 4 & 3 \end{vmatrix} + (-1) \begin{vmatrix} 2 & 2 \\ 4 & -1 \end{vmatrix}$$

$$= 6(6 + 1) - (-1)(6 - 4) + (-1)(-2 - 8) = 54$$

Next we compute D_x by expanding across the last row:

$$D_x = \begin{vmatrix} 7 & 2 & 1 \\ -1 & -1 & 3 \\ 0 & -1 & -1 \end{vmatrix} = 0 \begin{vmatrix} 2 & 1 \\ -1 & 3 \end{vmatrix} - (-1) \begin{vmatrix} 7 & 1 \\ -1 & 3 \end{vmatrix} + (-1) \begin{vmatrix} 7 & 2 \\ -1 & -1 \end{vmatrix}$$

$$= 0 - (-1)(21 + 1) + (-1)(-7 + 2) = 27$$

For variety, we compute D_y by expanding down the second column:

$$D_y = \begin{vmatrix} 2 & 7 & 1 \\ 4 & -1 & 3 \\ 6 & 0 & -1 \end{vmatrix} = -7 \begin{vmatrix} 4 & 3 \\ 6 & -1 \end{vmatrix} + (-1) \begin{vmatrix} 2 & 1 \\ 6 & -1 \end{vmatrix} - 0 \begin{vmatrix} 2 & 1 \\ 4 & 3 \end{vmatrix}$$

$$= -7(-4 - 18) + (-1)(-2 - 6) - 0 = 162$$

We compute D_z by expanding down the third column:

$$D_z = \begin{vmatrix} 2 & 2 & 7 \\ 4 & -1 & -1 \\ 6 & -1 & 0 \end{vmatrix} = 7 \begin{vmatrix} 4 & -1 \\ 6 & -1 \end{vmatrix} - (-1) \begin{vmatrix} 2 & 2 \\ 6 & -1 \end{vmatrix} + 0 \begin{vmatrix} 2 & 2 \\ 4 & -1 \end{vmatrix}$$

$$= 7(-4 + 6) - (-1)(-2 - 12) + 0 = 0$$

Hence $x = \dfrac{D_x}{D} = \dfrac{27}{54} = \dfrac{1}{2}$, $y = \dfrac{D_y}{D} = \dfrac{162}{54} = 3$, and $z = \dfrac{D_z}{D} = \dfrac{0}{54} = 0$.

The solution is $\left(\dfrac{1}{2}, 3, 0\right)$. The student should verify this solution.

7.6 Exercises

In Exercises 1–12, evaluate the determinant.

1. $\begin{vmatrix} 2 & 3 \\ 1 & 4 \end{vmatrix}$ **2.** $\begin{vmatrix} 1 & 0 \\ 5 & -2 \end{vmatrix}$

3. $\begin{vmatrix} 1 & -5 \\ 2 & 4 \end{vmatrix}$ **4.** $\begin{vmatrix} 3 & -1 \\ -1 & 3 \end{vmatrix}$

5. $\begin{vmatrix} 4 & -3 \\ -8 & 6 \end{vmatrix}$ **6.** $\begin{vmatrix} 1 & 0 \\ 0 & 1 \end{vmatrix}$

7. $\begin{vmatrix} 2 & 3 & -1 \\ 1 & 4 & 1 \\ 2 & 0 & 1 \end{vmatrix}$ **8.** $\begin{vmatrix} 1 & 0 & 2 \\ 5 & -2 & 1 \\ 3 & -1 & 0 \end{vmatrix}$

9. $\begin{vmatrix} 1 & 3 & -1 \\ 1 & 0 & 1 \\ 2 & 0 & 5 \end{vmatrix}$ **10.** $\begin{vmatrix} 1 & 0 & 0 \\ 3 & -3 & 1 \\ 2 & -1 & 5 \end{vmatrix}$

11. $\begin{vmatrix} 2 & 3 & 1 \\ 2 & 0 & 2 \\ 4 & 6 & 2 \end{vmatrix}$ **12.** $\begin{vmatrix} 3 & 1 & 0 \\ 3 & -3 & 1 \\ 2 & -1 & 0 \end{vmatrix}$

In Exercises 13–32, solve the system of equations using Cramer's rule.

13. $\begin{cases} x - 2y = 4 \\ 3x + 5y = 1 \end{cases}$ **14.** $\begin{cases} x + 2y = -1 \\ 6x - y = 4 \end{cases}$

15. $\begin{cases} 3x - y = 7 \\ 3x + 5y = -1 \end{cases}$ **16.** $\begin{cases} 2x + 3y = 9 \\ 6x - y = -3 \end{cases}$

17. $\begin{cases} 3x + 4y = 4 \\ 5x - 5y = 1 \end{cases}$ **18.** $\begin{cases} -x + 2y = 1 \\ 6x - 2y = 4 \end{cases}$

19. $\begin{cases} 2x + 4y = 4 \\ x + 2y = 2 \end{cases}$ **20.** $\begin{cases} -x + 2y = 1 \\ 2x - 4y = 3 \end{cases}$

21. $\begin{cases} 5x - 10y = 4 \\ x - 2y = 1 \end{cases}$ **22.** $\begin{cases} x + y = 1 \\ 3x + 3y = 3 \end{cases}$

23. $\begin{cases} x + 2y + z = 8 \\ x + 4y - z = 12 \\ x - 2y + z = -4 \end{cases}$ **24.** $\begin{cases} x - y + z = 6 \\ 2x - 2y - z = 3 \\ 3x - 2y - z = 5 \end{cases}$

25. $\begin{cases} 3x + 2y + z = 1 \\ x - 2y - z = 3 \\ x - 2y + z = 3 \end{cases}$ **26.** $\begin{cases} 2x + y + z = -1 \\ 2x - y - z = -5 \\ 3x - 2y - z = -9 \end{cases}$

27. $\begin{cases} 2x - 2y + z = 8 \\ x - 2y - z = 6 \\ x - 2y + z = 4 \end{cases}$ **28.** $\begin{cases} x - y + z = 1 \\ x - 2y + z = 0 \\ x - 2y - z = 5 \end{cases}$

29. $\begin{cases} x + 2y + z = 1 \\ 2x + 4y + z = 2 \\ x - 2y + z = -4 \end{cases}$ **30.** $\begin{cases} x - y + z = 3 \\ x - 2y - z = 0 \\ 3x - 3y + 3z = 9 \end{cases}$

31. $\begin{cases} x + z = 1 \\ 2y - z = 2 \\ x - z = 4 \end{cases}$ **32.** $\begin{cases} x - y + z = 6 \\ x - z = 0 \\ 3y + z = 8 \end{cases}$

33. Brian invests $10,000, part at 6% and part at 8.5%, earning a yearly interest equivalent to earning 7% on the entire amount. How much is invested at each rate?

34. A lawyer charges $110 an hour for her time and $50 an hour for her assistant's time. A client received a bill from the lawyer that included a $5280 charge for their combined time. If the amount of time spent by the lawyer and her assistant had been reversed, the bill for their time would have been $4320. How much time did the lawyer and her assistant spend on the client's case?

35. A manufacturer produces two types of computers: model A and model B. Model A requires 1.5 hours to manufacture and 1 hour to assemble; Model B requires 1 hour to manufacture and 30 minutes to assemble. If the company can allocate 100 hours for manufacture and 59 hours for assembly, how many of each type can be produced?

36. Raju splits up $20,000 into three investments. He has a bank account paying 5%, a certificate of deposit paying 6.5%, and a stock paying 9%. His annual interest from the three investments is $1365. If he invested $2000 more in his certificate than in the other two investments combined, how much did he invest at each rate?

37. A radio manufacturer produces three models, model A, model B, and model C. The company knows how much time is needed for production and assembly and how much raw materials cost for each model.

This information is found in the accompanying table. If the company has allocated 123 hours for production, 145 hours for assembly, and $1280 for raw materials, how many of each model can the company produce if it wants to use all the time allocated for each phase of the process and all the money for raw materials?

Model	Production hours	Assembly hours	Cost of raw materials
A	0.8	0.9	$9.80
B	0.7	0.5	$6.20
C	0.4	0.8	$3.10

38. A quadratic function is a function of the form $f(x) = ax^2 + bx + c$, where a, b, and c are constants. Find the quadratic function $f(x)$ such that $f(1) = 5, f(-1) = 7$, and $f(2) = 19$.

Question for Thought

39. Discuss the differences between $\begin{vmatrix} 5 & 0 \\ -1 & 2 \end{vmatrix}$ and $\begin{bmatrix} 5 & 0 \\ -1 & 2 \end{bmatrix}$.

7.7   Properties of Determinants

In the last section, we defined 2 × 2 and 3 × 3 determinants and discussed Cramer's rule applied to 2 × 2 and 3 × 3 systems. In this section we generalize our findings from the previous section to $n \times n$ determinants and discuss some of their properties. Then we generalize Cramer's rule to $n \times n$ systems of linear equations. To do this, we need a more formal way to refer to determinants, minors, and the sign array. This requires that we introduce some new notation and incorporate some of the ideas and notation used previously with matrices.

When we refer to a matrix, we usually refer to its size by specifying the number of rows and the number of columns. In this section we will restrict our discussion to square matrices. We will refer to an $n \times n$ square matrix as a *matrix of order n*.

If A is a matrix of order n, then we can represent the determinant of the matrix A by $|A|$. Shown next is a matrix A of order n and its determinant, $|A|$.

A Matrix of Order *n*

$$A = \begin{bmatrix} a_{11} & a_{12} & a_{13} & \cdots & a_{1n} \\ a_{21} & a_{22} & a_{23} & \cdots & a_{2n} \\ a_{31} & a_{32} & a_{33} & \cdots & a_{3n} \\ \vdots & \vdots & \vdots & & \vdots \\ a_{n1} & a_{n2} & a_{n3} & \cdots & a_{nn} \end{bmatrix}$$

The Determinant of *A*

$$|A| = \begin{vmatrix} a_{11} & a_{12} & a_{13} & \cdots & a_{1n} \\ a_{21} & a_{22} & a_{23} & \cdots & a_{2n} \\ a_{31} & a_{32} & a_{33} & \cdots & a_{3n} \\ \vdots & \vdots & \vdots & & \vdots \\ a_{n1} & a_{n2} & a_{n3} & \cdots & a_{nn} \end{vmatrix}$$

Again, the subscripts of a indicate the position of the entry in the matrix; in general, a_{ij} is the entry in the ith row and jth column.

Although a matrix is a rectangular array of numbers, *the determinant of a matrix is a single real number* defined only when the matrix is square.

In the last section, we needed minors to define the determinant of a matrix of order 3 (a 3×3 matrix). Recall that the minor of an entry of a 3×3 determinant is a 2×2 determinant arrived at by deleting the row and column containing the entry. For example, for the determinant

$$|A| = \begin{vmatrix} 3 & 7 & -1 \\ 0 & 8 & 1 \\ 2 & 5 & 3 \end{vmatrix},$$

the minor of $2 = a_{31}$ is the 2×2 determinant

$$\begin{vmatrix} 3 & 7 & -1 \\ 0 & 8 & 1 \\ 2 & 5 & 3 \end{vmatrix} = \begin{vmatrix} 7 & -1 \\ 8 & 1 \end{vmatrix}.$$

We use the symbol M_{ij} to designate the minor of a_{ij}. Hence, for A the minor of a_{31} is $M_{31} = \begin{vmatrix} 7 & -1 \\ 8 & 1 \end{vmatrix}$.

In general,

> If A is a matrix of order n (where $n \geq 2$), then M_{ij}, the **minor** of a_{ij}, is the determinant of the matrix of order $n - 1$ arrived at by deleting the row and column containing a_{ij}.

For the determinant $\begin{vmatrix} 6 & 3 & 7 & -1 \\ 0 & 0 & 8 & 1 \\ -3 & 2 & 5 & 3 \\ 9 & -1 & 4 & 1 \end{vmatrix}$, the minor of $8 = a_{23}$ is the determinant

$$M_{23} = \begin{vmatrix} 6 & 3 & 7 & -1 \\ 0 & 0 & 8 & 1 \\ -3 & 2 & 5 & 3 \\ 9 & -1 & 4 & 1 \end{vmatrix} = \begin{vmatrix} 6 & 3 & -1 \\ -3 & 2 & 3 \\ 9 & -1 & 1 \end{vmatrix}.$$

The sign array tells us the sign prefixed to the minor when the determinant is expanded. As with 3×3 determinants, the sign array for an $n \times n$ determinant is constructed by starting with a $+$ sign in the top left corner and alternating signs along the rows and columns:

$$\begin{vmatrix} + & - & + & - & + & \cdots \\ - & + & - & + & - & \cdots \\ + & - & + & - & + & \cdots \\ - & + & - & + & - & \cdots \\ + & - & + & - & + & \cdots \\ \vdots & \vdots & \vdots & \vdots & \vdots & \end{vmatrix}$$

Although the pattern is straightforward, it can be described as follows: In the sign array, the entry in the ith row and jth column is positive when the sum $i + j$ is an even number and negative when the sum $i + j$ is an odd number.

Recognizing that $(-1)^k = +1$ when k is even and $(-1)^k = -1$ when k is odd, we can express a minor prefixed with its sign from the sign array as $(-1)^{i+j} M_{ij}$; we call this term the **cofactor** of the element a_{ij} and designate it A_{ij}.

Definition of a Cofactor of a Matrix

The **cofactor** A_{ij} of the element a_{ij} is defined by

$$A_{ij} = (-1)^{i+j}M_{ij}$$

What is the difference between
a minor and a cofactor?

The cofactor A_{ij} is the minor M_{ij} prefixed with its sign. We can write the expansion of a 3×3 determinant down its first column as

$$|A| = \begin{vmatrix} a_{11} & a_{12} & a_{13} \\ a_{21} & a_{22} & a_{23} \\ a_{31} & a_{32} & a_{33} \end{vmatrix} = a_{11}A_{11} + a_{21}A_{21} + a_{31}A_{31} = a_{11}M_{11} - a_{21}M_{21} + a_{31}M_{31}$$

We can now define an $n \times n$ determinant by first-column expansion as we did with 3×3 determinants:

If A is a matrix of order $n \geq 2$, then

$$|A| = a_{11}A_{11} + a_{21}A_{21} + a_{31}A_{31} + \cdots + a_{n1}A_{n1}$$
$$= a_{11}M_{11} - a_{21}M_{21} + a_{31}M_{31} + \cdots + a_{n1}(-1)^{n+1}M_{n1}$$

Again, this definition is based on a first-column expansion. We can define a determinant in general as follows: If A is a matrix of order n ($n \geq 2$), the determinant $|A|$ is found by multiplying elements in any row (or column) by their respective cofactors and adding the resulting products.

Example 1 If $A = \begin{bmatrix} -2 & 3 & 0 & -1 \\ 9 & 3 & 1 & 2 \\ 2 & -1 & 0 & -4 \\ 0 & 1 & 4 & -7 \end{bmatrix}$, find $|A|$.

Solution Since the third column has two zeros, we can reduce the amount of computation by expanding down the third column to get

$$\begin{vmatrix} -2 & 3 & 0 & -1 \\ 9 & 3 & 1 & 2 \\ 2 & -1 & 0 & -4 \\ 0 & 1 & 4 & -7 \end{vmatrix} = 0A_{13} + 1A_{23} + 0A_{33} + 4A_{43}$$

$$= 0M_{13} - 1M_{23} + 0M_{33} - 4M_{43}$$
$$= -M_{23} - 4M_{43}$$

Now we have to expand M_{23} and M_{43} using 2×2 minors:

$$M_{23} = \begin{vmatrix} -2 & 3 & -1 \\ 2 & -1 & -4 \\ 0 & 1 & -7 \end{vmatrix} \qquad \text{Expand down the first column.}$$

$$= -2 \begin{vmatrix} -1 & -4 \\ 1 & -7 \end{vmatrix} - 2 \begin{vmatrix} 3 & -1 \\ 1 & -7 \end{vmatrix} + 0 \begin{vmatrix} 3 & -1 \\ -1 & -4 \end{vmatrix} = -2(11) - 2(-20) = 18$$

and

$$M_{43} = \begin{vmatrix} -2 & 3 & -1 \\ 9 & 3 & 2 \\ 2 & -1 & -4 \end{vmatrix} \qquad \text{Expand across the last row.}$$

```
[A]
  [[-2  3  0 -1]
   [9  3  1  2]
   [2 -1  0 -4]
   [0  1  4 -7]]
det([A])
              -638
```

$$= 2\begin{vmatrix} 3 & -1 \\ 3 & 2 \end{vmatrix} - (-1)\begin{vmatrix} -2 & -1 \\ 9 & 2 \end{vmatrix} + (-4)\begin{vmatrix} -2 & 3 \\ 9 & 3 \end{vmatrix} = 2(9) + (5) - 4(-33) = 155$$

Hence

$$\begin{vmatrix} -2 & 3 & 0 & -1 \\ 9 & 3 & 1 & 2 \\ 2 & -1 & 0 & -4 \\ 0 & 1 & 4 & -7 \end{vmatrix} = -M_{23} - 4M_{43}$$

$$= -18 - 4(155) = \boxed{-638}$$

If we had to evaluate a 7×7 determinant, then we would have to expand it to seven 6×6 cofactors, then expand *each* of the seven 6×6 cofactors to six 5×5 cofactors, and so on. As you can see, evaluating determinants of higher-order matrices can be quite tedious.

As we have noted in examples before, when we expand determinants down columns or across rows that contain zeros, computations can be significantly reduced. The following example illustrates one condition under which the computation of a determinant is straightforward.

Example 2   Prove that the determinant of a third-order matrix in triangular form is equal to the product of the entries in its main diagonal.

Solution   We begin with the determinant of a 3×3 matrix in triangular form:

$$\begin{vmatrix} a & b & c \\ 0 & d & e \\ 0 & 0 & f \end{vmatrix} \qquad \text{If we expand down column 1, we get}$$

$$= a\begin{vmatrix} d & e \\ 0 & f \end{vmatrix} - 0\begin{vmatrix} b & c \\ 0 & f \end{vmatrix} + 0\begin{vmatrix} b & c \\ d & e \end{vmatrix}$$

$$= a(df - 0e) = adf$$

which is the product of the entries in its main diagonal.

We can generalize this result:

The determinant of a matrix in triangular form is equal to the product of the entries in its main diagonal.

Hence, if we can somehow transform a determinant into one whose matrix is in triangular form, the computation of a determinant is greatly simplified.

As with matrices, we can define operations on rows that will help us to transform a determinant into one whose matrix is in triangular form. Unlike matrices, for determinants these operations apply to columns as well. Also unlike matrices, the row and column operations we define do not always produce equivalent forms of the determinant.

Elementary Row and Column Operations for Determinants

1. Interchanging any two rows (or columns) in a determinant changes the sign of the determinant.

For example: $\begin{vmatrix} a & b \\ c & d \end{vmatrix} = - \begin{vmatrix} c & d \\ a & b \end{vmatrix}$ Switching two rows

2. Multiplying a row (or column) by a nonzero constant k multiplies the value of the determinant by k. We can also view this operation as factoring a common factor from a row or column of a determinant.

For example: $\begin{vmatrix} a & b & kc \\ d & e & kf \\ g & h & ki \end{vmatrix} = k \begin{vmatrix} a & b & c \\ d & e & f \\ g & h & i \end{vmatrix}$ Factoring $k \neq 0$ from column 3

Exercises 7–10 are related to the elementary row operations.

3. Adding a multiple of one row to another does not affect the value of the determinant; adding a multiple of one column to another does not affect the value of the determinant.

For example: $\begin{vmatrix} a & b \\ c & d \end{vmatrix} = \begin{vmatrix} a & b \\ c + ka & d + kb \end{vmatrix}$ Adding a multiple of row 1 to row 2

In Section 7.5 we noted that if the coefficient matrix of a linear system of equations has an inverse, then we can solve the associated system. We also mentioned that a matrix has an inverse if and only if we can use the elementary row operations to transform it into the identity matrix. It is natural to ask, Is there some way to determine whether a matrix has an inverse?

If a matrix is transformed into an equivalent matrix with a row or column of zeros, then clearly it cannot be transformed into the identity matrix and therefore has no inverse. It turns out that if any row of a matrix is a multiple of any other row, then the matrix can be transformed into an equivalent matrix with a row of zeros and hence its determinant will be zero. We can summarize this discussion in the following very useful theorem.

Exercise 30 discusses when the determinant of a 2 × 2 matrix will be zero.

Theorem 7.3

A square matrix A has an inverse if and only if $|A| \neq 0$.

This agrees quite well with the observation we made in Section 7.6 that, according to Cramer's rule, if the determinant of a system is 0, then the system does not have a unique solution. The exercises ask you to generalize Cramer's rule (see Exercises 21 and 22) to an $n \times n$ system of linear equations.

7.7 Exercises

In Exercises 1–6, given

$$A = \begin{vmatrix} 2 & 5 & 3 & -1 \\ 7 & 4 & 0 & -3 \\ -2 & 1 & 8 & 9 \\ -6 & -4 & 6 & -5 \end{vmatrix}$$

find the following:

1. The minor of 6
2. The cofactor of -6
3. The cofactor of 0
4. The minor of 1
5. A_{24}
6. M_{32}
7. Use the determinant $\begin{vmatrix} 4 & 1 \\ -2 & 3 \end{vmatrix}$ to illustrate the elementary row and column operations for determinants.

That is,
(a) Show that interchanging rows or columns changes the sign of the determinant.
(b) Show that multiplying a row or column by a constant multiplies the value of the determinant by that constant.
(c) Show that adding a multiple of one row to another, or adding a multiple of one column to another, does not affect the value of the determinant.

8. Repeat Exercise 7 for the determinant

$$\begin{vmatrix} 2 & -1 & 0 \\ -1 & 3 & -2 \\ 4 & 1 & 3 \end{vmatrix}.$$

9. Prove each of the elementary row and column operations for 2×2 determinants.

10. Prove each of the elementary row and column operations for 3×3 determinants.

11. Evaluate the following determinant by using the elementary row and column operations to transform it into a triangular determinant.

$$\begin{vmatrix} 3 & -3 & 1 \\ 5 & 2 & 1 \\ 5 & 2 & 4 \end{vmatrix}$$

12. Evaluate the following determinant by using the elementary row and column operations to transform it into a triangular determinant.

$$\begin{vmatrix} 3 & -3 & 6 & 15 \\ 2 & 1 & -3 & 1 \\ -2 & 2 & -4 & -10 \\ 6 & 1 & -4 & 2 \end{vmatrix}$$

In Exercises 13–20, evaluate the determinants.

13. $\begin{vmatrix} 1 & 2 & -1 \\ 0 & 3 & 4 \\ 5 & 1 & 2 \end{vmatrix}$ 14. $\begin{vmatrix} 8 & -2 & 5 \\ 3 & 1 & 0 \\ 2 & -1 & 0 \end{vmatrix}$

15. $\begin{vmatrix} 1 & 1 & -3 & 4 \\ 3 & 1 & 0 & 2 \\ 3 & 1 & 0 & 4 \\ 2 & 2 & 0 & 5 \end{vmatrix}$ 16. $\begin{vmatrix} 2 & 3 & 1 & 5 \\ 0 & 0 & 0 & 0 \\ 5 & 3 & 1 & 1 \\ 2 & 9 & 5 & 4 \end{vmatrix}$

17. $\begin{vmatrix} 2 & 1 & 3 & 1 \\ -1 & 2 & 4 & 0 \\ 3 & 1 & 1 & -2 \\ 2 & -1 & 0 & 5 \end{vmatrix}$

18. $\begin{vmatrix} 1 & 1 & 3 & -1 & 2 \\ 2 & 1 & 0 & 3 & -1 \\ 0 & 1 & 3 & 1 & 2 \\ 0 & 4 & 1 & 1 & -2 \\ 1 & 1 & 2 & 1 & 3 \end{vmatrix}$

19. $\begin{vmatrix} 1 & 4 & 0 & 3 & 8 \\ 2 & 3 & 0 & 4 & 1 \\ 5 & 6 & 0 & 2 & 3 \\ 4 & 2 & 0 & 3 & 1 \\ 5 & 1 & 0 & 3 & 6 \end{vmatrix}$

20. $\begin{vmatrix} 2 & 3 & 0 & 0 & 1 & 2 \\ 1 & 3 & 4 & 0 & -1 & 3 \\ 2 & 3 & 1 & 0 & -1 & 2 \\ 1 & 3 & 0 & 1 & 1 & 3 \\ 2 & 3 & 0 & 0 & 1 & 4 \\ 3 & 0 & 5 & -1 & 1 & 0 \end{vmatrix}$

21. In Section 7.6 we discussed Cramer's rule as it applies to 2×2 and 3×3 systems of linear equations. Generalize Cramer's rule to the 4×4 system

$$\begin{cases} a_{11}x_1 + a_{12}x_2 + a_{13}x_3 + a_{14}x_4 = k_1 \\ a_{21}x_1 + a_{22}x_2 + a_{23}x_3 + a_{24}x_4 = k_2 \\ a_{31}x_1 + a_{32}x_2 + a_{33}x_3 + a_{34}x_4 = k_3 \\ a_{41}x_1 + a_{42}x_2 + a_{43}x_3 + a_{44}x_4 = k_4 \end{cases}$$

22. Generalize Cramer's rule to an $n \times n$ system of linear equations.

In Exercises 23–28, solve the system of equations using Cramer's rule.

23. $\begin{cases} w + x = -1 \\ w + y + z = 2 \\ x + y = 1 \\ y + z = 3 \end{cases}$ 24. $\begin{cases} w + x = 1 \\ w + y = 1 \\ x + y = 0 \\ y + z = -1 \end{cases}$

25. $\begin{cases} w + x + 2y - z = 1 \\ 2x + 3y + z = 1 \\ w + y - z = 0 \\ w + x - y = 4 \end{cases}$

26. $\begin{cases} w + x + y + z = 2 \\ w + x + y = -1 \\ w + x + z = 1 \\ x - y + z = 1 \end{cases}$

27. $\begin{cases} w + x + y + z = 0 \\ w + x + y = 0 \\ 2w + 2x + 2y = 0 \\ 2x - y + z = 0 \end{cases}$

28. $\begin{cases} w + 2x + y + z = 5 \\ w - x - y = 0 \\ 3w - 3x - 3z = 1 \\ x - y + z = 0 \end{cases}$

Questions for Thought

29. In a system of equations, two equations are equivalent. If you are trying to solve the system using Cramer's rule, explain what kinds of results you could expect.

30. Consider the general 2×2 matrix $A = \begin{bmatrix} a & b \\ c & d \end{bmatrix}$.

    (a) Show that if row 2 is a multiple of row 1, then the determinant of this matrix is 0. [HINT: If $kR_1 = R_2$, then $ka = c$ and $kb = d$, etc.)

    (b) Show that if $|A| = 0$ then row 2 is a multiple of row 1.

7.8   Systems of Linear Inequalities

Throughout this chapter we have been discussing systems of linear equations. Let's now turn our attention to systems of inequalities.

Graphing Linear Inequalities in Two Variables

On the number line, a one-dimensional coordinate system, the graph of a linear inequality such as $x \geq a$ is a *half-line* (or ray) starting at (or bounded by) the *point* $x = a$ (Figure 7.7).

Figure 7.7

The graph of $x \geq a$ on a number line

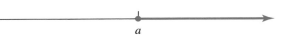

An analogous situation occurs with graphing linear inequalities in *two variables* on a two-dimensional rectangular coordinate plane: The graph of $Ax + By \leq C$ is a *half-plane* bounded by the *line* $Ax + By = C$. See Figure 7.8.

Figure 7.8

The graph of a typical solution set for $Ax + By \leq C$

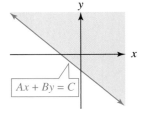

Consider the inequality $y - x \leq 5$. What do the solutions look like? If we isolate y and rewrite this equation as $y \leq x + 5$, then we can see that we are looking for all points on the rectangular coordinate system in which the y-coordinate is less than or equal to 5 more than the x-coordinate—all points on or below the line $y = x + 5$. This is pictured in Figure 7.9(a).

On the other hand, the same equation can be rewritten as $y - 5 \leq x$, which can be interpreted as all points on the plane such that the x-coordinate is greater than or equal to 5 less than the y-coordinate—all points on or to the right of the line $y - 5 = x$. This is pictured in Figure 7.9(b).

Figure 7.9

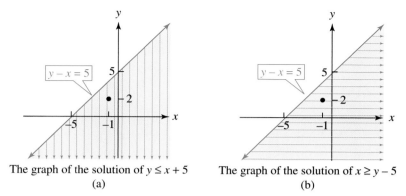

The graph of the solution of $y \le x + 5$
(a)

The graph of the solution of $x \ge y - 5$
(b)

Both views of the problem leave us with the same solution: the half-plane to the right of and bounded by the line $y - x = 5$. The shaded region is the solution. This means that any point lying in the shaded region is a solution to this inequality, and any point not in the region is not a solution. For example, the point $(-1, 2)$ is in the shaded region and $y - x \le 5$ is true for $x = -1$ and $y = 2$. The point $(-3, 6)$ is *not* in the shaded region and $y - x \le 5$ is *not* true for $x = -3$ and $y = 6$.

Example 1   Graph the solution of $2y - 3x > 12$.

Solution   We know that the solution will be a half-plane bounded by the *line $2y - 3x = 12$.* Hence, we first graph the line, as indicated in Figure 7.10(a). Notice that we draw a dashed line rather than a solid line. The dashed line indicates a strict inequality; that is, the points on the line itself are *not* part of the solution set.

Figure 7.10

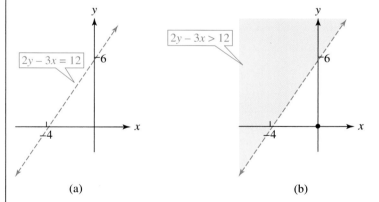

(a)

(b)

We know that all solutions lie on either one side of the line or the other side, so we need only test one point *not on the line.* If the coordinates of the test point we choose satisfy the inequality, then all points on that side of the line satisfy the inequality and, therefore, the solution is the half-plane that includes the test point.

If the test point does not satisfy the inequality, then no points on that side of the line satisfy the inequality, and the solution is the half-plane on the other side of the line. $(0, 0)$ is often a convenient test point (so long as it does not fall on the line):

$$2y - 3x > 12 \qquad \text{Substitute } x = 0 \text{ and } y = 0.$$
$$2(0) - 3(0) \overset{?}{>} 12 \qquad \text{This inequality is false.}$$

Since the inequality is false for the test point, $(0, 0)$, we shade the *other* side of the line, as indicated in Figure 7.10(b).   ▌

Graphing Linear Inequalities

In summary, we graph linear inequalities as follows:

1. Sketch a graph of the equation of the line that is the boundary of the solutions. The line should be dashed if it is a strict inequality ($<$ or $>$) and solid if it is a weak inequality ($\leq$ or $\geq$).

2. Choose a test point: Pick a point not on the line and determine whether its coordinates satisfy the inequality.

3. If the coordinates of the test point satisfy the inequality, shade in the half-plane on the side of the line that includes the test point; if the coordinates of the test point do not satisfy the inequality, shade the other side of the line. The solution to the inequality is the shaded side together with the line if it is solid.

Example 2 | Graph the following inequalities: **(a)** $3x \leq 2y$ **(b)** $x < -1$

Solution | **(a)** $3x \leq 2y$

Why can't we use (0, 0) as the test point?

1. Sketch the graph of $3x = 2y$ using a solid line. See Figure 7.11(a).

2. We will use the point $(1, 4)$ as the test point, since it does not lie on the line. We substitute $x = 1$ and $y = 4$ into $3x \leq 2y$. Because $3(1) \leq 2(4)$ is true, we shade in the half-plane on the side of the line that contains the point $(1, 4)$.

3. The graph of the solution set to $3x \leq 2y$ is shown in Figure 7.11(a).

(b) We could graph this inequality using the same approach as in part (a). However, we know that the graph of $x = -1$ is a vertical line one unit to the left of the y-axis. Hence we draw the dashed line in Figure 7.11(b). It then follows that to satisfy the inequality $x < -1$, a point must have its x-coordinate to the left of -1. Therefore, the graph of the solution set is the shaded region in Figure 7.11(b).

Figure 7.11

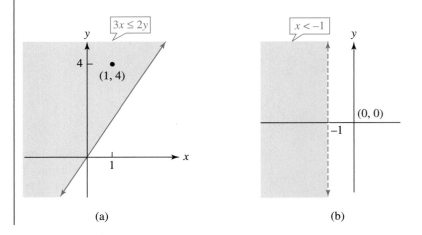

(a)   (b)

Now let's turn our attention to solving systems of linear inequalities in two variables by graphing the solution set of each inequality in the system and then determining where the solution sets intersect.

Example 3 | Solve the following system of linear inequalities: $\begin{cases} x - 2y < 14 \\ x + y < 5 \end{cases}$

Solution We begin by sketching the solution set to the first inequality, $x - 2y < 14$.

1. Graph the equation $x - 2y = 14$.

2. Draw a dashed line to indicate that the line is not included in the solution set.

3. Pick a test point not on the line and determine whether it satisfies the inequality $x - 2y < 14$. The point $(0, 0)$ is not on the line; because $0 - 2(0) < 14$ is true, $(0, 0)$ satisfies the inequality.

4. Shade the half-plane on the side of the line containing $(0, 0)$ to indicate that the half-plane is the solution set. The graph of $x - 2y < 14$ is shown in Figure 7.12(a).

Next we follow these same steps and graph the inequality $x + y < 5$, as shown in Figure 7.12(b). Notice that each individual inequality is a half-plane. The dashed lines indicate that the solutions do *not* include the lines.

Figure 7.12 (a), (b)

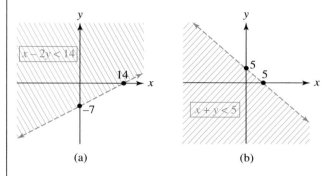

(a)   (b)

Because we are looking for ordered pairs that satisfy both inequalities, this means we want the points of intersection of the two half-planes. This is shown as the crosshatched region in Figure 7.12(c).

Figure 7.12(c)

The crosshatched region is the solution.

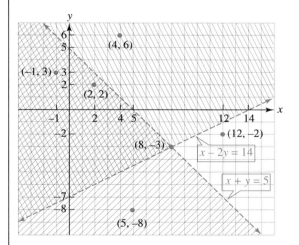

The crosshatched region is where the solutions lie; that is, every ordered pair representing a point in this region satisfies *both* inequalities. For example, $(-1, 3)$ and $(2, 2)$ satisfy both inequalities and lie in this region. The points are solutions to the system of inequalities. On the other hand, points such as $(4, 6)$, $(5, -8)$, and $(12, -2)$ do not satisfy *both* inequalities; they do not lie in this region, and therefore they are not solutions to the system.

To determine the points of intersection of the half-planes, we need to solve the system of equations represented by the dashed lines. By using one of the methods discussed in this chapter, we find that the two lines intersect at the point $(8, -3)$. ▌

Example 4 | Solve the following system of linear inequalities: $\begin{cases} 3x - 2y \le 12 \\ 5x + 2y > 4 \\ y \le 0 \end{cases}$

Solution | We graph each line and determine which half-plane to shade as described at the beginning of this section. See Figures 7.13(a)–(c).

Figure 7.13
(a), (b), (c)

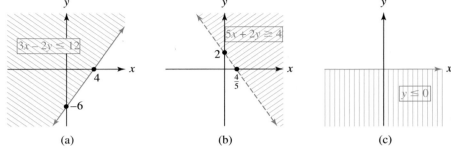

(a)   (b)   (c)

It's always a good idea to check the solution with a test point. Does (2, −2) work?

We use one of the methods discussed in this chapter to find the points of intersection of the lines: (2, −3), (4, 0), and $(\frac{4}{5}, 0)$. The solution is the region where all three half-planes intersect—the shaded region in Figure 7.13(d).

Figure 7.13(d)

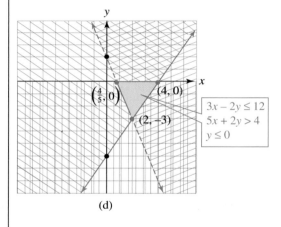

(d)

The points representing the corners of the shaded polygonal region (in this case, a triangular region) are called the *vertices* of the solution set. They are the points of intersection of each pair of lines in Example 4. ∎

Example 5 | Solve the following system of linear inequalities: $\begin{cases} x - 3y < 0 \\ 2x - 6y \ge 12 \end{cases}$

Solution | We graph each inequality by graphing the lines and shading the appropriate regions. Since we are looking for ordered pairs that satisfy both inequalities, we are looking for the points of intersection of the two half-planes. When we attempt to find the intersection of the two lines, we find that they have no points in common. As Figure 7.14 shows, the shaded regions never intersect, so there is   no solution .

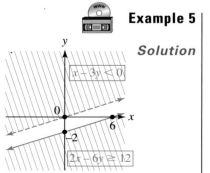

Figure 7.14

∎

Example 6 The Bedding Store needs to order a shipment of single and double mattresses. A single mattress costs $80 and takes up 16 cu ft of storage space, and a double mattress costs $120 and takes up 36 cu ft of storage space. If the store manager wants to order no more than $6000 worth of mattresses and has at most 1440 cu ft of storage space in the store's storage area, how many of each mattress can she order? Write a system of inequalities to describe this situation and sketch the solution set of the system.

Solution Let x = number of single mattresses and y = number of double mattresses. We can translate the information given into the following system of inequalities:

$$80x + 120y \leq 6000 \quad \text{Represents the total cost of the mattresses}$$
$$16x + 36y \leq 1440 \quad \text{Since the storage space is at most 1440 cu ft}$$
$$x \geq 0 \quad \text{Since we cannot have a negative number of mattresses}$$
$$y \geq 0$$

We have chosen to let the horizontal axis represent the number of single mattresses and the vertical axis to represent the number of double mattresses. Figure 7.15 is a sketch of the system.

Figure 7.15

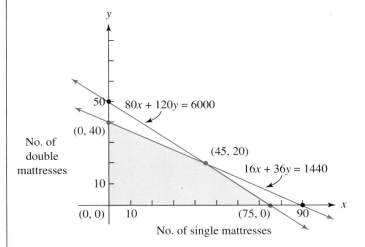

Noting the boundaries, we stay within the first quadrant: The solution set is the shaded region shown in Figure 7.15. The points of intersection are (0, 0), (0, 40), (45, 20), and (75, 0). Any point in the shaded region satisfies the restrictions. (Note that the point (75, 0) represents what would happen if the store manager ordered single mattresses only.) Keep in mind that since we cannot order a fraction of a mattress, only whole-number coordinates make sense. For example, (12.5, 18.2) is a point in the region but is not a realistic option. ∎

7.8 Exercises

In Exercises 1–8, sketch the graph of the given inequality in a rectangular coordinate system.

1. $2y - 3x \leq 12$
2. $5y + 4x + 20 > 0$
3. $4x + 7y > 10$
4. $y - 5x \geq 8$
5. $x + 3 < 0$
6. $y - 4 \leq 2$
7. $y \geq 4x$
8. $3x - 4y < 0$

In Exercises 9–28, sketch the solution set of the system of inequalities.

9. $\begin{cases} 2x + y \leq 12 \\ 4x + y \geq 8 \end{cases}$

10. $\begin{cases} -x + 2y \leq -4 \\ x - y < 3 \end{cases}$

11. $\begin{cases} 3x - 5y < 15 \\ 2x + y > 10 \end{cases}$

12. $\begin{cases} 3x - 5y < 15 \\ 3x - 5y > 10 \end{cases}$

13. $\begin{cases} 5x - 2y \leq 10 \\ 6x - 5y > -30 \end{cases}$

14. $\begin{cases} x \leq 3y - 12 \\ 2x + 5y > -10 \end{cases}$

15. $\begin{cases} 6x + 8y \leq 24 \\ 9x + 12y \geq 18 \end{cases}$

16. $\begin{cases} 3x - 12y \geq 24 \\ 12x \leq 8y + 24 \end{cases}$

17. $\begin{cases} 5y < 3x - 15 \\ 8x - 2y \leq 10 \end{cases}$

18. $\begin{cases} x + 7y < -14 \\ 3x + 21y > -21 \end{cases}$

19. $\begin{cases} 2x + y \leq 12 \\ 4x + y \geq 8 \\ y \geq 0 \end{cases}$

20. $\begin{cases} 5x + 2y \geq -10 \\ -2x + 5y \leq -10 \\ x \leq 1 \end{cases}$

21. $\begin{cases} x + y - 4 \leq 2 \\ 2x - 6y > 12 \\ x \geq 0 \end{cases}$

22. $\begin{cases} 2x - 5y < -20 \\ 4x + 2y > -8 \\ y \leq 8 \end{cases}$

23. $\begin{cases} x - y < 4 \\ 2x + y < 12 \\ x \geq 0 \\ y \geq 0 \end{cases}$

24. $\begin{cases} x + 2y > -1 \\ x - y < 4 \\ x \geq 0 \\ y \leq 1 \end{cases}$

25. $\begin{cases} x + y \geq -1 \\ 3x - 2y < 12 \\ x \geq -1 \\ y \leq 5 \end{cases}$

26. $\begin{cases} 5x - 2y > 0 \\ x - 2y \leq 8 \\ y > -1 \\ y \leq 0 \end{cases}$

27. $\begin{cases} 5x - 2y < 10 \\ 2x - 5y > -10 \\ x + y \geq -5 \\ x \geq -2 \end{cases}$

28. $\begin{cases} x < 3y - 1 \\ 2x - 5y \geq -10 \\ 2x - 2y < 1 \\ x \geq -1 \end{cases}$

For Exercises 29–36, write a system of inequalities to describe the conditions of the problem and sketch the solution set of the system.

29. An appliance store needs to order a shipment of VCRs and TVs. A VCR costs $200 and takes up 2 cu ft of storage space, and a TV costs $380 and takes up 6 cu ft of storage space. If the store wants to order at least $6000 worth of TVs and VCRs but has at most 120 cu ft of storage space in its storage area, how many of each can it order?

30. An electronics company makes two types of calculators—a scientific model and a graphing model. The scientific model requires $8 in materials and takes 1 hour to assemble and package. The graphing model requires $25 in materials and takes $\frac{1}{2}$ hour to assemble and package. The company decides to spend a maximum of $15,000 on materials and to allot a maximum of 1000 hours for packing and assembly. How many of each type can the company make?

31. Carol makes two kinds of jewelry boxes; one type is custom-made to specifications and the other type is a standard model made from precut parts. On the average, it takes her 3 hours to make a custom-made jewelry box and 1 hour to make the standard model. She can make a profit of $80 on the custom-made box and $20 on the standard model. If she can spend no more than 20 hours a week and wants to earn at least $200 per week from making jewelry boxes, how many of each type can she produce?

32. Joe wants to eat a more balanced breakfast. He reads the sides of two cereal boxes and finds out that 1 ounce of cereal X contains 8 grams of carbohydrates and 0.25 gram of sodium, whereas 1 ounce of cereal Y contains 12 grams of carbohydrates and 0.34 gram of sodium. He wants to create a mixture of cereals with at least 24 grams of carbohydrates and no more than 1 gram of sodium. How much of each cereal could he use?

33. Lana has no more than $60,000 available for investment. She can invest in two types of certificates of deposit: a 6-month certificate, which yields a 5% yearly return on investment, and a 1-year certificate, which yields a 6.5% yearly return. How much could she invest in each if she wants to earn at least $3300 interest in one year?

34. A judge is working on the weekly schedule of hearings in her courtroom, where she hears both minor criminal and civil complaints. She finds that the average civil complaint requires 45 minutes to settle; the average criminal complaint requires 30 minutes. Next week she wants to schedule at least 20 hours but no more than 35 hours of hearings. She prefers to adjudicate at least as many criminal complaints as civil complaints. How many of each type of case can she hear next week?

35. A lawyer handles both divorce cases and malpractice suits. The average divorce case takes 12 hours of his time and 20 hours of his assistants' time. The average malpractice suit takes 22 hours of his time and 18 hours of his assistants' time. The lawyer decided that he can put in no more than 60 hours per week and his assistants should put in at least 40 hours a week. Given these constraints, how many cases of each type could the lawyer handle weekly?

36. An orthodontist uses two types of appliances (A and B), both shown to be effective for overbites. Over a period of a year, each appliance A requires 6 hours of the doctor's labor and 12 hours of the dental assistants' labor; each appliance B requires 4 hours of the doctor's labor and 16 hours of the dental assistants' labor. Over the year, the doctor cannot put in more than 1200 hours of labor for appliances, and the dental assistants cannot put in more than 3200 hours. How many of each appliance could the orthodontist use?

7.9 An Introduction to Linear Programming: Geometric Solutions

Example 6 in Section 7.8 illustrates how we can take a set of restrictions, represent them mathematically, and then picture the restrictions so we may make the most appropriate decisions. Throughout this text we have shown how we can express relationships between two variables or functions of one variable in an effort to determine important points. For example, in Section 4.4, we expressed profit as a quadratic function of price and used what we knew about quadratic functions to determine the price that produces a maximum profit.

There are many problems in economics and the sciences where it is important to find the maximum or minimum value of a function of *several* variables, given a set of restrictions on these variables. In this section, we discuss how to find solutions to these problems when the function and the restrictions are linear; this is called **linear programming**. Although we can generalize most of our discussion to several variables, we will restrict our attention to linear functions of two variables. Before we continue, we will discuss some terminology.

The solution sets to some of the systems of linear inequalities we have seen so far have been polygonal regions, such as in Figures 7.13 and 7.15 of the previous section. These regions are called **convex sets**. A set is **convex** if for any two points in the set, the line segment joining them lies completely in the set. See Figure 7.16. In particular, a convex set is **bounded** if it can be enclosed within a circle.

With this terminology in hand, we can start by reexamining Example 6 in Section 7.8, which is restated here: The Bedding Store needs to order a shipment of single and double mattresses. A single mattress costs $80 and takes up 16 cu ft of storage space, and a double mattress costs $120 and takes up 36 cu ft of storage space. If the store manager wants to order no more than $6000 worth of mattresses and has at most 1440 cu ft of storage space in its storage area, how many of each mattress can she order?

Recall that the solution set for the shipment of mattresses was represented by the system of inequalities

$$\begin{cases} 80x + 120y \le 6000 \\ 16x + 36y \le 1440 \\ x \ge 0 \\ y \ge 0 \end{cases}$$

The solution set appears in Figure 7.17; we note that this set is a bounded convex set.

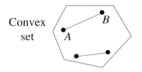

Convex set

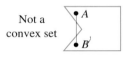

Not a convex set

Figure 7.16

Figure 7.17

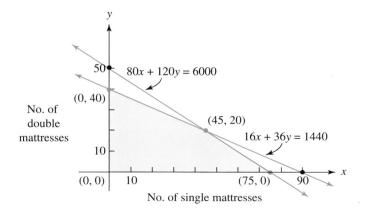

Suppose The Bedding Store makes a profit of $25 on the sale of each single mattress and $50 on the sale of each double mattress. Let's examine the amount of profit the store can make on the sale of all mattresses in this shipment.

Since x is the number of single mattresses and y is the number of double mattresses, the total profit the store would make on any shipment is

$$P = 25x + 50y \text{ dollars}$$

If x and y were unrestricted, then P could be any value. For example, if $x = 50$ and $y = 100$, then $P = \$6250$. But suppose we are subject to the conditions of the example, where we must restrict the values of x and y so that (x, y) must lie in the solution set given before; how does that restrict the values of P? Is there some maximum or minimum value for profit $P = 25x + 50y$ with this restriction? To answer this question, we examine P restricted to the given solution set. We will call this solution set (the shaded area) the set of **feasible solutions** for P.

With (x, y) confined to the feasible solution set, we see that since $(20, 15)$ and $(50, 10)$ lie in the solution set, then P can be $\$1250 = 25(20) + 50(15)$ or P can be $\$1750 = 25(50) + 50(10)$. To determine whether there is some maximum value for P, we start by examining the graph of the equation $P = 25x + 50y$.

Suppose $P = 4000$; that is, suppose $25x + 50y = 4000$. The graph of $25x + 50y = 4000$ is a line, and if we graph $25x + 50y = 4000$ on the same set of coordinate axes as the set of feasible solutions, then we find that the line $25x + 50y = 4000$ does not intersect this set. This means that given the restrictions on (x, y) (lying in the shaded area), P can never be 4000. (The store cannot make $\$4000$ from this shipment.) See Figure 7.18.

Figure 7.18

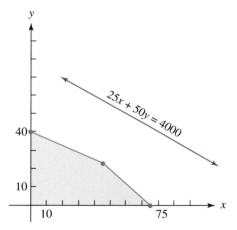

If we allowed P to take on consecutively lower values, such as 3000, 2000, 1500, 500, and −100, and graphed $P = 25x + 50y$ for each value of P, as illustrated in Figure 7.19, we would find that eventually the line would intersect the shaded region of feasible solutions. This means that P can take on those values where the line $P = 25x + 50y$ intersects the shaded region. For example, since $500 = 25x + 50y$ intersects the shaded region, P can be $\$500$.

Looking at Figure 7.19, we see that as P changes, the lines $P = 25x + 50y$ are parallel (since they have the same slope), and, as the line $P = 25x + 50y$ moves toward and then through the shaded area, the values of P get progressively smaller. Moving in this direction, we can see that the first point where the line $P = 25x + 50y$ touches the shaded region will give the maximum value of P for that region, and the last point where the line touches before leaving the shaded region will give the minimum value of P for that region.

Figure 7.19

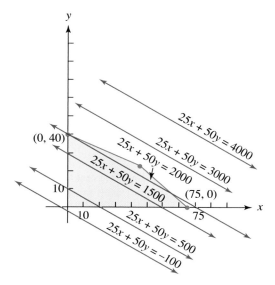

It seems intuitive that as P gets smaller, the *first point the line touches on the shaded region will be a vertex, and the last point will be a vertex.* Hence P has its maximum and minimum values in the region at one or more of the vertices. This will be true for any linear function $F = ax + by + c$ where the set of feasible solutions is a bounded convex set. We state the following without proof.

A Linear Programming Theorem

Given a linear function F,

$$F = ax + by + c$$

where a, b, and c are constants, and x and y are subject to the constraints of a system of linear inequalities,

1. If the set of feasible solutions is a bounded convex set, then F will attain a maximum and a minimum value.
2. If F does have a maximum or a minimum value, it will occur at one of the vertices of the set of feasible solutions.

We call the linear function F the **objective function**. We refer to the system of inequalities as the **constraints**.

Returning to the question of the maximum the store can make on this shipment of mattresses, we identify the function $P = 25x + 50y$ as the objective function, with the following constraints:

$$\begin{cases} 80x + 120y \le 6000 \\ 16x + 36y \le 1440 \\ x \ge 0 \\ y \ge 0 \end{cases}$$

By the preceding theorem, the maximum the store can profit on the sale of the mattresses ordered will occur at a vertex. Therefore, we compute P at each vertex and look for the largest value of P. The vertices are $(0, 0)$, $(0, 40)$, $(45, 20)$, and $(75, 0)$.

If we check the vertex $(0, 0)$, we get $P = 25x + 50y = 25(0) + 50(0) = \$0$.

The vertex $(0, 40)$ yields $P = 25x + 50y = 25(0) + 50(40) = \$2000$.

The vertex $(45, 20)$ yields $P = 25x + 50y = 25(45) + 50(20) = \$2125$.

Maximum value

The vertex $(75, 0)$ yields $P = 25x + 50y = 25\ (75) + 50\ (0) = \$1875$.

Hence the maximum the store could make from this shipment (given the constraints) is \$2125, which will occur if the store orders 45 single mattresses and 20 double mattresses.

Example 1

A bank has \$90 million to invest for business loans and personal loans. A state regulation requires that the bank must invest at least twice as much money in business loans as in personal loans. If the bank earns a yearly profit of 7% on business loans and a yearly profit of 8% on personal loans, what is the maximum profit the bank can make, given its constraints?

Solution

Let x = the amount of dollars (in millions) lent by the bank for business loans and let y = the amount of dollars (in millions) lent by the bank for personal loans. The yearly profit earned on loans can be expressed as a linear function of x and y, as follows:

$$P = 0.07x + 0.08y \qquad \text{This is the objective function.}$$

This objective function, P, is what we want to maximize. The constraints are as follows:

Note that under the given constraints, the bank could invest \$65 million in business loans and \$10 million in personal loans (the ordered pair $(65, 10)$ satisfies the constraints). The profit function would yield a profit of $P = 0.07x + 0.08y = 0.07(65) + 0.08(10) = 5.35$ million dollars.

$$\begin{cases} x + y \le 90 & \text{Represents the total amount the bank invests (in millions of dollars)} \\ 2y \le x & \text{Represents the state regulation (at least twice as much invested in business as in personal loans)} \\ x \ge 0 & \text{Since the bank cannot invest a negative amount of money} \\ y \ge 0 \end{cases}$$

Figure 7.20

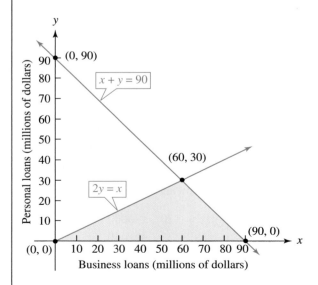

The graph of the constraints appears in Figure 7.20. Note the vertices are $(0, 0)$, $(60, 30)$, and $(90, 0)$. We note that the shaded area is a bounded convex set; therefore, by the linear programming theorem, there is a maximum, and this maximum occurs

The point (60, 30) is found by solving the system

$$\begin{cases} x + y = 90 \\ x = 2y \end{cases}$$

at a vertex. So we only have to check the vertices to determine how much of each type of loan will maximum the bank's profit. If we check the vertex (0, 0), we get

$$P = 0.07x + 0.08y = 0.07(0) + 0.08(0) = 0 \text{ dollars}$$

If we check the vertex (60, 30), we get

$$P = 0.07(60) + 0.08(30) = 6.6 \text{ million dollars}$$

If we check the vertex (90, 0), we get

$$P = 0.07(90) + 0.08(0) = 6.3 \text{ million dollars}$$

Hence the maximum profit of $6.6 million occurs if the bank can make $60 million in business loans and $30 million in personal loans.

7.9 Exercises

1. What are the maximum and minimum values of $f = 3x + 2y$ subject to the following constraints?

$$\begin{cases} x + 2y \le 6 \\ y \le x + 1 \\ x \ge 0 \\ y \ge 0 \end{cases}$$

2. What are the maximum and minimum values of $f = 3x + 9y$ subject to the same constraints as Exercise 1?

3. What are the maximum and minimum values of $f = 0.3x - y$ subject to the following constraints?

$$\begin{cases} x - 2y \le 4 \\ y + x \le 1 \\ x \ge 0 \end{cases}$$

4. What are the maximum and minimum values of $f = 3x - 0.5y$ subject to the constraints of Exercise 3?

5. What are the maximum and minimum values of $f = 2x + 5y - 1$ subject to the following constraints?

$$\begin{cases} 2x + 3y \ge 6 \\ x + y \le 3 \\ x \ge 0 \end{cases}$$

6. What are the maximum and minimum values of $f = 8x - 2y + 4$ subject to the constraints of Exercise 5?

7. A bank has a total of $75 million to invest for high-risk and conservative investments. A federal regulation requires that the bank can put no more than 30% of its total investment funds into high-risk ventures. If the rate of profit for conservative investments aver-

ages 10% yearly and the rate of profit for high-risk investments is 14% yearly, what is the maximum profit the bank can make given its constraints?

8. A computer store needs to order two types of computers: model A and model B. Model A costs $400 for the store to buy wholesale, and the wholesale price of model B is $200. Model A takes up 12 cu ft of space and model B takes up 16 cu ft of space. The store owner decides that he wants to buy at most $80,000 worth of the models wholesale, but he has only 6000 cu ft of storage space. If his profit is $300 on each model A he sells and $350 on each model B, how many of each model should he buy to maximize his profit?

9. A computer store needs to order two types of printers: model X and model Y. Model X costs $400 for the store to buy wholesale, and the wholesale price of model Y is $500. Model X takes up 16 cu ft of space and model Y takes up 12 cu ft of space. The store owner decides that she wants to buy at most $50,000 worth of the models wholesale, but she has only 1600 cu ft of storage space. Her profit is $90 on each model X she sells and $100 on each model Y she sells. How many of each model should she order to maximize her profit?

10. Farmer Jack has 100 acres available to plant crops. Crop A costs $8 an acre for seed and $80 an acre for labor. Crop B costs $10 an acre for seed and $60 an acre for labor. He does not want to spend more than $900 total for seed or more than $7000 for labor. If his profits are $120 per acre for crop A and $140 per acre for crop B, how many of each crop should he plant to maximize his profits?

11. An electronics company makes two types of calculators, a scientific model and a graphing model. The scientific model requires $8 in materials and takes 1 hour to assemble and package. The graphing model requires $25 in materials and takes $\frac{1}{2}$ hour to assemble and package. The company decides to spend a maximum of $15,000 on material and to allot a maximum of 1000 hours for packing and assembly. Under these constraints, if the electronics company profits $10 on each scientific model and $30 on each graphing model, how many of each should they produce if they want the maximum profit?

12. Under the constraints of Exercise 11, if the electronics company profits $8 on each scientific model and $40 on each graphing model, how many of each should they produce if they want the maximum profit?

13. Maria wants to eat a more balanced breakfast. She reads the sides of two cereal boxes and finds out that 1 ounce of cereal X contains 8 grams of carbohydrates and 0.25 gram of sodium, whereas 1 ounce of cereal Y contains 12 grams of carbohydrates and 0.34 gram of sodium. She wants to create a mixture of cereals that contains at least 24 grams of carbohydrates and no more than 1 gram of sodium. Under these constraints, if it costs Maria 25¢ an ounce for cereal X and 30¢ an ounce for cereal Y, what combination of cereals would be the least expensive?

14. Under the constraints of Exercise 13, if it costs Maria 25¢ an ounce for cereal X and 40¢ an ounce for cereal Y, what combination of cereals would be the least expensive?

15. A lawyer handles both divorce cases and malpractice suits. The average divorce case takes 12 hours of her time and 20 hours of her assistants' time. The average malpractice suit takes 22 hours of her time and 18 hours of her assistants' time. The lawyer decided that she can put in no more than 60 hours per week, and her assistants can put in no more than 80 hours a week. Under these constraints, if the lawyer averages $6000 on divorce cases and $8000 on malpractice suits, how many of each type of case should she take on weekly to maximize her income?

16. An orthodontist uses two types of appliances (A and B), both shown to be effective for overbites. Over a year, each appliance A requires 6 hours of the doctor's labor and 14 hours of the dental assistants' labor; each appliance B requires 4 hours of the doctor's labor and 16 hours of the dental assistants' labor. Over the year the doctor cannot put in more than 1200 hours of labor for appliances, and the assistants cannot put in more than 3200 hours. Under these constraints, if the orthodontist makes a profit of $2000 on appliance A and $1500 on appliance B, how many of each appliance should he use in order to maximize profit?

17. A pharmaceutical company produces two types of drugs, Zomine X and Zomine Y, which help to reduce the effects of migraine headaches. But each drug contains substances that produce negative side effects as well. Of each gram of Zomine X, 5 mg will directly relieve migraine pain; but 2 mg will produce a Type I negative side effect and 4 mg will produce a Type II negative side effect. On the other hand, of each gram of Zomine Y, 2 mg will provide pain relief; but 1 mg produces a Type I negative side effect and 1 mg will produce a Type II negative side effect. The body will not tolerate more than 6 mg daily of anything producing a Type I effect, nor will it tolerate more than 8 mg daily of anything producing a Type II side effect. For the drug to have any effect, the user must take at least 1 gram of either or both Zomines. How many grams of each type of Zomine should be administered daily to maximize the pain relief?

Chapter 7　*Summary*

After completing this chapter, you should be able to:

1. Solve linear systems by the elimination, Gaussian elimination, and substitution methods. (Sections 7.1 and 7.2)
The point of the elimination method is to combine equations to eliminate variables (and equations) until we have a single equation in one variable.
For example:

Solve the following system of equations.

$$\begin{cases} x - 2y + 3z = 13 & \textbf{(1)} \\ 2x + y - z = -5 & \textbf{(2)} \\ x - 2y + z = 7 & \textbf{(3)} \end{cases}$$

Solution:

The first step is to eliminate a variable and arrive at two equations in two variables. Look at equations (2) and (3). We will eliminate z.

$$2x + y - z = -5$$
$$\underline{x - 2y + z = \quad 7} \qquad \text{Add equations (2) and (3).}$$
$$3x - y \quad = \quad 2 \qquad \text{Call this equation (4).}$$

Now we have to find another equation in the same two variables, x and y. We will use equations (1) and (2):

$$\begin{array}{l} x - 2y + 3z = 13 \\ 2x + y - z = -5 \end{array} \quad \xrightarrow[\text{Multiply by 3}]{\text{As is}} \quad \begin{array}{l} x - 2y + 3z = \quad 13 \\ \underline{6x + 3y - 3z = -15} \\ 7x + y \quad = -2 \end{array} \quad \begin{array}{l}\text{Call this} \\ \text{equation (5).}\end{array}$$

$$\begin{cases} 3x - y = \quad 2 \\ \underline{7x + y = -2} \end{cases} \qquad \text{Add equations (4) and (5).}$$
$$\quad 10x \qquad = \quad 0 \qquad \text{Which yields } x = 0$$

Now substitute $x = 0$ back into equation (4) and solve for y:

$$3x \quad - y = 2$$
$$3(0) - y = 2 \quad \Rightarrow \quad y = -2$$

Finally, substitute $x = 0$ and $y = -2$ back into equation (1) and solve for z:

$$x - 2y \quad + 3z = 13$$
$$0 - 2(-2) + 3z = 13 \quad \Rightarrow \quad z = 3$$

Hence our solution is $(0, -2, 3)$. The student should check this solution.

2. Solve linear systems using augmented matrices, by either Gaussian elimination or the Gauss–Jordan method. (Section 7.3)
   Using matrix methods, we need only focus on the coefficients of the variables in a system of equations. We use row operations to transform the augmented matrix to triangular or reduced echelon form, which yields either the solution or a simpler system to solve.
   For example:
   Solve the following system using Gauss–Jordan elimination.

$$\begin{cases} x + y + 4z = 3 \\ 4x + 2y + 4z = 8 \\ 2x + 3y + 12z = 8 \end{cases}$$

Solution:

First, we write the system's associated matrix:

$$\begin{cases} x + y + 4z = 3 \\ 4x + 2y + 4z = 8 \\ 2x + 3y + 12z = 8 \end{cases} \Rightarrow \left[\begin{array}{ccc|c} 1 & 1 & 4 & 3 \\ 4 & 2 & 4 & 8 \\ 2 & 3 & 12 & 8 \end{array} \right]$$

Now use the elementary row operations to transform the matrix into reduced echelon form.

(a) Since 1 is in the upper left-hand corner, begin by sweeping out the rest of the first column.

$$\begin{bmatrix} 1 & 1 & 4 & | & 3 \\ 4 & 2 & 4 & | & 8 \\ 2 & 3 & 12 & | & 8 \end{bmatrix} \quad \begin{matrix} -4R_1 + R_2 \rightarrow R_2 \\ -2R_1 + R_3 \rightarrow R_3 \end{matrix} \quad \begin{bmatrix} 1 & 1 & 4 & | & 3 \\ 0 & -2 & -12 & | & -4 \\ 0 & 1 & 4 & | & 2 \end{bmatrix}$$

(b) Next, we want to get 1 in the middle of the main diagonal (in row 2, column 2):

$$\begin{bmatrix} 1 & 1 & 4 & | & 3 \\ 0 & -2 & -12 & | & -4 \\ 0 & 1 & 4 & | & 2 \end{bmatrix} \quad -\tfrac{1}{2}R_2 \rightarrow R_2 \quad \begin{bmatrix} 1 & 1 & 4 & | & 3 \\ 0 & 1 & 6 & | & 2 \\ 0 & 1 & 4 & | & 2 \end{bmatrix}$$

(c) Now we sweep out the rest of the second column:

$$\begin{bmatrix} 1 & 1 & 4 & | & 3 \\ 0 & 1 & 6 & | & 2 \\ 0 & 1 & 4 & | & 2 \end{bmatrix} \quad \begin{matrix} -R_2 + R_1 \rightarrow R_1 \\ -R_2 + R_3 \rightarrow R_3 \end{matrix} \quad \begin{bmatrix} 1 & 0 & -2 & | & 1 \\ 0 & 1 & 6 & | & 2 \\ 0 & 0 & -2 & | & 0 \end{bmatrix}$$

(d) Next, we want to get 1 in the lower right entry of the main diagonal:

$$\begin{bmatrix} 1 & 0 & -2 & | & 1 \\ 0 & 1 & 6 & | & 2 \\ 0 & 0 & -2 & | & 0 \end{bmatrix} \quad -\tfrac{1}{2}R_3 \rightarrow R_3 \quad \begin{bmatrix} 1 & 0 & -2 & | & 1 \\ 0 & 1 & 6 & | & 2 \\ 0 & 0 & 1 & | & 0 \end{bmatrix}$$

(e) Finally, sweep out the rest of the last column:

$$\begin{bmatrix} 1 & 0 & -2 & | & 1 \\ 0 & 1 & 6 & | & 2 \\ 0 & 0 & 1 & | & 0 \end{bmatrix} \quad \begin{matrix} 2R_3 + R_1 \rightarrow R_1 \\ -6R_3 + R_2 \rightarrow R_2 \end{matrix} \quad \begin{bmatrix} 1 & 0 & 0 & | & 1 \\ 0 & 1 & 0 & | & 2 \\ 0 & 0 & 1 & | & 0 \end{bmatrix}$$

Read off the solution to get $x = 1$, $y = 2$, $z = 0$.

3. Add and multiply matrices, including scalar multiplication. (Section 7.4)

(a) We can add two matrices of the same size by adding their corresponding elements. To multiply a matrix by a constant, called scalar multiplication, multiply each entry by the constant.

For example:

Given matrices

$$A = \begin{bmatrix} 1 & 4 & 2 \\ -1 & 0 & 3 \end{bmatrix} \quad B = \begin{bmatrix} 2 & 1 \\ -1 & 3 \\ 0 & 5 \end{bmatrix} \quad C = \begin{bmatrix} 1 & -1 \\ 4 & 0 \\ 3 & 1 \end{bmatrix}$$

to compute $2B - C$, notice that addition makes sense since B and C are the same size, 3×2.

$$2B - C = 2\begin{bmatrix} 2 & 1 \\ -1 & 3 \\ 0 & 5 \end{bmatrix} + (-1)\begin{bmatrix} 1 & -1 \\ 4 & 0 \\ 3 & 1 \end{bmatrix} \qquad \text{Find } 2B \text{ and } (-1)C.$$

$$= \begin{bmatrix} 4 & 2 \\ -2 & 6 \\ 0 & 10 \end{bmatrix} + \begin{bmatrix} -1 & 1 \\ -4 & 0 \\ -3 & -1 \end{bmatrix} = \begin{bmatrix} 3 & 3 \\ -6 & 6 \\ -3 & 9 \end{bmatrix}$$

(b) Two matrices can be multiplied if the number of rows of the matrix on the left is equal to the number of columns of the matrix on the right.

For example:

To compute the product AB using the matrices given above, we first note that, because A is a 2×3 matrix and B is a 3×2 matrix, AB makes sense and is a 2×2 matrix.

$$AB = \begin{bmatrix} 1 & 4 & 2 \\ -1 & 0 & 3 \end{bmatrix} \begin{bmatrix} 2 & 1 \\ -1 & 3 \\ 0 & 5 \end{bmatrix}$$

$$= \begin{bmatrix} 1(2) + 4(-1) + 2(0) & 1(1) + 4(3) + 2(5) \\ -1(2) + 0(-1) + 3(0) & -1(1) + 0(3) + 3(5) \end{bmatrix}$$

$$= \begin{bmatrix} -2 & 23 \\ -2 & 14 \end{bmatrix}$$

4. Find the inverse of a matrix and use it to solve a system of linear equations. (Section 7.5)

By using the inverse of a matrix, we can approach solving a system as though we were solving a simple linear equation. That is, if $AX = K$, then $X = A^{-1}K$.

For example:

Using a matrix inverse, find the solution to the following system of equations:

$$\begin{cases} x - 2y = 4 \\ x + 5y = -3 \end{cases}$$

Solution:

We identify matrices A, X, and K:

$$A = \begin{bmatrix} 1 & -2 \\ 1 & 5 \end{bmatrix} \quad X = \begin{bmatrix} x \\ y \end{bmatrix} \quad K = \begin{bmatrix} 4 \\ -3 \end{bmatrix}$$

Then $AX = K$. We find A^{-1} to be

$$A^{-1} = \begin{bmatrix} \frac{5}{7} & \frac{2}{7} \\ -\frac{1}{7} & \frac{1}{7} \end{bmatrix} = \frac{1}{7} \begin{bmatrix} 5 & 2 \\ -1 & 1 \end{bmatrix}$$

Now we can solve for X:

$$X = A^{-1}K$$

$$= \frac{1}{7} \begin{bmatrix} 5 & 2 \\ -1 & 1 \end{bmatrix} \begin{bmatrix} 4 \\ -3 \end{bmatrix} = \begin{bmatrix} 2 \\ -1 \end{bmatrix}$$

Hence $X = \begin{bmatrix} x \\ y \end{bmatrix} = \begin{bmatrix} 2 \\ -1 \end{bmatrix}$, which means $x = 2$ and $y = -1$.

5. Evaluate determinants. (Sections 7.6 and 7.7)

A 2×2 determinant is defined as

$$\begin{vmatrix} a & b \\ c & d \end{vmatrix} = ad - bc$$

A 3×3 determinant is evaluated by expansion of minors: We can expand down any column or across any row provided we prefix each entry with the appropriate sign from the sign array.

For example:

We evaluate determinants as follows:

(a) $\begin{vmatrix} 5 & 1 \\ 3 & -2 \end{vmatrix} = (5)(-2) - (3)(1) = -13$

(b) $\begin{vmatrix} 4 & 1 & -2 \\ -3 & 2 & 3 \\ 2 & 0 & -1 \end{vmatrix}$ We expand across the third row to get

$$= 2\begin{vmatrix} 1 & -2 \\ 2 & 3 \end{vmatrix} - 0\begin{vmatrix} 4 & -2 \\ -3 & 3 \end{vmatrix} + (-1)\begin{vmatrix} 4 & 1 \\ -3 & 2 \end{vmatrix}$$

$$= 2[3 - (-4)] - 0 + (-1)[8 - (-3)]$$

$$= 2(7) - 11 = 3$$

6. Solve systems of equations using Cramer's rule. (Sections 7.6 and 7.7)

We can solve $n \times n$ linear systems of equations by using Cramer's rule. Cramer's rule utilizes determinants of the system's coefficients.

For example:

Solve the following system of equations using Cramer's rule:

$$\begin{cases} 3x - y = -4 \\ 2x + y - 2z = 3 \\ y - 6z = 1 \end{cases}$$

Solution:

First, we compute D, because if $D = 0$, then there is no unique solution. We compute D by expanding across the third row:

$$D = \begin{vmatrix} 3 & -1 & 0 \\ 2 & 1 & -2 \\ 0 & 1 & -6 \end{vmatrix}$$

$$= 0\begin{vmatrix} -1 & 0 \\ 1 & -2 \end{vmatrix} - 1\begin{vmatrix} 3 & 0 \\ 2 & -2 \end{vmatrix} + (-6)\begin{vmatrix} 3 & -1 \\ 2 & 1 \end{vmatrix}$$

$$= 0 - 1(-6) - 6(3 + 2) = -24$$

Next, we compute D_x by expanding down the last column:

$$D_x = \begin{vmatrix} -4 & -1 & 0 \\ 3 & 1 & -2 \\ 1 & 1 & -6 \end{vmatrix}$$

$$= 0\begin{vmatrix} 3 & 1 \\ 1 & 1 \end{vmatrix} - (-2)\begin{vmatrix} -4 & -1 \\ 1 & 1 \end{vmatrix} + (-6)\begin{vmatrix} -4 & -1 \\ 3 & 1 \end{vmatrix}$$

$$= 0 + 2(-4 + 1) - 6(-4 + 3) = 0$$

We compute D_y by expanding down the first column:

$$D_y = \begin{vmatrix} 3 & -4 & 0 \\ 2 & 3 & -2 \\ 0 & 1 & -6 \end{vmatrix}$$

$$= 3 \begin{vmatrix} 3 & -2 \\ 1 & -6 \end{vmatrix} - 2 \begin{vmatrix} -4 & 0 \\ 1 & -6 \end{vmatrix} + 0 \begin{vmatrix} -4 & 0 \\ 3 & -2 \end{vmatrix}$$

$$= 3(-18 + 2) - 2(24) + 0 = -96$$

We compute D_z by expanding across the third row:

$$D_z = \begin{vmatrix} 3 & -1 & -4 \\ 2 & 1 & 3 \\ 0 & 1 & 1 \end{vmatrix}$$

$$= 0 \begin{vmatrix} -1 & -4 \\ 1 & 3 \end{vmatrix} - 1 \begin{vmatrix} 3 & -4 \\ 2 & 3 \end{vmatrix} + 1 \begin{vmatrix} 3 & -1 \\ 2 & 1 \end{vmatrix}$$

$$= 0 - 1(9 + 8) + 1(3 + 2) = -12$$

Hence $x = \dfrac{D_x}{D} = \dfrac{0}{-24} = 0$, $\quad y = \dfrac{D_y}{D} = \dfrac{-96}{-24} = 4$, and $z = \dfrac{D_z}{D} = \dfrac{-12}{-24} = \dfrac{1}{2}$.

7. Solve a variety of verbal problems that give rise to systems of linear equations. (Sections 7.1–7.7)

8. Sketch the solution set of a system of linear inequalities. (Section 7.8)

On the same coordinate plane, we graph each linear inequality as a half-plane. The intersection of the half-planes is the solution, which we represent as the darkest or crosshatched region. You can determine the vertices of the region by solving pairs of simultaneous equations.

For example:
Find the solution set for the following system of linear inequalities.

$$\begin{cases} 3x - 2y \le 12 \\ 2x + y > -2 \\ y \le 2 \end{cases}$$

Solution:
Graph each inequality. That is, graph the line and determine which half-plane to shade. The solution is the region where all three half-planes intersect, which is represented by the darkest shaded region in Figure 7.21.

Figure 7.21

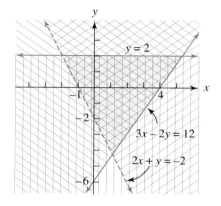

9. Use the linear programming theorem to solve applications. (Section 7.9)

Chapter 7   *Review Exercises*

In Exercises 1–12, solve the system of equations.

1. $\begin{cases} 2x + y = 9 \\ 3x - y = 11 \end{cases}$

2. $\begin{cases} -x + 2y = 8 \\ x - y = -5 \end{cases}$

3. $\begin{cases} 3x - 2y = 10 \\ 2x + y = -5 \end{cases}$

4. $\begin{cases} 5x + 2y = 34 \\ 3x + y = 20 \end{cases}$

5. $\begin{cases} 5x - 2y = 9 \\ x - 5y = -12 \end{cases}$

6. $\begin{cases} x = 3y - 1 \\ 2x - 5y = 1 \end{cases}$

7. $\begin{cases} x + y + z = 6 \\ 3x + y - z = 2 \\ 2x + y - z = 1 \end{cases}$

8. $\begin{cases} x - 2y + z = 6 \\ 3x - y + z = 14 \\ x + 2y - z = 2 \end{cases}$

9. $\begin{cases} x + y - z = 2 \\ 2x + 2y + 2z = 10 \\ x - y + z = 5 \end{cases}$

10. $\begin{cases} 3x + y + 2z = 1 \\ x - 2y - z = 1 \\ x + y + z = 1 \end{cases}$

11. $\begin{cases} 0.2a + 0.1b - c = 0.7 \\ a - 0.5b + 2c = 0.5 \\ a + 2b + c = 8 \end{cases}$

12. $\begin{cases} x - y = 2z + y \\ x - 3y = y - 1 \\ 2x - 3y = 3z \end{cases}$

In Exercises 13–16, put the augmented matrix into reduced echelon form.

13. $\left[\begin{array}{cc|c} -1 & 2 & 2 \\ 4 & 6 & 6 \end{array}\right]$

14. $\left[\begin{array}{cc|c} 3 & 0 & 6 \\ -6 & 1 & 3 \end{array}\right]$

15. $\left[\begin{array}{ccc|c} 1 & 1 & 2 & 2 \\ 2 & 0 & 2 & 0 \\ 3 & -1 & 1 & 1 \end{array}\right]$

16. $\left[\begin{array}{ccc|c} 1 & -1 & 0 & 5 \\ -2 & 2 & 1 & -1 \\ 4 & -1 & 0 & 2 \end{array}\right]$

In Exercises 17–20, solve the system of equations by Gaussian elimination.

17. $\begin{cases} x + y = 5 \\ 3x - y = 8 \end{cases}$

18. $\begin{cases} x + 2y = 9 \\ x - y = -6 \end{cases}$

19. $\begin{cases} x + y + z = -3 \\ 3x + y - z = 11 \\ 2x + y - z = 9 \end{cases}$

20. $\begin{cases} x + y - z = 1 \\ 2x + 2y + 2z = 1 \\ x - y + z = 3 \end{cases}$

In Exercises 21–24, solve the system of equations by Gauss–Jordan elimination.

21. $\begin{cases} x - 2y = 4 \\ 2x + y = 13 \end{cases}$

22. $\begin{cases} 3x + 2y = 3 \\ 3x + y = 0 \end{cases}$

23. $\begin{cases} x - 3y + z = 0 \\ x - y + z = 0 \\ x + 2y - z = 2 \end{cases}$

24. $\begin{cases} x + y + 2z = 7 \\ x - 2y - z = 1 \\ x + y + z = 4 \end{cases}$

In Exercises 25–34, perform the operations if possible.

25. $\begin{bmatrix} 3 & -2 \\ 5 & -8 \end{bmatrix} + \begin{bmatrix} 7 & -1 \\ 4 & 0 \end{bmatrix}$

26. $-2\begin{bmatrix} 2 & -3 & 8 \\ 5 & -1 & 0 \\ -2 & 7 & 18 \end{bmatrix}$

27. $2\begin{bmatrix} 1 & -3 \\ 5 & 2 \end{bmatrix} - \begin{bmatrix} 2 & 5 \\ 3 & 0 \end{bmatrix}$

28. $\begin{bmatrix} 5 & 3 & 6 \\ 3 & 9 & 0 \\ 2 & 12 & 6 \end{bmatrix} + \begin{bmatrix} 2 & -3 & 1 \\ 5 & 0 & 4 \end{bmatrix}$

29. $\begin{bmatrix} 1 & 0 & 2 & 1 \end{bmatrix} \begin{bmatrix} 1 \\ 0 \\ 2 \\ 1 \end{bmatrix}$

30. $\begin{bmatrix} 1 \\ 0 \\ 2 \\ 1 \end{bmatrix} \begin{bmatrix} 2 & 3 & 4 & 0 \\ 1 & 0 & 2 & 1 \end{bmatrix}$

31. $\begin{bmatrix} 1 & 0 \\ 0 & 1 \end{bmatrix} \begin{bmatrix} 5 & 8 \\ -4 & 1 \end{bmatrix}$

32. $\begin{bmatrix} -1 & 0 & 3 \\ 2 & 2 & 5 \end{bmatrix} \begin{bmatrix} 1 & 0 \\ -4 & 3 \\ 2 & -1 \end{bmatrix}$

33. $\begin{bmatrix} 1 & -1 & 0 \\ 2 & 3 & 1 \\ 5 & 1 & 3 \end{bmatrix} \begin{bmatrix} 1 & -2 & 1 \\ 2 & 0 & 1 \\ 3 & -1 & 4 \end{bmatrix}$

34. $\begin{bmatrix} 1 & -2 & 1 \\ 2 & 0 & 1 \\ 3 & -1 & 4 \end{bmatrix} \begin{bmatrix} 1 & -1 & 0 \\ 2 & 3 & 1 \\ 5 & 1 & 3 \end{bmatrix}$

In Exercises 35–36, find the inverse of the matrix if it exists.

35. $\begin{bmatrix} 1 & 3 \\ 2 & 2 \end{bmatrix}$

36. $\begin{bmatrix} 1 & -1 & 1 \\ 2 & -2 & 1 \\ 1 & 1 & 0 \end{bmatrix}$

In Exercises 37–40, solve the given system using matrix inverses.

37. $\begin{cases} x - 2y = -5 \\ 2x + y = 5 \end{cases}$

38. $\begin{cases} 3x + 2y = -4 \\ 3x + y = -2 \end{cases}$

39. $\begin{cases} x - y + z = 5 \\ x - 2y + z = 7 \\ x + 2y - z = -3 \end{cases}$

40. $\begin{cases} x - y + 2z = 1 \\ x - 2y + z = 2 \\ x - y + z = 2 \end{cases}$

In Exercises 41–44, evaluate the determinant.

41. $\begin{vmatrix} 1 & 0 \\ 5 & -2 \end{vmatrix}$

42. $\begin{vmatrix} 3 & 3 \\ 1 & 2 \end{vmatrix}$

43. $\begin{vmatrix} 1 & 0 & 2 \\ 5 & -2 & 1 \\ 3 & -1 & 0 \end{vmatrix}$

44. $\begin{vmatrix} 1 & 0 & 2 \\ 3 & -1 & 1 \\ 2 & -1 & 5 \end{vmatrix}$

In Exercises 45–48, solve the system of equations using Cramer's rule.

45. $\begin{cases} x - 2y = -5 \\ 3x + 5y = 7 \end{cases}$

46. $\begin{cases} x + 2y = 5 \\ 6x - 2y = 23 \end{cases}$

47. $\begin{cases} x + 2y + z = 5 \\ x + 4y - z = -1 \\ x - 2y + z = 5 \end{cases}$

48. $\begin{cases} x - y + z = 2 \\ 2x - 2y - z = -11 \\ 3x - 2y + z = -2 \end{cases}$

49. A manufacturer produces two types of cameras. The more expensive model requires 2 hours to manufacture and 20 minutes to assemble. The less expensive model requires 1 hour to manufacture and 15 minutes to assemble. If the company can allocate 280 hours for manufacture and 60 hours for assembly, how many of each type can be produced?

50. The Physics Department hires graders and tutors. For February the department budgets $855 for 20 hours of grading and 40 hours for tutoring. The next month the department budgets $630 for 40 hours of grading and 20 hours of tutoring. How much does the department pay for each hour of tutoring and for each hour of grading?

51. Brian invests $10,000, part at 6% and part at 8%, earning a yearly interest of $750 on the entire amount. How much is invested at each rate?

52. A lawyer charges $110 an hour for her time and $50 an hour for her assistant's time. A client received a bill from the lawyer that included a $2650 charge for their combined time. If the assistant worked 5 hours longer than the lawyer, how much time did the lawyer and her assistant spend on the client's case?

53. Josh has $40,000 in three investments. He has a bank account paying 5%, a certificate of deposit paying 6.5%, and a stock yielding 9%. His annual interest from the three investments is $2900. If he invested $6000 more in his stock than in his certificate of deposit, how much did he invest at each rate?

54. A radio manufacturer produces three models, model A, model B, and model C. The company knows how much time is needed for production and assembly and the cost of raw materials for each model. This information is found in the accompanying table. If the company has allocated 90 hours for product, 87 hours for assembly, and $895 for raw materials, how many of each model can the company produce if it wants to use all the time allocated for each phase of the process and all the money allocated for raw materials?

Model	Production hours	Assembly hours	Cost of raw materials
A	0.8	0.9	$9.80
B	0.7	0.5	$6.20
C	0.4	0.8	$4.20

In Exercises 55–58, sketch the solution set of the system of inequalities.

55. $\begin{cases} 2x + y \leq 12 \\ 3x + y \geq 12 \end{cases}$

56. $\begin{cases} -x + 2y \leq 8 \\ x - y < -3 \end{cases}$

57. $\begin{cases} y < 3x \\ 4x - y \leq 5 \\ y \leq 5 \end{cases}$

58. $\begin{cases} x + 3y > -1 \\ 3x - 6y > -3 \\ x \leq 4 \end{cases}$

In Exercises 59–62, write a system of inequalities to describe the conditions of the problem and sketch the solution set of the system.

59. An appliance store needs to order a shipment of VCRs and TVs. A VCR costs $200 and takes up 2 cu ft of storage space, and a TV costs $250 and takes up 6 cu ft of storage space. If the store wants to order at least $10,000 worth of TVs and VCRs but has at most 500 cu ft of storage space in its storage area, how many of each can it order?

60. Irena has at most $20,000 available for investment. She can invest in two types of certificates of deposit: a 6-month certificate, which yields a 5% yearly return on investment, and a 1-year certificate, which yields a 6.5% yearly return. She decides that she wants to put at least twice as much in the 1-year certificate as in the 6-month certificate and she wants at least $2000 but no more than $4000 invested in the 6-month certificate. How much could she invest in each certificate?

61. A judge is working on the weekly schedule of hearings in her courtroom, where she hears both minor criminal and civil complaints. She finds that the average minor civil complaint requires 1 hour to settle; the average minor criminal complaint requires 45 minutes. Next week she wants to schedule at least 20 hours of hearings but no more than 35 hours. She prefers to adjudicate more criminal complaints than civil complaints. How many of each type of case could she hear next week?

62. An orthodontist uses two types of appliances (A and B), both shown to be effective for overbites. Over a period of a year, each appliance A requires 6 hours of the doctor's labor and 14 hours of the dental assistants' labor; each appliance B requires 4 hours of the

doctor's labor and 16 hours of the dental assistants' labor. If the doctor wants to put in no more than 800 hours for appliances over the year and his assistants must put in at least 1000 hours labor on appliances, how many of each appliance could be used?

63. A bank has a total of $80 million to invest for both high-risk and conservative investments. A federal regulation requires that the bank have no more than 25% of its total investment funds in high-risk ventures. If the rate of return for conservative investments averages 8% yearly and the rate of return for high-risk investments is 14% yearly, what is the maximum profit the bank can make, given its constraints?

64. The owner of a computer store needs to order two types of printers, model X and model Y. Model X costs $300 wholesale, and the wholesale price of model Y is $400. Model X takes up 12 cu ft of space and model Y takes up 6 cu ft of space. The store owner decides that she wants to buy at least $48,000 worth of the models but has only 1080 cu ft of storage space. If her profit is $225 on each model X she sells and $100 on each model Y she sells, how many of each model should she buy to maximize profit?

Chapter 7   *Practice Test*

---

1. Solve the system of equations.

(a) $\begin{cases} 2x + y = 12 \\ 3x - y = 8 \end{cases}$

(b) $\begin{cases} -x + 2y = -4 \\ 5x + 3y = 7 \end{cases}$

(c) $\begin{cases} x - y - z = 4 \\ x + 2y + 2z = -2 \\ x - y + z = 3 \end{cases}$

2. Solve the system of equations using augmented matrices.

$\begin{cases} x + y - 2z = -7 \\ 2x + 2y + 2z = 10 \\ -x + y + z = -1 \end{cases}$

3. Perform the operations.

(a) $3\begin{bmatrix} 1 & 0 \\ -2 & 5 \end{bmatrix} - \begin{bmatrix} 2 & 5 \\ 1 & 3 \end{bmatrix}$

(b) $\begin{bmatrix} 3 & 5 & 1 \\ -4 & 6 & 0 \\ 2 & 1 & -1 \end{bmatrix}\begin{bmatrix} 1 & 0 & 4 \\ 0 & 1 & -1 \\ 2 & 1 & -2 \end{bmatrix}$

4. Find the inverse of the matrix if it exists.

$\begin{bmatrix} 2 & 8 \\ 1 & -4 \end{bmatrix}$

5. Solve the given system using matrix inverses.

$\begin{cases} x - 2y = 4 \\ 2x + y = 13 \end{cases}$

6. Evaluate the determinants.

(a) $\begin{vmatrix} 2 & 5 \\ -8 & 1 \end{vmatrix}$ (b) $\begin{vmatrix} 3 & 0 & 2 \\ 2 & -1 & 4 \\ 1 & 2 & -3 \end{vmatrix}$

7. Solve the system of equations using Cramer's rule.

$\begin{cases} x - 2y + 3z = 6 \\ -x + y - 2z = -4 \\ 2x + y - z = -1 \end{cases}$

8. A car rental agency charges a flat fee plus a mileage rate for a 1-day rental. If the charge for a 1-day rental with 120 miles is $65.70 and the charge for a 1-day rental with 80 miles is $51.30, find the flat fee and the charge per mile.

9. A horticulturist wants to mix three types of fertilizers, which contain 10%, 25%, and 30% nitrogen, respectively. If the final mixture should be 2000 lb of 21% nitrogen, all three types are used, and there is twice as much of the 30% type as of the 25% type, how much of each type is in the final mixture?

10. Sketch the solution set of the system of inequalities.

$$\begin{cases} 2y < 3x - 12 \\ 4x - y \le 16 \end{cases}$$

11. An appliance store needs to order two types of microwave ovens, model X and model Y. Model X costs $100, wholesale, and the wholesale price of

model Y is $200. Model X takes up 4 cu ft of space and model Y takes up 3 cu ft of space. The store owner decides that he wants to buy at most $10,000 worth of the models but has only 300 cu ft of storage space. Write a system of inequalities to describe this situation and sketch the solution set of the system.

12. The appliance store in Exercise 11 makes a profit of $60 on each model X microwave sold and a profit of $50 on each model Y microwave sold. How many of each type should be ordered to maximize the microwave profits within the given constraints? What is the maximum possible profit?

8

Conic Sections and Nonlinear Systems

In Section 3.1, we discussed graphs of equations of the form $Ax + By = C$: first-degree equations in two variables. In this chapter we discuss the graphs of the *general second-degree equation in two variables*,

$$Ax^2 + Bxy + Cy^2 + Dx + Ey + F = 0 \qquad (1)$$

If there are any ordered pairs satisfying such an equation, it can be shown that (with some exceptions) the graph will be one of the following four figures: a circle, a parabola, an ellipse, or a hyperbola. These figures are called **conic sections**, because they describe the intersection of a plane and a double-napped cone as illustrated below.

In this text we concentrate on the cases where $B = 0$ in the general second-degree equation. In other words, we focus on equations of the form (1) that do not contain an xy-term and where A and C are not both zero.

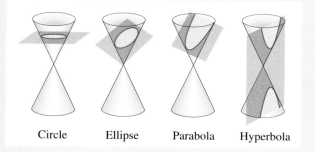

Circle   Ellipse   Parabola   Hyperbola

8.1   Conic Sections: Circles

In Section 3.1 we stated the geometric definition of a circle and derived the equation of a circle by applying the distance formula to this definition. We briefly review the discussion of the circle here, because we will take the same approach in deriving equations for the other conic sections.

Recall that the formula we use to find the distance between the points (x_1, y_1) and (x_2, y_2) in the Cartesian plane is $d = \sqrt{(x_2 - x_1)^2 + (y_2 - y_1)^2}$. The circle is defined to be the set of all points in a plane equidistant from a fixed point. We label the fixed point the center, $C(h, k)$, and the distance from the center to any point on the circle the radius, r. See Figure 8.1.

Figure 8.1

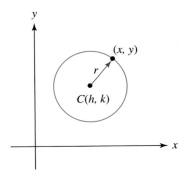

By the distance formula, any point (x, y) lies on the circle if and only if $\sqrt{(x - h)^2 + (y - k)^2} = r$, which, when we square each side, yields the standard form of a circle.

Standard Form for the Equation of a Circle with Center (*h, k*) and Radius *r*

The equation of a circle with center (h, k) and radius r is

$$(x - h)^2 + (y - k)^2 = r^2$$

To find an equation of the circle with center $(-1, 5)$ and radius 4, we simply use the standard form for the equation of a circle with $h = -1$, $k = 5$, and $r = 4$ to get

$$[x - (-1)]^2 + (y - 5)^2 = 4^2 \quad \text{or} \quad (x + 1)^2 + (y - 5)^2 = 16$$

Example 1   Find the center and radius of the circle

$$2x^2 + 2y^2 - 8x + 12y - 28 = 0$$

Solution   The equation $2x^2 + 2y^2 - 8x + 12y - 28 = 0$ is not in standard form, so we complete the square to put the equation in standard form. Before we do, however, we divide each side of the equation by 2 to make the coefficients of the second-degree terms equal to 1.

$2x^2 + 2y^2 - 8x + 12y - 28 = 0$ Divide both sides of the equation by 2.

$x^2 + y^2 - 4x + 6y - 14 = 0$ Add 14 to each side and group terms as shown.

$$(x^2 - 4x \quad) + (y^2 + 6y \quad) = 14$$

Complete the square for each quadratic expression

$$\left[\frac{1}{2}(-4)\right]^2 = 4, \quad \left[\frac{1}{2}(6)\right]^2 = 9$$

Add 4 and 9 to both sides of the equation.

$$(x^2 - 4x + 4) + (y^2 + 6y + 9) = 14 + 4 + 9$$

Rewrite the quadratic expressions in factored form.

$$(x - 2)^2 + (y + 3)^2 = 27$$

This is now in standard form.

We can now determine $h = 2$, $k = -3$, and $r^2 = 27 \Rightarrow r = \sqrt{27} = 3\sqrt{3} \approx 5.2$. Hence, the center is $(2, -3)$, and the radius is $3\sqrt{3}$. See Figure 8.2.

Figure 8.2

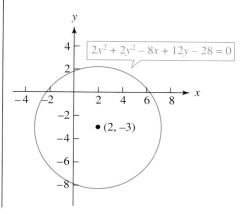

Calculator Exploration

To use a graphing calculator to graph a relation that is not a function, such as $x^2 + y^2 = 9$, you must first solve for y to get $y = \pm\sqrt{9 - x^2}$. Then graph both $y = \sqrt{9 - x^2}$ and $y = -\sqrt{9 - x^2}$ on the same set of coordinate axes.

1. Using this procedure, graph the circle $x^2 + y^2 = 12$ on your graphing calculator.

2. Using this procedure, graph the circle $(x + 5)^2 + (y - 3)^2 = 20$ on your graphing calculator.

NOTE: Use ZSquare on your calculator to ensure that the horizontal and vertical units are the same length. Otherwise, the circle will appear as an oval.

Will the graph of the equation $x^2 + y^2 = 0$ or $x^2 + y^2 = -25$ be a circle?

At this point you may have recognized that if the coefficients of the squared terms in the general second-degree equation (1) are identical (that is, if $A = C$ in the equation $Ax^2 + Cy^2 + Dx + Ey + F = 0$), then, by completing the square, we can always put the equation into the form $(x - h)^2 + (y - k)^2 = a$, where a is a constant. Although we expect this form to give us a circle, the graph will be a circle only if $a > 0$. We will have more to say about the exceptional cases in Section 8.5.

Example 2 Graph the solution to the inequality $(x - 3)^2 + (y - 2)^2 < 16$.

Solution We graph quadratic inequalities using the same approach we took with linear inequalities. We first graph the equation $(x - 3)^2 + (y - 2)^2 = 16$, which is a circle centered at $(3, 2)$ and with radius 4. This graph divides the plane into two regions: the

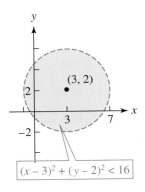

$(x - 3)^2 + (y - 2)^2 < 16$

Figure 8.3

How do we interpret
$(x - 3)^2 + (y - 2)^2 \geq 16$?

region inside the circle and the region outside the circle. The solution is one of the regions.

For the circle, we can determine which region is the solution by realizing that by the definition of the circle, $(x - 3)^2 + (y - 2)^2 < 16$ means all points *less than* 4 units from the center, $(3, 2)$. The solution is therefore the region inside the circle, and we shade in this region. We represent the circle by a dashed curve to indicate that the circle itself is not included in the solution. See Figure 8.3.

Alternatively, we can find the solution by using a test point—that is, by choosing a point in one of the regions and testing its coordinates to see whether it satisfies the inequality $(x - 3)^2 + (y - 2)^2 < 16$. If the coordinates satisfy the inequality, then we shade the region containing the test point; otherwise, we shade the region that does not contain the test point. ∎

8.1 Exercises

1. Find an equation of the circle with center $(3, -8)$ and radius 5.

2. Find an equation of the circle with center $\left(2, \dfrac{1}{3} \right)$ and radius 6.

3. Find an equation of the circle with center $(-3, 1)$ and radius 1. Put this equation into the general form of a second-degree equation.

4. Find an equation of the circle with center $\left(-\dfrac{1}{5}, \dfrac{1}{2} \right)$ and radius 3. Put this equation into the general form of a second-degree equation.

In Exercises 5–14, identify the center and radius of the circles described by the equation.

5. $(x - 2)^2 + (y + 3)^2 = 81$

6. $(x + 5)^2 + (y - 3)^2 = 16$

7. $\left(x - \dfrac{1}{2} \right)^2 + y^2 = 18$

8. $x^2 + \left(y - \dfrac{1}{3} \right)^2 = 12$

9. $x^2 + y^2 - 6x - 8y + 9 = 0$

10. $x^2 + y^2 + 10x - 6y + 10 = 0$

11. $x^2 + y^2 - 6x - 15 = 0$

12. $x^2 + y^2 - 14x + 31 = 0$

13. $4x^2 + 4y^2 - 4y - 47 = 0$

14. $9x^2 + 9y^2 - 6x - 6y - 79 = 0$

15. There is a theorem in geometry that any line tangent to a circle at a point, P, is perpendicular to the line passing through P and the center of the circle. (See the following figure.) Use this theorem to find an

equation of the line tangent to the circle $x^2 + y^2 = 5$ at the point $(1, 2)$.

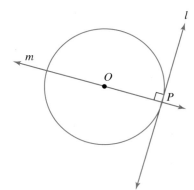

16. Use the theorem in Exercise 15 to find an equation of the line tangent to the circle $(x - 3)^2 + (y - 2)^2 = 40$ at the point $(5, -4)$.

17. Use the theorem in Exercise 15 to find an equation of the line tangent to the circle $x^2 + (y - 3)^2 = 8$ at the point $(2, 5)$.

18. Use the theorem in Exercise 15 to find an equation of the line tangent to the circle $(x - 1)^2 + y^2 = 32$ at the point $(5, -4)$.

In Exercises 19–22, sketch the solution for each inequality.

19. $x^2 + y^2 \leq 81$

20. $(x + 2)^2 + (y - 1)^2 \leq 25$

21. $x^2 + y^2 - 8x - 6y + 9 < 0$

22. $x^2 + y^2 - 2x + 10y > -10$

23. When we refer to the distance from a point to a line, we are referring to the shortest distance to the line, which is defined as the *perpendicular* distance: the length of the perpendicular line segment joining the point and the line. (See the following figure.)

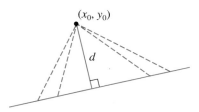

The (perpendicular) distance from the point (x_0, y_0) to the line $Ax + By + C = 0$ can be found by the formula

$$d = \frac{|Ax_0 + By_0 + C|}{\sqrt{A^2 + B^2}}$$

Use this formula to find the distance between the point $(3, -2)$ and the line $3x - 4y + 5 = 0$.

24. Use the formula in Exercise 23 to find the distance between the point $(0, -3)$ and the line $x = 5$.

25. Find an equation of the circle with center $(0, 0)$, if a tangent line to the circle has the equation $2x - 3y = 12$.
HINT: Use the formula in Exercise 23.

26. Find an equation of the circle with center $(2, 6)$, if a tangent line to the circle has the equation $x - 3y = 9$.
HINT: Use the formula in Exercise 23.

27. The (perpendicular) distance from the point (x_0, y_0) to the line $y = mx + b$ can be found by the formula

$$d = \frac{|mx_0 + b - y_0|}{\sqrt{1 + m^2}}$$

Use the formula to find the distance from the point $(2, 4)$ to the line $y = 3x - 4$.

28. Use the formula in Exercise 27 to find the distance from the point $(3, -1)$ to the line $y = 5x + 7$.

29. Use the formula in Exercise 27 to show that the (perpendicular) distance between two parallel lines $y = mx + b_1$ and $y = mx + b_2$ is found by the formula

$$d = \frac{|b_1 - b_2|}{\sqrt{1 + m^2}}$$

HINT: Choose any point on the line $y = mx + b_1$, and call it (x_0, y_0). Then use the formula for the distance from the point (x_0, y_0) to the line $y = mx + b_2$.

30. Use the formula in Exercise 29 to find the distance between the lines $y = 5x - 4$ and $y = 5x + 7$.

31. We will outline a way to demonstrate that the equation of the line tangent to the circle $x^2 + y^2 = r^2$ at the point (x_0, y_0) has the equation

$$xx_0 + yy_0 = r^2$$

(a) First draw a picture of the circle $x^2 + y^2 = r^2$, with tangent line passing through the point (x_0, y_0). Then draw a line passing through the center and through the point (x_0, y_0). (See the accompanying figure.)

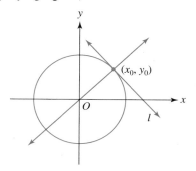

(b) To find the equation of a line, we need the point through which the line passes and the slope. We know that the tangent line passes through the point (x_0, y_0); we still need to find its slope.

(c) We know that the tangent line is perpendicular to the line segment joining the center and the point of tangency. Use this information to show that the tangent line has slope $-\dfrac{x_0}{y_0}$.

(d) Now we have the slope of the tangent line and a point through which the tangent line passes. Use the point–slope formula to show that the equation of the tangent line is $y - y_0 = -\dfrac{x_0}{y_0}(x - x_0)$.

(e) Multiplying both sides of the equation by y_0 and collecting the constant terms on the right-hand side of the equation, we get $xx_0 + yy_0 = (x_0)^2 + (y_0)^2$. How do we arrive at $xx_0 + yy_0 = r^2$?

32. Using the result of Exercise 31, find an equation of the line tangent to the circle $x^2 + y^2 = 29$ at the point $(2, 5)$.

In Exercises 33–36, use a graphing calculator to graph the equation.

33. $x^2 + y^2 = 36$

34. $2x^2 + 2y^2 = 18$

35. $(x - 3)^2 + (y + 4)^2 = 18$

36. $(x + 2)^2 + (y + 1)^2 = 2$

8.2   The Parabola

In Section 4.4 we examined the general quadratic function $f(x) = Ax^2 + Bx + C$ in detail and mentioned that the graph of this function is called a **parabola**. In this section we start with the geometric definition of a parabola and, as we did with the circle, derive the equation of a parabola by applying the distance formula to its definition.

Definition of a Parabola

A **parabola** is the set of all points in a plane whose distance from a fixed point is equal to its distance from a fixed line. We call the fixed point the **focus** and the fixed line the **directrix**. See Figure 8.4(a).

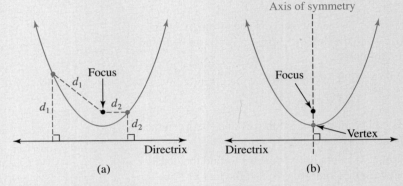

Figure 8.4   The parabola

The line passing through the focus and perpendicular to the directrix is called its **axis of symmetry**, and the point where the parabola intersects its axis of symmetry is called its **vertex**. See Figure 8.4(b).

We will concentrate on parabolas that have either a horizontal or a vertical axis of symmetry.

The Parabola with Vertex (0, 0)

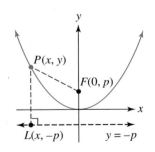

Figure 8.5

The parabola with focus $(0, p)$, $p > 0$, vertex $(0, 0)$, and directrix $y = -p$

Given the focus and directrix, we choose a coordinate system in such a way that the directrix is horizontal and the origin is midway between the focus and the directrix. Let the focus F have coordinates $(0, p)$, where $p > 0$, and let the directrix have equation $y = -p$. See Figure 8.5. Both the distance between the focus and the origin and the distance between the origin and the directrix are p.

By definition of a parabola, if we pick any point on the parabola, $P(x, y)$, the distance from $P(x, y)$ to its focus, $F(0, p)$, is equal to the distance from the point $P(x, y)$ to the point $L(x, -p)$. (Notice that $L(x, -p)$ is the point used to find the perpendicular distance to the line $y = -p$.)

$$\overline{|PF|} = \overline{|PL|} \qquad \text{Using the distance formula, we get}$$

$$\sqrt{(x - 0)^2 + (y - p)^2} = \sqrt{(x - x)^2 + (y + p)^2} \qquad \text{Squaring both sides, we get}$$

$$(x - 0)^2 + (y - p)^2 = (y + p)^2 \qquad \text{Then simplify.}$$

$$x^2 + (y - p)^2 = (y + p)^2$$

$$x^2 + y^2 - 2py + p^2 = y^2 + 2py + p^2 \qquad \text{Which yields}$$

$$x^2 = 4py$$

We have just derived the following:

Standard Form for the Equation of a Parabola with Focus (0, p) and Directrix y = −p

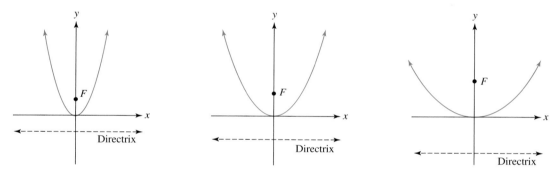

(for $p > 0$)

$x^2 = 4py$

$F(0, p)$

$y = -p$

The standard form for the equation of a parabola with focus $(0, p)$ and directrix $y = -p$ is

$$x^2 = 4py$$

This is a parabola with the vertex at the origin and having the y-axis as its axis of symmetry.

The "vertical" parabola with vertex at (0, 0)

Figure 8.6 shows the relationship between the shape of the parabola and the distance between its focus and directrix. Notice that the farther apart the focus and the directrix are, the wider the parabola.

Figure 8.6

Note the relationship between the shape of the parabola and the distance between its focus and directrix. The farther apart the focus and the directrix are, the wider the parabola.

We arrive at the same result with $p < 0$. In this case, the focus of the parabola, $(0, p)$, lies below the x-axis, and the horizontal directrix, $y = -p$, lies above the x-axis. The parabola still has its vertex at the origin and the y-axis as its axis of symmetry, but it opens downward, as indicated in Figure 8.7.

Keep in mind that if p is negative, then $-p$ is positive.

Figure 8.7

The parabola $x^2 = 4py$ with focus $(0, p)$, $p < 0$, vertex $(0, 0)$, and directrix $y = -p$

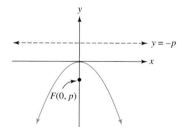

$y = -p$

$F(0, p)$

Example 1 Find the focus and directrix of the parabola $y = -\dfrac{1}{3}x^2$.

Solution For a parabola given in the form $x^2 = 4py$, we know that the equation of the directrix is $y = -p$ and that the focus is $(0, p)$, so we need to identify p. We can put the equation $y = -\dfrac{1}{3}x^2$ into the form $x^2 = 4py$ by solving for x^2:

$$y = -\frac{1}{3}x^2 \quad\Rightarrow\quad x^2 = -3y$$

We compare this with standard form to identify p:

$$x^2 = 4py$$

$$x^2 = -3y \qquad \text{Now we see } 4p = -3 \quad\Rightarrow\quad p = -\frac{3}{4}.$$

Hence the focus is $\left(0, -\frac{3}{4}\right)$ and the directrix is $y = \frac{3}{4}$. See Figure 8.8.

Figure 8.8

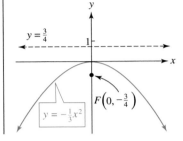

Why does this parabola open downward? If the focus is above the directrix, what can be said about the orientation of the parabola?

Next we look at the parabola with its vertex at the origin but symmetric with respect to the x-axis. The focus is $F(p, 0)$, and the directrix is $x = -p$, pictured in Figure 8.9.

Figure 8.9

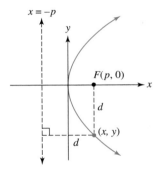

As with the parabola symmetric with respect to the y-axis, we can derive the equation of the parabola symmetric with respect to the x-axis using the distance formula. We arrive at the following:

Standard Form for the Equation of a Parabola with Focus $(p, 0)$ and Directrix $x = -p$

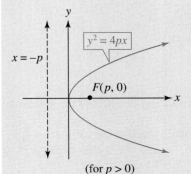

The "horizontal" parabola with vertex $(0, 0)$

The standard form for the equation of a parabola with focus $(p, 0)$ and directrix $x = -p$ is

$$y^2 = 4px$$

This is a parabola with the vertex at the origin and having the x-axis as its axis of symmetry.

Take a close look at the differences between the two standard forms: The key element for determining whether the parabola opens up (down) or right (left) is the second-degree (squared) term. There is only one second-degree term in the equation of a parabola; if there is an x^2-term, the parabola opens either up or down (y-axis symmetry), but if there is a y^2-term, the parabola opens either left or right (x-axis symmetry).

Example 2 Find the focus and directrix of the parabola $x = 2y^2$.

Solution We first determine which standard form to use. We note that since there is a y^2-term (and no x^2-term), we have a parabola symmetric with respect to the x-axis and use the form $y^2 = 4px$.

We can put the equation $x = 2y^2$ into the form $y^2 = 4px$ by solving for y^2:

$$x = 2y^2 \implies y^2 = \frac{1}{2}x$$

We compare this to the standard form for the parabola with the vertex at the origin and symmetric with respect to the x-axis to identify p:

$$y^2 = 4px$$
$$y^2 = \frac{1}{2}x \qquad \text{Now we have } 4p = \frac{1}{2} \implies p = \frac{1}{8}.$$

The focus is $\left(\frac{1}{8}, 0\right)$ and the directrix is $x = -\frac{1}{8}$. See Figure 8.10.

Figure 8.10

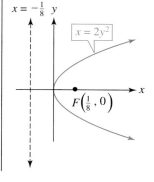

Why does this parabola open right? If the focus is to the right of the directrix, what can be said about the "orientation" of the parabola?

Example 3 Find the equation of the parabola with the vertex at the origin and symmetric with respect to the y-axis if its focus is $(0, -3)$.

Solution Given that the parabola has its vertex at the origin and is symmetric with respect to the y-axis, its equation will be of the form $x^2 = 4py$. Because we are given that the focus is $(0, -3)$, then $p = -3$. Hence the equation is $x^2 = 4(-3)y$, or $x^2 = -12y$. See Figure 8.11.

Figure 8.11

The parabola $x^2 = -12y$

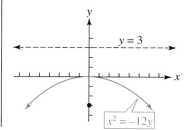

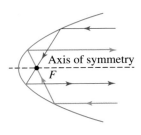

Figure 8.12

The reflecting property of a parabola.

As we have seen in previous chapters, there are many applications involving parabolas. If an object is fired or thrown in the air in a vertical direction and if other forces are negligible, gravity acts on the object in such a way that its height s above the ground at time t can be given by an equation such as $s = -16t^2 + 32t + 10$, which is a quadratic function of t whose graph we know is a parabola. We previously used this information to find the maximum height of an object and when it landed. In these cases, we found the vertex, the axis of symmetry, or intercepts to be important and ignored the focus.

Another important use of the parabola has to do with its reflecting property. Figure 8.12 illustrates that any ray entering the parabola parallel to its axis of symmetry will be reflected through the focus and back out parallel to the axis of symmetry. (See Exercise 55.)

This property has many applications. For example, if we were to create a mirror by rotating a parabola about its axis (called a paraboloid of revolution), then, if a source of light is placed at the focus, the beam would be reflected parallel to its axis. This concentrates the beam of light in one direction. You have probably noticed that flashlights, car headlights, and spotlights contain paraboloid-shaped reflecting surfaces (see Figure 8.13). The light source or bulb is placed at its focus.

Figure 8.13

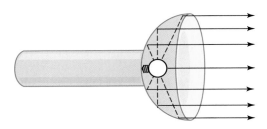

The same principle is used for receiving. For example, parabolic surfaces on telescopes, TV satellite dishes, and radiotelescopes use the principle that waves coming in parallel to the axis will be reflected into the focus.

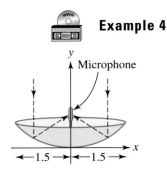

Figure 8.14

Example 4 A field microphone used at a football game (Figure 8.14) consists of a parabolic dish with the receiver placed at its focus. The dish can be described by rotating the parabola $y = \dfrac{1}{9}x^2$ about its axis of symmetry, where $-1.5 \le x \le 1.5$, and x is measured in feet. How deep is the dish, and where should the receiver be placed in relation to the bottom (vertex) of the dish?

Solution By Figure 8.14, we see that the cross section of the dish is a parabola. We can place the parabola such that its vertex is at the origin and the axis of symmetry is the y-axis. Hence we can use the form $x^2 = 4py$, which has focus $(0, p)$ as shown in Figure 8.15.

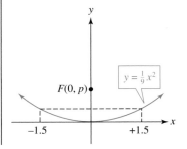

Figure 8.15

Looking at the equation of the cross section of the dish, $y = \dfrac{1}{9}x^2$, we can identify p by solving for x^2 to get $9y = x^2$. Hence $4p = 9$, and so $p = \dfrac{9}{4} = 2.25$ feet. This means that the receiver must be placed 2.25 feet above the vertex.

Notice that because $-1.5 \le x \le 1.5$, the dish has a diameter of 3 feet $[1.5 - (-1.5) = 3]$.

To find out how deep the dish is, we use the fact that $-1.5 \le x \le 1.5$. Looking at Figure 8.15, we see that if we substitute $x = 1.5$ (or $x = -1.5$) in the equation of the cross section, $y = \dfrac{1}{9}x^2$, we can find the "height," which will tell us how deep the dish is:

$$y = \frac{1}{9}x^2 \qquad \text{Let } x = 1.5 \text{ and find } y.$$

$$y = \frac{1}{9}(1.5)^2 = \boxed{0.25 \text{ foot, or 3 inches, deep}}$$

Calculator Exploration

Keep in mind that you must solve for y before sketching a graph with your graphing calculator.

1. Graph the following on the same set of coordinate axes:

$$x^2 = 12y \qquad (x - 2)^2 = 12y \qquad (x + 2)^2 = 12y$$

What can you conclude about the effect of h on the graph of $(x - h)^2 = 4py$?

2. Graph the folowing on the same set of coordinate axes:

$$x^2 = 12y \qquad x^2 = 12(y + 3) \qquad x^2 = 12(y - 3)$$

What can you conclude about the effect of k on the graph of $x^2 = 4p(y - k)$?

The Parabola Centered at (*h, k*)

Figure 8.16 illustrates what happens when we shift a parabola that is symmetric with respect to the *y*-axis and with vertex $(0, 0)$ *horizontally h* units and *vertically k* units. We arrive at a parabola with the following properties: Its vertex is (h, k); its axis of symmetry is the vertical line $x = h$; its focus is $F(h, k + p)$; and its directrix is $y = -p + k = k - p$.

Figure 8.16

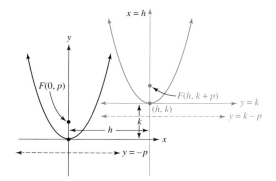

Horizontal and vertical shifting of the figure will produce a figure with axes of symmetry parallel to their original axes. We call this kind of shifting **translation of axes**. In the preceding case, the new axis of the parabola will be parallel to the y-axis.

Calculator Exploration

To graph a relation (that is not a function) such as $(y - 2)^2 = 20x$ using a graphing calculator, you must first solve explicitly for y, obtaining $y = 2 + \sqrt{20x}$ and $y = 2 - \sqrt{20x}$. Then graph both functions on the same set of coordinate axes.

On a graphing calculator, graph the following on the same set of coordinate axes:

$$y^2 = 20x \qquad (y - 2)^2 = 20x \qquad (y - 2)^2 = 20(x - 3).$$

What can you conclude about the effects of h and k on the graph of $(y - h)^2 = 4p(x - k)$?

If all three parabolas cannot be graphed on one set of axes, compare $y^2 = 20x$ with $(y - 2)^2 = 20x$ and then compare $(y - 2)^2 = 20x$ with $(y - 2)^2 = 20(x - 3)$.

In the same way we derived the equation for the parabola centered at the origin, we can derive the standard form of the equation for the parabola with vertex (h, k), focus $F(h, k + p)$, and axis of symmetry $x = h$. We arrive at the following standard form.

Standard Form for the Equation of a Parabola with Focus $(h, k + p)$ and Directrix $y = k - p$

The "vertical" parabola with vertex (h, k).

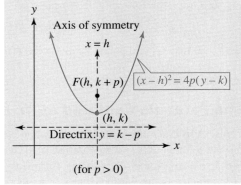

(for $p > 0$)

The standard form for the equation of a parabola with focus $(h, k + p)$ and directrix $y = k - p$ is

$$(x - h)^2 = 4p(y - k)$$

This is a parabola with vertex (h, k) and axis of symmetry $x = h$.

Compare the form of the parabola with its vertex at $(0, 0)$ and symmetric with respect to the y-axis with the form of the parabola with its vertex at (h, k) and symmetric with respect to a line parallel to the y-axis.

To change the graph of the parabola $x^2 = 4py$ into $(x - h)^2 = 4p(y - k)$:

Shift the vertex from	$(0, 0)$ to (h, k)
Shift the axis of symmetry from	$x = 0$ to $x = h$
Shift the focus from	$(0, p)$ to $(h, k + p)$

If we draw a new set of horizontal and vertical axes through the vertex, the new vertical axis is the axis of symmetry.

This suggests that if we have the parabola in the form $(x - h)^2 = 4p(y - k)$, then we can identify h, k (and p), *draw a new set of coordinate axes through the point* (h, k), *and graph the equation* $x^2 = 4py$ *on the new set of axes.*

Example 5 | Sketch the graph of the parabola $(x - 3)^2 = 8(y + 1)$. Identify the vertex, focus, axis of symmetry, and directrix.

Solution | We compare the graph of $(x - 3)^2 = 8(y + 1)$ with the standard form:

$$(x - h)^2 = 4p(y - k)$$

$$(x - 3)^2 = 8(y + 1) \qquad \text{Now we can identify } h = 3, \ k = -1, \text{ and } p = 2.$$

Hence the vertex is $(3, -1)$. We draw a new set of coordinate axes centered at $(3, -1)$ (and parallel to the x- and y-axes). See Figure 8.17(a). The axis of symmetry is the new vertical axis; it is parallel to and 3 units right of the original y-axis and, therefore, has equation $x = 3$. Since $p = 2$, if we move 2 units up from the vertex along the new vertical axis, we locate the focus at $(3, 1)$.

If we move vertically down 2 units from the vertex $(3, -1)$ and draw a line through this point parallel to the x-axis, we have the directrix. Since the directrix is parallel to and 3 units below the x-axis, its equation is $y = -3$. See Figure 8.17(b).

Figure 8.17

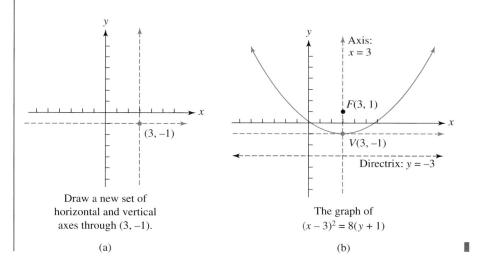

Draw a new set of horizontal and vertical axes through $(3, -1)$.

(a)

The graph of $(x - 3)^2 = 8(y + 1)$

(b)

The standard form for the equation of the parabola with vertex (h, k) and axis of symmetry parallel to the x-axis is summarized next.

Standard Form for the Equation of a Parabola with Focus $(h + p, k)$ and Directrix $x = h - p$

The "horizontal" parabola with vertex (h, k)

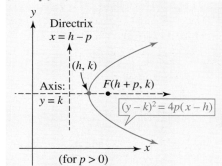

(for $p > 0$)

The standard form for the equation of a parabola with focus $(h + p, k)$ and directrix $x = h - p$ is

$$(y - k)^2 = 4p(x - h)$$

This is a parabola with vertex (h, k) and axis of symmetry $y = k$.

Example 6 Find the equation of the parabola with focus $(2, -1)$ and directrix $x = 6$.

Solution We plot the focus and sketch the directrix (Figure 8.18(a)). Because the directrix has equation $x = 6$, which is a vertical line, the parabola must have a horizontal axis of symmetry and, therefore, the form is $(y - k)^2 = 4p(x - h)$.

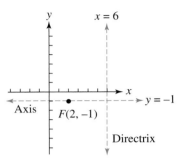 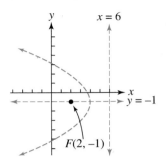

Figure 8.18(a)   **Figure 8.18(b)**

The directrix is $x = 6$ and the focus is $(2, -1)$.

The (horizontal) axis of symmetry passes through the focus $(2, -1)$ and therefore must have the equation $y = -1$. The vertex must lie on the axis of symmetry midway between the focus and directrix. The focus and directrix are 4 units apart, so you can count units and locate the vertex at $(4, -1)$: therefore, $h = 4$ and $k = -1$. Because $|p|$ equals the distance between the vertex and the focus, we can see that $|p| = 2$, and because the parabola opens left, $p < 0$, hence $p = -2$. We substitute the values $h = 4$, $k = -1$, and $p = -2$, into the general form:

$$(y - k)^2 = 4p(x - h)$$

$$[y - (-1)]^2 = 4(-2)(x - 4) \quad \text{or} \quad (y + 1)^2 = -8(x - 4)$$

See Figure 8.18(b). ∎

Example 7 Sketch the graph of the parabola $x^2 - 6x - 7 = 2y$, and identify its vertex, axis of symmetry, and focus.

Solution We recognize that there is an x^2-term, so the parabola has a vertical axis of symmetry. Thus we want the equation to be in the form $(x - h)^2 = 4p(y - k)$. We complete the square as follows:

$x^2 - 6x - 7 = 2y$	Isolate the terms containing powers of x on one side of the equation.
$x^2 - 6x \quad\quad = 2y + 7$	Complete the square for the left-hand side: Add $9 = \left[\frac{1}{2}(-6)\right]^2$ to each side of the equation.
$x^2 - 6x + 9 = 2y + 7 + 9$	Write the left-hand side in factored form.
$(x - 3)^2 = 2y + 16$	We want the right side to look like $4p(y - k)$, so we factor out 2, the coefficient of y, to get
$(x - 3)^2 = 2(y + 8)$	Now we can see $4p = 2$, so $p = \frac{1}{2}$.
$(x - 3)^2 = 4\left(\frac{1}{2}\right)[y - (-8)]$	Equation has the form $(x - h)^2 = 4p(y - k)$ and we can identify h, k, and p.

Thus $h = 3$, $k = -8$, and $p = \frac{1}{2}$. We draw a new set of coordinate axes through $(3, -8)$. The vertex is $(3, -8)$. The axis of symmetry is the line $x = 3$. If we move up $\frac{1}{2}$ unit above the vertex along the parabola's axis, we locate the focus at $(3, -\frac{15}{2})$. See Figure 8.19. ∎

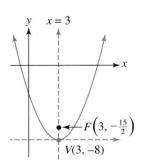

Figure 8.19

The graph of $x^2 - 6x - 7 = 2y$

At this point you may have recognized that if there is only one square term in the general equation $Ax^2 + Cy^2 + Dx + Ey + F = 0$ (that is, if $A = 0$ or $C = 0$ but not both), then, with the help of completing the square, we *may* be able to put the equation into the standard form of a parabola. This is a reasonable conjecture, but as with the circle, there are exceptions. We postpone discussion of the exceptional cases until Section 8.5.

8.2 Exercises

In Exercises 1–6, identify the focus and the directrix of the parabolas.

1. $x^2 = 12y$
2. $y^2 = -16x$
3. $x = -8y^2$
4. $y = -15x^2$
5. $6x^2 - 3y = 0$
6. $2y^2 + 10x = 0$

In Exercises 7–10, graph the parabola and identify the focus, axis, and directrix.

7. $4x^2 - 12y = 0$
8. $6y^2 - 24x = 0$
9. $x + y^2 = 0$
10. $x^2 + 16y = 0$

In Exercises 11–16, write the equation of the parabola using the given information.

11. The parabola has focus $(0, 4)$ and vertex at the origin.
12. The parabola has focus $(-4, 0)$ and vertex at the origin.
13. The parabola has focus $(0, -3)$ and directrix $y = 3$.
14. The parabola has focus $(3, 0)$ and directrix $x = -3$.
15. The parabola has its vertex at the origin and directrix $x = -5$.
16. The parabola has its vertex at the origin and directrix $y = 8$.
17. A TV satellite antenna consists of a parabolic dish with the receiver placed at its focus. The dish can be described by rotating the parabola $y = \frac{1}{12}x^2$ about its axis of symmetry, where $-6 \le x \le 6$ and x is measured in feet. How deep is the dish, and where should the receiver be placed in relation to the bottom (vertex) of the dish? (See the accompanying figure.)

18. A radar antenna consists of a parabolic dish with the receiver placed at its focus. The dish can be described by rotating the parabola $y = \frac{1}{20}x^2$ about its axis of symmetry, where $-8 \le x \le 8$ and x is measured in feet. How deep is the dish, and where should the receiver be placed in relation to the bottom (vertex) of the dish?

19. A spotlight has a parabolic reflecting mirror with the light source placed at its focus. The mirror can be described by rotating the parabola $y = \frac{1}{6}x^2$ about its axis of symmetry, where $-2 \le x \le 2$ and x is measured in feet. How deep is the mirror, and where should the light source be placed in relation to the bottom (vertex) of the mirror?

20. A reflecting telescope has a parabolic reflecting mirror with the eyepiece placed on the focus of the mirror. The mirror can be described by rotating the parabola $y = \frac{1}{10}x^2$ about its axis of symmetry, where $-2 \le x \le 2$ and x is measured in inches. How deep is the mirror, and where should the eyepiece be placed in relation to the bottom (vertex) of the mirror?

21. The arch of a building is constructed in the shape of a parabola. (See the accompanying figure.) The height of the arch is 30 feet, and the width of the base of the arch is 20 feet. How wide is the parabola 10 feet above the base of the arch? HINT: First draw the parabola with its vertex at $(0, 0)$, determine which form to use, identify p, and then write an equation of the parabola.

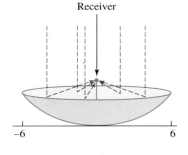

Receiver

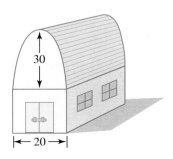

30

|← 20 →|

22. The cable of a suspension bridge is in the shape of a parabola. The cable extends over 200 feet of highway. The longest supporting wire, on either end of the cable, is 100 feet; the shortest, in the middle, is 30 feet. Find the length of the supporting wire that is 40 feet from the longest supporting wire. (See the figure below.)

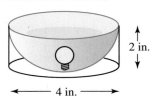

23. A flashlight is to be designed so that a cylindrical casing houses a parabolic reflecting mirror and the light source. (See the accompanying figure.) If the casing has a depth of 2 inches and a diameter of 4 inches, where should the light source be placed relative to the vertex of the mirror?

24. An engineer wants to design a parabolic microphone as described in Example 4. She wants the diameter of the dish to be 2 feet, but wants the receiver to be located below the lip of the dish. How deep can the dish be if the receiver is properly placed at the focus? HINT: Construct a parabola opening upward, with its vertex at the origin, and examine its equation in standard form. How would you translate the receiver being placed "below the lip of the dish" as an algebraic relationship between p and y?

In Exercises 25–30, identify the focus and the directrix of each parabola.

25. $(x - 2)^2 = 8y$ **26.** $(y + 5)^2 = -16x$
27. $8x = y^2 - 2y - 15$ **28.** $4y = x^2 - 6x + 5$
29. $-5y = x^2 - 4x + 19$ **30.** $2x = y^2 - 6y + 3$

In Exercises 31–34, graph the parabola and identify the focus, axis of symmetry, and directrix.

31. $8y = x^2 + 10x + 33$ **32.** $-12x = y^2 - 4y + 16$
33. $x = 3y^2 - 6y + 5$ **34.** $y = -2x^2 - 12x - 19$

In Exercises 35–40, write the equation of the parabola using the given information.

35. The parabola has focus (2, 4) and vertex (5, 4).
36. The parabola has focus (−2, 3) and vertex (−2, −1).
37. The parabola has focus (2, −3) and directrix $y = 4$.
38. The parabola has focus (3, 4) and directrix $x = 7$.
39. The parabola has its vertex at (2, 5) and directrix $x = -3$.
40. The parabola has its vertex at (3, −1), and directrix $y = 2$.

In Exercises 41–44, graph the solution to the inequality.

41. $y < 4x^2$
42. $-12x > y^2$
43. $x \geq y^2 - 6y + 5$
44. $y \geq -2x^2 - 12x - 8$

45. Use the distance formula to derive the equation for the parabola centered at the origin with focus $(p, 0)$ and directrix $x = -p$.

46. Use the distance formula to derive the equation for the parabola centered at (h, k) with focus $(h, k + p)$ and directrix $y = k - p$.

47. We define a line to be tangent to a parabola at point P if and only if it touches the parabola only at point P and the line is not parallel to its axis of symmetry. (See the accompanying figure.) The equation for the line tangent to the parabola $x^2 = 4py$ at the point (x_0, y_0) is given by

$$y = \frac{x_0}{2p}x - y_0$$

Use this formula to find the equation of the line tangent to the parabola $x^2 = 8y$ at the point (−4, 2).

Tangent line   Not a tangent line   Not a tangent line

48. Use the formula for the equation of the line tangent to the parabola in Exercise 47 to find the equation of the line tangent to the parabola $x^2 = 8y$ at the point (4, 2).

49. The *latus rectum* of a parabola is defined to be the line segment passing through the focus of the parabola that is perpendicular to its axis and has its endpoints on the parabola. (See the accompanying figure.) Show that the length of the latus rectum of the parabola $x^2 = 4py$ is $4p$.

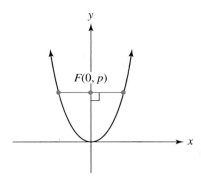

In Exercises 50–53, use a graphing calculator to graph the following.

50. $(x - 1)^2 = 18y$

51. $(y - 1)^2 = 24x$

52. $y^2 + 10y - x + 25 = 0$

53. $x^2 - 6x - 12y + 9 = 0$

Questions for Thought

54. The equations $x^2 + x - 6 = 0$ and $y^2 - 6y + 9 = 0$ are in the general form of a second-degree equation. If you graphed these equations on a rectangular coordinate system, what type of figure would you expect to have? Graph the equations on a set of coordinate axes. Is the graph what you expected?

55. We can state the reflecting property of the parabola in a more formal manner as follows: If l is a line tangent to a parabola at point P, then for any line q parallel to its axis of symmetry, the angle between l and q will equal the angle between l and the line joining P and the focus. (See the accompanying figure.)

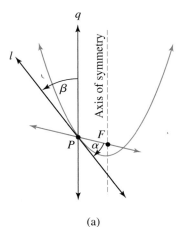

(a)

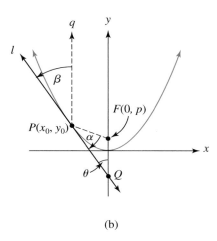

(b)

We outline a way to demonstrate the reflecting property of the parabola; that is, given the accompanying figure, where l is a line tangent to the parabola $x^2 = 4py$ at point $P(x_0, y_0)$ and q is a line parallel to the axis of symmetry, then $\alpha = \beta$. We use the formula of the tangent line given in Exercise 47.

(a) Exercise 47 states that the equation for the tangent line l is $y = \dfrac{x_0}{2p}x - y_0$. We designate the point where the tangent line crosses the y-axis as Q (forming the angle θ with the y-axis). Show that the coordinates of Q are $(0, -y_0)$.

(b) Show that the distance from the focus, $F(0, p)$ to Q is, therefore, $p + y_0$.

(c) The equation of the directrix is $y = -p$. Show that the perpendicular distance from the point $P(x_0, y_0)$ to the directrix is also $y_0 + p$.

(d) But the distance from the focus F to the point $P(x_0, y_0)$ must be equal to the distance from point $P(x_0, y_0)$ to the directrix. Explain why.

(e) Explain how we can now conclude that $|\overline{FQ}| = |\overline{FP}|$.

(f) But if $|\overline{FQ}| = |\overline{FP}|$, then $\alpha = \theta$. Why?

(g) Explain why $\theta = \beta$.

(h) Therefore, we conclude that $\alpha = \beta$.

8.3   The Ellipse

The **ellipse** is another conic section that is useful in providing a mathematical model for a variety of physical phenomena such as the orbits of planets.

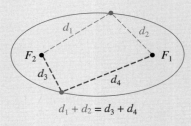

$$d_1 + d_2 = d_3 + d_4$$

Figure 8.20

Figure 8.21

We can actually use this definition to draw an ellipse by fastening the ends of a string at two points (see Figure 8.21). Obviously, the string must be longer than the distance between the two points. If you move the pencil while holding it taut against the string, you will trace out an ellipse. The points where the string is fastened are the foci of the ellipse, and the length of the string is the sum of the distances from the foci, which is constant.

As with the parabola, we use the definition of the ellipse to derive its equation. We start by choosing a coordinate system in such a way that the foci lie on a horizontal line with the origin midway between the two foci. See Figure 8.22.

Figure 8.22

Ellipse with foci $F_1(c, 0)$ and $F_2(-c, 0)$

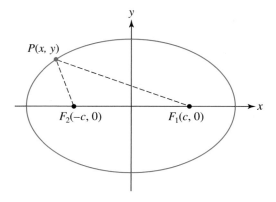

If we let $c > 0$ be the distance from the origin to the right focus F_1, then the distance to the left focus F_2, is also c. Hence the coordinates of F_1 are $(c, 0)$, and the coordinates of F_2 are $(-c, 0)$. The distance between the two foci, F_1 and F_2, is $2c$. We can derive an equation for this ellipse as follows.

By definition of an ellipse, if we pick any point on the ellipse, $P(x, y)$, the sum of the distance from $P(x, y)$ to $F_1(c, 0)$ and the distance from $P(x, y)$ to $F_2(-c, 0)$ is constant. To avoid working with fractions, we call this constant sum $2a$ and note that for any point on the ellipse, $2a > 2c$, or $a > c$. (This is equivalent to saying that the string is longer than the distance between the two foci.) Hence we have

$$|\overline{PF_1}| \qquad + \qquad |\overline{PF_2}| \qquad = 2a$$ Using the distance formula, we get

$$\sqrt{(x-c)^2 + (y-0)^2} + \sqrt{(x+c)^2 + (y-0)^2} = 2a$$ Or

$$\sqrt{(x-c)^2 + y^2} + \sqrt{(x+c)^2 + y^2} = 2a$$ Now isolate the first radical.

$$\sqrt{(x-c)^2 + y^2} = 2a - \sqrt{(x+c)^2 + y^2}$$ Square each side.

$$(x-c)^2 + y^2 = 4a^2 - 4a\sqrt{(x+c)^2 + y^2} + (x+c)^2 + y^2$$

$$x^2 - 2cx + c^2 + y^2 = 4a^2 - 4a\sqrt{(x+c)^2 + y^2} + x^2 + 2cx + c^2 + y^2$$

Simplify and isolate the radical term again.

$$4a\sqrt{(x+c)^2 + y^2} = 4a^2 + 4cx$$ Dividing by 4, we get

$$a\sqrt{(x+c)^2 + y^2} = a^2 + cx$$ Squaring each side again yields

$$a^2[(x+c)^2 + y^2] = (a^2 + cx)^2$$ Which yields

$$a^2x^2 + 2a^2cx + a^2c^2 + a^2y^2 = a^4 + 2a^2cx + c^2x^2$$ Which can be rewritten as

$$a^2x^2 - c^2x^2 + a^2y^2 = a^4 - a^2c^2$$ Or

$$x^2(a^2 - c^2) + a^2y^2 = a^2(a^2 - c^2)$$ We now divide both sides by $a^2(a^2 - c^2)$ to get

$$\frac{x^2}{a^2} + \frac{y^2}{a^2 - c^2} = 1$$ We label this equation (1).

We noted that a and c are positive and $a > c$; therefore, $a^2 - c^2 > 0$. If we let $b = \sqrt{a^2 - c^2}$, then $b > 0$ and $b^2 = a^2 - c^2$. Substituting into equation (1) gives us $\frac{x^2}{a^2} + \frac{y^2}{b^2} = 1$. Since $c > 0$ and $b^2 = a^2 - c^2$, we conclude $a^2 > b^2$, and hence $a > b$. Thus the ellipse centered at the origin, with focal points on the x-axis, has the equation

$$\frac{x^2}{a^2} + \frac{y^2}{b^2} = 1 \quad \text{(where } a > b\text{)}$$

To get the sense of how a, b, and c are related geometrically, consider the ellipse $\frac{x^2}{a^2} + \frac{y^2}{b^2} = 1$ in Figure 8.23.

Figure 8.23

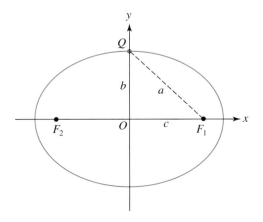

Let Q be a point where the ellipse intersects the y-axis. Q lies on the ellipse and therefore, the sum of the distances from the foci F_1 and F_2 is $2a$. Because Q is equidistant from F_1 and F_2, then $|\overline{QF_1}| = a$. Since $|\overline{OF_1}| = c$, by the Pythagorean Theorem, $|\overline{OQ}| = \sqrt{a^2 - c^2}$. We defined $\sqrt{a^2 - c^2}$ as b, therefore $b = |\overline{OQ}|$.

The x-intercepts of the graph of the equation $\dfrac{x^2}{a^2} + \dfrac{y^2}{b^2} = 1$ are found by setting $y = 0$:

$$\dfrac{x^2}{a^2} + \dfrac{y^2}{b^2} = 1 \qquad \text{Let } y = 0.$$

$$\dfrac{x^2}{a^2} + \dfrac{0^2}{b^2} = 1 \qquad \text{Solve for } x.$$

$$\dfrac{x^2}{a^2} = 1 \implies x^2 = a^2 \quad \text{or} \quad x = \pm a \qquad \text{Hence the } x\text{-intercepts are } \pm a.$$

We call the points $V_1(a, 0)$, and $V_2(-a, 0)$ the **vertices** of this ellipse, and the line segment $\overline{V_1V_2}$ is called the **major axis**. Similarly, if we let $x = 0$, then we find the y-intercepts to be $y = \pm b$. The line segment joining the points $(0, b)$ and $(0, -b)$ is called the **minor axis** of the ellipse. We summarize these facts as follows.

Standard Form of an Ellipse with Foci at $(c, 0)$ and $(-c, 0)$

The ellipse with horizontal major axis, centered at $(0, 0)$

How are the endpoints of the *major* axis related to the vertices?

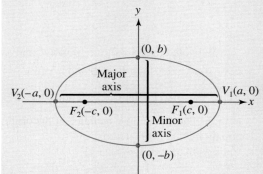

The graph of the equation

$$\dfrac{x^2}{a^2} + \dfrac{y^2}{b^2} = 1$$

for $a > b$ is an ellipse with center $(0, 0)$ and vertices $(\pm a, 0)$. The endpoints of the minor axis are $(0, \pm b)$. The foci are $(\pm c, 0)$, where $c^2 = a^2 - b^2$. The length of the major axis is $2a$; the length of the minor axis is $2b$.

Example 1   Sketch the graph of the following ellipse. Identify the foci.

$$\dfrac{x^2}{16} + \dfrac{y^2}{9} = 1$$

Solution   To sketch the graph of an ellipse centered at the origin, we compare our equation with the standard form of the ellipse $\dfrac{x^2}{a^2} + \dfrac{y^2}{b^2} = 1$ and we get

$$a^2 = 16 \implies a = 4 \quad \text{and} \quad b^2 = 9 \implies b = 3 \qquad \text{(Remember } a \text{ and } b \text{ are positive.)}$$

We plot the vertices $(\pm 4, 0)$ and the endpoints of the minor axis, $(0, \pm 3)$, and sketch the graph of the ellipse, as shown in Figure 8.24.

Figure 8.24

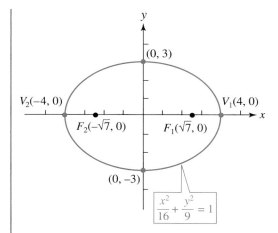

Notice that $a > b$ and $c^2 = a^2 - b^2 = 16 - 9 = 7 \implies c = \sqrt{7}$. Hence the foci are $(\pm\sqrt{7}, 0)$. ∎

Example 2 | Use a graphing calculator to graph the ellipse $3x^2 + 15y^2 = 45$. Identify the foci.

Solution | Since this equation is not a function, we first have to solve for y before we can enter it into the graphing calculator:

$$3x^2 + 15y^2 = 45 \qquad \text{Isolate } y^2$$
$$15y^2 = 45 - 3x^2 \qquad \text{Divide both sides by 15 to get}$$
$$y^2 = 3 - \frac{x^2}{5} \qquad \text{Take square roots.}$$
$$y = \pm\sqrt{3 - \frac{x^2}{5}}$$

We enter *both* equations: $y = \sqrt{3 - \frac{x^2}{5}}$ and $y = -\sqrt{3 - \frac{x^2}{5}}$. The graph of the ellipse appears in Figure 8.25.

Figure 8.25

The graph of $3x^2 + 15y^2 = 45$

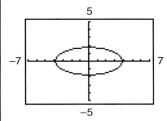

To find the foci, we first put the equation into standard form. We observe that standard form requires that a 1 be on one side. For us to have 1 on the right-hand side, we must divide both sides of the equation by 45.

$$3x^2 + 15y^2 = 45 \qquad \text{Divide both sides by 45.}$$
$$\frac{3x^2}{45} + \frac{15y^2}{45} = \frac{45}{45} \qquad \text{Simplify.}$$
$$\frac{x^2}{15} + \frac{y^2}{3} = 1 \qquad \text{The equation is now in standard form.}$$

Comparing our equation with the standard form of the ellipse, $\dfrac{x^2}{a^2} + \dfrac{y^2}{b^2} = 1$, we get

$$a^2 = 15 \implies a = \sqrt{15} \quad \text{and} \quad b^2 = 3 \implies b = \sqrt{3}$$

Again, $a > b$ and $c^2 = a^2 - b^2 = 15 - 3 = 12 \implies c = \sqrt{12} = 2\sqrt{3}$. Hence the foci are $(\pm 2\sqrt{3}, 0)$. ∎

Using the same analysis (again with $a > b$), we use the distance formula to derive the equation of an ellipse with center (0, 0) and *foci on the y-axis*.

Standard Form of an Ellipse with Foci at (0, c) and (0, −c)

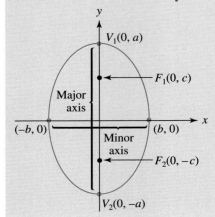

The ellipse with vertical major axis, centered at (0, 0)

The graph of the equation

$$\frac{x^2}{b^2} + \frac{y^2}{a^2} = 1$$

for $a > b$ is an ellipse with center (0, 0), and vertices $(0, \pm a)$. The endpoints of the minor axis are $(\pm b, 0)$. The foci are $(0, \pm c)$, where $c^2 = a^2 - b^2$. The length of the major axis is $2a$; the length of the minor axis is $2b$.

Examine the two standard forms of the ellipse; observe their similarities and differences. The first has its foci on the x-axis; the second has its foci on the y-axis. Notice that the two forms switch a and b as denominators of x^2 and y^2. This ensures that a is larger than b, so the foci found by $c^2 = a^2 - b^2$ will exist. Also, as a consequence, the major axis (the axis containing the foci) is always the longer axis (of length $2a$), and the vertices, the endpoints of the longer axis, are always determined by a.

When the denominator of the y^2-term is larger than the denominator of the x^2-term, is the major axis along the x-axis or y-axis? What distinguishes the major axis from the minor axis?

The choice of which form to use is dependent on which denominator is larger: If the denominator of x^2 is larger than the denominator of y^2, then the major axis lies on the x-axis, and we use the first form (since $a > b$). If the denominator of y^2 is greater than the denominator of x^2, then the major axis lies on the y-axis, and we use the second form (again, $a > b$). Thus a^2 is always the larger denominator.

Example 3   Sketch the graph of $16x^2 + 4y^2 = 16$. Identify its foci.

Solution   First we put the equation into standard form. In order for us to have 1 on the right-hand side, we must divide by 16.

$$16x^2 + 4y^2 = 16 \qquad \text{Divide both sides by 16.}$$

$$\frac{16x^2}{16} + \frac{4y^2}{16} = \frac{16}{16} \qquad \text{Simplify.}$$

$$\frac{x^2}{1} + \frac{y^2}{4} = 1 \qquad \text{The equation is now in standard form.}$$

Comparing our equation with both standard forms of the ellipse, we note that the denominator of y^2 is greater than the denominator of x^2; therefore, since a must be greater than b, we use the second form, $\dfrac{x^2}{b^2} + \dfrac{y^2}{a^2} = 1$, an ellipse with foci on the y-axis. Hence

$$a^2 = 4 \;\Rightarrow\; a = 2 \quad \text{and} \quad b^2 = 1 \;\Rightarrow\; b = 1$$

We plot the vertices $(0, \pm 2)$ and the endpoints of the minor axis $(\pm 1, 0)$, and sketch the graph of the ellipse. See Figure 8.26.

Figure 8.26

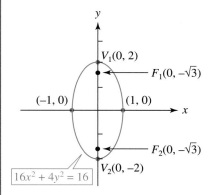

Notice that $a > b$ and $c^2 = a^2 - b^2 = 4 - 1 = 3 \;\Rightarrow\; c = \sqrt{3}$. The foci are $(0, \pm\sqrt{3})$.

Example 4 The arch of a bridge is semielliptical with a horizontal major axis. The base of the arch covers the entire 50-foot width of a two-lane road, and the highest part of the arch is 15 feet vertically above the centerline of the road. Can a truck that is 10 feet wide and 14 feet high pass through this bridge, staying to the right of the centerline? See Figure 8.27.

Figure 8.27

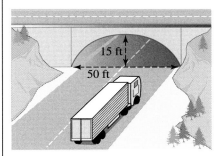

Solution

What do we need to find?

Whether the bridge is high enough for the truck to pass through

What information are we given?

We note that the bridge opening is a semiellipse, and we can draw this ellipse on a rectangular coordinate system centered at the origin, with major axis lying on the x-axis. We sketch the graph and label the given information on the graph shown in Figure 8.28 on page 520; the length of the major axis is 50 feet. Hence $a = 25$, and the vertices are $(\pm 25, 0)$. Since the highest part of the arch is 15 feet, we can label the top endpoint of the minor axis $(0, 15)$.

Figure 8.28

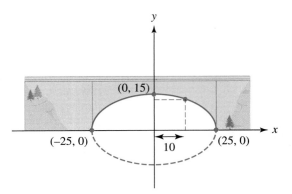

Can we restate what we need to find in terms of the figure?

The problem asks whether a truck that is 10 feet wide and 14 feet high can pass through the arch of the bridge staying to the right of the centerline. This means that we need to determine the height of the arch 10 feet from the center. Hence, on the graph in Figure 8.28, this means we need to find y when $x = 10$. Algebraically, if we have the equation of the ellipse, then all we need to do is substitute $x = 10$ in the equation and find y. Therefore our next step is to find the equation of the ellipse with the given information.

How do we find the equation of the ellipse?

We use the form where the major axis is horizontal. Because $a = 25$, and $b = 15$, our equation is, therefore,

$$\frac{x^2}{25^2} + \frac{y^2}{15^2} = 1$$

(Note the foci lie on the horizontal axis, which means that the larger of a and b determines the denominator of x.) Now we just substitute $x = 10$ in the equation, and solve for y:

$$\frac{x^2}{25^2} + \frac{y^2}{15^2} = 1 \qquad \text{Let } x = 10; \text{ solve for } y.$$

$$\frac{10^2}{25^2} + \frac{y^2}{15^2} = 1$$

$$\frac{y^2}{15^2} = 1 - \frac{10^2}{25^2} = \frac{21}{25}$$

$$y^2 = 15^2\left(\frac{21}{25}\right) = 189 \qquad \text{Hence } y = \pm\sqrt{189} = \pm 3\sqrt{21}.$$

Any suggestions as to how the truck can get through the tunnel without going into the left lane?

We ignore the negative value; the height of the arch 10 feet from the center is $3\sqrt{21} \approx 13.75$ feet. Hence a truck 10 feet wide and 14 feet high will not be able to make it through. ∎

Eccentricity

Ellipses can range in shape from almost circular to elongated. Figure 8.29 depicts two ellipses with the same vertices and centered at the origin, but with different foci.

Figure 8.29

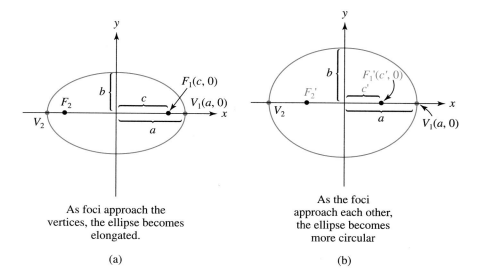

As foci approach the
vertices, the ellipse becomes
elongated.

(a)

As the foci
approach each other,
the ellipse becomes
more circular

(b)

To understand how the foci affect the shape of the ellipse, let us reexamine the relationship among the foci, major axes, and minor axes algebraically defined by the equation $b^2 = a^2 - c^2$. See Figure 8.29.

Looking at $b^2 = a^2 - c^2$, if we keep a constant and vary c, we find that as c gets closer to a, b becomes smaller. Geometrically, this means as the foci approach the vertices, the minor axis gets shorter. Note that the ellipse becomes elongated.

We use c and a to define the term *eccentricity,* denoted by e, which numerically describes the "roundness" of an ellipse.

Definition of Eccentricity of an Ellipse

The **eccentricity**, e, of an ellipse is defined by

$$e = \frac{c}{a} = \frac{\sqrt{a^2 - b^2}}{a}$$

The e being used here is **not** the base
for the natural exponential function or
for the natural log function.

We point out here that the eccentricity, e, is not the constant $e \approx 2.72$ that is the base of natural logarithms, but a variable value that describes the shape of an ellipse. Let's examine how the values of e are related to the shape of the ellipse.

Since $c > 0$ and $a > 0$, then $e = \dfrac{c}{a} > 0$. Also since $c < a$, then $e = \dfrac{c}{a} < 1$.

Putting these together, we have $0 < e < 1$.

Looking at the form of the eccentricity, $e = \dfrac{\sqrt{a^2 - b^2}}{a}$, keeping a constant, we see that as b gets smaller or closer to 0, algebraically e gets closer to 1. Hence, the flatter (or more elongated) the ellipse, the closer e gets to 1. (See the Different Perspectives box on page 522.)

On the other hand, as b gets closer to a (a must always be larger than b), algebraically $\sqrt{a^2 - b^2}$ gets closer to 0, and so e gets closer to 0. Hence, the ellipse becomes more circular as e approaches 0.

Different Perspectives: **Eccentricity**

Graphical Description

The eccentricity, e:

As e approaches 1, the ellipse becomes more elongated.

As e approaches 0, the ellipse becomes more circular.

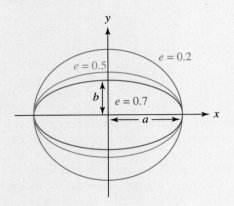

Algebraic Description

Consider eccentricity e defined by

$$e = \frac{\sqrt{a^2 - b^2}}{a}$$

As the value of b approaches 0, e approaches 1.

As the value of b approaches a, e approaches 0.

Example 5   Find the eccentricity, e, for the ellipse $\dfrac{x^2}{25} + \dfrac{y^2}{8} = 1$.

Solution   Since the eccentricity is $e = \dfrac{\sqrt{a^2 - b^2}}{a}$, we find e by first identifying a^2 and b^2. For the ellipse $\dfrac{x^2}{25} + \dfrac{y^2}{8} = 1$, we have $a^2 = 25$ (and, therefore, $a = 5$) and $b^2 = 8$. Hence

$$e = \frac{\sqrt{a^2 - b^2}}{a}$$

$$e = \frac{\sqrt{25 - 8}}{5} = \frac{\sqrt{17}}{5} \approx \boxed{0.825}$$

Figure 8.30

The solar system

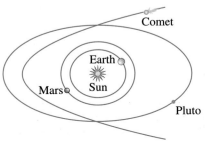

The orbits of the planets are elliptical with the sun at one of the foci. The orbits of Earth and Mars are almost circular (for Earth, $e \approx 0.017$; for Mars, $e \approx 0.093$), whereas the orbits of other planets are less circular, such as Pluto ($e \approx 0.25$) and Mercury ($e \approx 0.21$). See Figure 8.30. Many comets have elliptical orbits with the sun as the focus; for example, Haley's comet has eccentricity $e \approx 0.967$.

Regardless of the eccentricity of the ellipse, it can be shown that *the point on the ellipse closest to a focus is the nearest vertex; the point farthest from a focus is the farthest vertex.* We use this fact in the next example.

Example 6

Earth traces an elliptical path around the sun, with the sun at one of the foci. The length of half the major axis is 93 million miles, and the eccentricity is $e = 0.017$. Estimate the closest Earth is to the sun.

Solution

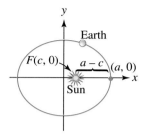

Figure 8.31

The earth traces an elliptical orbit with the sun at one focus.

Let's sketch the graph of an ellipse with its center at the origin and a horizontal major axis. By the fact stated above, Earth (on the ellipse) is closest to the focus (the sun) at the vertex closest to that focus. If we can find the distance between the focus and its vertex, then we have the distance we require. This distance is $a - c$ in Figure 8.31.

Because one-half the major axis is 93 million miles, we have $a = 9.3 \times 10^7$. Now we need to find c. We are given $e = 0.017$, so we use the definition of e to find c, as follows:

$$e = \frac{c}{a} \qquad \text{Substitute } e = 0.017 \text{ and } a = 9.3 \times 10^7 \text{ to get}$$

$$0.017 = \frac{c}{9.3 \times 10^7} \qquad \text{Solve for } c.$$

$$c = 0.017(9.3 \times 10^7)$$

$$= 1.581 \times 10^6 \quad \text{or approximately 1.6 million miles}$$

Hence, the nearest distance that Earth is to the sun is

$$a - c = 93 \text{ million miles} - 1.6 \text{ million miles} = 91.4 \text{ million miles} \qquad ∎$$

Figure 8.32

The reflecting properties of the ellipse.

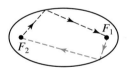

The ellipse also has a reflecting property similar to that of the parabola. In general, any ray emanating from one focus will be reflected off the ellipse to the other focus. This is illustrated in Figure 8.32.

This property is used in "whispering rooms," such as those found in St. Paul's Cathedral in London or the Capitol in Washington, D.C. In these rooms, the cross sections of the ceilings are made up of elliptical arcs. If you stand at one focus and whisper something quietly, sound is reflected off the ceiling so that whoever is standing at the other focus will be able to hear you.

Calculator Exploration

On a graphing calculator, graph the following on the same set of coordinate axes:

$$\frac{x^2}{36} + \frac{y^2}{4} = 1 \qquad \frac{(x-2)^2}{36} + \frac{y^2}{4} = 1 \qquad \frac{(x-2)^2}{36} + \frac{(y+5)^2}{4} = 1$$

(Recall that you must first solve explicitly for y.) What can you conclude about the effects of h and k on the graph of $\dfrac{(x-h)^2}{a^2} + \dfrac{(y-k)^2}{b^2} = 1$? If you cannot graph all three ellipses on one set of axes, compare the first with the second and then the second with the third.

The Ellipse Centered at (*h, k*)

As with the parabola, Figure 8.33 on page 524 demonstrates that the ellipse centered at $(0, 0)$ with foci on the x-axis can be shifted h units horizontally and k units vertically, to be centered at (h, k). This ellipse will have foci on the line $y = k$, a line

Figure 8.33

An ellipse shifted h units
horizontally and k units vertically

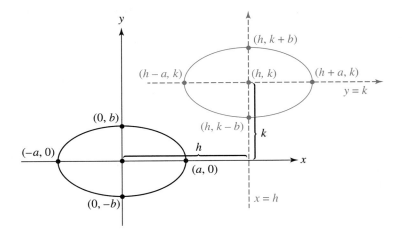

parallel to the x-axis. Its vertices are $(h \pm a, k)$; minor axis endpoints, $(h, k \pm b)$; and foci, $(h \pm c, k)$, where $c^2 = a^2 - b^2$.

Using the distance formula, we can derive the general form for this ellipse.

Standard Form of an Ellipse Centered at (h, k) with Foci at ($h + c$, k) and ($h - c$, k)

The graph of the equation

$$\frac{(x - h)^2}{a^2} + \frac{(y - k)^2}{b^2} = 1$$

for $a > b$ is an ellipse with center (h, k) and vertices $(h \pm a, k)$. The endpoints of the minor axis are $(h, k \pm b)$. The foci are $(h \pm c, k)$, where $c^2 = a^2 - b^2$. The length of the major axis is $2a$; the length of the minor axis is $2b$.

The ellipse with horizontal
major axis, centered at (h, k)

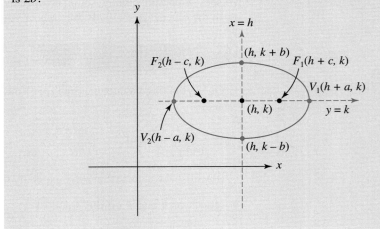

Comparing $\dfrac{x^2}{a^2} + \dfrac{y^2}{b^2} = 1$, the standard form of the ellipse centered at $(0, 0)$,

with $\dfrac{(x - h)^2}{a^2} + \dfrac{(y - k)^2}{b^2} = 1$, the standard form of the ellipse centered at (h, k),

we see that both ellipses have the same shape, but the center is moved h units horizontally and k units vertically.

Rather than memorizing new vertices, axes, and foci, if we have the ellipse in the form $\dfrac{(x-h)^2}{a^2} + \dfrac{(y-k)^2}{b^2} = 1$, we can identify h and k, draw a new set of coordinate axes through (h, k) (parallel to the x- and y-axes), and graph the ellipse on the new set of axes as though it were in the form $\dfrac{x^2}{a^2} + \dfrac{y^2}{b^2} = 1$.

Example 7 Sketch a graph of the following and identify its axes, center, vertices, and foci:

$$\frac{(x-3)^2}{16} + \frac{(y+4)^2}{4} = 1$$

Solution The ellipse $\dfrac{(x-3)^2}{16} + \dfrac{(y+4)^2}{4} = 1$ is in a standard form, so

$$h = 3 \quad k = -4 \quad a^2 = 16 \;\Rightarrow\; a = 4 \quad \text{and} \quad b^2 = 4 \;\Rightarrow\; b = 2$$

We draw a new set of coordinate axes centered at (h, k) which is $(3, -4)$. See Figure 8.34. The center of the ellipse is $(3, -4)$. The major axis lies on the horizontal line $y = -4$. Since $a = 4$, move 4 units horizontally left and right of the center $(3, -4)$ to find the vertices $(-1, -4)$ and $(7, -4)$. The minor axis lies on the vertical line $x = 3$. Since $b = 2$, move 2 units vertically above and below the center $(3, -4)$ to find the endpoints of the minor axis: $(3, -2)$ and $(3, -6)$.

Figure 8.34

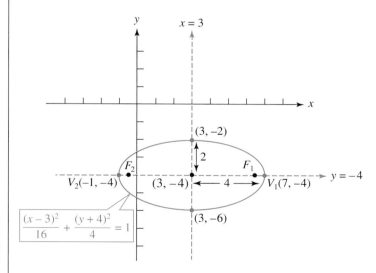

If we draw a new set of horizontal and vertical axes through the center, the new horizontal and vertical axes contain the axes of the ellipse.

Because $c^2 = a^2 - b^2 = 16 - 4 = 12 \;\Rightarrow\; c = \sqrt{12} = 2\sqrt{3}$, we can find the foci by moving $2\sqrt{3} \approx 3.46$ units horizontally left and right from the center $(3, -4)$ to get $F_2(3 - 2\sqrt{3}, -4)$ and $F_1(3 + 2\sqrt{3}, -4)$. The graph appears in Figure 8.34. ∎

Similarly, we can shift an ellipse centered at the origin with foci on the y-axis horizontally and vertically so that it is centered at (h, k): The new foci will lie on the line $x = h$, a line parallel to the y-axis. We can derive the following.

Standard Form of an Ellipse Centered at (*h*, *k*) with Foci at (*h*, *k* + *c*) and (*h*, *k* − *c*)

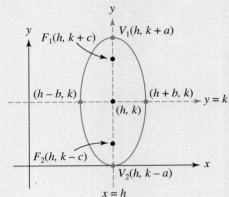

The ellipse with vertical major axis, centered at (*h*, *k*)

The graph of the equation

$$\frac{(x - h)^2}{b^2} + \frac{(y - k)^2}{a^2} = 1$$

for $a > b$ is an ellipse with center (h, k) and vertices $(h, k \pm a)$. The endpoints of the minor axis are $(h \pm b, k)$. The foci are $(h, k \pm c)$, where $c^2 = a^2 - b^2$. The length of the major axis is $2a$; the length of the minor axis is $2b$.

The ellipse $\dfrac{(x - h)^2}{b^2} + \dfrac{(y - k)^2}{a^2} = 1$ has the same shape as the ellipse $\dfrac{x^2}{b^2} + \dfrac{y^2}{a^2} = 1$, but its center is shifted h units horizontally and k units vertically.

Example 8   Sketch a graph of the ellipse $16x^2 + 25y^2 - 64x - 50y - 311 = 0$ and identify its axes, center, and vertices.

Solution   The equation $16x^2 + 25y^2 - 64x - 50y - 311 = 0$ is not in standard form, but we can put it in standard form by completing the square:

$$16x^2 + 25y^2 - 64x - 50y - 311 = 0$$ Add 311 to each side and group the terms as shown.

$$(16x^2 - 64x \quad) + (25y^2 - 50y \quad) = 311$$

The squared terms do not have coefficients equal to 1, and completing the square requires that the coefficients be 1. We can factor out the coefficient of each squared term, as follows:

$$(16x^2 - 64x \quad) + (25y^2 - 50y \quad) = 311$$ Factor 16 from the first group and 25 from the second group.

$$16(x^2 - 4x \quad) + 25(y^2 - 2y \quad) = 311$$ Complete the square for each expression within parentheses; add 4 and 1 within parentheses.

$$16(x^2 - 4x + 4) + 25(y^2 - 2y + 1) = 311 + 16 \cdot 4 + 25 \cdot 1$$ Note that we actually added 16 · 4 and 25 · 1 to both sides of the equation.

Now we write the left-hand expressions in factored form and simplify the numerical expression on the right-hand side:

$$16(x - 2)^2 + 25(y - 1)^2 = 400$$ Divide both sides by 400.

$$\frac{16(x - 2)^2}{400} + \frac{25(y - 1)^2}{400} = \frac{400}{400}$$ Which simplifies to

$$\frac{(x - 2)^2}{25} + \frac{(y - 1)^2}{16} = 1$$ This is in standard form.

Because 25 is the larger of the two denominators and is associated with the x^2-term, we have $a^2 = 25$, and the ellipse has a horizontal major axis. The ellipse is in standard form, so

$$h = 2 \quad k = 1 \quad a^2 = 25 \implies a = 5 \quad \text{and} \quad b^2 = 16 \implies b = 4$$

Note that the major axis for this form is parallel to the x-axis.

We draw a new set of coordinate axes centered at $(2, 1)$. The center of the ellipse is $(2, 1)$. The major axis lies on the horizontal line $y = 1$. Since $a = 5$, move 5 units horizontally right and left of $(2, 1)$ to find the vertices $(7, 1)$, and $(-3, 1)$.

The minor axis lies on the vertical line $x = 2$. Since $b = 4$, move 4 units vertically above and below $(2, 1)$, to find the endpoints of the minor axis: $(2, 5)$ and $(2, -3)$. The graph is sketched in Figure 8.35.

Figure 8.35

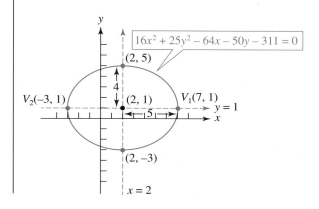

As with the circle and the parabola, we can manipulate the general equation $Ax^2 + Cy^2 + Dx + Ey + F = 0$ in such a way as to put it into one of the standard conic forms, depending on the coefficients of the squared terms. If the coefficients of the squared terms are equal ($A = C$), we expect the graph to be a circle; if there is only one squared term (either $A = 0$ or $C = 0$ but not both), we expect a parabola. We can draw a similar conclusion for the ellipse: We expect an ellipse if $A \neq C$ and A and C have the same sign. We return to this idea in Section 8.5.

Graph the figure $\dfrac{x^2}{4} + \dfrac{y^2}{9} = 0.$ Is it what you expected?

8.3 Exercises

In Exercises 1–10, identify the vertices and the foci of the ellipse.

1. $\dfrac{x^2}{25} + \dfrac{y^2}{9} = 1$ **2.** $\dfrac{x^2}{9} + \dfrac{y^2}{16} = 1$

3. $\dfrac{x^2}{12} + \dfrac{y^2}{18} = 1$ **4.** $\dfrac{x^2}{20} + \dfrac{y^2}{2} = 1$

5. $\dfrac{x^2}{24} + \dfrac{y^2}{9} = 1$ **6.** $\dfrac{y^2}{30} + \dfrac{x^2}{25} = 1$

7. $25x^2 + 9y^2 = 225$ **8.** $6x^2 + y^2 = 18$

9. $y^2 + 30x^2 = 30$ **10.** $x^2 + 2y^2 = 16$

In Exercises 11–20, graph the ellipse and identify the vertices and the foci.

11. $\dfrac{x^2}{49} + \dfrac{y^2}{9} = 1$ **12.** $\dfrac{y^2}{81} + \dfrac{x^2}{4} = 1$

13. $\dfrac{x^2}{36} + \dfrac{y^2}{25} = 1$ **14.** $\dfrac{x^2}{81} + \dfrac{y^2}{100} = 1$

15. $x^2 + 9y^2 = 36$ **16.** $9x^2 + 4y^2 = 144$

17. $12x^2 + y^2 = 24$ **18.** $18y^2 + 6x^2 = 36$

19. $3x^2 + 8y^2 = 12$ **20.** $5x^2 + 10y^2 = 5$

In Exercises 21–24, identify the eccentricity of the ellipse.

21. $\dfrac{x^2}{25} + \dfrac{y^2}{36} = 1$ **22.** $\dfrac{y^2}{16} + \dfrac{x^2}{4} = 1$

23. $8x^2 + 3y^2 = 12$ **24.** $15x^2 + 10y^2 = 5$

In Exercises 25–28, use the test point method discussed in Example 2 Section 8.1 on page 499 to sketch the solution set to the inequality.

25. $\dfrac{x^2}{36} + \dfrac{y^2}{24} > 1$ **26.** $\dfrac{y^2}{9} + \dfrac{x^2}{4} \leq 1$

27. $3x^2 + 4y^2 \leq 12$ **28.** $10x^2 + 15y^2 > 30$

In Exercises 29–32, write an equation of the ellipse using the given information.

29. The ellipse has foci $(2, 0)$ and $(-2, 0)$ and vertices $(4, 0)$ and $(-4, 0)$.

30. The ellipse has foci $(0, 3)$ and $(0, -3)$ and vertices $(0, 5)$ and $(0, -5)$.

31. The ellipse is centered at the origin; its major axis is horizontal, with length 8; the length of its minor axis is 4.

32. The ellipse is centered at the origin. The axes of the ellipse have lengths 5 and 9; the major axis is vertical.

33. The arch of a bridge is semielliptical with a horizontal major axis. If the base of the arch covers the entire 80-foot width of the road and the highest part of the bridge is 20 feet above the road, find the height of the arch 10 feet from the side of the road.

34. The arch of a bridge is semielliptical with a horizontal major axis. If the base of the arch covers the entire 80-foot width of the road and the highest part of the bridge is 20 feet above the road, find the height of the arch 10 feet from the center of the road.

35. The moon orbits Earth in an elliptical path, with Earth at one focus. The eccentricity of this orbit is $e \approx 0.055$, and the length of the major axis of this orbit is 468,972 miles. What is the minimum distance between the centers of Earth and Moon?

36. A satellite is launched from Earth and maintains an elliptical orbit with (the center of) Earth as one of its foci. The longest and shortest distances from Earth's surface are 800 miles and 300 miles, respectively. If Earth's radius is 4000 miles, find the equation of the orbit.

In Exercises 37–44, identify the center, the vertices, and the foci of the ellipse.

37. $\dfrac{(x - 2)^2}{16} + \dfrac{(y - 3)^2}{4} = 1$

38. $\dfrac{(x + 3)^2}{9} + \dfrac{(y - 1)^2}{49} = 1$

39. $\dfrac{(y - 3)^2}{18} + \dfrac{(x + 1)^2}{12} = 1$

40. $\dfrac{(x + 5)^2}{24} + \dfrac{y^2}{2} = 1$

41. $4x^2 + y^2 - 16x - 6y + 21 = 0$

42. $x^2 + 2y^2 + 4x + 12y + 6 = 0$

43. $6x^2 + 5y^2 - 60x - 10y + 65 = 0$

44. $x^2 + 2y^2 + 8x - 4y - 6 = 0$

In Exercises 45–52, graph the ellipse; identify the vertices and the foci.

45. $\dfrac{(x - 2)^2}{9} + \dfrac{(y + 3)^2}{8} = 1$

46. $\dfrac{(x + 3)^2}{8} + \dfrac{y^2}{12} = 1$

47. $\dfrac{(x + 2)^2}{16} + \dfrac{(y - 1)^2}{24} = 1$

48. $\dfrac{x^2}{6} + \dfrac{(y - 5)^2}{18} = 1$

49. $16x^2 + 9y^2 - 32x + 54y = 47$

50. $x^2 + 4y^2 + 4x + 40y + 100 = 0$

51. $x^2 + 4y^2 - 2x - 24y = -29$

52. $4x^2 + y^2 - 24x + 4y + 28 = 0$

In Exercises 53–58, identify and graph the figure, labeling its important aspects.

53. $\dfrac{x^2}{9} + \dfrac{y^2}{4} = 1$

54. $\dfrac{x}{9} + \dfrac{y}{4} = 1$

55. $\dfrac{(x + 2)^2}{16} + \dfrac{(y - 1)^2}{16} = 1$

56. $y^2 - 6y - 2x + 1 = 0$

57. $x^2 - 9y = 0$

58. $4x^2 + 3y^2 - 8x - 6y - 5 = 0$

In Exercises 59–62, write an equation of the ellipse using the given information.

59. The ellipse has foci $(1, 4)$ and $(5, 4)$ and vertices $(0, 4)$ and $(6, 4)$.

60. The ellipse has foci $(-1, 2)$ and $(-1, 10)$ and vertices $(-1, 0)$ and $(-1, 12)$.

61. The major axis of the ellipse lies on the line $y = -5$ and has length 4; the minor axis lies on the line $x = 3$ and has length 2.

62. The axes of the ellipse have lengths 1 and 6, the major axis is vertical, and the center of the ellipse is $(-3, 5)$.

63. Use the distance formula to derive the equation for the ellipse centered at the origin with foci $(0, \pm c)$ and vertices $(0, \pm a)$.

64. Use the distance formula to derive the equation for the ellipse centered at (h, k), with foci $(h \pm c, k)$ and vertices $(h \pm a, k)$.

65. We define a line to be tangent to an ellipse at point P if and only if it touches the ellipse only at point P. The equation for the line tangent to the ellipse $\frac{x^2}{a^2} + \frac{y^2}{b^2} = 1$ at the point (x_0, y_0) is given by

$$\frac{xx_0}{a^2} + \frac{yy_0}{b^2} = 1$$

(See the accompanying figure.) Use this formula to find an equation of the line tangent to the ellipse $\frac{x^2}{16} + \frac{y^2}{4} = 1$ at the point $(-2, \sqrt{3})$.

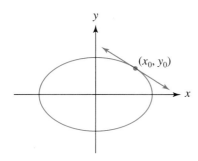

66. Use the formula for the equation of the line tangent to the ellipse in Exercise 65 to find the equation of the line tangent to the ellipse

$$\frac{x^2}{8} + \frac{y^2}{9} = 1$$

at the point $(0, 3)$.

67. The line segment with endpoints on an ellipse that passes through a focus and is perpendicular to the major axis is called a *latus rectum* of the ellipse. (See the accompanying figure.) Show that $\frac{2b^2}{a}$ is the length of each latus rectum of the ellipse $\frac{x^2}{a^2} + \frac{y^2}{b^2} = 1$.

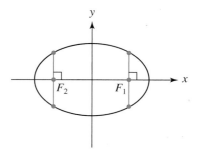

In Exercises 68–71, use your graphing calculator to graph the equations.

68. $\frac{x^2}{12} + \frac{y^2}{4} = 1$

69. $8x^2 + 6y^2 = 24$

70. $x^2 + 2y^2 - 2x = 1$

71. $\frac{(x - 3)^2}{12} + \frac{(y + 2)^2}{4} = 1$

Question for Thought

72. The equations $x^2 + 2y^2 + 3 = 0$ and $y^2 + 3x^2 = 0$ are in the general form of a second-degree equation. If you graphed these figures, what type of figures would you expect to have? Graph each figure on a set of coordinate axes. Is it what you expected?

8.4   The Hyperbola

As with the other conic sections, we begin with the geometric definition of the **hyperbola**.

Definition of a Hyperbola

A **hyperbola** is the set of all points in a plane such that the absolute value of the difference of their distances from two fixed points is a positive constant. The fixed points are called the **foci** of the hyperbola. (See Figure 8.36.)

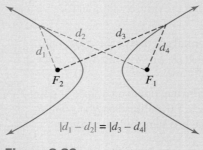

$$|d_1 - d_2| = |d_3 - d_4|$$

Figure 8.36

Whereas an ellipse is determined by the points whose sum of distances to the two foci is a constant, a hyperbola is determined by points whose *difference* of the distances to the two foci is constant.

As with the ellipse, given the definition of the hyperbola, we start by choosing a coordinate system in such a way that the foci lie on the horizontal axis with the origin midway between the two foci. Let $c > 0$ be the distance from the origin to each of the foci: F_1 and F_2. Hence the coordinates of F_1 are $(c, 0)$, and the coordinates of F_2 are $(-c, 0)$. The distance between the two foci, F_1 and F_2, is $2c$. (See Figure 8.37.) We will derive the equation for this hyperbola.

Figure 8.37

Hyperbola with foci $F_1(c, 0)$ and $F_2(-c, 0)$

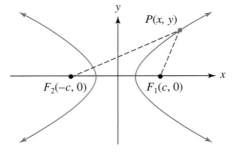

By the definition of a hyperbola, if we pick any point on the hyperbola, $P(x, y)$, the absolute value of the difference of the distance from $P(x, y)$ to $F_1(c, 0)$ and the distance from $P(x, y)$ to $F_2(-c, 0)$ is constant. To avoid working with fractions, we call this constant distance $2a$ and note that $2c > 2a$, or $c > a$. Hence, using the distance formula, we get

To show $2c > 2a$, consider $\triangle PF_1F_2$ in Figure 8.37. We have
$|\overline{PF_1}| + |\overline{F_1F_2}| > |\overline{PF_2}|$ implies that
$|\overline{F_1F_2}| > |\overline{PF_2}| - |\overline{PF_1}| = 2a$
But $|\overline{F_1F_2}| = 2c$. Hence $2c > 2a$. We can argue similarly for a point P on the left branch of the hyperbola.

$$\left| \sqrt{(x - c)^2 + (y - 0)^2} - \sqrt{(x + c)^2 + (y - 0)^2} \right| = 2a$$

If we manipulate this equation in the same way we did for the ellipse, we get

$$\frac{x^2}{a^2} - \frac{y^2}{c^2 - a^2} = 1$$

Because a and c are positive (they are distances), and $c > a$, we have $c^2 - a^2 > 0$. If we let $b = \sqrt{c^2 - a^2}$, then $b^2 = c^2 - a^2$; hence we have derived the following.

The hyperbola centered at the origin and with focal points on the x-axis has the equation

$$\frac{x^2}{a^2} - \frac{y^2}{b^2} = 1 \qquad (c > a)$$

The x-intercepts of the graph of this equation are found by setting $y = 0$:

$$\frac{x^2}{a^2} - \frac{y^2}{b^2} = 1 \qquad \text{Let } y = 0; \text{ solve for } x.$$

$$\frac{x^2}{a^2} - \frac{0^2}{b^2} = 1$$

$$\frac{x^2}{a^2} = 1 \implies x^2 = a^2 \implies x = \pm a \qquad \text{The } x\text{-intercepts are } \pm a.$$

We call the points $V_1(a, 0)$, and $V_2(-a, 0)$ the **vertices** of this hyperbola, and we call the line segment $\overline{V_1 V_2}$ the **transverse axis**. See figure in box on page 532.

Note that if we perform the same analysis to find the y-intercepts (let $x = 0$ and solve for y), we end up with $-\dfrac{y^2}{b^2} = 1 \implies y^2 = -b^2$, which is impossible (because $-b^2$ must be negative and y^2 cannot be negative). Therefore, there are no y-intercepts. The line segment joining the points $(0, b)$ and $(0, -b)$ is called the **conjugate axis** of the hyperbola. See figure in box on page 532.

For example, Figure 8.38 illustrates the graph of the hyperbola $\dfrac{x^2}{16} - \dfrac{y^2}{9} = 1$. Because $a^2 = 16 \implies a = 4$, the vertices are $(\pm 4, 0)$. Also, $b^2 = 9 \implies b = 3$, and since $c^2 = a^2 + b^2$, we have $c^2 = 16 + 9 = 25 \implies c = 5$. Hence the foci are $(\pm 5, 0)$.

Figure 8.38

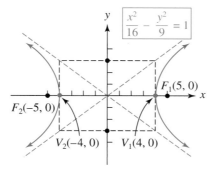

If we solve the equation $\dfrac{x^2}{16} - \dfrac{y^2}{9} = 1$ for y, we get

$$y = \pm \sqrt{\frac{9x^2 - 144}{16}} \qquad \text{Which can be rewritten as}$$

$$y = \pm \sqrt{\frac{9}{16}(x^2 - 16)} \quad \text{or} \quad y = \pm \frac{3}{4}\sqrt{x^2 - 16}$$

Why must we have $|x| \geq 4$?

Looking carefully at this form of the standard equation, we see that we must have $|x| \geq 4$. Notice that this is consistent with Figure 8.38; x never takes on values strictly between -4 and 4, and therefore there are no y-intercepts.

In our previous experience with graphs, we have worked with horizontal and vertical asymptotes. A distinguishing feature of the hyperbola is that it has asymptotes, although they are neither horizontal nor vertical. These asymptotes are shown in Figure 8.38. They determine the shape of the hyperbola and are helpful guides in sketching its graph. For the graph of the hyperbola $\dfrac{x^2}{16} - \dfrac{y^2}{9} = 1$ in Figure 8.38, the equations of the asymptotes are $y = \pm\dfrac{3}{4}x$.

In Exercise 55 we manipulate the standard form of the hyperbola to determine the asymptotes of the general hyperbola. We state this result here without proof.

The asymptotes of the hyperbola $\dfrac{x^2}{a^2} - \dfrac{y^2}{b^2} = 1$ are

$$y = \pm\frac{b}{a}x$$

We summarize this discussion as follows.

Standard Form of a Hyperbola with Foci at $(c, 0)$ and $(-c, 0)$

The "horizontal" hyperbola centered at $(0, 0)$

How are the endpoints of the *transverse* axis related to the vertices?

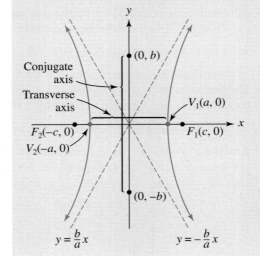

The graph of the equation

$$\frac{x^2}{a^2} - \frac{y^2}{b^2} = 1$$

is a hyperbola centered at $(0, 0)$ with vertices $(\pm a, 0)$. The foci are $(\pm c, 0)$, where $c^2 = a^2 + b^2$. The endpoints of the conjugate axis are $(0, \pm b)$. The length of the transverse axis is $2a$; the length of the conjugate axis is $2b$. The asymptotes of the hyperbola are $y = \pm\dfrac{b}{a}x$.

We refer to the two "arcs" of the hyperbola as the *branches of the hyperbola.*

An efficient way to draw the hyperbola is to first plot the vertices, $(\pm a, 0)$, and the endpoints of the conjugate axis, $(0, \pm b)$. See Figure 8.39. Draw a rectangle such

Figure 8.39

Draw the rectangle and the diagonals. The diagonals of the rectangle are contained in the asymptotes of the hyperbola.

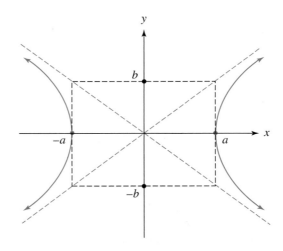

that these points are the midpoints of the sides of the rectangle. Then draw the diagonals of the rectangle. *The lines containing the diagonals of the rectangle are the asymptotes of the hyperbola.* The vertices and the asymptotes are the only guides we need to draw the hyperbola.

Note that one diagonal of the rectangle passes through the origin and the point (a, b). Using the slope–intercept form, you can verify that the equation of this diagonal is the asymptote $y = \dfrac{b}{a}x$.

Example 1   Sketch the graph of the hyperbola $9x^2 - 25y^2 = 225$. Identify the foci and the asymptotes.

Solution   First we put the equation into standard form, which requires that a 1 be on one side of the equation. To have 1 on the right-hand side, we must divide by 225:

$$9x^2 - 25y^2 = 225 \qquad \text{Divide both sides by 225.}$$

$$\frac{9x^2}{225} - \frac{25y^2}{225} = \frac{225}{225} \qquad \text{Simplify.}$$

$$\frac{x^2}{25} - \frac{y^2}{9} = 1 \qquad \text{This equation is in standard form.}$$

We compare our equation with the standard form of the hyperbola, and we get

$$a^2 = 25 \;\Rightarrow\; a = 5 \quad \text{and} \quad b^2 = 9 \;\Rightarrow\; b = 3 \;\;(a \text{ and } b \text{ are both positive})$$

We plot the vertices $(\pm 5, 0)$ and the endpoints of the conjugate axis: $(0, \pm 3)$. We draw a rectangle such that the points we just plotted are the *midpoints of the sides of the rectangle.* The lines containing the diagonal of the rectangle are the asymptotes

$$y = \pm \frac{b}{a}x \;\Rightarrow\; y = \pm \frac{3}{5}x$$

Because $c^2 = a^2 + b^2 = 25 + 9 = 34$, $c = \sqrt{34}$. Hence the foci are $(\pm\sqrt{34}, 0)$. (The length of the transverse axis is $2(5) = 10$, and the length of the conjugate axis is $2(3) = 6$.) We sketch the graph as shown in Figure 8.40.

Figure 8.40

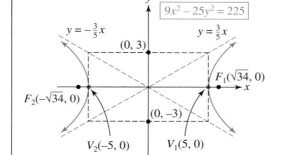

As with the hyperbola with foci on the x-axis, we can use the same analysis (again with $c > a$) to identify the standard form of the hyperbola with foci on the y-axis, which is summarized next.

Standard Form of a Hyperbola with Foci at (0, c) and (0, −c)

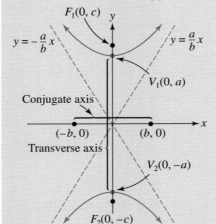

The "vertical" hyperbola centered at (0, 0)

The graph of the equation

$$\frac{y^2}{a^2} - \frac{x^2}{b^2} = 1$$

is a hyperbola centered at $(0, 0)$, with vertices $(0, \pm a)$. The foci are $(0, \pm c)$, where $c^2 = a^2 + b^2$. The endpoints of the conjugate axis are $(\pm b, 0)$. The length of the transverse axis is $2a$; the length of the conjugate axis is $2b$. The asymptotes of

the hyperbola are $y = \pm \dfrac{a}{b}x$.

For the hyperbola centered at the origin with foci on the y-axis, the vertices $(0, \pm a)$ are the y-intercepts. There are no x-intercepts.

Compare the two forms of the hyperbola, noting the similarities and the differences: The first form has its foci on the x-axis and the second has its foci on the y-axis. Notice that the two forms switch a and b as denominators of x^2 and y^2. The choice of form to use depends on which squared term has a positive coefficient: If the x^2-term has a positive coefficient, a^2 is the denominator of x^2, and the foci lie on the x-axis. On the other hand, if the y^2-term has a positive coefficient, a^2 is the denominator of y^2, and the foci lie on the y-axis.

Example 2 Using a graphing calculator, graph the hyperbola $36y^2 - x^2 = 9$. Identify its foci.

Solution This equation is not solved explicitly for y. We first have to solve for y before we can enter it into the graphing calculator.

$$36y^2 - x^2 = 9 \qquad \text{Isolate } y^2.$$

$$36y^2 = 9 + x^2 \qquad \text{Divide both sides by 36 to get}$$

$$y^2 = \frac{9 + x^2}{36} \qquad \text{Take square roots.}$$

$$y = \pm\sqrt{\frac{9 + x^2}{36}}$$

We enter *both* equations: $y = \dfrac{\sqrt{9 + x^2}}{6}$ and $y = -\dfrac{\sqrt{9 + x^2}}{6}$ The graph of the hyperbola is in Figure 8.41.

Figure 8.41

The graph of $36y^2 - x^2 = 9$

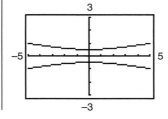

To find the foci, we first put the equation into standard form. For us to have 1 on the right-hand side, we must divide by 9.

$$36y^2 - x^2 = 9$$ Divide both sides by 9.

$$\frac{36y^2}{9} - \frac{x^2}{9} = \frac{9}{9}$$ Simplify.

$$4y^2 - \frac{x^2}{9} = 1$$ The equation is not quite in standard form;
however, we can rewrite $4y^2$ as $\frac{y^2}{\frac{1}{4}}$.

$$\frac{y^2}{\frac{1}{4}} - \frac{x^2}{9} = 1$$

Comparing our equation with both standard forms of the hyperbola, we note that the y^2-term has a positive coefficient; therefore a^2 is the denominator of y^2. We use the second form, $\frac{y^2}{a^2} - \frac{x^2}{b^2} = 1$, a hyperbola with foci on the y-axis. Hence

$$a^2 = \frac{1}{4} \implies a = \frac{1}{2} \quad \text{and} \quad b^2 = 9 \implies b = 3$$

Because $c^2 = a^2 + b^2 = 9 + \frac{1}{4} = \frac{37}{4} \implies c = \sqrt{\frac{37}{4}} = \frac{\sqrt{37}}{2}$, the foci are $\left(0, \pm \frac{\sqrt{37}}{2}\right)$. ∎

Hyperbolas are important in physics and engineering. The following example illustrates the hyperbola's use in navigation.

Example 3   An explosion is recorded by two microphones, M_1 and M_2, which are 2 miles apart. Microphone M_1 receives the sound 4 seconds before microphone M_2. If the speed of sound is 1100 ft/sec, determine the possible locations of the explosion in relation to the location of the microphones.

Solution   Let's start by putting the microphones in a coordinate system, where M_1 and M_2 lie on a horizontal axis with the origin midway between. Although we are not sure how far away the explosion was from either microphone, we do know that M_2 received the sound 4 seconds after M_1. Since the speed of sound is 1100 ft/sec, this means that the explosion, E, took place 4400 feet farther from M_2 than from M_1, or that *the difference between the distance from M_1 to E and the distance from M_2 to E is 4400 feet*. (See Figure 8.42.)

Figure 8.42

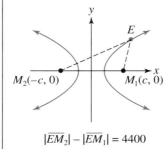

$$|\overline{EM_2}| - |\overline{EM_1}| = 4400$$

The possible points or locations of E that satisfy these conditions fit the definition of a hyperbola and therefore trace the path of a hyperbola, with the microphones at the foci.

To get the equation of the hyperbola, we note that the microphones are the foci, located 1 mile from the origin, and so $c = 5280$ feet. (We will express the equation in terms of feet.) The difference between the distances is 4400 feet; from the derivation of the hyperbola, this distance is $2a$. Hence $2a = 4400 \Rightarrow a = 2200 \Rightarrow a^2 = 4,840,000$.

Also,

$$b^2 = c^2 - a^2 = 5280^2 - 2200^2 = 23,038,400$$

How can we tell the explosion occurred on the branch with microphone M_1?

Hence the equation of the hyperbola (in feet) is $\dfrac{x^2}{4,840,000} - \dfrac{y^2}{23,038,400} = 1$. The explosion occurred somewhere on the right branch (the branch closest to M_1) of this hyperbola. ∎

The long-range navigational system known as LORAN locates ships and planes by applying the principle illustrated by Example 3. We have shown that using a pair of microphones, we can locate the source of a sound, such as an explosion, somewhere along one branch of a hyperbola.

Actually we can do the same thing with three microphones rather than four. How?

If another pair of microphones were placed at another location at a fixed distance from each other and also recorded the sound, we would find that the source of the sound could be located somewhere along one branch of another hyperbola (relative to the second pair of microphones). Since the sound source would lie on both hyperbolas, we could pinpoint the exact location of the sound by finding where the two hyperbolas intersect, as illustrated in Figure 8.43.

Figure 8.43

Pinpointing location L as the intersection of two hyperbolas

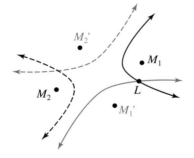

On a graphing calculator, graph the following on the same set of coordinate axes:

$$\frac{x^2}{9} - \frac{y^2}{4} = 1 \qquad \frac{(x-2)^2}{9} - \frac{y^2}{4} = 1 \qquad \frac{(x-2)^2}{9} - \frac{(y+1)^2}{4} = 1$$

What can you conclude about the effects of h and k on the graph of

$$\frac{(x-h)^2}{a^2} - \frac{(y-k)^2}{b^2} = 1?$$

The Hyperbola Centered at (*h, k*)

Figure 8.44 demonstrates that the hyperbola centered at $(0, 0)$ with foci on the x-axis can be shifted h units horizontally and k units vertically to be centered at (h, k).

This hyperbola will have foci on the line $y = k$, the line parallel to the x-axis and shifted k units from the x-axis. Its vertices will be $(h \pm a, k)$, conjugate axis endpoints, $(h, k \pm b)$, and foci, $(h \pm c, k)$, where $c^2 = a^2 + b^2$.

Figure 8.44

The hyperbola shifted h units horizontally and k units vertically

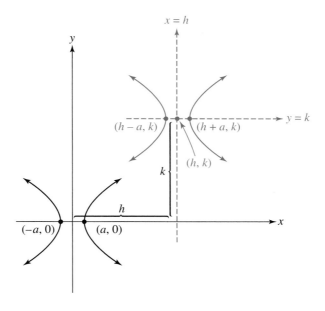

As with the other conics, we generalize the hyperbola centered at (h, k).

Standard Form of a Hyperbola Centered at (h, k) with Foci at $(h + c, k)$ and $(h - c, k)$

The "horizontal" hyperbola centered at (h, k)

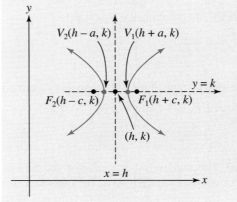

The graph of the equation

$$\frac{(x - h)^2}{a^2} - \frac{(y - k)^2}{b^2} = 1$$

is a hyperbola centered at (h, k) with vertices $(h \pm a, k)$. The endpoints of the conjugate axis are $(h, k \pm b)$. The foci are $(h \pm c, k)$, where $c^2 = a^2 + b^2$. The length of the transverse axis is $2a$; the length of the conjugate axis is $2b$.

The asymptotes of the hyperbola $\dfrac{(x - h)^2}{a^2} - \dfrac{(y - k)^2}{b^2} = 1$ are given by $y - k = \pm \dfrac{b}{a}(x - h)$.

Similarly, we can shift a hyperbola with foci on the y-axis horizontally and vertically to be centered at (h, k), and the new foci will lie on the line $x = h$, the line parallel to the y-axis, h units from the y-axis, with the following result.

Standard Form of a Hyperbola Centered at (h, k) with Foci at (h, $k + c$) and (h, $k - c$)

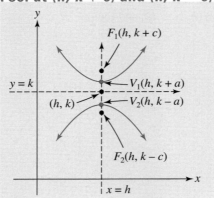

The "vertical" hyperbola centered at (h, k)

The graph of the equation

$$\frac{(y - k)^2}{a^2} - \frac{(x - h)^2}{b^2} = 1$$

is a hyperbola centered at (h, k) with vertices (h, $k \pm a$). The endpoints of the conjugate axis are ($h \pm b$, k). The foci are (h, $k \pm c$), where $c^2 = a^2 + b^2$. The length of the transverse axis is $2a$; the length of the conjugate axis is $2b$.

The asymptotes of the hyperbola $\dfrac{(y - k)^2}{a^2} - \dfrac{(x - h)^2}{b^2} = 1$ are given by $y - k = \pm\dfrac{a}{b}(x - h)$. The hyperbola $\dfrac{(y - k)^2}{a^2} - \dfrac{(x - h)^2}{b^2} = 1$ has the same shape as the hyperbola $\dfrac{y^2}{a^2} - \dfrac{x^2}{b^2} = 1$, but its center is shifted h units horizontally and k units vertically.

As with the ellipse, rather than memorizing the new set of vertices, foci, axes, and asymptotes of the hyperbola centered at (h, k), we use the fact that it has the same shape as the hyperbola centered at the origin and graph an equation of the form $\dfrac{(y - k)^2}{a^2} - \dfrac{(x - h)^2}{b^2} = 1$ by drawing a new set of coordinate axes through the point (h, k), and graphing the equation $\dfrac{y^2}{a^2} - \dfrac{x^2}{b^2} = 1$ on the new set of axes. This is demonstrated next.

Example 4   Sketch a graph of the following and identify its center, axes, vertices, and asymptotes.

(a) $\dfrac{(x + 2)^2}{9} - \dfrac{(y - 4)^2}{36} = 1$ (b) $25y^2 - 4x^2 - 250y + 24x = -489$

Solution   (a) This equation is already in a standard form. Because the x^2-term has a positive coefficient, we use the first standard form (the foci lie on a horizontal line) and the denominator of the x^2-term is a^2. Hence $a^2 = 9$. We can now read

$$h = -2 \quad k = 4 \quad a^2 = 9 \implies a = 3 \quad b^2 = 36 \implies b = 6$$

We draw a new set of coordinate axes centered at (h, k), which is (-2, 4). The hyperbola is centered at (-2, 4). The transverse axis lies on the horizontal line $y = 4$. Since $a = 3$, move 3 units horizontally left and right from its center (-2, 4) to find the vertices (-5, 4), and (1, 4).

The conjugate axis lies on the vertical line $x = -2$. Since $b = 6$, move 6 units vertically up and down from its center (-2, 4) to find the endpoints of the conjugate axis: (-2, 10) and (-2, -2). The rectangle is drawn such that the points plotted are the midpoints of the sides of the rectangle, and the asymptotes are the lines passing through the diagonals of the rectangle. The equations of the asymptotes are

$$y - k = \pm\frac{b}{a}(x - h) \quad \Rightarrow \quad y - 4 = \pm 2(x + 2)$$

or

$$y = 2x + 8 \quad \text{and} \quad y = -2x$$

The graph is shown in Figure 8.45.

Figure 8.45

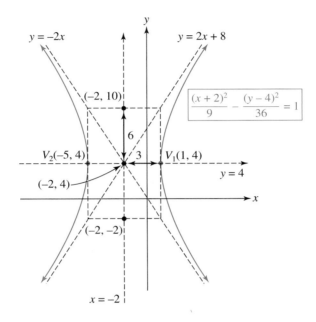

(b) This equation is not in standard form, but as with ellipses, we can put it in standard form by completing the square.

$$25y^2 - 4x^2 - 250y + 24x = -489$$

Group terms containing powers of x; group terms containing powers of y.

$$(25y^2 - 250y \quad) + (-4x^2 + 24x \quad) = -489$$

Factor 25 from the first group and -4 from the second group.

$$25(y^2 - 10y \quad) - 4(x^2 - 6x \quad) = -489$$

We complete the square for each expression within parentheses, to get

$$25(y^2 - 10y + 25) - 4(x^2 - 6x + 9) = -489 + 25 \cdot 25 - 4 \cdot 9$$

Note that we actually added $25 \cdot 25$ and $-4 \cdot 9$ to both sides of the equation.

Now we write the expression in factored form and simplify the numerical expressions on the right-hand side to get $25(y - 5)^2 - 4(x - 3)^2 = 100$, which when written in standard form is

$$\frac{(y - 5)^2}{4} - \frac{(x - 3)^2}{25} = 1$$

Because the "y^2-term" has a positive coefficient, we compare this form with the second form of the hyperbola centered at (h, k), where the foci lie on a vertical line and a^2 is the denominator of the y^2-term. We can now read

$$h = 3 \quad k = 5 \quad a^2 = 4 \quad \Rightarrow \quad a = 2 \quad b^2 = 25 \quad \Rightarrow \quad b = 5$$

We draw a new set of coordinate axes centered at $(3, 5)$. The hyperbola is centered at $(3, 5)$. The transverse axis lies on the vertical line $x = 3$. Because $a = 2$, move 2 units vertically up and down from its center $(3, 5)$ to find the vertices $(3, 7)$ and $(3, 3)$.

The conjugate axis lies on the horizontal lines $y = 5$. Since $b = 5$, move 5 units horizontally left and right from its center $(3, 5)$ to find the endpoints of the conjugate axis: $(-2, 5)$ and $(8, 5)$. The rectangle is drawn such that the points plotted are the midpoints of the sides of the rectangle, and the asymptotes are the lines passing through the diagonals of the rectangle. The equations of the asymptotes are

$$y - k = \pm\frac{a}{b}(x - h) \quad \Rightarrow \quad y - 5 = \pm\frac{2}{5}(x - 3)$$

or

$$y = \frac{2}{5}x + \frac{19}{5} \quad \text{and} \quad y = -\frac{2}{5}x + \frac{31}{5}$$

The graph appears in Figure 8.46.

Figure 8.46

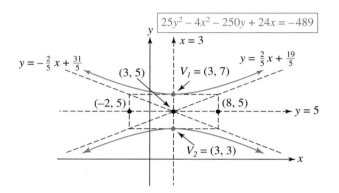

As with the other conic sections, we can manipulate the general equation $Ax^2 + Cy^2 + Dx + Ey + F = 0$ in such a way as to put it into one of the standard conic forms, depending on the coefficients of the squared terms. For example, if the coefficients of the squared terms have the same sign (and are not equal to each other or 0), then we would expect the graph to be an ellipse. We can draw a similar conclusion for the hyperbola: If A and C have opposite signs in the equation $Ax^2 + Cy^2 + Dx + Ey + F = 0$, then we would expect the graph of this equation to be a hyperbola. We will discuss these conditions and their exceptions in the next section.

8.4 Exercises

In Exercises 1–10, identify the vertices, foci, and equations of the asymptotes of the hyperbola.

1. $\dfrac{x^2}{9} - \dfrac{y^2}{16} = 1$

2. $\dfrac{y^2}{16} - \dfrac{x^2}{16} = 1$

3. $\dfrac{y^2}{12} - \dfrac{x^2}{18} = 1$

4. $\dfrac{x^2}{2} - \dfrac{y^2}{20} = 1$

5. $\dfrac{x^2}{9} - \dfrac{y^2}{18} = 1$

6. $\dfrac{y^2}{36} - \dfrac{x^2}{15} = 1$

7. $25x^2 - 9y^2 = 225$

8. $6y^2 - 3x^2 = 18$

9. $30y^2 - x^2 = 30$

10. $x^2 - 8y^2 = 16$

In Exercises 11–16, graph the hyperbola and identify the vertices, foci, and equations of the asymptotes.

11. $\dfrac{x^2}{9} - \dfrac{y^2}{49} = 1$

12. $\dfrac{y^2}{81} - \dfrac{x^2}{4} = 1$

13. $8x^2 - y^2 = 24$

14. $20x^2 - 5y^2 = 40$

15. $3y^2 - 8x^2 = 24$ **16.** $3y^2 - 15x^2 = 30$

17. Coast Guard station A is located 100 miles east of Coast Guard station B. Radio signals are sent simultaneously from stations A and B, traveling at a rate of 980 ft/μ sec (microsecond). A boat is sailing within range of both signals. If the boat receives the signal from station A 200 μ sec after receiving the signal from station B, express the location of the boat as an equation in relation to the two stations. HINT: Create a coordinate system where the two stations are located on the *x*-axis, each station on opposite sides of and equidistant from the *y*-axis.

18. Sketch the graph of the equation found in Exercise 17. If this boat is sailing parallel to and 20 miles north of the two Coast Guard stations, use the graph to show its exact location relative to the two Coast Guard stations.

19. Some comets travel in a hyperbolic orbit with the sun as a focus (and we never see them again). It can be shown that the vertex of a branch of a hyperbola is the point on the hyperbola closest to the focus associated with that branch. Given this fact and that the path of the comet is described by the hyperbola $4x^2 - 3y^2 - 12 = 0$, with the sun as a focus, find how close the comet gets to the sun. (The numbers given are in terms of *AU*, or *astronomical units*, where 1 AU $\approx$ 93,000,000 miles.) (See the accompanying figure.)

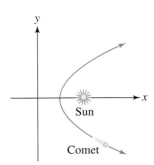

20. The impulse engines break down on the Starship Enterprise again, and the ship is traveling on momentum at a slow, constant rate of speed. Unfortunately, the ship is carrying medical supplies that a planet desperately needs within 24 hours; the shipboard computers calculate that at the current rate of speed, they would not arrive for another 4 days. The captain gets an idea: If they can approach a star (conveniently nearby) and maintain a hyperbolic orbit, the star's gravity will accelerate the ship and send it out at a rate of speed that will get them to the planet on time. Given their speed and the gravitational attraction of the star, the shipboard computer computes

the angle of approach and comes up with the equation $5x^2 - 9y^2 - 45 = 0$ as the path they must travel using the star as a focus. Noting the comments in Exercise 19, how close does the ship get to the star?

In Exercises 21–28, identify the center, vertices, foci, and equations of the asymptotes of the hyperbola.

21. $\dfrac{(x - 1)^2}{4} - \dfrac{(y - 3)^2}{4} = 1$

22. $\dfrac{(y - 3)^2}{9} - \dfrac{(x + 2)^2}{16} = 1$

23. $\dfrac{(y + 3)^2}{12} - \dfrac{(x + 1)^2}{18} = 1$

24. $\dfrac{(x + 5)^2}{2} - \dfrac{y^2}{24} = 1$

25. $x^2 - 4y^2 - 6x + 8y = 11$

26. $25y^2 - 9x^2 - 100y - 54x - 206 = 0$

27. $2y^2 - 3x^2 + 4y + 6x = 49$

28. $x^2 - 2y^2 - 2x - 12y = 35$

In Exercises 29–36, graph the hyperbola, identify the center, vertices, foci, and equations of the asymptotes.

29. $\dfrac{(x + 4)^2}{9} - \dfrac{(y + 3)^2}{16} = 1$

30. $\dfrac{(y - 1)^2}{4} - \dfrac{x^2}{36} = 1$

31. $\dfrac{x^2}{5} - \dfrac{(y - 3)^2}{15} = 1$

32. $\dfrac{(y - 3)^2}{20} - \dfrac{(x + 1)^2}{18} = 1$

33. $9x^2 - 16y^2 - 36x - 32y = 124$

34. $2y^2 - x^2 + 4y - 2x = 17$

35. $25y^2 - 36x^2 - 150y + 288x - 1251 = 0$

36. $18x^2 - y^2 - 72x + 6y - 45 = 0$

In Exercises 37–38, write the equation of the hyperbola using the given information.

37. The hyperbola has vertices $(2, -1)$ and $(10, -1)$ and foci $(0, -1)$ and $(12, -1)$.

38. The transverse axis of the hyperbola lies on the line $y = -2$ and has length 4; the conjugate axis lies on the line $x = 3$ and has length 6.

In Exercises 39–44, identify and graph the figure, labeling its important aspects.

39. $\dfrac{x^2}{12} + \dfrac{y^2}{4} = 1$

40. $\dfrac{x}{12} - \dfrac{y}{4} = 1$

41. $\dfrac{(x + 2)^2}{9} - \dfrac{(y - 1)^2}{25} = 1$

42. $3y^2 - 2x^2 - 12 = 0$

43. $x^2 - 2y - 12 = 0$

44. $4x^2 + 4y^2 - 8x - 16y + 4 = 0$

45. Use the distance formula to derive an equation for the hyperbola centered at the origin with foci $(\pm c, 0)$ and vertices $(\pm a, 0)$.

46. Use the distance formula to derive an equation for the hyperbola centered at (h, k) with foci $(h \pm c, k)$ and vertices $(h \pm a, k)$.

47. An equation for the line tangent to the hyperbola

$\dfrac{x^2}{a^2} - \dfrac{y^2}{b^2} = 1$ at the point (x_0, y_0) is given by

$$\dfrac{xx_0}{a^2} - \dfrac{yy_0}{b^2} = 1$$

Use this formula to find an equation of the line tangent to the hyperbola $\dfrac{x^2}{16} - \dfrac{y^2}{4} = 1$ at the point $(4, 0)$.

48. Use the formula for the equation of the line tangent to the hyperbola in Exercise 47 to find an equation of the line tangent to the hyperbola

$$\dfrac{x^2}{6} - \dfrac{y^2}{9} = 1$$

at the point $(6, 3\sqrt{5})$.

49. The line segment AB passing through a focus of a hyperbola and perpendicular to the transverse axis has its endpoints on the hyperbola. See the accompanying figure.

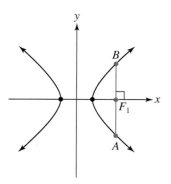

Show that, for the hyperbola $\dfrac{x^2}{a^2} - \dfrac{y^2}{b^2} = 1$, the length of the line segment AB is $\dfrac{2b^2}{a}$.

In Exercises 50–53, use your graphing calculator to graph the equations.

50. $\dfrac{x^2}{8} - \dfrac{y^2}{4} = 1$

51. $9x^2 - 3y^2 = 24$

52. $x^2 - 3y^2 - 2x = 2$

53. $\dfrac{(x + 3)^2}{8} - \dfrac{(y - 2)^2}{2} = 1$

Questions for Thought

54. (a) Show that if we start with the equation $\dfrac{x^2}{9} - \dfrac{y^2}{4} = 1$ and solve for y, we obtain

$$y = \pm\dfrac{2}{3}\sqrt{x^2 - 9}$$

(b) Compute y using the following four equations for these values of x: $x = 4, 10, 20, 100, 200, 1000, 2000$.

$$y = \pm\dfrac{2}{3}\sqrt{x^2 - 9} \qquad y = \pm\dfrac{2}{3}x$$

(c) What can you conclude about the relationship between the hyperbola and the lines

$$y = \pm\dfrac{2}{3}x$$

55. (a) Show that if we start with the equation $\dfrac{x^2}{a^2} - \dfrac{y^2}{b^2} = 1$ and solve for y, we obtain

$$y = \pm\dfrac{b}{a}\sqrt{x^2 - a^2}$$

(b) Show that we can factor x^2 in the radicand to get

$$y = \pm\dfrac{b}{a}\sqrt{x^2\left(1 - \dfrac{a^2}{x^2}\right)} \qquad \text{Which is}$$

$$y = \pm\dfrac{b}{a}x\sqrt{1 - \dfrac{a^2}{x^2}}$$

(c) What happens to the radical term in the last equation above as $x \to \infty$, or $x \to -\infty$?

(d) What can you conclude about y as $x \to \infty$, or $x \to -\infty$?

56. The *eccentricity* of the hyperbola $\dfrac{x^2}{a^2} - \dfrac{y^2}{b^2} = 1$ ($c > a$) is defined to be

$$e = \frac{c}{a} = \frac{\sqrt{a^2 + b^2}}{a}$$

(See the accompanying figure.)

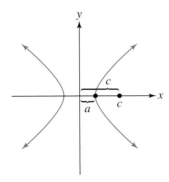

(a) What is the range of possible values for the eccentricity, e, of a hyperbola?

(b) What happens to the shape of the hyperbola as e gets larger or smaller?

57. The equation $x^2 - 4y^2 = 0$ is in the general form of a second-degree equation. If you graphed this figure, what type of figure would you expect to have? Graph the figure on a set of coordinate axes. Is it what you expected?

58. Finding the solutions of an inequality involving a hyperbola is not as obvious as for a circle or ellipse. The definition of the hyperbola is a bit more complicated, and the hyperbola splits the plane into three rather than two regions. The accompanying figure is an example of the graph of the second-degree inequality

$$\frac{(x - 1)^2}{49} - \frac{(y - 3)^2}{16} \le 1$$

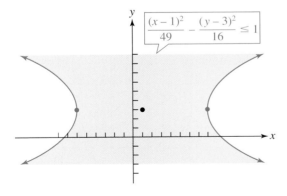

Sketch the graph of the inequality $\dfrac{x^2}{49} - \dfrac{y^2}{16} > 1$ using the following steps:

(a) Sketch the graph of the hyperbola $\dfrac{x^2}{49} - \dfrac{y^2}{16} = 1$. Should the hyperbola be dashed or solid?

(b) Choose a test point (not on the hyperbola) in each region (the regions within each branch of the hyperbola and the region between the two branches) and determine whether the coordinates of the point satisfy the inequality.

(c) Shade in the region(s) containing the points where the inequality is satisfied. This is the graph of the solution set.

(d) Suppose you choose the point (r, s) as a test point for the region "inside" the right branch of the hyperbola you just sketched. In which region does the point $(-r, s)$ lie?

(e) Looking at the inequality, if the test point (r, s) satisfies (or does not satisfy) the inequality, can you conclude anything about whether $(-r, s)$ satisfies the inequality?

(f) Putting together (d) and (e), if we find that solutions lie in the region inside one branch of the hyperbola, what can we conclude about the solutions with respect to the other branch?

(g) How many test points do we really need? Can we logically deduce the solution region(s) using one test point? Explain your answer.

In Exercises 59–62, use the results of Exercise 58 to graph the solutions of the inequality.

59. $\dfrac{x^2}{9} - \dfrac{y^2}{49} < 1$

60. $\dfrac{y^2}{81} - \dfrac{x^2}{4} \ge 1$

61. $8x^2 - y^2 \le 24$

62. $\dfrac{x^2}{5} + \dfrac{(y + 3)^2}{16} > 1$

8.5   Identifying Conic Sections: Degenerate Forms

We summarize the equations of conic sections derived in Sections 8.1–8.4.

Conic Sections

1. The circle:

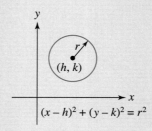

$$(x - h)^2 + (y - k)^2 = r^2$$

2. The parabola:

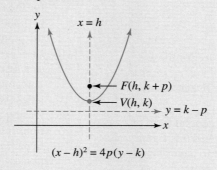

$$(x - h)^2 = 4p(y - k)$$

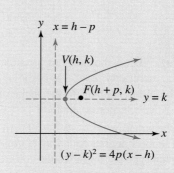

$$(y - k)^2 = 4p(x - h)$$

3. The ellipse:

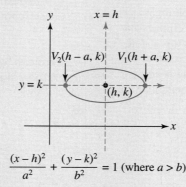

$$\frac{(x - h)^2}{a^2} + \frac{(y - k)^2}{b^2} = 1 \text{ (where } a > b)$$

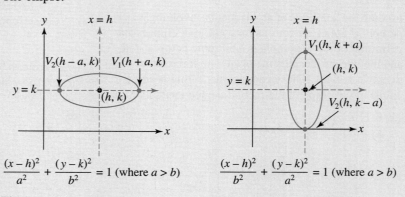

$$\frac{(x - h)^2}{b^2} + \frac{(y - k)^2}{a^2} = 1 \text{ (where } a > b)$$

4. The hyperbola:

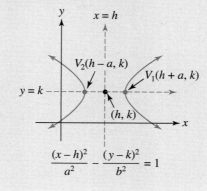

$$\frac{(x - h)^2}{a^2} - \frac{(y - k)^2}{b^2} = 1$$

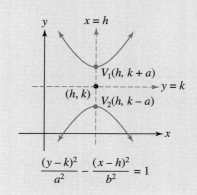

$$\frac{(y - k)^2}{a^2} - \frac{(x - h)^2}{b^2} = 1$$

As we moved through the sections in this chapter, we made mention of the general form

$$Ax^2 + Cy^2 + Dx + Ey + F = 0 \qquad (1)$$

where and A and C are not both zero, and its relationship to each standard form of the conic section being examined. It may have occurred to you that the general second-degree equation (with the help of completing the square) will always turn out to be a conic section.

However, consider the equation $x^2 + y^2 = -1$. This equation seems to be in the form of a circle, yet the equation cannot be graphed since there are no real numbers which, when squared and added, will yield a negative number. The equation $(x - 2)^2 + (y + 5)^2 = 0$ is also in the form of a circle, yet the graph of this equation is the single point $(2, -5)$. If the equation is of form (1), but its graph is not one of the four conic sections discussed in this chapter, we refer to the graph as a **degenerate form of a conic section**.

Recall our discussion at the beginning of this chapter about conic sections, in general, being the intersection of a double-napped cone and a plane.

If the plane is horizontal, we arrive at a circle, but if we move it down to the vertex of the double-napped cone, we end up with a point. So if the equation can be graphed, then the degenerate form of a circle is a point. (See Figure 8.47.)

Figure 8.47

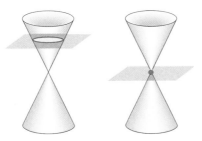

At the end of Section 8.1, we mentioned that if the coefficients of x^2 and y^2 in equation (1) are equal, then we would *expect* the graph of the equation to be a circle. Keeping in mind the possibility of a degenerate form, we can now modify this statement to read as follows: *If the coefficients of x^2 and y^2 in equation (1) are equal, then the graph of the equation (if it exists) is either a circle or its degenerate form, a point.*

The intersection of the plane and double-napped cone producing an ellipse would also produce a point if the plane were moved to the vertex of the double-napped cone, so if the ellipse can be graphed, the degenerate form is a point. At the end of Section 8.3, we mentioned that if the coefficients of x^2 and y^2 in equation (1) are not equal but have the same sign, then we would expect the graph of the equation to be an ellipse. Again, we can now modify this statement to read as follows: *If the coefficients of x^2 and y^2 in equation (1) have the same sign but are not equal, then the graph of the equation (if it exists) is either an ellipse, or its degenerate form, a point.*

Suppose we begin with the plane and double-napped cone used to produce the hyperbola and move the plane so that it remains vertical and intersects the vertex of the double-napped cone. The figure it traces is two intersecting lines. (See Figure 8.48.) This condition can be represented by the form

$$\frac{x^2}{a^2} - \frac{y^2}{b^2} = 0$$

Figure 8.48

If we tried to graph this equation, we would get two intersecting lines. We can see this by solving for y in this equation to get

$$y = \pm\frac{b}{a}x$$

the equations of two intersecting lines. Hence the degenerate form of a hyperbola is two intersecting lines.

As with the circle and the ellipse, we can modify the closing statement made in Section 8.4 as follows: *If the coefficients of x^2 and y^2 in equation (1) have different signs, then the graph of the equation (if its exists) is either a hyperbola or its degenerate form, two intersecting lines.*

At the end of Section 8.2, we stated that if one of the coefficients of x^2 or y^2 in equation (1) is 0, then we expect the graph of the equation to be a parabola. The equation $x^2 - 16 = 0$ is a second-degree equation that satisfies this condition, but if we solve for x, we get $x = -4$, $x = 4$, which are the equations of two vertical lines. On the other hand, the equation $x^2 - 6x + 9 = 0$, which is $(x - 3)^2 = 0$, is the equation of the single vertical line $x = 3$. We refer to these forms as degenerate forms of the parabola and modify the statement made at the end of Section 8.2 to read as follows: *If one of the coefficients of x^2 or y^2 in equation (1) is 0, then the graph of the equation (if its exists) is either a parabola, or one of its degenerate forms, two parallel lines, or a single line.*

Now we can conclude the following.

Suppose the equation $Ax^2 + Cy^2 + Dx + Ey + F = 0$ (where A and C are not both zero) can be graphed. The graph of this equation is a conic section or one of its degenerate forms:

1. If either $A = 0$ or $C = 0$, then the graph of the equation will be a parabola or one of its degenerate forms, a line or two parallel lines.

2. If the signs of A and C are the same, the graph of the equation will be a circle, if $A = C$; an ellipse, if $A \neq C$; or their degenerate form, a point.

3. If the signs of A and C are different, then the graph of the equation will be a hyperbola or its degenerate form, two intersecting lines.

Example 1   Identify and graph each of the following.

(a) $-3x^2 - 3y^2 - 6x + 12y - 15 = 0$
(b) $x^2 - 9y^2 - 4x + 18y - 14 = 0$
(c) $9x^2 - 16y^2 = 0$

Solution   (a) $-3x^2 - 3y^2 - 6x + 12y - 15 = 0$

We look at the equation and note that the coefficients of the squared terms are identical, so we can conclude that if it can be graphed, it will be a circle or a point (its degenerate form). Using the methods discussed earlier (the student should verify this), we put it into standard form to get $(x + 1)^2 + (y - 2)^2 = 0$, which is the point $(-1, 2)$. See Figure 8.49.

(b) $x^2 - 9y^2 - 4x + 18y - 14 = 0$

We look at the equation and note that the coefficients of the squared terms are opposite in sign, so we can conclude that if it can be graphed, it will be a hyperbola or its degenerate form, two interesting lines. Using the methods discussed

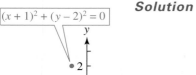

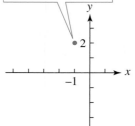

Figure 8.49

earlier (the student should verify this), we put it into standard form to get $\dfrac{(x-2)^2}{9} - \dfrac{(y-1)^2}{1} = 1$, which is the hyperbola graphed in Figure 8.50.

Figure 8.50

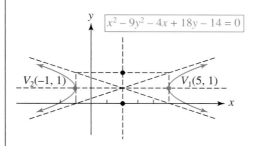

(c) $9x^2 - 16y^2 = 0$

Because the coefficients of the squared terms are opposite in sign, we conclude that if the equation can be graphed, it would describe a hyperbola or one of its degenerate forms. We cannot put this equation in standard form, but we note that we can solve for y to get two intersecting lines:

$$y = \pm\frac{3}{4}x$$

which is a degenerate form of the hyperbola, graphed in Figure 8.51.

Figure 8.51

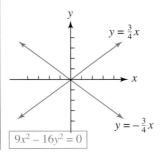

8.5 Exercises

In Exercises 1–10, identify the type of conic section if it can be graphed.

1. $3x^2 + 2y^2 - 18x = 0$

2. $4x^2 + 3y - 8x - 37 = 0$

3. $5x^2 - 4y^2 - 10x - 8y + 25 = 0$

4. $3x^2 - 3y^2 - 18y = 0$

5. $6x^2 + 6y^2 - 8x - 2y - 7 = 0$

6. $2y^2 - 5x + 3y = 0$

7. $-x^2 + y^2 - 50x - 6y - 16 = 0$

8. $-2x^2 - 2y^2 - 12x - 6y + 16 = 0$

9. $25x^2 + 8y^2 - 225 = 0$

10. $x^2 + y - 18 = 0$

In Exercises 11–40, identify the type of conic section if it can be graphed, and graph it.

11. $y^2 + 16x = 0$

12. $2x^2 + 3y^2 - 8x - 6y - 37 = 0$

13. $x^2 - 10x + 25 = 0$

14. $9y^2 + 18 = 0$

15. $2x^2 + y^2 - 8x - 2y - 7 = 0$

16. $y^2 - 5x + 4y + 24 = 0$

17. $25x^2 + y^2 - 50x - 6y - 16 = 0$

18. $2x^2 + 3y^2 - 12x - 6y + 24 = 0$

19. $25x^2 + 9y^2 - 225 = 0$

20. $x^2 + y^2 - 18 = 0$

21. $x^2 - y^2 = 18$

22. $8x^2 + 2y^2 - 24 = 0$

23. $y^2 - 6y = 16$

24. $-3x^2 - 3y^2 - 30x + 12y - 91 = 0$

25. $-10x^2 + y = 0$

26. $-20x^2 + 9y^2 - 18y - 171 = 0$

27. $x^2 + y^2 + 2x - 6y - 5 = 0$

28. $y^2 - 4y = -8x$

29. $x^2 + y^2 - 4x - 2y + 3 = 0$

30. $6x^2 - 8y^2 = -24$

31. $9x^2 - y^2 - 18x - 4y = 139$

32. $30x^2 + y^2 = 0$

33. $x^2 + y^2 + 36 = 0$

34. $3x^2 - 4y^2 - 8y = 52$

35. $9x^2 + 25y^2 + 18x + 9 = 0$

36. $x^2 + 14x + 49 = 0$

37. $16y^2 - x^2 = 0$

38. $2x^2 + 2y^2 + 12x - 4y + 20 = 0$

39. $x^2 - 2x + 12y = 47$

40. $9x^2 - 20y^2 - 225 = 0$

Question for Thought

41. Consider the general equation

$$Ax^2 + Cy^2 + Dx + Ey + F = 0$$

where A and C are not both 0.

(a) If $AC < 0$, what can you conclude about the graph of the general equation?

(b) If $AC > 0$, what can you conclude about the graph of the general equation?

(c) If $AC = 0$, what can you conclude about the graph of the general equation?

8.6 Nonlinear Systems of Equations and Inequalities

We have devoted much time and attention to graphing a wide variety of functions and equations. In the course of our discussions we solved various types of equations, particularly to find the x- and y-intercepts of a graph. In fact, in Sections 3.6, 5.1, and 6.1, we also saw how solutions to an equation can lead us to the points of intersection of two graphs.

In Chapter 7 we investigated linear systems of equations and inequalities. In this section we discuss nonlinear systems. In the course of this development we review many of the basic graphs in our catalog.

Up to this point we have been considering systems involving only first-degree equations (or inequalities). A **nonlinear system** is one in which at least one of the equations is not a linear equation. The following are two examples of nonlinear systems that we will consider in this section.

$$\begin{cases} x^2 + y^2 = 9 \\ \quad\quad y = x^2 - 3 \end{cases}$$

$$\begin{cases} y = \dfrac{7}{x} \\ x = y + 3 \end{cases}$$

When the system involves one linear equation and one quadratic equation or two quadratic equations, it is called a **quadratic system**.

Before we discuss the algebraic techniques for solving such systems, let's take a moment to use our knowledge of the graphs of the conic sections to analyze what we might expect from a quadratic system. If we examine the first system, we recognize that the graph of the first equation, $x^2 + y^2 = 9$, is a circle and the graph of the second equation, $y = x^2 - 3$, is a parabola. Keeping in mind that each point of intersection of the graphs of the equations corresponds to a solution to the system of equations, it would be useful to know how many points of intersection a circle and a parabola can have.

The graphs in Figure 8.52 illustrate a variety of possibilities for a circle and parabola. As the figure shows, a circle and a parabola can intersect in 0, 1, 2, 3, or 4 points. Can there be more than 4 points of intersection? If you play around with the graphs of a circle and a parabola, you will quickly convince yourself that there can be no more than 4 points of intersection.

Figure 8.52

Possible intersections of a parabola and a circle.

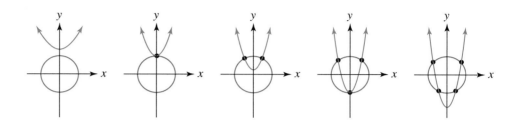

A similar analysis should convince us that any two distinct conic sections (straight line, parabola, circle, ellipse, and hyperbola) can intersect in *at most* 4 points. Exercise 59 suggests how we might prove this fact algebraically.

Knowing how many solutions a system of equations may have can be helpful in detecting errors or extraneous solutions.

Calculator Exploration

Using a graphing calculator.

1. Graph the equations $x^2 - y = 3$ and $x + y = 3$ on the same set of coordinate axes and find their points of intersection.
2. Graph the equations $x^2 + 3y^2 = 1$ and $x - y = 1$ on the same set of coordinate axes and find their points of intersection.

Let us now discuss algebraic techniques for solving nonlinear systems. As with linear systems, there are two basic methods: elimination and substitution. As the following example illustrates, we don't always have an option as to which method to use.

Example 1 | Solve the following system of equations.

$$\begin{cases} 2x^2 - y^2 = 7 \\ x - y = 1 \end{cases}$$

Solution | A moment's thought about using the elimination method should convince us that it will not be effective on this system. In order for the elimination method to work, we must eliminate *all* occurrences of one of the variables, which requires that we have like terms to eliminate. However, in this system the two equations have no like terms and so the elimination method will not work. Consequently, we will use the substitution method.

To make the substitution process as straightforward as possible, we try to use the "simpler" equation and explicitly solve for whichever variable appears to be easier to isolate.

In this example, the second equation is simpler, because it is linear in both variables, whereas the first is quadratic in both.

We solve the second equation for y, obtaining $y = x - 1$. Let's substitute $y = x - 1$ into the first equation.

$$2x^2 - y^2 = 7 \qquad \text{We substitute } y = x - 1 \text{ and solve for } x.$$
$$2x^2 - (x - 1)^2 = 7$$
$$2x^2 - (x^2 - 2x + 1) = 7$$
$$2x^2 - x^2 + 2x - 1 = 7$$
$$x^2 + 2x - 8 = 0 \qquad \text{We can solve this equation by factoring.}$$
$$(x + 4)(x - 2) = 0$$
$$x = -4 \quad \text{or} \quad x = 2$$

To get the corresponding y-values, we take advantage of the equation solved explicitly for y.

$y = x - 1 \qquad \text{Substitute } x = -4. \qquad\qquad y = x - 1 \qquad \text{Substitute } x = 2.$
$y = -4 - 1 \qquad\qquad\qquad\qquad\qquad\qquad y = 2 - 1 = 1$
$y = -5$

Therefore, the solutions to the system are $(-4, -5)$ and $(2, 1)$. It is left to the student to check that these ordered pairs satisfy *both* equations.

Note that these algebraic solutions agree quite well with the points of intersection that appear in Figure 8.53.

Figure 8.53

The graphs of $2x^2 - y^2 = 7$ and $x - y = 1$.

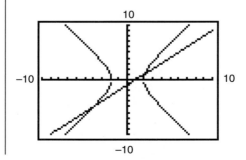

Example 2   Solve the following system of equations.

$$\begin{cases} x^2 + y^2 = 9 \\ \quad y = x^2 - 3 \end{cases}$$

We offer two possible approaches to solving this system.

Solution 1   *Using the Substitution Method*

The fact that the second equation is solved explicitly for y suggests that, perhaps, the substitution method will be the easiest. However, as we will see, it raises some additional problems.

$$x^2 + y^2 = 9$$

We use the second equation to substitute $y = x^2 - 3$.

$$x^2 + (x^2 - 3)^2 = 9$$
$$x^2 + x^4 - 6x^2 + 9 = 9$$
$$x^4 - 5x^2 = 0$$
$$x^2(x^2 - 5) = 0$$
$$x = 0 \quad \text{or} \quad x = \pm\sqrt{5}$$

We can solve for y by substituting into the second equation.

If $x = 0$, then $\qquad$ If $x = \pm\sqrt{5}$, then

$$y = x^2 - 3 = 0^2 - 3 = -3 \qquad y = (\pm\sqrt{5})^2 - 3 = 5 - 3 = 2$$

Therefore we have three solutions: $(0, -3)$, $(\sqrt{5}, 2)$, $(-\sqrt{5}, 2)$. It is left to the student to verify that these ordered pairs satisfy *both* equations.

Before we look at the second approach, two important points must be made.

1. When we substituted $y = x^2 - 3$ into the first equation, we obtained a *fourth-degree* equation. We know that when we encounter a fourth-degree equation, we may or may not be able to solve it, because we don't have a general method for solving fourth-degree equations. In this example, we were able to solve it because we could factor the fourth-degree polynomial. Thus it would seem to be to our advantage, as we manipulate the given system, to try to keep the degree of any resulting equations as small as possible.

2. In the past when we solved linear systems, once we obtained the value of one of the variables, it did not matter into which equation we substituted to obtain the other variable. However, in this example let's see what happens if we substitute the x-values we obtained into the first equation instead of the second equation. To solve for y, we substitute $x = 0$ and $x = \pm\sqrt{5}$ into the first equation:

$$x^2 + y^2 = 9 \qquad \text{Let } x = 0.$$
$$y^2 = 9$$
$$y = \pm 3$$

The "solutions" are $(0, 3)$, $(0, -3)$.

$$x^2 + y^2 = 9 \qquad \text{Let } x = \pm\sqrt{5}.$$
$$(\pm\sqrt{5})^2 + y^2 = 9$$
$$5 + y^2 = 9$$
$$y^2 = 4$$
$$y = \pm 2$$

This gives 4 "solutions": $(\sqrt{5}, 2)$, $(-\sqrt{5}, 2)$, $(\sqrt{5}, -2)$, $(-\sqrt{5}, -2)$.

Note that we have obtained a total of six apparent solutions, which, based on our previous discussion, we know is impossible for a quadratic system. In fact, if we substitute one of these 6 ordered pairs that we did not obtain earlier, such as $(0, 3)$, into the second equation, we get

$$y = x^2 - 3 \qquad \text{Substitute } (0, 3).$$
$$3 \overset{?}{=} (0)^2 - 3$$
$$3 \neq -3 \qquad \text{So } (0, 3) \text{ is not a solution to the system.}$$

The reason we obtained more apparent solutions is that by substituting into the first equation, which is quadratic in y, we generated more y-values than by substituting into the second equation, which is linear in y. Thus, to avoid extraneous solutions, once we have found the values of one of the variables and we want to find the values of the other variable, it is a good idea to substitute into the equation of lowest degree in the variable we are seeking.

Solution 2 | *Using the Elimination Method.*
We rewrite the second equation in the system to get

$$x^2 + y^2 = 9$$
$$\underline{-x^2 + y = -3} \qquad \text{Adding the two equations eliminates } x^2.$$
$$y^2 + y = 6$$
$$y^2 + y - 6 = 0$$
$$(y + 3)(y - 2) = 0$$
$$y = -3 \quad \text{or} \quad y = 2$$

We can now substitute these *y*-values into one of the original equations to obtain the corresponding *x*-values. Based on the comments made in point 2, we note that both of the original equations are of degree 2 in *x*, so it makes no difference which equation we use. We substitute into the second equation.

$y = x^2 - 3 \qquad \text{Substitute } y = -3.$	$y = x^2 - 3 \qquad \text{Substitute } y = 2.$
$-3 = x^2 - 3$	$2 = x^2 - 3$
$x^2 = 0$	$x^2 = 5$
$x = 0$	$x = \pm\sqrt{5}$

This result gives us the same three solutions obtained by the substitution method: $(0, -3), (-\sqrt{5}, 2), (\sqrt{5}, 2)$.

We include a sketch of the graph of the two equations, since the graph often gives us a fairly simple way to check whether we have the correct number of solutions. See Figure 8.54.

Finally, we note that by using the elimination method, we avoided the potential problem of dealing with a fourth-degree equation, which we encountered when we used the substitution method. ∎

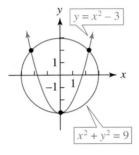

Figure 8.54

Example 3 | Solve and sketch a graph of the following system of equations.

$$\begin{cases} y = \dfrac{7}{x} \\ x = y + 3 \end{cases}$$

Round the answers to the nearest hundredth.

Solution | Since each equation is solved explicitly for one of the variables, we may substitute the second equation into the first equation or vice versa. We will use the second equation to substitute into the first equation.

$$y = \frac{7}{x} \qquad \text{Substitute } x = y + 3.$$
$$y = \frac{7}{y + 3} \qquad \text{Multiply both sides by } y + 3.$$
$$y^2 + 3y = 7$$
$$y^2 + 3y - 7 = 0 \qquad \text{We use the quadratic formula to solve for } y.$$
$$y = \frac{-3 \pm \sqrt{37}}{2} \qquad \text{Using a calculator, we get}$$
$$y = 1.54, \, -4.54 \qquad \text{Rounded to two decimal places}$$

We get the corresponding *x*-values by substituting into the second equation in the system:

$$x = 1.54 + 3 = 4.54 \quad \text{or} \quad x = -4.54 + 3 = -1.54$$

Thus the solutions are (4.54, 1.54) and (−1.54, −4.54). The graphs of the equations in this system are shown in Figure 8.55.

Figure 8.55

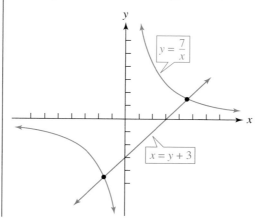

Example 4   Find the points of intersection (if any) of the graphs of the equations $16x^2 + 25y^2 = 400$ and $x^2 + y^2 = 9$.

Solution   The problem is asking us to solve the following system of equations.

$$\begin{cases} 16x^2 + 25y^2 = 400 \\ x^2 + y^2 = 9 \end{cases}$$

Since the system does have like terms, we can use the elimination method.

$$16x^2 + 25y^2 = 400 \quad \xrightarrow{\text{As is}} \quad 16x^2 + 25y^2 = 400$$
$$x^2 + y^2 = 9 \quad \xrightarrow{\text{Multiply by} -25} \quad \underline{-25x^2 - 25y^2 = -225}$$
$$-9x^2 = 175$$
$$x^2 = -\frac{175}{9}$$
$$x = \pm\sqrt{-\frac{175}{9}}$$

Thus there are no real solutions to this system, which means that the graphs of the two equations do not intersect.

We could have come to the same conclusion had we first graphed the two equations. The graph of $16x^2 + 25y^2 = 400$ is an ellipse and the graph of $x^2 + y^2 = 9$ is a circle. Both have centers at the origin. Their graphs appear in Figure 8.56.

Figure 8.56

The graphs of $16x^2 + 25y^2 = 400$ and $x^2 + y^2 = 9$ do not intersect.

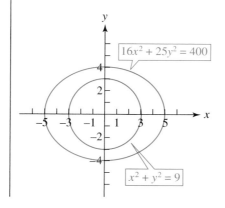

In Section 7.8 we saw that we could describe the solutions for a system of linear inequalities by graphing each of the inequalities separately. The same approach can be used for some nonlinear systems of inequalities.

Example 5   Sketch the solutions to the following system of inequalities.

$$\begin{cases} y \le 4 - x^2 \\ x - y \le 2 \end{cases}$$

Solution   As with a system of linear inequalities, we sketch the solutions of each inequality and see where those regions overlap.

To find the solutions to $y \le 4 - x^2$, we begin by sketching the graph of the boundary of the region, $y = 4 - x^2$. We see that the solution set will be the region on and below (inside) the parabola, since the inequality says that y is less than or equal to $4 - x^2$. Alternatively, we may choose a test point not on the graph of $y = 4 - x^2$ and obtain the same result. The solution set of the first inequality is indicated by the shaded region in Figure 8.57(a).

Figure 8.57

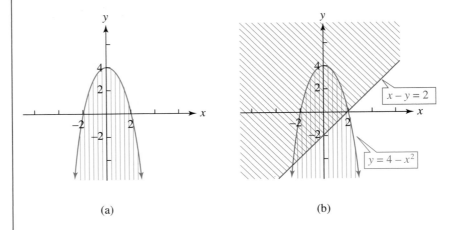

(a)   (b)

Proceeding in a similar manner, the solution to the second inequality $x - y \le 2$ is the shaded region on and above the line $x - y = 2$ in Figure 8.57(b). Thus the solutions of the system of inequalities consist of those points that are in *both* shaded regions. This region is indicated in Figure 8.57(b) by the crosshatching.   ∎

Real-life applications often give rise to nonlinear systems of equations.

Example 6   Suppose that a manufacturer wants to produce rectangular boxes with square tops and bottoms, such that the volume of a box is 200 cu cm. The company has on hand a supply of metal sheets, each of which measures 210 sq cm in area. Assuming no waste of material, to the nearest hundredth square centimeter, what should the dimensions of the box be so that each metal sheet produces one rectangular box?

Solution   We begin by drawing a diagram of the proposed box (see Figure 8.58). We are told that the top and bottom of the box are square, so we label each edge of the top and bottom x. We also label the height of the box h.

Figure 8.58

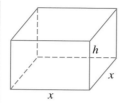

The fact that we want the volume of each box to be 200 cu cm translates into the equation $x^2h = 200$, and the fact that each metal sheet has an area of 210 sq cm means that the surface area of the box must be 210 sq cm. The area of both the top and bottom is x^2, whereas the area of each of the four sides is xh, so the surface area of the box is $2x^2 + 4xh$. We thus have a second equation, $2x^2 + 4xh = 210$.

Therefore, finding the required dimensions of the box is equivalent to solving the following system of equations.

$$\begin{cases} x^2h = 200 \\ 2x^2 + 4xh = 210 \end{cases}$$

We can solve the first equation for h, obtaining $h = \dfrac{200}{x^2}$, and substitute it into the second equation:

$$2x^2 + 4xh = 210 \qquad \text{We substitute } h = \frac{200}{x^2}.$$

$$2x^2 + 4x\left(\frac{200}{x^2}\right) = 210 \qquad \text{Simplify.}$$

$$2x^2 + \frac{800}{x} = 210 \qquad \text{Multiply both sides of the equation by } x.$$

$$2x^3 + 800 = 210x$$

$$2x^3 - 210x + 800 = 0 \qquad \text{Divide both sides of the equation by 2.}$$

$$x^3 - 105x + 400 = 0$$

A graphing calculator can be used to approximate the solutions to $x^3 - 105x + 400 = 0$.

By the Rational Root Theorem (Section 5.4), there are many possible rational roots that we may check by using either long division or synthetic division. We find that 5 is a root, so $(x - 5)$ is a factor of $x^3 - 105x + 400$ and we have

$$(x - 5)(x^2 + 5x - 80) = 0 \qquad \text{We obtain the other roots by using the quadratic formula.}$$

$$x = 5 \quad \text{or} \quad x = \frac{-5 \pm \sqrt{345}}{2} \qquad \text{Using a calculator, we get}$$

$$x = 5 \quad \text{or} \quad x = 6.79 \quad \text{or} \quad x = -11.79 \qquad \begin{array}{l}\text{Rounded to the nearest}\\ \text{hundredth}\end{array}$$

Since x represents the length of a side of the top and bottom of the box, it makes no sense for x to be negative. By substituting the two possible x-values into $h = \dfrac{200}{x^2}$ we get

$$h = \frac{200}{x^2} \quad \text{Substitute } x = 5. \qquad h = \frac{200}{x^2} \quad \text{Substitute } x = 6.79.$$

$$h = \frac{200}{25} = 8 \qquad h = \frac{200}{(6.79)^2} = 4.34 \quad \begin{array}{l}\text{Rounded to the nearest}\\ \text{hundredth}\end{array}$$

Therefore, we have two possible configurations for the box. We have either $x = 5$ cm and $h = 8$ cm or $x = 6.79$ cm and $h = 4.34$ cm. ∎

8.6 Exercises

In Exercises 1–22, solve the system of equations.

1. $\begin{cases} x^2 + y^2 = 10 \\ x + y = 2 \end{cases}$

2. $\begin{cases} x^2 + y^2 = 8 \\ x^2 + y = 6 \end{cases}$

3. $\begin{cases} 4x^2 + y^2 = 36 \\ 2x - y = 6 \end{cases}$

4. $\begin{cases} x^2 + y^2 = 4 \\ x^2 - y = 9 \end{cases}$

5. $\begin{cases} x^2 - y^2 = 16 \\ 3x - y = 1 \end{cases}$

6. $\begin{cases} y = x^2 + 8x - 10 \\ y = 3x + 4 \end{cases}$

7. $\begin{cases} x^2 - y^2 = 8 \\ x^2 + 2y^2 = 11 \end{cases}$

8. $\begin{cases} y = 2 - x^2 \\ x^2 = y^2 - 4 \end{cases}$

9. $\begin{cases} x^2 + y^2 = 1 \\ \dfrac{x^2}{4} + y^2 = 1 \end{cases}$

10. $\begin{cases} 4x^2 + 5y^2 = 10 \\ 5x^2 + 4y^2 = 10 \end{cases}$

11. $\begin{cases} y = x^2 - 2x - 4 \\ x - y = 2 \end{cases}$

12. $\begin{cases} y = x^2 + x - 1 \\ y = -x^2 - x + 1 \end{cases}$

13. $\begin{cases} y = \sqrt{x} \\ 2x + y = 10 \end{cases}$

14. $\begin{cases} y = \sqrt{2x - 1} \\ y = \sqrt{x - 1} \end{cases}$

15. $\begin{cases} (x - 2)^2 + y^2 = 13 \\ y = \sqrt{3 - x} \end{cases}$

16. $\begin{cases} x = y^2 - 4 \\ y = -\sqrt{x^2 + 4} \end{cases}$

17. $\begin{cases} y = (x - 3)^2 \\ (x - 3)^2 + y^2 = 12 \end{cases}$

18. $\begin{cases} x^2 + (y + 2)^2 = 2 \\ x = (y + 2)^2 \end{cases}$

19. $\begin{cases} xy = 2 \\ x + 3y = 7 \end{cases}$

20. $\begin{cases} xy = 5 \\ 2y - 3x = 2 \end{cases}$

21. $\begin{cases} y = \log_2(x - 1) \\ y = \log_2(x + 3) - 1 \end{cases}$

22. $\begin{cases} y = \log_3(x - 3) \\ y = 2 - \log_3(x + 5) \end{cases}$

In Exercises 23–34, solve the system of equations and sketch the graphs.

23. $\begin{cases} x^2 + y^2 = 1 \\ x^2 + (y - 1)^2 = 1 \end{cases}$

24. $\begin{cases} 9x^2 + 4y^2 = 36 \\ x^2 - y^2 = 16 \end{cases}$

25. $\begin{cases} y = 2x^2 \\ y = x^2 - 1 \end{cases}$

26. $\begin{cases} y = 2x^2 - 1 \\ y = x^2 \end{cases}$

27. $\begin{cases} x^2 + y^2 = 9 \\ x - y = 3 \end{cases}$

28. $\begin{cases} y = x^2 - 4 \\ y = 4 - x^2 \end{cases}$

29. $\begin{cases} y = x^2 - 6x \\ 2x + 3y + 13 = 0 \end{cases}$

30. $\begin{cases} y = x^2 - 4x + 1 \\ 3x - 2y = 7 \end{cases}$

31. $\begin{cases} x^2 + y^2 = 9 \\ 2y = x^2 - 1 \end{cases}$

32. $\begin{cases} x^2 + 9y^2 = 9 \\ x^2 - y^2 = 4 \end{cases}$

33. $\begin{cases} x^2 + y^2 = 16 \\ x + y = 2 \end{cases}$

34. $\begin{cases} y = x^2 - 6x + 9 \\ 2x - y = 1 \end{cases}$

35. Solve the following system of equations:

$$\begin{cases} \dfrac{8}{x^2} - \dfrac{4}{y^2} = 3 \\ \dfrac{6}{x^2} + \dfrac{8}{y^2} = 5 \end{cases}$$

HINT: Let $u = \dfrac{1}{x^2}$ and $v = \dfrac{1}{y^2}$, and rewrite the system in terms of u and v.

36. Solve the following system of equations:

$$\begin{cases} \dfrac{2}{\sqrt{x}} + \dfrac{5}{\sqrt{y}} = 6 \\ \dfrac{8}{\sqrt{x}} - \dfrac{3}{\sqrt{y}} = 1 \end{cases}$$

HINT: Let $s = \dfrac{1}{\sqrt{x}}$ and $t = \dfrac{1}{\sqrt{y}}$, and rewrite the system in terms of s and t.

37. Solve the following system of equations:

$$\begin{cases} y = 5^x \\ y = 5^{2x} - 12 \end{cases}$$

HINT: Note that $5^{2x} = (5^x)^2$. Estimate your answer to the nearest tenth and interpret your answer graphically.

38. Solve the following system of equations:

$$\begin{cases} y = 3^x \\ y = 9^x - 2 \end{cases}$$

HINT: Note that $9^x = (3^2)^x = (3^x)^2$. Estimate your answer to the nearest tenth and interpret your answer graphically.

39. Find the dimensions of a rectangle whose area is 35 sq cm and whose perimeter is 27 cm.

40. If the diagonal of a rectangle is 10 in. and the perimeter of the rectangle is 28 in., find the dimensions of the rectangle.

41. If the hypotenuse of a right triangle is 25 cm and the area of the right triangle is 84 sq cm, find the lengths of the legs.

42. Find the dimensions of a piece of pipe (an open right circular cylinder) whose volume is 72π cu cm and surface area is 48π sq cm.

43. (a) If we want the system $\begin{cases} y = x^2 - 2x - 3 \\ y = K \end{cases}$ to have exactly one solution, what must the value of K be?

   (b) If we want the system to have no real solutions, what can the value of K be?

44. Find the point(s) where the graph of $y = 9 - x^2$ intersects the graph of $2x + y = 6$.

45. Where do the circle $x^2 + y^2 = 16$ and the ellipse $4x^2 + y^2 = 64$ intersect?

46. What points are common to the graphs of $x^2 - 4y^2 = 12$ and $3y - x + 1 = 0$?

47. A farmer has 180 feet of fence with which to enclose a rectangular garden, one side of which will be bounded by the side of a barn. (See the accompanying figure.) If no fence is used along the barn, is it possible to enclose an area of 4000 sq ft? If so, how?

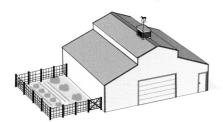

48. Is it possible for the farmer in Exercise 47 to enclose an area of 4200 sq ft with the 180 feet of fencing?

In Exercises 49–54, sketch the solution set to each system of inequalities.

49. $\begin{cases} x^2 + y^2 \le 4 \\ y \ge 1 - x^2 \end{cases}$ **50.** $\begin{cases} y \ge x^2 - 4 \\ x + y \le 2 \end{cases}$

51. $\begin{cases} 4x^2 + 9y^2 < 36 \\ x^2 + y^2 > 4 \end{cases}$ **52.** $\begin{cases} x \le \sqrt{9 - y^2} \\ y \le x \end{cases}$

53. $\begin{cases} 4x^2 - 25y^2 \le 100 \\ x^2 + y^2 < 100 \end{cases}$ **54.** $\begin{cases} y \ge x^2 - 4x + 1 \\ y \le -x^2 - 4x \end{cases}$

In Exercises 55–58, use your graphing calculator to find the solutions to each system of equations. Estimate the solutions to one decimal place.

55. $\begin{cases} y = x^2 \\ y = 3x^2 - 2 \end{cases}$ **56.** $\begin{cases} y = 5x^2 - 4 \\ y = 4x^2 \end{cases}$

57. $\begin{cases} x^2 + y^2 = 16 \\ x - y = 3 \end{cases}$ **58.** $\begin{cases} y = x^2 - 3 \\ y = 3 - x^2 \end{cases}$

Question for Thought

59. Consider a quadratic system consisting of two equations in two variables. Suppose we use either the substitution or elimination method to produce an equation involving only one variable. What is the maximum degree of such an equation? Why? What does this tell us about the maximum number of solutions to this system? Why?

Chapter 8   *Summary*

After completing this chapter you should be able to:

1. Graph a parabola, and identify its vertex, focus, axis of symmetry, and directrix. (Section 8.2)

For example:

Sketch the graph of the parabola

$$x^2 - 2x - 8y = 23$$

Solution:

We recognize that there is an x^2-term, so the parabola has a vertical axis of symmetry. Thus we want the equation to be in the form

$$(x - h)^2 = 4p(y - k)$$

To do this, we complete the square as follows:

$$x^2 - 2x - 8y = 23$$ Isolate the terms containing powers of x on one side.

$$x^2 - 2x = 8y + 23$$ Complete the square for the left-hand side.

$$x^2 - 2x + 1 = 8y + 23 + 1$$ Write in the form $(x - h)^2 = 4p(y - k)$.

$$(x - 1)^2 = 8(y + 3)$$ Now we can identify $h, k,$ and p.

Thus $h = 1$, $k = -3$, and $p = 2$. We draw a new set of coordinate axes through $(1, -3)$. The vertex is $(1, -3)$. The axis of symmetry is the line $x = 1$. If we move up 2 units above the vertex along the parabola's axis, we locate the focus at $(1, -1)$. The directrix is the horizontal line located 2 units below the vertex and has equation $y = -5$. See Figure 8.59.

Figure 8.59

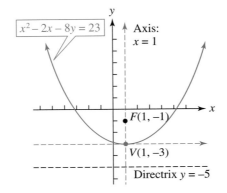

2. Write the equation of the parabola given certain conditions. (Section 8.2)
   For example:
   Find the equation of the parabola if its focus is $(2, 0)$ and its vertex is $(-4, 0)$.
   Solution:
   Given that the parabola has vertex $(-4, 0)$ and focus $(2, 0)$, we note that the vertex and focus lie on the horizontal line $y = 0$, the x-axis, which is also its axis of symmetry. The form we use is $(y - k)^2 = 4p(x - h)$. Since its vertex is $(-4, 0)$, we have $h = -4$ and $k = 0$. The focus for this form is $(h + p, k)$. Since the focus is $(2, 0)$, we have

   $$h + p = 2 \quad \Rightarrow \quad -4 + p = 2 \quad \Rightarrow \quad p = 6$$

   (Alternatively, we could have counted the number of units between the focus and vertex to get 6.) Hence the equation is

   $$(y - 0)^2 = 4(6)[x - (-4)] \quad \text{or} \quad y^2 = 24(x + 4)$$

   See Figure 8.60.

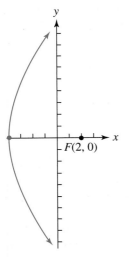

Figure 8.60
$y^2 = 24(x + 4)$

3. Graph an ellipse and identify its vertices, foci, and axes. (Section 8.3)
   For example:
   Sketch the graphs of the ellipses

(a) $\dfrac{x^2}{12} + \dfrac{y^2}{4} = 1$ (b) $\dfrac{(x+1)^2}{16} + \dfrac{(y-2)^2}{25} = 1$

Solution:

(a) The equation is in standard form with the x^2-term having the larger denominator, and therefore the major axis lies on the x-axis. Hence, we use the standard form of the ellipse $\dfrac{x^2}{a^2} + \dfrac{y^2}{b^2} = 1$. Comparing our equation with this general form, we get

$$a^2 = 12 \implies a = 2\sqrt{3} \quad \text{and}$$
$$b^2 = 4 \implies b = 2$$

We plot the vertices $(\pm 2\sqrt{3}, 0)$ and the endpoints of the minor axis $(0, \pm 2)$, and sketch the graph of the ellipse in Figure 8.61. Since $c^2 = a^2 - b^2 = 12 - 4 = 8$, we have $c = 2\sqrt{2}$. Hence the foci are $(\pm 2\sqrt{2}, 0)$.

Figure 8.61

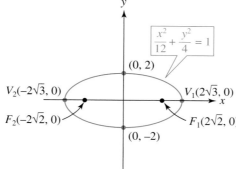

(b) This ellipse is in a standard form with the "y^2-term" having the larger denominator; therefore, the major axis is a line parallel to the y-axis, and we use the form $\dfrac{(x-h)^2}{b^2} + \dfrac{(y-k)^2}{a^2} = 1$. We compare $\dfrac{(x+1)^2}{16} + \dfrac{(y-2)^2}{25} = 1$ with this form and identify $h = -1$, $k = 2$,

$$a^2 = 25 \implies a = 5 \quad \text{and}$$
$$b^2 = 16 \implies b = 4$$

We draw a new set of coordinate axes centered at (h, k), which is $(-1, 2)$, the center of the ellipse. The major axis lies on the vertical line $x = -1$. Since $a = 5$, move 5 units vertically above and below the center, $(-1, 2)$, to find the vertices, $(-1, 7)$ and $(-1, -3)$.

The minor axis lies on the horizontal line $y = 2$. Since $b = 4$, move 4 units left and right of the center, $(-1, 2)$, to find the endpoints of the minor axis, $(-5, 2)$ and $(3, 2)$. Because

$$c^2 = a^2 - b^2 = 25 - 16 = 9 \implies c = 3$$

we find the foci by moving 3 units vertically above and below the center, $(-1, 2)$, to get $F_1(-1, 5)$ and $F_2(-1, -1)$. See Figure 8.62 on page 560.

Figure 8.62

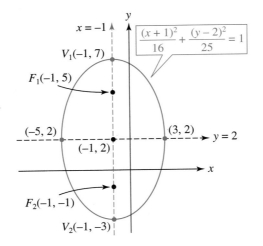

4. Graph a hyperbola and identify its vertices, foci, axes, and asymptotes. (Section 8.4)
   For example:
   Sketch the graph of the hyperbola $2y^2 - x^2 = 4$.
   Solution:
   First, we put the equation into standard form, which requires that 1 be on one side. For us to have 1 on the right-hand side, we must divide by 4.

   $2y^2 - x^2 = 4$ Divide both sides by 4 to get

   $$\frac{y^2}{2} - \frac{x^2}{4} = 1$$

   Comparing our equation with both standard forms of the hyperbola, we note that the y^2-term has a positive coefficient, and therefore a^2 is the denominator of y^2.

   We use the form $\dfrac{y^2}{a^2} - \dfrac{x^2}{b^2} = 1$, a hyperbola with foci on the y-axis. Hence

   $$a^2 = 2 \;\Rightarrow\; a = \sqrt{2} \quad \text{and} \quad b^2 = 4 \;\Rightarrow\; b = 2$$

   We plot the vertices $\left(0, \pm\sqrt{2}\right)$ and the endpoints of the conjugate axis $(\pm 2, 0)$, and draw the rectangle and the diagonals. We sketch the graph in Figure 8.63.

Figure 8.63

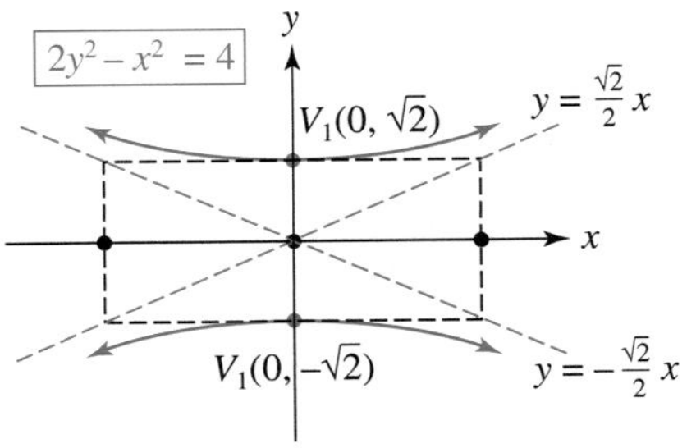

$2y^2 - x^2 = 4$

$V_1(0, \sqrt{2})$

$y = \frac{\sqrt{2}}{2}x$

$V_1(0, -\sqrt{2})$

$y = -\frac{\sqrt{2}}{2}x$

The asymptotes are $y = \pm\dfrac{a}{b}x \;\Rightarrow\; y = \pm\dfrac{\sqrt{2}}{2}x$. Since

$c^2 = a^2 + b^2 = 2 + 4 = 6$, $c = \sqrt{6}$. Hence, the foci are $\left(0, \pm\sqrt{6}\right)$.

5. Identify conic sections (or their degenerate forms) by their equations in the general form $Ax^2 + Cy^2 + Dx + Ey + F = 0$. (Section 8.5)

The coefficients of the square terms of the equation can help you to identify the conic section determined by that equation.

For example:

Assuming the equation

$$3x^2 - 5y^2 - 6x + 8y - 9 = 0$$

can be graphed, we identify its graph without putting the equation in standard form, by observing that the signs of the coefficients of the squared terms are opposite. Therefore, the equation will yield a hyperbola or its degenerate form.

6. Solve a system of two nonlinear equations in two unknowns using the elimination and substitution methods. (Section 8.6)

For example:

Solve the following system of equations.

$$\begin{cases} x^2 + y^2 = 5 \\ y = 3x - 1 \end{cases}$$

The fact that the second equation is solved explicitly for y suggests that the substitution method will be the easiest.

$$x^2 + y^2 = 5$$

We use the second equation to substitute $y = 3x - 1$.

$$x^2 + (3x - 1)^2 = 5$$
$$x^2 + 9x^2 - 6x + 1 = 5$$
$$10x^2 - 6x - 4 = 0$$
$$5x^2 - 3x - 2 = 0$$
$$(5x + 2)(x - 1) = 0$$
$$x = -\frac{2}{5} \quad \text{or} \quad x = 1$$

We can solve for y by substituting into the second equation in the system.

If $x = -\frac{2}{5}$, $y = -\frac{11}{5}$. If $x = 1$, $y = 2$. Therefore, we have two solutions:

$\left(-\frac{2}{5}, -\frac{11}{5}\right)$ and $(1, 2)$. It is left to the student to verify that these ordered pairs satisfy both equations.

7. Sketch the solution set to a system of nonlinear inequalities. (Section 8.6)

Chapter 8   *Review Exercises*

In Exercises 1–4, identify the center and radius of the circle described by the equation.

1. $(x - 3)^2 + (y + 4)^2 = 18$
2. $x^2 + (y - 1)^2 = 12$
3. $x^2 + y^2 - 8x - 8y - 4 = 0$
4. $x^2 + y^2 - 10x + 6y + 15 = 0$
5. There is a theorem in geometry that any line tangent to a circle at a point P is perpendicular to the line

passing through P and the center of the circle. Using this theorem, find the equation of the line tangent to the circle $x^2 + y^2 = 10$ at the point $(1, -3)$.

6. Find the equation of the line tangent to the circle

$$x^2 + (y - 5)^2 = 4$$

at the point $(2, 5)$. (See Exercise 5.)

In Exercises 7–12, graph the parabola and identify the focus, axis and directrix.

7. $8x - y^2 = 0$

8. $y + 5x^2 = 0$

9. $(x - 4)^2 = 4y$

10. $(y - 2)^2 = -16x$

11. $4x = y^2 - 10y + 21$

12. $6y = x^2 - 4x - 2$

In Exercises 13–14, write the equation of the parabola using the given information.

13. The parabola has its vertex at the origin and directrix $x = 3$.

14. The parabola has its vertex at $(1, -1)$ and directrix $y = 2$.

15. A spotlight has a parabolic reflecting mirror with the light source placed at its focus. The mirror can be described by rotating the parabola $y = \dfrac{1}{8}x^2$ about its axis of symmetry, where $-3 \le x \le 3$ and x is measured in feet. How deep is the mirror, and where should the light source be placed in relation to the bottom (vertex) of the mirror?

16. An arch of a building is constructed in the shape of a parabola. (See the accompanying figure.) The height of the arch is 60 feet, and the width of the base of the arch is 30 feet. How wide is the arch 10 feet above the base?

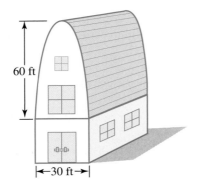

In Exercises 17–24, graph the ellipse and identify the vertices and foci.

17. $\dfrac{x^2}{9} + \dfrac{y^2}{36} = 1$

18. $\dfrac{x^2}{81} + \dfrac{y^2}{6} = 1$

19. $24x^2 + 2y^2 = 48$

20. $20y^2 + 10x^2 = 40$

21. $\dfrac{(x - 2)^2}{6} + \dfrac{(y + 4)^2}{4} = 1$

22. $\dfrac{x^2}{16} + \dfrac{(y + 5)^2}{12} = 1$

23. $x^2 + 4y^2 - 6x + 8y = 3$

24. $6x^2 + 5y^2 + 12x - 20y - 4 = 0$

In Exercises 25–26, write the equation of the ellipse using the given information.

25. The ellipse is centered at the origin; the axes of the ellipse have lengths 6 and 8; the major axis is vertical.

26. The ellipse has foci $(2, 5)$ and $(8, 5)$ and vertices $(0, 5)$ and $(10, 5)$.

In Exercises 27–34, graph the hyperbola and identify the foci, vertices, and asymptotes.

27. $\dfrac{x^2}{49} - \dfrac{y^2}{4} = 1$

28. $\dfrac{y^2}{81} - \dfrac{x^2}{8} = 1$

29. $x^2 - 8y^2 = 24$

30. $10y^2 - 5x^2 = 40$

31. $\dfrac{(y - 2)^2}{2} - \dfrac{(x - 1)^2}{8} = 1$

32. $\dfrac{(x + 3)^2}{12} - \dfrac{y^2}{24} = 1$

33. $x^2 - 4y^2 - 6x - 8y = 11$

34. $5y^2 - 6x^2 - 20y - 12x - 16 = 0$

In Exercises 35–38, identify the curve if it could be graphed.

35. $x^2 + 6y^2 - 8x - 2y - 7 = 0$

36. $2y^2 - 2x + 5y = 0$

37. $-x^2 + y^2 + 18x - 6y + 16 = 0$

38. $4x^2 + 4y^2 - 12x - 8y - 5 = 0$

In Exercises 39–52, identify the curve if it could be graphed and then graph it.

39. $\dfrac{x^2}{36} - \dfrac{y^2}{4} = 1$

40. $\dfrac{x^2}{49} + \dfrac{y^2}{4} = 1$

41. $\dfrac{x}{4} + \dfrac{y}{9} = 1$

42. $\dfrac{(x - 1)^2}{4} + \dfrac{(y + 1)^2}{8} = 1$

43. $\dfrac{(y - 2)^2}{2} + \dfrac{(x - 1)^2}{2} = 1$

44. $\dfrac{(y-2)^2}{2} - \dfrac{(x-1)^2}{2} = 1$

45. $x^2 + 64y = 0$

46. $x^2 + 4y^2 - 2x + 24y + 21 = 0$

47. $x^2 - 8x + 16 = 0$

48. $3y^2 - 3y - 18 = 0$

49. $x^2 + 2y^2 - 2x - 8y - 7 = 0$

50. $x^2 + 4x - 5y + 24 = 0$

51. $x^2 + 25y^2 - 6x - 50y + 34 = 0$

52. $3x^2 - y^2 - 6x - 6y - 6 = 0$

In Exercises 53–60, solve the system of equations.

53. $\begin{cases} x^2 - y^2 = 8 \\ \quad\; y = x - 2 \end{cases}$

54. $\begin{cases} y = 1 - x^2 \\ x^2 = y^2 - 5 \end{cases}$

55. $\begin{cases} x^2 - y^2 = 40 \\ \quad x + y = 10 \end{cases}$

56. $\begin{cases} 2x^2 + y^2 = 19 \\ 3x - y = 10 \end{cases}$

57. $\begin{cases} x^2 + y^2 = 4 \\ \dfrac{y^2}{4} - \dfrac{x^2}{4} = 1 \end{cases}$

58. $\begin{cases} 2x^2 + 3y^2 = 8 \\ 5x^2 + 4y^2 = 7 \end{cases}$

59. $\begin{cases} x^2 + y^2 = 13 \\ \quad\quad y = x^2 - 1 \end{cases}$

60. $\begin{cases} x^2 + 2y^2 = 6 \\ x^2 - y^2 = 3 \end{cases}$

61. If the hypotenuse of a right triangle is 4 cm and the area of the right triangle is 4 sq cm, find the lengths of the legs.

62. Find the dimensions of a piece of pipe (an open right circular cylinder) whose volume is 18π cu cm and surface area is 21π sq cm.

In Exercises 63–64, sketch the solution set to each system of inequalities.

63. $\begin{cases} 25x^2 + 4y^2 \le 100 \\ \quad x^2 + y^2 > 9 \end{cases}$

64. $\begin{cases} y \ge x^2 + 1 \\ y \le -x^2 + 4 \end{cases}$

Chapter 8   *Practice Test*

In Exercises 1–4, sketch the graph of the equation and label the important aspects of the graph.

1. $4x^2 + 4y = 0$

2. $4x^2 + 4y^2 = 36$

3. $\dfrac{x^2}{25} + \dfrac{y^2}{49} = 1$

4. $\dfrac{x^2}{25} - \dfrac{y^2}{49} = 1$

In Exercises 5–10, sketch the graph of the following equations and label the important aspects of the graph.

5. $(x-3)^2 + (y+5)^2 = 5$

6. $2(x-3)^2 + 8y + 24 = 0$

7. $\dfrac{(x-3)^2}{4} + \dfrac{(y-2)^2}{12} = 1$

8. $\dfrac{(y+2)^2}{25} - \dfrac{x^2}{18} = 1$

9. $4x^2 + 5y^2 - 24x + 30y + 61 = 0$

10. $3x^2 - 4y^2 - 6x = 45$

11. Find an equation of the parabola that has its vertex at the origin and directrix $x = -2$.

12. Find an equation of the ellipse with foci (3, 1) and (−1, 1) and vertices (5, 1) and (−3, 1).

13. Solve the system
$$\begin{cases} x^2 + 2y^2 = 3 \\ 2x^2 - y^2 = 1 \end{cases}$$

14. Sketch the solution set of the system of inequalities.
$$\begin{cases} x^2 + y^2 \le 4 \\ \quad\quad y < x^2 - 2 \end{cases}$$

Sequences, Series, and Related Topics

9

Recognizing and identifying patterns is a fundamental idea that appears throughout mathematics and serves as a fruitful starting point for analyzing a wide variety of problems. In the first four sections we discuss sequences and series, two particularly important types of patterns in mathematics that play a fundamental role in calculus and ultimately even in the way scientific calculators compute values of certain functions. In Section 9.5 we introduce the method of mathematical induction, a powerful method of proof that can be used to show that certain patterns hold for all natural numbers. In Section 9.6 we discuss some basic counting principles, which we apply in Section 9.7 on probability. In Section 9.8 we use a number of the ideas presented in this chapter to prove the binomial theorem.

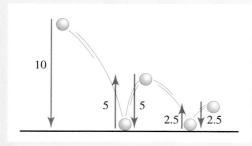

9.1   Sequences

The word *sequence,* used in everyday conversation, usually refers to a list of things occurring in a specific order. For example, if a person wants to build a house, he or she would probably have to arrange for the following sequence of events, which we have labeled with subscripts to indicate that they follow a particular order.

E_1:   Get plans drawn up and obtain all required building permits.

E_2:   Lay the foundation.

E_3:   Erect frame and roof.

E_4:   Install plumbing and wiring.

E_5:   Put up interior walls.

E_6:   Finish all interior surfaces.

Note that in order to construct the house, one is concerned not only with the events on the list, but also with the order in which they occur—which is first, which is second, and so on.

In mathematics, a sequence is viewed in much the same way: an "ordered" set of things. For example, we could have the sequence of numbers 1, 3, 5, 7, where 1 is the first, 3 is the second, 5 is the third, and 7 is the fourth number. The following are examples of sequences:

1, 4, 9, 16, 25, 36, 49, 64, 81, 100

2, 4, 6, 8, 10

2, 4, 6, 8, 10, . . .

5, 25, 125, 625, . . .

The first two sequences are examples of **finite sequences** because they terminate. The third and fourth sequences are examples of **infinite sequences** because, as the three-dot notation indicates, they continue on without end in the same established pattern. For the most part, we focus our attention on infinite sequences.

Although we may describe a sequence intuitively as a list of things in a particular order, we still need to give a precise mathematical definition of sequence. Since we are concerned with order, we can associate the first term in a sequence with the number 1, the second with the number 2, the third with the number 3, and so on. This approach lends itself quite naturally to using function notation to define a sequence.

Definition of a Sequence

A **sequence** is a function whose domain is a set of consecutive positive integers, usually beginning with 1.

The sequence 2, 4, 8, 16, 32, . . . can be viewed as a function, f, with domain $\{1, 2, 3, 4, 5, \ldots\}$ and range $\{2, 4, 8, 16, 32, \ldots\}$ where

$f(1) = 2$ *2 is the first number in the sequence.*

$f(2) = 4$ *4 is the second number in the sequence.*

$f(3) = 8$ *8 is the third number in the sequence.*

$f(4) = 16$ *16 is the fourth number in the sequence.*

$f(5) = 32$ *32 is the fifth number in the sequence.*

$\qquad \vdots$ *And so on*

In mathematics, rather than indicate the terms of the sequence as $f(1), f(2), f(3)$, and $f(4)$, we usually use subscript notation and write f_1, f_2, f_3, and f_4, respectively.

Note that the subscripts are the domain elements.

Thus $f_1 = 2$, $f_2 = 4$, $f_3 = 8$, $f_4 = 16$, and $f_5 = 32$, and so on. The entire sequence is denoted as $\{f_i\}$.

In a sequence $\{a_i\}$, i is called the **index**. Unless otherwise indicated, we will start our indices at 1. Thus we write $\{a_i\} = 1, \frac{1}{2}, \frac{1}{3}, \frac{1}{4}, \frac{1}{5}$ to mean $a_1 = 1$, $a_2 = \frac{1}{2}$, $a_3 = \frac{1}{3}$, $a_4 = \frac{1}{4}$, $a_5 = \frac{1}{5}$. The functional values a_1, a_2, a_3, a_4, a_5 are called the **terms** of the sequence.

Often, rather than writing out the terms of a sequence, we simply define a sequence by writing a formula for the general term of the sequence, $\{a_n\}$, in terms of n. For example, a general term for the sequence $\{a_n\} = 3, 6, 9, 12, 15, \ldots$ can be written as $a_n = 3n$, where it is understood that n can be any positive integer. *Unless otherwise indicated, we begin with $n = 1$ and find each term of the sequence by substituting successive positive integers for n.* Thus, for $a_n = 3n$,

$$a_1 = 3(1) = 3$$
$$a_2 = 3(2) = 6$$
$$a_3 = 3(3) = 9$$
$$a_4 = 3(4) = 12$$
$$\vdots$$

Example 1

Write the first four terms and the seventh term of the sequence whose nth term is given by:

(a) $a_n = 3^n$ **(b)** $u_n = n^2 - 1$ **(c)** $b_n = \dfrac{n-1}{n}$ **(d)** $y_n = \dfrac{(-1)^n}{n+2}$

Solution

Most graphing calculators can compute the terms of a sequence. The following screen illustrates the terms of the sequence in part (b)
$u_n = n^2 - 1$
computed simultaneously for the values $n = 1, 2, 3, 4$ and the value when $n = 7$.

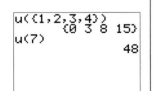

We can also graph a sequence and compute its values by using $\boxed{\text{TRACE}}$.

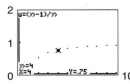

The graph of the sequence
$$b_n = \frac{n-1}{n}$$

(a) Because $a_n = 3^n$, the first term of the sequence, a_1, is found by substituting 1 for n. We get

$$a_1 = 3^1 = 3$$ The second term, a_2, is found by substituting 2 for n in a_n. Hence,
$$a_2 = 3^2 = 9$$ The third term is
$$a_3 = 3^3 = 27$$ And the fourth term is
$$a_4 = 3^4 = 81$$

So, the first four terms of the sequence are $3, 9, 27,$ and 81. The seventh term is $a_7 = 3^7 = 2187$.

(b) $u_n = n^2 - 1$ We substitute 1 for n in u_n to get the first term.
$$u_1 = (1)^2 - 1 = 0$$ For u_2, u_3, and u_4, we similarly substitute 2, 3, and 4, respectively, for n in u_n to get
$$u_2 = (2)^2 - 1 = 3$$
$$u_3 = (3)^2 - 1 = 8$$
$$u_4 = (4)^2 - 1 = 15$$

Hence, the first four terms of the sequence are $0, 3, 8, 15$. The seventh term is $u_7 = (7)^2 - 1 = 48$.

(c) $b_n = \dfrac{n-1}{n}$

$$b_1 = \frac{1-1}{1} = 0 \qquad\qquad b_2 = \frac{2-1}{2} = \frac{1}{2}$$

$$b_3 = \frac{3-1}{3} = \frac{2}{3} \qquad\qquad b_4 = \frac{4-1}{4} = \frac{3}{4}$$

Hence the first four terms are $0, \dfrac{1}{2}, \dfrac{2}{3}, \dfrac{3}{4}$, and the seventh term is given by

$$b_7 = \frac{7-1}{7} = \frac{6}{7}.$$

(d) Since $y_n = \dfrac{(-1)^n}{n+2}$,

$$y_1 = \frac{(-1)^1}{1+2} = \frac{-1}{3} \qquad y_2 = \frac{(-1)^2}{2+2} = \frac{1}{4}$$

$$y_3 = \frac{(-1)^3}{3+2} = \frac{-1}{5} \qquad y_4 = \frac{(-1)^4}{4+2} = \frac{1}{6}$$

Thus the first four terms of the sequence are $-\dfrac{1}{3}, \dfrac{1}{4}, -\dfrac{1}{5}, \dfrac{1}{6}$. The seventh term is $y_7 = \dfrac{(-1)^7}{7+2} = -\dfrac{1}{9}.$ ∎

Example 2 Suppose that a ball has the property that it always rebounds to a height equal to $\frac{1}{2}$ the distance from which it falls, and that such a ball is dropped from a height of 10 feet. Let d_n be the total vertical distance the ball has traveled when it hits the ground for the nth time. Find the first five terms of this sequence and graph them.

Solution Figure 9.1 illustrates the vertical distance traveled by the bouncing ball.

Figure 9.1

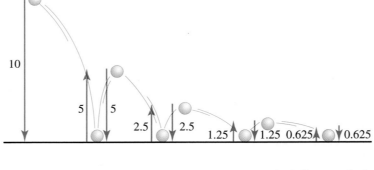

Using this figure and the fact that d_n is the total vertical distance the ball has traveled when it hits the ground for the nth time, we have

$$d_1 = 10$$
$$d_2 = 10 + 2(5) = 20$$
$$d_3 = 10 + 2(5) + 2(2.5) = 25$$
$$d_4 = 10 + 2(5) + 2(2.5) + 2(1.25) = 27.5$$
$$d_5 = 10 + 2(5) + 2(2.5) + 2(1.25) + 2(.625) = 28.75$$

as the first five terms of the sequence.

As defined previously, a sequence is a function whose domain is a subset of the natural numbers. Thus we are being asked to graph $d_n = f(n)$ for $n = 1, 2, 3, 4,$ and 5. As usual, we use the horizontal axis for the inputs (n) and the vertical axis for the outputs ($d_n = f(n)$). The graph appears in Figure 9.2. ∎

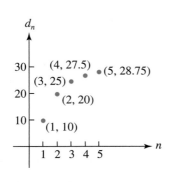

Figure 9.2

The graph of the sequence d_n for $n = 1, 2, 3, 4,$ and 5.

Do you think that the total distance the ball has traveled ever reaches 30 feet?

Examining a short sequence of numbers may allow us to identify a pattern and guess at the general term, but we must be careful: Two different sequences may start with the same terms. For example, if we examine the sequence $-1, 0, 1, \ldots$, we might guess that the general term of this sequence is $a_n = n - 2$. Then

$$a_1 = 1 - 2 = -1 \quad a_2 = 2 - 2 = 0 \quad a_3 = 3 - 2 = 1 \quad a_4 = 4 - 2 = 2$$

so that the first three terms of $\{a_n\}$ agree with the first three terms in $-1, 0, 1, \ldots$. However, if we consider the sequence with general term $b_n = (n - 2)^3$, then

$$b_1 = (1 - 2)^3 = -1 \quad b_2 = (2 - 2)^3 = 0 \quad b_3 = (3 - 2)^3 = 1$$
$$b_4 = (4 - 2)^3 = 8$$

Can you find any other possible formulas for the general term?

Even though the first three terms of $\{a_n\}$ agree with the first three terms of $\{b_n\}$, they are not the same sequence. (Look at the fourth term of each sequence.)

Thus when we examine the pattern in the terms of a sequence to look for a formula for the general term, we are only making a guess based on the available data.

Example 3 Find a possible general term for the following sequences:
(a) 6, 12, 18, 24, 30, ...   (b) 10; 100; 1000; 10,000; 100,000; ...
(c) $\dfrac{1}{3}, \dfrac{1}{5}, \dfrac{1}{7}, \dfrac{1}{9}, \ldots$

Solution We find the general term by examining the given terms and looking for a pattern.

(a) We can see that the first five terms are multiples of 6, so we can write $a_1 = 6 = 6 \cdot 1$, $a_2 = 12 = 6 \cdot 2$, $a_3 = 18 = 6 \cdot 3, \ldots$; therefore, a general term is possibly $a_n = 6n$.

(b) We can see that the first five terms are powers of 10, so we can write $a_1 = 10 = 10^1$, $a_2 = 100 = 10^2$, $a_3 = 1000 = 10^3, \ldots$; therefore, a general term is possibly $a_n = 10^n$.

(c) Given $a_1 = \dfrac{1}{3}$, $a_2 = \dfrac{1}{5}$, $a_3 = \dfrac{1}{7}$, $a_4 = \dfrac{1}{9}, \ldots$, we can see that the denominators are the odd numbers 3, 5, 7, 9. Thus a general term is possibly $a_n = \dfrac{1}{2n + 1}$. ∎

Describing certain real-life situations can lead us quite naturally to sequences.

Example 4 Elizabeth deposits \$2000 in a bank that pays 5% simple annual interest. Write a sequence showing the amount of money Elizabeth has in the bank at the end of each year for 4 years.

Solution In Section 6.5, we developed the formula $A = P(1 + r)^t$ for the amount of money, A, in an account if P dollars is invested at an annual interest rate of r for t years. (Remember that in the formula, r is the interest rate written as a decimal.) In this example, the interest rate is 5%, so $r = 0.05$, and $1 + r = 1.05$.

If we let $A_k =$ the amount in Elizabeth's account at the end of year k, then using this formula gives

$$A_1 = 2000(1.05) = \$2100$$
$$A_2 = 2000(1.05)^2 = \$2205$$
$$A_3 = 2000(1.05)^3 = \$2315.25$$
$$A_4 = 2000(1.05)^4 = \$2431.01$$

Therefore, the sequence of amounts in the bank at the end of each year for 4 years is

$$\$2100, \$2205, \$2315.25, \text{ and } \$2431.01$$

All the sequences we have discussed thus far have had an explicit formula for the nth term—for example, $a_n = 3n + 2$. The next example illustrates a different way to specify the terms of a sequence. A **recursively defined sequence** is a sequence in which later terms are defined by referring to earlier terms.

Example 5 Find the fourth term of each of the following recursively defined sequences:

(a) $a_1 = 5$ and $a_{k+1} = 2a_k + 3$ (b) $a_1 = 4$, $a_2 = 3$, and $a_{k+2} = a_{k+1} - 2a_k$

Solution (a) The given recursive relationship, $a_{k+1} = 2a_k + 3$, means that each term after the first is 3 more than twice the previous term. We are given that $a_1 = 5$, so to find a_2 we substitute a_1 into the recursive formula.

a_{k+1} is the term following a_k and a_{k+2} is the term following a_{k+1}.

$$a_{k+1} = 2a_k + 3 \qquad\qquad \text{For } k = 1 \text{ we have}$$
$$a_2 = 2a_1 + 3 = 2(5) + 3 = 13 \qquad \text{So } a_2 = 13$$
$$a_3 = 2a_2 + 3 = 2(13) + 3 = 29 \qquad \text{So } a_3 = 29$$
$$a_4 = 2a_3 + 3 = 2(29) + 3 = 61$$

Therefore, $a_4 = 61$.

(b) In the sequence defined by $a_{k+2} = a_{k+1} - 2a_k$, we are given the first two terms and the formula, which tells us that successive terms are obtained by taking the difference between the previous term and twice the term before that. Thus

$$a_1 = 4$$
$$a_2 = 3 \qquad \text{Given that } a_{k+2} = a_{k+1} - 2a_k, \text{ we have}$$
$$a_3 = a_2 - 2a_1 = 3 - 2(4) = -5 \qquad \text{So } a_3 = -5$$
$$a_4 = a_3 - 2a_2 = -5 - 2(3) = -11$$

Therefore, $a_4 = -11$.

Note that if we want to find the fifteenth term of a recursively defined sequence, we usually have to find the first fourteen terms of the sequence, whereas if we want to find the fifteenth term of a sequence for which we have a formula in terms of n, we can just substitute $n = 15$. In the next few sections, we discuss some special recursively defined sequences.

Some very important sequences in mathematics involve terms that are particular products, indicated with the following special notation.

Definition of $n!$

For n a positive integer, $n!$ (read "n factorial") is defined as

$$n! = n(n - 1)(n - 2) \cdots 3 \cdot 2 \cdot 1$$

In words, $n!$ is the *product* of all the positive integers less than or equal to n. We also define $0! = 1$.

Example 6 Compute each of the following:

(a) $5!$ (b) $\dfrac{8!}{4!}$ (c) $\dfrac{7! - 3!}{(7 - 3)!}$

Solution   **(a)** Using the definition of the factorial notation, we get

$$5! = 5 \cdot 4 \cdot 3 \cdot 2 \cdot 1 = \boxed{120}$$

(b) Again using the definition of $n!$, we get

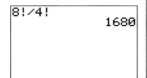

Most graphing calculators can compute factorials, as illustrated in the following display.

$$\frac{8!}{4!} = \frac{8 \cdot 7 \cdot 6 \cdot 5 \cdot \cancel{4} \cdot \cancel{3} \cdot \cancel{2} \cdot 1}{\cancel{4} \cdot \cancel{3} \cdot \cancel{2} \cdot 1} \qquad \text{Reduce.}$$

$$= 8 \cdot 7 \cdot 6 \cdot 5 = \boxed{1680}$$

(c) We must compute the factorials carefully.

$$\frac{7! - 3!}{(7 - 3)!} = \frac{7! - 3!}{4!} \qquad \begin{array}{l}\text{We could compute each factorial separately and then}\\ \text{evaluate the fraction; however, we offer a more}\\ \text{algebraic approach.}\end{array}$$

$$= \frac{7 \cdot 6 \cdot 5 \cdot 4 \cdot 3 \cdot 2 \cdot 1 - 3 \cdot 2 \cdot 1}{4 \cdot 3 \cdot 2 \cdot 1} \qquad \text{Factor and reduce.}$$

$$= \frac{\cancel{3} \cdot \cancel{2} \,(7 \cdot 6 \cdot 5 \cdot 4 - 1)}{4 \cdot \cancel{3} \cdot \cancel{2} \cdot 1} = \boxed{\frac{839}{4}}$$

Example 7   Find the fourth term of the sequence $a_n = \dfrac{2n!}{(2n)!}$.

Solution   We must be careful to clearly distinguish $2n!$ from $(2n)!$
For $n = 4$ we have

$$a_4 = \frac{2 \cdot 4!}{(2 \cdot 4)!} = \frac{2 \cdot 4!}{8!} = \frac{2 \cdot 4 \cdot 3 \cdot 2 \cdot 1}{8 \cdot 7 \cdot 6 \cdot 5 \cdot 4 \cdot 3 \cdot 2 \cdot 1} = \boxed{\frac{1}{840}}$$

We will have occasion to apply factorial notation in Sections 9.6–9.8.

9.1 Exercises

In Exercises 1–30, write the first four terms and the eighth term of the sequence whose general term is given.

1. $a_n = 5n - 3$

2. $a_n = 4n + 2$

3. $b_n = 3n + 2$

4. $b_n = 6n + 1$

5. $c_k = 3^k$

6. $c_k = 2^{k-1}$

7. $x_j = \dfrac{j + 1}{j}$

8. $x_j = \dfrac{j - 1}{j + 1}$

9. $x_n = \dfrac{(-1)^{n+1}}{n + 2}$

10. $y_n = \dfrac{(-1)^n}{2n}$

11. $a_n = \dfrac{(-1)^n n}{n + 1}$

12. $a_n = \dfrac{(-1)^n n^2}{3n - 2}$

13. $a_n = 2^n + |n - 3|$

14. $a_n = 3^n + |5 - n|$

15. $b_n = 1 + (-0.1)^n$

16. $b_n = 2 - (0.2)^n$

17. $x_n = \dfrac{2^n}{n^2}$

18. $y_n = \dfrac{n^3}{3^n}$

19. $a_n = 1 + (-1)^n$

20. $b_n = \dfrac{(-1)^n + (-1)^{n+1}}{2}$

21. $c_n = \left(1 + \dfrac{1}{n}\right)^n$

22. $d_n = \dfrac{1}{n + 1} - \dfrac{1}{n + 2}$

23. $a_n = \sqrt{n^2 + 1}$

24. $b_n = 5$

25. $a_n = \dfrac{1}{n!}$

26. $b_n = \dfrac{(-1)^n}{(n + 1)!}$

27. $c_n = \dfrac{n!}{n^n}$

28. $d_n = \dfrac{2^n}{(n - 1)!}$

29. $t_n = \dfrac{(3n)!}{3^n n!}$

30. $s_n = \dfrac{(2n - 1)!}{(2n + 1)!}$

In Exercises 31–42, find the first six terms of the given recursive sequence.

31. $a_1 = 3; a_{n+1} = 2a_n - 1$

32. $a_1 = 1; a_n = \dfrac{1}{a_{n-1} + 1}$ for $n > 1$

33. $a_1 = -1; a_n = (1 - a_{n-1})^2$ for $n > 1$

34. $a_1 = 2; a_n = \sqrt{4 + (a_{n-1})^2}$ for $n > 1$

35. $a_1 = 1; a_{n+1} = 2^{a_n}$

36. $a_1 = 3; a_{n+1} = na_n$

37. $a_1 = 2; a_n = \dfrac{n}{a_{n-1}}$ for $n > 1$

38. $a_1 = 2; a_n = 3a_{n-1} - n$ for $n > 1$

39. $a_1 = 0; a_2 = 1; a_{n+2} = \dfrac{1}{2}(a_{n+1} + a_n)$

40. $a_1 = 1; a_2 = 2; a_n = \dfrac{a_{n-2}}{a_{n-1}}$ for $n \geq 3$

41. $a_1 = -1; a_2 = 1; a_n = (a_{n-2})(a_{n-1})$ for $n \geq 3$

42. $a_1 = 1; a_2 = 1; a_{n+2} = na_{n+1} + (n + 1)a_n$

In Exercises 43–46, graph the first five terms of the sequence.

43. $y_n = 2^{n-1}$

44. $y_n = n^2 + 1$

45. $y_n = (-1)^n$

46. $y_n = \dfrac{(-1)^n}{n}$

47. A promotional firm begins a telemarketing campaign by making 100 phone calls on the first day. Each day the firm will make 20 more calls than were made the day before. Let c_n be the *total* number of calls made by the end of day n. Graph the first five terms of this sequence.

48. In a certain bacteria culture, the number of bacteria present triples every hour. Suppose that a culture begins with 10 bacteria. Let b_n be the number of bacteria present at the end of hour n. Graph the first 4 terms of this sequence.

In Exercises 49–58, find a possible formula for the general term of the given sequence.

49. 5, 10, 15, 20, . . .

50. 3, 8, 13, 18, . . .

51. 1, 3, 5, 7, 9, . . .

52. 5, 7, 9, 11, 13, . . .

53. $\dfrac{1}{2}, \dfrac{3}{4}, \dfrac{5}{6}, \dfrac{7}{8}, \ldots$

54. $\dfrac{2}{3}, \dfrac{4}{9}, \dfrac{8}{27}, \dfrac{16}{81}, \ldots$

55. $\dfrac{1}{2}, -\dfrac{1}{3}, \dfrac{1}{4}, -\dfrac{1}{5}, \ldots$

56. $-\dfrac{1}{2}, \dfrac{2}{3}, -\dfrac{3}{4}, \dfrac{4}{5}, \ldots$

57. $-0.4, 0.04, -0.004, 0.0004, \ldots$

58. $0.07, -0.0007, 0.000007, -0.00000007, \ldots$

59. Ann gets a raise of $650 at the end of each year she works for her company. If her starting salary is $24,500, write a sequence to show her income at the end of each year for the first 3 years.

60. Sean gets a raise of $600 every 6 months for the first 2 years he works for his firm. If his starting salary is $22,750, write a sequence to show his income at the end of each 6-month period for the first 2 years.

61. The population of Centerville increases at a rate of 2% per year. If the population is currently 60,000 people, write a sequence to show the population at the end of each year for 3 years.

62. The population of Rivertown is decreasing at a rate of 1.5% per year. If the population is currently 130,000 people, write a sequence to show the population at the end of each year for 4 years.

63. Juan deposits $1800 in a bank that is paying an effective annual yield of 3.5%. Write a sequence to show the amount of money Juan has in the bank at the end of each year for 5 years.

64. Elena deposits $2750 in a bank that is paying an effective annual yield of 4.2%. Write a sequence to show how much Elena has in the bank at the end of each year for 3 years.

65. A country is depleting its oil reserves at the rate of 1.35% per year. If it starts out with reserves of 2.3 billion barrels of oil in 1999, write a sequence to show the amount of oil reserves left at the end of each year for 5 years.

66. A city finds that its electric power consumption on peak summer days is increasing at the rate of 4.2% per year. If the peak electric power consumption in the summer of 1998 is 8.3 megawatts, write a sequence to show the peak electric power consumption during the next six summers.

67. A ball rebounds to one-half the distance from which it falls. If it is dropped from a height of 60 feet, write a sequence to show how high it bounces for the first five bounces.

68. A ball rebounds to two-thirds the distance from which it falls. If it is dropped from a height of 60 feet, write a sequence to show how high it bounces for the first five bounces.

69. George borrows $3000 and promises to pay back $200 per month plus 1% of the remaining balance each month. Write a sequence showing how much George must pay each month for the first 4 months.

70. Lena borrows \$8000 and promises to pay back \$300 per month plus 2% of the remaining balance each month. Write a sequence showing how much Lena owes each month for the first 6 months.

Questions for Thought

71. A very famous sequence is the Fibonacci sequence, F_n, where F_n is defined recursively as follows:

$$F_1 = 1; \quad F_2 = 1; \quad F_n = F_{n-2} + F_{n-1} \text{ for } n > 2$$

(a) Write the first eight terms of the Fibonacci sequence.

(b) Compute $F_1 + F_2 + F_3 + \cdots + F_n$ and $F_{n+2} - 1$ for $n = 5, 7$, and 10. What do you notice?

(c) Compute $(F_1)^2 + (F_2)^2 + (F_3)^2 + \cdots + (F_n)^2$ and $F_n \cdot F_{n+1}$ for $n = 4, 6$, and 9. What do you notice?

72. Suppose $\{a_n\}$ is a recursively defined sequence. Would you expect to be able to compute a_{16} directly? Explain.

73. Simplify each of the following factorial expressions.

(a) $\dfrac{(n+1)!}{n!}$ **(b)** $\dfrac{(2n+3)!}{(2n)!}$ **(c)** $\dfrac{(2n-1)!}{(2n+1)!}$

9.2 Series and Sigma Notation

In the previous section we were interested in the individual terms of a sequence. In this section we describe the process of adding the terms of a sequence. In the next two sections, we discuss some specific applications of this process.

Given a sequence, $\{a_k\}$, we can associate a sum with it, obtained by adding the first n terms of the sequence. For example, for the sequence

$$\{5, 10, 15, 20, 25, 30, \ldots, 5k, \ldots\}$$

we can associate the sum S_n, where S_n *is the sum of the first n terms of the sequence* $\{a_k\}$. Thus we have the following definition.

Definition of a Series

The series S_n associated with the sequence $\{a_k\}$ is defined by

$$S_n = a_1 + a_2 + a_3 + \cdots + a_n$$

In words, this says that S_n is the sum of the first n terms of the sequence $\{a_k\}$.

Note that we usually use uppercase letters to designate series and lowercase letters to designate sequences.

Thus for the sequence $\{5, 10, 15, 20, 25, 30, \ldots, 5k, \ldots\}$, we have

$S_1 = a_1 = 5$ — S_1 is the *first* term of the sequence.

$S_2 = a_1 + a_2 = 5 + 10$ — S_2 is the sum of the first *two* terms of the sequence.

$S_3 = a_1 + a_2 + a_3 = 5 + 10 + 15$ — S_3 is the sum of the first *three* terms of the sequence.

$S_4 = a_1 + a_2 + a_3 + a_4 = 5 + 10 + 15 + 20$ — S_4 is the sum of the first *four* terms of the sequence.

$\vdots$

$S_n = a_1 + a_2 + a_3 + a_4 + \cdots + a_n = 5 + 10 + 15 + 20 + \cdots + 5n$ — S_n is the sum of the first n terms of the sequence.

To find a series associated with the sequence $\{a_n\}$, where $a_n = 7n - 8$, we first write out some of the terms of the sequence: $-1, 6, 13, 20, 27, \ldots$. We can then find the indicated sums of the given sequence:

Verify that these are the first few terms of $\{a_n\}$.

$$\begin{aligned} S_1 &= a_1 & &= -1 \\ S_2 &= a_1 + a_2 &= -1 + 6 &= 5 \\ S_3 &= a_1 + a_2 + a_3 &= -1 + 6 + 13 &= 18 \\ S_4 &= a_1 + a_2 + a_3 + a_4 &= -1 + 6 + 13 + 20 &= 38 \end{aligned}$$

Example 1   Find S_5 for the sequence whose nth term is given by

(a) $a_n = 3n - 5$ (b) $b_n = 2^n - 1$ (c) $x_n = \dfrac{(-1)^n}{n}$

Solution   (a) S_5 is the sum of the first five terms of the sequence $\{a_n\}$, where $a_n = 3n - 5$. We find the first five terms by successively substituting $n = 1, 2, 3, 4, 5$ into the given formula for a_n. We find that $a_1 = -2$, $a_2 = 1$, $a_3 = 4$, $a_4 = 7$, and $a_5 = 10$. Thus

$$S_5 = a_1 + a_2 + a_3 + a_4 + a_5 = -2 + 1 + 4 + 7 + 10 = \boxed{20}$$

(b) S_5 is the sum of the first five terms of the sequence $\{b_n\}$, where $b_n = 2^n - 1$. Again, we find the first five terms: $b_1 = 1$, $b_2 = 3$, $b_3 = 7$, $b_4 = 15$, and $b_5 = 31$. Hence,

$$S_5 = b_1 + b_2 + b_3 + b_4 + b_5 = 1 + 3 + 7 + 15 + 31 = \boxed{57}$$

(c) We begin by finding the first five terms:

$$x_1 = \frac{(-1)^1}{1} = -1 \qquad x_2 = \frac{(-1)^2}{2} = \frac{1}{2} \qquad x_3 = \frac{(-1)^3}{3} = \frac{-1}{3},$$

$$x_4 = \frac{(-1)^4}{4} = \frac{1}{4} \qquad x_5 = \frac{(-1)^5}{5} = \frac{-1}{5}$$

Therefore, we have $S_5 = -1 + \dfrac{1}{2} + \dfrac{-1}{3} + \dfrac{1}{4} + \dfrac{-1}{5} = \boxed{-\dfrac{47}{60}}$.

Sigma Notation

When we have a formula for the general term of a sequence, we can express a series for the sequence in a more compact form by using a special notation for sums. The Greek (uppercase) letter sigma, Σ, often called the summation symbol, is used along with the general term of the sequence to indicate a series.

Definition of Sigma Notation

$$\sum_{k=1}^{N} a_k = a_1 + a_2 + a_3 + a_4 + \cdots + a_N$$

k is called the **index of summation**, or simply the **index**.

The Σ indicates that we are going to *add* the terms. The number k beneath the sigma is the number substituted for the index, k, to find the *first* term of the sum. Then, we increase the index by one and evaluate each subsequent term until the index reaches the number above the sigma, N, which is the last number used for k in the sum. In this example, 1 is called the **lower limit** of the summation and N is called the **upper limit** of the summation. In other words, we are adding the terms of the sequence $\{a_n\}$ starting with a_1 up to and including a_N.

Remember that sigma notation is used to express a *sum*.

Example 2   Evaluate each of the following.

(a) $\displaystyle\sum_{k=1}^{6} 4k$ (b) $\displaystyle\sum_{j=1}^{5} \frac{1}{j+1}$ (c) $\displaystyle\sum_{k=2}^{4} 10k^2$ (d) $\displaystyle\sum_{n=3}^{7} (4n - 3)$

Solution

(a) $\displaystyle\sum_{k=1}^{6} 4k$ Each term in the sum is found by substituting consecutive integers for k in the general term $4k$, starting with $k = 1$ and ending with $k = 6$. Remember: $\sum$ is telling us to add the terms.

$$= 4(1) + 4(2) + 4(3) + 4(4) + 4(5) + 4(6)$$

$$= \boxed{84}$$

The letters i, j, k, and n are popular choices for the index of summation.

(b) $\displaystyle\sum_{j=1}^{5} \frac{1}{j+1}$ Find each term in the sum by substituting consecutive integers for j in the general term $\dfrac{1}{j+1}$, starting with $j = 1$ and ending with $j = 5$.

$$\sum_{j=1}^{5} \frac{1}{j+1} = \frac{1}{1+1} + \frac{1}{2+1} + \frac{1}{3+1} + \frac{1}{4+1} + \frac{1}{5+1}$$

$$= \frac{1}{2} + \frac{1}{3} + \frac{1}{4} + \frac{1}{5} + \frac{1}{6}$$

$$= \frac{87}{60} = \boxed{\frac{29}{20}}$$

The index does not have to start with 1. $\displaystyle\sum_{k=2}^{4}$ means that we start with $k = 2$ and continue adding terms up to and including $k = 4$.

(c) $\displaystyle\sum_{k=2}^{4} 10k^2$ This time the first term is found by substituting $k = 2$ in $10k^2$. The next terms are found by substituting $k = 3$ and $k = 4$.

$$= 10(2)^2 + 10(3)^2 + 10(4)^2$$

$$= 40 + 90 + 160$$

$$= \boxed{290}$$

(d) $\displaystyle\sum_{n=3}^{7} (4n - 3)$ This time we are using n as the index of summation. The first term is found by evaluating the expression $4n - 3$ for $n = 3$. The next terms are found by evaluating the expression $4n - 3$ for $n = 4, 5, 6,$ and 7, respectively.

$$= [4(3) - 3] + [4(4) - 3] + [4(5) - 3] + [4(6) - 3] + [4(7) - 3]$$

$$= 9 + 13 + 17 + 21 + 25$$

$$= \boxed{85}$$

Note that the index is merely a "counter" and so can be replaced with any other letter. Hence

$$\sum_{i=1}^{10} a_i = \sum_{j=1}^{10} a_j = \sum_{h=1}^{10} a_h$$

Recall that we denote the sum of the first N terms of a sequence $\{a_i\}$ by S_N, so using sigma notation, we may write

$$S_N = a_1 + a_2 + a_3 + \cdots + a_N = \sum_{i=1}^{N} a_i$$

Example 3   Given a sequence for which $a_k = 2k^3$, evaluate S_4.

Solution $\displaystyle S_4 = \sum_{k=1}^{4} 2k^3 = 2(1^3) + 2(2^3) + 2(3^3) + 2(4^3) = 2 + 16 + 54 + 128 = \boxed{200}$

Example 4   Evaluate the following sum: $\displaystyle\sum_{n=1}^{5}\left(\frac{1}{n}-\frac{1}{n+1}\right)$

Solution   We replace n successively with the values 1, 2, 3, 4, and 5, in the expression $\dfrac{1}{n}-\dfrac{1}{n+1}$ and add the results.

$$\sum_{n=1}^{5}\left(\frac{1}{n}-\frac{1}{n+1}\right)=\left(\frac{1}{1}-\frac{1}{1+1}\right)+\left(\frac{1}{2}-\frac{1}{2+1}\right)+\left(\frac{1}{3}-\frac{1}{3+1}\right)+\left(\frac{1}{4}-\frac{1}{4+1}\right)+\left(\frac{1}{5}-\frac{1}{5+1}\right)$$

$$=\left(\frac{1}{1}-\frac{1}{2}\right)+\left(\frac{1}{2}-\frac{1}{3}\right)+\left(\frac{1}{3}-\frac{1}{4}\right)+\left(\frac{1}{4}-\frac{1}{5}\right)+\left(\frac{1}{5}-\frac{1}{6}\right)$$

$$=\frac{1}{1}-\frac{1}{2}+\frac{1}{2}-\frac{1}{3}+\frac{1}{3}-\frac{1}{4}+\frac{1}{4}-\frac{1}{5}+\frac{1}{5}-\frac{1}{6}=1-\frac{1}{6}=\boxed{\frac{5}{6}}$$

> Notice that all but the first and last terms drop out.

Such a sum, where many of the terms drop out, is called a **telescoping sum**.   ∎

Example 5   Given $f(x)=\displaystyle\sum_{n=1}^{4}nx^{n-1}$, find $f(2)$.

Solution   We first write out $f(x)$ without sigma notation, and then substitute $x=2$.

$$f(x)=\sum_{n=1}^{4}nx^{n-1}$$

$$=1x^{0}+2x^{1}+3x^{2}+4x^{3}\qquad\text{So}$$

$$f(x)=1+2x+3x^{2}+4x^{3}$$

> Now that $f(x)$ is in more familiar form, we compute $f(2)$.

$$f(2)=1+2(2)+3(2)^{2}+4(2)^{3}=1+4+12+32=\boxed{49}$$ ∎

Because sigma notation is merely a shorthand way of denoting a sum, we can restate some of the real number properties using sigma notation.

Properties of Σ Notation

1. $\displaystyle\sum_{k=1}^{n}ca_{k}=c\left(\sum_{k=1}^{n}a_{k}\right)$, where c is any constant.

2. $\displaystyle\sum_{k=1}^{n}(a_{k}+b_{k})=\sum_{k=1}^{n}a_{k}+\sum_{k=1}^{n}b_{k}$

3. $\displaystyle\sum_{k=1}^{n}(a_{k}-b_{k})=\sum_{k=1}^{n}a_{k}-\sum_{k=1}^{n}b_{k}$

Each of these properties is a direct consequence of the commutative, associative, and distributive properties of the real numbers. See Exercise 68.

In the next section we investigate a special type of sequence and its corresponding series.

9.2 Exercises

In Exercises 1–26, use the given sequence (or general term) to find the indicated sum, S_n.

1. $3, 7, 11, 15, 19, 23, 27, \ldots;$ S_5

2. $-8, -3, 2, 7, 12, 17, \ldots;$ S_4

3. $-2, 0, 2, 4, 6, 8, \ldots;$ S_6

4. $-1, 5, 11, 17, 23, \ldots;$ S_8

5. $10, 6, 2, -2, -6, -10, \ldots;$ S_7

6. $14, 11, 8, 5, 2, -1, \ldots;$ S_{10}

7. $1, 2, 4, 8, 16, \ldots;$ S_6

8. $1, \dfrac{1}{2}, \dfrac{1}{4}, \dfrac{1}{8}, \ldots;$ S_5

9. $a_n = 3n + 1;$ S_2 10. $a_n = 2n + 1;$ S_3

11. $b_m = 5m - 1;$ S_4 12. $b_m = 4m + 3;$ S_5

13. $x_j = 2^j;$ S_5 14. $x_j = 3^j;$ S_6

15. $x_k = 2^k + 1;$ S_6 16. $x_k = 2^{k+1};$ S_6

17. $x_k = 3^{k-2};$ S_5 18. $x_k = 3^k - 2;$ S_5

19. $a_n = \dfrac{n+1}{n};$ S_4

20. $a_n = \dfrac{n}{n+2};$ S_4

21. $b_j = \dfrac{(-1)^{j+1}}{j+2};$ S_5

22. $b_n = \dfrac{(-1)^n n}{n+1};$ S_3

23. $r_m = \dfrac{1}{m!};$ S_4

24. $t_n = \dfrac{2^n}{(n+1)!};$ S_3

25. $a_n = \dfrac{1}{n} - \dfrac{1}{n+1};$ S_7

26. $b_n = \dfrac{n}{n+1} - \dfrac{n+1}{n+2};$ S_6

In Exercises 27–42, rewrite each sum without using sigma notation; then compute each sum.

27. $\displaystyle\sum_{n=1}^{5} n$

28. $\displaystyle\sum_{j=1}^{8} 3j$

29. $\displaystyle\sum_{i=1}^{6} 5(i - 1)$

30. $\displaystyle\sum_{k=1}^{4} 4(k + 2)$

31. $\displaystyle\sum_{k=2}^{6} k^2$

32. $\displaystyle\sum_{n=3}^{5} n^3$

33. $\displaystyle\sum_{m=3}^{7} (2m^2 - 5)$

34. $\displaystyle\sum_{k=2}^{6} (3k^2 - 4)$

35. $\displaystyle\sum_{n=2}^{4} (n^2 - 3n + 1)$

36. $\displaystyle\sum_{j=3}^{5} (10j - j^2)$

37. $\displaystyle\sum_{j=1}^{5} \dfrac{1}{j+1}$

38. $\displaystyle\sum_{j=1}^{4} \left(\dfrac{1}{j} + 1\right)$

39. $\displaystyle\sum_{n=1}^{5} (\sqrt{n+1} - \sqrt{n})$ 40. $\displaystyle\sum_{m=1}^{4} \left(\dfrac{2}{m} - \dfrac{2}{m+1}\right)$

41. $\displaystyle\sum_{n=1}^{5} 3$

42. $\displaystyle\sum_{k=3}^{8} 5$

In Exercises 43–46, use sigma notation to represent the sum of the first n terms of the given sequence.

43. $4, 8, 12, \ldots, 4k, \ldots$ for $n = 5$

44. $2, 5, 8, \ldots, 3k - 1, \ldots$ for $n = 8$

45. $2, 8, 18, \ldots, 2k^2, \ldots$ for $n = 7$

46. $7, 9, 11, \ldots, (2k + 5), \ldots$ for $n = 8$

In Exercises 47–52, express the given sum using sigma notation. (The answers to these exercises are not unique.)

47. $2 + 6 + 10 + 14 + 18$

48. $5 + 25 + 125 + 625$

49. $1 + 3 + 5 + 7 + \cdots + 51$

50. $4 + 7 + 10 + 13 + \cdots + 52$

51. $\dfrac{1}{2} + \dfrac{1}{2 \cdot 3} + \dfrac{1}{3 \cdot 4} + \cdots + \dfrac{1}{99 \cdot 100}$

52. $\dfrac{2}{5} + \dfrac{4}{9} + \dfrac{6}{13} + \cdots + \dfrac{20}{41}$

In Exercises 53–58, find S_6 for the given recursively defined sequence.

53. $a_1 = 3; a_n = 2a_{n-1} + 3$ for $n > 1$

54. $a_1 = 64; a_{k+1} = \dfrac{1}{2}a_k$

55. $a_1 = -2; a_{n+1} = (a_n + 1)^2$

56. $a_1 = 2; a_{k+1} = \dfrac{1}{a_k}$

57. $a_1 = 2; a_2 = 3; a_{n+2} = a_{n+1} \cdot a_n$

58. $a_1 = 5; a_k = ka_{k-1}$ for $k > 1$

59. Write out $\displaystyle\sum_{n=1}^{8} \log_3 \frac{n+1}{n}$ without using sigma notation and simplify the sum. HINT: Use the properties of logarithms.

60. Write out $\displaystyle\sum_{n=1}^{6} \log_8 2^n$ without using sigma notation and simplify the sum. HINT: Use the properties of logarithms.

61. Given that $f(x) = \displaystyle\sum_{k=1}^{5} x^k$, find $f(1), f(-2)$, and $f\left(\dfrac{1}{2}\right)$.

62. Given that $g(x) = \displaystyle\sum_{k=1}^{4} \frac{e^x}{k}$, find $g(0)$, $g(1)$, and $g(2)$.

63. Given that $F(x) = \displaystyle\sum_{k=0}^{6} \frac{x^k}{k!}$:

  (a) Find $F(1)$, $F(2)$, and $F(-1)$ correct to four decimal places. (Remember that, by definition, $0! = 1$.)

  (b) Compare the values of $F(1)$, $F(2)$, and $F(-1)$ with the values of e, e^2, and e^{-1}, respectively (where e is the base for natural logarithms). What do you notice?

64. Given that $f(x) = \displaystyle\sum_{k=1}^{6} \frac{x^{k-1}}{(k-1)!}$:

  (a) Find $f(0), f(0.5), f(1)$, and $f(1.5)$ correct to four decimal places.

  (b) Compare the values of $f(0), f(0.5), f(1)$, and $f(1.5)$ with the values of e^0, $e^{0.5}$, e^1, and $e^{1.5}$, respectively. What do you notice?

Questions for Thought

65. Show that for c a constant, $\displaystyle\sum_{k=1}^{N} c = Nc$.

66. Show that $\displaystyle\sum_{i=1}^{n} a_i = \sum_{i=2}^{n+1} a_{i-1}$.

67. Write the polynomial

$$p(x) = a_n x^n + a_{n-1} x^{n-1} + a_{n-2} x^{n-2} + \cdots$$
$$\cdots + a_2 x^2 + a_1 x + a_0$$

using sigma notation.

68. Use the basic properties of the real numbers to prove the three properties of sigma notation listed in the box on page 576.

9.3   Arithmetic Sequences and Series

Arithmetic Sequences

In the last section we saw that to compute the sum of a series, we actually had to add all the terms in the series. Sometimes, however, the special structure of a sequence allows us to derive a formula for its associated series.

Let's examine the following sequences of numbers:

$$\{a_k\} = 2, 5, 8, 11, 14, \ldots$$
$$\{b_k\} = 1, 6, 11, 16, 21, \ldots$$
$$\{c_k\} = 6, 4, 2, 0, -2, \ldots$$

Can you find the next three terms in each sequence?

Rather than trying to find a formula immediately, we may find it fruitful to examine the relationship between successive terms of a sequence.

For example, in the sequence $\{a_k\} = 2, 5, 8, 11, 14, \ldots$, we observe that each term after the first is 3 more than the preceding term. Assuming that this same pattern continues for successive terms, the three terms after 14 would be 17, 20, 23. Symbolically, we can represent this relationship as

$$a_{k+1} = a_k + 3$$

In Section 9.1 we noted that different sequences may agree for a number of terms. Thus we cannot necessarily predict a formula for the sequence from the first few terms.

Note that this means that any term, (say, the $(k + 1)$st term), is 3 more than the previous term (the kth term). For $k = 10$, $a_{11} = a_{10} + 3$. Equivalently, we could write $a_{k+1} = a_k + 3$ as a difference:

$$a_{k+1} - a_k = 3$$

This says that the *difference* between any two successive terms is 3.

In the second sequence, $\{b_k\} = 1, 6, 11, 16, 21, \ldots$, we observe that each term after the first is 5 more than the previous term. Assuming that this same pattern continues, we conclude that adding 5 will produce the next term in the sequence. Algebraically we write $b_{k+1} = b_k + 5$, which says that any term after the first is found by adding 5 to the preceding term. Thus, the three terms following 21 would be 26, 31, 36. Again, we can write $b_{k+1} = b_k + 5$ as a difference:

$b_{k+1} - b_k = 5$ says that the difference between any two successive terms is 5.

$$b_{k+1} - b_k = 5$$

Notice that in $\{c_k\} = 6, 4, 2, 0, -2, \ldots$, the third sequence given, each term is 2 less than the previous term. Hence we could describe the relationship between successive terms in the sequence as

$$c_{k+1} = c_k - 2$$

So in $\{c_k\}$, the term -2 would be followed by $-4, -6, -8$, and so on. Rewriting this relationship as a difference, we have

$$c_{k+1} - c_k = -2$$

The difference between two consecutive terms is -2 because $4 - 6 = -2$ and $-2 - (0) = -2$, etc.

Sequences in which the difference between successive terms is constant are called *arithmetic sequences*. This can be stated as follows.

Definition of an Arithmetic Sequence

An **arithmetic sequence** is one in which the difference between successive terms (called the **common difference**) is constant. This common difference is usually denoted by d.

Algebraically, we write $a_{k+1} - a_k = d$ or equivalently $a_{k+1} = a_k + d$.

An arithmetic sequence is sometimes called an *arithmetic progression*.

Hence, referring back to the sequences given at the beginning of this section, the common difference in the first sequence, $\{a_k\}$, is 3; the common difference in the second sequence, $\{b_k\}$, is 5; and the common difference in the third sequence, $\{c_k\}$, is -2.

Why is an arithmetic sequence a recursive sequence?

In other words, if $a_1, a_2, a_3, a_4, \ldots, a_{k-1}, a_k, \ldots$ is an arithmetic sequence, then $a_{k+1} - a_k = d$; or equivalently $a_{k+1} = a_k + d$ for all k.

Similarly, if an arithmetic sequence has a common difference of d and we start with a_1, then the next term in the sequence, a_2, is

$$a_2 = a_1 + d \qquad \text{The common difference is added to the first term.}$$

The next term, a_3, is

$$a_3 = a_2 + d = (a_1 + d) + d = a_1 + 2d$$
$$a_4 = a_3 + d = (a_1 + 2d) + d = a_1 + 3d \qquad \text{Note the pattern.}$$
$$a_5 = a_4 + d = a_1 + 4d \qquad$$

To get the fifth term, we add *four* times the common difference to a_1.

$$a_6 = a_5 + d = a_1 + 5d$$

To get the sixth term, we add *five* times the common difference to a_1.

$$\vdots$$

In general, we have

$$a_n = a_{n-1} + d = a_1 + (n - 1)d$$

We have thus derived the following formula:

If $\{a_n\}$ is an arithmetic sequence with first term a_1 and common difference d, then the nth term is given by

$$a_n = a_1 + (n - 1)d \qquad \text{Formula 9.1}$$

Thus, for example, in any arithmetic sequence, $a_9 = a_1 + 8d$. In words, this says that to get the ninth term in an arithmetic sequence, we add eight times the common difference to the first term.

Example 1   Given an arithmetic sequence with first term 5 and common difference 4, find the first five terms and the twentieth term.

Solution   The first term of the arithmetic sequence is 5; hence

$$a_1 = 5$$

> In an arithmetic sequence with common difference d, each term is d more than the previous term. Since the common difference, d, is 4, a_2 is 4 more than a_1.

$$a_2 = 5 + 4 = 9$$
$$a_3 = 9 + 4 = 13 \qquad \text{a_3 is 4 more than a_2.}$$
$$a_4 = 13 + 4 = 17$$
$$a_5 = 17 + 4 = 21$$

Thus the first five terms are $5, 9, 13, 17,$ and 21 .

To find the twentieth term, we can use Formula 9.1:

$$a_n = a_1 + (n - 1)d \qquad \text{Substituting } n = 20, a_1 = 5, \text{ and } d = 4, \text{ we get}$$

Notice that to get a_{20} we add 19 times the common difference to a_1.

$$a_{20} = 5 + (20 - 1)4 = 5 + 19(4) = \boxed{81} \qquad \blacksquare$$

Example 2   Given an arithmetic sequence whose first two terms are -3 and 7, find the next three terms and the sixteenth term.

Solution   Since the first two terms of the sequence are -3 and 7, we have $a_1 = -3$ and $a_2 = 7$; because the sequence is *arithmetic,* the difference between the first two terms is also the common difference between any two successive terms. Therefore, the common difference is $d = a_2 - a_1 = 7 - (-3) = 10$.
Since $a_2 = 7$ and $d = 10$, we have

$$a_3 = 7 + 10 = 17 \qquad \text{Because } a_3 = a_2 + d$$
$$a_4 = 17 + 10 = 27 \qquad \text{Because } a_4 = a_3 + d$$
$$a_5 = 27 + 10 = 37$$

Therefore, the three terms following -3 and 7 are $17, 27,$ and 37 .

The sixteenth term is found by using Formula 9.1:

$$a_n = a_1 + (n - 1)d \qquad \text{Where } n = 16, a_1 = -3, \text{ and } d = 10$$

$$a_{16} = -3 + (15) \cdot 10 = \boxed{147} \qquad \blacksquare$$

Given an arithmetic sequence a_j, let's examine the difference between the fourth term, a_4, and the tenth term, a_{10}. If we start with a_4, we have to add d to get a_5, another d to get a_6, another d to get a_7, and so on, until we get to a_{10}. Hence to get from a_4 to a_{10} we have to add d six times, or

$$a_{10} = a_4 + 6d$$

We can derive a formula for the relationship between *any* two terms in an arithmetic sequence as follows: We begin with Formula 9.1, giving the nth term of an arithmetic sequence with first term a_1 and common difference d:

$$a_n = a_1 + (n - 1)d$$

Now suppose we have two terms in an arithmetic sequence, a_j and a_k. Let's apply this formula to a_j and a_k by substituting $n = k$ and $n = j$; we get

$$a_k = a_1 + (k - 1)d \qquad \text{And}$$
$$a_j = a_1 + (j - 1)d \qquad \text{Subtracting the second equation from the first,}$$
$$\text{we get}$$

$$
\begin{aligned}
a_k - a_j &= (k - 1)d - (j - 1)d \\
&= kd - d - jd + d \\
&= kd - jd \qquad \text{Factor out } d. \\
&= (k - j)d \qquad \text{Which gives us the following formula}
\end{aligned}
$$

The difference between any two terms, a_j and a_k, in an arithmetic sequence with common difference d is

$$a_k - a_j = (k - j)d \qquad\qquad \text{Formula 9.2}$$

Example 3 | Given an arithmetic sequence with $a_3 = 12$ and $a_9 = 14$, find a_1 and a_{30}.

Solution | The key feature of an arithmetic sequence is its common difference d. Formula 9.2 allows us to use any two terms in an arithmetic sequence to find d. We are given an arithmetic sequence with third term $a_3 = 12$ and ninth term $a_9 = 14$. Using Formula 9.2 with $j = 3$ and $k = 9$, the difference between these two terms is

$$
\begin{aligned}
a_9 - a_3 &= (9 - 3)d \qquad \text{Substitute for } a_9 \text{ and } a_3. \\
14 - 12 &= (9 - 3)d \\
2 &= 6d \qquad\qquad \text{So} \\
\frac{1}{3} &= d
\end{aligned}
$$

Once we know the value of d, we can find a_1 by substituting $d = \dfrac{1}{3}$ in Formula 9.1 for the nth term of an arithmetic sequence:

$$a_n = a_1 + (n - 1)d \qquad \text{Because we know the value of } a_3, \text{ we use } n = 3 \text{ to obtain}$$
$$a_3 = a_1 + (3 - 1)d \qquad \text{Substituting } a_3 = 12 \text{ and } d = \frac{1}{3}, \text{ we get}$$
$$12 = a_1 + 2\left(\frac{1}{3}\right)$$
$$12 = a_1 + \frac{2}{3}$$
$$11\frac{1}{3} = a_1$$

Knowing that $a_9 = 14$ and $d = \dfrac{1}{3}$, how can we find a_{30} without finding a_1?

Knowing $a_1 = 11\dfrac{1}{3}$ and $d = \dfrac{1}{3}$, we can use Formula 9.1 to find a_{30}.

$$a_n = a_1 + (n-1)d \qquad \text{Substituting } a_1 = 11\tfrac{1}{3}, \, d = \tfrac{1}{3}, \text{ and } n = 30, \text{ we get}$$

$$a_{30} = 11\frac{1}{3} + (30-1)\frac{1}{3} = \frac{34}{3} + \frac{29}{3} = \frac{63}{3} = 21$$

Therefore, $a_{30} = 21$. ∎

Example 4   Determine whether the sequence with the following general term is arithmetic.
   (a) $a_k = 4k - 7$ **(b)** $a_n = n^2 - n$

Solution   **(a)** To show that a sequence is arithmetic, we must show that the difference between *any* two consecutive terms is constant. Since $a_k = 4k - 7$, the first few terms of the sequence are $-3, 1, 5, 9, 13, \ldots$, which we can see are the terms of an arithmetic sequence with $a_1 = -3$ and $d = 4$.

More formally, we can prove that this is an arithmetic sequence by showing that the difference between *any* two consecutive terms is a constant. We have

Note that $a_{k-1} = 4(k-1) - 7$.

$$a_k - a_{k-1} = 4k - 7 - [4(k-1) - 7] = 4k - 7 - [4k - 4 - 7]$$
$$= 4k - 7 - 4k + 11 = 4$$

Thus the sequence has a common difference of 4 and is, therefore, arithmetic.
   (b) If we list the first few terms of the sequence with general term $a_n = n^2 - n$, we get $0, 2, 6, 12, \ldots$. We can immediately see that the difference between the first two terms, 0 and 2, is 2, whereas the difference between the next two terms, 2 and 6, is 4. Since these differences are not the same, the sequence is *not* arithmetic. ∎

Arithmetic Series

The particular structure of an arithmetic sequence allowed us to develop a formula for its nth term (Formula 9.1). This same structure allows us to develop formulas for S_n, the sum of the first n terms of an *arithmetic* sequence. The series obtained from an arithmetic sequence is called an **arithmetic series**.

We begin by examining a special arithmetic sequence and its associated series: Let $a_1 = 1$ and $d = 1$; then $a_2 = 2, a_3 = 3, a_4 = 4, \ldots, a_k = k$. In other words, this sequence is the sequence of positive integers. The series for this sequence is denoted by

$$A_n = 1 + 2 + 3 + \cdots + (n-2) + (n-1) + n$$

which is the sum of the first n positive integers. Let's see whether we can find a way of computing $A_{100} = 1 + 2 + 3 + \cdots + 98 + 99 + 100$ without actually adding 100 numbers. We write A_{100}; below it we rewrite A_{100} in reverse order.

$$A_{100} = \quad 1 + \quad 2 + \quad 3 + \cdots + \quad 98 + \quad 99 + 100$$

Note that there are 100 columns, and the sum in each column is 101.

$$A_{100} = 100 + \quad 99 + \quad 98 + \cdots + \quad 3 + \quad 2 + \quad 1$$

Add the two equations to get

$$2A_{100} = \underbrace{(101) + (101) + (101) + \cdots + (101) + (101) + (101)}_{100 \text{ times}}$$

Therefore, we have

$$2A_{100} = 100(101)$$ And so

$$A_{100} = \frac{100(101)}{2} = 5050$$

We can generalize this approach and derive a formula for the sum of the first k positive integers. We follow the same steps as we just did for A_{100}.

We write the series A_k; below it, we rewrite the same series in reverse order.

$$A_k = \quad 1 \quad + \quad 2 \quad + \quad 3 \quad + \cdots + (k-2) + (k-1) + \quad k$$

$$A_k = \quad k \quad + (k-1) + (k-2) + \cdots + \quad 3 \quad + \quad 2 \quad + \quad 1$$

$$2A_k = (k+1) + (k+1) + (k+1) + \cdots + (k+1) + (k+1) + (k+1)$$

$$\underbrace{\qquad\qquad\qquad\qquad\qquad\qquad\qquad\qquad\qquad\qquad}_{k \text{ times}}$$

Note that there are k columns and the sum in each column is $k + 1$. Add the two equations to get

Therefore, we have

$$2A_k = k(k+1)$$ And so

$$A_k = \frac{k(k+1)}{2}$$

Thus we have derived Formula 9.3.

The sum of the first k positive integers is given by

$$A_k = 1 + 2 + 3 + \cdots + k = \frac{k(k+1)}{2}$$ Formula 9.3

For example, using this formula for $k = 20$, we find that the sum of the first 20 positive integers is

$$A_{20} = 1 + 2 + 3 + \cdots + 20 = \frac{20(20+1)}{2} = \frac{20(21)}{2} = 210$$

We can now derive a formula for the general arithmetic series S_n.

Recall that $S_n = a_1 + a_2 + a_3 + a_4 + \cdots + a_{n-2} + a_{n-1} + a_n$ is called the arithmetic series associated with the sequence $\{a_n\}$. We can use the formula $a_n = a_1 + (n-1)d$ to rewrite each term in the sequence in terms of a_1, n, and d. In other words, since $a_2 = a_1 + d$, $a_3 = a_1 + 2d$, ..., $a_n = a_1 + (n-1)d$, we can write the entire sum S_n in terms of a_1, n, and d as follows:

$$S_n = a_1 + \quad a_2 \quad + \quad a_3 \quad + \quad a_4 \quad + \cdots + \quad a_{n-1} \quad + \quad a_n \qquad \text{Becomes}$$

$$\downarrow \qquad \downarrow \qquad \downarrow \qquad \downarrow \qquad \cdots \qquad \downarrow \qquad \downarrow$$

$$S_n = a_1 + (a_1 + d) + (a_1 + 2d) + (a_1 + 3d) + \cdots + [a_1 + (n-2)d] + [a_1 + (n-1)d]$$

We collect all the a_1 terms (there are n of them).

$$S_n = na_1 + [d + 2d + 3d + \cdots + (n-2)d + (n-1)d]$$

Now factor out d from within the brackets.

$$= na_1 + d[1 + 2 + 3 + \cdots + (n-2) + (n-1)] \qquad (1)$$

Inside the brackets we have the sum of the first $n - 1$ positive integers. We can use Formula 9.3 to find the sum of the first $n - 1$ positive integers by substituting $n - 1$ for k. From Formula 9.3 we have

$$1 + 2 + 3 + \cdots + k = \frac{k(k + 1)}{2} \qquad \text{Substitute } n - 1 \text{ for } k.$$

$$1 + 2 + 3 + \cdots + n - 1 = \frac{(n - 1)[(n - 1) + 1]}{2} = \frac{(n - 1)n}{2} \qquad (2)$$

Returning to equation (1), we have

$$S_n = na_1 + d[1 + 2 + 3 + \cdots + (n - 2) + (n - 1)] \qquad \begin{array}{l}\text{Substituting the result}\\\text{from (2), we get}\end{array}$$

$$= na_1 + d\frac{(n - 1)n}{2} \qquad \begin{array}{l}\text{Combine into a single}\\\text{fraction.}\end{array}$$

$$= \frac{2na_1}{2} + \frac{n(n - 1)d}{2} = \frac{2na_1 + n(n - 1)d}{2} \qquad \begin{array}{l}\text{Factor out } n \text{ in the}\\\text{numerator.}\end{array}$$

$$= \frac{n[2a_1 + (n - 1)d]}{2}$$

We finally rewrite this formula by separating out the factor of $\frac{n}{2}$.

Sum of an Arithmetic Series
The sum of the first n terms of an arithmetic sequence is given by

$$S_n = a_1 + a_2 + a_3 + \cdots + a_n = \sum_{i=1}^{n} a_i = \frac{n}{2}[2a_1 + (n - 1)d]$$

Formula 9.4

We can rewrite $2a_1$ as $a_1 + a_1$ and, since $a_n = a_1 + (n - 1)d$, Formula 9.4 for S_n becomes

$$S_n = \frac{n}{2}[2a_1 + (n - 1)d] = \frac{n}{2}[a_1 + \underbrace{a_1 + (n - 1)d}_{a_n}] = \frac{n}{2}(a_1 + a_n) = n\left(\frac{a_1 + a_n}{2}\right)$$

Hence, we have derived an alternative (and simpler) version for Formula 9.4.

Sum of an Arithmetic Series
An alternative form for the sum of the first n terms of an arithmetic sequence is given by

$$S_n = a_1 + a_2 + a_3 + \cdots + a_n = \frac{n}{2}(a_1 + a_n) = n\left(\frac{a_1 + a_n}{2}\right)$$

Formula 9.5

In words, Formula 9.5 says that the sum of the first n terms of an arithmetic sequence is equal to n times the average of the first term and the nth term.

We note that Formula 9.4 is most useful when we know n, a_1 (the first term), and d (the common difference), whereas Formula 9.5 is most useful when we know n, a_1, and a_n (the last term).

Example 5 | Given the arithmetic sequence $\{a_k\} = 3, 7, 11, 15, \ldots$, find

(a) S_{12} (b) $\displaystyle\sum_{k=1}^{40} a_k$

Solution

How do you find the common difference of an arithmetic sequence?

(a) We could calculate S_{12} by writing out the first 12 terms in the sequence and adding them together, or we could use one of the formulas for S_n. The sequence 3, 7, 11, 15, ... has first term $a_1 = 3$ and common difference $d = 4$ (why?), and for S_{12} we have $n = 12$. Thus we can substitute these values into Formula 9.4 to get

$$S_n = \frac{n}{2}[2a_1 + (n-1)d] \qquad \text{We substitute } n = 12, a_1 = 3, \text{ and } d = 4.$$

You may want to check this answer by writing out the first 12 terms in the sequence and finding their sum.

$$S_{12} = \frac{12}{2}[2 \cdot 3 + (12-1)4] = 6[6 + 11(4)] = \boxed{300}$$

(b) To find the sum of the first 40 terms of the sequence, we are looking for the series S_{40}. Thus we use Formula 9.4 again, this time with $n = 40$:

$$S_{40} = \frac{40}{2}[2 \cdot 3 + (40-1)4] = \boxed{3240} \qquad\blacksquare$$

Example 6 | Find the sum of the first 25 terms of the sequence whose general term is $a_n = 7n$.

Solution

The following display illustrates how we can compute the sum of the terms of this sequence on a graphing calculator.

```
sum(seq(7n,n,1,2
5))
              2275
```

This example is asking us to find the sum of the 25 terms 7, 14, 21, 28, 35, ... , 175. We notice that this is an arithmetic sequence with $a_1 = 7(1)$ and $a_{25} = 7(25) = 175$. (What is the common difference for this sequence?) Since we can easily identify the first and last terms, we use the alternative formula for an arithmetic series (Formula 9.5):

$$S_n = \frac{n}{2}(a_1 + a_n)$$

$$S_{25} = 25\left(\frac{7 + 175}{2}\right) = 25(91) = \boxed{2275}$$

Try finding S_{25} using Formula 9.4. Which formula is easier to use in this example? $\blacksquare$

Example 7 | A job applicant finds that firm A offers a starting annual salary of \$32,500 with a guaranteed raise of \$1400 per year, whereas firm B offers a higher starting salary of \$36,000 but will guarantee a raise of only \$900 per year.

(a) What would the annual salary be in the tenth year at each firm?

(b) Over the first 10 years, how much would be earned at each firm?

Solution

(a) The annual salaries at firm A form the arithmetic sequence 32,500; 33,900; 35,300; ... with first term \$32,500 and common difference \$1400. If we let $a_n =$ the annual salary at firm A in year n, then the annual salary in year 10 is

$$a_{10} = a_1 + 9d = 32,500 + 9(1400) = \$45,100$$

Similarly, the annual salaries at Firm B form the arithmetic sequence 36,000; 36,900; 37,800; ... with first term \$36,000 and common difference \$900. If we let $b_n =$ the annual salary at firm B in year n, then the annual salary in year 10 is

$$b_{10} = b_1 + 9d = 36,000 + 9(900) = \$44,100$$

Therefore, in the tenth year the annual salary would be \$45,100 at firm A and \$44,100 at firm B.

(b) To determine the amount that would be earned at each firm over the first 10 years we need to add the first 10 annual salaries; $a_1 + a_2 + a_3 + \cdots + a_{10}$ represents the amount of money earned over the first 10 years at firm A and $b_1 + b_2 + b_3 + \cdots + b_{10}$ represents the amount of money earned over the first 10 years at firm B. In other words, we want to find the sum of the first 10 terms of each arithmetic sequence.

$$a_1 + a_2 + a_3 + \cdots + a_{10} = \frac{10}{2}(a_1 + a_{10})$$ We know $a_1 = 32{,}500$ and, from part (a), $a_{10} = 45{,}100$.

$$= 5(32{,}500 + 45{,}100) = 5(77{,}600) = 388{,}000$$

$$b_1 + b_2 + b_3 + \cdots + b_{10} = \frac{10}{2}(b_1 + b_{10})$$ We know $b_1 = 36{,}000$ and, from part (a), $b_{10} = 44{,}100$.

$$= 5(36{,}000 + 44{,}100) = 5(80{,}100) = 400{,}500$$

Therefore, over the first 10 years a total of $388,000 would be earned at firm A, but a total of $400,500 would be earned at firm B.

Note that in year 10 the salary is higher at Firm A but the amount earned over the 10 year period is higher at Firm B. ∎

You might find it interesting to compute how much money would be earned at each firm during the next 10 years.

9.3 Exercises

In Exercises 1–6, determine whether the given sequence is arithmetic.

1. 4, 7, 10, 13, . . .

2. 1, 4, 9, 16, . . .

3. 2, 6, 10, 14, 20, 26, . . .

4. $\frac{1}{2}, \frac{5}{4}, 2, \frac{11}{4}, \ldots$

5. $a_n = 5 - 2n$

6. $a_k = \frac{1}{k}$

In Exercises 7–12, use the given information about an arithmetic sequence to find the common difference, d.

7. $a_1 = 3$ and $a_5 = 23$

8. $a_6 = -2$ and $a_{11} = 53$

9. $a_{20} - a_5 = 10$

10. $a_{12} - a_3 = -8$

11. $a_n = 4n + 1$

12. $a_k = \frac{k + 4}{5}$

In Exercises 13–24, use the given information about the arithmetic sequence to find the requested values.

13. $a_4 = 8$ and $a_8 = 10$. Find d and a_1.

14. $a_4 = 5$ and $d = 6$. Find a_1 and a_9.

15. $a_5 = 1$ and $d = \frac{2}{5}$. Find a_1 and a_{20}.

16. $a_7 - a_2 = 30$. Find d and $a_{12} - a_1$. Can you find a_1?

17. $\{a_n\} = 2, 5, 8, 11, \ldots$. Find d and a_{18}.

18. $\{a_n\} = -8, -1, 6, 13, \ldots$. Find a_8 and a_{100}.

19. $\{a_k\} = \frac{1}{2}, \frac{5}{6}, \frac{7}{6}, \ldots$. Find d and a_{10}.

20. $\{a_n\} = \frac{1}{4}, \frac{7}{12}, \frac{11}{12}, \ldots$. Find d and a_{21}.

21. $a_3 = -\frac{1}{3}, a_4 = -\frac{5}{6}$. Find a_1 and a_9.

22. $a_2 = -1, a_6 = -\frac{1}{2}$. Find a_4 and a_9.

23. $a_{10} = 11$ and $a_7 - a_4 = -\frac{5}{4}$. Find d and a_1.

24. $a_{61} = 102$ and $d = -\frac{5}{3}$. Find a_1.

25. Jim is scheduled to get a raise of $250 every 6 months during his first 5 years on the job. If his starting salary is $25,250 per year, what is his annual salary at the end of 3 years?

26. Rosa is scheduled to get a raise of $150 every three months during her first 3 years on the job. If her starting salary is $28,500, what is her yearly salary at the end of 2 years?

27. Olga is planning an exercise schedule in which she begins bench pressing 50 pounds and increases the weight by 2 pounds at the beginning of each subsequent week. How much will she be bench pressing during the twentieth week?

28. Janice begins a savings program in which she will save $1000 the first year, and each subsequent year will save $200 more than she did the previous year. How much will she save during the eighth year?

29. Suppose that when an object is dropped from a hot air balloon, it falls 16 feet the first second, 48 feet the second second, 80 feet the third second, and so on. The distance it falls each second form an arithmetic sequence. How many feet will the object fall during the fifth second? During the sixth second?

30. How many feet will the object in Exercise 29 fall during the first 10 seconds?

In Exercises 31–42, use the given information about each arithmetic sequence to find the indicated arithmetic series, S_n.

31. $a_1 = 4$ and the common difference is 5; S_6

32. $a_1 = 9$ and the common difference is -1; S_8

33. $b_4 = 2$ and $b_7 = 17$; S_7

34. $c_9 = 12$ and $c_{15} = 30$; S_5

35. $a_3 = 2$ and the common difference is $\frac{1}{2}$; S_7

36. $a_5 = 6$ and the common difference is $-\frac{1}{3}$; S_{12}

37. $\{a_n\} = 3, -2, -7, \ldots$; S_6

38. $\{a_j\} = 4, 13, 22, \ldots$; S_9

39. $\{a_k\} = -4, -2, 0, 2, 4, \ldots$; S_{10}

40. $\{b_m\} = 6, 7, 8, \ldots$; S_{17}

41. $\{x_n\} = \frac{1}{4}, \frac{1}{3}, \frac{5}{12}, \ldots$; S_8

42. $\{t_n\} = -\frac{3}{2}, -1, -\frac{1}{2}, \ldots$; S_7

43. Find $\sum_{j=1}^{8} 4j$.

44. Find $\sum_{k=1}^{6} 3k$.

45. Find $\sum_{m=1}^{10} 10m$.

46. Find $\sum_{n=1}^{9} 8n$.

47. Find $\sum_{i=1}^{7} (3i + 4)$.

48. Find $\sum_{k=1}^{5} (4k - 9)$.

49. Find $\sum_{i=40}^{70} i$.

50. Find $\sum_{n=30}^{100} n$.

51. Find $\frac{1}{\pi} + \frac{3}{\pi} + \frac{5}{\pi} + \cdots + \frac{15}{\pi}$.

52. Find $\frac{e}{2} + e + \frac{3e}{2} + \cdots + 10e$.

53. A pile of logs is in the shape of a pyramid with 40 logs in the bottom level. Each level above the first has two fewer logs than the level below. If the pile of logs has 12 levels, how many logs are there in the pile?

54. A theater has 38 rows of seats. The first row has 17 seats, the second row has 20 seats, the third row has 23 seats, and so on. What is the seating capacity of the theater?

55. A contest offers a total of 18 prizes. The first prize is worth $10,000, and each successive prize is worth $500 less than the next higher prize. Find the value of the eighteenth prize and the total value of all the prizes.

56. A contest offers 10 prizes with a total value of $13,250. If the difference in value between successive prizes is $250, what is the value of the first prize?

57. A grocer wants to empty 12 cases of breakfast cereal, with 48 boxes in each case. She wants to stack the boxes in a pyramid with each row containing two fewer boxes than the row below it and the top row containing one box.
 (a) Can such a pyramid be constructed so that it uses up all the boxes in the 12 cases?
 (b) How many boxes high must the pyramid be?
 (c) With how many boxes must she start in the bottom row of the pyramid?

58. Find the sum of the first 100 positive even integers.

59. Find the sum of the first 100 positive odd integers.

60. Find the sum of the first 30 numbers in the arithmetic sequence 8, 16, 24, ...

61. Find the sum of the first 25 numbers in the arithmetic sequence $-3, 7, 17, 27, \ldots$.

62. Given an arithmetic sequence with $S_{10} = 165$ and $a_1 = 3$, find a_{10}.

63. Given an arithmetic sequence with $S_{20} = 910$ and $a_{20} = 95$, find a_1.

64. Given an arithmetic sequence with $S_{16} = 368$ and $a_1 = 1$, find a_8.

65. Given an arithmetic sequence with $S_{15} = 390$ and $a_{15} = 56$, find S_5.

66. Given an arithmetic sequence with $a_1 = 9, d = 6$, and $S_N = 969$, find N.

67. Given an arithmetic sequence with $a_1 = -5$ and $S_{22} = 2431$, find d, the common difference.

Questions for Thought

68. Show that the terms $\frac{1}{1 + \sqrt{3}}, -\frac{1}{2}, \frac{1}{1 - \sqrt{3}}$ form an arithmetic sequence.

69. In Formula 9.2, does it matter whether $k > j$? Explain.

70. How does the fact that the sequence with $a_k = 4k - 7$ has a constant common difference of 4 relate to the fact that the line with equation $y = 4x - 7$ has a slope of 4?

9.4   Geometric Sequences and Series

Geometric Sequences

In the last section we saw that the structure of arithmetic sequences and series allows us to develop special formulas that pertain to them. In this section we analyze another type of special sequence and series.

Consider the following two sequences:

$$\{a_k\} = 4, 8, 16, 32, 64, \ldots \qquad \{b_j\} = 3, 15, 75, 375, 1875, \ldots$$

Neither of these sequences is arithmetic, because in each sequence the difference between consecutive terms is not constant. (For the sequence $\{a_k\}$, $8 - 4 = 4$, $16 - 8 = 8$, and $32 - 16 = 16$, and for the sequence $\{b_j\}$, $15 - 3 \neq 75 - 15$.)

However, upon closer inspection of each sequence, we find that there is a pattern: In $\{a_k\}$, each term after the first is 2 times the previous term,

$$8 = 2 \cdot 4, \qquad 16 = 2 \cdot 8, \qquad 32 = 2 \cdot 16, \ldots$$

and in $\{b_j\}$ each term after the first is 5 times the previous term:

$$15 = 5 \cdot 3, \qquad 75 = 5 \cdot 15, \qquad 375 = 5 \cdot 75, \ldots$$

Equivalently, we can describe these sequences by saying that the *quotient* of two consecutive terms (each term divided by its preceding term) is constant. In the case of $\{a_k\}$, this quotient is $\dfrac{a_{k+1}}{a_k} = 2$. In the case of $\{b_j\}$, this quotient is $\dfrac{b_{j+1}}{b_j} = 5$.

A sequence for which this type of relationship holds for every pair of consecutive terms is called a *geometric sequence*.

> ### Definition of a Geometric Sequence
>
> A sequence $\{a_k\}$ in which the ratio of consecutive terms is constant is called a **geometric sequence**. This **common ratio** is usually denoted by r. In other words, a geometric sequence is defined by the fact that
>
> $$\frac{a_{k+1}}{a_k} = r \qquad \text{or, equivalently,} \qquad a_{k+1} = ra_k$$

What are the similarities and differences between arithmetic and geometric sequences?

A geometric sequence is sometimes called a *geometric progression*.

In a geometric sequence, each successive term is found by *multiplying* the previous term by a constant. This makes a geometric sequence a recursively defined sequence.

In the sequence $\{a_k\}$ given at the beginning of this section, each successive term is found by multiplying the previous term by 2. Symbolically, we can represent this relationship as

$$a_{k+1} = 2a_k$$

If $k = 15$, we have $a_{15} = 2a_{14}$.

Note the similarities and differences between an arithmetic sequence and a geometric sequence: In an arithmetic sequence, the *differences* between successive terms are constant, whereas in a geometric sequence, the *quotients* of successive terms are constant.

Suppose $a_1, a_2, a_3, a_4, \ldots, a_{k-1}, a_k, \ldots$ is a geometric sequence with common ratio r. Then if we start with a_1, we have

$$a_2 = a_1 r$$

To get the second term, we multiplied the first term by the common ratio, r. The next term is

$$a_3 = ra_2 = r(a_1 r) = a_1 r^2$$

Because $a_2 = a_1 r$; similarly

$$a_4 = ra_3 = r(a_1 r^2) = a_1 r^3$$

Continuing the pattern, we have

$$a_5 = ra_4 = r(a_1 r^3) = a_1 r^4$$

$$a_6 = ra_5 = a_1 r^5$$

$$\vdots$$

In general we have

$$a_n = ra_{n-1} = a_1 r^{n-1}$$

This result gives us the following general formula.

The nth Term of a Geometric Sequence

The nth term of a geometric sequence with common ratio r and first term a_1 is

$$a_n = a_1 r^{n-1} \qquad \text{for } n \geq 1$$

Formula 9.6

In other words, the terms of a geometric sequence may be written as

$$a_1, \ a_1 r, \ a_1 r^2, \ a_1 r^3, \ldots, \ a_1 r^n, \ldots$$

Example 1 Given the geometric sequence $5, 15, 45, \ldots$, find the next two terms and the seventh term.

Solution We are given that the sequence is geometric, so we first find the common ratio, r, which is $\dfrac{15}{5} = 3$. Note that we can use *any* two consecutive terms to find the common ratio.

$$\frac{15}{5} = \frac{45}{15} = 3$$

Therefore, the term following 45 in the sequence is found by multiplying 45 by the common ratio 3. The term following 45 is $45 \cdot 3 = \boxed{135}$ and the term following 135 is $135 \cdot 3 = \boxed{405}$. The seventh term is found by using Formula 9.6 for the nth term of a geometric sequence.

$$a_n = a_1 r^{n-1} \qquad \text{Here we substitute } a_1 = 5, \ n = 7, \text{ and } r = 3.$$

$$a_7 = a_1 r^6$$

$$= 5 \cdot 3^6$$

$$= 5(729)$$

$$= \boxed{3645}$$

Example 2 Find the eighth term of the geometric sequence whose first term is 5 and whose fourth term is $\dfrac{1}{25}$.

Solution First we need to find the common ratio, r. Given $a_1 = 5$ and $a_4 = \dfrac{1}{25}$, we use Formula 9.6 for the nth term of a geometric sequence as follows:

$$a_n = a_1 r^{n-1} \qquad \text{Use the fact that we are given } a_1 \text{ and } a_4, \text{ and let } n = 4.$$

$$a_4 = a_1 r^3 \qquad \text{Substitute } a_1 = 5 \text{ and } a_4 = \dfrac{1}{25}.$$

$$\frac{1}{25} = 5r^3 \qquad \text{Now solve for } r.$$

$$\frac{1}{125} = r^3 \;\Rightarrow\; \frac{1}{5} = r \qquad \text{Now we use Formula 9.6 again to find } a_8.$$

$$a_8 = a_1 r^7 = 5\left(\frac{1}{5}\right)^7 = \boxed{\frac{1}{15{,}625}}$$

Example 3   Ramona deposits \$3500 in a bank account paying an annual interest rate of 6%. Show that the amounts she has in the account at the end of each year form a geometric sequence.

Solution   This example is very similar to Example 4 in Section 9.1. Recall that if P dollars is invested at an annual interest rate of r, then the amount of money A in the account after t years is given by the formula $A = P(1 + r)^t$. Therefore, if we let A_k be the amount of money in Ramona's account at the end of year k, we have

$$A_k = 3500(1 + 0.06)^k = 3500(1.06)^k \quad \text{and so} \quad \frac{A_{k+1}}{A_k} = \frac{3500(1.06)^{k+1}}{3500(1.06)^k} = 1.06$$

Since the ratio of any two successive terms is a constant, this sequence is geometric.

Example 4   Suppose a substance loses one-half its radioactive mass per year. If we start with 100 grams of radioactive substance, approximately how much is left after 10 years?

Solution   If we record how much is left after each year and note that each term is $\frac{1}{2}$ the previous term, we obtain the following geometric sequence:

Let A_j be the amount left at the end of year j. Then

$$A_1 = \frac{1}{2}(100) \qquad\qquad \text{is the amount left at the end of year 1}$$

$$A_2 = \frac{1}{2} \text{ of } A_1 = \frac{1}{2}\left[\frac{1}{2}(100)\right] = 100\left(\frac{1}{2}\right)^2 \qquad \text{is the amount left at the end of year 2}$$

$$A_3 = \frac{1}{2} \text{ of } A_2 = \frac{1}{2}\left[100\left(\frac{1}{2}\right)^2\right] = 100\left(\frac{1}{2}\right)^3 \qquad \text{is the amount left at the end of year 3}$$

In general we see that

Notice that A_j is a geometric sequence with ratio $r = 1/2$.

$$A_j = 100\left(\frac{1}{2}\right)^j \qquad \text{is the amount left at the end of year } j$$

Hence, at the end of 10 years there is

$$A_{10} = 100\left(\frac{1}{2}\right)^{10} = 100\left(\frac{1}{1024}\right) = \frac{100}{1024} \approx 0.09766$$

Therefore, there is approximately 0.1 gram of the radioactive substance remaining after 10 years.

Geometric Series

If $\{a_k\}$ is a geometric sequence, then its associated series, S_n, is

$$S_n = a_1 + a_2 + a_3 + \cdots + a_{n-1} + a_n$$

and S_n is called the **geometric series** associated with the geometric sequence. Throughout this discussion n is, as usual, a positive integer.

As with arithmetic sequences, we can find a formula to describe a geometric series associated with it. Before we begin to derive the formula, however, we first make note of the following product:

$$(1 - r)(1 + r + r^2 + r^3 + \cdots + r^{n-1}) = 1 - r^n \qquad (1)$$

We leave the proof as an exercise (see Exercise 89), but we demonstrate this product below for $n = 6$:

$(1 - r)(1 + r + r^2 + r^3 + r^4 + r^5)$ Use the distributive property to get

$$= 1(1 + r + r^2 + r^3 + r^4 + r^5) - r(1 + r + r^2 + r^3 + r^4 + r^5)$$

$$= 1 + r + r^2 + r^3 + r^4 + r^5 - r - r^2 - r^3 - r^4 - r^5 - r^6$$

All but the first and last terms drop out.

$$= 1 - r^6$$

Hence,

$$(1 - r)(1 + r + r^2 + r^3 + r^4 + r^5) = 1 - r^6 \qquad \text{As required}$$

If we divide both sides of equation (1) by $1 - r$ (assuming $r \neq 1$), we get

$$1 + r + r^2 + r^3 + \cdots + r^{n-2} + r^{n-1} = \frac{1 - r^n}{1 - r} \qquad (2)$$

We use this fact to find a formula for the sum of the first n terms of a geometric sequence as follows: Suppose $\{a_j\}$ is a geometric sequence. If we write out the terms of S_n, we get

$$S_n = a_1 + a_2 + a_3 + a_4 + \cdots + a_{n-1} + a_n$$

and since $\{a_j\}$ is a geometric sequence, we can use Formula 9.6, which says that $a_j = a_1 r^{j-1}$, to rewrite each term using a_1 and r:

$S_n = a_1 + a_1 r + a_1 r^2 + a_1 r^3 + \cdots + a_1 r^{n-2} + a_1 r^{n-1}$ Factoring out a_1, we get

$$= a_1(1 + r + r^2 + r^3 + \cdots + r^{n-2} + r^{n-1})$$

Using the result in equation (2) on the expression inside the parentheses, we get

$$= a_1\left(\frac{1 - r^n}{1 - r}\right)$$

Thus we have derived Formula 9.7.

The Sum of a Geometric Series

The formula for the sum of the first n terms of a geometric sequence is

$$S_n = a_1 + a_1 r + a_1 r^2 + \cdots + a_1 r^{n-1} = \frac{a_1(1 - r^n)}{1 - r} \qquad \text{for } r \neq 1$$

Formula 9.7

Equivalently, we can write the formula for S_n using sigma notation: If $\{a_k\}$ is a geometric sequence with $r \neq 1$, then

$$S_n = \sum_{k=1}^{n} a_k = \frac{a_1(1 - r^n)}{1 - r}$$

Example 5 | Given the geometric sequence $\{a_i\} = 6, 18, 54, \ldots$, find

(a) S_7 (b) $\displaystyle\sum_{i=1}^{12} a_i$

Solution | (a) To find S_7 we could simply find the next four terms of the sequence and add the first seven terms together. Instead, we use Formula 9.7, noting that the first term is $a_1 = 6$ and $r = 3$.

Find S_7 by finding the first seven terms of the sequence and adding them.

$$S_n = \frac{a_1(1 - r^n)}{1 - r}$$

$$S_7 = \frac{6(1 - 3^7)}{1 - 3}$$

$$= \frac{6(1 - 2187)}{-2} = \frac{6(-2186)}{-2} = \boxed{6558}$$

Does this answer agree with what you obtained by finding the first seven terms in the sequence and adding them together?

(b) Here we are looking for the sum of the first 12 terms in the sequence $\{a_i\}$. Again, $a_1 = 6$ and $r = 3$, but this time $n = 12$.

$$\sum_{i=1}^{12} a_i = S_{12} = \frac{6(1 - 3^{12})}{1 - 3}$$

$$= \frac{6(1 - 531{,}441)}{-2} = \frac{6(-531{,}440)}{-2} = \boxed{1{,}594{,}320}$$

To appreciate the usefulness of this formula, try to compute all 12 terms and find their sum. ∎

As the last example shows, the terms in a geometric sequence can grow very rapidly.

Infinite Geometric Series

Thus far we have discussed finite geometric series—that is, we have computed the sum of a finite number of terms in a geometric sequence. We would now like to try to make sense out of the sum of an **infinite geometric sequence**. In other words, how are we to understand the sum

$$a_1 + a_1 r + a_1 r^2 + \cdots + a_1 r^n + \cdots$$

where the intention is to add all the terms in the geometric sequence? Let's look at two examples of infinite geometric sequences:

$$\{a_n\} = 2^n \quad \text{and} \quad \{b_n\} = \frac{1}{2^n}$$

Let's write out a few terms of each sequence:

$$\{a_n\} = \{2^n\} = \{2, 4, 8, 16, 32, 64, \ldots\} \qquad \{b_n\} = \left\{\frac{1}{2^n}\right\} = \left\{\frac{1}{2}, \frac{1}{4}, \frac{1}{8}, \frac{1}{16}, \frac{1}{32}, \frac{1}{64}, \ldots\right\}$$

The following display illustrates the partial sums S_{10}, S_{20}, and S_{30} for both sequences.

The partial sums of $\{a_n\}$

```
sum(seq(2^n,n,1,
10))
              2046
sum(seq(2^n,n,1,
20))
           2097150
```

```
sum(seq(2^n,n,1,
30))
        2147483646
```

The partial sums of $\{b_n\}$

```
sum(seq(1/2^n,n,
1,10))
         .9990234375
sum(seq(1/2^n,n,
1,20))
         .9999990463
```

```
sum(seq(1/2^n,n,
1,30))
         .9999999991
```

Verify these partial sums by using Formula 9.7.

For the sequence $b_n = \dfrac{1}{2^n}$, we have $b_{10} \approx 0.001$, $b_{15} \approx 0.00003$, $b_{20} \approx 0.000001$.

We note that for the sequence $\{a_n\}$, the *terms* of the sequence are getting very large, and the *partial sums* are also getting very large. However, for the sequence $\{b_n\}$, the *terms* of the sequence are getting very small, and the *partial sums* seem to be getting closer and closer to 1. Thus for the sequence $\{b_n\}$ we can summarize this discussion as follows.

1. As n increases, it appears that b_n approaches, or gets close to, 0.

2. As n increases, it appears that S_n approaches, or gets close to, 1.

This suggests that as n gets larger and larger, S_n is getting closer and closer to 1. In fact, this is precisely what happens.

Based on this discussion, we say that the sum of the *infinite geometric series* $\dfrac{1}{2} + \dfrac{1}{4} + \dfrac{1}{8} + \cdots$ is 1.

Indeed, we find that there are some cases where we can determine the sum of an infinite series and some cases where we cannot. We now show that this is dependent on the size of r by generalizing the argument we made before for the sequence $\{b_k\}$.

If we start with a geometric sequence, $\{a_k\}$, we know that the associated series S_n can be found by Formula 9.7 (provided $r \neq 1$):

$$S_n = \frac{a_1(1 - r^n)}{1 - r}$$

Let's suppose $-1 < r < 1$ (usually written as $|r| < 1$) and take a closer look at what happens to the right-hand side of the last equation as n gets very large.

If $|r| < 1$, then r^n will get closer and closer to 0 as n gets larger and larger. The discussion above illustrates this for $r = \dfrac{1}{2}$. If n gets "large enough," then r^n gets so close to 0 that it becomes "negligible"—that is, so small that we can ignore the term r^n being subtracted from 1 in the numerator. Thus we have the following.

Sum of an Infinite Series

If $|r| < 1$, the sum S of an infinite geometric series $a_1 + a_1r + a_1r^2 + \cdots$ is given by

$$S = \frac{a_1}{1 - r} \qquad \text{Formula 9.8}$$

If $|r| \geq 1$, then the partial sums S_n increase (we say *increase without bound*) and thus we cannot find the sum S.

We can write S using sigma notation by writing the symbol ∞ above the Σ.

$$S = \sum_{n=1}^{\infty} a_1 r^{n-1} = \frac{a_1}{1 - r} \qquad \text{for } |r| < 1$$

The symbol ∞ means that we keep increasing the index n without ever reaching an upper limit.

Example 6   Find the sum of the given infinite geometric sequence, if possible.

(a) $3, \dfrac{9}{4}, \dfrac{27}{16}, \dfrac{81}{64}, \ldots$ (b) $\dfrac{1}{25}, \dfrac{3}{25}, \dfrac{9}{25}, \ldots$

Solution   (a) $3, \dfrac{9}{4}, \dfrac{27}{16}, \dfrac{81}{64}, \ldots$ is an infinite geometric sequence with first term 3, and common

ratio $r = \dfrac{\frac{9}{4}}{3} = \dfrac{3}{4}$. Since $|r| < 1$, the sum S can be found by using Formula 9.8:

$$S = \dfrac{3}{1 - \frac{3}{4}} = \dfrac{3}{\frac{1}{4}} = 12$$

Hence $S = 3 + \dfrac{9}{4} + \dfrac{27}{16} + \dfrac{81}{64} + \cdots = \boxed{12}$.

(b) $\dfrac{1}{25}, \dfrac{3}{25}, \dfrac{9}{25}, \ldots$ is an infinite sequence with first term $\dfrac{1}{25}$ and common ratio

$r = \dfrac{\frac{3}{25}}{\frac{1}{25}} = 3$. Since $|r| > 1$, the sum does not exist.

Compute S_{10}, S_{20}, and S_{30} for the sequences in Example 6.

You may find it instructive to compute S_{10}, S_{20}, S_{30} for the geometric sequences in this example to see that the sums in part (a) are getting closer and closer to 12, whereas the sums in part (b) are getting very large.   ∎

Example 7   Politicians often promote tax cuts by referring to the "trickle down" effect, which claims that people with extra income will spend more money, thus putting more money in the hands of others who will in turn spend more money. Suppose that a tax rebate of $1600 is proposed for each taxpayer, and assume that each person will spend 80% of any additional money resulting from this tax cut. Thus, the initial taxpayer will spend 80% of the $1600, and this "extra" spent money will then fall into the hands of people who will also spend 80% of what they receive, and so on. Under these assumptions, determine how much spending would be generated by the intitial $1600 rebate.

Solution   We will let A_k represent the amount of money spent by the kth person in the spending chain. Thus

$$A_1 = 0.80(1600) = \$1280$$
$$A_2 = (0.80)(0.80)(1600) = (0.80)^2(1600) = \$1024$$
$$A_3 = (0.80)(0.80)(0.80)(1600) = (0.80)^3(1600) = \$819.20$$
$$\vdots \qquad\qquad \vdots$$
$$A_k = (0.80)^k(1600)$$
$$\vdots \qquad\qquad \vdots$$

We want to find how much would be spent altogether, so assuming that this process continues indefinitely, we add the terms of an infinite geometric series with ratio $r = 0.8$.

Therefore, the total amount T spent would be

$$T = \dfrac{A_1}{1 - r} = \dfrac{1280}{1 - 0.8} = 6400$$

Thus, based on these assumptions, the $1600 tax rabate would generate a total of $6400 in spending.   ∎

Example 8 Express the repeating decimal $0.35\overline{35}$ as a fraction.

Solution In Example 1 of Section 1.1 we demonstrated one approach to solving this type of problem. Now we offer a different approach, using what we now know about infinite geometric series.

We can rewrite $0.35\overline{35}$ as an infinite sum in the following way:

$$0.35\overline{35} = 0.35 + 0.0035 + 0.000035 + \cdots$$

Note that the terms of this series, $0.35, 0.0035, 0.000035, \ldots$ form an infinite geometric sequence; the first term is 0.35 and $r = 0.01$ (each term is 0.01 times the preceding term). Since $|r| < 1$, we can find the sum using Formula 9.8:

$$0.35\overline{35} = 0.35 + 0.0035 + 0.000035 + \cdots = \sum_{k=1}^{\infty} 0.35(0.01)^{k-1} = \frac{0.35}{1 - 0.01}$$

$$= \frac{0.35}{0.99} = \frac{35}{99}$$

Thus, $0.35\overline{35} = \dfrac{35}{99}$.

Try converting 0.353535353535 into a fraction using a graphing calculator.

9.4 Exercises

In Exercises 1–8, use the given pattern or general term to determine whether the given sequence is arithmetic, geometric, or neither.

1. $3, 6, 9, 12, \ldots$

2. $3, 9, 27, 81, \ldots$

3. $\dfrac{2}{5}, \dfrac{1}{10}, \dfrac{1}{40}, \ldots$

4. $\dfrac{4}{3}, 8, 48, \ldots$

5. $a_n = 2 + \dfrac{1}{n}$

6. $b_n = \dfrac{1}{n^2}$

7. $s_n = \dfrac{4^n}{7^{n+2}}$

8. $t_n = 7n - 2$

In Exercises 9–22, determine whether the given sequence $\{a_n\}$ is arithmetic or geometric (assume it is one or the other) and find the indicated a_k.

9. $1, 5, 25, 125, \ldots$; find a_7.

10. $-2, 6, -18, 54, \ldots$; find a_6.

11. $\dfrac{1}{2}, -\dfrac{1}{3}, \dfrac{2}{9}, \ldots$; find a_5.

12. $\dfrac{5}{4}, 5, 20, \ldots$; find a_5.

13. $1, 8, 15, \ldots$; find a_{10}.

14. $1, 8, 64, \ldots$; find a_{10}.

15. $9, 3, 1, \ldots$; find a_6.

16. $9, 3, -3, \ldots$; find a_6.

17. $2, 1, \dfrac{1}{2}, \dfrac{1}{4}, \ldots$; find a_9.

18. $-5, \dfrac{5}{6}, -\dfrac{5}{36}, \ldots$; find a_5.

19. $8, 2, \dfrac{1}{2}, \dfrac{1}{8}, \ldots$; find a_9.

20. $9, \dfrac{9}{10}, \dfrac{9}{100}, \ldots$; find a_8.

21. $\sqrt{5}, 5, 5\sqrt{5}, 25, \ldots$; find a_7.

22. $1, -\dfrac{1}{\sqrt{2}}, \dfrac{1}{2}, \ldots$; find a_9.

In Exercises 23–28, use the given information about a geometric sequence to find the indicated a_n.

23. $a_1 = 10$ and $r = 2$; find a_4.

24. $a_1 = 4$ and $r = -3$; find a_6.

25. $a_3 = 1$ and $a_6 = 216$; find r and a_1.

26. $a_4 = -4$ and $a_9 = 128$; find a_7.

27. $a_2 = \dfrac{1}{\sqrt{3}}$ and $a_5 = -\dfrac{1}{9}$; find r and a_8.

28. $a_5 = 2$ and $a_9 = 4$; find a_7.

29. Find the fifth term of the geometric sequence whose first term is 1 and whose fourth term is 343.

30. Find the third term of the arithmetic sequence whose second term is 1 and whose fifth term is 64.

31. Find the eighth term of the arithmetic sequence whose first term is 5 and whose fourth term is -40.

32. Find the first term of the geometric sequence whose second term is 3 and whose seventh term is -96.

33. Find the seventh term of the geometric sequence whose first term is 1 and whose fifth term is $\frac{1}{81}$.

34. Find the sixth term of the geometric sequence whose first term is 2 and whose fourth term is $\frac{2}{27}$.

You may find the interest formula in Section 6.5 helpful in the next few exercises.

35. A certain luxury car loses one-tenth its value each year. If the car is worth $28,000 today, how much will it be worth 4 years from now?

36. A boat is now worth $34,000 and loses 12% of its value each year. What will it be worth in 5 years?

37. A machine depreciates in value by one-fifth each year. If the machine is now worth $29,000, how much will it be worth 3 years from now?

38. An antique chair is now worth $9500. If it appreciates in value at the rate of 8% per year, how much will it be worth 5 years from now?

39. The population of Porterville is increasing at a rate of 2.5% per year. If the population is currently 100,000, what will the population be 10 years from now?

40. The population of Hilltown is decreasing at a rate of 3% per year. If the population is currently 65,000, what will the population be 4 years from now?

41. Roberta invests $10,000 in a bank that pays 5% simple annual interest. How much will Roberta have in the bank at the end of 5 years?

42. How much would Roberta (Exercise 41) have in the bank at the end of 5 years if the bank paid 5% annual interest compounded quarterly?

43. Teresa invests $5000 in a bond that pays 6% interest compounded semiannually. How much is her bond worth at the end of 6 years?

44. How much will Teresa's bond be worth at the end of 10 years? See Exercise 43.

45. A ball bounces up one-half the distance from which it falls. How far does it bounce up on the fifth rebound if the ball is dropped from a height of 20 feet?

46. A ball bounces up one-third the distance from which it falls. How far does it bounce up on the fourth rebound if the ball is dropped from a height of 27 feet?

In Exercises 47–62, the given sequence is either arithmetic or geometric. Find the indicated S_n. Round to four decimal places where necessary.

47. $4, 12, 36, \ldots;\quad S_5$

48. $-3, 6, -12, \ldots;\quad S_8$

49. $\frac{2}{3}, -\frac{2}{5}, \frac{6}{25}, \ldots;\quad S_6$

50. $1, 4, 16, \ldots;\quad S_7$

51. $2, 6, 10, \ldots;\quad S_8$

52. $2, 6, 18, \ldots;\quad S_8$

53. $-\frac{3}{4}, \frac{1}{4}, -\frac{1}{12}, \ldots;\quad S_5$

54. $-\frac{3}{4}, \frac{1}{4}, \frac{5}{4}, \ldots;\quad S_5$

55. $\frac{1}{4}, \frac{1}{2}, 1, \ldots;\quad S_{10}$

56. $3, \frac{3}{10}, \frac{3}{100}, \ldots;\quad S_6$

57. $a_k = 2^k;\quad S_7$

58. $b_j = \left(\frac{2}{5}\right)^j;\quad S_9$

59. $a_k = (-1)^{k+1}\left(\frac{1}{3}\right)^k;\quad S_8$

60. $b_j = (-0.4)^k;\quad S_{20}$

61. $a_n = 2n + 1;\quad S_7$

62. $a_n = 2^{n+1};\quad S_7$

In Exercises 63–66, find each sum for the geometric sequence $\{a_i\} = 1, \frac{1}{10}, \frac{1}{100}, \ldots$.

63. $\displaystyle\sum_{i=1}^{6} a_i$

64. $\displaystyle\sum_{j=1}^{8} 5a_j$

65. $\displaystyle\sum_{k=3}^{10} a_k$

66. $\displaystyle\sum_{n=1}^{\infty} a_n$

67. A ball is dropped from a height of 60 feet and always rebounds one-third the distance from which it falls. Find the total distance the ball has traveled when it hits the ground for the fifth time.

68. Greg promises to pay Lamar a debt in the following way: Greg will pay 1¢ on the first day, 2¢ on the second day, 4¢ on the third day, and so on, where the payment on a particular day is double that of the previous day. At this rate, what is the total amount Greg will have paid Lamar at the end of 30 days?

69. Suppose a chain letter is started by sending a letter to five people and the letter requires each person who receives the letter to send it on to five additional people. Assume that no person receives more than one letter.
  (a) Show that the numbers of letters being sent at each stage form a geometric sequence.
  (b) How many letters are sent at the fourth stage?
  (c) How many letters are sent all together if the chain reaches the eighth stage?

70. Repeat Exercise 69 if the letter requires you to send the letter on to six people.

In Exercises 71–80, find S, the sum of the infinite geometric series (if it exists).

71. $2 + 1 + \dfrac{1}{2} + \dfrac{1}{4} + \cdots$

72. $1 + \dfrac{2}{3} + \dfrac{4}{9} + \cdots$

73. $\displaystyle\sum_{k=1}^{\infty} \dfrac{9}{3^k}$

74. $\displaystyle\sum_{k=1}^{\infty} 4^{3-k}$

75. $\dfrac{1}{5} + \dfrac{1}{10} + \dfrac{1}{20} + \cdots$

76. $\dfrac{1}{5} - \dfrac{1}{10} + \dfrac{1}{20} + \cdots$

77. $\displaystyle\sum_{n=1}^{\infty} 5\left(\dfrac{1}{3}\right)^{n-1}$

78. $\displaystyle\sum_{j=1}^{\infty} 2^{j-2}$

79. $\dfrac{1}{5} + \dfrac{4}{15} + \dfrac{16}{75} + \cdots$

80. $7 + \dfrac{7}{10} + \dfrac{7}{100} + \cdots$

81. Express $0.47\overline{47}$ as a fraction.

82. Express $3.139\overline{139}$ as a fraction.

83. Express $1.068\overline{1681}$ as a fraction.

84. Express $5.987\overline{171}$ as a fraction.

85. Suppose a ball is dropped from a height of 15 ft and that it always rebounds to a height equal to two-thirds of the distance from which it falls. If you let the ball bounce "forever," what is the total distance it would travel until it comes to rest?

86. A pendulum sweeps out a 9-ft arc on its first pass; each succeeding time it passes, it travels seven-eighths of the distance of the previous arc. How far would it travel before coming to rest?

87. Assuming the scenario described in Example 7, determine how much spending is generated by a tax rebate of $1850, assuming that each consumer in the chain will spend 70% of any additional money received.

88. Suppose that a scenario similar to that described in Example 7 holds when a tax surcharge is imposed. How much spending is inhibited by a tax surcharge of $1200, assuming that 75% of this money would have been spent by each person in the chain?

Questions for Thought

89. Show that
$$(1 - r)(1 + r + r^2 + r^3 + \cdots + r^{n-1}) = 1 - r^n.$$

90. Show that the terms of the sum $\displaystyle\sum_{i=1}^{N} cb^i$, where c is a constant, form a geometric sequence. What is the common ratio?

91. Use the result of Exercise 90, find the following sums.

   (a) $\displaystyle\sum_{i=1}^{10} 4 \cdot 5^i$ **(b)** $\displaystyle\sum_{i=1}^{10} 3 \cdot 2^i$

92. Explain what happens to the partial sums S_n for the series $\displaystyle\sum_{k=1}^{\infty} 2^k$ as n increases. Explain why the formula
$$S = \dfrac{a_1}{1 - r}$$
does not make any sense for this series.

93. Use a calculator to compute the partial sums S_{10}, S_{20}, and S_{30} for each of the following. Use these values to guess the sum of the infinite series and compare your guess to the answer you obtain by using Formula 9.8.

   (a) $\displaystyle\sum_{k=1}^{\infty} \left(\dfrac{2}{5}\right)^k$ **(b)** $\displaystyle\sum_{n=1}^{\infty} \dfrac{3^n}{4^{n+1}}$ **(c)** $\displaystyle\sum_{n=1}^{\infty} \dfrac{5^{n+2}}{6^n}$

9.5   Mathematical Induction

Throughout the text we have proven some of the facts (theorems) that we have stated and outlined the proofs of others. We have done algebraic proofs, geometric proofs, and trigonometric proofs. In this section we discuss a very different method for proving certain facts about natural numbers.

     Let n be a natural number and consider the following collection of statements, which we call S_n.

     S_n: The sum of the first n odd natural numbers is n^2.

S_8 is the statement that the sum of the first 8 odd natural number is $8^2 = 64$; S_{12} is the statement that the sum of the first 12 odd natural numbers is $12^2 = 144$. For each natural number n, we obtain a different statement S_n. In other words, the S_n's form a sequence of statements.

We can easily verify that S_8 and S_{12} are true by adding the first 8 or the first 12 odd natural numbers. How might we prove or disprove S_n for all natural numbers n?

Next, consider the following collection of statements, which we call T_n.

$$T_n: \quad n^2 - n + 41 \text{ is a prime number.}$$

T_5 is the statement that $5^2 - 5 + 41 = 61$ is a prime number. T_9 is the statement that $9^2 - 9 + 41 = 113$ is a prime number. For each natural number n, we obtain a different statement T_n. In other words, the T_n's also form a sequence of statements.

We can check that both 61 and 113 are prime numbers, so that T_5 and T_9 are true statements. How might we prove or disprove T_n for all natural numbers n?

Before we prove a statement to be true or false, it is often called a *conjecture*. Let's begin by examining the truth of S_n and T_n for some values of n. The calculations are listed in Table 9.1. (In effect, we are acting much like a scientist does by performing an experiment to help us decide whether to believe a particular statement.)

Table 9.1

	Statement			Statement	
n	$S_n: 1 + 3 + 5 + \cdots + (2n - 1) = n^2$	True?	$T_n: n^2 - n + 41$ is a prime number		True?
1	$S_1: 1$ $= 1^2$	True	$T_1: 1^2 - 1 + 41 = 41$ is prime.		True
2	$S_2: 1 + 3$ $= 2^2$	True	$T_2: 2^2 - 2 + 41 = 43$ is prime.		True
3	$S_3: 1 + 3 + 5$ $= 3^2$	True	$T_3: 3^2 - 3 + 41 = 47$ is prime.		True
4	$S_4: 1 + 3 + 5 + 7$ $= 4^2$	True	$T_4: 4^2 - 4 + 41 = 53$ is prime.		True
5	$S_5: 1 + 3 + 5 + 7 + 9$ $= 5^2$	True	$T_5: 5^2 - 5 + 41 = 61$ is prime		True

Based on this evidence, we might easily conclude that both statements, S_n and T_n, are true. Keep in mind that for one of these conjectures to be true, it must be true for *all* natural numbers n, whereas for one of these conjectures to be false, it must be shown false for only one natural number n. It turns out that S_n is true (which we will prove a bit later in this section), but T_n is false.

In fact, T_n is true for all natural numbers up to and including $n = 40$. However, for $n = 41$ we get $41^2 - 41 + 41 = 41^2$, which is not a prime. Thus despite the fact that T_n is true for the first 40 natural numbers, it is not true in general. This "prime-generating" polynomial $n^2 - n + 41$ is generally credited to Leonhard Euler.

We certainly cannot check a conjecture about natural numbers for every possible natural number, and regardless of how many values of n for which we verify the conjecture, the conjecture still must be *proven* true. How then can we prove that a conjecture is true for *all natural numbers n*?

One of the most powerful methods for proving a conjecture true for all natural numbers is to use the *principle of mathematical induction*, which can be stated as follows.

The Principle of Mathematical Induction

Suppose P_n is a statement involving the natural number n for which the following two conditions hold.

1. P_1 is true; and
2. For each natural number k, if P_k is true, then P_{k+1} is true. (We often write this as $P_k \implies P_{k+1}$.)

Then P_n is true for all natural numbers n.

The idea underlying mathematical induction is a fairly simple one. Condition 1 says P_1 must be true. Condition 2 says that whenever P_k is true, it must follow that P_{k+1} is true. Therefore, by condition 2 if P_1 is true, then P_2 must be true. But again by condition 2, if P_2 is true, then P_3 must be true and if P_3 is true, then P_4 must also be true, and so on. In this way we know that P_n is true for all n. These two conditions are often called hypotheses, because we do not know whether they are true in a particular case.

We can think of the statements P_n as a sequence of dominoes, as indicated in Figure 9.3.

Figure 9.3

The domino analogy for mathematical induction

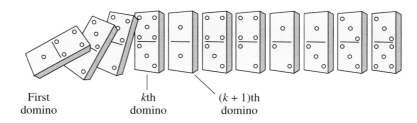

First domino kth domino $(k+1)$th domino

Here is how the steps in the principle of mathematical induction listed in the box correspond to the domino analogy.

Mathematical Induction

Hypotheses:
1. P_1 is true.
2. If P_k is true then P_{k+1} is true for any k.

Conclusion: $\{ P_n$ is true for all n.

Domino Analogy

Hypotheses:
1. The first domino is knocked over.
2. All the dominoes are set up in such a way that if any domino is knocked over, then the next domino is knocked over.

Conclusion: $\{$ All the dominoes are knocked over.

Note that in the domino analogy, to conclude that all the dominoes are knocked down, *both* hypotheses must be satisfied. If all we know is that the first domino is knocked over without knowing how closely the dominoes are set up, we cannot conclude that all the dominoes are knocked over. See Figure 9.4.

Figure 9.4

The domino analogy with hypothesis 2 not satisfied

Similarly, even if we know that the dominoes are set up closely enough so that if one is knocked over, the next one will fall also, unless we know that the first one is actually knocked over, we cannot conclude that they will all fall down. See Figure 9.5.

Figure 9.5

The domino analogy with hypothesis 1 not satisfied

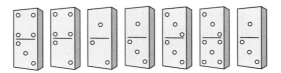

Let us return to the statement S_n discussed previously in this section.

Example 1   Prove that the sum of the first n odd natural numbers is equal to n^2.

Solution   We want to prove that

$$S_n: \quad 1 + 3 + 5 + \cdots + (2n - 1) = n^2$$

We use the principle of mathematical induction to prove the truth of S_n for all natural numbers n.

Step 1:   We must verify that S_1 is true. However, S_1 is just the statement that $1 = 1^2$, which is true.

Step 2:   *Assuming* that S_k is true, we must prove that S_{k+1} is true.

Let's examine S_k and S_{k+1}.

S_k: $1 + 3 + 5 + \cdots + (2k - 1) = k^2$ This is called the induction hypothesis.   (1)

S_{k+1} is obtained by substituting $k + 1$ for n in S_n.

S_{k+1}: $1 + 3 + 5 + \cdots + (2k - 1) + [2(k + 1) - 1] = (k + 1)^2$

   Simplify $2(k + 1) - 1$.

S_{k+1}: $1 + 3 + 5 + \cdots + (2k - 1) + (2k + 1) = (k + 1)^2$ (2)

Our goal in Step 2 is to show that S_{k+1} follows from S_k, or in other words, that we can derive equation (2) from equation (1). How do we derive equation (2) from equation (1)? We look at equations (1) and (2) and observe that the *left-hand sides* of equations (1) and (2) differ only by the quantity $(2k + 1)$, so we add $(2k + 1)$ to both sides of equation (1).

$$1 + 3 + 5 + \cdots + (2k - 1) \qquad\qquad = k^2$$

   This is equation (1). We add $2k + 1$ to both sides to obtain

$$1 + 3 + 5 + \cdots + (2k - 1) + (2k + 1) = k^2 + 2k + 1$$

   Factor the right-hand side to get

$$1 + 3 + 5 + \cdots + (2k - 1) + (2k + 1) = (k + 1)^2 \qquad \text{Which is equation (2)}$$

Because equation (2) follows from equation (1), we have demonstrated that $S_k \Rightarrow S_{k+1}$. Thus we have satisfied conditions (1) and (2) of the principle of mathematical induction, and so we have proven that S_n is true for all n.   ∎

Students are sometimes a bit confused by the induction hypothesis because they think that we are assuming what we are supposed to prove, but this is not so. We are not proving that either P_k or P_{k+1} is true. Rather, to carry out step 2 in an induction proof, we must prove that *if P_k is true, it must necessarily follow that P_{k+1} is true.*

Example 2 Prove that $1 + 4 + 9 + \cdots + n^2 = \dfrac{n(n + 1)(2n + 1)}{6}$ is true for all natural numbers n.

Solution Let P_n: $1 + 4 + 9 + \cdots + n^2 = \dfrac{n(n + 1)(2n + 1)}{6}$.

Step 1: We must verify that P_1 is true. P_1 is the statement that
$$1^2 = \frac{1(1 + 1)(2(1) + 1)}{6} = \frac{1(2)(3)}{6} = \frac{6}{6}, \text{ which is true.}$$

Step 2: *Assuming* that P_k is true, we must prove that P_{k+1} is true. Again, we begin by examining P_k and P_{k+1}.

P_k: $1^2 + 2^2 + 3^2 + \cdots + k^2 = \dfrac{k(k + 1)(2k + 1)}{6}$ This is the induction hypothesis.   (1)

P_{k+1}: $1^2 + 2^2 + 3^2 + \cdots + k^2 + (k + 1)^2 = \dfrac{(k + 1)[(k + 1) + 1][2(k + 1) + 1]}{6}$ Simplifying, we get

P_{k+1}: $1^2 + 2^2 + 3^2 + \cdots + k^2 + (k + 1)^2 = \dfrac{(k + 1)(k + 2)(2k + 3)}{6}$ (2)

Step 2 requires us to derive equation (2) from equation (1). By examining equations (1) and (2), we notice that their left-hand sides differ by $(k + 1)^2$, so we add $(k + 1)^2$ to both sides of equation (1).

$1^2 + 2^2 + 3^2 + \cdots + k^2 \qquad = \dfrac{k(k + 1)(2k + 1)}{6}$ This is equation (1). Add $(k + 1)^2$ to both sides.

$1^2 + 2^2 + 3^2 + \cdots + k^2 + (k + 1)^2 = \dfrac{k(k + 1)(2k + 1)}{6} + (k + 1)^2$

$\qquad = \dfrac{k(k + 1)(2k + 1) + 6(k + 1)^2}{6}$ Factor out $(k + 1)$.

$\qquad = \dfrac{(k + 1)[k(2k + 1) + 6(k + 1)]}{6}$

$\qquad = \dfrac{(k + 1)[2k^2 + 7k + 6]}{6}$ Factor again.

$1^2 + 2^2 + 3^2 + \cdots + k^2 + (k + 1)^2 = \dfrac{(k + 1)(k + 2)(2k + 3)}{6}$ This is equation (2).

Remember that in step 2 of an induction proof, we are not asserting that either P_k or P_{k+1} is true. We are proving only that
$$P_k \;\Rightarrow\; P_{k+1}.$$

Thus we have derived equation (2) from equation (1), and we have verified both conditions of mathematical induction, so we may conclude that the given formula for the sum of the squares of the first n natural numbers is true for all n.

Formulas such as the one in this example and similar ones in the exercise set come up in calculus. ∎

Example 3 Prove that the statement S_n: $(x + y)^n = x^n + y^n$ is not true for *all* natural numbers n.

Solution In order to prove that a statement is not true for all natural numbers n, it is sufficient to find *one* natural number for which the given statement is not true. For example, S_2 says that $(x + y)^2 = x^2 + y^2$, but we know that

Recall that such a number n is called a counterexample.

$$(x + y)^2 = x^2 + 2xy + y^2 \neq x^2 + y^2$$

Therefore, we have demonstrated that S_2 is false and so S_n is not true for all natural numbers n. ∎

Sometimes a statement P_n is not true for some initial values of n but is then true for all larger values of n. For example, consider the statement

$$P_n: \quad 2^n > n^2$$

We can easily check that this statement is true for $n = 1$, false for $n = 2, 3, 4$, and true for $n = 5, 6, 7$, and 8. However, the logic of mathematical induction applies just as well if we start with $n = 1$, $n = 5$, or any other value of n. In the domino analogy, this corresponds to skipping the first few dominoes and knocking over the fifth domino, for instance, and then causing all dominoes from the fifth one on to fall down.

Example 4   Prove that the statement $P_n: 2^n > n^2$ is true for all $n \geq 5$.

Solution   To prove P_n is true for $n \geq 5$, we will use the following fact, which you are asked to prove in Exercise 31.

$$n^2 > 2n + 1 \qquad \text{for } n \geq 3 \tag{$*$}$$

Step 1:   We must verify that P_5 is true. P_5 is the statement that $2^5 > 5^2$, which is true.

Step 2:   *Assuming* that P_k is true, we must prove that P_{k+1} is true; that is, we assume $2^k > k^2$ and we must prove $2^{k+1} > (k + 1)^2$, or, equivalently, that $2^{k+1} > k^2 + 2k + 1$.

We start with the induction hypothesis.

$$2^k > k^2 \qquad \text{Multiply both sides of this inequality by 2.}$$

$$2(2^k) > 2k^2$$

$$2^{k+1} > 2k^2 = k^2 + k^2 > k^2 + 2k + 1 \qquad \text{because by } (*), k^2 > 2k + 1. \text{ Therefore } 2^{k+1} > k^2 + 2k + 1, \text{ as required.}$$

We have thus satisfied both conditions of mathematical induction for $n \geq 5$ and so we have $2^n > n^2$ for $n \geq 5$.   ∎

We conclude this section with an example illustrating that mathematical induction can also be used to prove certain divisibility properties of natural numbers.

Example 5   Prove that for all natural numbers n, $4^n - 1$ is divisible by 3.

Solution   Before we proceed, let's keep in mind that, for a natural number N to be divisible by 3, we must have $N = 3M$ for some natural number M. (For example, 21 is divisible by 3 because $21 = 3 \cdot 7$.)

Step 1:   We must verify that P_1 is true. P_1 is the statement that $4^1 - 1 = 3$ is divisible by 3, which is true.

Step 2:   *Assuming* that P_k is true, we must prove that P_{k+1} is true. That is, we assume that $4^k - 1$ is divisible by 3 and we must prove that $4^{k+1} - 1$ is also divisible by 3. Let's look at $4^{k+1} - 1$.

$$4^{k+1} - 1 = 4(4^k) - 1 \qquad \text{We rewrite } -1 \text{ as } -4 + 3.$$

$$= 4(4^k) - 4 + 3 \qquad \text{Factor out 4 from the first two terms.}$$

$$= 4(4^k - 1) + 3 \qquad \text{By the induction hypothesis, } 4^k - 1 \text{ is divisible by 3.}$$
$$\qquad\qquad\qquad \text{Therefore, } 4^k - 1 = 3N \text{ for some natural number } N.$$

$$= 4(3N) + 3 \qquad \text{By factoring out 3, we see that the right-hand side is a multiple of 3.}$$

$$4^{k+1} - 1 = 3(4N + 1) \qquad \text{Which means that } 4^{k+1} - 1 \text{ is divisible by 3, as required.}$$

Therefore, by the principle of mathematical induction, $4^n - 1$ is divisible by 3 for all natural numbers n.   ∎

Throughout mathematical history there have been many conjectures about natural numbers. Here are just a few of the more famous ones and their current status.

1. We are all familiar with natural numbers a, b, and c such that $a^2 + b^2 = c^2$. For example, $3^2 + 4^2 = 5^2$. We have encountered many such *Pythagorean triples* as the sides of a right triangle.

   We may ask the following question: For $n > 2$, are there any natural numbers a, b, and c such that $a^n + b^n = c^n$? In other words, are there natural numbers a, b, and c such that $a^3 + b^3 = c^3$ or $a^8 + b^8 = c^8$, etc.? In 1637, the great mathematician Pierre de Fermat (1601–1665) wrote, in the margin of a book, that he had found a "marvelous" proof that there are no natural number solutions to $a^n + b^n = c^n$ for $n > 2$ but that "the margin was too narrow to contain it." This conjecture is known as "Fermat's Last Theorem."

   There is much controversy about whether Fermat actually had a proof, but for more than 350 years mathematicians of the first rank have attempted unsuccessfully to prove Fermat's Last Theorem. In 1994, Dr. Andrew Wiles announced that he had proved certain results that in turn prove Fermat's Last Theorem, and so, finally, Fermat's Last Theorem is proven.

2. It was conjectured from the times of the early Greek mathematicians that every natural number can be expressed as the sum of the squares of four integers. For example, $39 = 1^2 + 2^2 + 3^2 + 5^2$ and $11 = 0^2 + 1^2 + 1^2 + 3^2$. In 1770, Joseph-Louis Lagrange, probably the greatest mathematician of the eighteenth century, finally gave the first complete proof of this fact.

3. In 1742 the mathematician Christian Goldbach conjectured that every even natural number greater than 2 can be written as the sum of two prime numbers. For example, $20 = 3 + 17$ and $34 = 11 + 23$. Goldbach's conjecture still awaits a proof.

9.5 Exercises

In Exercises 1–20, use the principle of mathematical induction to prove the given statement is true for all natural numbers n.

1. P_n: $1 + 3 + 5 + \cdots + (2n - 1) = n^2$

2. S_n: $2 + 4 + 6 + \cdots + 2n = n^2 + n$

3. T_n: $3 + 7 + 11 + \cdots + (4n - 1) = n(2n + 1)$

4. R_n: $1 + 4 + 7 + \cdots + (3n - 2) = \dfrac{n(3n - 1)}{2}$

5. P_n: $2 + 5 + 8 + \cdots + (3n - 1) = \dfrac{n(3n + 1)}{2}$

6. S_n: $1^3 + 2^3 + 3^3 + \cdots + n^3 = \left(\dfrac{n(n + 1)}{2}\right)^2$

7. R_n: $2 + 9 + 16 + \cdots + (7n - 5) = \dfrac{n(7n - 3)}{2}$

8. T_n: $5 + 9 + 13 + \cdots + (4n + 1) = n(2n + 3)$

9. P_n: $2^2 + 4^2 + 6^2 + \cdots + (2n)^2 = \dfrac{2n(n + 1)(2n + 1)}{3}$

10. P_n: $\quad 1^2 + 3^2 + 5^2 + \cdots + (2n + 1)^2 = \dfrac{n(2n - 1)(2n + 1)}{3}$

11. P_n: $\quad 2 + 2^2 + 2^3 + \cdots + 2^n = 2^{n+1} - 2$

12. P_n: $\quad 3 + 3^2 + 3^3 + \cdots + 3^n = \dfrac{3^{n+1} - 3}{2}$

13. T_n: $\quad \dfrac{1}{1 \cdot 2} + \dfrac{1}{2 \cdot 3} + \dfrac{1}{3 \cdot 4} + \cdots + \dfrac{1}{n(n + 1)} = \dfrac{n}{n + 1}$

14. T_n: $\quad \dfrac{1}{1 \cdot 3} + \dfrac{1}{3 \cdot 5} + \dfrac{1}{5 \cdot 7} + \cdots + \dfrac{1}{(2n - 1)(2n + 1)} = \dfrac{n}{2n + 1}$

15. S_n: $\quad 1 \cdot 2 + 2 \cdot 3 + 3 \cdot 4 + \cdots + n(n + 1) = \dfrac{n(n + 1)(n + 2)}{3}$

16. S_n: $\quad 1 \cdot 2 + 3 \cdot 4 + 5 \cdot 6 + \cdots + (2n - 1)(2n) = \dfrac{n(n + 1)(4n - 1)}{3}$

17. T_n: $\quad \left(1 + \dfrac{1}{1}\right)\left(1 + \dfrac{1}{2}\right)\left(1 + \dfrac{1}{3}\right) \cdots \left(1 + \dfrac{1}{n}\right) = n + 1$

18. T_n: $\quad \left(1 - \dfrac{1}{2}\right)\left(1 - \dfrac{1}{3}\right)\left(1 - \dfrac{1}{4}\right) \cdots \left(1 - \dfrac{1}{n + 1}\right) = \dfrac{1}{n + 1}$

19. S_n: $\quad a + (a + d) + (a + 2d) + \cdots + [a + (n - 1)d] = n\left(a + \dfrac{n - 1}{2}d\right)$

20. S_n: $\quad a + ar + ar^2 + \cdots + ar^n = \dfrac{a(1 - r^n)}{1 - r} \quad$ for $r \neq 1$

21. Prove that $n^3 - n + 6$ is divisible by 3 for all natural numbers n.

22. Prove that 5 is a factor of $6^n - 1$ for all natural numbers n.

23. Use mathematical induction to prove that $x - 1$ is a factor of $x^n - 1$ for all natural numbers n. Can you think of another way to prove this fact?

24. Use mathematical induction to prove that $n^2 + n$ is divisible by 2 for all natural numbers n. Can you think of another way to prove this fact?

25. Prove that $x - y$ is a factor of $x^n - y^n$ for all natural numbers n.

26. Use the result of Exercise 6 and Example 2 to show that

$$\sum_{k=1}^{n} k^3 = \left(\sum_{k=1}^{n} k\right)^2$$

27. Prove that $9^n - 1$ is divisible by 4 for all natural numbers n.

28. Prove that $n^3 + 3n^2 + 2n$ is divisible by 6 for all natural numbers n. Can you think of another way to prove this fact?

29. Use mathematical induction to prove DeMoivre's theorem, which says that

$$[r(\cos \theta + i \sin \theta)]^n = r^n(\cos n\theta + i \sin n\theta)$$

30. Use mathematical induction to prove that, for all natural numbers n, $(a - bi)^n$ is the complex conjugate of $(a + bi)^n$.

In Exercises 31–40, use the principle of mathematical induction.

31. Prove that $n^2 > 2n + 1$ for $n \geq 3$.

32. Prove that $(n + 2)^2 < n^3$ for $n \geq 3$.

33. Prove that $\left(\dfrac{7}{5}\right)^n > n$ for $n \geq 5$.

34. Prove that $(1.1)^n > n$ for $n \geq 39$.

35. Prove that $n^2 > 10n$ for $n > 10$. Can you think of another way to prove this fact? HINT: Think about the graph of $y = x^2 - 10x$.

36. Prove that $n^2 > 3n + 10$ for $n > 5$. Can you think of another way to prove this fact? HINT: Think about the graph of $y = x^2 - 3x - 10$.

37. Prove that $n! > 2^n$ for $n \geq 4$.

38. Prove that $n^n > n!$ for $n > 1$.

39. Assuming the truth of the triangle inequality, which says that $|a + b| \leq |a| + |b|$, prove that

$$|a_1 + a_2 + \cdots + a_n| \leq |a_1| + |a_2| + \cdots + |a_n|$$

40. Prove each of the following inequalities is true for all natural numbers n.

(a) $\dfrac{1}{\sqrt{1}} + \dfrac{1}{\sqrt{2}} + \dfrac{1}{\sqrt{3}} + \cdots + \dfrac{1}{\sqrt{n}} \geq \sqrt{n}$

(b) $\dfrac{1}{\sqrt{1}} + \dfrac{1}{\sqrt{2}} + \dfrac{1}{\sqrt{3}} + \cdots + \dfrac{1}{\sqrt{n}} < 2\sqrt{n}$

(c) Based on the results of parts (a) and (b), what can you conclude about

$$\dfrac{1}{\sqrt{1}} + \dfrac{1}{\sqrt{2}} + \dfrac{1}{\sqrt{3}} + \cdots + \dfrac{1}{\sqrt{n}}?$$

In Exercises 41–46, find the smallest natural number $N > 1$ for which the statement is true, and then prove the statement is true for all natural numbers greater than or equal to N.

41. $\left(\dfrac{4}{3}\right)^n > \dfrac{4}{3}n$

42. $n + 20 \leq n^2$

43. $n^3 > (n + 5)^2$

44. $n^2 > 2n$

45. $2^n > n^2$. HINT: Use the result of Exercise 31.

46. $\dfrac{1}{n + 5} > \dfrac{1}{2n}$

In Exercises 47–58, determine what can be concluded from the given information about the sequence of statements P_n. For example, if P_6 is true and $P_k \Rightarrow P_{k+1}$ for all k, then we can conclude that P_n is true for all $n \geq 6$.

47. P_k is true for $1 \leq k < 50$.

48. P_1 is true but $P_k \nRightarrow P_{k+1}$.

49. P_{10} is true and $P_k \Rightarrow P_{k+1}$.

50. P_{10} is not true and $P_k \Rightarrow P_{k+1}$.

51. P_{10} is not true and $P_k \nRightarrow P_{k+1}$.

52. P_{10} is true and $P_k \nRightarrow P_{k+1}$.

53. P_1 is true and $P_k \Rightarrow P_{k+2}$.

54. P_1 and P_2 are true and P_k and P_{k+1} together imply P_{k+2}.

55. P_1 and P_2 are true and $P_k \Rightarrow P_{k+2}$.

56. P_{100} is true and $P_k \Rightarrow P_{k-1}$.

57. P_1 is true and $P_k \Rightarrow P_{3k}$.

58. P_1 is true and $P_k \Rightarrow P_{k+3}$.

Questions for Thought

59. True or false: $5^n + 3$ is divisible by 4 for all natural numbers n. Justify your answer.

60. True or false: $n^2 + n + 11$ is a prime number for all natural numbers n. Justify your answer.

9.6 Permutations and Combinations

In this section we discuss some basic counting techniques. Let's consider the following situation.

Example 1 Suppose Gerald wants to buy a computer, monitor, and printer. He is considering two choices for the computer, let's call them C_1 and C_2; two choices for the monitor, M_1 and M_2; and three choices for the printer, P_1, P_2, and P_3. Assuming that all the components are compatible, how many different computer systems of computer, monitor, and printer can Gerald put together?

Solution To count the number of possible configurations, we must consider the number of different ways Gerald can choose a computer, monitor, and printer. One way to analyze the number of possibilities is by using a **tree diagram**, as illustrated in Figure 9.6 on page 606. In the figure, $C_1 M_2 P_3$, for example, means that Gerald has chosen computer C_1, monitor M_2, and printer P_3.

From this tree diagram we can see that there are 12 possible configurations for the computer system. There are 2 choices for the computer, 2 choices for the monitor, and 3 choices for the printer, and we note that $12 = 2 \cdot 2 \cdot 3$.

Figure 9.6

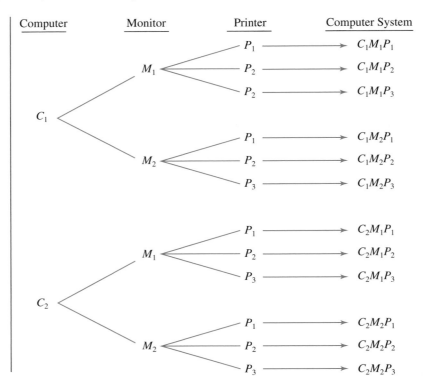

Similarly, if Nina has 3 skirts and 5 blouses, she can make $3 \cdot 5 = 15$ different outfits (assuming that she can wear any skirt with any blouse).

The situations we have just described are merely special cases of the **Fundamental Counting Principle**.

The Fundamental Counting Principle

Let $E_1, E_2, E_3, \ldots, E_k$ be a sequence of k events such that E_1 can occur in n_1 ways, E_2 can occur in n_2 ways, E_3 can occur in n_3 ways, and so on. Then the total number of ways all the events can occur is

$$n_1 \cdot n_2 \cdot n_3 \cdots n_k$$

Example 2 In a certain state a license plate number consists of 3 letters followed by 4 digits. How many different such license plates can there be?

Solution In Example 1 we were able to list all the possibilities with a tree diagram because there were only a relatively small number of possibilities. In this example, however, there are an enormous number of possibilities and it would be extremely impractical to try to list and count them all. Instead, we can view a license plate number as a series of seven blank spaces:

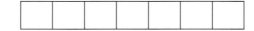

The first three spaces can be filled in with any of the 26 letters A–Z, and the last four spaces can be filled in with any of the ten digits 0–9. In the language of the Fundamental Counting Principle, E_1 is choosing the letter for the first space; E_2 is choosing the letter for the second space; and E_5 is choosing the digit for the fifth space. If we enter the number of ways each space can be filled in, we have

26	26	26	10	10	10	10

and so, by the Fundamental Counting Principle, there are

$$26 \cdot 26 \cdot 26 \cdot 10 \cdot 10 \cdot 10 \cdot 10 = \boxed{175,760,000}$$

different license plates. ∎

Example 3   If a fair coin is tossed ten times, how many possible different sequences of heads and tails are there?

Solution   Each time we toss the coin (each toss is an **event**) there are two possible outcomes: heads or tails. Thus by the Fundamental Counting Principle, there are a total of

$$2 \cdot 2 \cdot 2 \cdot 2 \cdot 2 \cdot 2 \cdot 2 \cdot 2 \cdot 2 \cdot 2 = 2^{10} = \boxed{1024}$$

possible sequences of heads and tails.

Notice that in formulating this example, we are concerned with the order of the heads and tails. Thus the sequence HHTHTHTHHT is different from HTHTTHHHTH, even though they both have 6 heads and 4 tails. ∎

Permutations

One of the important applications of the Fundamental Counting Principle is in determining the number of ways that n distinct objects can be arranged, where the order of the objects matters.

Example 4   Suppose an organization is having an election for president, vice president, and secretary, and the election is run so that the person receiving the most votes becomes president, second most becomes vice president, and third most becomes secretary. If there are 6 candidates, John (J), Kita (K), Debra (D), Tom (T), Gina (G), and Lamar (L), competing for the three positions, how many outcomes to the election are there?

Solution   Viewing the three positions to be filled as three vacant blanks, it is important to recognize that

Pres.	V. Pres.	Sec'y		Pres	V. Pres.	Sec'y
K	T	L		T	L	K

are different outcomes. The first means that Kita is president, Tom is vice president, and Lamar is secretary, whereas the second means Tom is president, Lamar is vice president, and Kita is secretary. In other words, the *order* in which the top three finish is important. There are 6 possible presidents. Once the president is chosen, there are now 5 possible people left to be vice president, and then there are 4 people left to be secretary. Using the Fundamental Counting Principle and the box format, we would describe the possibilities as

6	5	4

and so there are

$$6 \cdot 5 \cdot 4 = \boxed{120}$$

Note we are assuming that there are no ties. different possible outcomes of the election. ∎

In the license plate problem, we were allowed to use letters and digits more than once, but in the last example, once someone is elected to a position, he or she is no longer available for another position.

Definition of a Permutation

A **permutation** is an arrangement of distinct objects in a definite order.

When we speak of finding the number of permutations of n objects, we mean the number of ways in which the n objects can be rearranged.

Example 5   Suppose that the Kentucky Derby has 9 horses entered in the race.
(a) In how many different orders can the horses finish?
(b) How many different 1-2-3 finishes can there be?
(c) How many different 7-8-9 finishes can there be?

(Assume that there are no ties.)

Solution   **(a)** We are being asked to find the number of permutations of the 9 distinct horses. First place can be any of the 9 horses; second place can then be any of the remaining 8 horses; third place can then be any of the remaining 7 horses, and so on. We can view the problem as

9	8	7	6	5	4	3	2	1

and so by the Fundamental Counting Principle we have

$$9 \cdot 8 \cdot 7 \cdot 6 \cdot 5 \cdot 4 \cdot 3 \cdot 2 \cdot 1 = \boxed{362{,}880}$$

The 9 horses can finish in any of 362,880 different orders.

(b) A similar analysis gives

Number of possible horses in position 1	Number of possible horses in position 2	Number of possible horses in position 3
↓	↓	↓
9	8	7

and so there are $9 \cdot 8 \cdot 7 \cdot = \boxed{504}$ possible 1-2-3 finishes.

(c) Interestingly, the result for positions 7-8-9 is the same as for positions 1-2-3. After all, just as any of the 9 horses can finish first, so, too, any of the 9 horses can finish seventh.

Number of possible horses in position 7	Number of possible horses in position 8	Number of possible horses in position 9
↓	↓	↓
9	8	7

and so there are $9 \cdot 8 \cdot 7 = \boxed{504}$ possible 7-8-9 finishes. ∎

Note that in analyzing the number of permutations of the 9 horses in Example 5(a), we encountered $9 \cdot 8 \cdot 7 \cdot 6 \cdot 5 \cdot 4 \cdot 3 \cdot 2 \cdot 1$. Recall from Section 9.1 that this product can be denoted as 9!.

In Example 5, we were interested in the number of permutations of 3 distinct objects taken from a set of 9 distinct objects. In such a situation, we say that we are permuting 9 objects taken 3 at a time, and we denote this number as $_9P_3$. We will use the notation $_nP_r$ to mean the number of permutations of n distinct objects taken r at a time. Using this notation, the results of Example 5 can be generalized to the following theorem.

Another notation for $_nP_r$ is $P(n, r)$.

Theorem 9.1

(a) The number of permutations of n distinct objects is $_nP_n = n!$.

(b) The number of permutations of n distinct objects taken r at a time is
$$_nP_r = n(n - 1)(n - 2) \cdots (n - r + 1).$$

Proof   (a) We want to fill n boxes with n choices for the first, $n - 1$ for the second, $n - 2$ for the third, and so on.

n	$n - 1$	$n - 2$	$\cdots$	3	2	1

and so by the Fundamental Counting Principle the number of permutations is
$$n(n - 1)(n - 2) \cdots 3 \cdot 2 \cdot 1 = n! \qquad \text{As required}$$

(b) In part (b) we are arranging only r out of the n objects. Thus we are filling r boxes. Note that in the second box we subtract 1, in the third box we subtract 2, in the fourth box, 3, and so on, until the rth box, where we subtract $r - 1$. So the entry in the rth box is $n - (r - 1) = n - r + 1$. Therefore, the number of permutation is
$$_nP_r = n(n - 1)(n - 2) \cdots (n - r + 1) \qquad \text{As required} \qquad \blacksquare$$

Example 6   Suppose that the police are arranging a lineup of six suspects.
(a) How many such lineups are possible?
(b) How many lineups are possible if two specific suspects must be next to each other?

Solution   (a) The number of permutations of the six suspects is
$$_6P_6 = 6! = 6 \cdot 5 \cdot 4 \cdot 3 \cdot 2 \cdot 1 = \boxed{720}$$

(b) Let's refer to the six suspects as A, B, C, D, E, and F and let's suppose that A and B are the two suspects who must be next to each other. We can then view the two suspects who must be next to each other as one object (AB). We are then counting the number of permutations of the five objects (AB), C, D, E, F, which we know is $5! = 120$. However, we also get 120 arrangements with suspects A and B reversed. In other words, we also have 120 permutations of (BA), C, D, E, F.
   Therefore, there are $120 + 120 = \boxed{240}$ lineups with two specific suspects next to each other. $\blacksquare$

Example 7   How many different permutations are there of the letters of each word?
(a) DADDY   (b) AARDVARK

Solution   (a) Up to this point we have been examining the permutations of distinct objects. The word DADDY has 5 letters, but they are not distinct. Let's temporarily distinguish the D's in the word as $D_1AD_2D_3Y$, which Theorem 9.1 tells us has $_5P_5 = 5! = 120$ permutations. Among these permutations we are counting

$$AYD_1D_2D_3 \qquad AYD_1D_3D_2 \qquad AYD_2D_1D_3$$
$$AYD_2D_3D_1 \qquad AYD_3D_1D_2 \qquad AYD_3D_2D_1$$

However, these six permutations are actually indistinguishable, because they all appear as *AYDDD*. Thus we see that for each permutation of $D_1AD_2D_3Y$, there are 6 permutations that contain the same letters in the same order. Note that $6 = 3!$, which is the number of permutations of the 3 *D*'s that are repeated. To obtain the actual number of distinct permutations, we must divide the total number of permutations by 6.

Number of distinct permutations of DADDY

$$= \frac{\textit{number of permutations}}{\textit{number of permutations of the 3 D's}} = \frac{{}_5P_5}{{}_3P_3} = \frac{5!}{3!} = \frac{120}{6} = \boxed{20}$$

(b) Applying the same type of analysis to the word AARDVARK, we can view it as $A_1A_2R_1DVA_3R_2K$, which has a total of $8! = 40{,}320$ permutations. Due to the fact that there are 3 *A*'s, *each* permutation of AARDVARK is counted $6 = 3!$ times in the list of permutations of $A_1A_2R_1DVA_3R_2K$. But because of the fact that there are 2 *R*'s, *each* of these 6 repetitions is counted $2 = 2!$ times. Thus each permutation of AARDVARK appears $12 = 6 \times 2 = 3! \times 2!$ times in the list of permutations of $A_1A_2R_1DVA_3R_2K$. Therefore, we have

Number of distinct permutations of AARDVARK

$$= \frac{\textit{total no. of permutations}}{\textit{no. of permutations of the 3 A's} \times \textit{no. of permutations of the 2 R's}}$$

$$= \frac{{}_8P_8}{{}_3P_3 \cdot {}_2P_2} = \frac{8!}{3! \times 2!} = \frac{8 \cdot 7 \cdot 6 \cdot 5 \cdot 4 \cdot 3 \cdot 2 \cdot 1}{6 \cdot 2} = \boxed{3360}$$

The previous example generalizes to the following theorem.

Theorem 9.2

Suppose a set of n objects has r_1 identical objects of one type, r_2 identical objects of a second type, r_3 identical objects of a third type, . . . , and r_k identical objects of a kth type. The number of *distinct* permutations of the n objects is

$$\frac{n!}{r_1! \cdot r_2! \cdot r_3! \cdots r_k!}$$

Example 8   How many different 7-digit numbers can be made using all the digits 1, 1, 5, 5, 8, 8, and 8?

Solution   Because we are using 7 digits (1, 1, 5, 5, 8, 8, 8), there are 7! permutations of these digits. Since there are two 1's, two 5's, and three 8's, Theorem 9.2 gives us

$$\frac{7!}{2! \cdot 2! \cdot 3!} = \frac{7 \cdot 6 \cdot 5 \cdot 4 \cdot 3 \cdot 2 \cdot 1}{2 \cdot 2 \cdot 6} = \boxed{210 \text{ distinct 7-digit numbers}}$$

Combinations

In Example 4 we analyzed the number of ways in which a president, vice president, and secretary can be chosen from among a group of candidates. If there were 10 candidates for the 3 positions, then there would be $10 \cdot 9 \cdot 8 = 720$ different possible outcomes of the election. In this analysis we recognized that the *order* in which the candidates are chosen is important. The next example, which deals with a situation in which the order of the objects is *not* important, illustrates a particularly important application of Theorem 9.2.

Example 9 A committee of 3 people is to be chosen from among 10 members of a club. How many such committees are there?

Solution This example is asking for the number of ways we can choose 3 distinct objects (people) out of 10 distinct objects *without regard to their order.* Suppose we think of the 10 people being numbered 1 through 10. If persons 3, 7, and 8 are chosen for the committee, it makes no difference if they were chosen in the order 3, 7, 8 or 8, 3, 7. The committee still consists of the same people. Therefore we take the total number of 3-person committees $_{10}P_3$ (that is, the number of permutations of 10 objects taken 3 at a time, which we know is $10 \cdot 9 \cdot 8$) and divide that by $_3P_3$ (the number of permutations of the 3 members of the committee, which is 3!), and so the number of 3-person committees is

$$\frac{_{10}P_3}{_3P_3} = \frac{10 \cdot 9 \cdot 8}{3!} = 120$$

∎

Let's examine Example 9 from a slightly different perspective. We can think of the number of ways of choosing a 3-person committee as the number of permutations of a set of 10 people in which one subset consists of 3 "identical" people (the people who *are* chosen for the committee) and another subset of 7 identical people (the people who *are not* chosen for the committee). Therefore, according to Theorem 9.2, the number of such permutations is

$$\frac{10!}{3! \cdot 7!} = \frac{10 \cdot 9 \cdot 8 \cdot 7 \cdot 6 \cdot 5 \cdot 4 \cdot 3 \cdot 2 \cdot 1}{(7 \cdot 6 \cdot 5 \cdot 4 \cdot 3 \cdot 2 \cdot 1)(3 \cdot 2 \cdot 1)} = \frac{10 \cdot 9 \cdot 8}{3!} = \frac{_{10}P_3}{_3P_3} = 120$$

as we saw before.

The key feature of this example is that we were choosing a subset of 3 elements from a set of 10 elements *without regard to order.* A selection of a subset from a larger set where the order of the selection is not important is called a **combination.**

We will denote the number of ways in which r elements can be chosen from a set of n elements as

$_nC_r$ is also sometimes written as $C(n, r)$.

$$_nC_r \quad \text{or} \quad \binom{n}{r}, \qquad \text{which are often read as "} n \text{ choose } r\text{"}$$

Generalizing the discussion of the last example we have Theorem 9.3.

Theorem 9.3

The number of combinations of n distinct objects taken r at a time is

$$_nC_r = \frac{n!}{r! \cdot (n - r)!} = \frac{n(n - 1)(n - 2) \cdots (n - r + 1)}{r!} = \frac{_nP_r}{_rP_r}$$

Recall that in Section 9.1, we defined $0! = 1$, so that based on this theorem we have $_nC_0 = \dfrac{n!}{0! \cdot (n - 0)!} = \dfrac{n!}{n!} = 1$.

How would we interpret $_nC_0$? Based upon this discussion, we have n items and we are choosing 0 of them. There is only one way to choose 0 items: Do not choose any of them. Thus we would expect $_nC_0$ to equal 1, which agrees with the result obtained using the theorem.

The following example puts to use a number of the ideas we have developed in this section.

Example 10 A poker hand consists of 5 cards dealt from a standard deck of 52 playing cards.
(a) How many different 5-card poker hands are there?
(b) How many different 5-card flushes are there? (A flush means 5 cards of the same suit.)

Solution

Most graphing calculators can compute permutations and combinations. The following display illustrates $_{52}C_5$.

```
52 nCr 5
             2598960
```

(a) Once the cards are dealt, the player can put them in any order that he or she wishes to, so the order in which the cards are dealt is irrelevant. The number of possible 5-card poker hands is the number of ways we can choose 5 distinct elements from a set of 52 distinct elements, which is precisely $_{52}C_5$. By Theorem 9.3 we have

$$_{52}C_5 = \frac{52 \cdot 51 \cdot 50 \cdot 49 \cdot 48}{5 \cdot 4 \cdot 3 \cdot 2 \cdot 1} = \boxed{2{,}598{,}960}$$

(b) We can think of dealing a flush as a two-step process: First, choose one of the four suits (spades, hearts, clubs, or diamonds) and second, choose 5 out of the 13 cards of the suit. Therefore, by the Fundamental Counting Principle we have

$$= \underbrace{(\textit{no. of ways to pick a suit})}_{_4C_1} \cdot \underbrace{(\textit{no. of ways to pick 5 cards of that suit})}_{_{13}C_5}$$

$$= {_4C_1} \cdot {_{13}C_5} = (4) \cdot \left(\frac{13 \cdot 12 \cdot 11 \cdot 10 \cdot 9}{5 \cdot 4 \cdot 3 \cdot 2 \cdot 1}\right) = \boxed{5148}$$

We offer the following suggestions to the reader.

Guidelines for Solving Counting Problems

When solving a counting problem, read the problem and ask yourself, "Is order important in the selection process?"

1. If the answer is yes, then the problem calls for permutations.
2. If the answer is no, then the problem calls for combinations.
3. Keep in mind that the Fundamental Counting Principle can often be used in conjunction with the various theorems in this section.

Example 11 Show that $\binom{n}{k} = \binom{n}{n-k}$.

Solution

Remember $\binom{n}{k}$ means the same thing as $_nC_k$.

According to Theorem 9.3 we have

$$\binom{n}{k} = \frac{n!}{k! \cdot (n-k)!}$$

And we also have

$$\binom{n}{n-k} = \frac{n!}{(n-k)! \cdot [n-(n-k)]!} = \frac{n!}{(n-k)! \cdot k!}$$

From these two equations we have

$$\binom{n}{k} = \binom{n}{n-k}$$

As required

We will find the result established in the last example useful in the next section, as we develop the Binomial Theorem.

9.6 Exercises

In Exercises 1–10, evaluate the given quantity and describe it verbally.

1. $_8P_3$

2. $_{10}C_3$

3. $\binom{12}{7}$

4. $\binom{12}{5}$

5. $_6P_5$

6. $_6P_1$

7. $\binom{18}{12}$

8. $\binom{18}{6}$

9. $\dfrac{_8P_4}{_4P_4}$

10. $\dfrac{12!}{3!} \cdot \dfrac{5!}{4!}$

11. If $_nC_2 = 105$, find n.

12. If $_nP_2 = 90$, find n.

13. A student has 5 pairs of slacks, 8 shirts, and 3 sweaters. Assuming that the student is a nonconformist and does not care about how the clothes match, how many different outfits consisting of slacks, shirt, and sweater are possible?

14. A car manufacturer offers a particular car model with 12 choices of paint color, 3 choices of body style, and 4 choices of interior trim. How many different configurations of color, body style, and interior trim are possible?

15. A mattress manufacturer offers 8 different choices in covering fabric (called the *ticking*), 5 different choices in firmness, and 4 different sizes. How many different types of mattresses are available?

16. A travel agent is offering a special discount on vacations to 12 different destinations with 9 different departure dates and 3 different categories of hotel accommodations. How many different possible discount vacations are available?

17. In how many ways can a student arrange 8 different books on a shelf?

18. In how many ways can a student arrange 9 different books consisting of 3 science books, 2 history books, and 4 mathematics books if
   (a) The books can be arranged in any order?
   (b) All books of the same type must be next to each other?

19. A CD player has a "shuffle" feature, which will play the selections in a random order. If a particular CD has 15 selections, in how many different orders can the CD be played?

20. A horse race has 10 entries. How many possible 1-2-3 finishes are there?

21. A jewelry manufacturer produces "initial" rings, which contain the first initials of the wearer's first and last names. If the manufacturer wants to produce every possible initial ring, how many rings must be made?

22. A company wants to buy initial rings for all its employees from the manufacturer of Exercise 21. If the company has 680 employees, explain why the company will have to order more than one copy of at least one ring.

23. Suppose a state's license plates each consist of 4 letters followed by 3 digits, using only the digits 1 through 9.
   (a) How many such license plates are possible?
   (b) How many such license plates are possible if the first digit must be odd?
   (c) How many such license plates are possible if the first digit must be even and the first letter must be a consonant? (Consider Y to be a consonant.)

24. A state lottery requires participants to choose 6 numbers from the numbers 1 through 40. If the order of the numbers is unimportant, how many ways are there to choose the 6 numbers?

25. A multiple-choice examination consists of 20 questions. If each question has 4 choices, how many possible ways are there to answer all the questions? What if there are 5 choices for each question?

26. A true–false examination has 15 questions. How many possible ways are there to answer the questions?

27. A student is required to answer 10 out of 12 questions on an examination. How many different sets of 10 questions can the student answer?

28. An audio component store has a special sale on 3 different amplifiers, 4 different tuners, 2 different cassette decks, 5 different CD players, and 8 different sets of speakers. How many different stereo systems can be made choosing one of each type of component?

29. A bag contains 9 red marbles and 7 blue marbles. In how many ways can you choose
   (a) 5 blue marbles?
   (b) 4 red marbles?
   (c) 9 marbles of any color?
   (d) 3 blue and 6 red marbles?

30. A scrabble tournament has 16 participants. If the tournament committee wants each player to meet every other player exactly once, how many games must be scheduled?

31. A club has 25 members. A committee of 6 is to be chosen, and then this committee is to select two co-chairpersons. In how many ways can this be done?

32. Repeat Exercise 31 if the committee is to select one chairperson and a secretary.

33. Two teams are playing a best-of-seven series to determine the champion. How many different ways are there for the series to go the full seven games?

34. Four people are to be selected from a group of four couples. In how many ways can the four people be selected if
(a) There are no restrictions?
(b) One member from each couple is to be chosen?
(c) At least one couple is to be included?
(d) Exactly one couple is to be included?

35. An attache case has a three-digit number combination lock. Each of the digits can be chosen from 1, 2, 3, 4, 5.
(a) How many such combinations are there?
(b) How many of the combinations are even numbers?

36. Using the digits 3, 4, 5, 6, 7, 8, 9 and assuming that digits can be repeated, determine how many four-digit numbers can be made if
(a) There are no restrictions.
(b) The number is greater than 5000.
(c) The number is less than 4000.
(d) The number is not divisible by 5.

37. Five cards are dealt from an ordinary deck of playing cards. In how many ways can the cards be dealt so that
(a) All the cards are spades?
(b) All the cards are the same suit?
(c) Exactly 3 are aces?
(d) The hand is a full house (3 of one kind and 2 of another)?

38. If a 5-card poker hand is dealt from a deck of 52 playing cards, how many hands will contain
(a) Exactly one pair?
(b) Two pairs?
(c) A straight (five cards in sequence; e.g., 4 5 6 7 8)?

39. Suppose we have 10 points in the plane such that no 3 of them fall on the same line.
(a) How many different lines are determined by these 10 points?
(b) How many different triangles are determined by these 10 points?

40. Find the number of diagonals of (a) a pentagon; (b) an octagon.

41. How many different arrangements are there of the letters of each word?
(a) COMPUTER
(b) ENCYCLOPEDIA
(c) EFFERVESCENT

42. In a family of 6 children, how many possible boy–girl orders of birth are there?

43. Suppose there are 3 routes connecting city A to city B, 5 routes connecting city B to city C, and 2 routes connecting city C to city D. How many routes are there from city A to city D?

44. A group of 20 basketball players wants to divide itself up into 4 teams of 5 players each. In how many ways can this be done?

In Exercises 45–46, prove the given identity.

45. $\dbinom{n}{r} + \dbinom{n}{r-1} = \dbinom{n+1}{r}$

46. Use mathematical induction to prove that
$$\dbinom{n}{0} + \dbinom{n}{1} + \dbinom{n}{2} + \cdots + \dbinom{n}{n} = 2^n$$

Questions for Thought

47. How are the ideas of a permutation and a combination related?

48. Use the method of mathematical induction to prove that the number of permutations of n distinct objects is $n!$.

9.7 Probability

If two dice are tossed, what are the chances of rolling a 7? If you purchase a lottery ticket, what are your chances of winning? If a five-card poker hand is dealt from a standard deck of 52 playing cards, how likely is it that you will be dealt exactly three aces? Questions such as these are the subject of *probability*. The concepts and methods of probability are used to predict results in a wide variety of real-life situations.

In order to discuss probability, we must introduce some vocabulary. Any process such as rolling a die, tossing a coin, giving a patient medication, or manufacturing an item, is an *experiment*. Any one of the possibilities that can result from performing an experiment is called an *outcome*.

We will restrict our attention to situations where all the outcomes of an experiment are "equally likely." For example, if we toss a coin (this is the experiment), then we expect the two possible outcomes—heads and tails—are equally likely. By this we mean that, if we toss the coin many times, we expect that about half the time the result will be heads and about half the time the result will be tails. In such a case we often say that we are tossing a *fair* coin.

Similarly, if we roll a fair die many times, then we expect that each of the six possible outcomes, 1, 2, 3, 4, 5, and 6, will come up about one-sixth of the time.

The set of all possible outcomes of an experiment is called the *sample space*. Thus if we toss a coin once the sample space is $\{H, T\}$, because these are the only two possible outcomes. If we roll a die once the sample space is $\{1, 2, 3, 4, 5, 6\}$, because these are the only possible outcomes.

We will be dealing with finite sample spaces only.

Example 1 | If an experiment consists of tossing a coin 3 times, write out the sample space S.

Solution | In order to know that we have written out a complete sample space, it is usually a good idea to know how many outcomes are possible. The counting principles we learned in the previous section prove extremely useful here.

We are tossing a coin 3 times. Each time we toss the coin we have two possibilities: heads (H) or tails (T). Thus, according to the Fundamental Counting Principle, there are $2 \cdot 2 \cdot 2 = 8$ possible outcomes. One way of obtaining the 8 possible outcomes is by using a tree diagram, as shown in Figure 9.7.

Reading the possible outcomes from the tree diagram, we see that the sample space consists of 8 possible outcomes:

$$S = \{HHH, HHT, HTH, THH, TTH, THT, HTT, TTT\}$$

where HTH means: first toss—heads, second toss—tails, and third toss—heads. ∎

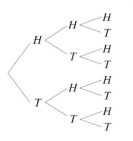

Figure 9.7

The tree diagram for tossing a coin 3 times

In any given experiment we are usually interested in a particular result—that is, a particular set of outcomes. For example, in the coin-tossing situation of Example 1, we might be interested in getting exactly 2 heads. Since the sample space is the set of *all* possible outcomes, a particular set of outcomes is a subset of the sample space. The following definition makes this idea of obtaining particular outcomes of an experiment precise.

Definition

Any subset of a sample space associated with a particular experiment is called an **event**. If E is an event we will write $n(E)$ to represent the number of elements in E.

Referring back to Example 1, if we toss a coin 3 times and E is the event "we get exactly two heads," then we have

Note that E is a subset of the sample space S.

$$E = \{HHT, HTH, THH\}$$

and so we have $n(E) = 3$. In other words, if you toss a coin 3 times then there are 3 possible ways to get exactly 2 heads.

We are now prepared to define the notion of probability. Intuitively, we know that if we roll a die then there are six equally likely outcomes, so that the chance of rolling a 5 is $\frac{1}{6}$. Of the six equally likely outcomes, three are odd. So it seems reasonable to say that the chance of rolling a die and getting an odd number is $\frac{3}{6} = \frac{1}{2}$. This type of reasoning is the basis for the following definition.

Definition

Let S be the sample space of an experiment and E an event. The **probability** of the event E, denoted as $P(E)$, is

$$P(E) = \frac{n(E)}{n(S)} = \frac{\text{number of elements in } E}{\text{number of elements in } S}$$

We note that since the number of elements in any event is greater than or equal to 0 and is less than or equal to the number of elements in the sample space, we have

$$0 \le n(E) \le n(S) \qquad \text{Divide each member of the inequality by } n(S).$$

$$\frac{0}{n(S)} \le \frac{n(E)}{n(S)} \le \frac{n(S)}{n(S)}$$

$$0 \le \frac{n(E)}{n(S)} \le 1 \qquad \text{But } \frac{n(E)}{n(S)} = P(E)$$

$$0 \le P(E) \le 1$$

Thus the probability of an event must fall between 0 and 1 (inclusive). The closer a probability is to 0, the less likely the event is to happen; the closer to 1, the more likely it is to happen. If $P(E) = 0$, then the event is called *impossible*; if $P(E) = 1$, then the event is called *certain*.

Example 2

Returning to the situation of Example 1, a coin is tossed 3 times. Find:
(a) the probability of getting exactly 2 heads
(b) the probability of getting at least 2 heads
(c) the probability of getting no heads

Solution

As before, the sample space is

When writing out a sample space, it is usually most efficient to write down the outcomes in a systematic order. We recorded the outcomes as follows: 3 heads, 2 heads and 1 tail, 2 tails and 1 head, and 3 tails.

$S = \{HHH, HHT, HTH, THH, TTH, THT, HTT, TTT\}$

Let A be the event of getting *exactly* 2 heads.

Let B be the event of getting *at least* 2 heads.

Let C be the event of getting *no* heads.

We have $A = \{HHT, HTH, THH\}$, $B = \{HHT, HTH, THH, HHH\}$, $C = \{TTT\}$. Therefore we have

(a) $P(A) = \dfrac{n(A)}{n(S)} = \dfrac{3}{8}$ Remember that $n(A)$ means the number of outcomes in A.

(b) $P(B) = \dfrac{n(B)}{n(S)} = \dfrac{4}{8} = \dfrac{1}{2}$

(c) $P(C) = \dfrac{n(C)}{n(S)} = \dfrac{1}{8}$

Keep in mind what we mean by a probability. If we toss a coin we know that the probability of getting a head is $\frac{1}{2}$. This means that if we toss a coin 100 times, we would expect to get approximately 50 heads. In fact, it is unlikely that we would get *exactly* 50 heads, but very likely that we would get between 45 and 55 heads.

There are many occasions when it is impractical to list the entire sample space. In such situations we often use the counting principles developed in the last section, as illustrated in the next few examples.

Example 3 | Suppose a state lottery game requires a player to choose 5 distinct numbers from 1 to 40. Find the probability of winning this lottery game.

Solution | The sample space for this game is the total number of ways of choosing five numbers out of the forty numbers 1, 2, 3, . . . , 40. Recall from the previous section that this is exactly what we have denoted as $_{40}C_5$—the number of ways of choosing 5 objects out of 40 objects without regard to their order. Thus the sample space contains

$$_{40}C_5 = \frac{40!}{(40-5)! \cdot 5!} = \frac{40!}{35! \cdot 5!} = 658{,}008 \text{ elements}$$

This means there are 658,008 ways of choosing 5 distinct numbers from 1 to 40. Since there is only one winning set of numbers, the probability of winning is $\frac{1}{658{,}008} \approx 0.0000015$.

■

Example 4 | A five-card poker hand is drawn from a standard deck of 52 playing cards. What is the probability of drawing a flush (all 5 cards of the same suit)?

Solution | Let F be the event of drawing a flush and S be the sample space. The experiment in this example consists of drawing five cards, and so the sample space consists of all possible ways in which we can choose 5 cards out of 52 cards. Thus the number of elements in the sample space is

$$n(S) = {}_{52}C_5 = \frac{52!}{5!(52-5)!} = \frac{52!}{5!47!} = 2{,}598{,}960$$

The event of getting a flush can be thought of as a two-step process: choose one suit (hearts, spades, clubs, or diamonds) out of four, and then choose 5 of the 13 cards in that suit. According to the Fundamental Counting Principle, we must multiply the number of ways we can choose the suit, $_4C_1$, by the number of ways we can choose the 5 cards in that suit, $_{13}C_5$. Thus the number of elements in F is

$$n(F) = {}_4C_1 \cdot {}_{13}C_5 = 4 \cdot 1287 = 5148$$

Thus the probability of drawing a flush is

$$P(F) = \frac{n(F)}{n(S)} = \frac{\text{number of ways to draw a flush}}{\text{number of ways to draw a 5-card poker hand}} = \frac{5{,}148}{2{,}598{,}960} \approx 0.002$$

What is the significance of this answer? Since 0.002 is $\frac{2}{1000} = \frac{1}{500}$, this answer means that, on average, if you are dealt 5 cards many, many times, you will get a flush about once in every 500 hands.

■

Example 5   A box of computer disks contains 40 disks, of which six are defective. If two disks are chosen from the box at random, what is the probability that both are not defective?

Solution   The experiment in this example involves choosing two disks out of 40. The sample space consists of the number of ways we can choose two disks out of 40, which is $_{40}C_2$. Since there are six defective disks, there are 34 nondefective disks, and the number of ways of picking two nondefective disks is $_{34}C_2$. Thus the probability of the event D of choosing two nondefective disks is

$$P(D) = \frac{n(D)}{n(S)} = \frac{_{34}C_2}{_{40}C_2} = \frac{561}{780} \approx 0.72$$

Thus we have an approximately 72% chance of choosing two nondefective disks. ∎

Sometimes probabilities are easier to find indirectly. We will find the following idea very useful. The *complement* of an event E, denoted as E', is the set of all outcomes not in E. In other words, we are dividing the sample space into two parts: the outcomes in E, and those not in E (and hence in E'). For example, consider the case of tossing a coin three times. Let E be the event of tossing exactly two heads; then E' is the event of not tossing exactly two heads (which means tossing 0, 1, or 3 heads). See Figure 9.8.

Figure 9.8

Events E and E' always make up the entire sample space.

$$E \qquad\qquad E'$$
$$\{\overline{(THH, HTH, HHT)}\ \overline{(TTT, TTH, THT, HTT, TTT)}\}$$

We recognize that, together, events E and E' always make up the entire sample space S, and therefore the number of outcomes in E plus the number of outcomes in E' equals the total number of outcomes in the sample space. Thus we have

$$n(E) + n(E') = n(S). \qquad \text{Divide both sides by } n(S).$$

$$\frac{n(E)}{n(S)} + \frac{n(E')}{n(S)} = \frac{n(S)}{n(S)} \qquad \text{Using the definition of probability, we have}$$

$$P(E) + P(E') = 1 \qquad \text{and therefore}$$

$$P(E) = 1 - P(E')$$

Thus we have proved the following.

Probability of the Complement of an Event

Let S be the sample space associated with an experiment, and E an event. Then

$$P(E) = 1 - P(E')$$

This says that the probability an event occurs is 1 minus the probability the event does not occur.

Example 6   Suppose we roll a die four times. What is the probability that at least one of the rolls comes up 3?

Solution   Each time we roll a die, there are 6 possible outcomes. Therefore, if we roll the die 4 times, there are $6 \cdot 6 \cdot 6 \cdot 6 = 6^4 = 1296$ outcomes in the sample space S. In other words, $n(S) = 1296$.

If we let E be the event of rolling at least one 3, then E consists of rolling one, two, three, or four 3's. Computing this probability directly is rather complicated. However, using the idea of the complementary event makes this computation significantly easier.

The complementary event E' would mean that none of the rolls come up 3. In order for none of the rolls to come up 3, we must roll a number other than 3 on each of the four rolls. This means that we have five choices (1, 2, 4, 5, 6) on each of the four rolls, and so there are $5^4 = 625$ outcomes comprising event E'. In other words, $n(E') = 625$.

Therefore, using the result about the relationship between the probability of an event and its complement, we get

Note that even though the probability of rolling a three on any particular roll is $\frac{1}{6}$, the probability of rolling at least one 3 on four rolls of a die is slightly greater than $\frac{1}{2}$.

$$P(E) = 1 - P(E') = 1 - \frac{n(E')}{n(S)} = 1 - \frac{625}{1296} \approx 0.52$$

Thus we have a bit more than a 50% chance of rolling at least one 3 when we roll a die four times. ∎

Example 7   Suppose we roll a pair of dice. Compute the probability that we roll a 7 (that is, that the sum of the two dice is 7), and the probability that we roll a total other than 7.

Solution   Let's begin with the sample space. Since there are 6 possibilities on each throw, we know there are $6 \cdot 6 = 36$ outcomes in the sample space.

When we throw a pair of dice we will call one of them the first die and the other one the second die. We will record a throw as an ordered pair such as (3, 5) to mean that the first die came up 3 and the second die came up 5. In this way we can exhibit the sample space as follows:

(1, 1)	(1, 2)	(1, 3)	(1, 4)	(1, 5)	(1, 6)
(2, 1)	(2, 2)	(2, 3)	(2, 4)	(2, 5)	(2, 6)
(3, 1)	(3, 2)	(3, 3)	(3, 4)	(3, 5)	(3, 6)
(4, 1)	(4, 2)	(4, 3)	(4, 4)	(4, 5)	(4, 6)
(5, 1)	(5, 2)	(5, 3)	(5, 4)	(5, 5)	(5, 6)
(6, 1)	(6, 2)	(6, 3)	(6, 4)	(6, 5)	(6, 6)

We can see that there are 6 outcomes for which the roll is a 7. (They are highlighted in the sample space.) Therefore the probability of rolling a 7 is $\frac{6}{36} = \frac{1}{6}$, and the probability of rolling any number other than 7 (which is the complementary event) is $1 - \frac{1}{6} = \frac{5}{6}$. ∎

Many questions about probabilities involve events that are described by more than one condition. For example, we may ask: In drawing a card from a deck, what is the probability of drawing an ace *or* a heart? If two dice are thrown, what is the probability that the first die is even *and* the sum of the two dice is 7? The ideas we introduce here will allow us to answer questions such as these rather directly.

Mutually Exclusive Events

Two events that have no outcomes in common are called *mutually exclusive* events. In other words, it is impossible for mutually exclusive events to occur at the same time. For example, if we toss three coins, then the events

A: We get at least two heads   and B: We get three tails

are mutually exclusive since it is impossible to get three tails and also get at least two heads when you toss a coin three times.

Suppose A and B are mutually exclusive events. Then since A and B have no outcomes in common, we know that

Recall that $A \cup B$ is read as A union B and means the set of elements in A or B or both.

$$n(A \cup B) = n(A) + n(B)$$

This means the number of outcomes in A or B is equal to the number of outcomes in A plus the number of outcomes in B. We can represent this in Figure 9.9.

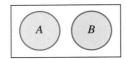

$$\frac{\text{Number of outcomes}}{\text{in } A \text{ or } B} = \text{Number of outcomes in } A + \text{Number of outcomes in } B$$

Figure 9.9

These events A and B are mutually exclusive.

Consequently

$$P(A \cup B) = \frac{n(A \cup B)}{n(S)} = \frac{n(A) + n(B)}{n(S)} = \frac{n(A)}{n(s)} + \frac{n(B)}{n(S)} = P(A) + P(B)$$

and we have proved the following result.

The Probability of Mutually Exclusive Events

If A and B are mutually exclusive events, then the probability of A or B is

$$P(A \cup B) = P(A) + P(B)$$

Example 8   Suppose we toss three coins. What is the probability that we get at least two heads or we get three tails?

Solution   Let's define the events A and B as follows:

A: We get at least two heads   and B: We get three tails

Since these events are mutually exclusive, we know that

$$P(A \text{ or } B) = P(A) + P(B) = \frac{4}{8} + \frac{1}{8} = \frac{5}{8} \qquad \text{See Figure 9.10.}$$

Figure 9.10

The sample space for tossing a coin three times

In an experiment such as the one in Example 8, which has a small sample space, we could just as easily have counted up those outcomes for which there are at least two heads or exactly three tails without separating the events A and B. However, the next example illustrates the usefulness of our result about the probability of mutually exclusive events.

Example 9   Suppose that each question on a 10-question multiple-choice test has 4 choices. If you guess randomly, what is the probability that you will score 90% or higher?

Solution   The sample space consists of all possible ways to answer this 10-question multiple-choice test. Since we have four choices for each question, there are $4^{10} = 1,048,576$ possible outcomes in the sample space (that is, different ways to answer the 10 questions). In order to score at least 90% on the exam, you must answer exactly 9 or exactly 10 questions correctly. Since these are mutually exclusive events, we have

It would be highly impractical to write out the 1,048,576 possible outcomes in the sample space.

$$P(\text{at least } 90\%) = P(9 \text{ correct}) + P(10 \text{ correct})$$

We multiply $_{10}C_9$ by 3 because there are 3 possible choices for the one question we get wrong.

To compute these probabilities, we recognize that the number of ways we can choose the 9 questions answered correctly, which we will denote as $n(9\text{ correct})$, is exactly $_{10}C_9(3)$. Similarly, $n(10\text{ correct}) = {}_{10}C_{10}$. Therefore we have

$$P(\text{at least } 90\%) = P(9\text{ correct}) + P(10\text{ correct})$$

$$= \frac{n(9\text{ correct})}{n(S)} + \frac{n(10\text{ correct})}{n(S)}$$

$$= \frac{{}_{10}C_9(3)}{n(S)} + \frac{{}_{10}C_{10}}{n(S)}$$

$$= \frac{30}{1,048,576} + \frac{1}{1,048,576} \approx 0.0000296$$

This means that if you guess randomly, you have about 3 chances in 100,000 of getting at least 9 correct.

The Probability of the Union of Two Events

Let's consider the following situation of two events that are not mutually exclusive.

Example 10   What is the probability that a card drawn from a standard deck will be either a heart or a picture card?

Solution   Let A be the event of drawing a heart and B be the event of drawing a picture card. The example is asking us to compute $P(A \cup B)$; however, we note that these events are not mutually exclusive. It is possible for the card to be both a heart and a picture card. Figure 9.11 shows a diagram of the sets A and B.

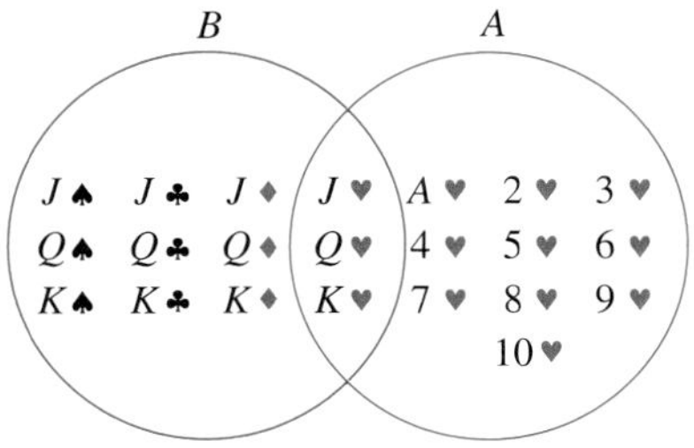

Looking at the figure, we can see that $n(A \cup B) = 22$ and therefore

$$P(A \cup B) = \frac{n(A \cup B)}{n(S)} = \frac{22}{52} = \frac{11}{26}$$

Figure 9.11

Examining the figure again, we see that there are 13 outcomes in event A [that is, $n(A) = 13$], and there are 12 outcomes in event B [that is, $n(B) = 12$]. Therefore, we have

$$P(A) + P(B) = \frac{n(A)}{n(S)} + \frac{n(B)}{n(S)} = \frac{13}{52} + \frac{12}{52} = \frac{25}{52}$$

However, this result is not $P(A \cup B)$ because as Figure 9.11 illustrates, we have counted those outcomes that are common to both A and B twice. The outcomes common to both A and B are denoted as $A \cap B$. Since $n(A \cap B) = 3$, we have $P(A \cap B) = \frac{3}{52}$.

Therefore, to compute $P(A \cup B)$, we must subtract what we have counted twice; that is, we must subtract $P(A \cap B) = \frac{3}{52}$. Thus we have

$$P(A \cup B) = \frac{13}{52} + \frac{12}{52} - \frac{3}{52} = \frac{22}{52} = \frac{11}{26}$$

The previous example generalizes to the following result.

The Probability of the Union of Two Events
If A and B are two events, then the probability of A *or* B is

$$P(A \cup B) = P(A) + P(B) - P(A \cap B)$$

What happens to this formula if A and B are mutually exclusive?

Example 11

Suppose we roll a pair of fair dice. Find
(a) The probability that the first die is 6, or the sum of the dice is 7 or 8.
(b) The probability that the sum of the dice is at least 10.

Solution

Let's begin with the sample space. As we did with Example 7, we record a throw as an ordered pair such as (3, 5) to mean that the first die came up 3 and the second die came up 5. We exhibit the sample space as follows.

B

(1, 1)	(1, 2)	(1, 3)	(1, 4)	(1, 5)	(1, 6)
(2, 1)	(2, 2)	(2, 3)	(2, 4)	(2, 5)	(2, 6)
(3, 1)	(3, 2)	(3, 3)	(3, 4)	(3, 5)	(3, 6)
(4, 1)	(4, 2)	(4, 3)	(4, 4)	(4, 5)	(4, 6)
(5, 1)	(5, 2)	(5, 3)	(5, 4)	(5, 5)	(5, 6)
(6, 1)	(6, 2)	(6, 3)	(6, 4)	(6, 5)	(6, 6)

A

Compute the probability of $A \cup B$ as $\dfrac{n(A \cup B)}{n(S)}$.

(a) In the diagram of the sample space, we have highlighted event *A*, which is that the first die is 6, and event *B*, which is that the sum of the dice is 7 or 8. We can see that *A* has 6 elements and *B* has 11 elements, and that *A* and *B* are not mutually exclusive. Therefore,

$$P(A) = \frac{6}{36} \qquad P(B) = \frac{11}{36} \qquad P(A \cap B) = \frac{2}{36}$$

and so

$$P(A \cup B) = P(A) + P(B) - P(A \cap B) = \frac{6}{36} + \frac{11}{36} - \frac{2}{36} = \frac{15}{36} = \frac{5}{12}$$

(b) If the sum of the two dice is at least 10, then the sum must be 10 or 11 or 12. Since these events are mutually exclusive, we have

$$P(\text{at least } 10) = P(10) + P(11) + P(12) \qquad \text{Using the diagram of the sample space, we get}$$

$$= \frac{3}{36} + \frac{2}{36} + \frac{1}{36} = \frac{6}{36} = \frac{1}{6}$$

Conditional Probability

Sometimes how we compute a probability is affected by the information we have. For example, suppose we pick a person at random and want to know the probability that he or she owns stock. If we know additionally that this person earns an annual income of $100,000, we might very well revise our estimate of the probability to a higher value.

Let's consider another example. Suppose we draw a card from a standard 52-card deck. Consider the following events:

A: the card is a jack

B: the card is a picture card

The probability that the card is a jack is $\frac{4}{52} = \frac{1}{13}$. However, suppose we know that the card is a picture card, and then ask about the probability that the card is a jack. Since there are only 12 picture cards, the probability that the card is a jack is $\frac{4}{12} = \frac{1}{3}$.

By knowing that the card is a picture card, we are in effect considering the probability that the card is a jack in a *restricted sample space*. That is why we are dividing by 12 (the number of picture cards) rather than by 52 (the total number of cards).

In general, we are describing the probability of an event A *given* that event B has occurred. We denote this as $P(A \mid B)$, and it is called the *conditional probability of A given B*. Based upon this discussion, we have the following result.

The Conditional Probability of *A* Given *B*

If A and B are two events, then the probability of A given that B has occurred is

$$P(A \mid B) = \frac{n(A \cap B)}{n(B)}$$

Example 12 | In a class of 20 students, 6 have no siblings, 8 have a brother and a sister, 4 have a brother only, and 2 have a sister only. If a student is chosen at random, find the probability of each of the following events.

(a) The student has a brother.
(b) The student has a brother, given that the student has a sister.
(c) The student has a brother, given that the student has a sibling.

Solution | Let's define the following events:

B: The student has a brother.   We have $n(B) = 8 + 4 = 12$.
F: The student has a sister.   We have $n(F) = 8 + 2 = 10$.
S: The student has a sibling.   We have $n(S) = 8 + 4 + 2 = 14$.

(a) There are 20 students in the class, 12 of whom have a brother (8 have a brother and sister, and 4 have a brother only). Therefore,

$$P(B) = \frac{12}{20} = \frac{3}{5}$$

(b) We are being asked to find the probability that the student has a brother, given that the student has a sister. Since 10 students have a sister and, of these, 8 have a brother, clearly this probability is $\frac{8}{10}$. The conditional probability formula gives us

$$P(B \mid F) = \frac{n(B \cap F)}{n(F)} = \frac{8}{10} = \frac{4}{5}$$

(c) We are being asked to find the probability that the student has a brother, given that the student has a sibling. Thus we want to find

$$P(B \mid S) = \frac{n(B \cap S)}{n(S)} = \frac{12}{14} = \frac{6}{7}$$

The Probability of the Intersection of Two Events

It turns out that the definition of conditional probability allows us to derive a formula for the probability that event A *and* event B both occur:

$$P(B \mid A) = \frac{n(A \cap B)}{n(A)} \qquad \text{Divide numerator and denominator by } n(S).$$

$$= \frac{\dfrac{n(A \cap B)}{n(S)}}{\dfrac{n(A)}{n(S)}} \qquad \text{By the definition of probability this becomes}$$

$$P(B \mid A) = \frac{P(A \cap B)}{P(A)} \qquad \text{Multiply both sides by } P(A).$$

$$P(B \mid A) \cdot P(A) = P(A \cap B)$$

Thus we have derived the following formula.

The Probability of the Intersection of Two Events
If A and B are two events, then the probability of A *and* B is

$$P(A \cap B) = P(A) \cdot P(B \mid A)$$

Example 13 Suppose two cards are drawn, without replacement, from a standard deck of 52 playing cards. Find the probability that the first card is an ace and the second card is a picture card.

Solution Let A be the event that the first card is an ace, and B be the event that the second card is a picture card. The probability of drawing an ace on the first card is $P(A) = \frac{4}{52} = \frac{1}{13}$. Since there are now 51 cards remaining in the deck, the probability that the second card is a picture card given that the first card drawn is an ace is $P(B \mid A) = \frac{12}{51}$ (there are 12 picture cards in the remaining deck of 51 cards). Therefore, we have

$$P(A \cap B) = P(A) \cdot P(B \mid A) = \frac{1}{13} \cdot \frac{12}{51} = \frac{12}{663} \approx 0.018$$

Independent Events

Suppose we toss a coin and then throw a die. Let A be the event that the coin comes up heads, and B be the event that the die comes up 5. If we want to know the probability that the coin comes up a head *and* the die comes up 5, then we are seeking $P(A \cap B)$.

We exhibit the sample space S for this experiment in Figure 9.12.

Figure 9.12.

The sample space S for tossing a coin and rolling a die

$(H, 1)$	$(H, 2)$	$(H, 3)$	$(H, 4)$	$(H, 5)$	$(H, 6)$
$(T, 1)$	$(T, 2)$	$(T, 3)$	$(T, 4)$	$(T, 5)$	$(T, 6)$

We can see that the sample space has 12 outcomes and that only one satisfies the condition of tossing a head and rolling a 5. Therefore the probability of tossing a head and rolling a 5 is $\frac{1}{12}$.

Note that $P(A) = P(\text{Head}) = \frac{1}{2}$ and $P(B) = P(\text{roll a 5}) = \frac{1}{6}$. If we multiply these two probabilities we get $\frac{1}{2} \cdot \frac{1}{6} = \frac{1}{12}$, which is $P(A \cap B)$.

This is not a coincidence. If events A and B are *independent*, which means that the occurrence of A does not affect the probability of B (and vice versa), then the probability that events A and B both occur is the product of the probabilities that A and B occur individually.

In other words, events A and B are independent if and only if $P(A \mid B) = P(A)$ and $P(B \mid A) = P(B)$. Consequently, for *independent events*, the formula for the probability of A *and* B that we stated above becomes the following.

The Probability of the Intersection of Independent Events

If A and B are two independent events, then the probability of A *and* B is

$$P(A \cap B) = P(A) \cdot P(B)$$

Example 14

A state lottery has a daily "numbers" game that requires you to choose a three-digit number. What is the probability that the winning number will have three different digits?

Solution

The first digit can be any of the digits from 0 to 9, so the probability of the first digit is $\frac{10}{10} = 1$. In order for the second digit to be different from the first, it must be one of the remaining 9 digits, so the probability the second digit is different is $\frac{9}{10}$. Similarly, in order for the third digit to be different from the first two, it must be one of the remaining 8 digits; thus the probability the third digit is different is $\frac{8}{10}$. Since the digits are chosen independently, the probability the three digits are different is

$$1 \cdot \frac{9}{10} \cdot \frac{8}{10} = \frac{72}{100} = 0.72$$

This means that the winning number will have three different digits about 72% of the time. ∎

Example 15

Suppose you are rolling a pair of dice. Find the probability that you roll two consecutive 8's.

Solution

Each time we roll a pair of dice, the result is independent of what we have rolled previously. As we see from the illustration of the sample space in Example 11, the probability of rolling an 8 is $\frac{5}{36}$. Therefore, the probability of rolling two consecutive 8's is

$$P(\text{rolling two 8's}) = \frac{5}{36} \cdot \frac{5}{36} = \frac{25}{1296} \approx 0.02$$

$0.02 = \frac{2}{100} = \frac{1}{50}$ In other words, there is about 1 chance in 50 to roll two consecutive 8's. ∎

The probability ideas we have presented apply to a wide variety of situations as illustrated in the exercises. As you do the exercises, make sure you understand the situation described in terms of whether there is replacement or not, whether order matters or not, whether the events are independent or not, and whether you are seeking A *and* B or A *or* B.

9.7 Exercises

In Exercises 1–10, write out the sample space for each experiment.

1. The following spinner is spun twice.

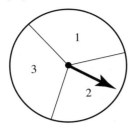

2. The spinner of Exercise 1 is spun three times.

3. A die is rolled and then a coin is tossed.

4. A die is rolled and then a coin is tossed twice.

5. A bag contains five balls numbered 1, 2, 3, 4, and 5. A ball is selected and then a second ball is chosen.

6. Repeat Exercise 5 if the first ball is replaced before the second ball is chosen.

7. A student answers four true/false questions randomly.

8. A department chairperson chooses one of three classrooms, 101, 102, or 103, and assigns one of three teachers, Prof. Lu, Prof. Johnson, or Prof. Gold, to each of the rooms.

9. A committee of two people is to be selected from six county legislators, *A*, *B*, *C*, *D*, *E*, and *F*.

10. A salesperson must decide which two cities to visit from a list of five cities, *A*, *B*, *C*, *D*, and *E*.

11. A coin is tossed three times.
   (a) Find the probability of getting exactly one tail.
   (b) Find the probability of at least one tail.
   (c) Find the probability of no tails.

12. A die is rolled and then a coin is tossed.
   (a) Find the probability of getting an even number *and* a head.
   (b) Find the probability of getting an even number *or* a head.
   (c) Find the probability of getting a number less than 3 and a tail.

13. A die is rolled twice.
   (a) Find the probability of getting at least one 6.
   (b) Find the probability the sum of the two dice is 8.
   (c) Find the probability that both rolls are not the same.

14. A die is rolled twice.
   (a) Find the probability the sum of the two dice is odd.

(b) Find the probability one of the dice is odd and the other is even.
   (c) Find the probability that the first roll is greater than the second.

15. A card is drawn at random from a standard 52-card deck. Find the probability of the given event.
   (a) The card is an ace.
   (b) The numerical value of the card is less than 5.
   (c) The numerical value of the card is not less than 5.

16. A card is drawn at random from a standard 52-card deck. Find the probability of the given event.
   (a) The card is a queen or a king.
   (b) The card is a heart or a king.
   (c) The card is neither a heart nor a king.

17. A jar contains 8 red marbles, 7 black marbles, and 5 white marbles. A marble is drawn randomly from the jar. Find the probability of the given event.
   (a) The marble is black.
   (b) The marble is not white.
   (c) The marble is red or white.

18. A suitcase contains 16 loose socks: 4 pairs are identical black socks, 3 pairs are identical blue socks, and 1 pair is brown.
   (a) Find the probability a sock drawn randomly is blue.
   (b) If one sock is drawn and it is black, what is the probability that a second sock drawn randomly will be brown?
   (c) If one sock is drawn and it is blue, what is the probability that a second sock drawn randomly will make a matching blue pair?

19. The letters of the word PROBABILITY are written on individual slips of paper.
   (a) If one slip of paper is chosen randomly, what is the probability that it is the letter *L*?
   (b) If one slip of paper is chosen randomly, what is the probability that it is the letter *P* or *I*?
   (c) If one slip of paper is chosen randomly, what is the probability that it is a consonant?

20. The letters of the word PROFESSOR are written on individual slips of paper.
   (a) If one slip of paper is chosen randomly, what is the probability that it is the letter *F*?
   (b) If one slip of paper is chosen randomly, what is the probability that it is the letter *P* or *R*?
   (c) If one slip of paper is chosen randomly, what is the probability that it is a vowel?

21. The accompanying spinner is used in a board game. Find the probability of the given event.

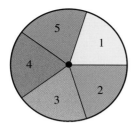

(a) The spinner stops on an odd number.
(b) The spinner stops on a number other than 1 or 2.
(c) The spinner stops on an even number or a number less than 5.

22. The accompanying spinner is used in a board game. Find the probability of the given event.

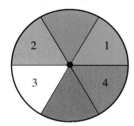

(a) The spinner stops on a color other than red.
(b) The spinner stops on a number other than 3.
(c) The spinner stops on a number or a color.

23. A state lottery game requires you to choose a four-digit number.
(a) What is the probability that the winning number will have four different digits?
(b) What is the probability that the winning number will have the same digit repeated 4 times? Is this the complementary event to part (a)?

24. A state lottery game requires you to choose a three-digit number.
(a) What is the probability that the winning number will be even?
(b) What is the probability that the winning number will start with 5 and be odd?
(c) What is the probability that the winning number does not have an 8 in it?

In Exercises 25–36, a five-card poker hand is dealt from a standard 52-card deck. Find the probability that the poker hand contains the indicated cards. Round off your answer to seven decimal places.

25. Five clubs
26. A straight (five cards in consecutive order; e.g., 3, 4, 5, 6, and 7 regardless of suit)

27. A flush (five cards of the same suit)
28. A royal flush (10, jack, queen, king, and ace in the same suit)
29. Four of a kind (four jacks or four 7's, etc.)
30. Five picture cards
31. Exactly three of a kind
32. A full house (three of one kind and two of another kind)
33. Two pair (two of one kind and two of a different kind)
34. Exactly one pair
35. At least one spade
36. At most four clubs [*Hint*: See Exercise 25.]
37. The New York State Lotto game requires the winner to choose six different numbers from 1 to 56. What is the probability of winning the lottery? Give your answer as "one chance in how many."
38. The New York State Lotto game formerly required the winner to choose six different numbers from 1 to 48. What was the probability of winning this lottery? Give your answer as "one chance in how many."
39. A company categorizes its employees by salary in the following table.

Salary	Number of Employees
$19,999 or less	26
Between $20,000 and $29,999	43
Between $30,000 and $39,999	17
Between $40,000 and $49,999	14
$50,000 or above	28

(a) What is the probability that a random employee earns less than $20,000?
(b) What is the probability that a random employee earns less than $40,000?
(c) What is the probability that a random employee earns at least $50,000?
40. A company categorizes its employees by years of service in the following table.

Years of Service	Number of Employees
0–5	34
6–10	57
11–15	60
16–20	48
More than 20	26

(a) What is the probability that a random employee has worked for the company less than 6 years?

(b) What is the probability that a random employee has worked for the company at least 6 years?

(c) What is the probability that a random employee has been employed by the company between 11 and 20 years?

41. A psychology experiment was designed to test a subject for ESP (extrasensory perception). Four cards—a jack, queen, king, and ace—are shuffled and placed face down in front of the subject, who is then asked to identify each card.

(a) If the subject simply guesses randomly, what is the probability of guessing all the cards correctly?

(b) If the subject simply guesses randomly, what is the probability of guessing only one card incorrectly?

(c) If the subject simply guesses randomly, what is the probability of guessing two of the four incorrectly?

42. Repeat Exercise 41 if the experiment is done with five different cards.

43. A study of the effectiveness of a "heartburn" medication was conducted with 600 subjects who were asked to eat food that normally gives them heartburn. Some in the study were given one dose of the medication, some were given 2 doses, and some were given no medication. The following table contains the results of the study.

	No Medication	One Dose	Two Doses	Total
Heartburn	102	43	28	173
No Heartburn	18	197	212	427
Total	120	240	240	600

(a) What is the probability that a randomly chosen subject had one dose of medication and no heartburn?

(b) What is the probability that a randomly chosen subject had no medication and no heartburn?

(c) What is the probability that a randomly chosen subject had two doses of medication and had heartburn?

44. A political polling service interviews 500 voters and asks them whether they think that the method of funding political campaigns needs to be reformed. The following table contains the results of the survey.

	Democrats	Republicans	Independents	Total
Reform	115	93	61	269
No Reform	106	99	26	231
Total	221	192	87	500

(a) What is the probability that a randomly chosen person surveyed is a Republican who favors campaign reform?

(b) What is the probability that a randomly chosen person surveyed favors reform?

(c) What is the probability that a randomly chosen person surveyed is a Democrat or Independent who opposes reform?

45. A jar contains 20 white poker chips, 15 blue poker chips, and 10 red poker chips. If you randomly choose 5 chips, find the following probabilities.
(a) All 5 chips are blue.
(b) At least one chip is blue.
(c) Exactly one chip is red.

46. A jar contains 20 white poker chips, 15 blue poker chips, and 10 red poker chips. If you randomly choose 6 chips, find the following probabilities.
(a) 3 chips are white and 3 chips are red.
(b) All 6 chips are white.
(c) None of the chips are blue.

47. Suppose we interpret a baseball player's lifetime batting average of 0.289 as his probability of getting a hit during any at bat.
(a) Find the probability the player will go "4 for 4"—that is, get 4 hits in 4 at bats.
(b) Find the probability the player will get no hits in 4 at bats.
(c) Find the probability the player will get at least 1 hit in 4 at bats.

48. Suppose we interpret a basketball player's lifetime free-throw percentage of 0.81 as her probability of making any particular free throw.

(a) Find the probability the player will make all 5 free throws she takes in a particular game.

(b) Find the probability the player will make none of the 5 free throws she takes during a particular game.

(c) Find the probability the player will make at least 4 out of the 5 free throws she takes during a particular game.

49. Suppose you toss a coin 4 times. Find the probability of each of the following events.
(a) You get all heads or all tails.
(b) You get at least 3 heads.
(c) You get exactly 1 head or exactly 1 tail.

50. Suppose you toss a coin 4 times. Find the probability of each of the following events.
(a) You get exactly 3 heads or at least 1 tail.
(b) You get 4 heads or at most 2 tails.
(c) You get exactly 2 heads in a row or 2 tails in a row.

51. Suppose you roll a pair of dice. Find the probability of each of the following events.
(a) You roll a 10, 11, or 12.
(b) You roll less than 5.
(c) You roll a 7 or 11.

52. Suppose you roll a pair of dice. Find the probability of each of the following events.
(a) You roll an odd number or at least a 10.
(b) You roll a double (both dice the same) or at least a 7.
(c) You roll a double or an even number.

53. Suppose you are drawing cards from a standard deck of 52 cards. Find the probability of each of the following events.
(a) You draw two picture cards.
(b) The second card you draw is a picture card given that the first card is a picture card.
(c) The second card you draw is a picture card given that the first card is not a picture card.

54. Suppose you are drawing cards from a standard deck of 52 cards. Find the probability of each of the following events.
(a) You draw two aces.
(b) The second card you draw is an ace given that the first card is an ace.
(c) The second card you draw is an ace given that the first card is not an ace.

55. A card is drawn from a standard deck of 52 cards. Find the probability of each of the following events.
(a) The card is a picture card or a diamond.
(b) The card is a king or a spade.
(c) The card is an odd-numbered card or a black card. (Count an ace as 1.)

56. A card is drawn from a standard deck of 52 cards. Find the probability of each of the following events.
(a) The card is a non-picture card or a heart.
(b) The card is an ace or a club.
(c) The card is a picture card or a red card.

57. If you randomly arrange the letters C, E, I, P, R, T, U, what is the probability that you will arrange them into the word PICTURE?

58. If you randomly arrange the letters A, A, C, D, E, L, N, R, what is the probability that you will arrange them into the word CALENDAR?

59. A jar contains 18 red marbles, 12 black marbles, and 10 green marbles. Find the probability of each of the following events.
(a) You select 3 marbles and get one of each color.
(b) You select 3 marbles and get all 3 of the same color.
(c) You select a marble and replace it 3 times, and get one of each color.

60. A jar contains 10 white balls numbered 1 through 10, and 7 blue balls numbered from 1 to 7. Find the probability of each of the following events.
(a) You select two balls and they are both white and even.
(b) You select two balls, and both are white or both are even.
(c) You select two balls and both are blue, but one is odd and one is even.

61. In American roulette, a ball is spun on a wheel and has an equal chance of landing in any of the 38 slots numbered 0, 00, 1, 2, . . . , 36. The slots numbered 0 and 00 are green; odd-numbered slots are red; even-numbered slots are black. Find the probability of each of the following.
(a) The ball lands red.
(b) The ball lands on an odd number.
(c) The ball comes up 18 or 36.
(d) The wheel is spun twice and the ball comes up green twice.

62. Suppose the roulette wheel is spun 10 times. What is the probability that the numbers 12, 13, 14, 15, 16, and 17 do not come up on any of the spins?

63. A compact disk manufacturer knows that the probability a CD will be defective is 0.003.
   (a) If you select a batch of 10 CDs, what is the probability that at most 1 will be defective?
   (b) If you select a batch of 100 CDs, what is the probability they will all be nondefective?

64. A pack of 20 compact disks is known to have 3 defective disks.
   (a) If a customer buys 5 of these disks, what is the probability that none is defective?
   (b) If a customer buys 5 of these disks, what is the probability that she has selected all 3 defective disks?

65. Four letters addressed to four different people are randomly placed in their mailboxes, one letter per mailbox. What is the probability that each of the letters gets to its intended recipient?

66. A teacher gives her class a study sheet containing 10 exam problems, from which 6 exam questions will be chosen. If Noreen knows how to solve 8 of the 10 study problems, what is the probability that she will solve 5 out of the 6 exam problems correctly?

Questions for Thought

67. Explain why the closer a probability is to 0, the less likely the event is to occur, and the closer a probability is to 1, the more likely it is to occur.

68. In Example 5 we found that the probability of choosing two nondefective disks is approximately 72%. Given the fact that we are starting with $\frac{34}{40} = 0.85 = 85\%$ nondefective disks, do you find this answer of 72% surprising? Explain.

69. In Example 6 we computed the probability of rolling at least one 3 when rolling a die 4 times by using the probability of the complementary event. Compute this same probability directly and compare the amount of computation involved.

70. Compare the answers to Exercises 37 and 38. Describe how your chances of winning have changed under the new lottery format.

71. Probabilities are often given as *odds*. If your odds of winning are 2 to 1, this means that your probability of winning is $\frac{2}{3}$ and of losing is $\frac{1}{3}$. In general, if the odds of an event happening are a to b, then the probability of the event happening is $\frac{a}{a+b}$. Using our definition of probability, how would you explain this process of changing odds into probability?

72. If A and B are mutually exclusive events, then what is $P(A \mid B)$?

73. If A and B are mutually exclusive events, then what is $P(B \mid A)$?

74. Can mutually exclusive events be independent? Explain.

75. Each time we toss a coin, the probability of getting a head is $\frac{1}{2}$. Each toss is independent of the previous tosses. If you were to toss a fair coin five times and got five heads, what is the probability of getting a head on the sixth toss?

9.8 The Binomial Theorem

Recall that a sum of the form $A + B$ is called a binomial. In Chapter 1, we discussed special products of the form $(A + B)^2$ and $(A + B)^3$. In this section we develop a general formula called the **Binomial Theorem** for raising a binomial to a power.

We begin by actually multiplying out, or *expanding,* $(A + B)^n$ for $n = 0, 1, 2, 3, 4,$ and 5.

$(A + B)^n$	Expanded form
For $n = 0$:	$(A + B)^0 = 1$
For $n = 1$:	$(A + B)^1 = A + B$
For $n = 2$:	$(A + B)^2 = A^2 + 2AB + B^2$
For $n = 3$:	$(A + B)^3 = A^3 + 3A^2B + 3AB^2 + B^3$
For $n = 4$:	$(A + B)^4 = A^4 + 4A^3B + 6A^2B^2 + 4AB^3 + B^4$
For $n = 5$:	$(A + B)^5 = A^5 + 5A^4B + 10A^3B^2 + 10A^2B^3 + 5AB^4 + B^5$

Let's examine the bottom row of this table. We note that the terms (as we have arranged them) begin with A^5, which we can view as A^5B^0. In the subsequent terms, the exponent of A decreases by one each time, and the exponent of B increases by one each time. Also note that the sum of the exponents in each term is 5. We see this same pattern in the other rows of the table as well. It seems reasonable to believe that this pattern will hold in general; in fact, this is the case.

Let's now turn our attention to the coefficients and see whether we can understand how they can be computed. Keep in mind that

$$(A + B)^5 = (A + B)(A + B)(A + B)(A + B)(A + B)$$

There are five binomials and the terms in the expanded form are the result of all possible products of the terms in the binomials. For example, to get A^3B^2 in the expansion of $(A + B)^5$, we must multiply the A-term from three of the binomials with the B-term from the other two binomials. This can be done in exactly $\binom{5}{3}$ ways—the number of ways we can choose 3 out of the 5 binomials from which to get the factor A. We could equally well reason that this can be done in exactly $\binom{5}{2}$ ways—the number of ways we can choose 2 out of the 5 binomials from which to get the factor B. We leave it to the reader to verify that $\binom{5}{3} = \binom{5}{2} = 10$.

Generalizing this argument to $(A + B)^n$, we conclude that the coefficient of the $A^{n-k}B^k$-term is $\binom{n}{k}$ or $\binom{n}{n-k}$. Recall that in Example 11 of Section 9.6 we proved algebraically that $\binom{n}{k} = \binom{n}{n-k}$.

Summarizing this discussion leads to the next theorem.

The Binomial Theorem

$$(A + B)^n = A^n + \binom{n}{1}A^{n-1}B + \binom{n}{2}A^{n-2}B^2 + \cdots$$

$$+ \binom{n}{k}A^{n-k}B^k + \cdots + \binom{n}{n-1}AB^{n-1} + B^n$$

Note that we normally don't write the first and last coefficients in $\binom{n}{0}$ or $\binom{n}{n}$ form, but rather simply write them as 1. A formal proof of the Binomial Theorem is outlined in Exercise 53.

It is because of the Binomial Theorem that $\binom{n}{k}$ is often called a *binomial coefficient.*

Example 1   Use the Binomial Theorem to expand $(x + y)^8$.

Solution   By the Binomial Theorem we have

$$(x + y)^8 = x^8 + \binom{8}{1}x^7y + \binom{8}{2}x^6y^2 + \binom{8}{3}x^5y^3 + \binom{8}{4}x^4y^4$$
$$+ \binom{8}{5}x^3y^5 + \binom{8}{6}x^2y^6 + \binom{8}{7}xy^7 + y^8$$

Computing the binomial coefficients we have

$$\binom{8}{1} = \frac{8}{1} = 8 \qquad\qquad \binom{8}{5} = \binom{8}{3} = 56$$

$$\binom{8}{2} = \frac{8 \cdot 7}{2 \cdot 1} = 28 \qquad\qquad \binom{8}{6} = \binom{8}{2} = 28$$

$$\binom{8}{3} = \frac{8 \cdot 7 \cdot 6}{3 \cdot 2 \cdot 1} = 56 \qquad\qquad \binom{8}{7} = \binom{8}{1} = 8$$

$$\binom{8}{4} = \frac{8 \cdot 7 \cdot 6 \cdot 5}{4 \cdot 3 \cdot 2 \cdot 1} = 70$$

Note that we have made use of the fact that $\binom{n}{k} = \binom{n}{n-k}$. Our final answer is

$$(x + y)^8 = x^8 + 8x^7y + 28x^6y^2 + 56x^5y^3 + 70x^4y^4 + 56x^3y^5 + 28x^2y^6 + 8xy^7 + y^8 \quad\blacksquare$$

Example 2   Expand $(2x - 3)^4$ using the Binomial Theorem.

Solution   We view $(2x - 3)^4$ as $(2x + (-3))^4$, so according to the Binomial Theorem we have

$$(2x - 3)^4 = (2x)^4 + \binom{4}{1}(2x)^3(-3) + \binom{4}{2}(2x)^2(-3)^2 + \binom{4}{3}(2x)(-3)^3 + (-3)^4$$

$$= 16x^4 + 4(8)(-3)x^3 + 6(4)(9)x^2 + 4(2)(-27)x + 81$$

$$= 16x^4 - 96x^3 + 216x^2 - 216x + 81$$

Notice that because of the fact that the exponents alternate from even to odd, the signs in the expansion of $(A - B)^n$ will alternate from positive to negative. $\quad\blacksquare$

There is another way to compute binomial coefficients. We write out the binomial coefficients of $(A + B)^n$ for $n = 0, 1, 2, 3, 4, 5, 6$, and tabulate them in the following triangular array.

$$
\begin{array}{c}
n = 0 \qquad\qquad\qquad\qquad\quad 1 \\
n = 1 \qquad\qquad\qquad\qquad 1 \quad 1 \\
n = 2 \qquad\qquad\qquad 1 \quad 2 \quad 1 \\
n = 3 \qquad\qquad 1 \quad 3 \quad 3 \quad 1 \\
n = 4 \qquad 1 \quad 4 \quad 6 \quad 4 \quad 1 \\
n = 5 \quad 1 \quad 5 \quad 10 \quad 10 \quad 5 \quad 1 \\
n = 6 \quad 1 \quad 6 \quad 15 \quad 20 \quad 15 \quad 6 \quad 1
\end{array}
$$

If we examine the table carefully we can see that each row begins and ends with a 1 and that each of the other entries in a row can be obtained by adding the two numbers diagonally above it. Thus $10 = 6 + 4$ and $15 = 5 + 10$, as indicated by the triangles in the table.

The triangular array is called *Pascal's triangle,* named for the famous seventeenth-century French mathematician Blaise Pascal (1623–1662). We can use Pascal's triangle to generate the binomial coefficients for higher values of n by simply adding rows to the table, as described.

Example 3 | Expand $(x^2 + y^3)^7$.

Solution | From the Binomial Theorem we know the pattern of the powers. We may compute the binomial coefficients by using the formula for $\binom{n}{k}$ or we may generate the line for $n = 7$ in Pascal's triangle. We choose the latter approach.

Using the row for $n = 6$ in the table, we have

$$
\begin{array}{ccccccccc}
n = 6 & 1 & & 6 & & 15 & & 20 & & 15 & & 6 & & 1 \\
n = 7 & 1 & & 7 & & 21 & & 35 & & 35 & & 21 & & 7 & & 1
\end{array}
$$

Using these coefficients, we get

$$
\begin{aligned}
(x^2 + y^3)^7 &= (x^2)^7 + 7(x^2)^6 y^3 + 21(x^2)^5(y^3)^2 + 35(x^2)^4(y^3)^3 \\
&\quad + 35(x^2)^3(y^3)^4 + 21(x^2)^2(y^3)^5 + 7(x^2)(y^3)^6 + (y^3)^7 \\
&= x^{14} + 7x^{12}y^3 + 21x^{10}y^6 + 35x^8 y^9 + 35x^6 y^{12} + 21x^4 y^{15} + 7x^2 y^{18} + y^{21}
\end{aligned}
$$

For relatively small values of n, Pascal's triangle offers a convenient way to generate the binomial coefficients. However, if we want to compute the coefficient of a specific term in the binomial expansion for a fairly large value of n, it can be rather time-consuming to generate all the rows of Pascal's triangle until we get to the row we need, as the following example illustrates.

Example 4 | Find the coefficient of x^8 in the expansion of $(\sqrt{x} - 3)^{20}$.

Solution | Since we are dealing with a question about the exponent of x, it seems reasonable to expect that writing $\sqrt{x}$ as $x^{1/2}$ may be helpful. A typical term in the expansion of $(x^{1/2} - 3)^{20}$ will be

$$
\binom{n}{k}(x^{1/2})^k (-3)^{n-k}
$$

If we are looking for the term containing x^8, then we need

$$(x^{1/2})^k \doteq x^8$$
$$x^{k/2} = x^8 \Rightarrow k = 16$$

Using the formula for the general term of the binomial expansion with $n = 20$ and $k = 16$, we get

$$\binom{20}{16}(x^{1/2})^{16}(-3)^4$$

To find the binomial coefficient $\binom{20}{16}$ or its equivalent, $\binom{20}{4}$, from Pascal's triangle would require us to complete the table all the way down to the row for $n = 20$. It seems much more efficient to directly compute the binomial coefficient we need:

$$\binom{20}{16} = \binom{20}{4} = \frac{20 \cdot 19 \cdot 18 \cdot 17}{4 \cdot 3 \cdot 2 \cdot 1} = 4845$$

Therefore, the term corresponding to $k = 16$ is

$$4845(x^{1/2})^{16}(-3)^4 = 4845(81)x^8 = 392{,}445x^8$$

and so the coefficient of x^8 is $\boxed{392{,}445}$.

9.8 Exercises

In Exercises 1–30, expand the given binomial and simplify.

1. $(x - 4)^3$

2. $(t + r)^5$

3. $(a + 2)^7$

4. $(y - z)^4$

5. $(2x + y)^6$

6. $(x + 2y)^6$

7. $(t - 3r)^5$

8. $(3t - r)^5$

9. $(w^2 + 1)^8$

10. $(1 - w^2)^7$

11. $(x^3 - y^3)^4$

12. $(x^2 + y^2)^3$

13. $(a^2 + 2b)^9$

14. $(c - 3d^2)^8$

15. $\left(x - \dfrac{1}{2}\right)^4$

16. $\left(y + \dfrac{1}{3}\right)^3$

17. $\left(\dfrac{x}{3} + 4\right)^3$

18. $\left(\dfrac{a}{4} - 3\right)^4$

19. $(x - \sqrt{x})^6$

20. $(2x - \sqrt{5})^4$

21. $(\sqrt{x} + \sqrt{y})^6$

22. $(\sqrt{a} - \sqrt{3})^5$

23. $\left(x - \dfrac{1}{x}\right)^5$

24. $\left(x - \dfrac{1}{\sqrt{x}}\right)^6$

25. $\left(\dfrac{x}{y} + \dfrac{y}{x}\right)^4$

26. $\left(\dfrac{2}{x} - \dfrac{x}{2}\right)^3$

27. $(x^{-1} - y^2)^5$

28. $(x + y^{-2})^4$

29. $(2r^{-1} + s^{-2})^6$

30. $(a^2 + 5b^{-2})^3$

31. What is the coefficient of x^6 in the expansion of $(x + 5)^{10}$?

32. What is the coefficient of x^6 in the expansion of $(x^2 + 5)^{10}$?

33. What is the coefficient of the term containing a^5 in the expansion of $(a - 3b)^7$?

34. What is the coefficient of term containing s^4 in the expansion of $(2r - 3s)^{10}$?

35. What is the coefficient of z^5 in the expansion of $(2\sqrt{z} + 1)^{12}$?

36. What is the coefficient of z^5 in the expansion of $(\sqrt{2z} + 1)^{12}$?

37. What is the coefficient of the term containing c^3 in the expansion of $(\sqrt{c} - \sqrt{d})^8$?

38. What is the coefficient of x in the expansion of $\left(2x^2 - \dfrac{3}{\sqrt{x}}\right)^8$?

39. Use the Binomial Theorem to write out the expansion of $(a + b + c)^4$. HINT: Think of $(a + b + c)^4$ as $[(a + b) + c]^4$.

40. Use the Binomial Theorem to write out the expansion of $(x - 2 + y)^3$. HINT: Think of $(x - 2 + y)^3$ as $[(x - 2) + y]^3$.

In Exercises 41–44, simplify the given expression.

41. $\dbinom{9}{4}\dbinom{4}{2}$

42. $\dbinom{6}{3}+\dbinom{4}{3}-\dbinom{10}{3}$

43. $\dfrac{(n+1)!}{n!}$

44. $\dfrac{(n+1)!(n-1)!}{(n!)^2}$

45. Is it true that $n! = n(n-1)!$ for $n \geq 1$?

46. Is it true that $(n+m)! = n! + m!$?

In Exercises 47–50, expand the power of the given complex number. Simplify your answers by using the fact that $i^2 = -1$.

47. $(1+i)^5$

48. $(2-i)^4$

49. $(2+\sqrt{-9})^3$

50. $(5-\sqrt{-6})^6$

Questions for Thought

51. In general, which do you think is greater, $(n!)^2$ or $(n^2)!$?

52. Show that $\dbinom{n}{0}+\dbinom{n}{1}+\dbinom{n}{2}+\cdots+\dbinom{n}{n}=2^n$ by using the Binomial Theorem on $(1+1)^n$.

53. This exercise outlines a proof of the Binomial Theorem using the principle of mathematical induction.

Outline of Proof of Binomial Theorem   Let P_n be the statement that

$$(A+B)^n = A^n + \binom{n}{1}A^{n-1}B + \binom{n}{2}A^{n-2}B^2 + \cdots + \binom{n}{n-1}AB^{n-1} + B^n$$

Step 1:   First, we verify that P_1 is true; that is, we need to verify that

$$(A+B)^1 = A^1 + \binom{1}{1}A^{1-1}B^1 = A^1 + 1A^0B^1 = A^1 + B^1 \qquad \text{This is true.}$$

Step 2:   We want to prove that $P_k \Rightarrow P_{k+1}$. We assume that P_k is true—that is, we assume

$$(A+B)^k = \binom{k}{0}A^k + \binom{k}{1}A^{k-1}B + \binom{k}{2}A^{k-2}B^2 + \cdots + \binom{k}{k-1}AB^{k-1} + \binom{k}{k}B^k \qquad \text{This is the induction hypothesis.}$$

Multiply both sides of this equation by $(A+B)$ to obtain

$$(A+B)^{k+1} = (A+B)\left[\binom{k}{0}A^k + \binom{k}{1}A^{k-1}B + \binom{k}{2}A^{k-2}B^2 + \cdots + \binom{k}{k-1}AB^{k-1} + \binom{k}{k}B^k\right]$$

On the right-hand side, distribute the $(A+B)$, giving

$$(A+B)^{k+1} = A\left[\binom{k}{0}A^k + \binom{k}{1}A^{k-1}B + \binom{k}{2}A^{k-2}B^2 + \cdots + \binom{k}{k-1}AB^{k-1} + \binom{k}{k}B^k\right]$$
$$+ B\left[\binom{k}{0}A^k + \binom{k}{1}A^{k-1}B + \binom{k}{2}A^{k-2}B^2 + \cdots + \binom{k}{k-1}AB^{k-1} + \binom{k}{k}B^k\right]$$

This equation becomes

$$(A+B)^{k+1} = \left[\binom{k}{0}A^{k+1} + \binom{k}{1}A^kB + \binom{k}{2}A^{k-1}B^2 + \cdots + \binom{k}{k-1}A^2B^{k-1} + \binom{k}{k}AB^k\right]$$
$$+ \left[\binom{k}{0}A^kB + \binom{k}{1}A^{k-1}B^2 + \binom{k}{2}A^{k-2}B^3 + \cdots + \binom{k}{k-1}AB^k + \binom{k}{k}B^{k+1}\right]$$

Regroup the like terms to get

$$(A+B)^{k+1} = \binom{k}{0}A^{k+1} + \left[\binom{k}{1}+\binom{k}{0}\right]A^kB + \left[\binom{k}{2}+\binom{k}{1}\right]A^{k-1}B^2 + \cdots + \left[\binom{k}{k}+\binom{k}{k-1}\right]AB^k + \binom{k}{k}B^{k+1}$$

Now use the fact that $\dbinom{k}{k}=\dbinom{k}{0}=1$ and the result of Section 9.6, Exercise 45, that $\dbinom{n}{r-1}+\dbinom{n}{r}=\dbinom{n+1}{r}$

to rewrite the last equation as $(A+B)^{k+1} = A^{k+1} + \dbinom{k+1}{1}A^kB + \dbinom{k+1}{2}A^{k-1}B^2 + \cdots + \dbinom{k+1}{k}AB^k + B^{k+1}$

which proves that P_{k+1} is true. Thus, by the principle of mathematical induction, P_n is true for all values of n, which proves the Binomial Theorem.

After having completed this chapter, you should:

1. Be able to find a specific term in a sequence. (Section 9.1)
   For example:
   Given a sequence with general term

   $$a_n = \frac{3n - 1}{n^2} \qquad \text{Then}$$

   $$a_7 = \frac{3(7) - 1}{7^2} = \frac{21 - 1}{49} = \frac{20}{49}$$

2. Understand and be able to use sigma (Σ) notation. (Section 9.2)
   For example:

   $$\sum_{k=1}^{5} (2k^2 - 3) = [2(1)^2 - 3] + [2(2)^2 - 3] + [2(3)^2 - 3] + [2(4)^2 - 3] + [2(5)^2 - 3]$$

   $$= -1 + 5 + 15 + 29 + 47$$
   $$= 95$$

3. Recognize arithmetic and geometric sequences. (Sections 9.3, 9.4)
   For example:
   (a) An *arithmetic* sequence is one in which the difference between successive terms is constant. Thus the numbers 4, 9, 14, 19, 24 are the first five terms of an arithmetic sequence with first term 4 and a common difference of 5.
   (b) A *geometric* sequence is one in which the quotient of successive terms is constant. Thus the numbers 4, 12, 36, 108, 324 are the first five terms of a geometric sequence with first term 4 and a common ratio of 3.

4. Be able to find a specific term of a given arithmetic or geometric sequence. (Sections 9.3, 9.4)
   For example:
   (a) To find the tenth term of the arithmetic sequence whose first term is 3 and whose common difference is 8, we use Formula 9.1 for the *n*th term of an arithmetic sequence, which is

   $$a_n = a_1 + (n - 1)d \qquad \text{We substitute } a_1 = 3, n = 10, \text{ and } d = 8.$$
   $$a_{10} = 3 + (10 - 1)8 = 75$$

   (b) To find the seventh term of the geometric sequence whose first term is 2 and whose fourth term is -16, we can use Formula 9.6 for the *n*th term of a geometric sequence, which is

   $$a_n = a_1 r^{n-1} \qquad \text{We substitute } n = 4, a_1 = 2$$
   $$\text{and } a_n = a_4 = -16 \text{ to find } r.$$

   $$a_4 = a_1 r^3$$

   $$-16 = 2r^3 \implies r^3 = -8 \implies r = -2$$

   Now we can use the formula again to find a_7.

   $$a_7 = a_1 r^6 = 2(-2)^6 = 2(64) = 128$$

5. Be able to find the sum of an arithmetic or geometric series. (Sections 9.3, 9.4)
   For example:
   (a) To find the sum of the first eight terms of an *arithmetic* sequence whose first term is 5 and whose common difference is 4, we use Formula 9.4:

$$S_n = \frac{n}{2}[2a_1 + (n-1)d] \quad \text{We substitute } a_1 = 5, n = 8, \text{ and } d = 4 \text{ to get}$$

$$S_8 = \frac{8}{2}[2(5) + (8-1)4] = \frac{8}{2}[10 + 28]$$

$$= 152$$

(b) To find the sum of the first eight terms of a *geometric* sequence whose first term is 5 and whose common ratio is 4, we use Formula 9.7:

$$S_n = \frac{a_1(1 - r^n)}{1 - r} \quad \begin{array}{l} \text{We substitute } a_1 = 5, n = 8, \\ \text{and } r = 4 \text{ to get} \end{array}$$

$$S_8 = \frac{5(1 - 4^8)}{1 - 4} = 5\left(\frac{1 - 65{,}536}{1 - 4}\right)$$

$$= 5\left(\frac{-65{,}535}{-3}\right)$$

$$= 109{,}225$$

(c) To find the sum of an infinite geometric sequence $\{a_k\} = 2, \frac{2}{3}, \frac{2}{9}, \frac{2}{27}, \ldots$ (if it exists), we first identify a_1 and r. We are given $a_1 = 2$ and we find that $r = \frac{\frac{2}{3}}{2} = \frac{1}{3}$. Since $|r| < 1$, we know that the sum S exists and, by Formula 9.8, $S = \frac{a_1}{1 - r} = \frac{2}{1 - \frac{1}{3}} = 3.$

6. Understand the Fundamental Counting Principle and be able to apply it. (Section 9.6)

For example:

Marge is making up her class schedule, which is to contain 1 biology course, 1 math course, and 1 English course. If she chooses from 2 biology courses, 4 math courses, and 5 English courses, how many different possible schedules are there? (Assume none of the courses conflict.) According to the Fundamental Counting Principle, she can make $2 \cdot 4 \cdot 5 = 40$ different schedules.

7. Understand the difference between permutations and combinations and be able to compute the number of permutations or combinations in a variety of situations. (Section 9.6)

When you are counting the number of ways that a selection can be made where order matters, you are dealing with permutations. When you are counting the number of ways that a selection can be made where order does not matter, you are dealing with combinations.

For example:

(a) How many 3-letter arrangements can be made using the letters of the word PENCIL?

Because the order of the letters clearly matters, we are looking for the number of *permutations* of 3 out of 6 distinct objects, which is $_6P_3 = 6 \cdot 5 \cdot 4 = 120$. So there are 120 different 3-letter arrangements using the letters of PENCIL.

(b) A union local has 40 members, 5 of whom are to be chosen to attend a meeting of the national union. In how many ways can this 5-member group be chosen?

Because the order in which the 5-member group is chosen is not important, we want the number of *combinations* of 5 out of 40 distinct objects, which is

$$_{40}C_5 = \frac{40 \cdot 39 \cdot 38 \cdot 37 \cdot 36}{5 \cdot 4 \cdot 3 \cdot 2 \cdot 1} = 658{,}008$$

Thus there are 658,008 different 5-member groups that can go to the meeting.

8. Understand the principle of mathematical induction and be able to apply it to prove results about sets of natural numbers. (Section 9.5)

9. Be able to compute the probability of an event by writing out a sample space or using the counting methods discussed in Section 9.6. (Section 9.7)
   For example:
   (a) Find the probability of getting at least one head in tossing two coins:
   The sample space is $S = \{HH, HT, TH, TT\}$, hence $n(S) = 4$.
   If E is the event of getting at least one head in two coin tosses, then $E = \{HH, HT, TH\}$, and hence $n(E) = 3$. Hence, the probability of getting at least one head is $P(E) = \dfrac{n(E)}{n(S)} = \dfrac{3}{4}$.

   (b) A five-card poker hand is drawn from a standard deck of 52 playing cards. What is the probability of drawing three of a kind in a five-card hand?
   Let F be the event of drawing three of a kind and S be the sample space. The experiment in this example consists of drawing five cards, and so the sample space consists of all possible ways in which we can choose 5 cards out of 52 cards. Thus the number of elements in the sample space is

   $$n(S) = {}_{52}C_5 = \tfrac{52!}{5!(52-5)!} = \tfrac{52!}{5!47!} = 2{,}598{,}960$$

   The event of getting three of a kind can be thought of as a two-step process: Choose one rank (A, 2, 3, 4, etc.) out of 13, and then choose 3 of the 4 cards of that rank. According to the Fundamental Counting Principle, we must multiply the number of ways we can choose the rank, $_{13}C_1$, by the number of ways we can choose the 3 cards in that rank, $_4C_3$.

10. Be able to compute the probability of the union of two events. (Section 9.7)
    For example:
    Suppose we roll two dice. Let A be the event of rolling a 9 or a 10. Let B be the event of rolling a 5 on the first die. Since A and B are not mutually exclusive (you can roll a 5 on the first die and roll a total of 9 or 10), we have

    $$P(A \cup B) = P(A) + P(B) - P(A \cap B) = \frac{7}{36} + \frac{6}{36} - \frac{2}{36} = \frac{11}{36},$$

    whereas if we let E be the event of rolling an even total, and F the event of rolling an 11, these events are mutually exclusive and so we have

    $$P(E \cup F) = P(E) + P(F) = \frac{18}{36} + \frac{2}{36} = \frac{20}{36} = \frac{5}{9}.$$

11. Given two events, be able to compute the conditional probability of one event given the other. (Section 9.7)

For example:
Suppose we roll two dice. Let A be the event of rolling at least an 8. Let B be the event of rolling a 4 on the first die. The probability of rolling at least an 8 given a 4 was rolled on the first die is

$$P(A|B) = \frac{n(A \cap B)}{n(B)} = \frac{3}{6} = \frac{1}{2}$$

12. Understand and be able to apply the Binomial Theorem. (Section 9.7)
For example:
Use the Binomial Theorem to expand $(3x - 2)^5$.
According to the Binomial Theorem, we have

$$(3x - 2)^5 = (3x)^5 + \binom{5}{1}(3x)^4(-2) + \binom{5}{2}(3x)^3(-2)^2$$

$$+ \binom{5}{3}(3x)^2(-2)^3 + \binom{5}{4}(3x)(-2)^4 + (-2)^5$$

We have $\binom{5}{1} = \binom{5}{4} = \frac{5}{1} = 5$ and $\binom{5}{2} = \binom{5}{3} = \frac{5 \cdot 4}{2 \cdot 1} = 10$

Thus

$$(3x - 2)^5 = 243x^5 + 5(81)x^4(-2) + 10(27)x^3(4) + 10(9)x^2(-8) + 5(3)x(16) + (-32)$$
$$= 243x^5 - 810x^4 + 1080x^3 - 720x^2 + 240x - 32$$

Chapter 9   *Review Exercises*

In Exercises 1–8, write the first four terms and the twelfth term of the sequence whose nth term is given.

1. $a_n = 3n - 5$ **2.** $b_n = 7n + 1$
3. $x_n = 5n^2$ **4.** $y_n = 4n$
5. $a_n = \dfrac{(-1)^n}{n + 2}$ **6.** $b_n = \dfrac{(-1)^{n+2}}{n + 1}$
7. $x_n = 4 + (-1)^n$ **8.** $y_n = 2^n + |n - 5|$

In Exercises 9–12, find a possible general term for the given sequences.

9. $-2, 3, 8, 13, \ldots$ **10.** $3, 15, 75, 375, \ldots$
11. $\dfrac{1}{3}, -\dfrac{1}{6}, \dfrac{1}{9}, -\dfrac{1}{12}, \ldots$ **12.** $\dfrac{1}{2}, \dfrac{1}{4}, \dfrac{1}{8}, \dfrac{1}{16}, \ldots$

13. Byron gets a raise of $8 in his weekly salary at the end of each month during his first two years on a job. Let p_n be his weekly salary at the beginning of month

n. If his starting weekly salary is $250 per week, write a formula for p_n during his first 2 years on the job.

14. The population of Sudyville increases at a rate of 3% per year. If the current population is 22,500, write a sequence to show how the population changes over the next 4 years.

15. Soojin invests $5000 in a bond that increases in value by 5% each year. How much will her bond be worth in 5 years?

16. Jenna invests $5000 in a bond that increases in value by 6% each year. How much is the bond worth in 5 years?

In Exercises 17–22, assume that the given sequence is either arithmetic or geometric and find the indicated sum, S_n.

17. $3, 6, 9, 12, \ldots;$ S_7 **18.** $3, 7, 11, 15, \ldots;$ S_5

19. $2, 4, 8, 16, \ldots;\quad S_6$

20. $3, 9, 27, 81, 243, \ldots;\quad S_8$

21. $a_k = 5k - 2;\quad S_6$ **22.** $b_j = 3j + 4;\quad S_8$

In Exercises 23–28, rewrite each sum without using sigma notation; then compute each sum.

23. $\sum\limits_{m=1}^{6} 2m$ **24.** $\sum\limits_{j=4}^{9} 4j$

25. $\sum\limits_{n=1}^{5} (n - 2)^2$ **26.** $\sum\limits_{k=1}^{4} k^3$

27. $\sum\limits_{i=1}^{7} 5i^2$ **28.** $\sum\limits_{k=2}^{4} (k^2 - 6k + 1)$

In Exercises 29–34, use the given information to find the next two terms and the tenth term of the given *arithmetic* sequence.

29. $a_1 = 7, a_2 = 13$ **30.** $a_3 = 4, a_4 = -1$

31. $6, 11, 16, 21, \ldots$ **32.** $-4, -2, 0, 2, \ldots$

33. $2, 0, -2, -4, \ldots$ **34.** $\dfrac{1}{3}, 0, -\dfrac{1}{3}, \ldots$

35. Given the arithmetic sequence $\{x_i\}$ with $x_4 = 10$ and $x_8 = 22$, find x_1 and d.

36. Given the arithmetic sequence $\{y_i\}$ with $y_4 = -9$ and $y_6 = -13$, find y_2 and d.

37. Suppose that an object dropped from the top of a building falls 16 feet the first second, 48 feet the second second, and so on, the distances it falls each second forming an arithmetic sequence. How many feet will the object fall during the eighth second?

38. How many feet will the object described in Exercise 37 fall during the tenth second?

In Exercises 39–42, use a formula for S_n to find the indicated S_n for the given *arithmetic* sequence.

39. $\{a_k\}$, where the first term is 8 and the common difference is 6; find S_{12}.

40. $\{b_k\}$, where the first term is -9 and the common difference is -2; find S_9.

41. $4, 11, 18, \ldots;\quad S_7$ **42.** $-6, -2, 2, \ldots;\quad S_8$

43. Find $\sum\limits_{i=1}^{40} 4i$.   **44.** Find $\sum\limits_{i=10}^{20} 2i$.

45. Find the sum of the first 25 positive odd numbers.

46. Find the sum of the first 25 positive even numbers.

47. Given an arithmetic sequence with $a_5 = 5$ and $a_9 = 8$, find S_{12}.

48. Given an arithmetic sequence with $a_2 = 2$ and $a_7 = 5$, find S_{10}.

In Exercises 49–54, find the required a_n for the given *geometric* sequence.

49. $2, -6, 18, \ldots;\quad a_5$ **50.** $1, 4, 16, \ldots;\quad a_6$

51. $1, \dfrac{1}{2}, \dfrac{1}{4}, \ldots;\quad a_7$ **52.** $4, -\dfrac{4}{3}, \dfrac{4}{9}, \ldots;\quad a_8$

53. $\{a_k\}$ with $a_1 = 4$ and $r = 3$; find a_5.

54. $\{a_j\}$ with $a_2 = 4$ and $r = \dfrac{1}{2}$; find a_6.

55. Jim promises to pay Cindy a debt in the following way: He will pay her $3 the first month, $6 the second month, $12 the third month, and so on, where the payment in any month is double that of the previous month. How much would Cindy receive in the tenth month?

56. How much will Jim have paid Cindy of Exercise 55 all together at the end of 1 year?

For Exercises 57–60, find the sum, S (if it exists), of the infinite geometric sequence.

57. $5, 1, \dfrac{1}{5}, \ldots$ **58.** $1, -\dfrac{1}{3}, \dfrac{1}{9}, \ldots$

59. $\dfrac{1}{4}, \dfrac{1}{2}, 1, \ldots$ **60.** $\dfrac{3}{7}, 1, \dfrac{7}{3}, \ldots$

61. Express $0.39\overline{39}$ as a fraction.

62. Express $2.1\overline{745745745}$ as a fraction.

63. A ball bounces back three-fourths the distance it falls on a rebound. If you could let the ball bounce forever, what is the total distance it would travel if it is dropped from a height of 80 feet?

64. Use mathematical induction to prove that
$$1 \cdot 2 + 2 \cdot 3 + 3 \cdot 4 + \cdots + (n - 1)n$$
$$= \dfrac{(n - 1)n(n + 1)}{3}$$

65. Use mathematical induction to prove that $2^n < n!$ for $n \geq 4$.

66. Use mathematical induction to prove that $n! > n^3$ for $n \geq 6$.

67. If there are 12 horses entered in a race, how many possible 1-2-3 finishes can there be?

68. The first race at a racetrack has 8 horses entered, and the second race has 10 horses entered. In how many ways can you pick the first 3 horses in the first two races?

69. How many different arrangements are there of the letters of the word HYPERBOLA?

70. How many different arrangements are there of the letters of the word PARABOLA?

71. How many different arrangements are there of the letters of the word BANANAS?

72. A briefcase has a 3-digit combination lock, where each digit can be any of 0, 1, 2, 3, 4, 5, 6, 7, 8.
   (a) How many combinations are there?
   (b) How many of these combinations are even numbers?
   (c) How many of these combinations are greater than 500?

73. If a 5-card poker hand is dealt from a standard deck of 52 cards, in how many ways can you be dealt a
   (a) Spade flush?   **(b)** Straight?
   (c) Straight flush?

74. How many boy–girl combinations are there in a family with 5 children?

75. How many boy–girl birth sequences are there in a family with 5 children?

76. A die is rolled twice.
   (a) Find the probability of getting at least one 3.
   (b) Find the probability that the sum of the two dice is 10.
   (c) Find the probability that both rolls are the same.

77. A card is drawn at random from a standard 52-card deck. Find the probability of the given event.
   (a) The card is a jack, a queen, or a king.
   (b) The card is a club or a king.
   (c) The card is neither a club nor a king.

78. A state lottery game requires you to choose a five-digit number.
   (a) What is the probability that the winning number will have five different digits?
   (b) What is the probability that the winning number will have the same digit repeated 5 times?

79. A five-card poker hand is dealt from a standard 52-card deck. Find the probability that the poker hand contains a flush (5 cards of the same suit). Round your answer to 5 decimal places.

80. A psychology experiment was designed to test a subject for ESP (extrasensory perception). Three cards—a star, a circle, and a square—are shuffled and placed face down in front of the subject, who is then asked to identify each card.
   (a) If the subject simply guesses randomly, what is the probability of guessing all the cards correctly?

(b) If the subject simply guesses randomly, what is the probability of guessing only one card incorrectly?
   (c) If the subject simply guesses randomly, what is the probability of guessing two of the three incorrectly?

81. Suppose we interpret a basketball player's lifetime free-throw percentage of 0.78 as his probability of making any particular free throw.
   (a) Find the probability the player will make all 6 free throws he takes in a particular game.
   (b) Find the probability the player will make none of the 6 free throws he takes during a particular game.
   (c) Find the probability the player will make at least 1 out of the 6 free throws he takes during a particular game.

82. Suppose you roll a pair of dice. Find the probability of each of the following events.
   (a) You roll an even number or at least a 9.
   (b) You roll a double (both dice the same) or at least a 10.
   (c) You roll a double or an even number.

83. A jar contains 8 white balls numbered 1 through 8, and 5 blue balls numbered from 1 to 5. Find the probability of each of the following events.
   (a) You select two balls and they are both white and even.
   (b) You select two balls, and both are white or both are even.
   (c) You select two balls and both are blue, but one is odd and one is even.

84. Expand $(2x - y)^5$ and simplify.

85. Expand $\left(\dfrac{x}{3} + \dfrac{y}{4}\right)^3$ and simplify.

86. Expand $(x^{-2} + y^{-1})^6$ and simplify.

87. What is the coefficient of x^6 in the expansion of $(x + 5\sqrt{x})^8$?

88. What is the coefficient of a^7b^3 in the expansion of $(a - 3b)^{10}$?

1. Identify the first three terms and the eighth term of the following sequences:
   (a) $\{x_k\}$, where $x_k = k^2 - 3k$
   (b) $\{y_j\}$, where $y_j = 5j^3 - 2$

2. Rewrite the following as a sum without using sigma notation; then compute the sum.

$$\sum_{k=4}^{7} (3k^2 - 5k + 1)$$

3. Find the 112th term of the arithmetic sequence with first term -4 and common difference 6.

4. Find the sixth term of a geometric sequence with first term 3 and common ratio -2.

5. Find S_{20} for the arithmetic sequence 4, 8, 12, ...

6. Find S_5 for the geometric sequence 3, 9, 27, ...

7. Use the fact that each of the following sequences is either arithmetic or geometric and the given information to find the requested values.
   (a) Given the sequence 5, 9, 13, 17, . . . , find a_7.
   (b) Given the sequence 1, 5, 25, 125, . . . , find a_7.
   (c) Given $a_1 = \dfrac{1}{3}$ and $r = 6$, find a_5.
   (d) Given $a_3 = 1$, $a_5 = 2$, and $a_7 = 3$, find S_{10}.

8. Find the sum of the first 20 positive multiples of 10.

9. A pendulum sweeps out an 8-foot arc on its first pass, and each time it passes, it covers three-fourths the distance of the previous arc. How far would it travel before coming to rest?

10. Use the principle of mathematical induction to prove that $x - 2$ is a factor of $x^n - 2^n$.

11. Suppose a jar contains 10 yellow marbles, 8 blue marbles, and 5 green marbles.
    (a) In how many ways can you choose 6 marbles from the jar?
    (b) In how many ways can you select the 6 marbles choosing first 2 yellow, then 2 blue, and finally 2 green?

12. How many different arrangements are there of the letters in the word ABRACADABRA?

13. A state lottery game requires you to choose a seven-digit number.
    (a) What is the probability that the winning number will be even?
    (b) What is the probability that the winning number will start with 5 and be odd?
    (c) What is the probability that the winning number does not have an 8 in it?

14. A card is drawn from a standard deck of 52 cards. Find the probability of each of the following events.
    (a) The card is a picture card or a diamond.
    (b) The card is an ace or a club.
    (c) The card is an odd-numbered card (count an ace as 1) or a red card.

15. Use the Binomial Theorem to expand $(2x^2 - 3y)^8$.

Appendix: Using Technology to Model Data

To help us in identifying trends or making decisions, it is often helpful to collect real data and arrange them in an organized fashion. For example, the Heathcliff Electric Motor Company manufactures home transformers and has collected data over the years. Table A.1 shows the company's profit (in millions of dollars) for various prices (in dollars) of its smallest home transformer.

Table A.1

Price	20.00	20.50	21.50	22.00	23.10	24.75	25.60	26.40	27.20
Profit	12.3	13.5	14.6	15.8	16.2	16.2	15.6	13.5	11.9

We can plot the profit (as the dependent variable) vs. the price of the transformer (as the independent variable) as shown in Figure A.1. The graph that results from plotting data points this way is called a **scatterplot**.

Figure A.1

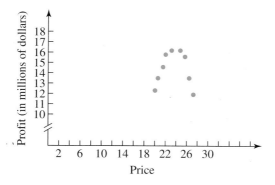

Notice that what we graphed looks as if it might be part of a parabola—a quadratic function. If it is, then we can use our knowledge of how a parabola behaves to *estimate* that the highest profit would occur if the price is somewhere between 23 and 24 dollars.

Let's take advantage of the technology available in the graphing calculator to obtain a scatterplot of this relationship. Open the STAT menu by pressing ⌗STAT⌗ (Figure A.2a, page 644). Then choose ⌗1:Edit⌗ and, using the arrow and ⌗ENTER⌗ keys, enter the item price as L1 and the profit as L2, being careful to pair the appropriate two values together (Figure A.2b).

Figure A.2

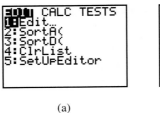

(a)

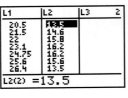

(b)

To obtain a scatterplot of the data, enter the STAT PLOT menu. First press ⌷2nd⌷ ⌷Y=⌷, and then choose ⌷1:Plot 1⌷ (Figure A.3a). Highlight the ON button and press ⌷ENTER⌷. We will plot the data using Type 1, with L1 being the price list and L2 being the profit list, as shown on the screen in Figure A.3b.

Figure A.3

(a)

(b)

Next press ⌷ZOOM⌷ ⌷9:ZoomStat⌷ (Figure A.4a), and you will see the scatterplot for the data (Figure A.4b).

Figure A.4

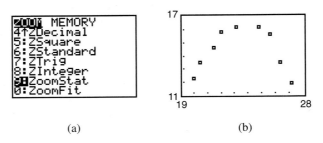

(a)   (b)

Let's consider another example. Suppose Table A.2 gives the National Hospital Care Expenditures data (in billions of dollars) from 1980 to 1995.

Table A.2

Year	1980	1985	1990	1992	1993	1994	1995
National Hospital Care Expenditures (in billions)	102.7	168.3	256.4	305.3	323.0	335.7	347.2

Figure A.5 gives a scatterplot of these data. Notice that the data seem to follow a linear trend (the data fall almost on a line).

Figure A.5

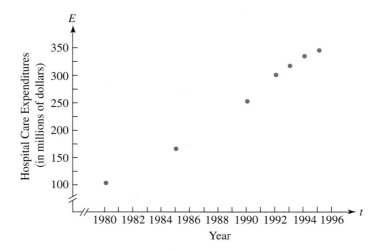

Figure A.6a shows the data for the National Hospital Care Expenditures entered in the calculator; the scatterplot of this data appears in Figure A.6b.

Figure A.6

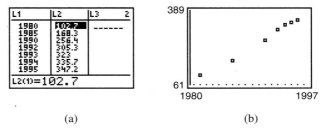

(a)   (b)

We noted that the data in the Heathcliff Company example appear to have the shape of a parabola. If we knew more about this approximate parabola, then we could better predict what price to charge in order to maximize our profit. The Hospital Expenditures trend looks almost like a line; if we knew more about this line, we could better predict future values. While we can make decisions based on the scatterplots we created, it would be more convenient if we had an equation. When we *model data*, we mean that we are seeking equations whose values agree well with the given data values.

Let's consider the Hospital Expenditures example shown in Figure A.7.

Figure A.7

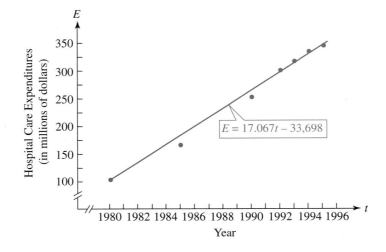

$E = 17.067t - 33,698$

We could use the scatterplot to get a very rough estimate of the national expenditures for hospitals in, say, 1996—it seems to be somewhere between 360 and 375 billion dollars. If we knew the equation of the line "closest to all the data points," which in this case is $E = 17.067t - 33{,}698$, then we could substitute $t = 1996$ and get $y = 367.7$. Although this is still an estimate, note that having an equation makes it more convenient to obtain an estimate than using a scatterplot.

Finding the line or curve that "fits" the data is called **regression analysis**. In the first section of this appendix we will discuss **linear regression**—finding an equation of the *line* that best "fits" the data.

A.1   Linear Regression

In Section 1.7 we discussed linear equations. We mentioned that if we know there is a (perfect) linear relationship between two variables (that is, the graph of the relationship is a line), and if we are able to identify two points on the line, then we can find its equation. Once we have its equation, we can find values of one variable given values of the other.

In real-life applications, although we may not have a perfect linear relationship, the relationship may "approximate" a line. In other words, the plotted points may not all fall exactly on a line, but they may fall close enough to a line that we can get an estimate sufficient for our needs. Consider the scatterplot of the National Hospital Expenditures example in Figure A.8.

Figure A.8

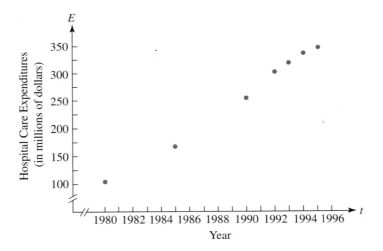

The relationship looks almost linear (E is growing at an almost constant rate), but points do not fall exactly in a straight line. Even so, we can draw a line "close enough" to the data points. The equation of this line is called the *mathematical model of these data* and gives us an algebraic description of them. The line that "comes closest" to all the data points is called the **regression line** or **line of best fit**. It is computed (using calculus) by finding the line that minimizes the sum of the squares of the (vertical) distances between the line and the data points.

Figure A.9 depicts some lines of best fit for various sets of data.

Figure A.9

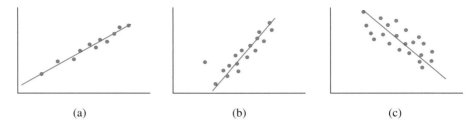

(a)   (b)   (c)

Let's stay with the Hospital Expenditures data given in Table A.2. The scatterplot of the data appears in Figure A.10.

Figure A.10

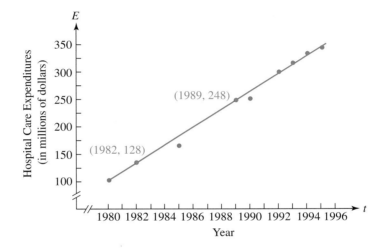

To estimate the line of best fit, we first draw in a line that "comes close" to the data points (see Figure A.10). Then we estimate any two points ON THE LINE, and then find its equation. For this example, we estimate that the points (1982, 128) and (1989, 248) are on this proposed line of best fit. Note we emphasized ON THE LINE: Picking two data points not on the line only gives you the line that goes through those two data points, which may not be the line of best fit.

For this example, letting t be the year and E the hospital expenditure, we find the equation by using our selected points (1982,128) and (1989, 248):

The slope is $m = \dfrac{248 - 128}{1989 - 1982} = \dfrac{120}{7}$

Hence the equation is $E - 128 = \frac{120}{7}(t - 1982)$ or $E = 17.14t - 33{,}849$.

Assuming the linear trend will continue, we can use the equation to predict that in 1996, there will be $17.14(1996) - 33{,}849 = 362$ billion dollars spent on hospitals, which seems to agree with the scatterplot. Note that since we estimated the regression line and then estimated two points on the line, our answer is a *rough* estimate.

Alternatively, we can take advantage of the technology available in the graphing calculator, which can give us the exact equation of the line of best fit after we enter the data. For example, let's return to the scatterplot for the National Health Care Expenditures we made earlier. We enter the data, L1 and L2, as we did previously (see Figures A.11a and b on page 648).

Figure A.11

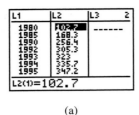

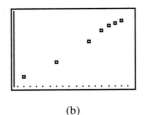

(a)   (b)

To find the equation of the "line of best fit" or the *regression equation*, press $\boxed{\text{STAT}}$, then use the arrow key to highlight the CALC menu, and then choose $\boxed{4:\text{LinReg(ax+b)}}$ (see Figures A.12a and b). Press $\boxed{\text{ENTER}}$ and the screen gives the linear regression equation (see Figure A.13).

Figure A.12

(a)   (b)

Figure A.13 gives us the regression equation $y = ax + b$ where $a = 17.06701278$ and $b = -33,698.26014$. Hence the equation is $y = 17.06701278x - 33,698.26014$. If we use *this* equation to predict the expenditures in 1996, we would get $y = 17.06701278(1996) - 33,698.26014 \approx 367.5$ billion dollars.

Figure A.13

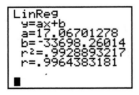

Having computed the regression equation, we can graph this line on the scatterplot by doing the following: First make sure the scatterplot is turned on as shown in Figure A.3. Next, go into the $\boxed{Y=}$ menu (Figure A.14 a), highlight Y1 , and then press $\boxed{\text{VARS}}$. Then choose $\boxed{5:\text{STATISTICS}}$ (Figure A.14b). Use the arrow key to

Figure A.14

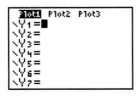

(a)   (b)

Figure A.15

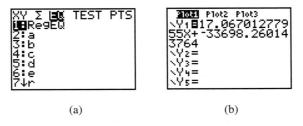

(a)   (b)

highlight EQ. Choose $\boxed{1 : RegEQ}$ (Figure A.15a); this takes the regression equation and stores it in Y1 for graphing (Figure A.15b). Press $\boxed{GRAPH}$, and the regression line will appear on the scatterplot (Figure A.16).

Figure A.16

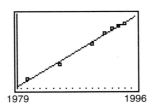

The Correlation Coefficient

The calculator uses a formula to compute the equation of the line of best fit. This formula does not take into account whether there actually is a linear relationship between variables. For example, consider the four scatterplots shown in Figure A.17.

Figure A.17

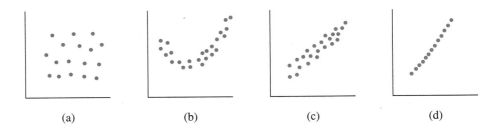

(a)   (b)   (c)   (d)

The first scatterplot exhibits no apparent relationship between variables. The second scatterplot exhibits what could be a quadratic relationship. The third scatterplot seems to exhibit a linear trend, and the fourth shows an almost perfect linear relationship. The calculator may generate a linear equation for each of these situations, whether or not there actually is a linear relationship. Be careful! Because you *can* compute a linear equation does not mean that there *is* a linear relationship. So the next question is, "When is it appropriate to generate a regression line for data? How can we tell whether there is a linear relationship?"

There is a way to quantify how close the data come to a line. We use a measure called the *correlation coefficient*.

The **correlation coefficient**, designated r, is a number that tells us how close the data come to fitting on a line. The correlation coefficient, r, ranges from -1 to $+1$, or equivalently, $|r| \leq 1$. When $|r| = 1$, all the data points are *on* the line. As $|r|$ deviates from 1, the data are less and less "linear." When $r = 0$, there is no linear relationship whatsoever between the variables.

The sign of r is the sign of the slope of the line. Hence, a negative r means the slope of the line is negative, or as one variable increases, the other variable decreases; a positive r indicates the slope of the line is positive, or as one variable increases, the other increases as well. Figure A.18 shows some examples of data distributions along with their correlation coefficients.

Figure A.18

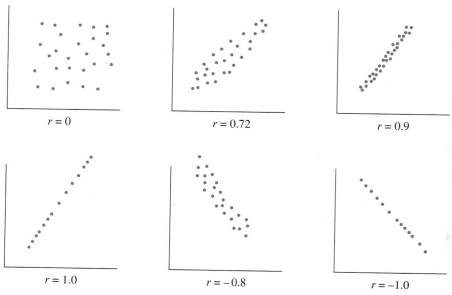

$r = 0$ $r = 0.72$ $r = 0.9$

$r = 1.0$ $r = -0.8$ $r = -1.0$

The correlation coefficient tells how close the data come to fitting on a line. Again, even though a calculator can compute a linear regression equation for the data, it does not mean that the data actually form or come close to forming a line. We first look at the value of r to see whether it is large enough to conclude that a line is the best way to describe the data: The further away r is from 1, the greater the error in the estimate using the linear regression equation.

Example 1

Table A.3 shows the percentage of adults who were defined as smokers from 1970 to 1987. Use a graphing calculator to create a scatterplot. Is there a linear trend in the data? If so, compute the correlation coefficient and the regression equation, and use this equation to estimate the percent of adults who were smokers in 1997.

Table A.3

Year	1970	1974	1978	1979	1980	1983	1985	1987
% of Adult Smokers (in millions)	37.4	37.1	34.1	33.5	33.2	32.1	30.1	28.8

Solution

First we enter the data in the STAT Menu using $\boxed{1:\text{Edit}}$ and entering the year and the percentage of adult smokers (Figure A.19a). Make sure the scatterplot (Plot1) is turned on. In the ZOOM Menu, choose $\boxed{9:\text{Zoom Stat}}$ to see the data plotted (Figure A.19b).

Figure A.19

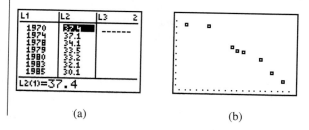

(a)   (b)

The trend seems somewhat linear. To determine the strength of the linear relationship, compute the correlation coefficient and the regression equation.[*] To do this, press $\boxed{\text{STAT}}$, then highlight the CALC menu, choose $\boxed{\text{4:LinReg(ax+b)}}$, and press $\boxed{\text{ENTER}}$ (see Figures 20a and b).

Figure A.20

(a)

(b)

We see that the correlation coefficient is $r \approx -0.98$. This is a strong linear relationship, and since r is negative, the slope is negative, which means the later the year, the smaller the percentage of adult smokers.

The regression equation is $y = -0.5259x + 1074.31$. Using this equation to predict the percent of adult smokers in 1997 yields $y = -0.5259(1997) + 1074.31 \approx 24.1\%$. The actual value was 24.7%. After getting the linear regression equation, go into the $\boxed{\text{Y=}}$ menu, highlight Y1=, press $\boxed{\text{VARS}}$, then $\boxed{\text{5:Statistics}}$, highlight EQ, and choose $\boxed{\text{1:RegEQ}}$ to graph the regression equation with the data, as shown in Figure A.21.

1969   1988

Figure A.21

In the next section, we will examine other regression models.

---

[*]Note that for the TI-83 Plus, the correlation coefficient, r, is computed if the "diagnostic" is turned on. To do this, press $\boxed{\text{2nd}}$ $\boxed{\text{CATALOG}}$ scroll down to DiagnosticOn, and then press $\boxed{\text{ENTER}}$.

A.1 Exercises

1. The accompanying table shows the gross domestic product of the United States (GDP) in billions of dollars for the period 1980–1998. Use a graphing calculator to create a scatterplot. Is there a linear trend in the data? If so, compute the correlation coefficient and the regression equation, and use the regression equation to estimate what the GDP was for 1999.

Year	1980	1985	1990	1992
GDP (billions of dollars)	2784.2	4180.7	5743.8	6244.4

Year	1994	1996	1998
GDP (billions of dollars)	6947.0	7661.6	8511.0

2. The table shows the total value, in millions of dollars, of timber cut in the United States from 1990 to 1996. Use a graphing calculator to create a scatterplot. Is there a linear trend in the data? If so, compute the correlation coefficient and the regression equation, and use this equation to estimate the value of timber that was cut in 1997.

Year	1990	1991	1992	1993
Value of Timber Cut (in millions of dollars)	1,191	1,012	938	918

Year	1994	1995	1996
Value of Timber Cut (in millions of dollars)	786	616	544

3. The table shows the size, in square feet, and price, in dollars, of real estate lots in a small town in the Midwest in 2003. Use a graphing calculator to create a scatterplot. Is there a linear trend in the data? If so, compute the correlation coefficient and the regression equation, and use this equation to estimate what the market price should be of a 10,000-sq-foot lot.

Lot Size	6,000	7,200	4,500	12,600
Market Value	2,450	3,000	1,500	5,500

Lot Size	15,500	18,800	8,600	5,800
Market Value	6,800	8,000	3,000	2,000

4. The table shows the percent of the U.S. civilian labor force that was unemployed from 1992 to 1998. Use a graphing calculator to create a scatterplot. Is there a linear trend in the data? If so, compute the correlation coefficient and the regression equation, and use the regression equation to estimate the percent of the civilian labor force that was unemployed in 1999.

Year	92	93	94	95	96	97	98
Percent Unemployed	7.5	6.9	6.1	5.6	5.4	4.9	4.5

5. The table shows the amount (in billions of dollars) public and private agencies spent on higher education from 1980 to 1996. Use a graphing calculator to create a scatterplot. Is there a linear trend in the data? If so, compute the correlation coefficient and the regression equation, and use this equation to estimate how much the agencies spent in 1997.

Year	80	85	90	93
Funding for Higher Education (in billions of dollars)	130.2	150.3	191.8	207.3

Year	94	95	96
Funding for Higher Education (in billions of dollars)	209.9	216.7	221.2

6. The JC Giles Company wants to determine how often it should run its television commercials. The table shows the number of times its commercials are run during a week along with the amount of money in sales (in thousands of dollars) during that same week.

Use a graphing calculator to create a scatterplot. Is there a linear trend in the data? If so, compute the correlation coefficient and the regression equation, and use this equation to estimate the amount of sales the company would get if the commercial were to run 16 times.

Number of Commercials	3	5	8	9
Sales (in thousands of dollars)	179.5	215.5	270	285

Number of Commercials	10	12	15
Sales (in thousands of dollars)	300	340	400

7. The table shows the cumulative grade point average (GPA) of 10 students at a certain college, along with their scores on the Graduate Record Exam (GRE). Use a graphing calculator to create a scatterplot. Is there a linear trend in the data? Compute the correlation coefficient and the regression equation, and discuss the results.

GPA	2.5	2.6	2.8	3.0	3.22
GRE	550	530	580	610	640

GPA	3.28	3.3	3.45	3.56	3.62
GRE	620	640	680	700	710

8. The table shows the final high school cumulative grade point average, HGPA, of 10 students, along with their first-year cumulative grade point average at a certain college, CGPA. Use a graphing calculator to create a scatterplot. Is there a linear trend in the data? Compute the correlation coefficient and the regression equation, and discuss the results.

HGPA	2.85	2.9	3.0	3.25	3.30
CGPA	1.92	1.80	2.1	2.55	2.85

HGPA	3.28	3.45	3.58	3.68	3.80
CGPA	2.9	3.1	2.8	2.95	3.2

9. The table shows the preprimary school enrollment (in thousands) in the United States from 1980 to 1997. Use a graphing calculator to create a scatterplot. Is there a linear trend in the data? Compute the correlation coefficient and the regression equation, and discuss the results.

Year	1980	1985	1990	1994
Number of Students (in thousands)	9,284	10,733	11,207	12,328

Year	1995	1996	1997
Number of Students (in thousands)	12,518	12,378	12,121

10. The table shows the number of secondary school teachers (in thousands) from 1980 to 1996. Use a graphing calculator to create a scatterplot. Is there a linear trend in the data? Compute the correlation coefficient and the regression equation, and discuss the results.

Year	1980	1985	1990	1991	1992
Number of Teachers (in thousands)	1,084	1,066	1,073	1,074	1,070

Year	1993	1994	1995	1996
Number of Teachers (in thousands)	1,095	1,132	1,164	1,197

11. The table shows the average salary (in thousands of dollars) of secondary-school teachers from 1985 to 1997. Use a graphing calculator to create a scatterplot. Is there a linear trend in the data? If so, compute the correlation coefficient and the regression equation, and use the regression equation to find the average secondary-school teacher salary in 1998.

Year	85	90	91	92	93
Average Salary (in thousands of dollars)	24.2	32.0	33.9	34.8	35.9

Year	94	95	96	97
Average Salary (in thousands of dollars)	36.6	37.5	38.4	39.1

12. The table shows the national public school per capita expenses (in dollars) from 1980 to 1997. Use a graphing calculator to create a scatterplot. Is there a linear trend in the data? If so, compute the correlation coefficient and the regression equation, and use this equation to estimate the per capita expenses in 1998.

Year	80	85	90	93
Per Capita Expenses (in dollars)	428	591	850	976

Year	94	95	96	97
Per Capita Expenses (in dollars)	1,018	1,062	1,112	1,162

13. The table shows the percent of schools with Internet access from 1994 to 1998. Use a graphing calculator to create a scatterplot. Is there a linear trend in the data? If so, compute the correlation coefficient and the regression equation, and use the regression equation to estimate the percent for 1999.

Year	1994	1995	1996	1997	1998
Percent of Instructional Classrooms with Internet Access	35	50	65	78	89

14. The table at the top of page 654 shows the percent of students enrolled in foreign language courses in public schools from 1976 to 1994. Use a graphing calculator to create a scatterplot. Is there a linear trend in the data? If so, compute the correlation coefficient and the regression equation, and use this equation to estimate the percent in 1996.

(continued)

Year	1976	1978	1982
Percent of Students Enrolled in Foreign Languages	22.7	23.0	22.6

Year	1985	1990	1994
Percent of Students Enrolled in Foreign Languages	32.3	38.4	42.2

15. The table shows per capita consumer expenditures (in dollars) on dentists' services in the United States from 1980 to 1997. Use a graphing calculator to create a scatterplot. Is there a linear trend in the data? If so, compute the correlation coefficient and the regression equation, and use the regression equation to estimate the per capita expenditure for 1998.

Year	1980	1985	1990	1992	1993
Dentist Service Expenditure (in dollars)	57	88	121	139	147

Year	1994	1995	1996	1997
Dentist Service Expenditure (in dollars)	157	165	172	182

16. The table shows the number of violent crimes (in thousands) in the United States from 1991 to 1997. Use a graphing calculator to create a scatterplot. Is there a linear trend in the data? If so, compute the correlation coefficient and the regression equation, and use this equation to estimate how many violent crimes occurred in 1998.

Year	1991	1992	1993	1994
Number of Violent Crimes (in thousands)	14,873	14,438	14,145	13,990

Year	1995	1996	1997
Number of Violent Crimes (in thousands)	13,863	13,494	13,175

17. The table shows the number of motor vehicle thefts (in thousands) in the United States from 1987 to 1997. Use a graphing calculator to create a scatterplot. Is there a linear trend in the data? If so, compute the correlation coefficient and the regression equation.

Year	1987	1988	1989	1990	1991	1992
Number of Thefts (in thousands)	1,289	1,433	1,565	1,636	1,662	1,611

Year	1993	1994	1995	1996	1997
Number of Thefts (in thousands)	1,563	1,539	1,472	1,394	1,354

18. The table shows the number of motor vehicle thefts (in thousands) in the United States from 1991 to 1997.

Year	1991	1992	1993	1994
Number of Thefts (in thousands)	1,662	1,611	1,563	1,539

Year	1995	1996	1997
Number of Thefts (in thousands)	1,472	1,394	1,354

(a) Use a graphing calculator to create a scatterplot. Is there a linear trend in the data? If so, compute the correlation coefficient and the regression equation, and use this equation to estimate the number of motor vehicle thefts that occurred in 1998.

(b) In Exercise 17, we extend the table to include data for 1987–1990. Use a graphing calculator to create a scatterplot. Is there now a linear trend in the data? Compute the correlation coefficient and the regression equation. Comment on the differences between the scatterplots in Exercises 17 and 18.

19. The table shows the number of tornadoes in the United States from 1987 to 1996. Use a graphing calculator to create a scatterplot. Is there a linear trend in the data? Compute the correlation coefficient and the regression equation, and discuss the results.

Year	1987	1988	1989	1990	1991
Number of Tornadoes	656	702	856	1,133	1,132

Year	1992	1993	1994	1995	1996
Number of Tornadoes	1,298	1,176	1,082	1,235	1,170

20. The table shows federal budget outlay for national defense from 1992 to 1998, in billions of dollars. Use a graphing calculator to create a scatterplot. Is there a linear trend in the data? If so, compute the correlation coefficient and the regression equation, and use the regression equation to estimate the budget outlay for defense in 1999.

Year	92	93	94	95
National Defense (in billions of dollars)	298.4	291.1	281.6	272.1

Year	96	97	98
National Defense (in billions of dollars)	265.8	270.5	268.5

21. The table shows the United States military expenditures (in billions of dollars) from 1990 to 1995.
   (a) Use a graphing calculator to create a scatterplot. Is there a linear trend in the data? If so, compute the correlation coefficient and the regression equation, and use this equation to estimate the military expenditures in 1996.
   (b) In Exercise 22, we extend the table to include data for 1987–1989. Use a graphing calculator to create a scatterplot. Is there now a linear trend in the data? Compute the correlation coefficient and the regression equation. Comment on the differences between the scatterplots in Exercise 21 and 22.

Year	1990	1991	1992
Military Expenditures (in billions of dollars)	1,106	1,049	974

Year	1993	1994	1995
Military Expenditures (in billions of dollars)	912	879	865

22. The table shows the United States military expenditures (in billions of dollars) from 1987 to 1995. Use a graphing calculator to create a scatterplot. Is there a linear trend in the data? If so, compute the correlation coefficient and the regression equation, and use this equation to estimate the military expenditures in 1996.

Year	1987	1988	1989	1990	1991
Military Expenditures (in billions of dollars)	1,051	1,080	1,089	1,106	1,049

Year	1992	1993	1994	1995
Military Expenditures (in billions of dollars)	974	912	879	865

23. The table shows the total credit market debt outstanding in the United States (in billions of dollars) from 1980 to 1997. Use a graphing calculator to create a scatterplot. Is there a linear trend in the data? If so, compute the correlation coefficient and the regression equation, and use this equation to estimate the credit market debt in 1998.

Year	1980	1985	1990	1992	1993
Credit Market Debt (in billions of dollars)	4,734	8,628	13,745	15,195	16,167

Year	1994	1995	1996	1997
Credit Market Debt (in billions of dollars)	17,212	18,445	19,794	21,200

24. The table at the top of page 656 shows the total household assets (in billions of dollars) of U.S. households from 1980 to 1998. Use a graphing calculator to create a scatterplot. Is there a linear trend in the data? If so, compute the correlation coefficient and the regression equation, and use the regression equation to find the household assets for 1999.

Year	1980	1985	1990	1995
Household Assets (in billions of dollars)	6,584	10,149	14,985	21,751

Year	1996	1997	1998
Household Assets (in billions of dollars)	23,908	27,020	30,121

25. The table shows outstanding mortgage debt (in billions of dollars) from 1980 to 1997. Use a graphing calculator to create a scatterplot. Is there a linear trend in the data? If so, compute the correlation coefficient and the regression equation, and use this equation to estimate the mortgage debt outstanding in 1998.

Year	1980	1985	1990	1991	1992
Mortgage Debt (in billions of dollars)	1,465	2,374	3,794	3,948	4,063

Year	1993	1994	1995	1996	1997
Mortgage Debt (in billions of dollars)	4,206	4,392	4,610	4,928	5,257

26. The table shows the number of mortgage loans outstanding from 1980 to 1998 (in thousands). Use a graphing calculator to create a scatterplot. Is there a linear trend in the data? If so, compute the correlation coefficient and the regression equation, and use this equation to estimate the number of mortgage loans outstanding in 1999.

Year	1980	1985	1990	1993	1994
Number of mortgages (in thousands)	30,033	34,004	40,638	44,562	47,462

Year	1995	1996	1997	1998
Number of mortgages (in thousands)	49,111	50,064	51,279	52,508

Questions for Thought

27. A student recorded the heights and weights of all the male students in his class and created the scatterplot shown here. Draw in a line of best fit and find the equation of the line. Use the equation to estimate the weight of a person who is 6′2″. How accurate do you think is your estimation using your equation?

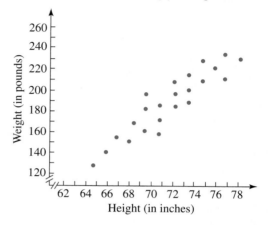

28. A student recorded the heights and weights of all the female students in his class and created the scatterplot shown here. Draw in a line of best fit and find the equation of the line. Use the equation to estimate the weight of a person who is 5′3″. How accurate do you think is your estimation using your equation?

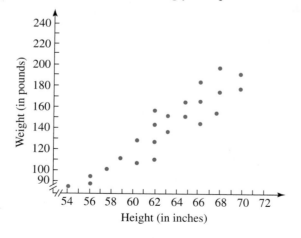

29. A math teacher recorded her students' scores on their first and second exams for one section and created the upper scatterplot shown at the top of page 657: The vertical axis is the first exam score; the horizontal axis is the second exam score. The teacher did the same thing for her other math section; the results are shown in the lower scatterplot.

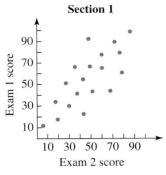

Section 1

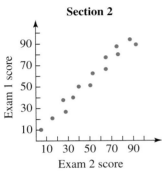

Section 2

(a) Discuss the trend in the data for each section and what it means.
(b) For each scatterplot, draw in a line of best fit, and find the equation of the line.
(c) For each section, use the equation to estimate the score a student would get on Exam 2 if they scored a 75 on Exam 1.
(d) Discuss the accuracy of the estimations you found in part (c). Compare the two scatterplots. Which equation do you think would be more accurate?

A.2 Other Regression Models

We saw in the previous section that some relationships can be described or approximated by a line. Throughout this text, we have seen naturally occurring phenomena related in other ways. In Section 5.5, we saw that some relationships could be expressed as quadratic functions. In Chapter 6 we saw that some relationships, such as populations, can be expressed using exponential models; other relationships can be expressed using logarithmic functions. In this section we will look at these other models.

Table A.4 shows the amount of carbon dioxide (as measured by carbon content in millions of metric tons) emitted in the United States from 1990 to 1996.

Table A.4

Year	1990	1992	1993	1994	1995	1996
Carbon Dioxide (in millions of tons)	1,355.9	1,360.6	1,393.6	1,413.8	1,428.1	1,478.8

Figures A.22a and b show the entered data and the scatterplot.

Figure A.22

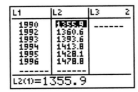

(a)

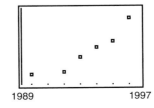

(b)

We can compute the linear regression equation (shown in Figure A.23a). The linear regression equation is $y = 20.05x - 38,561.2$ where x is the year and y is the amount of carbon dioxide in millions of tons. The line is graphed in Figure A.23b.

Figure A.23

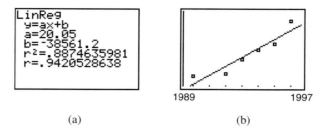

(a)   (b)

Not all data fit a linear model well. The graphing calculator gives us the capability to construct other models. Just as we did for linear regression, we can find a *curve* of best fit, provided we indicate what type of curve we would like to fit to the data. Let's see whether we can get a better fit using a parabola, or a quadratic model. If we press STAT, highlight the CALC menu, and choose 5:QuadReg, we get the **quadratic regression** equation, or the quadratic curve of best fit (see Figures A.24a and b).

Figure A.24

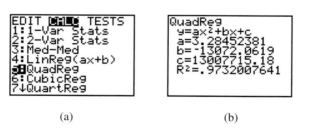

(a)   (b)

The screen in Figure A.24b shows that if we were to fit a quadratic equation $y = ax^2 + bx + c$ to the data, the resulting quadratic curve of best fit would be $y = 3.284523811x^2 - 13,072.0619x + 13,007,715.18$. (Note that you are not given r, but you are given $R^2 \approx .973$. It is beyond the scope of this text to discuss the derivation of R^2. We can, however, interpret it similarly to r^2, the square of the correlation coefficient we found in linear regression: The closer R^2 gets to 1, the closer the data fit the curve.) We see that the quadratic model fits the data better; that is, the data points come closer to the quadratic curve than to the line. Figures A.25a and b show the equation and the quadratic curve graphed with the scatterplot.

Figure A.25

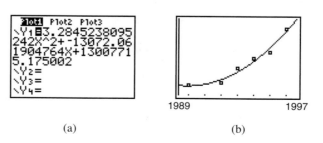

(a)   (b)

We point out here that the best fit may not necessarily be our ultimate goal. It really depends upon what we are looking to do with the model once we have found it. If we are using the model to make predictions, we should have a theoretical basis to determine the type of curve we want the data to fit. The next example illustrates this point.

Example 1   Table A.5 shows the total public and private school enrollment (kindergarten through college, in thousands) in the United States between 1980 and 1995. Create a scatterplot and guess what type of curve would fit the data best. Try linear, quadratic, and quartic regression models. Which model fits the data best? Which model makes the most sense to use for these data?

Table A.5

Year	1980	1985	1988	1990	1991	1992	1993	1994	1995
Total School Enrollment (in thousands)	58,305	57,226	58,485	60,267	61,605	62,686	63,241	63,986	64,803

Solution   Figures A. 26a and b show the entered data and the scatterplot. The regression equations and the curves are shown for the linear model (Figure A.27a), the quadratic model (Figure A.27b), and the quartic model (Figure A.27c). The quartic model seems to fit the data best. However, if we extend the quartic (see Figure A.28 on page 824) we see that there is a peak and then a downturn. We have no reason to predict such a radical downturn, so although the data may fit the quartic model better, it may make more sense to use the quadratic model.

Figure A.26

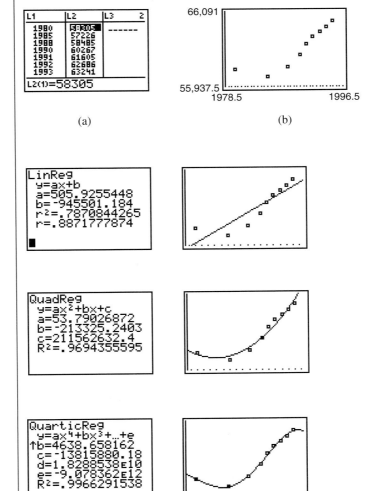

(a)

(b)

Figure A.27a
Linear regression

Figure A.27b
Quadratic regression

Figure A.27c
Quartic regression

Figure A.28

The quartic model extended to year 2002

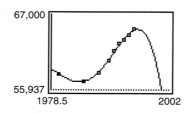

Exponential, Logistic, and Logarithmic Regression

Table A.6 shows the resident population of the United States from 1790 to 1840.

Table A.6

Year	1790	1800	1810	1820	1830	1840
Population (in millions)	3.9	5.3	7.2	9.6	13.0	17.1

We create a scatterplot (Figure A.29a). The trend almost looks linear; if we perform a linear regression, we arrive at the equation and r in Figure A.29b. However, in Chapter 5 we saw that many populations follow an exponential model; hence, we might expect an exponential model to be more appropriate.

Figure A.29

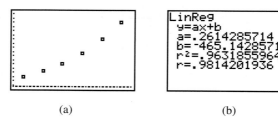

(a)   (b)

If we press $\boxed{\text{STAT}}$, highlight the CALC menu, and choose $\boxed{\text{0:ExpReg}}$, we get the exponential regression equation, or the exponential curve of best fit (see Figures A.30a, b, and c).

Figure A.30

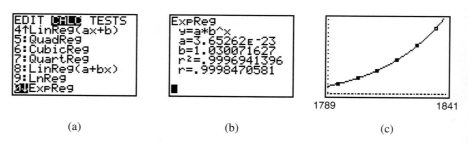

(a)   (b)   (c)

The screen shows that if we were to fit an exponential curve to the data, the exponential curve of best fit would be $y = a \cdot b^x$ where $a = 3.6526 \times 10^{-23}$ and $b = 1.03007$. Hence, $y = (3.6526 \times 10^{-23}) 1.03007^x$ is the exponential model of best fit.

The carbon dioxide emission example we discussed at the beginning of this section was a polynomial model—in particular, a quadratic model. The exponential model looks like it could work as well. Is there a way to look at the data to determine which model to try? Actually, there is:

If the ratio of y values for successive values of x is approximately the same, then the exponential model may be the most appropriate model for fitting the data.

Consider the exponential function $f(x) = Ce^{kx}$. Pick any two values of x, say x_1 and x_2. Then $\dfrac{f(x_2)}{f(x_1)} = \dfrac{Ce^{kx_2}}{Ce^{kx_1}} = e^{k(x_2 - x_1)}$. What this says is that for an exponential model, if the difference between any two x values is the same (if $x_2 - x_1$ is a constant), then the ratio of the corresponding y values will be the same $\left(\dfrac{f(x_2)}{f(x_1)}\right.$ will be constant $\left.\right)$.

Note that if $x_2 - x_1$ is constant, then $\dfrac{f(x_2)}{f(x_1)} = e^{k(x_2 - x_1)}$ is constant.

Note that for the population example, we compute the ratio $\dfrac{y_2}{y_1}$ for every 10 years ($x_2 - x_1 = 10$) and get

$$\frac{y_2}{y_1} = \frac{5.3}{3.9} \approx 1.36, \quad \frac{7.2}{5.3} \approx 1.36, \quad \frac{9.6}{7.2} \approx 1.33, \quad \frac{13.0}{9.6} \approx 1.35, \quad \frac{17.1}{13.0} \approx 1.32$$

The ratios are similar and the exponential model is appropriate. We will return to this later.

The logistic model is another model useful in predicting population growth where there are factors that eventually limit growth. The model for logistic growth is

$$y = \frac{C}{1 + ae^{-bt}}$$

Notice that as time goes on (as t gets bigger), ae^{-bt} gets closer to 0 and hence $y \to C$. The graph will have a horizontal asymptote of $y = C$. The curve for the general logistic is shown in Figure A.31.

Figure A.31

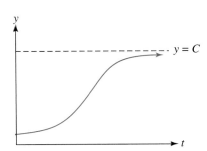

Consider a population starting anew at a particular location. The population starts increasing exponentially (accelerated growth)—slow at first, then faster and faster. But then, at some point, the curve is marked by decelerated growth. Growth is limited—perhaps limited by the food or shelter available, or by predators. This ceiling is called the *carrying capacity*.

Example 2   Table A.7 shows the deer population in New Zealand from 1979 to 1996. Try to fit quadratic, exponential, and logistic curves to the data. Which equation should you use?

Table A.7

Year	79	80	82	84	85	87	88	89
No. of deer	42,080	104,359	151,020	258,707	319,908	500,397	606,042	780,066

Year	90	91	92	93	94	95	96
No. of deer	976,290	1,129,503	1,135,242	1,078,479	1,231,109	1,178,704	1,192,138

Solution | The scatterplot is shown in Figure A.32b.

Figure A.32

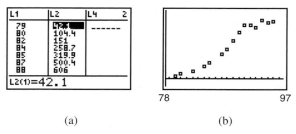

(a)   (b)

We see the quartic regression equation (Figure A.33a), and the logistic regression equation (Figure A.33b) graphed on the scatterplot.

Figure A.33

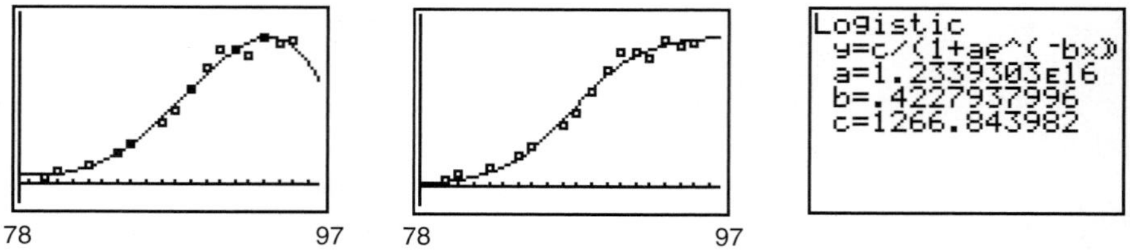

(a)   (b)   (c)

We try fitting to other curves and we see that the logistic is the closest; even though the quartic seems to hit most of the data points (it appears to be a better fit), theory supports use of the logistic model. By Figure A.33c, we see the model is

$$y = \frac{1266.844}{1 + (1.23393 \times 10^{16})e^{-0.4227938x}}$$

Note that $C \approx 1,267$, which means the horizontal asymptote is $y = 1,267$, or that 1,267,000 may be the carrying capacity of the population. If conditions that limit the population have not changed, we would expect the population to fluctuate around 1,267,000. ∎

Example 3 | Suppose we want to determine the height above the ground of an object using a barometer, an instrument that measures atmospheric pressure. The data in Table A.8 show the atmospheric pressure in pounds per square inch (ppsi) taken at various heights in feet. Find the curve of best fit for these data.

Table A.8

Atmos. pressure (ppsi)	14.7	14.16	13.66	13.17	12.69	12.23
Height (in feet)	0	1,000	2,000	3,000	4,000	5,000

Solution | The scatterplot (Figure A.34b) certainly looks linear; the linear regression model is given in Figure A.34c.

Figure A.34

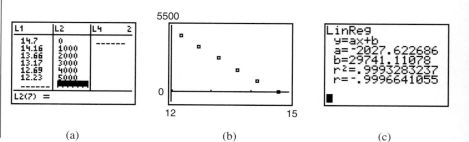

(a)   (b)   (c)

However, notice that for this example, the ratios of the x values for successive y values are close to being constant: $\frac{14.7}{14.16} \approx 1.038$, $\frac{14.16}{13.66} \approx 1.037$, $\frac{13.66}{13.17} \approx 1.037$, etc. This indicates that the x values might be growing exponentially.

Recall from Chapter 5 that the exponential and logarithmic functions are inverses of each other; that is, we can derive one from the other by reversing the assignment or switching x and y. Since we found that the x values are growing exponentially (for successive values of x), this implies that the y values must be growing logarithmically (for successive values of y). This suggests we use a logarithmic model.

We can compute a natural log regression as shown in Figure A.35 and get $y = 73{,}196.58\text{–}27{,}233.96 \ln x$ as a model that fits the data very well.

Figure A.35

(a)　　　　　　　　　　(b)　　　　　　　　　　(c)

Which Model to Choose? Linearizing Data

If we are simply fitting curves, sometimes we can easily see that a data set is linear; sometimes it is obvious if it is exponential. But as we have seen, it is not always easy to tell. A power curve and an exponential may be hard to discern. Unless there is some theoretical basis to use a particular curve, we give the following guidelines.

In general, the line is the easiest to see. If growth is rapid, you should check successive ratios: If the ratio of corresponding y values of successive terms is near constant, the exponential model is the most appropriate model to use. If the ratios are not near constant, then perhaps the power or polynomial model, which has less rapid growth, or the logistic model is more appropriate.

Since a line is easiest to identify, it may be helpful to transform the data to help determine which is a more appropriate model. This process is called *linearizing the data*.

Consider the exponential function $y = Ce^{kx}$. If we take the natural log of each side, we get $\ln y = \ln (Ce^{kx}) = \ln C + kx$. Now $\ln C$ is a constant (call it A), so the function is now $\ln y = A + kx$. This says that the natural log of y, $\ln y$, is a linear function of x. Hence, *if (x, y) are data points on a scatterplot, and if the points $(x, \ln y)$ form a line, then the data are best modelled by an exponential function.*

Consider the power function $y = ax^n$. If we take the natural log of each side, we get $\ln y = \ln(ax^n) = \ln a + n \ln x$. Now $\ln a$ is a constant (call it k), so the function is now $\ln y = k + n \ln x$. This says that the natural log of y, $\ln y$, is a linear function of the natural log of x, $\ln x$. Hence, *if (x, y) are data points on a scatterplot, and if the points $(\ln x, \ln y)$ form a line, then the data are best modelled by a power function.*

Finally, consider the function $y = a + b \ln x$. Notice that y is a linear function of the natural log of x, $\ln x$. Hence, *if (x, y) are data points on a scatterplot, and if the points $(\ln x, y)$ form a line, then the data are best modelled by a logarithmic function.*

Example 4 | Table A.9 shows the yearly population growth of alpacas if we begin with 2 males and 5 pregnant females at year 0.

Table A.9

End of Year	1	2	3	4	5	6	7	8	9	10
Number of Alpacas	11	15	20	27	36	46	60	77	98	126

Determine which model is appropriate by linearizing the data.

Solution | We begin by entering the data in the table as shown in Figure A.36. We suspect that the data are exponential rather than power, so we want to determine whether $(x, \ln y)$ falls on a line. Since L1 is x and L2 is y, we want to define L3 as the ln of each y value (L2). Move the cursor to highlight L3 and press [ln] [2nd] [L2] [)] [ENTER] (Figure A.37a). This will have the effect of assigning L3 to the ln of each L2 (see Figure A.37b). Next enter the STAT PLOT menu and define the y variable now as L3 by moving down to Ylist and pressing [2nd] [L3]. See Figure A.38a.

Figure A.36

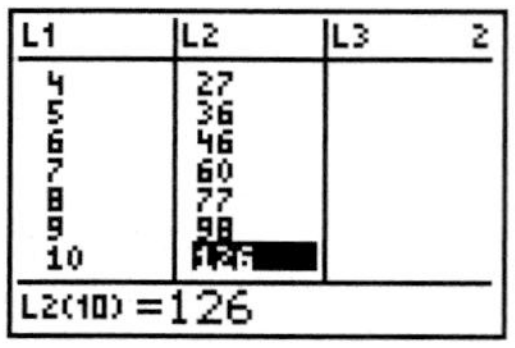

Figure A.37

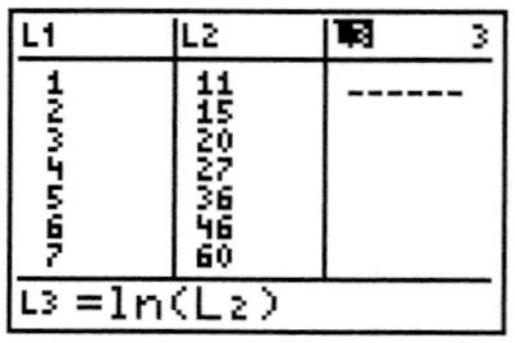

(a)

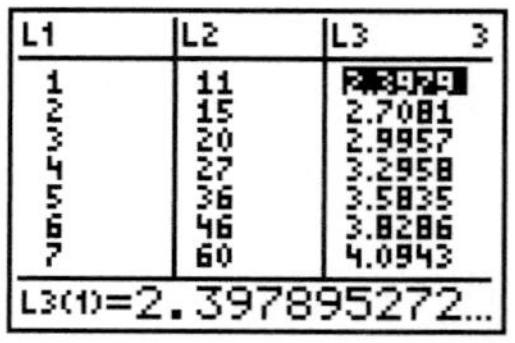

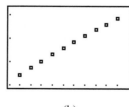

(b)

Figure A.38

(a)

(b)

Go to the ZOOM menu and choose [9:ZoomStat]. We can see in Figure A.38b that the data $(L1, L3)$ seem to form a line and since these data are in the form of $(x, \ln y)$, the guidelines tell us that an exponential model would be the most appropriate.

Figure A.39

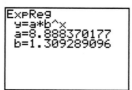

If we press [STAT], highlight the CALC menu, and choose [0:ExpReg], we get the exponential regression equation: $y = 8.88837 \cdot 1.30929^x$. (See Figure A.39.)

A.2 Exercises

1. The Bremit Company wants to determine how much to pay for advertising on television. The table shows the costs of weekly advertising (in hundreds of dollars) along with the company net profit (after paying for the commercials, in thousands of dollars) during that same week. Use a graphing calculator to create a scatterplot. Is there a quadratic trend in the data? If so, compute the regression equation, and use this equation to determine how much Bremit should pay in advertising in order to maximize its net profits.

Cost for Advertising (in hundreds of dollars)	25.2	38.5	18.2	48.2
Net Profit (in thousands of dollars)	352.1	438.4	235	426.1

Cost for Advertising (in hundreds of dollars)	60.2	54.3	62.0
Net Profit (in thousands of dollars)	300	378	275

2. A ball was thrown upward from an initial height of 20 feet, and the height of the ball was recorded every 0.5 second. The table shows the height of the ball (in feet) at various times t (seconds after the ball was thrown). Use a graphing calculator to create a scatterplot. Is there a quadratic trend in the data? If so, compute the regression equation, and use this equation to determine the maximum height of the ball.

Time (sec)	0.0	0.5	1.0	1.5	2.0
Height (feet)	20.0	45.8	64.3	73.4	76.5

Time (sec)	2.5	3.0	3.5	4.0
Height (feet)	71.1	55.4	32.6	4.2

3. The table shows the number of motor vehicle thefts in the United States from 1987 to 1996. Use a graphing calculator to create a scatterplot. Is there a quadratic trend in the data? If so, compute the regression equation, and use this equation to estimate how many motor vehicle thefts occurred in 1997.

Year	1987	1988	1989	1990	1991
Number of Thefts (in thousands)	1,289	1,433	1,565	1,636	1,662

Year	1992	1993	1994	1995	1996
Number of Thefts (in thousands)	1,611	1,563	1,539	1,472	1,394

4. The table shows credit market debt outstanding (in billions of dollars) in the United States from 1980 to 1997. Use a graphing calculator to create a scatterplot. Compute a linear model and a quadratic model. Which is a better fit? The credit market debt in 1998 was actually $23,291,000,000,000. Compute the credit market debt in 1998 predicted by the linear and then the quadratic model. Which was a better predictor for that year?

Year	1980	1985	1990	1992	1993
Debt (billions of dollars)	4,734	8,628	13,745	15,195	16,167

Year	1994	1995	1996	1997
Debt (billions of dollars)	17,212	18,445	19,794	21,200

5. The table at the top of page 832 shows the number of minutes (in millions) billed in the United States for international telephone service from 1985 to 1996. Use a graphing calculator to create a scatterplot. Compute a linear model and a quadratic model. Which is a better fit? The number of minutes billed in 1997 was actually 22,586,000,000 minutes. Compute the number of minutes in 1997 predicted by the linear and then the quadratic model. Which was a better predictor for that year?

Year	85	90	91	92
Minutes (in millions)	3,446	8,030	8,986	10,156

Year	93	94	95	96
Minutes (in millions)	11,393	13,393	15,387	19,119

Year	1990	1992	1993	1994
Number of Dealerships	24,825	23,500	22,950	22,850

Year	1995	1996	1997
Number of Dealerships	22,800	22,750	22,700

6. The accompanying table shows the population of 50-to-54-year-olds (in thousands) in the United States from 1988 to 1996. Use a graphing calculator to create a scatterplot. Compute a linear model and a quadratic model. Which is a better fit? The population of 50-to-54-year-olds in 1998 was actually 15,726,000. Compute the population predicted by the linear and then the quadratic model. Which was a better predictor for that year?

Year	1988	1990	1992	1994	1996
Population	10,995	11,314	12,054	13,194	13,927

7. The table shows the total assets (in billions of dollars) of U.S. households from 1980 to 1997. Use a graphing calculator to create a scatterplot. Compute a linear model and a quadratic model. Which is a better fit? The total assets in 1998 were actually $30,121,000,000,000. Compute the total assets predicted by the linear and then the quadratic model. Which was a better predictor for that year?

Year	1980	1985	1990
Household Assets (in billions of dollars)	6,584	10,149	14,985

Year	1995	1996	1997
Household Assets (in billions of dollars)	21,751	23,908	27,020

8. The table shows the number of franchised new car dealerships in the United States from 1990 to 1997. Use a graphing calculator to create a scatterplot. Compute a quadratic, a cubic, and a quartic model. The number of franchised new car dealerships in 1998 was actually 22,600. Compute the number of dealerships predicted by the three models. Which was the best predictor for that year?

9. The table shows the United States military expenditures (in billions of dollars) from 1987 to 1994. Use a graphing calculator to create a scatterplot. What type of trend do you see? Compute a quadratic, a cubic, and a quartic model. The total expenditures in 1995 were actually $865,000,000,000. Compute the expenditures predicted by the three models. Which was the best predictor for that year?

Year	1987	1988	1989	1990
Military Expenditures (in billions of dollars)	1,051	1,080	1,089	1,106

Year	1991	1992	1993	1994
Military Expenditures (in billions of dollars)	1,049	974	912	879

10. The table shows the total personal consumption expenditures for recreation (in billions of dollars) from 1990 to 1996. Use a graphing calculator to create a scatterplot. Compute a quadratic and a quartic model. Compute the amount spent on recreation predicted by the two models for 1997.

Year	1990	1991	1992	1993
Amount Spent on Recreation (in billions)	281.6	292.0	310.8	340.2

Year	1994	1995	1996
Amount Spent on Recreation (in billions)	370.2	402.5	431.1

11. Use a graphing calculator to create a scatterplot for the data in Exercise 4. Compute a linear, a quadratic, and an exponential model. Which is the best fit? The credit market debt in 1998 was actually $23,291,000,000,000. Compute the credit market debt in 1998 predicted by all three models. Which was the best predictor for that year?

12. Use a graphing calculator to create a scatterplot for the data in Exercise 7. Compute a linear, a quadratic, and an exponential model. Which is the best fit? The total assets in 1998 were actually $30,121,000,000,000. Compute the total assets predicted by all three models. Which is the best predictor for that year?

13. The table shows the per consumer expenditures (in dollars) for fees and admissions for recreation from 1985 to 1996. Use a graphing calculator to create a scatterplot. Compute a linear, a quadratic, and an exponential model. In 1997, $471 was the actual per consumer expenditure on fees and admissions. Compute the amount predicted by all three models. Which is the best predictor for that year?

Year	1985	1988	1990	1992
Amount Spent on Fees and Admissions (in dollars)	320	353	371	379

Year	1993	1994	1995	1996
Amount Spent on Fees and Admissions (in dollars)	414	439	433	459

14. The table shows the percent of instructional classrooms with Internet access from 1994 to 1997. Use a graphing calculator to create a scatterplot. Is there an exponential trend in the data? If so, compute the regression equation, and use the regression equation to predict the percent of instructional classrooms with Internet access in 1998.

Year	94	95	96	97
Percent of Instructional Classrooms with Internet Access	3	8	14	27

15. David makes an investment. The table shows how much his investment is worth (in dollars) at the end of each given year. Use a graphing calculator to create a scatterplot. Compute the exponential regression model. Assuming this model, determine how much was invested, and the yearly (simple) interest rate. [*Hint*: Compare the model you computed with the formula for simple interest found on page 397.]

(End of) Year	2	4	6	8	10
Value of Investment	5,539	6,136	6,796	7,529	8,340

16. In his lab, Manoj cultured a certain bacterium and measured its volume every few hours to determine its growth rate. The table shows the volume (in cubic centimeters) at the end of the given hours. Use a graphing calculator to create a scatterplot. Compute the exponential regression model.

(End of) Hour	1	2	5	8
Volume (in cc)	526	553	642	746

(End of) Hour	10	16	20	24
Volume (in cc)	824	1,113	1,360	1,660

17. The table shows the U.S. population (in millions) from 1900 to 1998. Use a graphing calculator to create a scatterplot. What type of trend do you see? Compute an exponential and a logistic model. In 2000, the population was 281,422,000. Compute the 2000 population predicted by both models. Which is the better predictor for that year?

Year	Population (in millions)
1900	75.99
1910	91.97
1920	105.17
1930	122.78
1940	131.67
1950	151.33
1960	179.32
1970	203.30
1980	226.52
1990	248.72
1998	270.56

18. The table shows the total number of endangered and threatened species in the United States at the end of selected calendar years from 1980 to 2000. Use a graphing calculator to create a scatterplot. You may have to enter the year 1980 as 0, 1982 as 2, etc. What type of trend do you see? Compute the exponential and the logistic regression equations. In 2001, the number of endangered and threatened species totalled 1,254. Compute the 2001 number predicted by both models. Which is the better predictor for that year?

Year	No. of endangered species
1980	281
1982	294
1984	326
1986	422
1988	526
1990	596
1992	753
1994	941
1996	1,053
1998	1,194
2000	1,244

19. The table shows the accumulated weekly gross of the movie *Gladiator* during the first eleven weeks of its showing in the United States in 2000. Use a graphing calculator to create a scatterplot. What type of trend do you see? Compute a logarithmic and a logistic model. Which model do you think is more appropriate for these data?

Week	Accumulated Gross (in millions)
1	49.02
2	83.39
3	109.94
4	130.58
5	143.10
6	153.63
7	161.59
8	167.33
9	171.94
10	175.01
11	177.44

20. The table shows the accumulated weekly gross of the movie *Monsters, Inc.*, in millions of dollars, during the first seven weeks of its showing in the United States in 2001. Use a graphing calculator to create scatterplot. What type of trend do you see? Compute a logarithmic and a logistic model. Which model do you think is more appropriate for these data?

Week	1	2	3	4
Accumulated Gross (in millions)	76.60	133.62	168.18	194.92

Week	5	6	7
Accumulated Gross (in millions)	205.81	213.85	220.65

Question for Thought

21. Although the exponential regression model used in the TI-83 Plus calculator is of the form $y = a \cdot b^x$, the exponential model we have been using in this text is $y = Ce^{kx}$, where C and k are constants. We can convert the $a \cdot b^x$ model to the Ce^{kx} model as follows:

> To convert from $y = a \cdot b^x$ to $y = Ce^{kx}$, we first assume the constants a and C are equal, and therefore we need to find the value of k that will make b^x equal to e^{kx}. We note that if $b^x = e^{kx}$ for all x, then, in particular, for $x = 1$ we get $b^1 = e^k$ or $b = e^k$, which means $\ln b = k$.
> Hence, to convert the model $y = a \cdot b^x$ to the model $y = Ce^{kx}$, we let $a = C$ and $k = \ln b$.

   (a) Convert the equation $y = 5 \cdot 3^x$ into an equation in the form $y = Ce^{kx}$.
   (b) On page 660, we found an exponential regression equation to describe the resident population of the United States from 1790 to 1840; this model was $y = (3.6526 \times 10^{-23})1.03007^x$. Express this model in $y = Ce^{kx}$ form.

ANSWERS
TO SELECTED EXERCISES

Chapter 1

Exercises 1.1

1. $\dfrac{2}{9}$ **3.** $\dfrac{41}{9}$ **5.** $\dfrac{8230}{999}$

7. Associative property of addition   **9.** False

11. Associative property of multiplication

13. Distributive property

15. Multiplicative inverse property

19. No, $\sqrt{2}$ and $-\sqrt{2}$ are examples of two irrational numbers whose sum (0) is not irrational.

21. ⟵──○──⟶ 4   **23.** ⟵──○──⟶ 5

25. ⟵──○────○──⟶ -8 -2

27. $(5, \infty)$ ⟵────○──⟶ 5

29. $[-5, \infty)$ ⟵────●──⟶ -5

31. $[-8, -5)$ ⟵──●────○──⟶ -8 -5

33. $(-\infty, -2]$ ⟵────●──⟶ -2

35. $(-9, -2]$ ⟵──○────●──⟶ -9 -2

37. 2   **39.** $\sqrt{2} - 1$ **41.** $\begin{cases} x - 5 & \text{if } x \geq 5 \\ 5 - x & \text{if } x < 5 \end{cases}$

43. $x^2 + 1$ **45.** 1 unit   **47.** 11 units

Exercises 1.2

1. -5 **3.** -36 **5.** -5.552 **7.** 3.29   **9.** -12

11. -7 **13.** 11   **15.** 10   **17.** $\dfrac{7}{2}$ **19.** $\dfrac{7}{12}$

21. $-\dfrac{59}{100}$ **23.** 0   **25.** $-\dfrac{82}{25}$ **27.** $\dfrac{48}{65}$ **29.** $\dfrac{20}{63}$

31. -14 **33.** 1   **35.** 7.0278   **37.** 1.15   **39.** 1.06

41. 0.087   **43. (a)** 240 million   **(b)** 8.96 million

45. (a) 1994, 330 billion; 1995, 347 billion; 1996, 363 billion; 1997, 380 billion

(b) 94–95, 5.04% increase; 95–96, 4.8% increase; 96–97, 4.6% increase

Exercises 1.3

1. $-6x^3y^3$ **3.** $45x^3y^3$ **5.** $6x^2 - 5x - 6$

7. $6x^2 - 19x - 7$ **9.** $6x^3 - 13x^2 + 9x - 2$

11. $125x^3 + 27$ **13.** $x^3 - 4x^2 + x + 6$

15. $x^2 - 6xy + 9y^2$ **17.** $x^4 - 49$

19. $4x^2 - 12xy + 9y^2$ **21.** $x^2 - 2xy + y^2 - 49$

23. $x^2 - 2xy + y^2 + 4x - 4y + 4$

25. $-4x + 8$ **27.** $5y^3 - 51y^2 + 125y + 25$

29. $-12x + 18$ **31.** $4xh + 2h^2$

33. (a) $192 - 5x - 2x^2$ **(c)** $\pi(10x + 25)$

(b) $102 + 12x$ **(d)** $138 + 36x$

35. $6 - 4x^2$ sq feet   **37.** $A = 2x^2 + \frac{1}{2}\pi x^2; P = 4x + x\pi$

39. $2x^2 + 20x + 16$ **41.** $(x + 5)(x - 4)$

43. $2(3x - 2)(2x + 3)$ **45.** $(3a - 1)(b - 2)$

47. $5(3x - 2)(x - 1)^2$ **49.** $3(2x^2 + 1)(x^2 - 4x - 6)$

51. $(x - 2)(a + b)$ **53.** $(x - 3)(x^2 - 5)$

55. $(x - 3)(x + 3)$ **57.** $(x - 4)^2$

59. $(2x - 1)(4x^2 + 2x + 1)$

61. $3(3x - 2)(9x^2 + 6x + 4)$

63. $(x + 3)(x + 4)(x - 4)$

65. $(x^2 - 6)(x - 2)(x + 2)$

67. $(x - y - 4)(x - y + 4)$

69. $(a + 1)(a - 1)^2(a^2 + a + 1)$ **71.** $n(n - 2)(n - 1)$

Exercises 1.4

1. $\dfrac{x - y}{x + y}$ **3.** $\dfrac{x + 1}{2x}$ **5.** $\dfrac{x + y}{2x(x - y)}$ **7.** $\dfrac{8x}{(x + 2)^3}$

9. 3   **11.** $\dfrac{12x^2 + 16}{(x^2 - 4)^3}$ **13.** $\dfrac{1}{x - 2}$

15. $5x - 1$ **17.** $\dfrac{1}{3x + 2}$ **19.** $\dfrac{25 - 6x}{45x^2y^3}$

21. $\dfrac{2xy - x^2}{y(x - y)}$ **23.** $\dfrac{5a + 3}{a + 2}$

25. $\dfrac{2x^3 - 5x^2 + 9x - 9}{(x - 2)(x - 1)}$ **27.** $-\dfrac{1}{x - 3}$ **29.** $\dfrac{x^2 - 4x - 1}{(x - 2)^2}$

31. $\dfrac{x + 1}{x}$ **33.** $\dfrac{-1}{(x + 1)(x + 3)}$ **35.** $\dfrac{-5}{x(x + h)}$

37. $\dfrac{2}{(x + h + 2)(x + 2)}$ **39.** $\dfrac{1}{x}$ **41.** $\dfrac{3x^2 + 7x + 2}{x^2 - 9x - 4}$

Exercises 1.5

1. x^6 **3.** $3x^5y^5$ **5.** $\dfrac{1+2x}{x^2+2x}$ **7.** $\dfrac{y+x}{x^2}$

9. 7.5×10^9

11. Approximately 8 minutes 20 seconds

13. $\dfrac{1}{8}$ **15.** $\dfrac{9}{4}$ **17.** $x^{5/6}y^{2/3}$ **19.** $(x^2+1)^{1/10}$

21. $x + 4x^{1/2} + 4$ **23.** $x - 4$ **25.** $\dfrac{3}{(x+1)^2}$

27. $\dfrac{4x(2x^2+1)}{(x^2+1)^{1/2}}$ **29.** $\dfrac{5x^3-6}{2(x^3-3)^{1/2}}$

31. $\dfrac{3}{x(x^3-3)^{1/2}} = 3x^{-1}(x^3-3)^{-1/2}$ **33.** $2x^2y^3\sqrt{6}$

35. $\dfrac{4x^4}{3}$ **37.** $\sqrt[4]{x^4-y^4}$ **39.** $6y\sqrt[4]{2y^2}$ **41.** $\dfrac{y^4}{x^3}$

43. $3x^3\sqrt[6]{3^5x}$ **45.** $\dfrac{5\sqrt{3x}}{3x^3}$ **47.** $\dfrac{\sqrt{x}+4}{x-16}$

49. $\dfrac{x-2}{\sqrt{(x-1)(x-2)}}$ **51.** $\dfrac{1}{\sqrt{x+3}}$

53. $\dfrac{1}{\sqrt{x+h}+\sqrt{x}}$ **55.** $18x^3y^3\sqrt{2xy}$

57. $\dfrac{3y\sqrt[3]{12x^2}}{2x}$ **59.** $\dfrac{y\sqrt{3xy}}{x^2}$

61. $\sqrt{x-1}$ **63.** $15\sqrt{3}-9\sqrt{2}$ **65.** $2-6\sqrt{3}$

67. $a - 25$ **69.** $12 - 6\sqrt{x}$ **71.** $\dfrac{\sqrt{14}+\sqrt{10}}{2}$ **73.** 12

75. $\dfrac{2x\sqrt{x^2-2}-1}{\sqrt{x^2-2}}$ **77.** $\dfrac{3x^2-8}{\sqrt{x^2-2}}$

79. $\dfrac{\sqrt{2x}-\sqrt{2(x+h)}}{h\sqrt{x(x+h)}}$

Exercises 1.6

1. 1   **3.** 1   **5.** $-i$ **7.** $3+4i$ **9.** $5-\dfrac{\sqrt{3}}{6}i$

11. $9-6i$ **13.** $17+0i=17$

15. $13-13i$ **17.** 61   **19.** $-5-12i$

21. $1+4i$ **23.** $3-2i$ **25.** $\dfrac{6}{5}-\dfrac{3}{5}i$ **27.** $-\dfrac{1}{5}+\dfrac{2}{5}i$

29. $0-1i=-i$ **31.** $\dfrac{26}{29}-\dfrac{22}{29}i$ **33.** -5

35. $4-6i$ **39.** $0-1i=-i$

Chapter 1 Review Exercises

1. $\dfrac{817}{99}$ **2.** $\dfrac{72,400}{999}$

3. Commutative property of addition
4. Distributive property
5. $[-4, \infty)$ ⟵————▸
                          -4

6. $(-8, -1]$ ⟵○————●▸
                    -8 -1

7. $(-\infty, 3)$ ◀————○——▸
                              3

8. $[0, \infty)$ ⟵——●————▸
                        0

9. $\sqrt{5}-1$ **10.** $7x^2+3$

11. 19   **12.** -18 **13.** $\dfrac{38}{9}$ **14.** $-\dfrac{25}{9}$ **15.** $-\dfrac{25}{6}$

16. $-\dfrac{28}{15}$ **17.** $108x^5y^5$ **18.** $6x^2y^4-9xy^4$

19. $2x^2+x-21$ **20.** $6x^2-11x+4$ **21.** $25a^6-b^2$
22. $25a^6-10a^3b+b^2$ **23.** $27x^3-1$
24. $32x^3+8x^2-2x+1$ **25.** $2x^3-9x^2+8x+4$
26. $x^4-10x^3+37x^2-60x+36$
27. $(x-5)(x+3)$ **28.** $(3x+2)(2x-5)$
29. $(x-2)(3xy-6y+5)$
30. $2x(x+8)^2(1-8x-x^2)$
31. $(5x-3)^2$ **32.** $(a^2+9)(a+3)(a-3)$
33. $(a+4)(a-4)(a+1)$ **34.** $2(x-3)(x+3)^2$
35. $(a-1)(a^2-a+1)(a+1)^2$

36. $(a+2+x)(a+2-x)$ **37.** $\dfrac{1}{x+y}$ **38.** $\dfrac{x}{x+3}$

39. $\dfrac{3x^2+x-2}{x(x-2)}$ **40.** $\dfrac{x^2-11x+12}{(x+3)(x-3)}$

41. $\dfrac{x+2}{(x-1)^2}$ **42.** $\dfrac{2x^2+2}{(x-1)^2(x+1)}$

43. $\dfrac{xy}{y-6x}$ **44.** $\dfrac{6y^2-3y+1}{6y(2y+1)}$ **45.** $\dfrac{y^2}{x^5}$ **46.** $\dfrac{x^7}{24y}$

47. $\dfrac{1}{x^2}$ **48.** $\dfrac{y^2+x}{x^2}$ **49.** $\dfrac{3}{xy^3}$ **50.** $\dfrac{3y^2-x}{y^2+xy^3}$

51. $x^{49/60}$ **52.** $\dfrac{2x}{x^2+4}$ **53.** $\dfrac{1}{16}$ **54.** $\dfrac{1}{27}$

55. $\dfrac{4x^5+8x^3+x^2+4x}{2(x^2+1)^{3/2}}$ **56.** $\dfrac{5x^2-3}{3(x^2-1)^{2/3}}$

57. $3x^2(x^2+1)^{1/2}$ **58.** $4x^5(x^2-1)^{1/3}$

59. $2x\sqrt{x+1}+\dfrac{x^2}{\sqrt{(x^2+1)^3}}$

60. $\sqrt[3]{x^2-1}+\dfrac{2x^2}{3\sqrt[3]{(x^2-1)^2}}$

61. $2x^4y^5\sqrt[3]{6}$ **62.** $x^4y^3\sqrt{3y}$ **63.** $\dfrac{5\sqrt{x}}{x^2}$

64. $\sqrt{3}+1$ **65.** $4\sqrt[3]{9a^2b^2}$ **66.** $6x^7y^2$

67. $\dfrac{2}{3}$ **68.** $\dfrac{3xy^2\sqrt{2}}{2}$ **69.** $2x-4\sqrt{2x}+4$ **70.** $2x-4$

71. $8\sqrt{a}-20$ **72.** $\sqrt{x-1}$ **73.** $\dfrac{\sqrt{22}+\sqrt{6}}{4}$

74. $x\sqrt{x}-3x+2\sqrt{x}-6$ **75.** $\dfrac{2\sqrt{x^2-4}-2}{\sqrt{x^2-4}}$

76. $\dfrac{15x^2-10}{\sqrt{3x^2-1}}$ **77.** $10^4\,\text{Å}=1\mu$ **78.** 5.916×10^9 km

79. $5-i$ **80.** $4-8i$ **81.** $24+10i$

82. 13   **83.** $\dfrac{3}{2}-i$ **84.** $\dfrac{3}{13}+\dfrac{24}{13}i$

Chapter 1 Practice Test

1. Multiplicative inverse property

2. (a) ; $(-\infty, -2]$

    (b) ; $(-6, -2]$

3. $\sqrt{5} - 1$ **4.** $\dfrac{26}{5}$

5. (a) $9a^6 - 4b^2$ **(b)** $3x^3 - 31x^2 + 75x + 25$

6. (a) $(5x - 2)(2x + 1)$ **(b)** $3(y + 3)(y^2 + 3y + 5)$

    (c) $(x - 3)(x + 3)(x + 2)$

7. (a) $\dfrac{x + 3}{(x + 1)^2}$ **(b)** $\dfrac{ab}{a + 5b}$

    (c) $\dfrac{(x^2 + y^2)(x + y)^2}{x^2 y^2}$ **(d)** $\dfrac{3x^2}{x^3 - 2}$

8. (a) $\dfrac{4x^2 \sqrt{xy}}{3y}$ **(b)** $\dfrac{5x\sqrt{2x}}{2}$

    (c) $9x^2 - 6x\sqrt{2} + 2$ **(d)** $\sqrt{10} + \sqrt{7}$

9. (a) $-16 - 30i$ **(b)** $\dfrac{3}{10} - \dfrac{11}{10}i$

Chapter 2

Exercises 2.1

1. $x = \dfrac{16}{5}$ **3.** $(-\infty, -\frac{1}{3}]$ **5.** $(\frac{23}{9}, \infty)$ **7.** $x = -\dfrac{6}{29}$

9. $y = \dfrac{1}{2}$ **11.** $(-\infty, \frac{5}{9})$ **13.** No solution

15. $x = -\dfrac{9}{5}$ **17.** $a = 4$ **19.** $x = \dfrac{3}{2}$

21. No solution   **23.** $y = \dfrac{2x + 6}{5}$

25. $W = \dfrac{S - 2LH}{2L + 2H}$ **27.** $h = \dfrac{2A}{b_1 + b_2}$

29. $y = \dfrac{1}{2}$ for $x \neq \dfrac{2}{3}$ **31.** $s > \dfrac{x - \mu}{1.96}$

33. $f = \dfrac{f_1 f_2}{f_1 + f_2}$ **35.** $x = \dfrac{2}{3 - y}$

37.   **39.**

41.   **43.**

45. s is multiplied by 9   **47.** E is divided by 9

49. (a) \$2,100   **(b)** \$0.75x **51.** \$0.8702x; \$261.06

53. \$1.4388x; \$575.52

55. Let x = no. of 20-lb boxes; then $50 - x$ = no. of 25-lb boxes. $20x + 25(50 - x) = 1175$; $x = 15$; 35 25-lb boxes, 15 20-lb boxes

57. Let x = no. of orchestra seats; then $56 - x$ = no. of balcony seats. $48x + 28(56 - x) = 2328$; $x = 38$; 38 orchestra seats

59. Let t = no. of minutes older machine works; then $110 - t$ = no. of minutes new machine works. $50t + 70(110 - t) = 7200$, $t = 25$; 1250 copies

61. Let x = no. of ounces of 20% solution. $0.20x + 0.50(5) = 0.30(x + 5)$, $x = 10$; 10 oz

63. Let x = amount invested in the high-risk certificate; then $\$8000 - x$ = amount invested in the lower-risk certificate. $0.14x + 0.09(8000 - x) \geq 890$, $x \geq \$3400$; at least \$3400.

65. Let t = amount of time it takes for both pipes to fill the pool together. $\dfrac{t}{3} + \dfrac{t}{2} = 1$, $t = \dfrac{6}{5}$; $1\frac{1}{5}$ days.

67. Let t = amount of time it takes to fill the tub with the drain open. $\dfrac{t}{10} - \dfrac{t}{15} = 1$, $t = 30$; 30 minutes.

Exercises 2.2

1. $x = 12, x = -12$ **3.** $(-9, 9)$ **5.** $[6, 10]$

7. $x = \dfrac{11}{2}, x = -\dfrac{3}{2}$ **9.** $(-4, -1)$

11. $(-\infty, 0) \cup (\frac{4}{3}, \infty)$ **13.** $(0, 3)$

15. $x = -\dfrac{3}{4}, x = -\dfrac{21}{4}$ **17.** $(-\infty, \frac{1}{3}) \cup (\frac{7}{3}, \infty)$

19. $(-\infty, -3] \cup [3, \infty)$ **21.** $x = -\dfrac{5}{2}, x = \dfrac{11}{2}$

23. $x = \dfrac{5}{2}, x = \dfrac{15}{2}$ **25.** $x = \dfrac{7}{3}, x = -1$

27. $x = 2, x = 1$ **29.** $x = -\dfrac{1}{3}, x = \dfrac{5}{3}$

31. $(-\frac{7}{3}, \frac{17}{3})$ **33.** No solution   **35.** $x^2 + 8$

37. $4x^2 + 9$ **39.** Not possible   **41.** $\dfrac{5 + x^6}{4}$

45. $74°F–84°F$ **47.** $5.52\%–6.48\%$

Exercises 2.3

1. $y = \pm\sqrt{65}$ **3.** $x = 5, -\dfrac{3}{2}$ **5.** $x = \pm\sqrt{10}$

7. $x = -1 \pm \sqrt{5}$ **9.** $a = \dfrac{4 \pm \sqrt{43}}{2}$

11. $x = 1 \pm \sqrt{39}$ **13.** $a = 1, \dfrac{1}{2}$ **15.** $z = \dfrac{7}{6}$

17. $x = 3, -1$ **19.** $x = -\dfrac{4}{3}$ **21.** $x = -3 \pm i\sqrt{11}$

23. $x \approx 4.7114, -2.4257$ **25.** 1.60   **27.** 0.65   **29.** 68.2

31. Distinct and complex   **33.** Distinct and real

35. Distinct and complex

37. $k(x - 5)(x - 3)$, k is any real no.

39. $kx(x - 7)$, k is any real no.

41. $k(x - i)(x + i)$, k is any real no.

43. Let x = the number. $x + \dfrac{1}{x} = \dfrac{13}{6}$; $x = \dfrac{2}{3}, \dfrac{3}{2}$

45. Let x = the number. $x\left(\dfrac{9}{2} - x\right) = 5$; $x = 2, 5/2$

47. Let w = width; then length = $2w + 5$. $w(2w + 5) = 75$, $w = 5$; 5 by 15

49. $2\sqrt{2} \approx 2.83$ ft   **51.** 32 sq in.

53. Let x = width of the path. $(21 + 2x)^2 - (21)^2 = 184$, $x = 2$; 2 feet

55. Let x = width of the path. $\pi(x + 8)^2 - \pi 8^2 = 57\pi$, $x = 3$ ft

57. (a) 96 ft **(b)** $\sqrt{10} \approx 3.16$ sec

Exercises 2.4

1. 0.50 **3.** 6 **5.** $a = 484$ **7.** $x = 25$ **9.** $x = 4$
11. $x = 9$ **13.** $x = \pm 1; \pm 4$ **15.** $x = -1; 343$
17. No solution **19.** $a = 16, 1/256$ **21.** $a = 81$
23. $x = 2, 3, 4, 6$ **25.** $a = 3^{4/3}$ **27.** $x = 30$
29. $y = \dfrac{1}{3^{4/3}}$ **31.** $y = \pm 3, \pm i\sqrt{2}$ **33.** $a = \pm 6, \pm i$
35. $x = \pm 3, -4$ **37.** $x = 1$ **39.** $x = 3$
41. $s = \sqrt{\dfrac{2gm}{K}} = \dfrac{\sqrt{2gmK}}{K}$

43. $a = \pm\sqrt{\dfrac{3b-2}{2x+3}}$
$= \dfrac{\pm\sqrt{(3b-2)(2x+3)}}{2x+3}\left(x \neq -\dfrac{3}{2}\right)$

45. $\rho = \sqrt{1 - \sigma_r^2(n-1)}$

Exercises 2.5

1. $(-\infty, -6) \cup (4, \infty)$ **3.** $(-\frac{1}{2}, 2)$ **5.** $[-3, 8]$
7. $(-\infty, -1] \cup [\frac{5}{2}, \infty)$ **9.** $\left[-\dfrac{1}{5}, \dfrac{2}{3}\right]$
11. $(-\infty, -\frac{3}{2}) \cup (\frac{1}{2}, \infty)$ **13.** $(-\infty, \infty)$
15. $(-\infty, 1) \cup (1, \infty)$ **17.** $(-\infty, -1) \cup (1, \infty)$
19. $(-\infty, -4) \cup (-1, 1)$ **21.** $\{-2\} \cup [2, \infty)$
23. $\left(\dfrac{3 - \sqrt{21}}{2}, \dfrac{3 + \sqrt{21}}{2}\right)$
25. $\left(-\infty, \dfrac{1 - \sqrt{17}}{4}\right] \cup \left[\dfrac{1 + \sqrt{17}}{4}, \infty\right)$ **27.** $(-\infty, -1)$
29. $(-\infty, -5) \cup [\frac{1}{2}, \infty)$ **31.** $(-1, 2)$
33. $[\frac{1}{5}, 2)$ **35.** $(-\infty, 0] \cup (1, \infty)$
37. $(-\frac{1}{2}, 1)$ **39.** $[-3, -1)$ **41.** $(-1, 2] \cup (3, \infty)$
43. $(-\infty, -1) \cup [0, 1)$ **45.** $(-5, -3) \cup (3, \infty)$
47. $[0, \infty)$ **49.** $(-\infty, -4) \cup (-1, 1) \cup (4, \infty)$
51. $\left(\dfrac{1}{3}, 1\right)$ **53.** $\left(\dfrac{1}{2}, \infty\right)$
55. Between (and including) 16°C and 26°C

Exercises 2.6

1. (a) Let s = length of a side. $P = 84 = 4s \Rightarrow$
$s = 21$ in.; $A = s^2 = 441$ sq. in.
(b) Let s = the length of a side. $P = x = 4s \Rightarrow$
$s = \frac{x}{4}; A = s^2 = \frac{x^2}{16}$ sq. in.
3. Let s = the length of a side. $s^2 + s^2 = 5^2 \Rightarrow s = \frac{5}{\sqrt{2}};$
$A = s^2 = \frac{25}{2} = 12\frac{1}{2}$ sq ft
5. Let s = the length of a side and d = the length of a diagonal. $A = 84 = s^2 \Rightarrow s = \sqrt{84}; d^2 = s^2 + s^2 \Rightarrow$
$d = \sqrt{168} \approx 12.96$ cm

7. Let s = the length of a side and d = the length of a diagonal: $P = 4s \Rightarrow s = \dfrac{P}{4}; d^2 = s^2 + s^2 \Rightarrow$
$d = s\sqrt{2} = \dfrac{P\sqrt{2}}{4}$
9. Let s = the length of a side and d = the length of a diagonal. $d^2 = s^2 + s^2 \Rightarrow s = \dfrac{d}{\sqrt{2}}; P = 4s = 2d\sqrt{2}$
11. Let d = the diameter and r = the radius.
$r = \dfrac{d}{2} = \dfrac{8}{2} = 4; V = \dfrac{4}{3}\pi r^3 \Rightarrow V = \dfrac{256\pi}{3} \approx 268.08$ cu ft
13. Let d = the diameter and r = the radius. $r = \dfrac{d}{2};$
$S = 4\pi r^2 \Rightarrow S = \pi d^2$
15. $A = 100$ sq in. **17.** $A = 4r^2$ sq in.
19. (a) $A = 36\pi \approx 113.10$ sq in.
(b) $A = \pi(3t)^2 = 9\pi t^2$ sq in.
21. $4500\pi \approx 14,137.17$ cu in.
23. $V = \dfrac{4}{3}\pi(3t)^3 = 36\pi t^3$ cu in.
25. Cost = $8(500) = \$4000$
27. Let s = the length of a side. $200 = s^2 \Rightarrow s = 10\sqrt{2};$
cost $= 8(4)(10\sqrt{2}) = 320\sqrt{2} \approx \$452.55$
29. Let C = the circumference and r = the radius.
$C = 2\pi r = 2\pi(5) = 10\pi$; cost = $2(10\pi) = 20\pi \approx$
$\$62.83$
31. Let A = the area, r = the radius, and C = the circumference. $A = 90 = \pi r^2 \Rightarrow r = \sqrt{\dfrac{90}{\pi}} = \dfrac{3\sqrt{10\pi}}{\pi};$
$C = 2\pi r = 6\sqrt{10\pi}$; cost $= 3(6\sqrt{10\pi})$
$= 18\sqrt{10\pi} \approx \$100.89$
33. 100 miles

Chapter 2 Review Exercises

1. $x = -\dfrac{2}{7}$ **2.** No solution **3.** $\left[\dfrac{15}{38}, \infty\right)$
4. $\left(-\dfrac{9}{4}, \infty\right)$ **5.** $x = \dfrac{38}{13}$ **6.** No solution **7.** $\left[\dfrac{8}{9}, 1\right)$
8. $\left[-2, \dfrac{2}{5}\right]$ **9.** $F = \dfrac{9C + 160}{5}$
10. $x = \dfrac{y+9}{1-3y}$
11. Let x = amount of pure water;
$0x + 0.60(2) < 0.40(x + 2); x > 1$; more than 1 liter
12. Let t = the number of hours it takes them to paint the room together; $\dfrac{t}{3} + \dfrac{t}{5/2} = 1; 1\dfrac{4}{11}$ hours
13. $x = \dfrac{11}{7}, -1$ **14.** $x = -\dfrac{1}{2}, \dfrac{7}{2}$ **15.** $\left(-\dfrac{1}{3}, \dfrac{11}{3}\right)$
16. $\left(-\infty, -\dfrac{1}{3}\right] \cup \left[\dfrac{11}{3}, \infty\right)$ **17.** $\left(-\infty, -\dfrac{7}{2}\right] \cup [4, \infty)$
18. $\left(-\dfrac{3}{2}, 3\right)$ **19.** $x = -\dfrac{1}{3}, -1$ **20.** $x = 0$

21. $x = \dfrac{1 \pm \sqrt{33}}{2}$ **22.** $t = 5, -1$ **23.** $x = \pm\sqrt{5}$

24. $x = -\dfrac{2}{3}, \dfrac{1}{2}$ **25.** $x = \dfrac{4}{3}, 5$ **26.** $x = 1$

27. $x = -\dfrac{3}{2}, 5$ **28.** $x = 3, -1$ **29.** $x = 5$

30. $x = \dfrac{193}{64}$ **31.** $x = \dfrac{27}{8}, -1$ **32.** $x = 0, 3$

33. $x = \pm\sqrt{\dfrac{9 - 7y^2}{5}} = \pm\dfrac{\sqrt{5(9 - 7y^2)}}{5}$

34. $a = \dfrac{b}{5}, a = -\dfrac{2b}{3}$ **35.** $\left(-\infty, -\dfrac{1}{3}\right) \cup (5, \infty)$

36. $\left(-\infty, -\dfrac{3}{2}\right] \cup [-1, \infty)$ **37.** No solution

38. $(-\infty, -2] \cup [-1, 1]$ **39.** $(-\infty, -1) \cup \left(\dfrac{2}{3}, \infty\right)$

40. $\left(-\infty, \dfrac{3}{2}\right] \cup (5, \infty)$ **41.** $(0, 1]$

42. $\left(-\infty, -\dfrac{5}{14}\right) \cup \left(-\dfrac{1}{3}, \infty\right)$

43. (a) 140 feet   (b) 5.5 seconds
44. Let $x =$ width of the path;
$\pi(x + 10)^2 - \pi(10)^2 = 44\pi$; $x = 2$ feet
45. $25\pi t^2$ sq in.
46. Let $C =$ the circumference;
$C = 2\pi r$; cost $= 3(2\pi r) = 6\pi r$ dollars

Chapter 2 Practice Test

1. $x = \dfrac{5}{2}$ **2.** $(-17, \infty)$ **3.** $x = 0, -7$

4. $\left[-\dfrac{9}{2}, \dfrac{21}{2}\right]$ **5.** $(-\infty, \frac{3}{2}) \cup (\frac{11}{2}, \infty)$

6. $(-\infty, \infty)$ **7.** $x = 2, -1$

8. $x = \pm\sqrt{\dfrac{5}{3}} = \pm\dfrac{\sqrt{15}}{3}$ **9.** $x = \frac{7}{5}$ **10.** $x = 6, 5$

11. $[-1, \frac{3}{2}]$ **12.** $(-\infty, 1) \cup (4, \infty)$

13. $(-\infty, -\frac{1}{2}) \cup (2, \infty)$ **14.** $[-5, 0)$

15. Let $t =$ time to process 200 forms. $\dfrac{t}{3} + \dfrac{t}{2\frac{1}{3}} = 1$,

$t = \dfrac{21}{16}$; $1\dfrac{5}{16}$ hours

16. Let $s = 0$; then $t = 4$: 4 seconds.
17. Let $A =$ the area, $r =$ the radius, and $C =$ the

circumference. $A = \pi r^2 \Rightarrow r = \sqrt{\dfrac{A}{\pi}} = \dfrac{\sqrt{A\pi}}{\pi}$,

$C = 2\pi r = 2\sqrt{A\pi}$, cost $= 6(2\sqrt{A\pi}) = 12\sqrt{A\pi}$
dollars.

Chapter 3

Exercises 3.1

1. x-int: -5; y-int: 4   **3.** x-int: 6; y-int: 2

5. x-int: $\dfrac{16}{3}$; y-int: -2 **7.** x-int: -3; y-int: $-\dfrac{2}{3}$

9. x-int: -3; no y-int   **11.** x-int: $-\dfrac{8}{5}$; y-int: $\dfrac{4}{3}$

13. x-int: 0; y-int: 0

15. 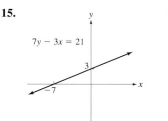   **17.**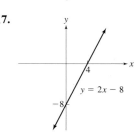

19. 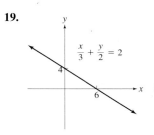   **21.**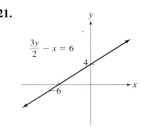

23. 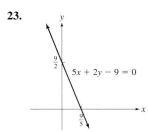   **25.**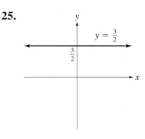

27. 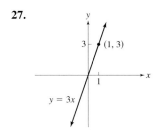   **29.**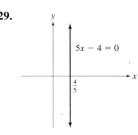

31.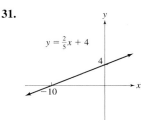

33. x-int $= 3.5$, y-int $= 7$
35. x-int $= -250$, y-int $= 1.67$
37. x-int $= -0.41$, y-int $= 2$
39. x-int $= 1.91$, y-int $= 2.8$
41.

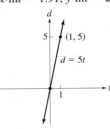

43.

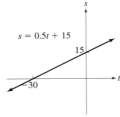

45. (a) $95v + 125t = 9110$
(b)    **(c)** 50

47. (a) $9a + 15c = 360$
(b) 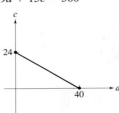   **(c)** 6

49. Length: $\sqrt{41}$; mp $= \left(-1, \dfrac{3}{2}\right)$

51. Length: $\sqrt{85}$; mp $= \left(0, \dfrac{1}{2}\right)$

53. Length: $|a|\sqrt{2}$; mp $= \left(\dfrac{a}{2}, \dfrac{a}{2}\right)$

55. $A = 15$ **57.** $A = \dfrac{13}{2}$ **59.** $w = 11, w = -5$

61. No   **63.** $(x - 2)^2 + (y - 3)^2 = 9$

65. $\left(x - \dfrac{1}{2}\right)^2 + (y - 4)^2 = 36$

67. Center: $(3, 2)$; $r = 4$ **69.** Center: $(0, 0)$; $r = 4$
71. Center: $(3, 5)$; $r = 5$ **73.** Center: $(0, -3)$; $r = 3$
75.

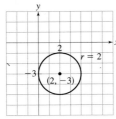

$(x - 2)^2 + (y + 3)^2 = 4$

77. $(x - 1)^2 + \left(y - \dfrac{3}{2}\right)^2 = \dfrac{205}{4}$

79. $(x - 3)^2 + (y + 5)^2 = 122$ **81.** $C = 4\pi\sqrt{17}$
83. $(x - 3)^2 + (y + 2)^2 = 4$
85. $(x - 3)^2 + (y + 3)^2 = 9$

87. (a) $x^2 + y^2 = 1$ **(b)** $\left(\dfrac{3}{5}, -\dfrac{4}{5}\right)$ and $\left(-\dfrac{\sqrt{3}}{2}, \dfrac{1}{2}\right)$

Exercises 3.2

1. $m = \dfrac{5}{3}$ **3.** $m = -1$ **5.** $m = -\dfrac{2}{3}$ **7.** $m = \dfrac{3}{2}$

9. $m = \dfrac{2}{3}$ **11.** $m = 0$ **13.** $m = \dfrac{4}{15}$ **15.** $m = \sqrt{3}$

17. $m = a + b$ **19.** $m = 1$
21.

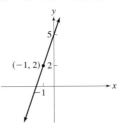

23.

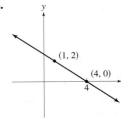

25.

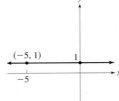

27. (a) 120 **(b)** 6
(c) That you burn 6 calories per minute during a brisk walk
29. (a) 18 **(b)** 280 **(c)** 20; Yes   **(d)** $\frac{1}{20}$
(e) For each 20 miles driven, the car uses 1 gallon of gasoline.
31. (a) $A = 20d + 65$
(b)

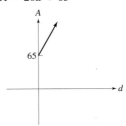

(c) $m = 20$; the slope is the amount earned per delivery.
33. Parallel   **35.** Neither   **37.** Perpendicular
39. $c = 8$ **41.** $t = \dfrac{1}{4}$ **43.** $h = -\dfrac{1}{2}, h = 5$

51. $m_4 < m_2 < m_3 < m_1$

Exercises 3.3

1. $y = 3x + 5$ **3.** $y = \dfrac{2}{5}x - \dfrac{11}{5}$

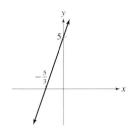
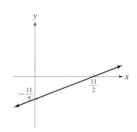

5. $y = 4x + 16$ **7.** $y = -\dfrac{1}{2}x + 7$ **9.** $y = 4x$

11. $y = -\dfrac{2}{3}x + 4$ **13.** $y = \dfrac{1}{5}x - 1$ **15.** $x = -3$

17. $y = -\dfrac{1}{3}x + 5$ **19.** $y = 5$ **21.** $y = -5$

23. $m = -2$ **25.** $m = \dfrac{2}{3}$ **27.** $m = \dfrac{1}{5}$ **29.** $m = -\dfrac{4}{3}$

31. $y = 4x - 9$ **33.** $y = \dfrac{4}{3}x + \dfrac{5}{3}$ **35.** $y = \dfrac{4}{5}x - 4$

37. $y = -\dfrac{7}{6}x + \dfrac{73}{6}$ **39.** $y = -x$ **41.** $y = 5$

43. $y = \dfrac{7}{2}x + 2$ **57.** $y = \dfrac{5}{3}x + 4$

59. $y = -\dfrac{5}{3}x - \dfrac{16}{45}$ **61.** $C = 0.30m + 29$

63. (a) $P = 10x + 100$ **(b)** $P = \$1000$

65. (a) $N = \dfrac{5}{3}s + 55$ **(b)** Approximately 97

67. (a) $P - 14.5 = \dfrac{9}{50}(y - 1990)$ **(b)** Approx. 18.1 million

69. (a) $N - 34.5 = -2.2(y - 1998)$
   (b) Approx. 19.1 million

71. $V = -\dfrac{2125}{3}t + 8500$

73. (a) $C = 8.25d + 12{,}375$ **(b)** $I = 49.95d$
   (c) 297 days
   (d) The break-even point is where the graphs intersect.

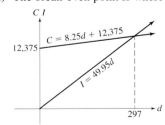

75. (a) $N = 275 - 10\left(\dfrac{P - 62.50}{2.50}\right)$ or $N = -4P + 525$
   (b) 225   **(c)** \$71.25

77. $C = \dfrac{5}{9}(F - 32)$; 37°C

79. (a) $h = 3.75f - 0.325$ **(b)** 194.7 cm

81. $y = \dfrac{5}{12}x - \dfrac{169}{12}$ **83.** All three answers are $x < -\dfrac{3}{2}$.

Exercises 3.4

1. $\{(A, 10), (B, 20), (C, 30)\}$; Domain $= \{A, B, C\}$;
   Range $= \{10, 20, 30\}$; it is a function.
3. $\{(6, A), (6, B), (10, B), (15, C), (19, D)\}$;
   Domain $= \{6, 10, 15, 19\}$; Range $= \{A, B, C, D\}$; it is not
   a function.
5. $\{(A, 7), (B, 7), (C, 7), (D, 7)\}$; Domain $= \{A, B, C, D\}$;
   Range $= \{7\}$; it is a function.
7. Domain $= \{-4, 2, 4\}$; Range $= \{-5, -3, 0, 6\}$; it is not a
   function.
9. Domain $= \{-\sqrt{3}, -1, 0, 1, \sqrt{3}\}$; Range $= \{0, 1, 3\}$; it is
   a function.
11. Domain $= \{0, 1, 2, 3, 4, 5\}$; Range $= \{5\}$; it is a function.

The equations in Exercises 13–42 *all* define y as a function of x.

13. All real numbers   **15.** $\{x \mid x \ne 2\}$ **17.** $(-\infty, \infty)$

19. $[-2, \infty)$ **21.** $\left\{x \mid x \ne -\dfrac{4}{3}\right\}$ **23.** $\{x \mid x \ne 0\}$

25. $(-\infty, -3) \cup (3, \infty)$ **27.** All real numbers

29. All real numbers   **31.** $\left\{x \mid x \le \dfrac{8}{5}\right\}$

33. All real numbers   **35.** $\{x \mid x \ge -5, x \ne 3\}$

37. $\left(0, \dfrac{3}{2}\right)$ **39.** $\{x \mid x \ne 5\}$ **41.** $\{x \mid x \ne \pm 4\}$

43. Function   **45.** Not a function
47. Function
49. Domain $= (-\infty, \infty)$; Range $= [-3, \infty)$
51. Domain $= (-\infty, \infty)$; Range $= (-\infty, 6]$
53. Domain $= [-5, 5]$; Range $= [-3, 3]$
55. $C = 0.11m + 110$ **57.** $d = 60 - 2t$
59. (a) 1996: $C = 167.2$, 1997: $C = 170.1$, 1998: $C = 173.6$,
   1999: $C = 177.6$, 2000: $C = 182.1$
   (b) 2001: $C = 187.2$
61. (a) 1996: $C = 152.7$, 1997: $C = 156.3$, 1998: $C = 160$,
   1999: $C = 163.7$, 2000: $C = 167.3$
   (b) 2001: $C = 171.0$
63. (a) 1996: $L = 5061.3$, 1997: $L = 6277.5$, 1998:
   $L = 7706.7$, 1999: $L = 9001.5$, 2000: $L = 9814.5$
   (b) Closest in 2000; farthest in 1998   **(c)** 9798.3
   (d) 1996: H $= 6625.1$,
   1997: H $= 8002.3$,
   1998: H $= 9759.3$,
   1999: H $= 11{,}239.1$,
   2000: H $= 11{,}784.7$
   (e) Closest in 2000; farthest in 1998
   (f) 10,739.1

Exercises 3.5

1. 28   **3.** $\sqrt{37}$ **5.** 161   **7.** $-\dfrac{1}{3}$ **9.** $5x + 8$

11. $3x^2 - 10x + 8$ **13.** $\sqrt{4x^2 - 3}$

15. $27x^2 - 12x + 1$ **17.** $20x + 33$

19. $3x^2 + 6xh + 3h^2 - 4x - 4h + 1$

21. $\dfrac{9}{10}$ **23.** $\dfrac{1}{\sqrt{14}} = \dfrac{\sqrt{14}}{14}$ **25.** $\dfrac{1}{3}$

27. $\dfrac{1}{\sqrt{t^2 - 1}}$ **29.** Undefined   **31.** $\dfrac{x + 1}{x + 2}$

33. $\sqrt{5}$ **35.** $\dfrac{40}{41}$ **37.** $\dfrac{a}{a + 3}$ **39.** $\dfrac{x - 1}{x}$

41. $20x + 25$ **43.** $\dfrac{1}{x - a}$ **45.** $\dfrac{1}{x} - \dfrac{1}{a}$

47. $\dfrac{1}{x} - a$ **49.** 156 **51.** 45 **53.** $\dfrac{1}{x^2 - x}$

55. $k^2 x^2 - kx$ **57.** $-\dfrac{7}{3}$ **59.** $\dfrac{1}{x^2} - \dfrac{1}{x} = \dfrac{1 - x}{x^2}$

61. $x + 2$ **63.** -3 **65.** $2x + h - 3$

67. (a) -14 (b) 0 (c) 35

69. (a) 1 (b) Undefined   (c) $\dfrac{3}{2}$ (d) Undefined

71. (a) $h(1) = 34; h(2) = 36; h(3) = 6$
  (b) 3.125 seconds

73. (a) 1 (b) -1 (c) -2 (d) $-\frac{1}{2}$ (e) 2
  (f) $x = -6, -3,$ and 2 (g) 3 solutions

75. $A = \begin{cases} 7 \text{ if } 0 \le h \le 35 \\ 7 + 0.12(h - 35) \text{ if } h > 35 \end{cases}$

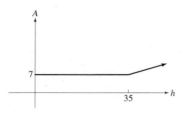

77. 1500 cans

Exercises 3.6

1. (a) -2 (b) 1 (c) 3 (d) 4 (e) $x = 3$
  (f) Domain: $[-3, 5]$; range: $[-2, 4]$
3. (a) Undefined   (b) 4 (c) Undefined
  (d) -3 (e) 1 (f) $(-7, -3] \cup [1, 6]$
  (g) $[-3, 2] \cup [3, 5)$
5. (a) -1.2 and 3.3 (b) -1.2 and 3.2
  (c) Domain: $(-\infty, \infty)$, range: $[-5, \infty)$
7. (a) -0.4 and 3.4 (b) -0.3 and 3.3
9. (a) $-0.6, 0,$ and 1.6 (b) $-0.6, 0,$ and 1.6
11. (a) $F(x) \to 1$ (b) $F(x) \to -4$ (c) $F(x) \to \infty$
13. (a) $(-\infty, -6] \cup [-2, \infty)$ (b) $[-6, -2]$
  (c) $[-7, -3] \cup [-1, \infty)$
  (d) $(-7, -3) \cup (-1, \infty)$
  (e) $(-\infty, -7) \cup (-3, -1)$
  (f) rel. max $= 3$; rel. min $= -1$
15. (a) $[3, \infty)$ (b) $(-\infty, 3]$ (c) $(-\infty, \infty)$
  (d) $(-\infty, 3) \cup (3, \infty)$ (e) Nowhere
17. (a) $(-\infty, 0) \cup [0, 2]$ (b) $[5, \infty)$ (c) $[2, 5]$
  (d) $(-\infty, 0) \cup (0, 7)$ (e) $(7, \infty)$

19. (a) 3 (b) 5 (c) 0 (d) $-5, -3, 5$ (e) 2
  (f) -3
21. (a) 3 (b) 3 (c) None   (d) 5
23. (a) $[-7, -6) \cup (-2, 4)$ (b) $(-6, -2) \cup (4, 6]$
  (c) $[-7, -6] \cup [-2, 4]$
25. (a) 4 (b) $[-7, -6) \cup (-2, 4)$
  (c) $[-6, -2] \cup [4, 6]$
27. (a) 3 (b) 4 (c) 5 (d) $-9, -2, 4$
  (e) $-7, -3, 7$ (f) -2
29. (a) 3 (b) 3 (c) 1 (d) 5
31. (a) $(-\infty, -7) \cup (-3, 7)$ (b) $(-7, -3) \cup (7, \infty)$
  (c) $(-\infty, -7] \cup [-3, 7]$
33. (a) -5 (b) $(-\infty, -7) \cup (-2, 7)$
  (c) $[-7, -2] \cup [7, \infty)$
35. (a) The number of each type of paramecium increases and
      then levels off.
  (b) The number of *aurelia* increases, whereas the *caudatum*
      die out.
37. (a) 13 (b) $\$700{,}000$
39.

39 graph x-axis from -10 to 10, y-axis from -10 to 10.

  (a) Zeros $= -2, 3$ (b) $(-\infty, -2) \cup (3, \infty)$
  (c) $(-2, 3)$ (d) No rel. max; rel. min $= -6.25$
41.

41 graph x-axis from -10 to 10, y-axis from -5 to 5.

  (a) Zeros $= -1, 1$ (b) $(-\infty, -1) \cup (1, \infty)$
  (c) $(-1, 1)$ (d) No rel. max; rel. min $= -1$
43.

43 graph x-axis from -10 to 10, y-axis from -10 to 10.

  (a) Zeros $= -1.88, 0.35, 1.53$
  (b) $(-1.88, 0.35) \cup (1.53, \infty)$
  (c) $(-\infty, -1.88) \cup (0.35, 1.53)$
  (d) Rel max. $= 3$; rel. min $= -1$
45. $x = -2.09$ **47.** $x = -1.62, 0.62, 1$
49. $x = -0.79, 0.79$ **51.** $[-1.73, 1.73]$
53. $(4.88, \infty)$

Exercises 3.7

1. y-axis symmetry   **3.** x-axis symmetry
5. Origin symmetry   **7.** None

9. (a)

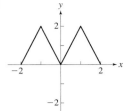

(b)

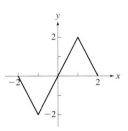

11. (a)

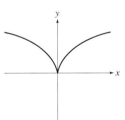

(b)

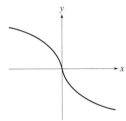

13. (a) *y*-axis reflection

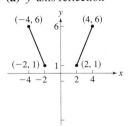

(b) *x*-axis reflection

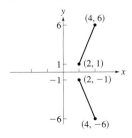

(c) Origin reflection

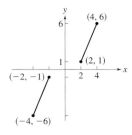

15. (a) *y*-axis reflection

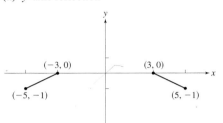

(b) *x*-axis reflection

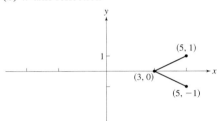

(c) Origin reflection

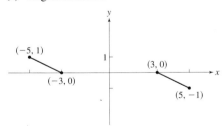

17. *y*-axis symmetry   **19.** Origin symmetry
21. Neither   **23.** Origin symmetry
25. *y*-axis symmetry   **27.** Origin symmetry
29. *y*-axis symmetry   **31.** Origin symmetry
33. Neither   **35.** Origin symmetry
37. *y*-axis symmetry   **39.** Neither

Chapter 3 Review Exercises

1.

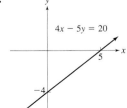

2.

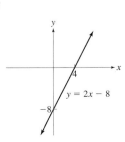

3.

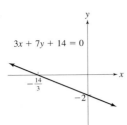

$3x + 7y + 14 = 0$

4.

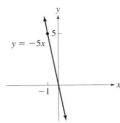

$y = -5x$

15.

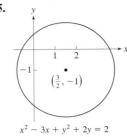

$x^2 - 3x + y^2 + 2y = 2$

16.

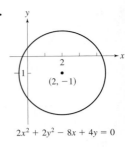

$2x^2 + 2y^2 - 8x + 4y = 0$

5.

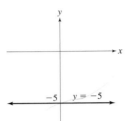

$y = -5$

6.

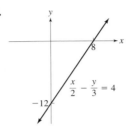

$\dfrac{x}{2} - \dfrac{y}{3} = 4$

17.

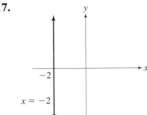

$x = -2$

18.

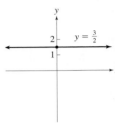

$y = \dfrac{3}{2}$

7.

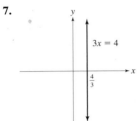

$3x = 4$

8.

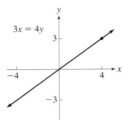

$3x = 4y$

19. **(a)** Yes **(b)** Domain $= (-\infty, \infty)$
(c) Range $= [-5, \infty)$
20. **(a)** Not a function **(b)** Domain $= (-\infty, 6]$
(c) Range $= (-\infty, \infty)$
21. **(a)** Yes
(b) Domain $= \{x \mid -5 \le x \le -2 \text{ or } 3 \le x \le 5\}$
(c) Range $= \{y \mid -5 \le y \le 4\}$
22. **(a)** Yes **(b)** Domain $= (-\infty, \infty)$
(c) Range $= (-\infty, \infty)$
23. When $x = 2$, $y = 16$; when $x = -2$, $y = 16$; it is a function
24. If $x = 16$, then $y = \pm 2$. It is not a function because $x = 16$
is assigned two y-values.

9.

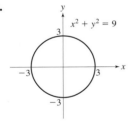

$x^2 + y^2 = 9$

10.

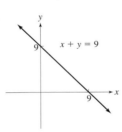

$x + y = 9$

25. $y = -\dfrac{9}{5}x - \dfrac{7}{5}$ **26.** $y = -\dfrac{7}{4}x + \dfrac{7}{2}$ **27.** $t = \pm 9$
28. Since $m_{\overline{PQ}} = -\dfrac{5}{4}$ and $m_{\overline{QR}} = \dfrac{4}{5}$, $\overline{PQ}$ is perpendicular to $\overline{QR}$,
and therefore $\triangle PQR$ forms a right triangle.
29. $(x + 4)^2 + (y - 1)^2 = 49$
30. Center: $(0, -2)$; $r = 2\sqrt{5}$
31. Center: $(-4, 1)$; $r = 5$
32.

11.

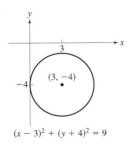

$(x - 3)^2 + (y + 4)^2 = 9$

12.

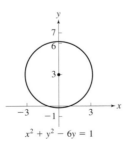

$x^2 + y^2 - 6y = 1$

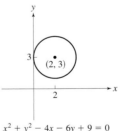

$x^2 + y^2 - 4x - 6y + 9 = 0$

33. $(x + 5)^2 + (y - 4)^2 = \dfrac{1}{4}$

34. $y = \dfrac{7}{3}x - \dfrac{23}{3}$ **35.** $\left(x - \dfrac{1}{2}\right)^2 + \left(y - \dfrac{1}{2}\right)^2 = \dfrac{41}{2}$

13.

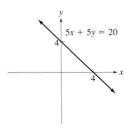

$5x + 5y = 20$

14.

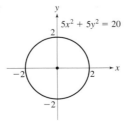

$5x^2 + 5y^2 = 20$

36. $x^2 + \left(y - \dfrac{4}{3}\right)^2 = 49$ **37.** $\left\{x \mid x \le \dfrac{5}{3}\right\}$

38. Domain $= (-\infty, \infty)$ **39.** $\{x \mid x \ne -1, 4\}$
40. Domain $= (-\infty, \infty)$ **41.** $[0, 6]$
42. Domain $= \{x \mid x \ge -6, x \ne 2\}$ **43.** $\{x \mid x > -4\}$

44. Domain $= (-\infty, \infty)$
45. $f(-3) = -22; f(2x) = -4x^2 + 8x - 1;$
$2f(x) = -2x^2 + 8x - 2$
46. $g(-9) = 5; g(0) = \sqrt{7}; g\left(\dfrac{1}{2}\right) = \sqrt{6}$
47. $h(2) = h(10) = h(-3) = \sqrt{5}$
48. $f(x + 3) = -4x^2 - 24x - 27; f(x) + 3 = -4x^2 + 12$
49. $g\left(\dfrac{1}{2}\right) = 0; g(-5)$ is undefined; $g(t + 1) = \dfrac{2t + 1}{t + 6}$
50. $H(-\tfrac{1}{2}) = -\tfrac{1}{4}, H(\tfrac{1}{r}) = \dfrac{r}{4 + r^2},$

$H(r + 1) = \dfrac{r + 1}{4r^2 + 8r + 5}$
51. $f(0) = -1; f(-1) = 0; f(-5) = 17$
52. $f(3) = 9; f(4) = 2; f(-2) = 2$
53. $-10x + 26$ **54.** -7 **55.** $\dfrac{3}{t(t + h)}$
56. $\dfrac{2}{(x - 4)(x - 1)}$
57. **(a)** $f(-4) = -2; f(-2) = 2; f(0) = 3; f(1) = 3$
  (b) $-5, -3, 4$ **(c)** $(-\infty, -5) \cup (-3, 4)$
  (d) $(-5, -3) \cup (4, \infty)$ **(e)** $[-4, -1]$
  (f) $(-\infty, -4] \cup [2, \infty)$ **(g)** $[-1, 2]$
  (h) Rel. min $= -2$
58. None   **59.** y-axis symmetry
60. Origin symmetry   **61.** x-axis symmetry
62. Origin symmetry   **63.** None
64. None   **65.** Origin symmetry
66. y-axis symmetry   **67.** Origin symmetry
68. $C(g) = \begin{cases} 1.32g & \text{if } 0 \le g \le 150 \\ 1.21(g - 150) + 198 & \text{if } g > 150 \end{cases}$
69. 89.5
70. $C(h) = \begin{cases} 75 & \text{if } 0 < h \le 1 \\ 75 + 95(h - 1) & \text{if } h > 1 \end{cases}$
71. $A = (6 + 2.5t)^2$ **72.** $x = 0, 1.52$
73. $x = -2.15, -0.52, 2.67$
74. $(-0.69, 0.73) \cup (1.71, \infty)$ **75.** $[-0.46, 1.95]$

Chapter 3 Practice Test

1. **(a)** -9 **(b)** $-2x^2 - 21x - 49$
  (c) $-32x^2 - 20x + 3$ **(d)** $-2x^4 - 5x^2 + 3$
  (e) $-4x - 2h - 5$
2. **(a)** $-\dfrac{7}{4}$ **(b)** $\dfrac{2 + t}{1 - 2t}$ **3.** $\dfrac{-3}{(x + 2)(x - 3)}$
4.

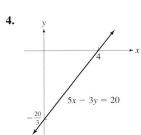

5. $y = \dfrac{3}{4}x - 3$

6. $y = -\dfrac{7}{3}x - \dfrac{29}{3}$ **7.** It is not a function.

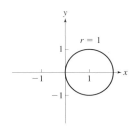

8. $(x + 1)^2 + \left(y - \dfrac{1}{2}\right)^2 = \dfrac{65}{4}$
9. **(a)** $f(-5) = -2; f(-3) = 3; f(0) = -1; f(4) = 4$
  (b) $-6, -4, -2, 2, 7$
  (c) $(-\infty, -6) \cup (-4, -2) \cup (2, 7)$
  (d) $(-6, -4) \cup (-2, 2) \cup (7, \infty)$
  (e) $[-5, -3] \cup [0, 3]$
  (f) $(-\infty, -5] \cup [-3, 0] \cup [5, \infty)$
  (g) $[3, 5]$ **(h)** Rel. max $= 3$
10. **(a)**

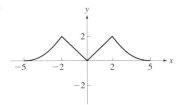

  (b)

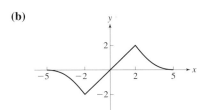

  (c)

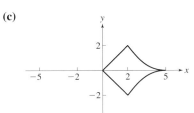

11. **(a)** Neither   **(b)** y-axis symmetry
  (c) Origin symmetry
12. $C = 3.5[2w + 2(4w - 3)] = 3.5(10w - 6)$
13. $x = -1, -0.41, 2.41$
14. $(-\infty, -1.60) \cup (1.60, \infty)$

Chapter 4

Exercises 4.1

1.

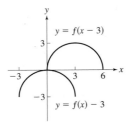

3.

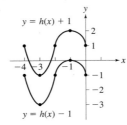

5.

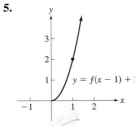

7.

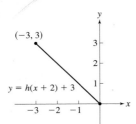

9.

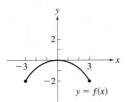

11.

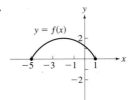

13.

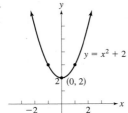

15.

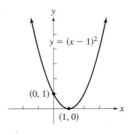

17.

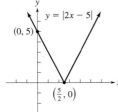

19.

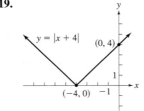

21.

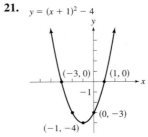

23.

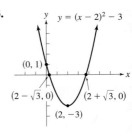

25.

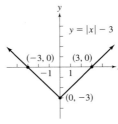

27.

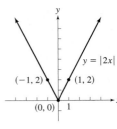

29.

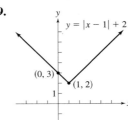

31.

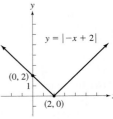

33.

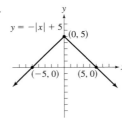

35.

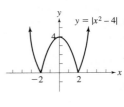

37. (a) 2   **(b)** 2   **(c)** 1   **(d)** -1 **(e)** -1
(f) $x < 0$ **(g)** $x = 0$ **(h)** $x > 0$ **(i)** None

39.

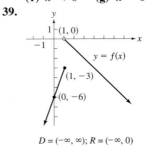

$D = (-\infty, \infty); R = (-\infty, 0)$

41.

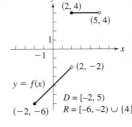

$D = [-2, 5)$
$R = [-6, -2) \cup \{4\}$

43.

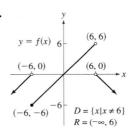

$D = \{x | x \neq 6\}$
$R = (-\infty, 6)$

45.

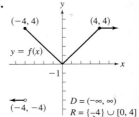

$D = (-\infty, \infty)$
$R = \{-4\} \cup [0, 4]$

47.
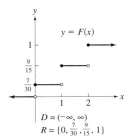

$D = (-\infty, \infty)$
$R = \{0, \frac{7}{30}, \frac{9}{15}, 1\}$

49.
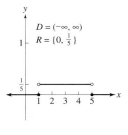

$D = (-\infty, \infty)$
$R = \{0, \frac{1}{5}\}$

Exercises 4.2

1.
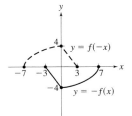

3.
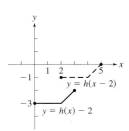

5.
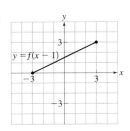

7.
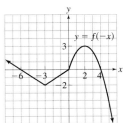

9.
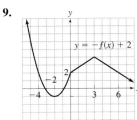

11.
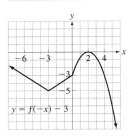

13.
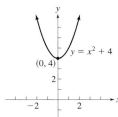

15.
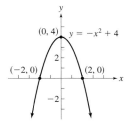

17.
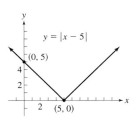

19.
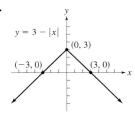

21.
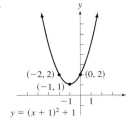

23.
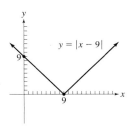

25.
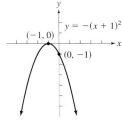

27.
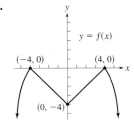

29.
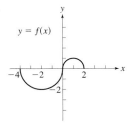

31.
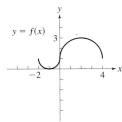

33. Shift $y = f(x)$ four units down.

35. Reflect $y = f(x)$ about the y-axis.

37. Reflect $y = f(x)$ about the x-axis and then shift 2 units up.

39. Reflect $y = f(x)$ about the x-axis and shift it 3 units up.

41. Shift $y = f(x)$ left 3 units and up 4 units.

43. Shift the graph of $y = f(x)$ two units to the left and then reflect it about the y-axis.

45. 6

47. -3

49.

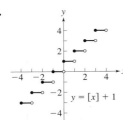

$y = [x] + 1$

51.

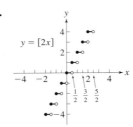

$y = [2x]$

53.

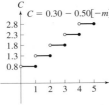

$C = 0.30 - 0.50[-m]$

55. (a) 7   **(b)** 2.4   **(c)** 5.7
(d) 8.9
(e) -6.5; $f(x)$ rounds the given number to the nearest tenth

Exercises 4.3

1. (a) $y = \dfrac{6(9 + x)}{x}, x > 0$ **(b)** $A = \dfrac{3(9 + x)^2}{x}, x > 0$

3. $A = \dfrac{1}{2}\pi r^2 - \dfrac{1}{2}(5x) =$

$\dfrac{1}{2}(\pi r^2 - 5\sqrt{4r^2 - 25}), r \geq \dfrac{5}{2}$

5. $V = 80 = x^2 h \;\Rightarrow\; h = \dfrac{80}{x^2}. A = 2x^2 + 4xh =$

$2x^2 + 4x\left(\dfrac{80}{x^2}\right) = \dfrac{2x^3 + 320}{x}, x > 0$

7. Since $C = 2\pi r, r = \dfrac{C}{2\pi}$. Hence

$A = \pi r^2 = \pi\left(\dfrac{C}{2\pi}\right)^2 = \dfrac{C^2}{4\pi}, C > 0.$

9. Since $\dfrac{20}{x + s} = \dfrac{6}{s}, s = \dfrac{3x}{7}, x > 0.$

11. Since $\dfrac{1 - 0}{4 - a} = \dfrac{1 - b}{4 - 0}, b = \dfrac{a}{a - 4}$. Hence

$A = \dfrac{1}{2}ab = \dfrac{1}{2}a\left(\dfrac{a}{a - 4}\right) = \dfrac{a^2}{2(a - 4)}, a > 4.$

13. $A =$ the area of the rectangle $+$ the area of the two semicircles

$= \dfrac{(200 - 2\pi r)(2r)}{2} + 2\left(\dfrac{1}{2}\right)\pi r^2$

$= 200r - \pi r^2, 0 < r < \dfrac{100}{\pi}.$

15. If h is the height of the balloon, then $h = rt = 5t$.

$d = \sqrt{120^2 + h^2} = \sqrt{120^2 + (5t)^2}$
$= \sqrt{14{,}400 + 25t^2}, \quad t \geq 0$

17. (a) If $s =$ the length of the side of the square, then the length of the wire for the square is $4s$. Thus, the length of the wire for the circle is $30 - 4s$. Hence

$2\pi r = 30 - 4s$ and so $r = \dfrac{30 - 4s}{2\pi}$. The area of the circle is $\pi\left(\dfrac{30 - 4s}{2\pi}\right)^2 = \dfrac{(15 - 2s)^2}{\pi}$. The area of the circle and the square is $s^2 + \dfrac{(15 - 2s)^2}{\pi}, 0 \leq s \leq \dfrac{15}{2}$.

(b) Using the same reasoning as in part (a), if $r =$ the radius of the circle, then the area of the square and the circle is $\pi r^2 + \dfrac{(15 - \pi r)^2}{4}, 0 \leq r \leq \dfrac{15}{\pi}$.

19. $R(n) = (24 - .50n)(3500 + 80n), 0 \leq n \leq 48$
21. $Y(n) = (800 + n)(750 - 30n) =$
$-30n^2 - 23{,}250n + 600{,}000$. Since $750 - 30n \geq 0$, the domain is $0 \leq n \leq 25$.
23. $R = R(n) = (9 + 0.5n)(8500 - 225n) =$
$-112.5n^2 + 2225n + 76{,}500$, for $0 \leq n \leq 37$
25. $R(r) = (1200 - 75r)(750 + 35r) =$
$-2625r^2 - 12{,}500r + 937{,}500$. Since $1250 - 75r \geq 0$, the domain is $0 \leq r \leq 16$.
27. Let $x =$ the side of the fence adjacent to the house; then the side opposite the house $= 80 - 2x. A = x(80 - 2x)$, $0 < x < 40$.
29. Let $x =$ the length of the vertical sides of the fence. If we let $a =$ the length of one entire horizontal side, then the cost, C, is $C = 3(2x) + 5(2x) + 5(2a)$, or $240 = 16x + 10a$.
Hence $a = \dfrac{120 - 8x}{5}$. The area enclosed is

$A = xa = x\left(\dfrac{120 - 8x}{5}\right) = -\dfrac{8}{5}x^2 + 24x, 0 < x < 15.$

31. (a) $L = L(x) = 4x + \dfrac{1600}{3x}$, for $x > 0$.

(b) $C = C(x) = 72x + \dfrac{9600}{x}$, for $x > 0$.

33. $L = \sqrt{x^2 + 1} + (6 - x)$, for $0 \leq x \leq 6$.

35. $I = \begin{cases} 0.0165b & \text{for } 0 \leq b \leq 2000 \\ 33 + 0.0125(b - 2000) & \text{for } b > 2000 \end{cases}$

37. $V = 18x(1 - 2x) = 18x - 36x^2$, for $0 < x < \dfrac{1}{2}$

39. Since $S = 4\pi r^2, r = \sqrt{\dfrac{S}{4\pi}}$. Hence $V = \dfrac{4}{3}\pi r^3 =$

$\dfrac{4}{3}\pi\left(\sqrt{\dfrac{S}{4\pi}}\right)^3 = \dfrac{S\sqrt{\pi S}}{6\pi}, S \geq 0.$

41. $V = \pi r^2 h = \pi r^2(10) \Rightarrow r^2 = \dfrac{V}{10\pi}$. Therefore,

$S = 2\pi rh + 2\pi r^2 = 20\pi\sqrt{\dfrac{V}{10\pi}} + 2\pi\left(\dfrac{V}{10\pi}\right) =$

$2\sqrt{10\pi V} + \dfrac{V}{5}$, for $V > 0$.

43. $D = \sqrt{x^4 - 13x^2 + 2x + 50}$, for all real numbers x
45. $D = \sqrt{y^2 - 7y + 12}$, for $y \leq 3$
47. $D = \sqrt{x^2 + x}$, for $x \geq 0$.

Exercises 4.4

1.
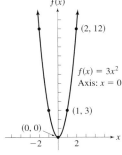
$f(x) = 3x^2$
Axis: $x = 0$
(2, 12)
(1, 3)
(0, 0)

3.
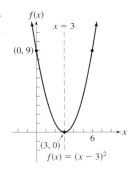
$x = 3$
(0, 9)
(3, 0)
$f(x) = (x - 3)^2$

5.
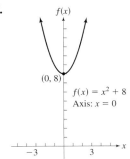
(0, 8)
$f(x) = x^2 + 8$
Axis: $x = 0$

7.
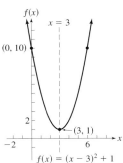
$x = 3$
(0, 10)
(3, 1)
$f(x) = (x - 3)^2 + 1$

9.
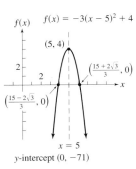
$f(x) = -3(x - 5)^2 + 4$
(5, 4)
$\left(\frac{15 + 2\sqrt{3}}{3}, 0\right)$
$\left(\frac{15 - 2\sqrt{3}}{3}, 0\right)$
$x = 5$
y-intercept (0, −71)

11.
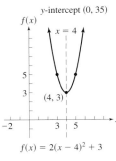
y-intercept (0, 35)
$x = 4$
(4, 3)
$f(x) = 2(x - 4)^2 + 3$

13.
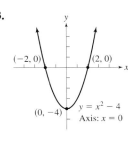
(−2, 0)   (2, 0)
(0, −4)
$y = x^2 - 4$
Axis: $x = 0$

15.
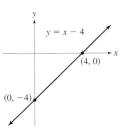
$y = x - 4$
(4, 0)
(0, −4)

17.
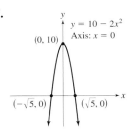
$y = 10 - 2x^2$
Axis: $x = 0$
(0, 10)
$(-\sqrt{5}, 0)$ $(\sqrt{5}, 0)$

19.
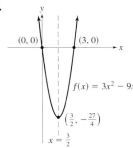
(0, 0)   (3, 0)
$f(x) = 3x^2 - 9x$
$\left(\frac{3}{2}, -\frac{27}{4}\right)$
$x = \frac{3}{2}$

21.
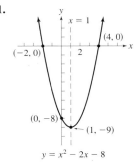
$x = 1$
(4, 0)
(−2, 0)
(0, −8)
(1, −9)
$y = x^2 - 2x - 8$

23.
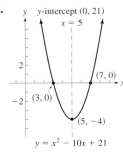
y-intercept (0, 21)
$x = 5$
(7, 0)
(3, 0)
(5, −4)
$y = x^2 - 10x + 21$

25.
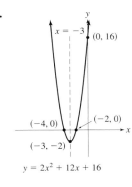
$x = -3$
(0, 16)
(−4, 0)   (−2, 0)
(−3, −2)
$y = 2x^2 + 12x + 16$

27.
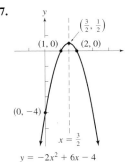
$\left(\frac{3}{2}, \frac{1}{2}\right)$
(1, 0)   (2, 0)
(0, −4)
$x = \frac{3}{2}$
$y = -2x^2 + 6x - 4$

29.
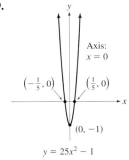
Axis: $x = 0$
$\left(-\frac{1}{5}, 0\right)$ $\left(\frac{1}{5}, 0\right)$
(0, −1)
$y = 25x^2 - 1$

31. Minimum is −9   **33.** Maximum is 9

35. Maximum is 36   **37.** Maximum is $-\dfrac{27}{2}$

39. Minimum is −36   **41.** Minimum is $\dfrac{27}{2}$

43. (a)
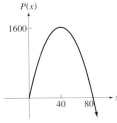
$P(x)$
1600
40   80
(b) $x = 40$ items
(c) Profit: \$1600 daily

45. $x = 12$ tables; \$425.92
47. $t = 27$ seconds; 11,664 feet
49. Let $x =$ one side; then the adjacent side is
$\dfrac{100 - 2x}{2} = 50 - x.$ $A = x(50 - x) = -x^2 + 50x.$

The maximum area is at $x = 25$: 25×25 feet.

51. Let $x =$ one number; then the other number is $104 - x$. $P = x(104 - x) = -x^2 + 104x$. The maximum product is at $x = 52$: 52 and 52.

53. Since $800 = 3x + 2y$, then $y = \dfrac{800 - 3x}{2}$. Hence

$A = xy = x\left(\dfrac{800 - 3x}{2}\right) = -\dfrac{3}{2}x^2 + 400x$. Maximum

area at $x = \dfrac{400}{3}$ feet: $x = 133\dfrac{1}{3}$ feet, $y = 200$ feet.

55. (a) 15   (b) \$4   (c)

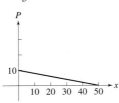

   (d) $R(x) = 10x - \dfrac{x^2}{5}$ (e)

   (f) 25 items; \$5

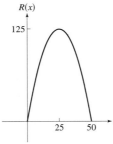

57. (a) 585   (b) \$40
    (c)

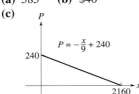

    (d)

$R = -\dfrac{1}{9}x^2 + 240x$
129,600

    (e) 1080 items; unit price is \$120

59. $\dfrac{72}{\pi + 4}$ **61.** $n = 2.125$ **63.** $n = 0$ **65.** $n \approx 9.89$

67. $r = 0$

69. When the ball has travelled 400 ft it will be at a height of 20 feet and so it will clear an 18-foot high fence.

71. \$1075

73. (a) \$3.50   (b) \$3.67

75. 10.5 lb   **77.** $f\left(-\dfrac{B}{2A}\right) = \dfrac{4AC - B^2}{4A}$

79. (a)   (b)

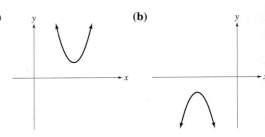

   (c) No because A $>$ 0 means the parabola opens up, whereas $B^2 - 4AC < 0$ means that there are no x-intercepts.

81. (a) At $x = -1$ and $x = 4$ (b) $(-\infty, -1)\cup(4, \infty)$
   (c) $(-1, 4)$

83. (a) 4   (b) 2   **85.** (a) 3   (b) $\dfrac{1}{3}$

Exercises 4.5

1. 11   **3.** $-\dfrac{5}{6}$ **5.** $\dfrac{2}{5}$ **7.** $-\dfrac{5}{2}$ **9.** 187   **11.** 5

13. $2x^2 + 2x - 5$

15. $2x^2 - 4x - 1$; Domain $=$ all real numbers

17. $\dfrac{1}{5x}, x \neq 0$

19. $x^3 - 2x^2 + x + 2$; Domain $=$ all real numbers

21. $\dfrac{4(3x - 2)}{x + 2}, x \neq -2$

23. $18x^2 - 27x + 7$; Domain $=$ all real numbers

25. $\dfrac{1 - x^3}{x^3}, x \neq 0$

27. $5, x \neq -2$

29. $27x^3 - 54x^2 + 36x - 9$; Domain $=$ all real numbers

31. $\dfrac{x + 2}{4}, x \neq -2$

33. (a) $\sqrt{2x - 6}; [3, \infty)$ (b) $2\sqrt{x + 1} - 7; [-1, \infty)$

35. (a) $\dfrac{18 + 6t}{(t - 3)^2}, t \neq 3$

   (b) $\dfrac{6}{t^2 + t - 3}, t \neq \dfrac{-1 \pm \sqrt{13}}{2}$

   (c) $t^4 + 2t^3 + 2t^2 + t$ (d) $\dfrac{2t - 6}{5 - t}, t \neq 5, 3$

37. (a) $\dfrac{1}{\sqrt{2x + 3}}, x > -\dfrac{3}{2}$ (b) $\dfrac{3x + 2\sqrt{x}}{x}, x > 0$

   (c) $\dfrac{\sqrt{3x^2 + 2x}}{x}, x \leq -\dfrac{2}{3}$ or $x > 0$

39. $f(x) = x^3$; $g(x) = x + 3$

41. $f(x) = x^2$; $g(x) = \dfrac{x + 4}{x - 1}$

43. $f(x) = x + 6$; $g(x) = \dfrac{1}{x}$

45. $g(x) = \dfrac{2}{5}x + 2$ **47.** $g(x) = \dfrac{8 - x}{5}$

49. (a) $N(h) = 75h^2 + 1700h + 14{,}625$ (b) 22,625
   (c) Beyond the given domain of N

51. (a) \$522.60   (b) \$89.70   (c) $P = 41.6 + \dfrac{481}{n}$

Exercises 4.6

1. Yes   **3.** No   **5.** Yes   **17.** $f^{-1}(x) = \dfrac{x + 9}{5}$

19. $F^{-1}(x) = \sqrt[3]{\dfrac{x-1}{2}} = \dfrac{\sqrt[3]{4x-4}}{2}$ **21.** $h^{-1}(x) = \dfrac{1-4x}{x}$

23. $f^{-1}(x) = \dfrac{(x+7)^2}{4}, x \geq -7$

25. $h^{-1}(x) = \sqrt{x+1}$ **27.** $g^{-1}(x) = \dfrac{2x+5}{3x-2}$

29. $f^{-1}(x) = x^{5/3}$ **31.** $f^{-1}(x) = (x-1)^{-7/5}, x \neq 1$

33. $f^{-1}(x) = \dfrac{x+6}{7}$ **35.** $f^{-1}(x) = \dfrac{3}{1-x}$

37. $f^{-1}(x) = \dfrac{x-3}{2x+5}$

39. $f^{-1}(x) = \sqrt{x}; D_f = D_{f^{-1}} = \{x \mid x \geq 0\}$;
$R_f = R_{f^{-1}} = \{y \mid y \geq 0\}$

41. 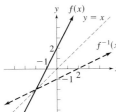   **43.**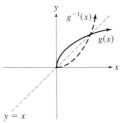

45. 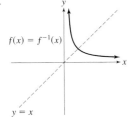   **47.**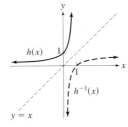

49. $f(2) = 3; f^{-1}(2) = \dfrac{5}{2}$ **51.** $f(2) = -6; f^{-1}(2) = 4$

53. $f(2) = 4; f^{-1}(2) = -3$

55. (a) The graph satisfies the horizontal line test; $F^{-1}(C) = \frac{5}{9}(C - 32)$.
   (b) $F(100) = 212; F^{-1}(212) = 100$
   (d) $F(100) = 212$ means that a Celsius temperature of 100° is equivalent to a Fahrenheit temperature of 212°; $F^{-1}(100) \approx 37.8$ means that a Fahrenheit temperature of 37.8° is equivalent to a Celsius temperature of 100°.

57. (a) $g = f(p) = 453.6p$ (b) $f^{-1}(p) = \frac{p}{453.6}$
   (c) $f(50) = 22,680$ is the number of grams in 50 pounds; $f^{-1}(50) \approx 0.11$ is the number of pounds in 50 grams.

59. (a) 13 15 20 8 5 18   (b) FATHER

61. $f(2) = 17; f^{-1}(2) = -0.32$

63. $f(2) = 31; f^{-1}(2) = 1.18$

65. $f^{-1}(x) = \dfrac{x+6}{3}$

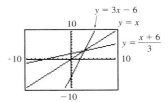

67. $f^{-1}(x) = x^3 - 4$

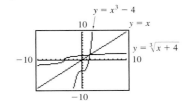

Chapter 4 Review Exercises

1. 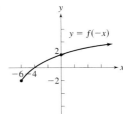   **2.**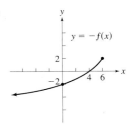

3. 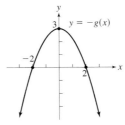   **4.**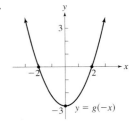

5. 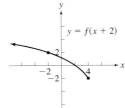   **6.**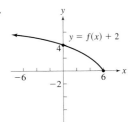

7. 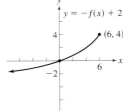   **8.**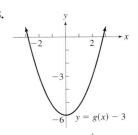

9.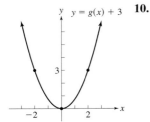

10.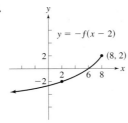

11.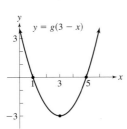

12.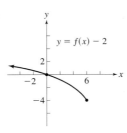

13.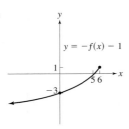

14.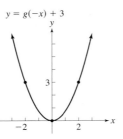

15.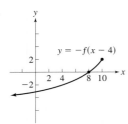

16.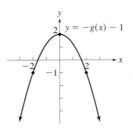

17. $y = f(-x)$

18. $y = -f(x)$

19. $y = f(x - 2)$

20. $y = -f(x) - 1$

21. Vertex: $(4, 6)$; axis: $x = 4$

22. Vertex $(3, -4)$; axis of symmetry: $x = 3$

23. $\dfrac{81}{8}$ **24.** $\dfrac{2}{3}$

25.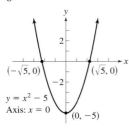

26.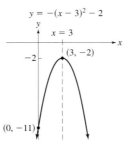

27.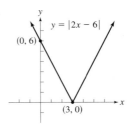

28.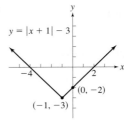

29. $f(x) = 2x^2 - 3x - 2$

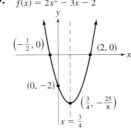

30.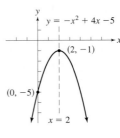

31.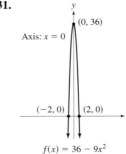

32.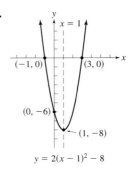

33.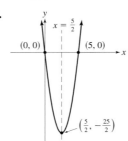

34.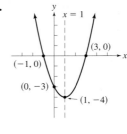

35.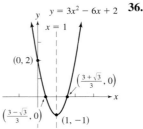

36.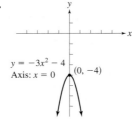

37. -2 **38.** $-\dfrac{130}{7}$ **39.** $\dfrac{133}{3}$ **40.** -48 **41.** $-\dfrac{1}{11}$

42. $\dfrac{s}{t}$ is not defined at $x = 4$ **43.** $-\dfrac{7}{3}$ **44.** -1

45. 40 **46.** -61 **47.** -3 **48.** $\dfrac{2}{3}$

49. $-8a^3 + 60a^2 - 150a + 133$ **50.** $\dfrac{2}{3}$

51. $3x^2 - 2x - 4$; Domain = all real numbers

52. $\dfrac{-2x^3 - 16}{x}, x \neq 0$ **53.** $\dfrac{2}{3x}, x \neq 0$

54. $\dfrac{3x}{2}, x \neq 0$

55. $x^3 - 3x^2 + 4x + 7$; Domain = all real numbers

56. $\dfrac{x + 8}{x(x - 4)}, x \neq 0, 4$ **57.** $\dfrac{15 - 6x}{x - 4}, x \neq 4$

58. $\dfrac{-2x^2 + 13x - 20}{3}, x \neq 4$

59. $-6x^2 + 8x + 3$; Domain = all real numbers
60. $12x^2 - 52x + 56$; Domain = $(-\infty, \infty)$

61. $-\dfrac{2}{x^3 + 8}, x \neq -2$ **62.** $\dfrac{8x^3 - 8}{x^3}, x \neq 0$

63. $-\dfrac{3}{7}$; Domain = all real numbers

64. $-3, x \neq 4$

65. $4x - 5$; Domain = all real numbers

66. $\dfrac{3x - 12}{19 - 4x}, x \neq 4, \frac{19}{4}$ **67.** $\dfrac{3x}{-2 - 4x}, x \neq -\dfrac{1}{2}, 0$

68. $\dfrac{-2x + 8}{3}, x \neq 4$ **69.** $\dfrac{8 - 2x}{5x - 26}, x \neq \dfrac{26}{5}, 4$

70. $\dfrac{15 - 6x}{8x - 22}, x \neq \frac{5}{2}, \frac{11}{4}$ **71.** $f(x) = \dfrac{1}{\sqrt{x}}; g(x) = x + 2$

72. $f(x) = x^4; g(x) = 3x + 2$
73. $f(x) = x^{-3}; g(x) = 5x - 7$
74. $f(x) = \sqrt[3]{x}; g(x) = \dfrac{x}{x + 1}$

75. (a) $N(h) = 16h^2 + 580h + 2350$ (b) 4926
   (c) Value is out of the domain of $N(h)$.
76. (a) $A = \pi(64 - t^2)^2$, for $0 \le t \le 8$
   (b) $A = 2304\pi \approx 7238$ square units
77. $A = 8x$ **78.** $V = 10\pi r - \pi r^3, r > 0$
79. Since $S = 180 = 4x^2 + 6xh, h = \dfrac{90 - 2x^2}{3x}$.
   Thus $V = 2x^2\left(\dfrac{90 - 2x^2}{3x}\right) = \dfrac{2x}{3}(90 - 2x^2)$.

80. (a) Let w be the width (or vertical height) of the garden,
   and l the length; then the cost C of the fence is
   $C = 6(2l + 2w) + 4(l + 3w) = 240 \Rightarrow$
   $l = \dfrac{30 - 3w}{2}$. Since $A = lw$, we have
   $A = \left(\dfrac{30 - 3w}{2}\right)w = 15w - \tfrac{3}{2}w^2$.
   (b) The maximum area occurs at the vertex of the parabola
   $A = 15w - \tfrac{3}{2}w^2$, which is 37.5 sq feet.
81. No   **82.** f has an inverse.
83. $D_f = R_f$ = all real numbers
   $D_{f^{-1}} = R_{f^{-1}}$ = all real numbers

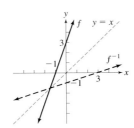

84. $D_f = \{x|x \ge 0\}$
   $R_f = \{y|y \le 0\}$
   $D_{f^{-1}} = \{x|x \le 0\}$
   $R_{f^{-1}} = \{y|y \ge 0\}$

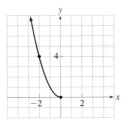

85. $D_f = R_f$ = all real numbers
   $D_{f^{-1}} = R_{f^{-1}}$ = all real numbers

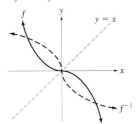

86. $D_f = \{x|x > 0\}$
   R_f = all real numbers
   $D_{f^{-1}}$ = all real numbers
   $R_{f^{-1}} = \{y|y > 0\}$

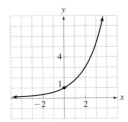

87. $f^{-1}(x) = 4x - 20$ **88.** $f^{-1}(x) = \dfrac{8 - x}{5}$

89. $f^{-1}(x) = \dfrac{3x + 4}{x - 1}$ **90.** $f^{-1}(x) = x^2 + 5, x \ge 0$

91. $f^{-1}(x) = \dfrac{3 - 6x}{x}$ **92.** $f^{-1}(x) = \dfrac{1}{x + 2}, x \neq -2$

93. $f(3) = \frac{5}{3}, f^{-1}(3) = \frac{7}{5}$
94. $f(3) = 34, f^{-1}(3) \approx -0.68$
95. $f(3) = 1, f^{-1}(3) = -1$

Chapter 4 Practice Test

1. (a)

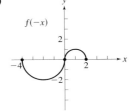

(b)

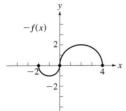

$-f(x)$

(c)

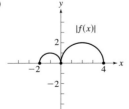

$|f(x)|$

(d)

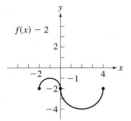

$f(x) - 2$

(e)

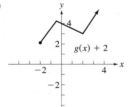

$g(x) + 2$

(f)

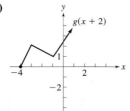

$g(x + 2)$

(g)

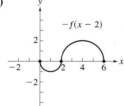

$-f(x - 2)$

2. (a) Vertex: $(-5, -3)$; axis: $x = -5$

  (b) Vertex: $\left(-\dfrac{5}{4}, \dfrac{33}{8}\right)$; axis: $x = -\dfrac{5}{4}$

3. (a)

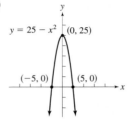

$y = 25 - x^2$ $(0, 25)$
$(-5, 0)$ $(5, 0)$

(b) $f(x) = |4x + 8|$

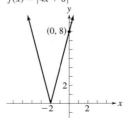

$(0, 8)$

(c) $y = -2(x + 3)^2 + 8$

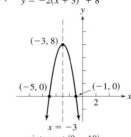

$(-3, 8)$
$(-5, 0)$ $(-1, 0)$
$x = -3$
y-intercept $(0, -10)$

(d) $f(x) = x^2 + 4x - 5$

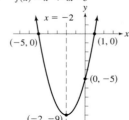

$x = -2$
$(-5, 0)$ $(1, 0)$
$(0, -5)$
$(-2, -9)$

(e) $y = 3x^2 - 48x$

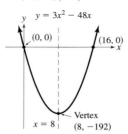

$(0, 0)$ $(16, 0)$
Vertex
$x = 8$ $(8, -192)$

(f) $f(x) = -x^2 + 3x - 4$

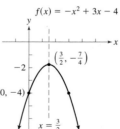

$\left(\dfrac{3}{2}, -\dfrac{7}{4}\right)$
$(0, -4)$
$x = \dfrac{3}{2}$

(g)

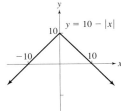

(h)

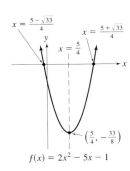

$f(x) = 2x^2 - 5x - 1$

4. (a) $-2\sqrt{2}$ **(b)** $\dfrac{13}{4}$ **(c)** $\dfrac{2}{3}$

(d) $\dfrac{7x^2 + 28x + 19}{(x + 2)^2}$ **(e)** $\dfrac{3}{9 - x^2}$

(f) $\dfrac{3x + 6}{7 + 2x}$ **(g)** $\dfrac{3}{10 - x}, x \geq 1, x \neq 10$

5. (a) $x \neq \dfrac{1}{2}, 1$ **(b)** $x \neq \pm 1$

6. Since $x^2 + 12^2 = (2r)^2, x = 2\sqrt{r^2 - 36}$. Thus
$A = \pi r^2 - 12x = \pi r^2 - 12(2\sqrt{r^2 - 36}) = \pi r^2 - 24\sqrt{r^2 - 36}$.

7.

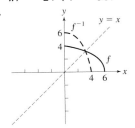

8. (a) $f^{-1}(x) = \dfrac{3x + 4}{5}$ **(b)** $f^{-1}(x) = \dfrac{x^3 - 9}{2}$

(c) $f^{-1}(x) = \dfrac{2}{x + 5}$

9. $f^{-1}(3) = -\dfrac{3}{2}$

10. (a) $d = f(p) = 1.56p$

(b) $f^{-1}(p) = \frac{p}{1.56}$; the number of pounds in p dollars
(c) $f(1000) = \$1560; f^{-1}(1000) = 641$ pounds

Chapter 5

Exercises 5.1

1. Degree is at least 5 and odd; leading coefficient is positive

3. Not a polynomial; has a break
5. Degree is 1; leading coefficient is positive
7. Degree is at least 4 and even; leading coefficient is negative
9. 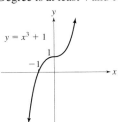 **11.**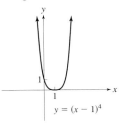

13. 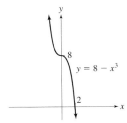 **15.**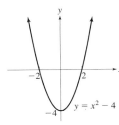

17. 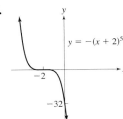 **19.**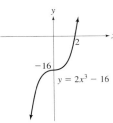

21. 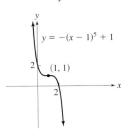 **23.**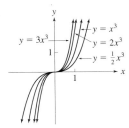

25. 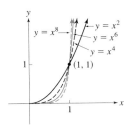 **27.** One possible graph: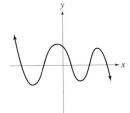

29. One possible graph: 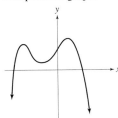 **31.** $y = x^3 - x^2 - 2x$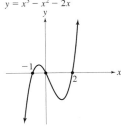

33. $y = x^2 - 6x + 8$

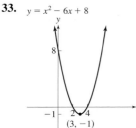

35.

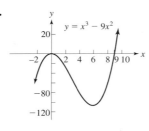

37.

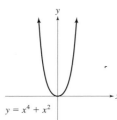

39. $y = x(x - 1)(x + 1)(x + 2)$

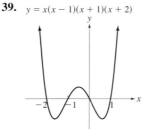

41. $y = (x + 1)(x - 2)(x + 3)(x - 4)$

43. $y = x^3 + x^2 - 2x - 2$

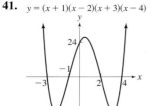
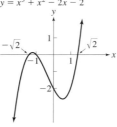

45. The polynomial appears to have one real zero and two turning points (one relative minimum and one relative maximum). As $x \to -\infty$, $y \to -\infty$ and as $x \to \infty$, $y \to \infty$.

47. The polynomial appears to have three distinct real zeros and three turning points (two relative minima and one relative maximum). As $x \to -\infty$, $y \to \infty$ and as $x \to \infty$, $y \to \infty$.

49. The polynomial appears to have two distinct real zeros and two turning points (one relative minimum and one relative maximum). As $x \to -\infty$, $y \to \infty$ and as $x \to \infty$, $y \to -\infty$.

51. x-intercepts: -3.34, -1.47, 0.81; turning points: relative maximum at $(-2.54, 2.88)$, relative maximum at $(-0.13, -4.06)$

53. x-intercepts: -3.09, -2.14, -0.11, 1.35; turning points: relative minima at $(-2.69, -2.29)$ and $(0.77, -5.75)$, relative maximum at $(-1.08, 5.04)$

55. x-intercepts: -1.17, 1.33; turning point: relative minimum at $(-0.62, -4.71)$

57.

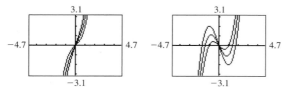

When $k = 1, 2, 3$, the graphs are all similar to the graph of $y = x^3$. When $k = -1, -2, -3$, the graphs are similar to each other with larger values of $|k|$ giving more pronounced turning points—that is, a larger maximum and a smaller minimum.

59.

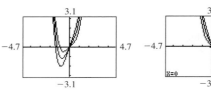

The graphs are similar for all values of k. When k is positive, the relative minimum is shifted down and to the left; when k is negative, the relative minimum is shifted down and to the right.

Exercises 5.2

1.

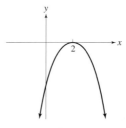

3.

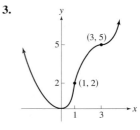

5.

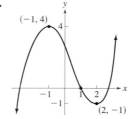

7. Between 1.2 and 1.3 **9.** Between -3.8 and -3.7

11. Between -2.2 and -2.1; between 0.7 and 0.8

13. $(0, 0)$, $(3, 27)$, $(-3, -27)$

15. $(0, 0)$, $(3, 243)$, $(-1, -1)$

17. $x^2 + 2x + 4$ **19.** $4x^3 + 6x^2h + 4xh^2 + h^3$

21. $n = 8$ **23.** $|x| < 0.182$

25. (a) $R = 430d - d^3$ **(b)** 11.97 per pound
(c) 3432

27. (a) $R = 2n + 0.45n^2 - 0.001n^3$ **(b)** 46.70 per item
(c) $14,102$

29. $S(w) = 6w^2 + 8w(w - 6)$; $S(20) = 4640$ sq in., $S(26) = 8216$ sq in., $S(42) = 22,680$ sq in.

31. (a) $A(0) = 0$ mg, $A(5) = 9.9$ mg, $A(10) = 31.8$ mg, $A(15) = 58.2$ mg, $A(20) = 81.6$ mg, $A(25) = 94.5$ mg, $A(30) = 89.4$ mg
(b) 26 minutes

33. (a) $V(x) = x(20 - 2x)^2$, $D_V = \{x \mid 0 < x < 10\}$
(b) 3.33 in. by 13.34 in. by 13.34 in.

35. Choice **(b)** gives values closest to the table values for all of the years listed.

37. $x = 0.30, 6.70$ **39.** $x = -0.82, 0.82$

41. $x < 1.77$ **43.** $x = 1.46$ **45.** $(-6.46, 0.46)$

47. $(-\infty, 1.61)$ **49.** $(-5.25, -1.29) \cup (1.03, \infty)$

Exercises 5.3

1. $x - 1 - \dfrac{18}{x - 4}$ **3.** $x - 2$ **5.** $2t - 5 - \dfrac{5}{3t - 4}$

7. $1 + \dfrac{1}{y^2 + 1}$ **9.** $2x^2 - 3x + 1$

11. $3x^3 + 2x^2 + 2x + 2 + \dfrac{8}{x - 1}$

13. $4c^2 - 2c + 1$ **15.** $2w^2 - w - 4 + \dfrac{8}{w^2 + w + 1}$

17. $x^3 + 2x^2 + 4x + 8$ **19.** $x^2 + 5x + 9 + \dfrac{24}{x - 2}$

21. $x - 6; R = -5$ **23.** $2x^2 - 3x + 3; R = 9$

25. $2x^3 + 3x^2 - 3x - 13$ **27.** $5x - 20; R = 57$

29. $2x^2 - 2x + 6; R = -12$

31. $x^3 + 4x^2 + 16x + 64; R = 192$ **33.** $6x^2 - 8$

35. $2x^2 + 4$ **37.** $x^2 - ax + a^2$

39. $x^3 + ax^2 + a^2 x + a^3$

41. $q(x) = x^2 + 6x + 34; R(x) = 165$

43. $k = 6$ **45.** $k = -3$

47. $x - 9; p(-4) = 44$ **49.** $2x^2 + x + 5; p(4) = 24$

51. $-x^3 - x^2 - x - 1; p(1) = 0$

53. $x^4 - x^3 + x^2 - 3x + 3; p(-1) = -6$

55. $p(4) = -1$ **57.** $p(-2) = -27$ **59.** $p(2) = 0$

61. $p(-3) = -220$ **63.** $p\left(\dfrac{1}{2}\right) = 1$

65. $p(x) = (x - 4)(2x + 1)(x - 2)$

67. 3 and -3

69. $3, -5, 2$ **71.** $\pm i$

73. $p(x) = x(x - 4)(x - (-3))$; the roots are $0, 4, -3$, all with multiplicity 1.

In Exercises 75 and 77, k is a constant.

75. **(a)** $p(x) = k(x - 1)^3 (x - 2)^2 (x - 3)$; deg. $= 6$

  (b) Same as **(a)** with $k = \dfrac{1}{2}$

77. $p(x) = kx^2 (x + 2)^2 (x - 1)$; deg. $= 5$

79. $2 + 3i, -\dfrac{3}{2}$ **81.** $3 - 4i, \dfrac{1}{2} \pm i$

83.

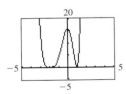

The graph has a turning point at each zero.

85.

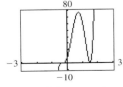

The graph has a turning point at the zero of even multiplicity, and crosses the x-axis at the zero of odd multiplicity.

Exercises 5.4

1. $-2, 1, 5$ **3.** $-3, \dfrac{1}{2}, 5$ **5.** $1, 2$ **7.** $-\dfrac{1}{2}, 1$

9. $-2, -1, 2$ **11.** $\dfrac{1}{3}, \dfrac{1}{2}, -1$ **13.** No rational roots

15. $-\dfrac{3}{4}$ **17.** $-1, -\dfrac{1}{3}, \dfrac{1}{3}, 2$

19. No positive zeros; 1 negative zero

21. No real zeros

23. 0 or 2 positive zeros; 1 negative zero

25. 1, 3, or 5 positive zeros; no negative zeros

27. 5 is upper bound; -1 is lower bound.

29. 6 is upper bound; -4 is lower bound.

31. 1 is upper bound; -2 is lower bound.

33. 2 is upper bound; -2 is lower bound.

35. 2 is upper bound; -6 is lower bound.

37. 1 (actually 0) is upper bound; -4 is lower bound.

39. 5 is upper bound; -3 is lower bound.

41. $x = 1, \dfrac{1}{3}$ **43.** $x = -3, \dfrac{-1 \pm \sqrt{5}}{2}, \dfrac{1}{2}$

45. $x = \dfrac{3}{2}, \dfrac{5}{3}, \pm i$ **47.** $x = \pm 3, \pm \dfrac{\sqrt{3}}{3}$

49. Positive for: $x < -4, -2 < x < 1, x > 3$; Negative for: $-4 < x < -2, 1 < x < 3$

51. Positive for: $-\sqrt{5} < x < 1, x > \sqrt{5}$; negative for: $x < -\sqrt{5}, 1 < x < \sqrt{5}$

53.

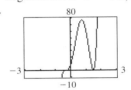

  (a) Zeros are $-0.33, 1.5,$ and 4.

  (b) Zeros are $-\dfrac{1}{3}, \dfrac{3}{2},$ and 4.

55. Zeros are $-\dfrac{3}{2}, -\dfrac{1}{2}, \dfrac{1}{4},$ and $\dfrac{5}{3}$.

Exercises 5.5

1. $D_f = \{x \mid x \neq \pm 5\}$; no zeros

3. $D_h = \left\{x \mid x \neq -4, \dfrac{5}{2}\right\}$; zeros: $x = 0, 4$

5. $D_g = \{x \mid x \neq 0\}$; no real zeros

7. **(a)**

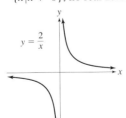

  (b)

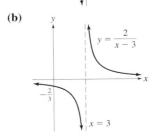

(c)

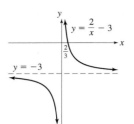

$y = \dfrac{2}{x} - 3$

$y = -3$

9. (a)

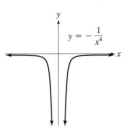

$y = -\dfrac{1}{x^4}$

(b)

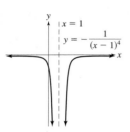

$x = 1$

$y = -\dfrac{1}{(x-1)^4}$

(c)

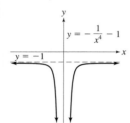

$y = -\dfrac{1}{x^4} - 1$

$y = -1$

11.

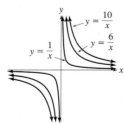

$y = \dfrac{10}{x}$

$y = \dfrac{6}{x}$

$y = \dfrac{1}{x}$

13.

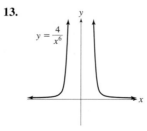

$y = \dfrac{4}{x^6}$

15.

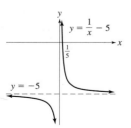

$y = \dfrac{1}{x} - 5$

$y = -5$

17.

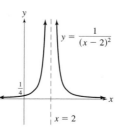

$y = \dfrac{1}{(x-2)^2}$

$x = 2$

19.

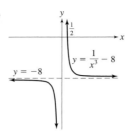

$y = \dfrac{1}{x^3} - 8$

$y = -8$

21.

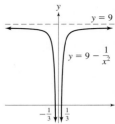

$y = 9$

$y = 9 - \dfrac{1}{x^2}$

23. $y = \dfrac{x}{x+5}$

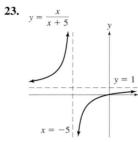

$y = 1$

$x = -5$

25. $y = \dfrac{3x-5}{x}$

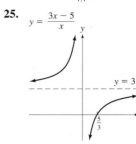

$y = 3$

$\dfrac{5}{3}$

27. $y = \dfrac{x+2}{x-2}$

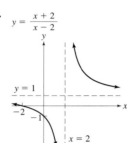

$y = 1$

$x = 2$

29. $y = \dfrac{5-x}{x+4}$

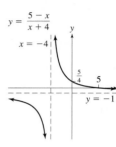

$x = -4$

$\dfrac{5}{4}$

$y = -1$

31. $y = \dfrac{2x}{x^2-1}$

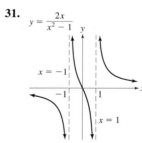

$x = -1$

$x = 1$

33. $y = \dfrac{x^2}{x^2-4}$

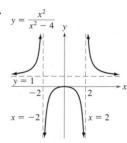

$y = 1$

$x = -2$

$x = 2$

35. $y = \dfrac{x-1}{x^2-2x-3}$

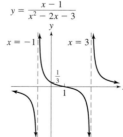

$x = -1$

$x = 3$

37. $y = \dfrac{2-x}{2x^2-x-3}$

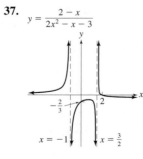

$x = -1$

$x = \dfrac{3}{2}$

39.
$$y = \frac{x^2 - 4x + 3}{x^2 - 2x}$$

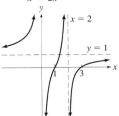

41. $-\dfrac{1}{5x}$ **43.** $\dfrac{-2x - h}{x^2(x + h)^2}$

45. The value of the expression is near 19.

47. $L = 3x + \dfrac{120,000}{x}$

53.

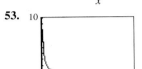

The graph seems to level off as the number of items x increases. By using Trace, it appears that the horizontal asymptote is $A = 0.4$, which means that as the number of items produced increases, the average price per item approaches 0.4 dollar (or 40 cents).

55.

The graph seems to level off as you move out to the right. This means that as the number of days increases, the number of insects levels off.
(a) $n(30) \approx 1214$, $n(60) \approx 1231$, $n(90) \approx 1237$
(b) By using Trace, it appears that the horizontal asymptote is $n = 1250$, which means that as the number of days increases, the number of insects increases toward 1250.

57. (a) $R(1) \approx 0.857$, $R(5) \approx 2.73$, $R(10) = 3.75$. For example, $R(5)$ gives the total resistance of a circuit with resistors of 5 ohms and 6 ohms connected in parallel.

(b)

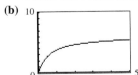

By using Trace, it appears that the horizontal asymptote is $R = 6$, which means that as the number of ohms x increases, the total resistance of the circuit increases toward 6 ohms.

59. (a) The domain is all real numbers greater than 0.

(b)

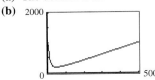

(c) As g increases, the values of C first decrease for a while and then steadily increase.
(d) The minimum cost is 200, which is the cost required to store 50 gallons.

Exercises 5.6

1.

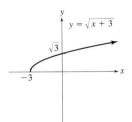

3.

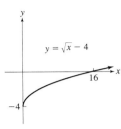

5.

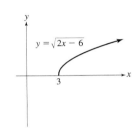

7.

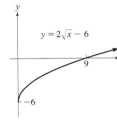

9.

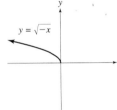

11.

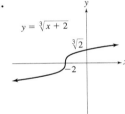

13.

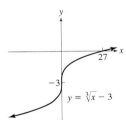

15.

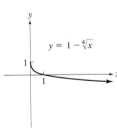

17.

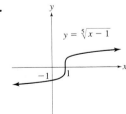

19.

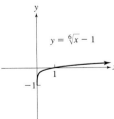

21. $y = \sqrt{16 - x^2}$

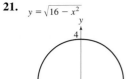

23. $y = -\sqrt{16 - x^2}$

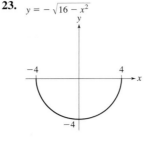

3. $y = (x - 1)^4 - 16$

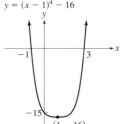

4. $y = (x + 2)^5 - 1$

y-intercept is at (0, 31)

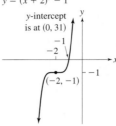

29. (4, 2) **31.** (1, −1) **33.** (0, 3)

35. $\dfrac{\sqrt{x} - 2}{x - 4} = \dfrac{1}{\sqrt{x} + 2}$, which is near $\dfrac{1}{4}$ when x is near 4

37. (a) $\sqrt{(x - 2)^2 + (y - 3)^2} = \sqrt{(x - 5)^2 + (y - 1)^2}$
(b) Both equations become $6x - 4y = 13$.

39. (a) $L = \sqrt{x^2 + 9} + 8 - x$
(b) $C = 2D\sqrt{x^2 + 9} + D(8 - x)$
(c) Approximately $14.7D$ dollars

41. $V = \dfrac{8}{3}\pi(s^2 - 64)$ **43.** $A = h\sqrt{1296 - h^2}$

45. (a) 0.7 sec (b) 24.8 cm

Exercises 5.7

1. $y = \dfrac{375}{8}$ **3.** $u = \dfrac{64}{25}$ **5.** $z = -\dfrac{80}{21}$ **7.** $z = \dfrac{243}{32}$

9. y is multiplied by a factor of $\dfrac{1}{4}$.

11. y is multiplied by a factor of 3.
13. s is multiplied by a factor of 6.
15. 15 pounds **17.** 9.1 inches **19.** 12,500 lumens
21. 133.33 lb/sq in. **23.** 32 ohms
25. 55 mi/hr
27. (a) Force is quadrupled. (b) Force is doubled.
(c) Force is multiplied by a factor of 8.
29. 266.67 pounds
31. (a) The force varies jointly as the two masses and inversely as the square of the distance.
(b) 6.67×10^{-11} (c) 8626 newtons
33. S varies jointly as r and h.
35. V varies directly as the cube of r.
37. z varies jointly as the square of x and the cube of y and inversely as w.

Chapter 5 Review Exercises

1.

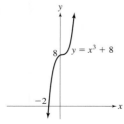

$y = x^3 + 8$

2.

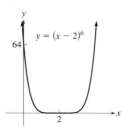

$y = (x - 2)^6$

5. $y = x^3 - x^2 - 6x$

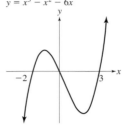

6. $y = x^4 - 4x^2$

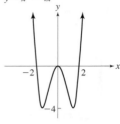

7. $y = x^2 - 6x + 9$

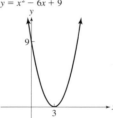

8. $y = 1 - x^4$

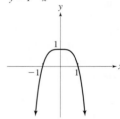

9. $y = \dfrac{1}{(x - 1)^2}$

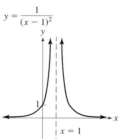

10. $y = -\dfrac{1}{x} + 2$

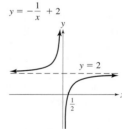

11. $y = \dfrac{1}{x + 2} - 3$

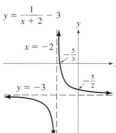

12. $y = \dfrac{1}{(x - 3)^4} + 2$

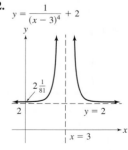

13. $y = \dfrac{x+3}{x+2}$

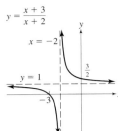

14. $y = \dfrac{x}{x+1}$

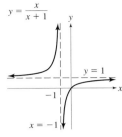

15. $y = \dfrac{x}{x^2-1}$

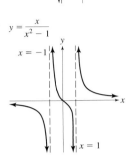

16. $y = \dfrac{x^2+2}{4-x^2}$

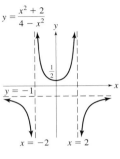

17.

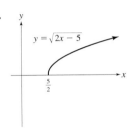

$y = \sqrt{2x-5}$

18.

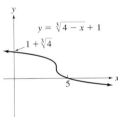

$y = \sqrt[3]{4-x} + 1$

19.

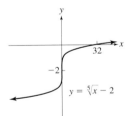

$y = \sqrt[5]{x} - 2$

20.

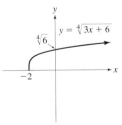

$y = \sqrt[4]{3x+6}$

21.

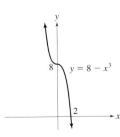

$y = 8 - x^3$

22. $y = (x-2)^3 + 8$

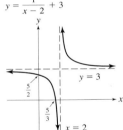

23. $x^2 + 2; R = -1$ **24.** $x^2 - 1; R = -x + 2$
25. $x^3 + 1$ **26.** $x^3 + 2x^2 - x - 1; R = 1$
27. $2x^2 + 3x + 5$ **28.** $x^3 - 2x^2 + 4x - 13; R = 29$
29. $x^4 + x^2 - x + 3; R = 2$
30. $x^5 - x^4 + x^3 - x^2 + x - 2; R = -1$

31. $-3, -1, 4$ **32.** $-2, \pm\sqrt{3}$ **33.** $5, \dfrac{1 \pm i\sqrt{7}}{2}$

34. $\dfrac{2}{3}, \pm 3$ **35.** $-4, \dfrac{1}{2}, \pm i$

36. -1 and 2 both of multiplicity 2
37. 1 more; $5 + 2i$; $p(x) = k(x^3 - 8x^2 + 9x + 58)$ where k is any real number
38. 2 more; $2 - 3i$ and $1 + i$;
$p(x) = k(x^4 - 6x^3 + 23x^2 - 34x + 26)$ where k is any real number
39. 0   **40.** 2 or 0   **41.** 3 or 1   **42.** 0
43. Upper bound: 4; lower bound: -1
44. Upper bound: 4; lower bound: -3
45. Upper bound: 1; lower bound: -1
46. Upper bound: 4; lower bound: -3

47. $A = \dfrac{1}{2}x\sqrt{25 - x^2}$ **48.** $S = 2x^2 + \dfrac{400}{x}$

49. $(0, 2)$ **50.** $(9, 0)$ **51.** $(10, 3)$ **52.** $(1, 3)$

53. $x = \dfrac{81}{2}$ **54.** $x = \dfrac{20}{27}$

55. Approximately 5.1 seconds   **56.** 6π cu cm

Chapter 5 Practice Test

1. (a) $y = \dfrac{1}{x-2} + 3$

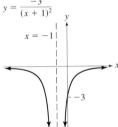

(b) $y = \dfrac{-3}{(x+1)^2}$

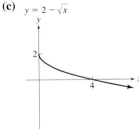

(c) $y = 2 - \sqrt{x}$

(d) $y = \dfrac{x+3}{x-2}$

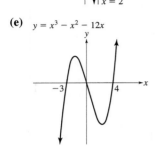

(e) $y = x^3 - x^2 - 12x$

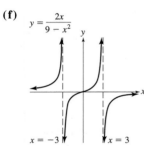

(f) $y = \dfrac{2x}{9 - x^2}$

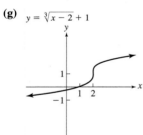

(g) $y = \sqrt[3]{x - 2} + 1$

(h) $y = (x + 1)^2(x - 2)^2$

2. $x^3 - 2x^2 - 5x - 1 - \dfrac{1}{3x - 4}$ **3.** $x - 2$ and $x + 2$

4. (a) $x = 3, 4, -\dfrac{1}{3}$ **(b)** $x = 0, \dfrac{3}{2}, \dfrac{3 \pm i\sqrt{11}}{2}$

5. The blood pressure would increase by a factor of approximately 5.

Chapter 6

Exercises 6.1

1. $x = 4$ **3.** $x = \dfrac{5}{2}$ **5.** $t = \dfrac{3}{2}$ **7.** $x = -6$

9. $x = \dfrac{1}{6}$ **11.** D_f: all real numbers   **13.** D_h: $[3, \infty)$

15. D_g: $x \neq \dfrac{1}{3}$ **17.**

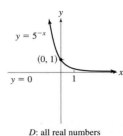

D: all real numbers
R: $\{y \mid y > 0\}$

19.

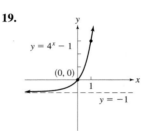

D: all real numbers
R: $\{y \mid y > -1\}$

21.

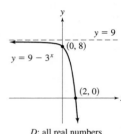

D: all real numbers
R: $\{y \mid y < 9\}$

23.

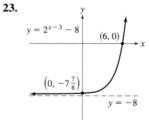

D: all real numbers
R: $\{y \mid y > -8\}$

25.

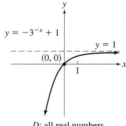

D: all real numbers
R: $\{y \mid y < 1\}$

27.

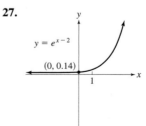

D: all real numbers
R: $\{y \mid y > 0\}$

29.

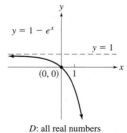

D: all real numbers
R: $\{y \mid y < 1\}$

31. $5^{x^2 - 6x + 2}$ **33.** $5^{\sqrt{x}}$ for $x \geq 0$ **35.** $\dfrac{1}{5^x + 1}$

37. $x = 0$ **39.** $x = \pm 3$ **41.** None   **43.** $x = 0$

49.

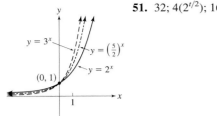

$y = 3^x$

$y = \left(\frac{5}{2}\right)^x$

$y = 2^x$

$(0, 1)$

51. $32; 4(2^{t/2}); 16,384$

53. (a)

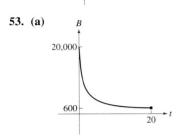

B

$20,000$

600

20 t

(b) Approx. \$11,819
55. (a) \$578.16   **(b)** \$438.63
57. (a) 1 atmosphere; 14.69 lb/sq in.
   (b) 0.33 atmosphere; 4.81 lb/sq in.
   (c) 1.05 atmosphere, 15.44 lb/sq in.
59. (a) 13,784; 14,270   **(b)** 23,182; 24,000
   (c) 1.04; 1.04
   (d) The population in year $t + 1$ is $2^{0.05}$ times the population in year t.
61. (b) Approx. 139
63. (a) $A = A_0 2^{-t/4}$;   **(b)** 7.07 mg; 0.16 mg
65. Approx. 3.1 min
67. (a)

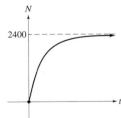

N

2400

t

(b) 1322; 1678, 1916   **(c)** 55%, 70%, 80%
69. (a) I is the initial concentration.
   (b) F is the concentration level that is approached as more
   and more time has elapsed.
71. 0.22 lumen   **73. (a)** 28%   **(b)** 29 days

Exercises 6.2

1. $7^2 = 49$ **3.** $\log_3 \frac{1}{81} = -4$ **5.** $\left(\frac{1}{4}\right)^{-3} = 64$

7. $\log_{27}\frac{1}{3} = -\frac{1}{3}$ **9.** $8^{-1/3} = \frac{1}{2}$ **11.** $\log_5 \sqrt[4]{5} = \frac{1}{4}$

13. $2^{-1} = \frac{1}{2}$ **15.** $\left(\frac{2}{3}\right)^{-3} = \frac{27}{8}$ **17.** $\log_4 1024 = 5$

19. $\log_{1/9} 3 = -\frac{1}{2}$ **21.** 5   **23.** -1 **25.** $\frac{4}{3}$

27. $-\frac{3}{4}$ **29.** Not defined   **31.** $-\frac{3}{2}$ **33.** -4 **35.** $\frac{2}{3}$

37. 1   **39.** -1 **41.** 7   **43.** $-\frac{1}{2}$ **45.** 6

47. $t = 81$ **49.** $t = 2$ **51.** $t = \frac{4}{5}$

53. $(-\infty, -2)\cup(2, \infty)$; $F(6) = 5$
55.

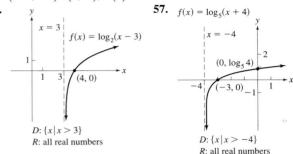

$x = 3$

$f(x) = \log_2(x - 3)$

1

1 3 $(4, 0)$ x

$D: \{x \mid x > 3\}$
$R:$ all real numbers

57. $f(x) = \log_5(x + 4)$

$x = -4$

2

$(0, \log_5 4)$

-4 $(-3, 0)$ 1 x

-1

$D: \{x \mid x > -4\}$
$R:$ all real numbers

59.

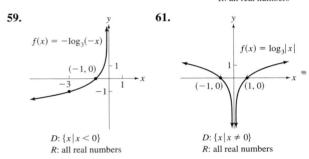

$f(x) = -\log_3(-x)$

1

$(-1, 0)$

-3 1 x

-1

$D: \{x \mid x < 0\}$
$R:$ all real numbers

61.

$f(x) = \log_3 |x|$

1

$(-1, 0)$ $(1, 0)$ x

$D: \{x \mid x \neq 0\}$
$R:$ all real numbers

63. $y = \log_6 x$ **65.** $y = \log_{16} x$

Exercises 6.3

1. $2 \log_4 x + \log_4 y + 3 \log_4 z$
3. $3 \log_b x - \log_b y - 4 \log_b z$
5. $\frac{1}{2} + \frac{3}{2} \log_5 x$ **7.** Cannot be simplified
9. $\log_b(x - 2) + \log_b(x + 2)$
11. $\frac{1}{3} \log_4 (x - 3) - \frac{1}{6} - \frac{2}{3} \log_4 x$
13. $\frac{1}{2} \log_b(x + 4) - \frac{1}{2} \log_b(x + 2)$ where $x \neq 4$
15. False   **17.** True   **19.** $\log_b 48$ **21.** $\log_b \frac{s^{1/2}}{t^{3/2}}$

23. 1   **25.** 3   **27.** $\log_b\left(\frac{x^3}{y^4 z^2}\right)$ **29.** $\log_b\left(\frac{x^{1/4}y^{1/3}}{z^{1/2}}\right)$

37. 2.47   **39.** -2.87 **41.** $x = \frac{9}{4}$ **43.** $x = 7$

45. $t = \frac{16}{7}$ **47.** $x = 5$ **49.** $x = 10$ **51.** $x = 8$

53. $x = 9$ **55.** 25

Exercises 6.4

1. 5   **3.** $\frac{1}{2}$ **5.** $x + 1$

7.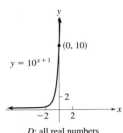

$y = 10^{x+1}$

(0, 10)

D: all real numbers
R: $\{y \mid y > 0\}$

9.

$y = 2 - \ln x$

$(e^2, 0)$

D: $\{x \mid x > 0\}$
R: all real numbers

11.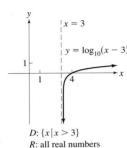

$x = 3$

$y = \log_{10}(x - 3)$

D: $\{x \mid x > 3\}$
R: all real numbers

13.

$y = \log x^2$

$(-1, 0)$ $(1, 0)$

D: $\{x \mid x \neq 0\}$
R: all real numbers

15. $f^{-1}(x) = \dfrac{1 + \ln x}{3}$ **17.** $g(x) = (\ln x)^2$

19. 0.71   **21.** -2.32 **23.** -0.18 **25.** $2 \log_9 x$

27. $\dfrac{\ln x}{\ln 10} = \dfrac{\ln x}{2.303}$ **29.** $x = \dfrac{1 - \ln 5}{2}$

31. $x = \dfrac{-2 + \ln 20}{3}$ **33.** $x = \dfrac{4 \ln 3 - \ln 7}{\ln 3}$

35. $x = \dfrac{2 \ln 2}{2 \ln 2 + \ln 6}$ **37.** $x = \ln 2$ **39.** $x = \dfrac{\ln 3}{4 \ln 2}$

41. $x = 64$ or $x = \dfrac{1}{2}$ **43.** $x = 0.235$ **45.** $x = 1.765, x = 5$

47. $x = -2.032$ **49.** $x = -1, x = 1.501$

53. $y = \log_4 x = \dfrac{\log_{10} x}{\log_{10} 4}$

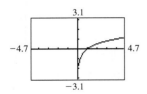

Exercises 6.5

1. No   **3.** $A = A_0 e^{-0.00002773t}$ **5.** 88.6 grams
7. $A = A_0 e^{-0.30944t}$; .00372 gram   **9.** 559 years
11. 18.9 hours; 30 hours   **13.** 0.94%
15. 9.78 billion   **17.** 1.7%; lower   **19.** 5.39 years
21. 6.91%; no   **23.** 6.49%   **25.** $4489.58
27. 11.552 years compounded continuously; 11.553 years compounded daily
29. (a) $199.08   (b) $9555.84
31. (a) $598.20   (b) $215,352
33. (a) 3
    (b) $10^{8.3} = 199,526,231.5$ times as intense.

35. E_3 is 10,000 times as intense as E_2, which is 10,000 times as intense as E_1.
37. 4.2   **39.** 5.6
41. (a) 2.51   (b) 2.51   (c) 2.5   (d) 4
43. $N = 0$ decibels
45. (a) 4.38%   (b) 5.22%
47. (d) Approx. 3560 years old
49. Approx. 2268 years old
51. (a) 70 grams   (b) 81.8 grams   (c) 89.0 grams
    (d) 99.999 grams
53. (a) $78.20   (b) 14.09 years
55. Stirling's formula; $2.422786847 \times 10^{18}$; actual value: $2.432902008 \times 10^{18}$

Chapter 6 Review Exercises

1.

$y = 2^{x-1}$
$(0, \frac{1}{2})$
$y = 0$

2. $f(x) = \left(\frac{1}{4}\right)^x - 1$

(0, 0)
$y = -1$

3. $f(x) = \log_{1/3}(x + 3)$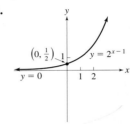
$(-2, 0)$
$(0, -1)$
$x = -3$

4.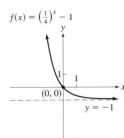
$y = 5 + \log_5 x$
$(0.00032, 0)$

5. $y = e^{x+2} - 1$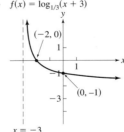
$(0, e^2 - 1)$
$(-2, 0)$
$y = -1$

6. $y = 2 + \ln x^2$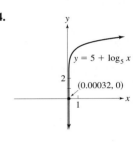
$(-0.368, 0)$ $(0.368, 0)$

7.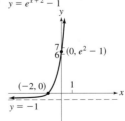
$y = 8 - 2^{x+3}$

8.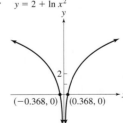
$y = -\ln x$

9. $6^{-1} = \dfrac{1}{6}$ **10.** $\log_9 3 = \dfrac{1}{2}$ **11.** $\log_8 \dfrac{1}{4} = -\dfrac{2}{3}$

12. $3^4 = 81$ **13.** $b^6 = b^6$ **14.** $\log_b t = \log_b t$

15. 3   **16.** -6 **17.** -2 **18.** -3 **19.** $\dfrac{2}{3}$ **20.** $\dfrac{4}{5}$

21. 7   **22.** 5   **23.** 20   **24.** 17   **25.** $\dfrac{3}{2}$ **26.** 0

27. $3 \log_b x + 4 \log_b y + 2 \log_b z$

28. $3 - \dfrac{1}{2} \log_2 x - \dfrac{1}{2} \log_2 y$

29. $\log_5 2 + \dfrac{1}{3} \log_5 x - \dfrac{1}{3} - \dfrac{4}{3} \log_5 y$

30. Cannot be simplified   **31.** Cannot be simplified

32. $\dfrac{1}{2} \log_b x - \dfrac{1}{3} \log_b y$ **33.** $x = -2$ **34.** $x = -\dfrac{3}{2}$

35. $x = \dfrac{\ln 29}{\ln(\frac{1}{9})} \approx -1.533$ **36.** $x = -1, x = 2$

37. $x = 1$ **38.** $x = 1$ **39.** $x = \dfrac{\ln 3 + \ln 7}{\ln 7} \approx 1.565$

40. $t = 3$ **41.** $x = 10$ **42.** $x = \dfrac{2 \ln 5}{\ln 3 - \ln 5} \approx -6.301$

43. $x = \dfrac{\ln 5}{\ln 3 - \ln 2} \approx 3.969$ **44.** $x = 8$ **45.** $x = 9$

46. $x = 1, 5$ **47.** $x = \dfrac{2 \ln 8 + 2 \ln 9}{3 \ln 8 - \ln 9} \approx 2.117$

48. $x = \dfrac{100}{99}$ **49.** $b = 5$ **50.** $x = \dfrac{7}{3}$

52. $f^{-1}(x) = -3 + \log_2 x$ **53.** $f^{-1}(x) = 1 + 3^{x-5}$

54. $\log_6 x = \dfrac{\log_3 x}{\log_3 6}$ **55.** 1.23   **56.** \$2953.86

57. \$5621.42   **58.** 11.6%

59. (a) 1418 bacteria   (b) 23 hours

60. 66.67 hours

61. $A = A_0 e^{-0.013078t}$ (t in days)

62. $A = A_0 e^{-0.138629t}$ (t in minutes)

63. The half-life is halved from 6.9 to 3.45 years.

64. 9.0 (to the nearest tenth)

65. 10^{-6} watt per sq cm

Chapter 6 Practice Test

1. (a)

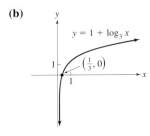

$$y = \left(\tfrac{1}{2}\right)^{x+1} - 4$$

(b)

$$y = 1 + \log_3 x$$

$\left(\tfrac{1}{3}, 0\right)$

2.

$f(x) = 2^{x-1}$

$(0, 2)$

$y = x$

$(2, 0)$

$f^{-1}(x) = -1 + \log_2 x$

3. (a) $-\dfrac{1}{2}$ **(b)** 0.71   **(c)** $\dfrac{4}{3}$ **(d)** 2

4. (a) $x = \dfrac{3}{2}$ **(b)** $x = 9.83$ **(c)** $x = 4$

   (d) $x = \dfrac{\ln 10 - \ln 5}{\ln 5} \approx 0.43$

5. (a) 949   **(b)** 50.26 hours

6. $A = A_0 e^{-0.000924196t}$ for t in years; 83.1 grams

Chapter 7

Exercises 7.1

1. $\left(\dfrac{1}{2}, \dfrac{1}{3}\right)$ **3.** $\left(\dfrac{32}{13}, \dfrac{18}{13}\right)$

5. $(4, 4)$ **7.** $(5, 0)$ **9.** $(3, 7)$ **11.** $(-10, -45)$

13. $(0, 3)$ **15.** $\left(2, -\dfrac{1}{2}\right)$ **17.** $(10, 50)$

19. $\left(\dfrac{4}{3}, 0\right)$ **21.** $a = 3, b = 2$ **23.** $\left(\dfrac{-17}{5}, \dfrac{48}{5}\right)$

25. $\left(\dfrac{39}{95}, \dfrac{-250}{19}\right)$

27. 10 liters of the 30% alcohol solution: 20 liters of the 45% alcohol solution

29. \$6000 at 9.5%: \$14,000 at 8%

31. Flat fee = \$29.50; mileage fee = 32¢ per mile

33. Speed of the plane = 600 mi/hr; speed of wind = 30 mi/hr

35. 60 of the more expensive model; 120 of the less expensive model

37. Let m = no. of miles and C = the daily cost. For the Speedy Agency, $C = 18 + 0.28m$; for the Hirsch Agency, $C = 14 + 0.32m$. For mileage below 100 miles, the Hirsch Agency is less expensive; for mileage above 100 miles, the Speedy Agency is less expensive.

39. For air time of more than 100 minutes per month, plan B is less expensive; for air time of less than 100 minutes, plan A is less expensive.

41. At $t = 1, 64 = -16 + v_0 + s_0$; at $t = 2$, $96 = -64 + 2v_0 + s_0$; $v_0 = 80$ ft/sec; $s_0 = 0$ feet

43. (a) $A = 5$ **(b)** $A = 45$

45. $A = 2, B = -4$ **47.** $A = 2, B = -3$

49. $(-4, -7)$

Exercises 7.2

1. $(2, 2, 2)$ **3.** $\left(2, -\dfrac{3}{2}, -\dfrac{1}{2}\right)$ **5.** Inconsistent

7. $(5, 0.5, 0.3)$ **9.** $(-8, 12, -3)$ **11.** $(2, 3, -2)$
13. $(-1, 0, -2)$ **15.** Inconsistent
17. Let b = the amount in the bank account, c = the amount invested in the certificate of deposit, and s = the amount invested in the stock. The three equations are $b + c + s = 19,750$, $0.06b + 0.08c + 0.105s = 1680$, and $0.06b + 0.08c = 0.105s$. $5000 in the bank account, $6750 in the certificate, and $8000 in the stock.
19. 150 of model A, 110 of model B, and 140 of model C
21. 4 oz of food A; 3.1 oz of food B; 3.8 oz of food C (rounded to the nearest tenth)
23. 20 liters from tank X; 18 liters from tank Y; 16 liters from tank Z
25. $15\frac{5}{9} \approx 15.56$ liters of the 10% solution, $35\frac{5}{9} \approx 35.56$ liters of the 25% solution, and $48\frac{8}{9} \approx 48.89$ liters of the 40% solution
27. $f(x) = 3x^2 - 2x + 1$ **29.** $A = 5, B = -5, C = 7$

Exercises 7.3

1. $\left[\begin{array}{cc|c} 7 & 1 & 11 \\ 1 & -3 & 4 \end{array}\right]$ **3.** $\left[\begin{array}{ccc|c} 3 & -2 & 1 & 15 \\ 2 & 1 & -3 & 10 \\ 5 & -3 & 1 & -2 \end{array}\right]$

5. $\left[\begin{array}{cc|c} 1 & 3 & 2 \\ 0 & -11 & -10 \end{array}\right]$ **7.** $\left[\begin{array}{ccc|c} 1 & 2 & 1 & 0 \\ 0 & 2 & 1 & 1 \\ 0 & 0 & 3 & -9 \end{array}\right]$

9. $(5, 1)$ **11.** $(2, -3)$ **13.** $(3, 7)$ **15.** $(-10, -45)$
17. $(0, 1)$ **19.** $(2, 2, 2)$ **21.** $(2, -1, 3)$

23. Inconsistent   **25.** $a = \frac{1}{2}, b = 2, c = 3$

27. $(2, 1, 0, -1)$ **29.** $\left[\begin{array}{cc|c} 1 & 0 & 2 \\ 0 & 1 & -1 \end{array}\right]$

31. $\left[\begin{array}{cc|c} 1 & 0 & -4 \\ 0 & 1 & 1 \end{array}\right]$ **33.** $\left[\begin{array}{ccc|c} 1 & 0 & 0 & 3 \\ 0 & 1 & 0 & 1 \\ 0 & 0 & 1 & -2 \end{array}\right]$

35. $(2, -1)$ **37.** $(1, 3, 1)$ **39.** $\left(-\frac{5}{2}, \frac{5}{2}, \frac{1}{2}\right)$
41. $A = 25, B = 13, C = -20, D = -10$

Exercises 7.4

1. $\left[\begin{array}{cc} 3 & 2 \\ 8 & 2 \end{array}\right]$ **3.** $\left[\begin{array}{cc} 0 & 0 \\ 0 & 0 \\ 0 & 0 \end{array}\right]$ **5.** Not defined

7. $\left[\begin{array}{cc} 0 & 9 \\ 51 & 0 \end{array}\right]$ **9.** $\left[\begin{array}{cc} -2 & 3 \\ 1 & 0 \end{array}\right]$ **11.** $\left[\begin{array}{cc} \frac{10}{3} & \frac{2}{3} \\ -\frac{4}{3} & 2 \end{array}\right]$

13. $\left[\begin{array}{cc} -3 & -4 \\ 1 & -3 \end{array}\right]$ **15.** $\left[\begin{array}{cc} -1 & -7 \\ 0 & -3 \end{array}\right]$ **17.** $\left[\begin{array}{cc} \frac{5}{2} & \frac{1}{2} \\ -1 & \frac{3}{2} \end{array}\right]$

19. $\left[\begin{array}{cc} \frac{9}{2} & -\frac{5}{2} \\ -2 & \frac{3}{2} \end{array}\right]$ **21.** $[10]$ **23.** $[3]$ **25.** $\left[\begin{array}{cc} 0 & 2 \\ -3 & 5 \end{array}\right]$

27. Not defined   **29.** $\left[\begin{array}{cc} -30 & -2 \\ 39 & -11 \end{array}\right]$

31. $\left[\begin{array}{ccc} 13 & -3 & -10 \\ 9 & -7 & -7 \\ -4 & -6 & 5 \end{array}\right]$ **33.** $\left[\begin{array}{cc} 5 & -7 \\ 3 & -14 \end{array}\right]$

35. $\left[\begin{array}{ccc} 3 & 1 & 0 \\ -2 & 4 & 6 \\ 3 & 8 & -10 \end{array}\right]$ **37.** False   **39.** True

43. AB is not defined; BA is a 3×4 matrix.
47. Yes. The additive inverse for an $n \times n$ matrix A is the $n \times n$ matrix $-A = -1(A)$.
51. Brand A is less expensive when ordered from Raju (62.8¢); brands B and C are less expensive when ordered from George (48.6¢ and 41¢, respectively). If all three brands must be ordered from one company, then the less expensive company is Raju ($1.536).

53. (a) $\left[\begin{array}{cc} 87,552 & 94,065 \\ 122,503 & 131,616 \\ 151,546 & 162,801 \end{array}\right]$

(b) The first column contains the 2001 total knapsack sales in the store, by catalog, and by Internet. The second column contains the same information for 2002.
(c) 2001: $361,601; 2002: $388,482
55. The inner product is not defined. The matrix product is

$\left[\begin{array}{ccc} 2 & -3 & 1 \\ 0 & 0 & 0 \\ 8 & -12 & 4 \end{array}\right]$.

Exercises 7.5

1. $\left[\begin{array}{cc} 3 & -1 \\ 3 & -2 \end{array}\right]$ **5.** $\left[\begin{array}{cc} 0 & \frac{1}{2} \\ \frac{1}{2} & -\frac{1}{4} \end{array}\right]$

7. $\left[\begin{array}{cc} -\frac{1}{16} & \frac{5}{16} \\ \frac{3}{16} & \frac{1}{16} \end{array}\right] = \frac{1}{16}\left[\begin{array}{cc} -1 & 5 \\ 3 & 1 \end{array}\right]$ **9.** No inverse

11. No inverse   **13.** $\frac{1}{2}\left[\begin{array}{ccc} -1 & 3 & -1 \\ -1 & 5 & -3 \\ 2 & -4 & 2 \end{array}\right]$

15. $\left[\begin{array}{ccc} 2 & -1 & -\frac{3}{2} \\ 2 & -1 & -2 \\ -1 & 1 & \frac{3}{2} \end{array}\right]$ **17.** No inverse

19. $\frac{1}{6}\left[\begin{array}{ccc} 1 & -2 & 3 \\ 3 & -6 & 3 \\ 4 & -2 & 0 \end{array}\right]$ **21.** $(5, -1)$ **23.** $(6, 1)$

25. $(3, 2)$ **27.** $(-10, -45)$ **29.** $(0, 3)$ **31.** $(2, 2, 2)$
33. $(1, 0, -1)$ **35.** Inconsistent   **37.** $(1, 0.1, -1)$
39. (a) $16,666.67 invested in the certificate; $3333.33 invested in the stock.
(b) $13,333.33 invested in the certificate; $6666.67 invested in the stock.
(c) $10,000 invested in the certificate; $10,000 invested in the stock.
41. Mice: 0.015 gm brand A, 0.003 gm brand B, 0.0112 gm brand C; rats: 0.021875 gm brand A, 0.003125 gm brand B, 0.0175 gm brand C

43. (a) $\begin{bmatrix} \frac{1}{2} & 0 & 0 \\ 0 & \frac{1}{4} & 0 \\ 0 & 0 & \frac{1}{10} \end{bmatrix}$ **(b)** $\begin{bmatrix} 5 & 0 & 0 \\ 0 & 3 & 0 \\ 0 & 0 & 6 \end{bmatrix}$

(c) $\begin{bmatrix} \frac{1}{a} & 0 & 0 \\ 0 & \frac{1}{b} & 0 \\ 0 & 0 & \frac{1}{c} \end{bmatrix}$

Exercises 7.6

1. 5   **3.** 14   **5.** 0   **7.** 19   **9.** -9 **11.** 0

13. $(2, -1)$ **15.** $\left(\frac{17}{9}, -\frac{4}{3}\right)$ **17.** $\left(\frac{24}{35}, \frac{17}{35}\right)$

19. Dependent: $\left(a, \frac{2-a}{2}\right)$ **21.** Inconsistent

23. $(1, 3, 1)$ **25.** $(1, -1, 0)$ **27.** $\left(4, -\frac{1}{2}, -1\right)$

29. $\left(-\frac{3}{2}, \frac{5}{4}, 0\right)$ **31.** $\left(\frac{5}{2}, \frac{1}{4}, -\frac{3}{2}\right)$

33. \$6000 invested at 6%; \$4000 invested at 8.5%

35. 36 of model A; 46 of model B

37. 80 of model A; 50 of model B; 60 of model C

Exercises 7.7

1. $\begin{vmatrix} 2 & 5 & -1 \\ 7 & 4 & -3 \\ -2 & 1 & 9 \end{vmatrix}$ **3.** $-\begin{vmatrix} 2 & 5 & -1 \\ -2 & 1 & 9 \\ -6 & -4 & -5 \end{vmatrix}$

5. $\begin{vmatrix} 2 & 5 & 3 \\ -2 & 1 & 8 \\ -6 & -4 & 6 \end{vmatrix}$ **11.** 63   **13.** 57   **15.** 24

17. -29 **19.** 0

23. $w = -1, x = 0, y = 1, z = 2$

25. $w = 1, x = 2, y = -1, z = 0$

27. Dependent: Solutions (w, x, y, z) are of the form

$\left(a, -\frac{a}{3}, -\frac{2a}{3}, 0\right)$

Exercises 7.8

1.
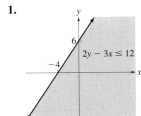

3.
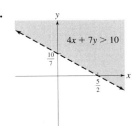

5.
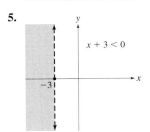

7.
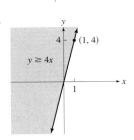

9.
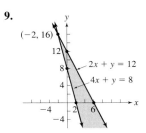

11.
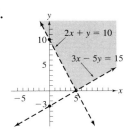

13.
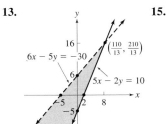

15.
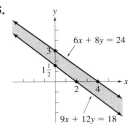

17.
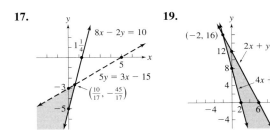

19.

21.
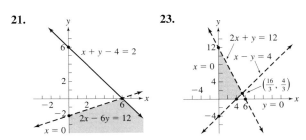

23.

25.
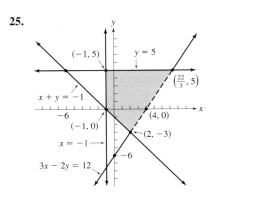

27.

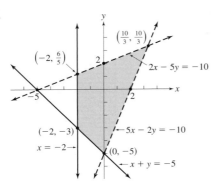

29. Let $V =$ no. of VCRs and $T =$ no. of TVs.

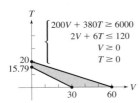

31. Let $s =$ no. of standard models and $c =$ no. of custom models.

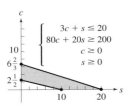

33. Let $s =$ amount invested in the 6-month certificate and $y =$ amount invested in the 1-year certificate.

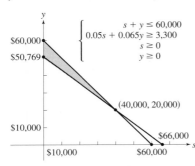

35. Let $d =$ no. of divorce cases and $m =$ no. of malpractice suits.

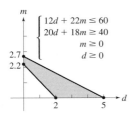

Exercises 7.9

1. The vertices are $(0, 0)$, $(0, 1)$, $\left(\dfrac{4}{3}, \dfrac{7}{3}\right)$, and $(6, 0)$. The minimum value of f is 0, which occurs at $(0, 0)$; the maximum value of f is 18, which occurs at $(6, 0)$.

3. The vertices are $(0, 1)$, $(0, -2)$, and $(2, -1)$. The minimum value of f is -1, which occurs at $(0, 1)$; the maximum value of f is 2, which occurs at $(0, -2)$.

5. The vertices are $(0, 2)$, $(0, 3)$, and $(3, 0)$. The minimum value of f is 5, which occurs at $(3, 0)$; the maximum value of f is 14, which occurs at $(0, 3)$.

7. Let $h =$ amount invested in high-risk and $c =$ amount invested in conservative investments (in millions). Maximum profit is \$8.4 million, where \$52.5 million is invested in conservative investments and \$22.5 million is invested in high-risk investments.

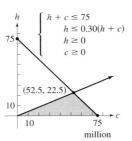

Vertices: $(0, 0)$, $(75, 0)$, and $(52.5, 22.5)$; $P = 0.10c + 0.14h$

9. Let $x =$ no. of model X and $y =$ no. of model Y. Maximum value of P is at $(62.5, 50)$, where $P = 10,625$. The closest integral value within constraints is $(63, 49)$. Maximum profit is \$10,570 for 63 of model X and 49 of model Y.

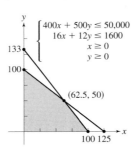

Vertices: $(0, 0)$, $(0, 100)$, $(62.5, 50)$, and $(100, 0)$; $P = 90x + 100y$

11. The maximum profit would be \$18,333.20 for g (graphing calculators) $= 333.33$ and s (scientific calculators) $= 833.33$; however, constrained to integral values, the maximum profit is \$18,320 for 333 graphing calculators and 833 scientific calculators.

13. The least expensive combination of cereals is 60¢ for 0 oz of cereal X and 2 oz of cereal Y.

15. The maximum profit would be \$26,800 for 3.04 divorce cases and 1.07 malpractice cases; however, constrained to integral values, the maximum profit is \$26,000 for 3 divorce cases and 1 malpractice case.

17. Let x = no. of grams of Zomine X and y = no. of grams of Zomine Y. Vertices are $(0, 1)$, $(0, 6)$, $(1, 4)$, $(2, 0)$, and $(1, 0)$. $R = 0.005x + 0.002y$. The *best* dosage is 1 gram of Zomine X and 4 grams of Zomine Y, yielding 0.013 gram of pain relief, given the constraints.

Chapter 7 Review Exercises

1. $(4, 1)$ **2.** $(-2, 3)$ **3.** $(0, -5)$ **4.** $(6, 2)$
5. $(3, 3)$ **6.** $(8, 3)$ **7.** $(1, 2, 3)$ **8.** $(4, 0, 2)$

9. $\left(\frac{17}{2}, -5, \frac{3}{2}\right)$ **10.** $(3, 4, -6)$ **11.** $(2, 3, 0)$

12. $\left(0, \frac{1}{4}, -\frac{1}{4}\right)$ **13.** $\begin{bmatrix} 1 & 0 & | & 0 \\ 0 & 1 & | & 1 \end{bmatrix}$

14. $\begin{bmatrix} 1 & 0 & | & 2 \\ 0 & 1 & | & 15 \end{bmatrix}$ **15.** $\begin{bmatrix} 1 & 0 & 0 & | & 3 \\ 0 & 1 & 0 & | & 5 \\ 0 & 0 & 1 & | & -3 \end{bmatrix}$

16. $\begin{bmatrix} 1 & 0 & 0 & | & -1 \\ 0 & 1 & 0 & | & -6 \\ 0 & 0 & 1 & | & 9 \end{bmatrix}$ **17.** $\left(\frac{13}{4}, \frac{7}{4}\right)$ **18.** $(-1, 5)$

19. $(2, 0, -5)$ **20.** $\left(2, -\frac{5}{4}, -\frac{1}{4}\right)$ **21.** $(6, 1)$

22. $(-1, 3)$ **23.** $(1, 0, -1)$ **24.** $(2, -1, 3)$

25. $\begin{bmatrix} 10 & -3 \\ 9 & -8 \end{bmatrix}$ **26.** $\begin{bmatrix} -4 & 6 & -16 \\ -10 & 2 & 0 \\ 4 & -14 & -36 \end{bmatrix}$

27. $\begin{bmatrix} 0 & -11 \\ 7 & 4 \end{bmatrix}$
28. Addition is not defined for matrices of different sizes.
29. [6]
30. Multiplication is not defined for a 4×1 matrix times a 2×4 matrix.

31. $\begin{bmatrix} 5 & 8 \\ -4 & 1 \end{bmatrix}$ **32.** $\begin{bmatrix} 5 & -3 \\ 4 & 1 \end{bmatrix}$

33. $\begin{bmatrix} -1 & -2 & 0 \\ 11 & -5 & 9 \\ 16 & -13 & 18 \end{bmatrix}$ **34.** $\begin{bmatrix} 2 & -6 & 1 \\ 7 & -1 & 3 \\ 21 & -2 & 11 \end{bmatrix}$

35. $\frac{1}{4}\begin{bmatrix} -2 & 3 \\ 2 & -1 \end{bmatrix}$ **36.** $\frac{1}{2}\begin{bmatrix} -1 & 1 & 1 \\ 1 & -1 & 1 \\ 4 & -2 & 0 \end{bmatrix}$

37. $(1, 3)$ **38.** $(0, -2)$ **39.** $(2, -2, 1)$
40. $(3, 0, -1)$ **41.** -2 **42.** 3   **43.** 3   **44.** -6

45. $(-1, 2)$ **46.** $\left(4, \frac{1}{2}\right)$ **47.** $(2, 0, 3)$

48. $(-1, 2, 5)$
49. 60 of the more expensive cameras; 160 of the less expensive cameras
50. $6.75 per hour for grading; $18 per hour for tutoring
51. $2500 invested at 6%; $7500 invested at 8%
52. The lawyer worked 15 hours; the assistant worked 20 hours.
53. $10,000 in the bank account; $12,000 in the certificate of deposit; $18,000 in the stock

54. 30 of model A; 80 of model B; 25 of model C
55.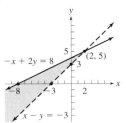
56.

57.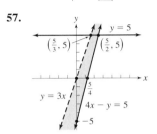

58.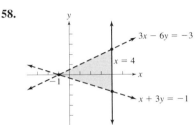

59. Let V = no. of VCRs and T = no. of TVs.
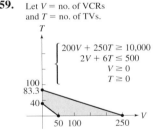

60. Let x = amount in 6-mo. certificate and y = amount in 1-year certificate.
$$\begin{cases} x + y \le 20{,}000 \\ y \ge 2x \\ 2000 \le x \le 4000 \end{cases}$$
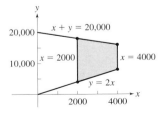

61. Let x = no. of civil complaints and y = no. of criminal complaints.

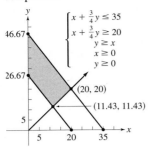

$$\begin{cases} x + \frac{3}{4}y \le 35 \\ x + \frac{3}{4}y \ge 20 \\ y \ge x \\ x \ge 0 \\ y \ge 0 \end{cases}$$

62. Let A = no. of appliance A and B = no. of appliance B.

$$\begin{aligned} 6A + 4B &\le 800 \\ 14A + 16B &\ge 1000 \\ A &\ge 0 \\ B &\ge 0 \end{aligned}$$

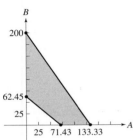

63. The maximum profit is $7.6 million for $60 million in conservative investments and $20 million in high-risk investments.

64. Let x = no. of model X and y = no. of model Y.

$$f = 225x + 100y$$
$$\begin{cases} 300x + 400y \ge 48000 \\ 12x + 6y \le 1080 \\ x \ge 0 \\ y \ge 0 \end{cases}$$

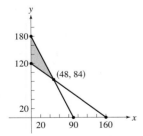

At $(0, 120)$, $f = 225(0) + 100(120)$
$= 12{,}000$
At $(0, 180)$, $f = 225(0) + 100(180)$
$= 18{,}000$
At $(48, 84)$, $f = 225(48) + 100(84)$
$= 19{,}200$

The maximum profit is $19,200 for 48 of model X and 84 of model Y.

Chapter 7 Practice Test

1. (a) $(4, 4)$ **(b)** $(2, -1)$ **(c)** $\left(2, -\frac{3}{2}, -\frac{1}{2}\right)$

2. $(3, -2, 4)$

3. (a) $\begin{bmatrix} 1 & -5 \\ -7 & 12 \end{bmatrix}$ **(b)** $\begin{bmatrix} 5 & 6 & 5 \\ -4 & 6 & -22 \\ 0 & 0 & 9 \end{bmatrix}$

4. $\frac{1}{16}\begin{bmatrix} 4 & 8 \\ 1 & -2 \end{bmatrix}$ **5.** $(6, 1)$

6. (a) 42 **(b)** -5

7. $\left(\frac{1}{2}, -\frac{1}{2}, \frac{3}{2}\right)$

8. The flat fee is $22.50 per day; mileage fee is 36¢ per mile.

9. 800 lb of the 10% mixture, 400 lb of the 25% mixture, and 800 lb of the 30% mixture

10.

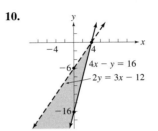

11. Let x = no. of model X and y = no. of model Y.

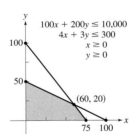

12. The maximum profit is $4600 for 60 model X and 20 model Y.

Chapter 8

Exercises 8.1

1. $(x - 3)^2 + (y + 8)^2 = 25$

3. $x^2 + y^2 + 6x - 2y + 9 = 0$

5. $C(2, -3); r = 9$ **7.** $C\left(\frac{1}{2}, 0\right); r = 3\sqrt{2}$

9. $C(3, 4); r = 4$ **11.** $C(3, 0); r = 2\sqrt{6}$

13. $C\left(0, \frac{1}{2}\right); r = 2\sqrt{3}$ **15.** $y = -\frac{1}{2}x + \frac{5}{2}$

17. $y = -x + 7$

19.

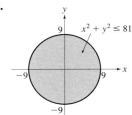

$x^2 + y^2 \leq 81$

21.

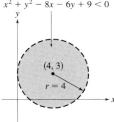

$x^2 + y^2 - 8x - 6y + 9 < 0$

$(4, 3)$

$r = 4$

25. $F(2, 2)$; directrix: $y = -2$
27. $F(0, 1)$; directrix: $x = -4$
29. $F\left(2, -\dfrac{17}{4}\right)$; directrix: $y = -\dfrac{7}{4}$
31. $8y = x^2 + 10x + 33$ **33.** $x = 3y^2 - 6y + 5$

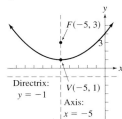

$F(-5, 3)$

Directrix: $y = -1$

$V(-5, 1)$

Axis: $x = -5$

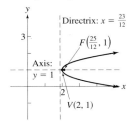

Directrix: $x = \frac{23}{12}$

$F\left(\frac{25}{12}, 1\right)$

Axis: $y = 1$

$V(2, 1)$

23. $\dfrac{22}{5}$ **25.** $x^2 + y^2 = \dfrac{144}{13}$ **27.** $\dfrac{\sqrt{10}}{5}$

33. $y_1 = \sqrt{36 - x^2}$; $y_2 = -\sqrt{36 - x^2}$

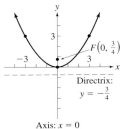

35. $y_1 = -4 + \sqrt{18 - (x - 3)^2}$;
$y_2 = -4 - \sqrt{18 - (x - 3)^2}$

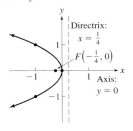

35. $(y - 4)^2 = -12(x - 5)$
37. $(x - 2)^2 = -14\left(y - \dfrac{1}{2}\right)$
39. $(y - 5)^2 = 20(x - 2)$
41. $y < 4x^2$ **43.** $x \geq y^2 - 6y + 5$

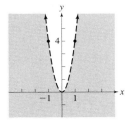

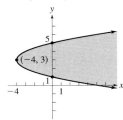

$(-4, 3)$

Exercises 8.2

1. $F(0, 3)$; directrix: $y = -3$
3. $F\left(-\dfrac{1}{32}, 0\right)$; directrix: $x = \dfrac{1}{32}$
5. $F\left(0, \dfrac{1}{8}\right)$; directrix: $y = -\dfrac{1}{8}$
7. $4x^2 - 12y = 0$ **9.** $x + y^2 = 0$

47. $y = -x - 2$
51. $y_1 = 1 + \sqrt{24x}$, $y_2 = 1 - \sqrt{24x}$

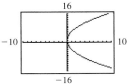

$F\left(0, \dfrac{3}{4}\right)$

Directrix: $y = -\dfrac{3}{4}$

Axis: $x = 0$

Directrix: $x = \dfrac{1}{4}$

$F\left(-\dfrac{1}{4}, 0\right)$

Axis: $y = 0$

53. $y = \dfrac{(x - 3)^2}{12}$

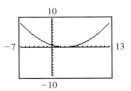

11. $x^2 = 16y$ **13.** $x^2 = -12y$ **15.** $y^2 = 20x$
17. The dish is 3 feet deep. The receiver should be placed 3 feet directly above the vertex.
19. The mirror is 8 inches deep. The light source should be placed 18 inches directly above the vertex.
21. $\dfrac{20\sqrt{6}}{3} \approx 16.33$ feet
23. $\dfrac{1}{2}$ inch above the vertex

Exercises 8.3

1. $V_1(5, 0)$, $V_2(-5, 0)$, $F_1(4, 0)$, $F_2(-4, 0)$
3. $V_1(0, 3\sqrt{2})$, $V_2(0, -3\sqrt{2})$, $F_1(0, \sqrt{6})$, $F_2(0, -\sqrt{6})$
5. $V_1(2\sqrt{6}, 0)$, $V_2(-2\sqrt{6}, 0)$, $F_1(\sqrt{15}, 0)$, $F_2(-\sqrt{15}, 0)$
7. $V_1(0, 5)$, $V_2(0, -5)$, $F_1(0, 4)$, $F_2(0, -4)$
9. $V_1(0, \sqrt{30})$, $V_2(0, -\sqrt{30})$, $F_1(0, \sqrt{29})$, $F_2(0, -\sqrt{29})$

11. $\dfrac{x^2}{49} + \dfrac{y^2}{9} = 1$

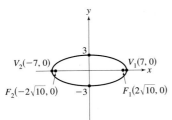

13. $\dfrac{x^2}{36} + \dfrac{y^2}{25} = 1$

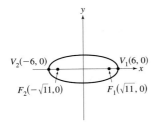

15. $x^2 + 9y^2 = 36$

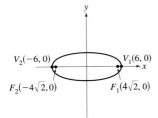

17. $12x^2 + y^2 = 24$

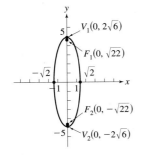

19. $3x^2 + 8y^2 = 12$

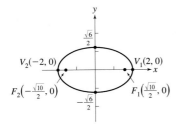

21. $e = \dfrac{\sqrt{11}}{6}$ **23.** $e = \dfrac{\sqrt{10}}{4}$

25. $\dfrac{x^2}{36} + \dfrac{y^2}{24} > 1$ **27.** $3x^2 + 4y^2 \le 12$

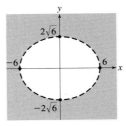

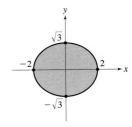

29. $\dfrac{x^2}{16} + \dfrac{y^2}{12} = 1$ **31.** $\dfrac{x^2}{16} + \dfrac{y^2}{4} = 1$

33. $5\sqrt{7} \approx 13.23$ feet   **35.** 221,589 miles

37. $C(2, 3),\ V_1(6, 3),\ V_2(-2, 3),\ F_1(2 + 2\sqrt{3}, 3),$
$F_2(2 - 2\sqrt{3}, 3)$

39. $C(-1, 3),\ V_1(-1, 3 + 3\sqrt{2}),\ V_2(-1, 3 - 3\sqrt{2}),$
$F_1(-1, 3 + \sqrt{6}),\ F_2(-1, 3 - \sqrt{6})$

41. $C(2, 3),\ V_1(2, 5),\ V_2(2, 1),\ F_1(2, 3 + \sqrt{3}),\ F_2(2, 3 - \sqrt{3})$

43. $C(5, 1),\ V_1(5, 1 + 3\sqrt{2}),\ V_2(5, 1 - 3\sqrt{2}),$
$F_1(5, 1 + \sqrt{3}),\ F_2(5, 1 - \sqrt{3})$

45. $\dfrac{(x - 2)^2}{9} + \dfrac{(y + 3)^2}{8} = 1$

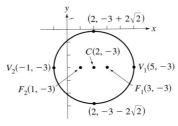

47. $\dfrac{(x + 2)^2}{16} + \dfrac{(y - 1)^2}{24} = 1$

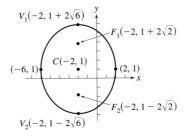

49. $16x^2 + 9y^2 - 32x + 54y = 47$

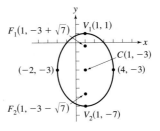

51. $x^2 + 4y^2 - 2x - 24y = -29$

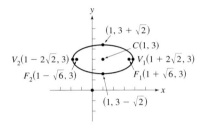

53. $\dfrac{x^2}{9} + \dfrac{y^2}{4} = 1$; Ellipse

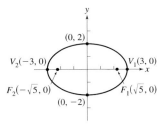

55. $\dfrac{(x + 2)^2}{16} + \dfrac{(y - 1)^2}{16} = 1$; Circle

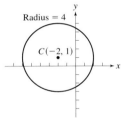

57. $x^2 - 9y = 0$; Parabola

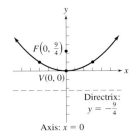

59. $\dfrac{(x - 3)^2}{9} + \dfrac{(y - 4)^2}{5} = 1$

61. $\dfrac{(x - 3)^2}{4} + (y + 5)^2 = 1$ **65.** $-x + 2\sqrt{3}\,y = 8$

69. $y_1 = \sqrt{4 - \dfrac{4x^2}{3}}$

$y_2 = -\sqrt{4 - \dfrac{4x^2}{3}}$

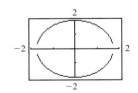

71. $y_1 = -2 + \sqrt{4 - \dfrac{(x - 3)^2}{3}}$

$y_2 = -2 - \sqrt{4 - \dfrac{(x - 3)^2}{3}}$

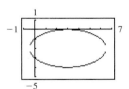

Exercises 8.4

1. $V_1(3, 0), V_2(-3, 0), F_1(5, 0), F_2(-5, 0)$; asymptotes: $y = \pm\dfrac{4}{3}x$

3. $V_1(0, 2\sqrt{3}), V_2(0, -2\sqrt{3}), F_1(0, \sqrt{30}), F_2(0, -\sqrt{30})$; asymptotes: $y = \pm\dfrac{\sqrt{6}}{3}x$

5. $V_1(3, 0), V_2(-3, 0), F_1(3\sqrt{3}, 0), F_2(-3\sqrt{3}, 0)$; asymptotes: $y = \pm\sqrt{2}x$

7. $V_1(3, 0), V_2(-3, 0), F_1(\sqrt{34}, 0), F_2(-\sqrt{34}, 0)$; asymptotes: $y = \pm\dfrac{5}{3}x$

9. $V_1(0, 1), V_2(0, -1), F_1(0, \sqrt{31}), F_2(0, -\sqrt{31})$; asymptotes: $y = \pm\dfrac{\sqrt{30}}{30}x$

11. $\dfrac{x^2}{9} - \dfrac{y^2}{49} = 1$

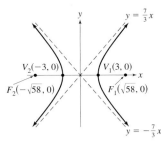

13. $8x^2 - y^2 = 24$

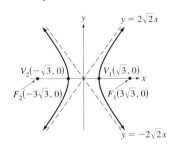

15. $3y^2 - 8x^2 = 24$

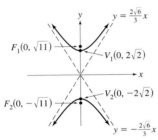

17. In miles (to one decimal place): $\dfrac{x^2}{344.5} - \dfrac{y^2}{2155.5} = 1$

19. $\sqrt{7} - \sqrt{3} \approx 0.9137$ AU

21. $C(1, 3)$, $V_1(3, 3)$, $V_2(-1, 3)$, $F_1(1 + 2\sqrt{2}, 3)$, $F_2(1 - 2\sqrt{2}, 3)$; asymptotes: $y = x + 2$, $y = -x + 4$

23. $C(-1, -3)$, $V_1(-1, -3 + 2\sqrt{3})$, $V_2(-1, -3 - 2\sqrt{3})$, $F_1(-1, -3 + \sqrt{30})$, $F_2(-1, -3 - \sqrt{30})$; asymptotes:
$y = \dfrac{\sqrt{6}}{3}x + \dfrac{\sqrt{6} - 9}{3}$, $y = -\dfrac{\sqrt{6}}{3}x - \dfrac{\sqrt{6} + 9}{3}$

25. $C(3, 1)$, $V_1(7, 1)$, $V_2(-1, 1)$, $F_1(3 + 2\sqrt{5}, 1)$, $F_2(3 - 2\sqrt{5}, 1)$; asymptotes: $y = \dfrac{1}{2}x - \dfrac{1}{2}$,
$y = -\dfrac{1}{2}x + \dfrac{5}{2}$

27. $C(1, -1)$, $V_1(1, -1 + 2\sqrt{6})$, $V_2(1, -1 - 2\sqrt{6})$, $F_1(1, -1 + 2\sqrt{10})$, $F_2(1, -1 - 2\sqrt{10})$; asymptotes:
$y = \dfrac{\sqrt{6}}{2}x - \dfrac{\sqrt{6} + 2}{2}$, $y = -\dfrac{\sqrt{6}}{2}x + \dfrac{\sqrt{6} - 2}{2}$

29. $\dfrac{(x + 4)^2}{9} - \dfrac{(y + 3)^2}{16} = 1$

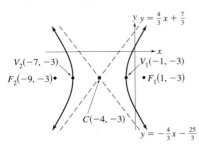

31. $\dfrac{x^2}{5} - \dfrac{(y - 3)^2}{15} = 1$

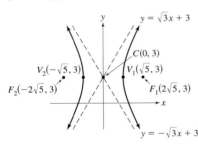

33. $9x^2 - 16y^2 - 36x - 32y = 124$

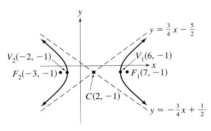

35. $25y^2 - 36x^2 - 150y + 288x - 1251 = 0$

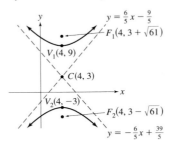

37. $\dfrac{(x - 6)^2}{16} - \dfrac{(y + 1)^2}{20} = 1$

39. $\dfrac{x^2}{12} + \dfrac{y^2}{4} = 1$; Ellipse

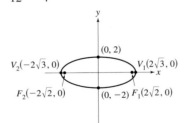

41. $\dfrac{(x + 2)^2}{9} - \dfrac{(y - 1)^2}{25} = 1$; Hyperbola

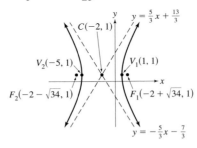

43. $x^2 - 2y - 12 = 0$; Parabola

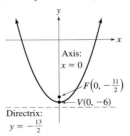

47. $x = 4$

51. $y_1 = \sqrt{3x^2 - 8}$
$y_2 = -\sqrt{3x^2 - 8}$

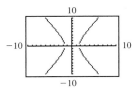

53. $y_1 = 2 + \sqrt{\dfrac{(x+3)^2}{4} - 2}$

$y_2 = 2 - \sqrt{\dfrac{(x+3)^2}{4} - 2}$

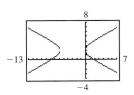

Exercises 8.5

1. An ellipse or its degenerate form
3. A hyperbola or its degenerate form
5. A circle or its degenerate form
7. A hyperbola or its degenerate form
9. An ellipse or its degenerate form
11. $y^2 + 16x = 0$; Parabola

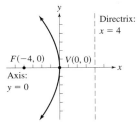

13. $x^2 - 10x + 25 = 0$;
Degenerate parabola: a line $x = 5$

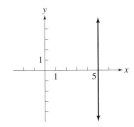

15. $2x^2 + y^2 - 8x - 2y - 7 = 0$; Ellipse

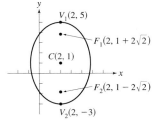

17. $25x^2 + y^2 - 50x - 6y - 16 = 0$; Ellipse

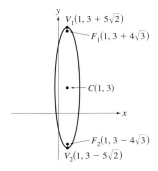

19. $25x^2 + 9y^2 - 225 = 0$; Ellipse

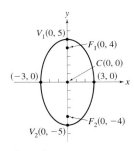

21. $x^2 - y^2 - 18 = 0$; Hyperbola

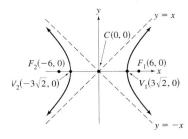

23. $y^2 - 6y - 16 = 0$; Degenerate parabola: 2 parallel lines,
$y = 8$ and $y = -2$

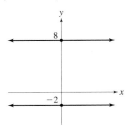

25. $-10x^2 + y = 0$; Parabola

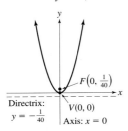

27. $x^2 + y^2 + 2x - 6y - 5 = 0$; Circle

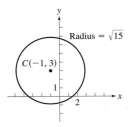

29. $x^2 + y^2 - 4x - 2y + 3 = 0$; Circle

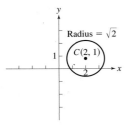

31. $9x^2 - y^2 - 18x - 4y - 139 = 0$; Hyperbola

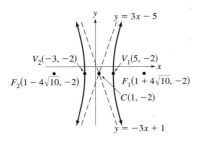

33. $x^2 + y^2 + 36 = 0$; Degenerate circle (not graphable)
35. $9x^2 + 25y^2 + 18x + 9 = 0$; Degenerate ellipse: the point $(-1, 0)$

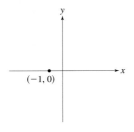

37. $16y^2 - x^2 = 0$; Degenerate hyperbola: two intersecting lines

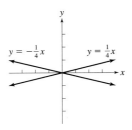

39. $x^2 - 2x + 12y - 47 = 0$; Parabola

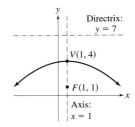

Exercises 8.6

1. $(3, -1), (-1, 3)$ **3.** $(0, -6), (3, 0)$
5. No real solutions
7. $(3, 1), (3, -1), (-3, 1), (-3, -1)$
9. $(0, 1), (0, -1)$
11. $\left(\dfrac{3 + \sqrt{17}}{2}, \dfrac{-1 + \sqrt{17}}{2}\right), \left(\dfrac{3 - \sqrt{17}}{2}, \dfrac{-1 - \sqrt{17}}{2}\right)$
13. $(4, 2)$ **15.** $(-1, 2)$
17. $(3 + \sqrt{3}, 3), (3 - \sqrt{3}, 3)$ **19.** $\left(6, \dfrac{1}{3}\right), (1, 2)$
21. $(5, 2)$ **23.** $\left(\dfrac{\sqrt{3}}{2}, \dfrac{1}{2}\right), \left(-\dfrac{\sqrt{3}}{2}, \dfrac{1}{2}\right)$

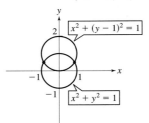

25. No real solutions

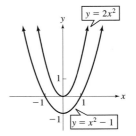

27. $(0, -3), (3, 0)$

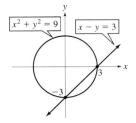

29. $\left(\dfrac{13}{3}, -\dfrac{65}{9}\right), (1, -5)$

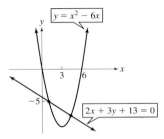

31. $(\sqrt{5}, 2), (-\sqrt{5}, 2)$

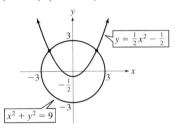

33. $(1 + \sqrt{7}, 1 - \sqrt{7}), (1 - \sqrt{7}, 1 + \sqrt{7})$

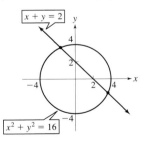

35. $(\sqrt{2}, 2), (\sqrt{2}, -2), (-\sqrt{2}, 2), (-\sqrt{2}, -2)$
37. $(\log_5 4, 4) \approx (0.9, 4)$ **39.** 3.5 cm by 10 cm
41. 7 cm by 24 cm
43. **(a)** $K = -4$ **(b)** $K < -4$
45. $(4, 0), (-4, 0)$
47. Yes; 40×100 or 50×80 ft
49.

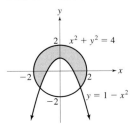

51.

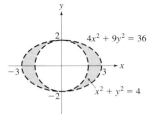

53.

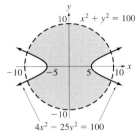

55. $(1, 1), (-1, 1)$ **57.** $(3.9, 0.9), (-0.9, -3.9)$

Chapter 8 Review Exercises

1. $C(3, -4); r = 3\sqrt{2}$ **2.** $C(0, 1); r = 2\sqrt{3}$
3. $C(4, 4); r = 6$ **4.** $C(5, -3); r = \sqrt{19}$
5. $y = \dfrac{1}{3}x - \dfrac{10}{3}$ **6.** $x = 2$
7. $8x - y^2 = 0$

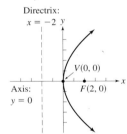

8. $y + 5x^2 = 0$

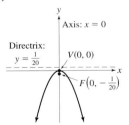

9. $(x - 4)^2 = 4y$

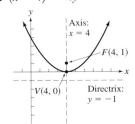

10. $(y - 2)^2 = -16x$

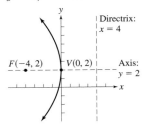

11. $4x = y^2 - 10y + 21$

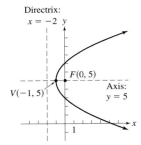

12. $6y = x^2 - 4x - 2$

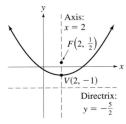

13. $y^2 = -12x$ **14.** $(x - 1)^2 = -12(y + 1)$

15. The mirror is $1\frac{1}{8}$ ft deep. The light source should be placed 2 ft above the vertex.

16. Width $= 5\sqrt{30} \approx 27.39$ ft

17. $\dfrac{x^2}{9} + \dfrac{y^2}{36} = 1$

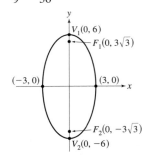

18. $\dfrac{x^2}{81} + \dfrac{y^2}{6} = 1$

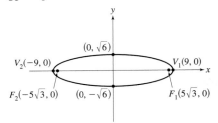

19. $24x^2 + 2y^2 = 48$

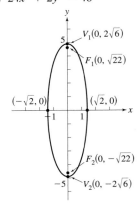

20. $20y^2 + 10x^2 = 40$

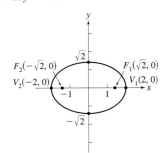

21. $\dfrac{(x - 2)^2}{6} + \dfrac{(y + 4)^2}{4} = 1$

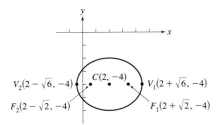

22. $\dfrac{x^2}{16} + \dfrac{(y+5)^2}{12} = 1$

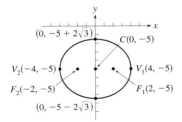

23. $x^2 + 4y^2 - 6x + 8y = 3$

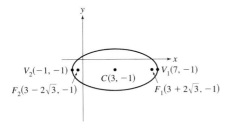

24. $6x^2 + 5y^2 + 12x - 20y - 4 = 0$

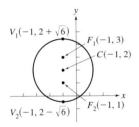

25. $\dfrac{x^2}{9} + \dfrac{y^2}{16} = 1$ **26.** $\dfrac{(x-5)^2}{25} + \dfrac{(y-5)^2}{16} = 1$

27. $\dfrac{x^2}{49} - \dfrac{y^2}{4} = 1$

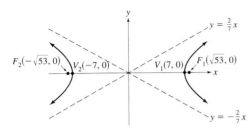

28. $\dfrac{y^2}{81} - \dfrac{x^2}{8} = 1$

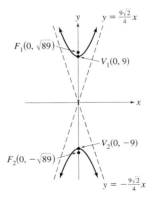

29. $x^2 - 8y^2 = 24$

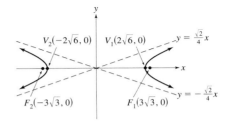

30. $10y^2 - 5x^2 = 40$

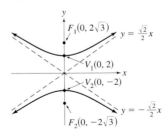

31. $\dfrac{(y-2)^2}{2} - \dfrac{(x-1)^2}{8} = 1$

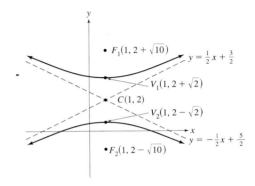

32. $\dfrac{(x+3)^2}{12} - \dfrac{y^2}{24} = 1$

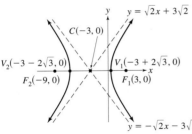

33. $x^2 - 4y^2 - 6x - 8y = 11$

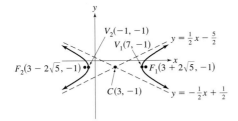

34. $5y^2 - 6x^2 - 20y - 12x - 16 = 0$

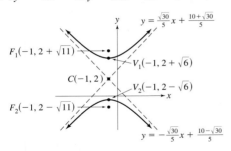

35. An ellipse or its degenerate form
36. A parabola or one of its degenerate forms
37. A hyperbola or its degenerate form
38. A circle or its degenerate form
39. $\dfrac{x^2}{36} - \dfrac{y^2}{4} = 1$; Hyperbola

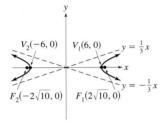

40. $\dfrac{x^2}{49} + \dfrac{y^2}{4} = 1$; Ellipse

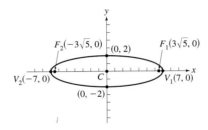

41. $\dfrac{x}{4} + \dfrac{y}{9} = 1$; Line

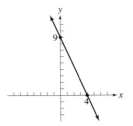

42. $\dfrac{(x-1)^2}{4} + \dfrac{(y+1)^2}{8} = 1$; Ellipse

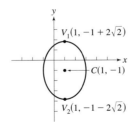

43. $\dfrac{(y-2)^2}{2} + \dfrac{(x-1)^2}{2} = 1$; Circle

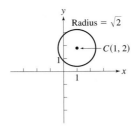

44. $\dfrac{(y-2)^2}{2} - \dfrac{(x-1)^2}{2} = 1$; Hyperbola

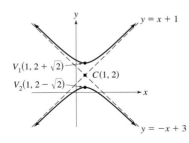

45. $x^2 = -64y$; Parabola

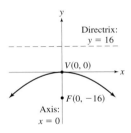

46. $x^2 + 4y^2 - 2x + 24y + 21 = 0$; Ellipse

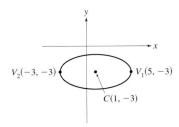

47. $x^2 - 8x + 16 = 0$; Degenerate parabola: a line

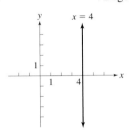

48. $3y^2 - 3y - 18 = 0$; Degenerate parabola: two lines

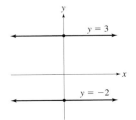

49. $x^2 + 2y^2 - 2x - 8y - 7 = 0$; Ellipse

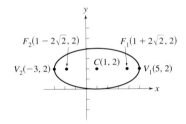

50. $x^2 + 4x - 5y + 24 = 0$; Parabola

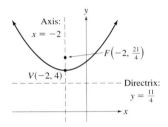

51. $x^2 + 25y^2 - 6x - 50y + 34 = 0$;
Degenerate ellipse: a point

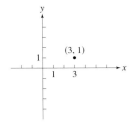

52. $3x^2 - y^2 - 6x - 6y - 6 = 0$;
Degenerate hyperbola: two intersecting lines

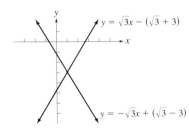

53. $(3, 1)$ **54.** $(2, -3), (-2, -3)$ **55.** $(7, 3)$
56. $\left(\dfrac{27}{11}, \dfrac{-29}{11} \right), (3, -1)$ **57.** $(0, 2), (0, -2)$
58. No real solutions   **59.** $(2, 3), (-2, 3)$
60. $(-2, -1), (-2, 1), (2, -1), (2, 1)$
61. Both legs are $2\sqrt{2} \approx 2.83$ cm
62. Dimensions: $r = 1\dfrac{5}{7}$ cm, $h = 6\dfrac{1}{8}$ cm
63. $\begin{cases} 25x^2 + 4y^2 \le 100 \\ x^2 + y^2 > 9 \end{cases}$

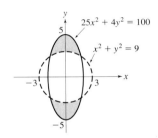

64. $\begin{cases} y \geq x^2 + 1 \\ y \leq -x^2 + 4 \end{cases}$

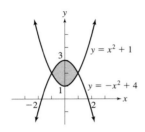

Chapter 8 Practice Test

1. $4x^2 + 4y = 0$; Parabola

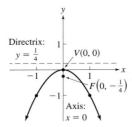

2. $4x^2 + 4y^2 = 36$; Circle   **3.** $\dfrac{x^2}{25} + \dfrac{y^2}{49} = 1$; Ellipse

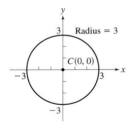
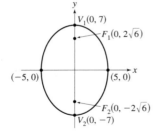

4. $\dfrac{x^2}{25} - \dfrac{y^2}{49} = 1$; Hyperbola

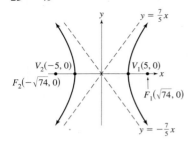

5. $(x - 3)^2 + (y + 5)^2 = 5$; Circle

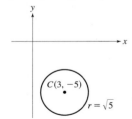

6. $2(x - 3)^2 + 8y + 24 = 0$; Parabola

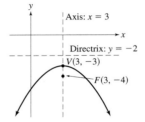

7. $\dfrac{(x - 3)^2}{4} + \dfrac{(y - 2)^2}{12} = 1$; Ellipse

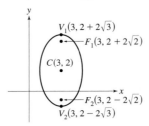

8. $\dfrac{(y + 2)^2}{25} - \dfrac{x^2}{18} = 1$; Hyperbola

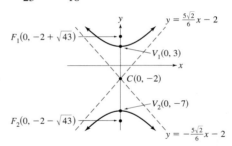

9. $4x^2 + 5y^2 - 24x + 30y + 61 = 0$; Ellipse

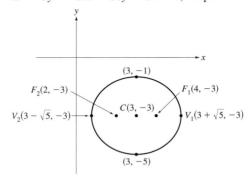

10. $3x^2 - 4y^2 - 6x = 45$; Hyperbola

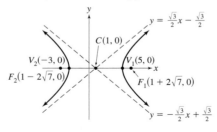

11. $y^2 = 8x$ **12.** $\dfrac{(x-1)^2}{16} + \dfrac{(y-1)^2}{12} = 1$

13. $(1,1), (1,-1), (-1,1), (-1,-1)$

14.

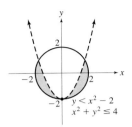

Chapter 9

Exercises 9.1

1. $a_1 = 2; a_2 = 7; a_3 = 12; a_4 = 17; a_8 = 37$

3. $b_1 = 5; b_2 = 8; b_3 = 11; b_4 = 14; b_8 = 26$

5. $c_1 = 3; c_2 = 9; c_3 = 27; c_4 = 81; c_8 = 6561$

7. $x_1 = 2; x_2 = \dfrac{3}{2}; x_3 = \dfrac{4}{3}; x_4 = \dfrac{5}{4}; x_8 = \dfrac{9}{8}$

9. $x_1 = \dfrac{1}{3}; x_2 = -\dfrac{1}{4}; x_3 = \dfrac{1}{5}; x_4 = -\dfrac{1}{6}; x_8 = -\dfrac{1}{10}$

11. $a_1 = -\dfrac{1}{2}; a_2 = \dfrac{2}{3}; a_3 = -\dfrac{3}{4}; a_4 = \dfrac{4}{5}; a_8 = \dfrac{8}{9}$

13. $a_1 = 4; a_2 = 5; a_3 = 8; a_4 = 17; a_8 = 261$

15. $b_1 = 0.9; b_2 = 1.01; b_3 = 0.999; b_4 = 1.0001;$
$b_8 = 1.00000001$

17. $x_1 = 2; x_2 = 1; x_3 = \dfrac{8}{9}; x_4 = 1; x_8 = 4$

19. $a_1 = 0; a_2 = 2; a_3 = 0; a_4 = 2; a_8 = 2$

21. $c_1 = 2; c_2 = \dfrac{9}{4}; c_3 = \dfrac{64}{27}; c_4 = \dfrac{625}{256}; c_8 = \dfrac{43,046,721}{16,777,216}$

23. $a_1 = \sqrt{2}; a_2 = \sqrt{5}; a_3 = \sqrt{10}; a_4 = \sqrt{17}; a_8 = \sqrt{65}$

25. $a_1 = 1; a_2 = \dfrac{1}{2}; a_3 = \dfrac{1}{6}; a_4 = \dfrac{1}{24}; a_8 = \dfrac{1}{40,320}$

27. $c_1 = 1; c_2 = \dfrac{1}{2}; c_3 = \dfrac{2}{9}; c_4 = \dfrac{3}{32}; c_8 = \dfrac{315}{131,072}$

29. $t_1 = 2; t_2 = 40; t_3 = 2240; t_4 = 246,400;$
$t_8 \approx 2.35 \times 10^{15}$

31. $3, 5, 9, 17, 33, 65$

33. $-1; 4; 9; 64; 3969; 15,745,024$

35. $1; 2; 4; 16; 65,536; 2^{65,536}$

37. $2, 1, 3, \dfrac{4}{3}, \dfrac{15}{4}, \dfrac{8}{5}$ **39.** $0, 1, \dfrac{1}{2}, \dfrac{3}{4}, \dfrac{5}{8}, \dfrac{11}{16}$

41. $-1, 1, -1, -1, 1, -1$

43.

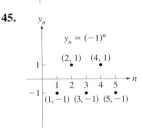

45.

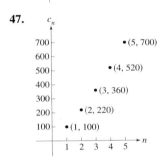

47.

49. $a_n = 5n$ **51.** $a_n = 2n - 1$ **53.** $a_n = \dfrac{2n-1}{2n}$

55. $a_n = \dfrac{(-1)^{n+1}}{n+1}$ **57.** $a_n = (-1)^n 4(10^{-n})$

59. $25,150; $25,800; $26,450$

61. $61,200; 62,424; 63,672$

63. $1863; $1928.21; $1995.70; $2065.55; 2137.84

65. In billions: $2.27; 2.24; 2.21; 2.18; 2.15$

67. In feet: $30; 15; 7.5; 3.75; 1.875$

69. $230; $227.70; $225.42; 223.17

Exercises 9.2

1. 55   **3.** 18   **5.** -14 **7.** 63   **9.** 11   **11.** 46

13. 62   **15.** 132   **17.** $\dfrac{121}{3}$ **19.** $\dfrac{73}{12}$ **21.** $\dfrac{109}{420}$

23. $\dfrac{41}{24}$ **25.** $\dfrac{7}{8}$ **27.** $1 + 2 + 3 + 4 + 5 = 15$

29. $5(1-1) + 5(2-1) + 5(3-1) +$
$\qquad 5(4-1) + 5(5-1) + 5(6-1) = 75$

31. $2^2 + 3^2 + 4^2 + 5^2 + 6^2 = 90$

33. $(2(3)^2 - 5) + (2(4)^2 - 5) + (2(5)^2 - 5) +$
$$(2(6)^2 - 5) + (2(7)^2 - 5) = 245$$

35. $(2^2 - 3(2) + 1) + (3^2 - 3(3) + 1) +$
$$(4^2 - 3(4) + 1) = 5$$

37. $\dfrac{1}{1 + 1} + \dfrac{1}{2 + 1} + \dfrac{1}{3 + 1} + \dfrac{1}{4 + 1} + \dfrac{1}{5 + 1} = \dfrac{29}{20}$

39. $(\sqrt{1 + 1} - \sqrt{1}) + (\sqrt{2 + 1} - \sqrt{2}) +$
$$(\sqrt{3 + 1} - \sqrt{3}) + (\sqrt{4 + 1} - \sqrt{4}) +$$
$$(\sqrt{5 + 1} - \sqrt{5}) = \sqrt{6} - 1$$

41. $3 + 3 + 3 + 3 + 3 = 15$ **43.** $\displaystyle\sum_{k=1}^{5} 4k$

45. $\displaystyle\sum_{k=1}^{7} 2k^2$ **47.** $\displaystyle\sum_{n=1}^{5} (4n - 2)$ **49.** $\displaystyle\sum_{k=1}^{26} (2k - 1)$

51. $\displaystyle\sum_{n=1}^{99} \dfrac{1}{n(n + 1)}$ **53.** 360   **55.** 459,033   **57.** 2081

59. 2   **61.** $f(1) = 5; f(-2) = -22; f\left(\dfrac{1}{2}\right) = \dfrac{31}{32}$

63. (a) $F(1) = 2.7181; F(2) = 7.3556; F(-1) = 0.3681$

Exercises 9.3

1. Arithmetic   **3.** Not arithmetic   **5.** Arithmetic

7. $d = 5$ **9.** $d = \dfrac{2}{3}$ **11.** $d = 4$

13. $d = \dfrac{1}{2}; a_1 = \dfrac{13}{2}$ **15.** $a_1 = -\dfrac{3}{5}; a_{20} = 7$

17. $d = 3; a_{18} = 53$ **19.** $d = \dfrac{1}{3}; a_{10} = \dfrac{7}{2}$

21. $a_1 = \dfrac{2}{3}; a_9 = -\dfrac{10}{3}$ **23.** $d = -\dfrac{5}{12}; a_1 = \dfrac{59}{4}$

25. \$26,750   **27.** 88 lb   **29.** 144 ft.; 176 ft.   **31.** 99

33. 14   **35.** $\dfrac{35}{2}$ **37.** -57 **39.** 50   **41.** $\dfrac{13}{3}$

43. 144   **45.** 550   **47.** 112   **49.** 1705   **51.** $\dfrac{64}{\pi}$

53. 348   **55.** \$1500; \$103,500
57. (a) Yes   **(b)** 24   **(c)** 47
59. 10,000   **61.** 2925   **63.** $a_1 = -4$ **65.** $S_5 = \dfrac{160}{7}$
67. $d = 11$

Exercises 9.4

1. Arithmetic   **3.** Geometric   **5.** Neither

7. Geometric   **9.** 15,625   **11.** $\dfrac{8}{81}$ **13.** 64

15. $\dfrac{1}{27}$ **17.** $\dfrac{1}{128}$ **19.** $\dfrac{1}{8192}$ **21.** $125\sqrt{5}$ **23.** 80

25. $r = 6; a_1 = \dfrac{1}{36}$ **27.** $r = -\dfrac{1}{\sqrt{3}}; a_8 = \dfrac{\sqrt{3}}{81}$

29. 2401   **31.** -100 **33.** $\dfrac{1}{729}$ **35.** \$18,370.80

37. \$14,848   **39.** 128,008   **41.** \$12,762.82
43. \$7128.80   **45.** 0.625 ft   **47.** 484

49. $\dfrac{3724}{9375} \approx 0.3972$ **51.** 128   **53.** $-\dfrac{61}{108} \approx -0.5648$

55. $\dfrac{1023}{4} = 255.75$ **57.** 254   **59.** $\dfrac{1640}{6561} \approx 0.2500$

61. 63   **63.** 1.11111   **65.** 0.011111111
67. 119.26 feet
69. (a) $5, 5^2, 5^3, 5^4, \ldots$ **(b)** 625   **(c)** 488,280

71. 4   **73.** $\dfrac{9}{2}$ **75.** $\dfrac{2}{5}$ **77.** $\dfrac{15}{2}$ **79.** S does not exist

81. $\dfrac{47}{99}$ **83.** $\dfrac{10,671}{9990}$ **85.** 75 feet   **87.** \$6166.67

Exercises 9.5

41. $N = 9$ **43.** $N = 5$ **45.** $N = 5$
47. $P_1, P_2, P_3, \ldots, P_{49}$ are true.
49. P_n is true for all $n \geq 10$.   **51.** P_{10} is not true.
53. P_n is true for all odd n.   **55.** P_n is true for all n.
57. P_n is true for all n that are powers of 3.

Exercises 9.6

1. 336   **3.** 792   **5.** 720   **7.** 18,564   **9.** 70   **11.** 15
13. 120   **15.** 160   **17.** 40,320
19. Approximately $1.3 \times 10^{12} = 1.3$ trillion   **21.** 676
23. (a) 333,135,504   **(b)** 185,075,280
   (c) 119,587,104
25. Approximately $1.1 \times 10^{12} = 1.1$ trillion; approximately
   $9.5 \times 10^{13} = 95$ trillion
27. 66
29. (a) 21   **(b)** 126   **(c)** 11,440   **(d)** 2940
31. 2,656,500   **33.** 40
35. (a) 125   **(b)** 50
37. (a) 1287   **(b)** 5148   **(c)** 4512   **(d)** 3744
39. (a) 45   **(b)** 120
41. (a) 40,320   **(b)** 119,750,400   **(c)** 9,979,200
43. 30

Exercises 9.7

1. $\{(1, 1), (1, 2), (1, 3), (2, 1), (2, 2), (2, 3), (3, 1), (3, 2), (3, 3)\}$
3. $\{(1, H), (2, H), (3, H), (4, H), (5, H), (6, H), (1, T), (2, T),$
   $(3, T), (4, T), (5, T), (6, T)\}$
5. $\{(1, 2), (1, 3), (1, 4), (1, 5), (2, 1), (2, 3), (2, 4), (2, 5),$
   $(3, 1), (3, 2), (3, 4), (3, 5), (4, 1), (4, 2), (4, 3), (4, 5), (5, 1),$
   $(5, 2), (5, 3), (5, 4)\}$
7. $\{TTTT, TTTF, TTFT, TFTT, FTTT, TTFF, TFFT, TFTF,$
   $FTFT, FFTT, FTTF, TFFF, FTFF, FFTF, FFFT, FFFF\}$
9. $\{A, B\}, \{A, C\}, \{A, D\}, \{A, E\}, \{A, F\}, \{B, C\}, \{B, D\},$
   $\{B, E\}, \{B, F\}, \{C, D\}, \{C, E\}, \{C, F\}, \{D, E\}, \{D, F\},$
   $\{E, F\}$
11. (a) 3/8   **(b)** 7/8   **(c)** 1/8
13. (a) 11/36   **(b)** 5/36   **(c)** 15/18
15. (a) 1/13   **(b)** 3/13   **(c)** 10/13
17. (a) 7/20   **(b)** 3/4   **(c)** 13/20
19. (a) 1/11   **(b)** 3/11   **(c)** 6/11
21. (a) 3/5   **(b)** 3/5   **(c)** 4/5
23. (a) 0.504
   (b) 1/1,000; it is not the complement of part **(a)**
25. 0.0004952   **27.** 0.0019808   **29.** 0.0002401
31. 0.0211285   **33.** 0.0475390   **35.** 0.7784663

37. One chance in 32,468,436

39. (a) 13/64   **(b)** 43/64   **(c)** 7/32

41. (a) 1/24   **(b)** 0   **(c)** 1/4

43. (a) 197/600   **(b)** 3/100   **(c)** 7/150

45. (a) $\frac{{}_{15}C_5}{{}_{45}C_5} \approx 0.002458$ **(b)** $1 - \frac{{}_{30}C_5}{{}_{45}C_5} \approx 0.88336$

   (c) $\frac{({}_{10}C_1)({}_{35}C_4)}{{}_{45}C_5} \approx 0.4286$

47. (a) $0.289^4 \approx 0.00698$ **(b)** $(1 - 0.289)^4 \approx 0.256$
   (c) $1 - (1 - 0.289)^4 \approx 0.744$

49. (a) 1/8   **(b)** 5/16   **(c)** 1/2

51. (a) 1/6   **(b)** 1/6   **(c)** 2/9

53. (a) $\frac{12 \cdot 11}{52 \cdot 51} \approx 0.0498$ **(b)** $11/51 \approx 0.216$
   (c) $12/51 \approx 0.235$

55. (a) $11/26 \approx 0.4231$ **(b)** $4/13 \approx 0.3077$
   (c) $9/13 \approx 0.6923$

57. $1/5040 \approx 0.000198$

59. (a) $\frac{6 \cdot 18 \cdot 12 \cdot 10}{40 \cdot 39 \cdot 38} \approx 0.2186$ **(b)** ≈ 0.117

   (c) $\frac{6 \cdot 18 \cdot 12 \cdot 10}{40 \cdot 40 \cdot 40} = 0.2025$

61. (a) $9/19 \approx 0.4737$ **(b)** $9/19 \approx 0.4737$
   (c) $1/19 \approx 0.0526$ **(d)** $(1/19)^2 \approx 0.0028$

63. (a) ≈ 0.9996 **(b)** $(0.997)^{100} \approx 0.7405$

65. $1/24 \approx 0.0417$

Exercises 9.8

1. $x^3 - 12x^2 + 48x - 64$

3. $a^7 + 14a^6 + 84a^5 + 280a^4 + 560a^3 +$
$$672a^2 + 448a + 128$$

5. $64x^6 + 192x^5y + 240x^4y^2 + 160x^3y^3 +$
$$60x^2y^4 + 12xy^5 + y^6$$

7. $t^5 - 15t^4r + 90t^3r^2 - 270t^2r^3 + 405tr^4 - 243r^5$

9. $w^{16} + 8w^{14} + 28w^{12} + 56w^{10} + 70w^8 +$
$$56w^6 + 28w^4 + 8w^2 + 1$$

11. $x^{12} - 4x^9y^3 + 6x^6y^6 - 4x^3y^9 + y^{12}$

13. $a^{18} + 18a^{16}b + 144a^{14}b^2 + 672a^{12}b^3 +$
$$2016a^{10}b^4 + 4032a^8b^5 + 5376a^6b^6 +$$
$$4608a^4b^7 + 2304a^2b^8 + 512b^9$$

15. $x^4 - 2x^3 + \frac{3}{2}x^2 - \frac{1}{2}x + \frac{1}{16}$

17. $\frac{1}{27}x^3 + \frac{4}{3}x^2 + 16x + 64$

19. $x^6 - 6x^5\sqrt{x} + 15x^5 - 20x^4\sqrt{x} + 15x^4 - 6x^3\sqrt{x} + x^3$

21. $x^3 + 6x^2\sqrt{xy} + 15x^2y + 20xy\sqrt{xy} +$
$$15xy^2 + 6y^2\sqrt{xy} + y^3$$

23. $x^5 - 5x^3 + 10x - \frac{10}{x} + \frac{5}{x^3} - \frac{1}{x^5}$

25. $\frac{x^4}{y^4} + \frac{4x^2}{y^2} + 6 + \frac{4y^2}{x^2} + \frac{y^4}{x^4}$

27. $x^{-5} - 5x^{-4}y^2 + 10x^{-3}y^4 - 10x^{-2}y^6 + 5x^{-1}y^8 - y^{10}$

29. $64r^{-6} + 192r^{-5}s^{-2} + 240r^{-4}s^{-4} + 160r^{-3}s^{-6} +$
$$60r^{-2}s^{-8} + 12r^{-1}s^{-10} + s^{-12}$$

31. 131,250   **33.** 189   **35.** 67,584   **37.** 28

39. $a^4 + 4a^3b + 6a^2b^2 + 4ab^3 + b^4 + 4a^3c +$
$$12a^2bc + 12ab^2c + 4b^3c + 6a^2c^2 + 12abc^2 +$$
$$6b^2c^2 + 4ac^3 + 4bc^3 + c^4$$

41. 756   **43.** $n + 1$ **45.** Yes   **47.** $-4 - 4i$

49. $-46 + 9i$

Chapter 9 Review Exercises

1. $a_1 = -2; a_2 = 1; a_3 = 4; a_4 = 7; a_{12} = 31$

2. $b_1 = 8; b_2 = 15; b_3 = 22; b_4 = 29; b_{12} = 85$

3. $x_1 = 5; x_2 = 20; x_3 = 45; x_4 = 80; x_{12} = 720$

4. $y_1 = 4; y_2 = 8; y_3 = 12; y_4 = 16; y_{12} = 48$

5. $a_1 = \frac{-1}{3}; a_2 = \frac{1}{4}; a_3 = \frac{-1}{5}; a_4 = \frac{1}{6}; a_{12} = \frac{1}{14}$

6. $b_1 = \frac{-1}{2}; b_2 = \frac{1}{3}; b_3 = \frac{-1}{4}; b_4 = \frac{1}{5}; b_{12} = \frac{1}{13}$

7. $x_1 = 3; x_2 = 5; x_3 = 3; x_4 = 5; x_{12} = 5$

8. $y_1 = 6; y_2 = 7; y_3 = 10; y_4 = 17; y_{12} = 4103$

9. $a_n = 5n - 7$ **10.** $a_n = 3(5^{n-1})$

11. $a_n = \frac{(-1)^{n+1}}{3n}$ **12.** $a_n = \frac{1}{2^n}$

13. $p_n = 250 + 8(n - 1)$

14. $a_1 = 22,500; a_2 = 23,175; a_3 = 23,870;$
$a_4 = 24,586; a_5 = 25,323$

15. \$6381.41   **16.** \$6691.13   **17.** 84   **18.** 55

19. 126   **20.** 9840   **21.** 93   **22.** 140

23. $2(1) + 2(2) + 2(3) + 2(4) + 2(5) + 2(6) = 42$

24. $4(4) + 4(5) + 4(6) + 4(7) + 4(8) + 4(9) = 156$

25. $(1 - 2)^2 + (2 - 2)^2 + (3 - 2)^2 +$
$$(4 - 2)^2 + (5 - 2)^2 = 15$$

26. $1^3 + 2^3 + 3^3 + 4^3 = 100$

27. $5(1)^2 + 5(2)^2 + 5(3)^2 + 5(4)^2 + 5(5)^2 +$
$$5(6)^2 + 5(7)^2 = 700$$

28. $(2^2 - 6(2) + 1) + (3^2 - 6(3) + 1) +$
$$(4^2 - 6(4) + 1) = -22$$

29. $a_3 = 19; a_4 = 25; a_{10} = 61$

30. $a_5 = -6; a_6 = -11; a_{10} = -31$

31. $a_5 = 26; a_6 = 31; a_{10} = 51$

32. $a_5 = 4; a_6 = 6; a_{10} = 14$

33. $a_5 = -6; a_6 = -8; a_{10} = -16$

34. $a_4 = -\frac{2}{3}; a_5 = -1; a_{10} = -\frac{8}{3}$

35. $d = 3; x_1 = 1$ **36.** $d = -2; y_2 = -5$ **37.** 240 ft

38. 304 ft   **39.** 492   **40.** -153 **41.** 175   **42.** 64

43. 3280   **44.** 330   **45.** 625   **46.** 650   **47.** 73.5

48. 41   **49.** 162   **50.** 1024   **51.** $\frac{1}{64}$ **52.** $-\frac{4}{2187}$

53. 324   **54.** $\frac{1}{4}$ **55.** \$1536   **56.** \$12,285   **57.** $\frac{25}{4}$

58. $\frac{3}{4}$ **59.** S does not exist.   **60.** S does not exist.

61. $\frac{13}{33}$ **62.** $\frac{21,724}{9990}$ **63.** 560 ft   **67.** 1320

68. 6720   **69.** 362,880   **70.** 6720   **71.** 420

72. (a) 729   **(b)** 405   **(c)** 323

73. (a) 1287   **(b)** 10,240   **(c)** 40
74. 6 possible boy–girl combinations   **75.** 32
76. (a) 11/36   **(b)** 1/12   **(c)** 1/6
77. (a) 3/13   **(b)** 4/13   **(c)** 9/13
78. (a) $\frac{10 \cdot 9 \cdot 8 \cdot 7 \cdot 6}{10 \cdot 10 \cdot 10 \cdot 10 \cdot 10} \approx 0.3024$ **(b)** $\frac{10}{10^5} = 0.0001$
79. $\frac{(_4C_1)(_{13}C_5)}{_{52}C_5} \approx 0.00198$
80. (a) 1/6   **(b)** 0   **(c)** 1/2
81. (a) $0.78^6 \approx 0.2252$ **(b)** $(1 - 0.78)^6 \approx 0.00011$
   (c) $1 - (1 - 0.78)^6 \approx 0.999887$
82. (a) 2/3   **(b)** 5/18   **(c)** 1/2
83. (a) $\frac{_4C_2}{_{13}C_2} \approx 0.769$ **(b)** 0.474359
   (c) $\frac{6}{_{13}C_2} \approx 0.769$
84. $32x^5 - 80x^4y + 80x^3y^2 - 40x^2y^3 + 10xy^4 - y^5$
85. $\frac{x^3}{27} + \frac{x^2y}{12} + \frac{xy^2}{16} + \frac{y^3}{64}$
86. $x^{-12} + 6x^{-10}y^{-1} + 15x^{-8}y^{-2} + 20x^{-6}y^{-3} +$
                 $15x^{-4}y^{-4} + 6x^{-2}y^{-5} + y^{-6}$
87. 43,750   **88.** -3240

Chapter 9 Practice Test

1. (a) $x_1 = -2; x_2 = -2; x_3 = 0; x_8 = 40$
   (b) $y_1 = 3; y_2 = 38; y_3 = 133; y_8 = 2558$
2. $(3(4)^2 - 5(4) + 1) + (3(5)^2 - 5(5) + 1)$
   $+ (3(6)^2 - 5(6) + 1) + (3(7)^2 - 5(7) + 1) = 272$
3. 662   **4.** -96 **5.** 840   **6.** 363
7. (a) 29   **(b)** 15,625   **(c)** 432   **(d)** $\dfrac{45}{2}$
8. 2100   **9.** 32 ft
11. (a) 100,947   **(b)** 12,600
12. 83,160
13. (a) 1/2   **(b)** 1/20   **(c)** $0.9^7 \approx 0.478$
14. (a) 11/26   **(b)** 4/13   **(c)** 9/13
15. $256x^{16} - 3072x^{14}y + 16,128x^{12}y^2 -$
   $48,384x^{10}y^3 + 90,720x^8y^4 - 108,864x^6y^5 +$
            $81,648x^4y^6 - 34,992x^2y^7 + 6561y^8$

Appendix

Exercises A.1

1.

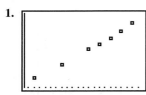

Linear trend: $y = 312.87503x - 616834.40533$;
$r \approx 0.9978$; for 1999, \$8602.8 billion

3.

Linear trend: $y = 0.46316x - 542.46669$;

$r \approx 0.9551$; \$4089

5.

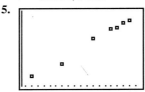

Linear trend: $y = 5.95960x - 349.28999$; $r \approx 0.9942$;
for 97, \$228.8 billion

7.

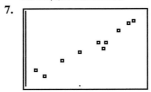

Linear trend: $r \approx 0.9723$, indicating for these students,
there is a strong correlation between their college GPA and
GRE scores.

9.

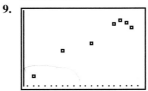

There seems to be a linear trend:
$y = 180.09426x - 347,057.8191$; $r \approx 0.9670$. Although
it looks as if the enrollment is falling off during the last
2–3 years, a look at the entire graph still seems to indicate a
linear trend.

11.

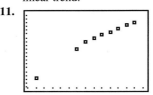

Linear trend: $y = 1.22238x - 78.42741$; $r \approx 0.9834$; for
1998, the average annual salary should be \$41,400.

13.

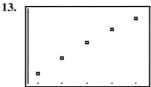

Linear trend: $y = 13.6x - 27082.2$; $r \approx 0.9980$. For 1999,
104.2% is predicted, which is impossible. Probably somewhere close to 100%.

15.

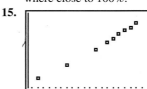

Linear trend: $y = 7.38172x - 14,563.02151$; $r \approx 0.9971$.
For 1998, the predicted per capita dentist expense would

be $186.

17.

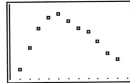

Not a linear trend—the scatterplot appears quadratic. Note that $r \approx -0.0995$, which indicates the relationship is not linear.

19.

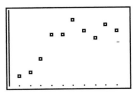

There seems to be a moderate linear trend:
$y = 59.28485x - 117,021.7758$; $r \approx 0.7989$.

21.

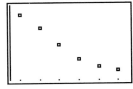

(a) Linear trend: $y = -50.77143x + 102,126.2381$; $r \approx -0.9774$. For this model, the 1996 expenditure predicted would be $786 billion.

(b) If the years 87–90 are added to the scatterplot, the trend seems less linear: $y = -30.55x + 61,825.61$, but $r \approx -0.8829$. The expenditures are still falling, but the linear trend is not as strong.

23.

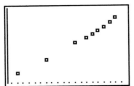

Linear trend: $y = 953.58065x - 1,883,883.591$; $r \approx 0.9964$. For 1998, the predicted credit market debt would be $21,371 billion.

25.

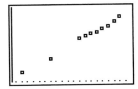

Linear trend: $y = 219.3023x - 432,792.9649$; $r \approx 0.9957$. For 1998, the predicted mortgage debt would be $5,373 billion.

Exercises A.2

1.

Quadratic trend:
$y = -0.39408x^2 + 32.40586x - 220.7993$. To maximize their net profit, they would have to spend $4,112 in advertising. Their maximum profit would be $445,390.

3.

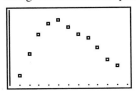

The trend seems quadratic:
$y = -14.71591x^2 + 58,618.42348x - 58,372,593.88$. For 1997, the number of thefts predicted by this equation would be 1,220,000.

5.

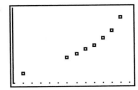

The linear model is $y = 1329.14286x - 111,042.3929$; the quadratic model is
$y = 97.17901x^2 - 16,260.25838x + 683,638.9806$. The quadratic model seems a better fit. In 1997, the linear model predicts 17,884 million minutes, while the quadratic model predicts 20,751 million minutes; the quadratic model is a better predictor for 1997.

7.

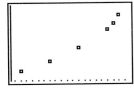

The linear model is $y = 1167.27837x - 2,306,068.101$; the quadratic model is
$y = 46.20184x^2 + 182,605.5224x + 180,436,047.9$. The quadratic model seems a better fit. In 1998, the linear model predicts $26,154 billion, while the quadratic model predicts $28,130 billion: the quadratic model is a better predictor for 1998.

9.

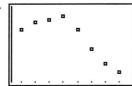

The quadratic model is
$y = -8.702381x^2 + 34,615.0595x - 34,420,594.07$;
the cubic model is
$y = 1.5454545x^3 - 9237.3842x^2 + 18,404,291.92x - 1.222268 \times 10^{10}$; and the quartic model is
$y = 0.776515152x^4 - 6181.06818x^3 + 18,450,491.35x^2 - (2.447763 \times 10^{10})x + 1.2177596 \times 10^{13}$. The quartic seems a better fit. In 1995, the quadratic model predicts expenditures of \$756 billion; the cubic model, \$832 billion; and the quartic model, \$967 billion. The cubic model is the best predictor for 1995.

11.

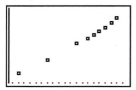

The exponential model is
$y = (1.596187 \times 10^{-70})1.089255^x$. The quadratic model still seems to be the best fit. In 1998, the linear model predicts \$21,371 billion, the quadratic model predicts \$22,145 billion, and the exponential model predicts \$24,434 billion, so the exponential model is a better predictor for 1998.

13.

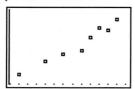

The linear model is $y = 12.475096x - 24,449.7126$, the quadratic model is
$y = 0.479303x^2 - 1895.70661x + 1,874,738.508$, and the exponential model is $y = (3.710416 \times 10^{-26})1.032931764^x$. The quadratic and exponential models seem very close, and fit better than the linear model. The exponential model seems slightly better. For 1997, the linear model predicts expenditures of \$463, the quadratic model predicts \$476, and the exponential model predicts \$468. The exponential model is the best predictor for 1997.

15.

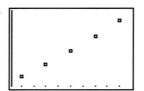

The exponential regression model is
$y = (5000.32568)1.052485^x$. This implies the principal was about \$5,000 invested at about 5.25% simple annual interest.

17.

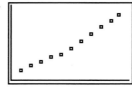

The exponential model is

$y = (2.1855877 \times 10^{-9})1.012886148^x$ and the logistic model is $\dfrac{641.87894}{1 + (9.37035 \times 10^{14})e^{-0.01709779x}}$. In the year 2000, the exponential model predicts a population of

288,960,000, while the logistic model predicts 276,590,000. The logistic seems to be a better predictor for the year 2000.

19.

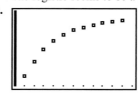

The logarithmic model is $y = 49.437187 + 55.944676\ln x$ and the logistic model is $\dfrac{174.4658}{1 + 3.912938e^{-0.59843x}}$. The logistic model seems to make more sense to use in this situation: We would not expect significant increases in income (in comparison to the cumulative total) from U.S. theaters after the eleventh week.

LIST OF APPLICATIONS

Application areas are followed by section numbers (with exercise numbers in parentheses).

INDEX